978 9401793759 -1

D1618172

The Changing World Religion Map

Stanley D. Brunn
Editor

The Changing World Religion Map

Sacred Places, Identities, Practices and Politics

Volume I

Donna A. Gilbreath
Assistant Editor

Editor
Stanley D. Brunn
Department of Geography
University of Kentucky
Lexington, KY, USA

Assistant Editor
Donna A. Gilbreath
UK Markey Cancer Center
Research Communications Office
Lexington, KY, USA

ISBN 978-94-017-9375-9   ISBN 978-94-017-9376-6   (eBook)
DOI 10.1007/978-94-017-9376-6

Library of Congress Control Number: 2014960060

Springer Dordrecht Heidelberg New York London
© Springer Science+Business Media Dordrecht 2015
This work is subject to copyright. All rights are reserved by the Publisher, whether the whole or part of the material is concerned, specifically the rights of translation, reprinting, reuse of illustrations, recitation, broadcasting, reproduction on microfilms or in any other physical way, and transmission or information storage and retrieval, electronic adaptation, computer software, or by similar or dissimilar methodology now known or hereafter developed.
The use of general descriptive names, registered names, trademarks, service marks, etc. in this publication does not imply, even in the absence of a specific statement, that such names are exempt from the relevant protective laws and regulations and therefore free for general use.
The publisher, the authors and the editors are safe to assume that the advice and information in this book are believed to be true and accurate at the date of publication. Neither the publisher nor the authors or the editors give a warranty, express or implied, with respect to the material contained herein or for any errors or omissions that may have been made.

Printed on acid-free paper

Springer Science+Business Media B.V. Dordrecht is part of Springer Science+Business Media (www.springer.com)

Preface: A Continuing Journey

Religion has always been a part of my life. I am a Presbyterian PK (preacher's kid). From my father I inherited not only an interest in the histories and geographies of religions, not just Christianity, but also a strong sense of social justice, a thread that has been part of my personal and professional (teaching, research, service) life. My mother was raised as a Quaker and from her I also learned much about social justice, peace and reconciliation and being a part of an effective voice calling for ends to war, social discrimination of various types, and other injustices that seem to be a continual part of daily life on the planet. My father had churches mostly in the rural Upper Middle West. These were open country and small town congregations in Illinois, Wisconsin, Minnesota, South Dakota, Nebraska, and Missouri. The members of these congregations were Germans, Czech, Scandinavians (Norwegians, and Swedes), and English. Perhaps or probably because of these experiences, I had friendships with many young people who comprised the mosaic of the rural Middle West. Our family moved frequently when I was living at home, primarily because my father's views on social issues were often not popular with the rural farming communities. (He lost his church in northwest Missouri in 1953 because he supported the Supreme Court's decision on desegregation of schools. By the time I graduated from high school in a small town in southeastern Illinois, I had attended schools in a half-dozen states; these include one-room school house experiences as well as those in small towns.

During my childhood days my interests in religion were, of course, important in the views I had about many subjects about those of different faiths and many places on the planet. I was born in a Catholic hospital, which I always attribute to the beginning of my ecumenical experiences. The schools I attended mixes of Catholics and Protestants; I had few experiences with Native Americans, Jews and African, and Asian Americans before entering college. But that background changed, as I will explain below. My father was always interested in missionaries and foreign missions and once I considered training for a missionary work. What fascinated me most about missionaries were that they were living in distant lands, places that I just longed to know about; an atlas was always my favorite childhood book, next to a dictionary. I was always glad when missionaries visited our churches and stayed in our homes. The fascination extended to my corresponding with missionaries in Africa, Asia, and Latin America. I was curious what kind of work they did. I also found them a source for stamps, a hobby that I have pursued since primary school. Also I collected the call letters of radio stations, some which were missionary stations, especially in Latin America. (Some of these radio stations are still broadcasting.)

When I enrolled as an undergraduate student at Eastern Illinois University, a small regional university in east central Illinois, I immediately requested roommates from different countries. I very much wanted to make friends with students from outside the United States and learn about their culture. During my 3 years at EIU, I had roommates from Jordan, Samoa, Costa Rica, Ethiopia, and South Korea; these were very formative years in helping me understand cross-cultural, and especially, religious diversity. On reflection, I think that most of the Sunday services I attended were mostly Presbyterian and Methodist, not Catholic, Lutheran, or Baptist. When I entered the University of Wisconsin, Madison, for the M.A. degree, I was again exposed to some different views about religion. The Madison church that fascinated me the most was the Unitarian church, a building designed by Frank Lloyd Wright. I remember how different the services and sermons were from Protestant churches, but intellectually I felt at home. My father was not exactly pleased I found the Unitarian church a good worship experience. The UW-Madison experience also introduced me to the study of geography and religion. This was brought home especially in conversations with my longtime and good friend, Dan Gade, but also a cultural geography course I audited with Fred Simoons, whose new book on religion and food prejudices just appeared and I found fascinating. Also I had conversations with John Alexander, who eventually left the department to continue in his own ministry with the Inter Varsity Christian Fellowship. A seminar on Cultural Plant Geography co-taught with Fred Simoons, Jonathan Sauer, and Clarence Olmstead provided some opportunities to explore cultural and historical dimensions of religion, which were the major fields where geographers could study religion. The geographers I knew who were writing about religion were Pierre Deffontaines, Eric Issac, and Xavier de Phanol. That narrow focus, has, of course, changed in the past several decades, as I will discuss below.

The move to Ohio State University for my doctoral work did not have the strong religious threads that had emerged before. I attended a variety of Protestant churches, especially Presbyterian, Congregational, and Methodist. I took no formal courses in geography that dealt with religion, although I was very interested when Wilbur Zelinsky's lengthy article on church membership patterns appeared in the *Annals of the Association of American Geographers* in 1961. I felt then that this was, and would be, a landmark study in American human geography, as the many maps of denominational membership patterns plus extensive references would form the basis for future scholars interested in religion questions, apart from historical and cultural foci which were the norm at that time. My first article on religion was on religious town names; I wrote it when I was at Ohio State with another longtime friend, Jim Wheeler, who had little interest in religion. I can still remember using my knowledge of biblical place names and going through a Rand McNally atlas with Jim identifying these town names. This study appeared in *Names*, which cultural geographers acknowledge is one of the premier journals concerned with names and naming processes. Even though my dissertation on changes in the central place functions in small towns in northwest Ohio and southeast Ohio (Appalachia) did not look specifically at churches, I did tabulate the number and variety during extensive fieldwork in both areas.

My first teaching job was at the University of Florida in fall 1966. I decided once I graduated from OSU that I wanted to live in a different part of the United States where I could learn about different regional cultures and politics. I was discouraged by some former teachers about teaching in Florida, especially about the region's segregation history, recent civil rights struggles in the South and also the John Birch Society (which was also active in Columbus when I lived there). The 3 years (1966–1969) in Gainesville were also very rewarding years. These were also very formative years in developing my interests in the social geography, a new field that was just beginning to be studied in the mid-1960s. Included in the forefront of this emerging field of social geographers were Anne Buttimer, Paul Claval, Yi-Fu Tuan, Dick Morrill, Richard Peet, Bill Bunge, Wilbur Zelinsky, David Harvey, and David Smith, all who were challenging geographers to study the social geographies of race, employment, school and housing discrimination, but also poverty, environmental injustice, inequities in federal and state programs promoting human welfare, the privileges of whiteness and the minorities' participation in the voting/political process. Living in northern Florida in the late 1960s or "Wallace years" could not help but alert one to the role that religion was playing in rural and urban areas in the South. Gainesville had distinct racial landscapes. I was definitely a "northerner" and carpetbagger who was an outcast in many ways in southern culture. One vivid memory is attending a University of Florida football game (a good example of regional pride and nationalism) and being about the only person seated while the band played "Dixie." I joined a Congregational/United Church of Christ church which was attended by a small number of "northern faculty" who were supportive of initiatives to end discriminatory practices at local, university, and broader levels. At this time I also was learning about the role of the Southern Baptist Church, a bastion of segregation that was very slow to accommodate to the wishes of those seeking ends to all kinds of overt and subtle discrimination (gender, race, class) practices. The term Bible Belt was also a label that rang true; it represented, as it still does, those who adhere to a literal interpretation of the Bible, a theological position I have never felt comfortable. I soon realized that if one really wanted to make a difference in the lives of those living with discrimination, poverty, and ending racial disenfranchisement in voting, religion was a good arena to express one's feelings and work with others on coordinated efforts. Published research that emerged from my Florida experiences included studies on poverty in the United States (with Jim Wheeler), the geographies of federal outlays to states, an open housing referendum in Flint, Michigan (with Wayne Hoffman), and school levies in cities that illustrated social inequities (with Wayne Hoffman and Gerald Romsa). My Florida years also provided me the first opportunity to travel in the developing world; that was made possible with a summer grant where I visited nearly 15 different Caribbean capitals where I witnessed housing, social, and infrastructure gaps. This experience provided my first experiences with the developing world and led to a Cities of the World class I taught at Michigan State University and also co-edited several editions with the same title of a book with Jack Williams.

The 11 years at Michigan State University did not result in any major research initiatives related to religion, although it did broaden my horizons about faiths other than Christianity. I began to learn about Islam, especially from graduate students in the department from Saudi Arabia, Libya, Kuwait, and Iran. Many of these I advised on religion topics about their own cultures, especially those dealing with pilgrimages and sacred sites. Probably the main gain from living in Michigan was support for and interest in an emerging secular society. The religious "flavors" of Michigan's religious landscape ran the full gamut from those who were very traditional and conservative to those who were globally ecumenical, interfaith, and even agnostic. I continued to be active in Presbyterian and United Churches of Christ, both which were intellectually and spiritually challenging places for adult classes and singing in a choir.

When I moved to the University of Kentucky in 1980, I knew that living in the Bluegrass State would be different from Michigan in at least two respects. One is that Kentucky was considered a moderate to progressive state with many strong traditional and conservative churches, especially the Southern Baptist denomination. Zelinsky's map accurately portrayed this region as having a dominance of conservative and evangelical Protestantism. Second, I realized that for anyone interested in advancing social issues related to race and gender equality or environmental quality (especially strip mining in eastern Kentucky), there would likely be some conflicts. I also understood before coming to Kentucky that alcoholic beverage consumption was a big issue in some countries; that fact was evident in an innovative regional map Fraser Hart prepared in a small book about the South. And then there was the issue about science and religion in school curricula. With this foreknowledge, I was looking forward to living in a region where the cross-currents of religion and politics meshed, not only experiencing some of these social issues or schisms firsthand, but also having an opportunity to study them, as I did.

I realized when I moved to Lexington, it was in many ways and still is a slowly progressing socially conscious city. Southern Baptist churches, Christian churches, and Churches of Christ were dominant in the landscape and in their influences on social issues. One could not purchase alcoholic beverages on Sunday in restaurants until a couple referenda were passed in the mid-1980s that permitted sales. I think 90 of the state's 120 counties were officially dry, although everyone living in a dry county knew where to purchase liquor. One could not see the then-controversial "The Last Temptation of Christ" movie when it appeared unless one would drive three hours to Dayton. "Get Right with God" signs were prominent along rural highways. The University of Kentucky chimes in Memorial Hall on campus played religious hymns until this practice stopped sometime in the middle of the decade; I am not exactly sure why. Public schools had prayers before athletic events; some still do. Teachers in some public schools could lose their jobs if they taught evolution. Creationism was (and still is) alive and well. I was informed by university advisors that the five most "dangerous" subjects to new UK students were biology, anthropology, astronomy, geology, and physical geography. Students not used to other than literal biblical interpretations were confused and confounded by evolutional science. Betting on horses was legal, even though gambling was frowned on by some religious leaders. Cock fighting and snake handling still existed (and still do) in pockets in rural eastern Kentucky. In many ways living in Kentucky was like living "on the dark side of gray." Lexington in many ways was and still is an island or outlier. Desegregation was a slow moving process in a city with a strong southern white traditional heritage. Athletic programs were also rather slow to integrate, especially UK basketball. In short, how could one not study religion in such an atmosphere. Living in Kentucky is sort of the antipode to living in agnostic-thriving New England and Pacific Northwest. I would expect that within 100 miles of Lexington one would discover one of the most diverse religious denominational and faith belief landscapes in the United States. There are the old regular mainline denominations, new faiths that have come into the Bluegrass and also many one-of-a-kind churches, especially in rural eastern, southern, and southeastern Kentucky.

I have undertaken a number of studies related to religion in Kentucky and the South in the past three decades. Some of these have been single-authored projects, others with students and faculty at UK and elsewhere. Some were presentations at professional meetings; some resulted in publications. The topics that fascinated me were ones that I learned from my geography colleagues and those in other disciplines that were understudied. These include the history and current patterns of wet/dry counties in Kentucky, a topic that appears in local and statewide media with communities deciding whether to approve the sale of alcoholic beverages. This study I conducted with historian Tom Appleton. With regularity there were clergy of some fundamentalist denominations who decried the sale of such drinks; opposing these clergy and their supporters were often those interested in promoting tourism and attracting out-of-state traffic on interstates. Also I looked into legislation that focused on science/education interfaces in the public schools and on the types of religious books (or avoidance of such, such as dealing with Marx, Darwin, and interfaith relations) in county libraries. Craig Campbell and I published an article in *Political Geography* on Cristo Redentor (Christ of the Andes statue) as an example of differential locational harmony. At the regional level I investigated with Esther Long the mission statements of seminaries in the South, a study that led to some interesting variations not only in their statements, but course offerings and visual materials on websites. I published with Holly Barcus two articles in *Great Plains Research* about denominational changes in the Great Plains. Missionaries have also been relatively neglected in geography, so I embarked on a study with Elizabeth Leppman that looked at the contents of a leading Quaker journal in the early part of the past century. Religions magazines, as we acknowledged in our study, were (and probably still are) a very important medium for educating the public about places and cultures, especially those where most Americans would have limited first-hand knowledge. The music/religion interface has long fascinated me, not only as a regular choir member, but for the words used to convey messages about spirituality, human welfare and justice, religious traditions and promises of peace and hope.

After 11 September 2001, I collected information from a number of churches in eastern Kentucky about how that somber event was celebrated and also what hymns they sung on the tenth anniversary. As expected, some were very somber and dignified, others had words about hope, healing, and reaching across traditional religious boundaries that separate us. I also co-authored an article (mostly photos) in *Focus* on the Shankill-Falls divided between Catholic and Protestant areas of Belfast with three students in my geography of religion class at the National University of Ireland in Maynooth. The visualization theme was integral to a paper published in *Geographica Slovenica* on ecumenical spaces and the web pages of the World Council of Churches and papers I delivered how cartoonists depicted the controversial construction of a mosque at Ground Zero. How cartoonists depicted God-Nature themes (the 2011 Haitian earthquake and Icelandic volcanic eruption) were the focus of an article in *Mitteilungen der Österreichsten Geographischen Gesellschaft*. I published in *Geographical Review* an article how the renaissance of religion in Russia is depicted on stamp issues since 1991. A major change in my thinking about the subject of religion in the South was the study that I worked on with Jerry Webster and Clark Archer, a study that appeared in the *Southeastern Geographer* in late 2011. We looked at the definition and concept of the Bible Belt as first discussed by Charles Heatwole (who was in my classes when I taught at Michigan State University) in 1978 in the *Journal of Geography*. We wanted to update his study and learn what has happened to the Bible Belt (or Belts) since this pioneering effort. What we learned using the Glenmary Research Center's county data on adherents for the past several decades was that the "buckle" has relocated. As our maps illustrated, the decline in those counties with denominations adhering to a literal interpretation of the Bible in western North Carolina and eastern Tennessee and a shift to the high concentration of Bible Belt counties in western Oklahoma and panhandle Texas. In this study using Glenmary data for 2000, we also looked at the demographic and political/voting characteristics of these counties. (In this volume we look at the same phenomenon using 2010 data and also discuss some of the visual features of the Bible Belt landscapes.)

What also was instrumental in my thinking about religion and geography interfaces were activities outside my own research agenda. As someone who has long standing interests in working with others at community levels on peace and justice issues, I worked with three other similarly committed adults in Lexington to organize the Central Kentucky Council for Peace and Justice. CKCPJ emerged in 1983 as an interfaith and interdenominational group committed to working on peace and justice issues within Lexington, in Central Kentucky especially, but also with national and global interests. The other three who were active in this initiative were Betsy Neale (from the Friends), Marylynne Flowers (active in a local Presbyterian church) and Ernie Yanarella (political scientist, Episcopalian, longtime friend, and also contributor to a very thoughtful essay on Weber in this volume). This organization is a key agent in peace/justice issues in the Bluegrass; it hosts meetings, fairs, conferences, and other events for people of all ages, plans annual marches on Martin Luther King Jr. holiday, and is an active voice on issues related to capital punishment, gun control, gay/lesbian issues, fair trade and employment, environmental responsibility and stewardship, and the rights of women, children, and minorities.

I also led adult classes at Maxwell Street Presbyterian Church where we discussed major theologians and religious writers, including William Spong, Marcus Borg, Joseph Campbell, Philip Jenkins, Diane Eck, Kathleen Norris, Diana Butler Bass, Francis Collins, Sam Harris, Paul Alan Laughlin, James Kugel, Dorothy Bass, and Garry Wills. We discuss issues about science, secularism, death and dying, interfaith dialogue, Christianity in the twenty-first century, images of God, missions and missionaries, and more. I also benefitted from attending church services in the many countries I have traveled, lived, and taught classes in the past three decades. These include services in elaborate, formal, and distinguished cathedrals in Europe, Russian Orthodox services in Central Asia, and services in a black township and white and interracial mainline churches in Cape Town. Often I would attend services where I understood nothing or little, but that did not diminish the opportunity to worship with youth and elders (many more) on Sunday mornings and listen to choirs sing in multiple languages. These personal experiences also became part of my religion pilgrimage.

While religion has been an important part of my personal life, it was less important as part of my teaching program. Teaching classes on the geography of religion are few and far between in the United States; I think the subject was accepted much more in the instructional and research arenas among geographers in Europe. I think that part of my reluctance to pursue a major book project on religion was that for a long time I considered the subject too narrowly focused, especially on cultural and historical geography. From my reading of the geography and religion literature, there were actually few studies done before 1970s. (See the bibliography at the end of Chap. 1). I took some renewed interest in the subject in the mid-1980s when a number of geographers began to examine religion/nature/environment issues. The pioneering works of Yi-Fu Tuan and Anne Buttimer were instrumental in steering the study of values, ethics, spirituality, and religion into some new and productive directions. These studies paved the way for a number of other studies by social geographers (a field that was not among the major fields until the 1970s and early 1980s). The steady stream of studies on geography and religion continued with the emergence of GORABS (Geography of Religion and Belief Systems) as a Specialty Group of the Association of American Geographers. The publication of more articles and special journal issues devoted to the geography of religion continued into the last decade of the twentieth century and first decade of this century. The synthetic works of Lily Kong that have regularly appeared in *Progress in Human Geography* further supported those who wanted to look at religion from human/environmental perspectives. These reviews not only introduced the study of religion within geography, but also to those in related scholarly disciplines.

As more and more research appeared in professional journals and more conferences included presentations on religion from different fields and subfields, it became increasing apparent that the time was propitious for a volume that looked at religion/geography interfaces from a number of different perspectives. From my own vantage point, the study of religion was one that could, should, might, and would benefit from those who have theoretical and conceptual training in many of the discipline's major subfields. The same applied to those who were regional specialists; there were topics meriting study from those who looked a political/religion issues in Southeast Asia or Central America as well as cultural/historical themes in southern Africa and continental Europe and symbolic/architectural features and built environments of religions landscapes in California, southeast Australia, and southwest Asia. Studying religions topics would not have to be limited to those in human geography, but could be seen as opportunities for those studying religion/natural disaster issues in East Asia and southeast United States as well as the spiritual roots of early and contemporary religious thinking in Central Asia, East Asia, Russia, and indigenous groups in South America. For those engaged in the study of gender, law, multicultural education, and media disciplines, there were also opportunities to contribute to the study of this emerging field. In short, there were literally "gold mines" of potential research topics in rural and urban areas everywhere on the continent.

About 7 years ago I decided to offer a class on the geography of religion in the Department of Geography at the University of Kentucky. The numbers were never larger (less than 15), but these were always enlightening and interesting, because of the views expressed by students. Their views about religion ran the gamut from very conservative to very liberal and also agnostic and atheist, which made, as one would expect, some very interesting exchanges. Students were strongly encouraged (not required, as I could not do this in a public university) to attend a half dozen different worship services during the semester. This did not mean attending First Baptist, Second Baptist, Third Baptist, etc., but different kinds of experiences. For some this course component was the first some had ever attended a Jewish synagogue, Catholic mass, Baptist service, an African American church, Unitarian church, or visited a mosque. Some students used this opportunity to attend Wiccan services, or visit a Buddhist and Hindu temple. Their write-ups about these experiences and the ensuing discussion were one of the high spots of the weekly class. In addition, the classes discussed chapters in various books and articles from the geography literature about the state of studying religion. And we always discussed current news items, using materials from the RNS (Religion News Service) website.

Another ingredient that stimulated my decision to edit a book on the geography of religion emerged from geography of religion conferences held in Europe in recent years. These were organized by my good friends Ceri Peach (Oxford), Reinhard Henkel (Heidelberg), and also Martin Baumann and Andreas Tunger-Zanetti (University of Lucerne). These miniconferences, held in Oxford, Lucerne, and Gottingen, usually attracted 20–40 junior and senior geographers and other religious scholars, and were a rich source of ideas for topics that might be studied. The opportunities for small group discussions, the field trips, and special events were conducive to learning about historical and contemporary changes in the religious landscapes of the European continent and beyond. A number of authors contributing to this volume presented papers at one or more of these conferences. Additional names came from those attending sessions at annual meetings of the Association of American Geographers.

Some of my initial thoughts and inspiration about a book came from the course I taught, conversations with friends who studied and did not study religion, and also the book I edited on megaengineering projects. This three-volume, 126-chapter book, *Engineering Earth: The Impacts of Megaengineering Projects*, was published in 2011 by Springer. There were only a few chapters in this book that had a religious content, one on megachurches, another on liberation theologians fighting megadevelopment projects in the Philippines and Guatemala. When I approached Evelien Bakker and Bernadette Deelen-Mans, my first geography editors at Springer, about a religion book, they were excited and supportive, as they have been since day one. They gave me the encouragement, certainly the latitude (and probably the longitude) to pursue the idea, knowing that I would identify significant cutting-edge topics about religion and culture and society in all major world regions. The prospectus I developed was for an innovative book that would include the contributions of scholars from the social sciences and humanities, those from different counties and those from different faiths. For their confidence and support, I am very grateful. The reviews they obtained of the prospectus were encouraging and acknowledging that there was a definite need for a major international, interdisciplinary, and interfaith volume. Springer also saw this book as an opportunity to emphasize its new directions in the social sciences and humanities. I also want to thank Stefan Einarson who came on board late in the project and shepherded the project to its completion with the usual Springer traits of professionalism, kindness, and commitment to the project's publication. And I wish to thank Chitra Sundarajan and her staff for helpful professionalism in preparing the final manuscript for publication.

The organization of the book, which is discussed in Chapter One, basically reflects the way I look at religion from a geographical perspective. I look at the subject as more than simply investigations into human geography's fields and subfields, including cultural and historical, but also economic, social, and political geography, but also human/environmental geography (dealing with human values, ethics, behavior, disasters, etc.). I also look at the study of religious topics and phenomena with respects to major concepts we use in geography; these include landscapes, networks, hierarchies, scales, regions, organization of space, the delivery of services, and virtual religion. I started contacting potential authors in September 2010. Since then I have sent or received over 15,000 emails related the volume.

I am deeply indebted to many friends for providing names of potential authors. I relied on my global network of geography colleagues in colleges and universities around the world, who not only recommended specific individuals, but also topics they deemed worthy of inclusion. Some were geographers, but many were not; some taught in universities, others in divinity schools and departments of religion around the world. Those I specifically want to acknowledge include: Barbara Ambrose, Martin Checa Artasu, Martin Baumann, John Benson, Gary Bouma, John Benson, Dwight Billings, Marion Bowman, John D. Brewer, David Brunn, David Butler, Ron Byars, Heidi Campbell, Caroline Creamer, Janel Curry, David Eicher, Elizabeth Ferris, Richard Gale, Don Gross, Wayne Gnatuk, Martin Haigh, Dan Hofrenning, Wil Holden, Hannah Holtschneider, Monica Ingalls, Nicole Karapanagiotis, Aharon Kellerman, Judith Kenny, Jean Kilde, Ted Levin, James

Munder, Alec Murphy, Tad Mutersbuagh, Garth Myers, Lionel Obidah, Sam Otterstrom, Francis Owusu, Maria Paradiso, Ron Pen, Ivan Petrella, Adam Possamai, Leonard Primiano, Craig Revels, Heinz Scheifinger, Anna Secor, Ira Sheskin, Doug Slaymaker, Patricia Solis, Anita Stasulne, Jill Stevenson, Robert Strauss, Tristan Sturm, Greg Stump, Karen Till, Andreas Tunger-Zenetti, Gary Vachicouras, Viera Vlčkova, Herman van der Wusten, Stanley Waterman, Mike Whine, Don Zeigler, Shangyi Zhou, and Matt Zook.

And I want to thank John Kostelnick who provided the GORABS Working Bibliography; most of the entries, except dissertations and theses, are included in Chap. 1 bibliography. Others who helped him prepare this valuable bibliography also need to be acknowledged: John Bauer, Ed Davis, Michael Ferber, Julian Holloway, Lily Kong, Elizabeth Leppman, Carolyn Prorock, Simon Potter, Thomas Rumney, Rana P.B. Singh, and Robert Stoddard. These are scholars who devoted their lifetimes to advancing research on geography and religion.

Finally I want to thank Donna Gilbreath for another splendid effort preparing all the chapters for Springer. She formatted the chapters and prepared all the tables and illustrations per the publisher's guidelines. Donna is an invaluable and skilled professional who deserves much credit for working with multiple authors and the publisher to ensure that all text materials were correct and in order. Also I am indebted to her husband, Richard Gilbreath, for helping prepare some of the maps and graphics for authors without cartographic services and making changes on others. As Director of the Gyula Pauer Center for Cartography and GIS, Dick's work is always first class. And, finally, thanks are much in order to Natalya Tyutenkova for her interest, support, patience, and endurance in the past several years working on this megaproject, thinking and believing it would never end.

The journey continues.

February 2014   Stanley D. Brunn

Contents of Volume I

Part I Introduction

1 Changing World Religion Map: Status, Literature and Challenges 3
   Stanley D. Brunn

Part II Nature, Ethics and Environmental Change

2 Nature, Culture and the Quest of the Sacred 71
   Anne Buttimer

3 Church, Politics, Faceless Men and the Face of God in Early Twenty-First Century Australia 83
   Mary C. Tehan

4 The Island Mystic/que: Seeking Spiritual Connection in a Postmodern World 97
   Laurie Brinklow

5 The Spatial Turn in Planetary Theologies: Ambiguity, Hope and Ethical Imposters 115
   Whitney A. Bauman

6 The Age of the World Motion Picture: Cosmic Visions in the Post-*Earthrise* Era 129
   Adrian Ivakhiv

7 Weber's Protestant Ethic Thesis and Ecological Modernization: The Continuing Influence of Calvin's Doctrine on Twenty-First Century Debates over Capitalism, Nature and Sustainability 145
   Ernest J. Yanarella

8	Exploring the Green Dimensions of Islam Mohammad Aslam Parvaiz	175
9	Making Oneself at Home in Climate Change: Religion as a Skill of Creative Adaptation Sigurd Bergmann	187
10	Scale-Jumping and Climate Change in the Geography of Religion Michael P. Ferber and Randolph Haluza-DeLay	203
11	All My Holy Mountain: Imaginations of Appalachia in Christian Responses to Mountaintop Removal Mining Andrew R.H. Thompson	217
12	God, Nature and Society: Views of the Tragedies of Hurricane Katrina and the Asian Tsunami Janel Curry	237
13	Japanese Buddhism and Its Responses to Natural Disasters: Past and Present Yukio Yotsumoto	255
14	Reshaping the Worldview: Case Studies of Faith Groups' Approaches to a New Australian Land Ethic Justin Lawson, Kelly Miller, and Geoff Wescott	273
15	"Let My People Grow." The Jewish Farming Movement: A Bottom-Up Approach to Ecological and Social Sustainability Rachel Berndtson and Martha Geores	297
16	Religious and Moral Hybridity of Vegetarian Activism at Farm Animal Sanctuaries Timothy Joseph Fargo	323

Part III Sacred Spaces and Places

17	Religions and Ideologies Paul Claval	349
18	Sacred Space and Globalization Alyson L. Greiner	363
19	Dark Green Religion: Advocating for the Sacredness of Nature in a Changing World Joseph Witt	381
20	Reinventing Agency, Sacred Geography and Community Formation: The Case of Displaced Kashmiri Pandits in India Devinder Singh	397

Contents of Volume I · xvii

21	Symbiosis in Diversity: The Specific Character of Slovakia's Religious Landscape	415
	Juraj Majo	
22	Religion Inscribed in the Landscape: Sacred Sites, Local Deities and Natural Resource Use in the Himalayas	439
	Elizabeth Allison	
23	Suppression of Tibetan Religious Heritage	461
	P.P. Karan	
24	Archaeological Approaches to Sacred Landscapes and Rituals of Place Making	477
	Edward Swenson	
25	Sacred Caves of the World: Illuminating the Darkness	503
	Leslie E. Sponsel	
26	Space, Time and Heritage on a Japanese Sacred Site: The Religious Geography of Kōyasan	523
	Ian Astley	
27	Greening the Goddess: Sacred Landscape, History and Legislation on the Cāmuṇḍī Hills of Mysore	545
	Caleb Simmons	
28	Pollution and the Renegotiation of River Goddess Worship and Water Use Practices Among the Hindu Devotees of India's Ganges/Ganga River	557
	Sya Buryn Kedzior	
29	Privileged Places of Marian Piety in South America	577
	David Pereyra	
30	Sacred Space Reborn: Protestant Monasteries in Twentieth Century Europe	593
	Linda Pittman	
31	Cemeteries as a Template of Religion, Non-religion and Culture	623
	Daniel W. Gade	
32	Visualizing the Dead: Contemporary Cemetery Landscapes	649
	Donald J. Zeigler	
33	Sacred, Separate Places: African American Cemeteries in the Jim Crow South	669
	Carroll West	

Contributors

Jamaine Abidogun Department of History, Missouri State University, Springfield, MO, USA

Afe Adogame School of Divinity, University of Edinburgh, Edinburgh, Scotland, UK

Christopher A. Airriess Department of Geography, Ball State University, Muncie, IN, USA

Kaarina Aitamurto Aleksanteri Institute, University of Helsinki, Helsinki, Finland

Mikael Aktor Institute of History, Study of Religions, University of Southern Denmark, Odense, Denmark

Elizabeth Allison Department of Philosophy and Religion, California Institute of Integral Studies, San Francisco, CA, USA

Johan Andersson Department of Geography, King's College London, London, UK

Stephen W. Angell Earlham College, School of Religion, Richmond, IN, USA

J. Clark Archer Department of Geography, School of Natural Resources, University of Nebraska-Lincoln, Lincoln, NE, USA

Ian Astley Asian Studies, University of Edinburgh, Edinburgh, UK

Steven M. Avella Professor of History, Marquette University, Milwaukee, WI, USA

Yulier Avello COPEXTEL S.A., Ministry of Informatics and Communications, Havana, Cuba

Erica Baffelli School of Arts, Languages and Cultures, University of Manchester, Manchester, UK

Bakama BakamaNume Division of Social Work, Behavioral and Political Science, Prairie View A&M University, Prairie View, TX, USA

Josiah R. Baker Methodist University, Fayetteville, NC, USA

Economics and Geography, Methodist University, Fayetteville, USA

Holly R. Barcus Department of Geography, Macalester College, St. Paul, MN, USA

David Bassens Department of Geography, Free University Brussels, Brussels, Belgium

Ramon Bauer Wittgenstein Centre for Demography and Global Human Capital (IIASA, VID/ÖAW, WU), Vienna Institute of Demography/Austrian Academy of Sciences, Vienna, Austria

Whitney A. Bauman Department of Religious Studies, Florida International University, Miami, FL, USA

Gwilym Beckerlegge Department of Religious Studies, The Open University, Milton Keynes, UK

Michael Bégin Department of Global Studies, Pusan National University, Pusan, Republic of Korea

Demyan Belyaev Collegium de Lyon/Institute of Advanced Studies, Lyon, France

Alexandre Benod Research Division, Department of Japanese Studies, Université de Lyon, Lyon, France

John Benson School of Teaching and Learning, Minnesota State University, Moorhead, MN, USA

Sigurd Bergmann Department of Philosophy and Religious Studies, Norwegian University of Science and Technology, Trondheim, Norway

Rachel Berndtson Department of Geographical Sciences, University of Maryland, College Park, MD, USA

Martha Bettis Gee Compassion, Peace and Justice, Peace and Justice Ministries, Presbyterian Mission Agency, Presbyterian Church (USA), Louisville, KY, USA

Warren Bird Research Division, Leadership Network, Dallas, TX, USA

Andrew Boulton Department of Geography, University of Kentucky, Lexington, KY, USA

Humana, Inc., Louisville, KY, USA

Kathleen Braden Department of Political Science and Geography, Seattle Pacific University, Seattle, WA, USA

Namara Brede Department of Geography, Macalester College, St. Paul, MN, USA

John D. Brewer Institute for the Study of Conflict Transformation and Social Justice, Queen's University Belfast, Belfast, UK

Laurie Brinklow School of Geography and Environmental Studies, University of Tasmania, Hobart, Australia

Interim Co-ordinator, Master of Arts in Island Studies Program, University of Prince Edward Island, Charlottetown, PE Canada

Dave Brunn Language and Linguistics Department, New Tribes Missionary Training Center, Camdenton, MO, USA

Stanley D. Brunn Department of Geography, University of Kentucky, Lexington, KY, USA

David J. Butler Department of Geography, University of Ireland, Cork, Ireland

Anne Buttimer Department of Geography, University College Dublin, Dublin, Ireland

Éric Caron Malenfant Demography Division, Statistics Canada, Ottawa, Canada

Lori Carter-Edwards Gillings School of Global Public Health, Public Health Leadership Program, University of North Carolina, Chapel Hill, NC, USA

Clemens Cavallin Department of Literature, History of Ideas and Religion, University of Gothenburg, Göteborg, Sweden

Martin M. Checa-Artasu Department of Sociology, Universidad Autónoma Metropolitana, Unidad Iztapalapa, Mexico, DF, Mexico

Richard Cimino Department of Anthropology and Sociology, University of Richmond, Richmond, VA, USA

Paul Claval Department of Geography, University of Paris-Sorbonne, Paris, France

Paul Cloke Department of Geography, Exeter University, Exeter, UK

Kevin Coe Department of Communication, University of Utah, Salt Lake City, UT, USA

Noga Collins-Kreiner Department of Geography and Environmental Studies, Centre for Tourism, Pilgrimage and Recreation, University of Haifa, Haifa, Israel

Louise Connelly Institute for Academic Development, University of Edinburgh, Edinburgh, Scotland, UK

Thia Cooper Department of Religion, Gustavus Adolphus College, St. Peter, MN, USA

Catherine Cottrell Department of Geography and Earth Sciences, Aberystwyth University, Aberystwyth, UK

Thomas W. Crawford Department of Geography, East Carolina University, Greenville, NC, USA

Janel Curry Provost, Gordon College, Wenham, MA, USA

Seif Da'Na Sociology and Anthropology Department, University of Wisconsin-Parkside, Kenosha, WI, USA

Erik Davis Department of Religious Studies, Rice University, Houston, TX, USA

Jenny L. Davis Department of American Indian Studies, University of Illinois, Urbana-Champaign, Urbana, USA

Kiku Day Department of Ethnomusicology, Aarhus University, Aarhus, Denmark

Renée de la Torre Castellanos Centro de Investigaciones y Estudios Superiores en Antropologia Social-Occidente, Guadalajara, Jalisco, Mexico

Frédéric Dejean Institut de recherche sur l'intégration professionnelle des immigrants, Collège de Maisonneuve, Montréal (Québec), Canada

Veronica della Dora Department of Geography, Royal Holloway University of London, UK

Sergio DellaPergola The Avraham Harman Institute of Contemporary Jewry, The Hebrew University of Jerusalem, Mt. Scopus, Jerusalem, Israel

Antoinette E. DeNapoli Religious Studies Department, University of Wyoming, Laramie, WY, USA

Matthew A. Derrick Department of Geography, Humboldt State University, Arcata, CA, USA

C. Nathan DeWall Department of Psychology, University of Kentucky, Lexington, KY, USA

Jualynne Dodson Department of Sociology, American and African Studies Program, Michigan State University, East Lansing, MI, USA

David Domke Department of Communication, University of Washington, Seattle, WA, USA

Katherine Donohue M.A. Diplomacy and International Commerce, Patterson School of Diplomacy and International Commerce, University of Kentucky, Lexington, KY, USA

Lizanne Dowds Northern Ireland Life and Times Survey, University of Ulster, Belfast, UK

Kevin M. Dunn School of Social Sciences and Psychology, University of Western Sydney, Penrith, NSW, Australia

Claire Dwyer Department of Geography, University College London, London, UK

Patricia Ehrkamp Department of Geography, University of Kentucky, Lexington, KY, USA

Paul Emerson Teusner School of Media and Communication, RMIT University, Melbourne, VIC, Australia

Chad F. Emmett Department of Geography, Brigham Young University, Provo, UT, USA

Ghazi-Walid Falah Department of Public Administration and Urban Studies, University of Akron, Akron, OH, USA

Yasser Farrés Department of Philosophy, University of Zaragoza, Pedro Cerbuna, Zaragoza, Spain

Timothy Joseph Fargo Department of City Planning, City of Los Angeles, Los Angeles, CA, USA

Michael P. Ferber Department of Geography, The King's University College, Edmonton, AB, Canada

Tatiana V. Filosofova Department of World Languages, Literatures, and Cultures, University of North Texas, Denton, TX, USA

John T. Fitzgerald Department of Theology, University of Notre Dame, Notre Dame, IN, USA

Colin Flint Department of Political Science, Utah State University, Logan, UT, USA

Daniel W. Gade Department of Geography, University of Vermont, Burlington, VT, USA

Armando Garcia Chiang Department of Sociology, Universidad Autónoma Metropolitana Iztapalapa, Iztapalapa, Mexico

Jeff Garmany King's Brazil Institute, King's College London, London, UK

Martha Geores Department of Geographical Sciences, University of Maryland, College Park, MD, USA

Hannes Gerhardt Department of Geosciences, University of West Georgia, Carrolton, GA, USA

Christina Ghanbarpour History Department, Saddleback College, Mission Viejo, CA, USA

Danilo Giambra Department of Theology and Religion, University of Otago-Te Whare Wānanga o Otāgo, Dunedin, New Zealand/Aotearoa

Banu Gökarıksel Department of Geography, University of North Carolina, Chapel Hill, NC, USA

Margaret M. Gold London Guildhall Faculty of Business and Law, London Metropolitan University, London, UK

Anton Gosar Faculty of Tourism Studies, University of Primorska, Portorož, Slovenia

Anne Goujon Wittgenstein Centre for Demography and Global Human Capital (IIASA, VID/ÖAW, WU), International Institute for Applied Systems Analysis (IIASA), Laxenburg, Austria

Vienna Institute of Demography/Austrian Academy of Sciences, Vienna, Austria

Alyson L. Greiner Department of Geography, Oklahoma State University, Stillwater, OK, USA

Daniel Jay Grimminger Faith Lutheran Church, Kent State University, Millersburg, OH, USA

School of Music, Kent State University, Kent, OH, USA

Zeynep B. Gürtin Department of Sociology, University of Cambridge, Cambridge, UK

Cristina Gutiérrez Zúñiga Centro Universitario de Ciencias Sociales y Humanidades, El Colegio de Jalisco, Zapopan, Jalisco, Mexico

Martin J. Haigh Department of Social Sciences, Oxford Brookes University, Oxford, UK

Anna Halafoff Centre for Citizenship and Globalisation, Deakin University, Burwood, VIC, Australia

Airen Hall Department of Theology, Georgetown University, Washington, DC, USA

Randolph Haluza-DeLay Department of Sociology, The Kings University, Edmonton, AB, Canada

Tomáš Havlíček Faculty of Science, Department of Social Geography and Regional Development, Charles University, Prague 2, Czechia

C. Michael Hawn Sacred Music Program, Perkins School of Theology, Southern Methodist University, Dallas, TX, USA

Bernadette C. Hayes Department of Sociology, University of Aberdeen, Aberdeen, Scotland, UK

Peter J. Hemming School of Social Sciences, Cardiff University, Cardiff, Wales, UK

William Holden Department of Geography, University of Calgary, Calgary, AB, Canada

Edward C. Holland Havighurst Center for Russian and Post-Soviet Studies, Miami University, Oxford, OH, USA

Beverly A. Howard School of Music, California Baptist University, Riverside, CA, USA

Martina Hupková Faculty of Science, Department of Social Geography and Regional Development, Charles University, Prague 2, Czechia

Tim Hutchings Post Doc, St. John's College, Durham University, Durham, UK

Ronald Inglehart Institute of Social Research, University of Michigan, Ann Arbor, MI, USA

World Values Survey Association, Madrid, Spain

Contributors

Marcia C. Inhorn Anthropology and International Affairs, Yale University, New Haven, CT, USA

Adrian Ivakhiv Environmental Program, University of Vermont, Burlington, VT, USA

Maria Cristina Ivaldi Dipartimento di Scienze Politiche "Jean Monnet", Seconda Università degli Studi di Napoli, Caserta, Italy

Thomas Jablonsky Professor of History, Marquette University, Milwaukee, WI, USA

Maria Jaschok International Gender Studies Centre, Lady Margaret Hall, Oxford University, Norham Gardens, UK

Philip Jenkins Institute for the Study of Religion, Baylor University, Waco, TX, USA

Wesley Jetton Student, University of Kentucky, Lexington, KY, USA

Shui Jingjun Henan Academy of Social Sciences, Zhengzhou, Henan Province, China

Mark D. Johns Department of Communication, Luther College, Decorah, IA, USA

James H. Johnson Jr. Kenan-Flagler Business School and Urban Investment Strategies Center, University of North Carolina, Chapel Hill, NC, USA

Lucas F. Johnston Department of Religion and Environmental Studies, Wake Forest University, Winston-Salem, NC, USA

Peter Jordan Austrian Academy of Sciences, Institute of Urban and Regional Research, Wien, Austria

Yakubu Joseph Geographisches Institut, University of Tübingen, Tübingen, Germany

Deborah Justice Yale Institute of Sacred Music, Yale University, New Haven, CT, USA

Akel Ismail Kahera College of Architecture, Art and Humanities, Clemson University, Clemson, SC, USA

P.P. Karan Department of Geography, University of Kentucky, Lexington, KY, USA

Sya Buryn Kedzior Department of Geography and Environmental Planning, Towson State University, Towson, MD, USA

Kevin D. Kehrberg Department of Music, Warren Wilson College, Asheville, NC, USA

Laura J. Khoury Department of Sociology, Birzeit University, West Bank, Palestine

Hans Knippenberg Department of Geography, Planning and International Development Studies, University of Amsterdam, Velserbroek, The Netherlands

Katherine Knutson Department of Political Science, Gustavus Adolphus College, St. Peter, MN, USA

Miha Koderman Science and Research Centre of Koper, University of Primorska, Koper-Capodistria, Slovenia

Lily Kong Department of Geography, National University of Singapore, Singapore, Singapore

Igor Kotin Museum of Anthropology and Ethnography, Russian Academy of Sciences, St. Petersburg, Russia

Katharina Kunter Faculty of Theology, University of Bochum, Bochum, Germany

Lisa La George International Studies, The Master's College, Santa Clarita, CA, USA

Shirley Lal Wijesinghe Faculty of Humanities, University of Kelaniya, Kelaniya, Sri Lanka

Ibrahim Badamasi Lambu Department of Geography, Faculty of Earth and Environmental Sciences, Bayero University Kano, Kano, Nigeria

Michelle Gezentsvey Lamy Comparative Education Research Unit, Ministry of Education, Wellington, New Zealand

Justin Lawson Health, Nature and Sustainability Research Group, School of Health and Social Development, Deakin University, Burwood, VIC, Australia

Deborah Lee Department of Geography, National University of Singapore, Singapore, Singapore

Karsten Lehmann Senior Lecturer, Science des Religions, Bayreuth University, Fribourg, Switzerland

Reina Lewis London College of Fashion, University of the Arts, London, UK

Micah Liben Judaic Studies, Kellman Brown Academy, Voorhees, NJ, USA

Edmund B. Lingan Department of Theater, University of Toledo, Toledo, OH, USA

Rubén C. Lois-González Departamento de Xeografía, Universidade de Santiago de Compostela, Galiza, Spain

Naomi Ludeman Smith Learning and Women's Initiatives, St. Paul, MN, USA

Katrín Anna Lund Department of Geography and Tourism, Faculty of Life and Environmental Sciences, University of Iceland, Reykjavik, Iceland

Avril Maddrell Department of Geography and Environmental Sciences, University of West England, Bristol, UK

Juraj Majo Department of Human Geography and Demography, Faculty of Sciences, Comenius University in Bratislava, Bratislava, Slovak Republic

Virginie Mamadouh Department of Geography, Planning and International Development Studies, University of Amsterdam, Amsterdam, The Netherlands

Mariana Mastagar Department of Theology, Trinity College, University of Toronto, Toronto, Canada

Alberto Matarán Department of Urban and Spatial Planning, University of Granada, Granada, Spain

René Matlovič Department of Geography and Applied Geoinformatics, Faculty of Humanities and Natural Sciences, University of Prešov, Prešov, Slovakia

Kvetoslava Matlovičová Department of Geography and Applied Geoinformatics, Faculty of Humanities and Natural Sciences, University of Prešov, Prešov, Slovakia

Hannah Mayne Department of Anthropology, University of Florida, Gainesville, FL, USA

Shampa Mazumdar Department of Sociology, University of California, Irvine, CA, USA

Sanjoy Mazumdar Department of Planning, Policy and Design, University of California, Irvine, CA, USA

Andrew M. McCoy Center for Ministry Studies, Hope College, Holland, MI, USA

Daniel McGowin Department of Geology and Geography, Auburn University, Auburn, AL, USA

James F. McGrath Department of Philosophy and Religion, Butler University, Indianapolis, IN, USA

Nick Megoran Department of Geography, University of Newcastle-upon-Tyne, Newcastle, UK

Amy Messer Department of Sociology, University of Kentucky, Lexington, KY, USA

Sarah Ann Deardorff Miller Researcher, Refugee Studies Centre, Oxford, UK

Kelly Miller Centre for Integrative Ecology, School of Life and Environmental Sciences, Deakin University, Burwood, VIC, Australia

Nathan A. Mosurinjohn Center for Social Research, Calvin College, Grand Rapids, MI, USA

Sven Müller Institute for Transport Economics, University of Hamburg, Hamburg, Germany

Erik Munder Institut für Vergleichende Kulturforschung - Kultur- u. Sozialanthropologie und Religionswissenschaft, Universität Marburg, Marburg, Germany

David W. Music School of Music, Baylor University, Waco, TX, USA

Kathleen Nadeau Department of Anthropology, California State University, San Bernadino, CA, USA

Caroline Nagel Department of Geography, University of South Carolina, Columbia, SC, USA

Pippa Norris John F. Kennedy School of Government, Harvard University, Cambridge, MA, USA

Government and International Relations, University of Sydney, Sydney, Australia

Orville Nyblade Makumira University College, Usa River, Tanzania

Lionel Obadia Department of Anthropology, Université de Lyon, Lyon, France

Daniel H. Olsen Department of Geography, Brigham Young University, Provo, UT, USA

Samuel M. Otterstrom Department of Geography, Brigham Young University, Provo, UT, USA

Barbara Palmquist Department of Geography, University of Kentucky, Lexington, KY, USA

Grigorios D. Papathomas Faculty of Theology, University of Athens, Athens, Greece

Nikos Pappas Musicology, University of Alabama, Tuscaloosa, AL, USA

Mohammad Aslam Parvaiz Islamic Foundation for Science and Environment (IFSE), New Delhi, India

Valerià Paül Departamento de Xeografía, Universidade de Santiago de Compostela, Galiza, Spain

Miguel Pazos-Otón Departamento de Xeografía, Universidade de Santiago de Compostela, Galiza, Spain

David Pereyra Toronto School of Theology, University of Toronto, Toronto, Canada

Bruce Phillips Loucheim School of Judaic Studies at the University of Southern California, Hebrew Union College-Jewish Institute of Religion, Los Angeles, CA, USA

Awais Piracha School of Social Sciences and Psychology, University of Western Sydney, Penrith, NSW, Australia

Linda Pittman Department of Geography, Richard Bland College of the College of William and Mary, Petersburg, VA, USA

Richard S. Pond Department of Psychology, University of North Carolina, Wilmington, NC, USA

Carolyn V. Prorok Independent Scholar, Slippery Rock, PA, USA

Steven M. Radil Department of Geography, University of Idaho, Moscow, ID, USA

Esther Long Ratajeski Independent Scholar, Lexington, KY, USA

Daniel Reeves Faculty of Science, Department of Social Geography and Regional Development, Charles University, Prague 2, Czechia

Arthur Remillard Department of Religious Studies, St. Francis University, Loretto, PA, USA

Claire M. Renzetti Department of Sociology, University of Kentucky, Lexington, KY, USA

Friedlind Riedel Department of Musicology, Georg-August-University of Göttingen, Göttingen, Germany

Sandra Milena Rios Oyola Department of Sociology and the Compromise after Conflict Research Programme, University of Aberdeen, Aberdeen, Scotland, UK

C.K. Robertson Presiding Bishop, The Episcopal Church, New York, NY, USA

Arsenio Rodrigues School of Architecture, Prairie View A&M University, Prairie View, TX, USA

Andrea Rota Institute for the Study of Religion, University of Bern, Bern, Switzerland

Rainer Rothfuss Geographisches Institut, University of Tübingen, Tübingen, Germany

Jeanmarie Rouhier-Willoughby Department of Modern and Classical Languages, Literatures and Cultures, University of Kentucky, Lexington, KY, USA

Rex J. Rowley Department of Geography-Geology, Illinois State University, Normal, IL, USA

Bradley C. Rundquist Department of Geography, University of North Dakota, Grand Forks, ND, USA

Simon Runkel Department of Geography, University of Bonn, Bonn, Germany

Joanna Sadgrove Research Staff, United Society, London, UK

Michael Samers Department of Geography, University of Kentucky, Lexington, KY, USA

Åke Sander Department of Literature, History of Ideas and Religion, University of Gothenburg, Göteborg, Sweden

Xosé M. Santos Departamento de Xeografía, Universidade de Santiago de Compostela, Galiza, Spain

Alessandro Scafi Medieval and Renaissance Cultural History, The Warburg Institute, University of London, London, UK

Anthony Schmidt Department of Communication Studies, Edmonds Community College, Edmonds, WA, USA

Mallory Schneuwly Purdie Institut de sciences sociales des religions contemporaines, Observatoire des religions en Suisse, Université de Lausanne – Anthropole, Lausanne, Switzerland

Anna J. Secor Department of Geography, University of Kentucky, Lexington, KY, USA

Hafid Setiadi Department of Geography, University of Indonesia, Depok, West Java, Indonesia

Fred M. Shelley Department of Geography and Environmental Sustainability, University of Oklahoma, Norman, OK, USA

Ira M. Sheskin Department of Geography and Regional Studies, University of Miami, Coral Gables, FL, USA

Lia Dong Shimada Conflict Mediator, Methodist Church in Britain, London, UK

Caleb Kwang-Eun Shin ABD, Korea Baptist Theological Seminary, Daejeon, Republic of Korea

J. Matthew Shumway Department of Geography, Brigham Young University, Provo, UT, USA

Dmitrii Sidorov Department of Geography, California State University, Long Beach, Long Beach, CA, USA

Caleb Simmons Religious Studies Program, University of Arizona, Tucson, AZ, USA

Devinder Singh Department of Geography, University of Jammu, Jammu, Jammu and Kashmir, India

Rana P.B. Singh Department of Geography, Banaras Hindu University, Varanasi, UP, India

Nkosinathi Sithole Department of English, University of Zululand, KwaZulu-Natal, South Africa

Vegard Skirbekk Wittgenstein Centre for Demography and Global Human Capital (IIASA, VID/ÖAW, WU), International Institute for Applied Systems Analysis, Laxenburg, Austria

Alexander Thomas T. Smith Department of Sociology, University of Warwick, Coventry, UK

Christopher Smith Independent Scholar, Tecumseh, OK, USA

Ryan D. Smith Compassion, Peace and Justice Ministries, Presbyterian Ministry at the U.N., Presbyterian Mission Agency, Presbyterian Church (USA), New York, NY, USA

Sara Smith Department of Geography, University of North Carolina, Chapel Hill, NC, USA

Leslie E. Sponsel Department of Anthropology, University of Hawaii, Honolulu, HI, USA

Chloë Starr Asian Christianity and Theology, Yale Divinity School, New Haven, CT, USA

Jeffrey Steller Public History, Northern Kentucky University, Highland Heights, KY, USA

Christopher Stephens Southlands College, University of Roehampton, London, UK

Jill Stevenson Department of Theater Arts, Marymount Manhattan College, New York, NY, USA

Anna Rose Stewart Department of Religious Studies, University of Kent, Canterbury, UK

Nancy Palmer Stockwell Senior Contract Administrator, Enerfin Resources, Houston, TX, USA

Robert Strauss President and CEO, Worldview Resource Group, Colorado Springs, CO, USA

Tristan Sturm School of Geography, Archaeology and Palaeoecology, Queen's University Belfast, Belfast, UK

Edward Swenson Department of Anthropology, University of Toronto, Toronto, ON, Canada

Anna Swynford Duke Divinity School, Duke University, Durham, NC, USA

Jonathan Taylor Department of Geography, California State University, Fullerton, CA, USA

Francis Teeney Institute for the Study of Conflict Transformation and Social Justice, Queen's University Belfast, Belfast, UK

Mary C. Tehan Stirling College, University of Divinity, Melbourne, Australia

Andrew R.H. Thompson The School of Theology, The University of the South, Sewanee, TN, USA

Scott L. Thumma Professor, Department of Sociology, Hartford Seminary, Hartford, CT, USA

Meagan Todd Department of Geography, University of Colorado, Boulder, CO, USA

Soraya Tremayne Fertility and Reproduction Studies Group, Institute of Social and Cultural Anthropology, University of Oxford, Oxford, UK

Gill Valentine Faculty of Social Sciences, University of Sheffield, Sheffield, UK

Inge van der Welle Department of Geography, Planning and International Development Studies, University of Amsterdam, Amsterdam, The Netherlands

Herman van der Wusten Department of Geography, Planning and International Development Studies, University of Amsterdam, Amsterdam, The Netherlands

Robert M. Vanderbeck Department of Geography, University of Leeds, West Yorkshire, UK

Jason E. VanHorn Department of Geography, Calvin College, Grand Rapids, MI, USA

Viera Vlčková Department of Public Administration and Regional Development, Faculty of National Economy, University of Economics in Bratislava, Bratislava, Slovakia

Geoffrey Wall Department of Geography and Environmental Management, University of Waterloo, Waterloo, ON, Canada

Robert H. Wall Counsel, Spilman Thomas & Battle, Winston-Salem, NC, USA

Kevin Ward School of Theology and Religious Studies, University of Leeds, West Yorkshire, UK

Barney Warf Department of Geography, University of Kansas, Lawrence, KS, USA

Stanley Waterman Department of Geography and Environmental Studies, University of Haifa, Haifa, Israel

Robert H. Watrel Department of Geography, South Dakota State University, Brookings, SD, USA

Gerald R. Webster Department of Geography, University of Wyoming, Laramie, WY, USA

Paul G. Weller Research, Innovation and Academic Enterprise, University of Derby, Derby, UK

Oxford Centre for Christianity and Culture, University of Oxford, Oxford, UK

Cynthia Werner Department of Anthropology, Texas A&M University, College Station, TX, USA

Geoff Wescott Centre for Integrative Ecology, School of Life and Environmental Sciences, Deakin University, Burwood, VIC, Australia

Carroll West Center for Historic Preservation, Middle Tennessee State University, Murfreesboro, TN, USA

Gerald West School of Religion, Philosophy and Classics, University of KwaZulu-Natal, Scottsville, South Africa

Mark Whitaker Department of Anthropology, University of Kentucky, Lexington, KY, USA

Thomas A. Wikle Department of Geography, Oklahoma State University, Stillwater, OK, USA

Justin Wilford Department of Geography, University of California, Los Angeles, CA, USA

Joseph Witt Department of Philosophy and Religion, Mississippi State University, Mississippi State, MS, USA

John D. Witvliet Calvin Institute of Christian Worship, Calvin College and Calvin Theological Seminary, Grand Rapids, MI, USA

Teresa Wright Department of Political Science, California State University, Long Beach, CA, USA

Ernest J. Yanarella Department of Political Science, University of Kentucky, Lexington, KY, USA

Yukio Yotsumoto College of Asia Pacific Studies, Ritsumeikan Asia Pacific University, Beppu, Oita, Japan

Samuel Zalanga Department of Anthropology, Sociology and Reconciliation Studies, Bethel University, St. Paul, MN, USA

Donald J. Zeigler Department of Geography, Old Dominion University, Virginia Beach, VA, USA

Shangyi Zhou School of Geography, Beijing Normal University, Beijing, China

Teresa Zimmerman-Liu Departments of Asian/Asian-American Studies and Sociology, California State University, Long Beach, CA, USA

Part I
Introduction

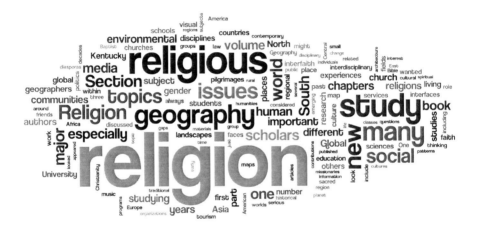

Chapter 1
Changing World Religion Map: Status, Literature and Challenges

Stanley D. Brunn

The Pew Research Center's Forum on Religion and Public Life (www.pewforum.org/global-religious-landscape.aspx) published a report on 18 December 2012 that reported eight of ten people (84 % of 6.9 billion in 2010) worldwide identify with some religious group. This study was based on polling results in more than 230 countries and territories in which adults were asked questions about their religious affiliation. Other highlights were:

- A number of religious groups are heavily concentrated in the Asia-Pacific region: Hindus (90 %), Buddhists (99 %), folk or traditional religions (90 %) and members of other religions (89 %).
- Seventy-five percent of those unaffiliated also live in the Asia-Pacific region. This region also has 75 % of the Muslims (also 20 % live in the Middle East; 16 % in Sub Saharan Africa).
- Forty-four percent of the Jews live in North America and 41 % in the Middle East and North Africa. Israel is the only country with a Jewish majority.
- Seventy-three percent of the world's population live in countries where their religion groups makes up a majority of the population while only 27 % live as religions minorities.
- Hindus and Christians tend to live in countries where they are majorities.
- Most Muslims (73 %) and religious unaffiliated groups (71 %) live in countries where they are the predominant religious group. The largest number of unaffiliated groups lives in China.
- The median age of Muslims (23 years) and Hindus (26 years) is younger than the median age of the world population (28 years). The median ages for other groups are: Christians (30), other religions (32), folk or traditional (33), religiously unaffiliated (34), Buddhists (34) and Jewish (36).

S.D. Brunn (✉)
Department of Geography, University of Kentucky, Lexington, KY 40506, USA
e-mail: brunn@uky.edu

This report illustrates a finding that has been known to scholars of religion for some time, viz., that religion is an important part of the fabric of most people's daily lives on the planet, regardless of the primary faith belief system, ideological histories and legacies, the stage of economic development and political system. After all, the quest for a meaning to life, a hope to improve the human condition and the role that humans play in their associations and interactions with others and their spiritual links to environmental worlds around them have been topics of interest from the earliest humans in local settings to those living in global cultures today. Those same curiosity yearnings and spiritualities are raised by new believers and searchers among those living within established religious traditions, those in new-found worlds of religions/spiritual experiences as well as those firmly embedded in secular and postmodern societies. What is important for those who study those with religious beliefs, doctrines and religious establishments and or those with spiritual quests for the divine are that these traits are just as relevant today as they were 10, 25 or even 50 years ago.

The formal study of religion as a specific and serious line of inquiry has not always been widely recognized in the academic world as one that is important or needs to be taken seriously. Philosophers and historians took the lead in our thinking about the subject, less important were those in medicine, education, law and the natural and social sciences. Not studying the many dimensions of religion or faith and a religious or spiritual life does not mean they are not deserving of serious study. Rather as we have seen in many western societies, including public universities, the study of religion was not considered a serious topic for systematic and critical scholarship, including the science/religion interfaces. For whatever reasons, including the wish to "separate the daily living from the holy" or the blind rejection or serious study of religion by unquestioning "prevailing isms" (historical and contemporary), studying religion or the many dimensions of religion became less than popular or fashionable. It was and is a subject that requires interdisciplinary, interfaith and international levels of serious inquiry. Studying religion seriously and critically also needed to move scholars beyond their own traditional and often personal "theological and religious comfort zones," which are not always easy. This often meant going beyond the study of familiar topics, issues and themes, to explore new dimensions, interfaces or fusions of religion with subjects, disciplines, paradigms, and fields hitherto considerable unimaginable. Medicine, science, law, the media and evolutionary sciences are just a few of these fields. Many religious thinkers in our past, even recent past, have had limited training and expertise in these arenas; many serious studies are now coming from scholars in these fields.

My reading of the contemporary religious landscape is that there are some (probably many) deep, profound, fascinating, exciting and important subjects that await scholarly inquiry, many that were not even on the intellectual horizons two decades ago. There are new fields of interdisciplinary and transdisciplinary study, including those dealing with the human/environment interfaces (megadisasters, climate change, loss of biodiversity, etc.), diasporas, gender, holistic health care, the body, human sexuality, neurosciences, mass media and social media, virtual worlds, tourism, and popular culture that are emerging as suitable arenas within which to study religion. These include a study of places, beliefs, identities, experiences, the sacred, virtual worlds, organizations, institutions and belongingness. Some of these new fields are looking at religion through new lens, including evolutionary biology, postmodernism, visualization, environmental and planetary sciences and cyberspace. The emergence of these arenas for study by scholars from the social and life sciences, humanities and fine arts have ushered in new college/university degree programs, electronic listservs and access to new and mega databases, new journals (paper and electronic) and new publishing houses, social media networking and novel ways of looking at old/familiar topics. In these emerging areas there are junior and senior scholars from around the world and in all major world religions. Some are working within their disciplines, others with those outside on some truly innovative, interesting and intriguing ways of looking at religion and culture, society, and politics. Many examples of these new topics appear in the contents of this book.

Another significant and recognized component in the study of religion that has evolved in the past decade is the importance of place, a concept of longstanding importance to those in the human and environmental geography communities. It is also a concept that has been introduced with some fresh insights by those studying religion and history, literature, drama, film studies, and the media (print and visual), not to mention its role in an understanding of identity, gender, the body, art (music, drama, photography, architecture etc.) and popular culture. Place has long been studied as a theme by those looking a pilgrimages, sacred places and landscapes and even the built environment, but even those studying the concept have been invigorated by scholars from outside geography who look at places and sacred worlds, human-nature experiences, diasporas, the media and religious communities with fresh insights and conceptual frameworks. That the origins of all major world religions are associated with specific place and regional attributes, that old and new diasporas mix in different place settings and that new age, animistic, agnostic and secular movements have specific place features attest to the importance of studying this overriding concept with some fresh and new insights.

1.1 Religion in Daily Life

If there was a global newspaper, electronic or paper, that was devoted to the subject of religion in today's world, it would (should) always be reporting events on all continents and human interest stories from many cultures. Reporters would not lack for coverage of major and minor events. Major events with religious/spiritual dimensions would include: interpreting the reinterpreting prehistoric and historic sacred texts, adherence to practices (strong or lenient), legal church/state issues, backgrounds to wars and conflicts, gender and faith issues, religious education, social media and religion, research by neurosciences on "the religious brain," and humanitarian efforts with a religions face to end poverty, racism, discrimination, social and environmental injustices.

The closest we have to such an information source is the daily (Monday through Friday) website of Religion News Service: http://www.religionnews.com with summarizes 15–20 stories about global events around the world and also provides links for each for additional information. This English site is a "must" for anyone wanting to keep apprised of global religion issues. It presents stories and accounts about the importance of religion in the personal and institutional fabric of the daily life about most residents on the planet, not just the English-speaking world. It illustrates and documents that religion as a topic is more important than western social scientists have long believed, including many who thought (not hoped) it would disappear as a major cultural fabric of a nation or people because of increased secularization, wealth, education and democracy. Such predictions or projections by western social scientists have been shown, and by other scholars especially recently, to be far off base. Religions issues, religious identity and religion in society and politics have, if anything, increased in importance in the past two decades. This observation is evident in the volume of recent research on religion and its interfaces with issues about globalization, conflict, development and environmental awareness.

1.2 Major Themes

The study of religion within the geography discipline, which forms the basis for this book project, has roots that go back to the mid-1960s. The first serious recognized effort to acknowledge religion in American geography is traced to the annual meeting of the AAG (Association of American Geographers), the largest organization of geography professionals, in 1976 or 1977 when a number of papers on religion topics were presented (Stoddard 2003). Mannfred Bűttner was an early organizer and leader, but so was Gisbert Rinschede. Bűttner founded the International Working Group on the Geography of Belief Systems, but it was never recognized officially by the IGU (International Geographical Union). In subsequent years sessions were organized along with the small Geography of the Bible Specialty Group. The first newsletter entitled "Geography of Religion/Belief Systems" appeared in 1977; the title of the organizing group and newsletter were changed later to the "Geography of Religion and Belief Systems" or GORABS as it is known today (Stoddard 2003; Stoddard and Prorok 2003). The group's newsletter is http://www.gorabs.org/ which publishes articles as well as contains information on sessions and presentations at the AAG annual meetings. The founders of GORABS, the newsletter and organizers of sessions included Robert Stoddard, Charles Heatwole, Edwin Gaustad, Jack Licate, Jane Ratcliffe, Klaus Gurgel, Manfred Bűttner, Carolyn Prorok, Robert Reed, Mary Lee Nolan, Richard Jackson, Barbara Weightman, Surinder Bhardwaj, Mohammed Hemmasi, Jeffrey Maddux, Thomas McCormick, Gregory Stein and Gisbert Rinschede. These individuals might be considered the first generation of geographers who pioneered the study of religion in geography. One of the early international conferences was held at Catholic University in Eichstätt, Germany in 1988. The names of many second and third generation geographers are listed in the bibliography that accompanies this introduction. GORABS became an official AAG Specialty Group in 1991 in large part because of these individuals' collective efforts. GORABS is a successful group that organizes sessions, has a keynote speaker and gives awards to deserving individuals. A similar organization that emerged in the past few years within the Royal Geographical Society and Institute of British Geographers is the Geographies of Religion, Spirituality and Faith Working Group (GRSFWG). The early organizers and leaders were Jenny Lund, Justin Beaumont, Richard Gale and David Butler.

In the early decades most of the sessions organized and papers presented at annual geography conferences dealt with cultural and historical topics. Topics such as pilgrimages, holy sites, rituals and sacred spaces were preferred; many of the scholars writing about these and related topics were trained in cultural and cultural-historical geography. Few were the studies that looked at membership patterns, missions and missionaries, territory and territoriality, geopolitics and religion, education, law, the media, and the intersections of nationalism, identity and culture. A shift in geographical thinking about religion occurred in the mid-to-late 1980s with studies about religion and images, identity and organizations and concerns about social justice, gender and equity, politics and law, environmental awareness and postmodern discourses.

My own thinking about the geography/religion interfaces has evolved over the years. While I have undertaken a number of studies on religion topics, sometimes with others, it was only one of several foci in my human and human/environmental geography research career. However, as I was observing what was transpiring with human and human/environmental geography in the 1990s and first decade of this century, I was discovering or rediscovering that there were openings for those interested in religion questions within the context of space, place, identity, landscape, territory, networking, cyberspace and human/nature interfaces.

In large part this volume represents two initiatives on my part. First, it is meant to expand significantly the study of religion topics for those in the humanities and in the social, environmental and policy sciences. That is, to discover openings or lacunae where the study of religion can and might be considered a legitimate focus for serious scholarly disciplinary and transdisciplinary inquiry. The second is to welcome and include as potential authors scholars whose work relates closely to questions about space, place, region, territory, networks, and human/nature inquiries. Maps and mapping are also to be part of this emerging scholarly discourse. To accomplish both initiatives, I reached out to friends and friends of friends for ideas about subject matter that fit within the title of this book in a very broad way and for scholars who might share both visions, that is, expand our thinking about religion in academic circles and offer original and meaningful contributors.

Early in the project's development (mid-2010), I had several major objectives in regards to content. I wanted to volume to be more than a disciplinary, or geographic, perspective. But I did wish to retain the importance of place, space, region, territory, human/nature interfaces and also maps and mapping in studying and understanding contemporary religious worlds. These, in my mind, are conceptual frameworks that scholars in various disciplines can and might use in studying religious phenomena, practices, processes, and events. This conviction has been more than supported as I observed the content of chapters coming in during the past couple years. I also wanted to be sure the volume was (a) interdisciplinary and transdisciplinary in topics and themes (there are differences between these two terms and each contributes significantly to what we can and might know and study), (b) international in topics as well as authors, and (c) multifaith and interfaith. From the outset I deliberately sought topics and authors from a wide variety of disciplines, fields and subfields and from those with interests and personal backgrounds in many of the world's major faiths.

As I proceeded with the above mental outline, I also thought of major thematic areas or subjects that would be important to include in such a volume. I wanted to be sure there were some fresh, new and innovative perspectives in looking at familiar topics that the social sciences and humanities have looked at previously. And I wanted these studies to have a solid theoretical or conceptual framework; it did not have to be grounded in geography, as other disciplines also provide foundations that are valuable in studying changes in and on the world religion map. I sought contributions about sacred places, pilgrimages, human/nature interfaces, missions and missionaries. But I wanted especially to see the study of religion "in and on the changing world map" address some new topics or little studied topics from religious scholarly communities around the world. For example, I wanted to look at issues about the impacts (positive and negative) of globalization, new diasporas, tourism, virtual religion, mass media and new social media, and questions about religion and gender, education, architecture, sports, music, food, fashion, theater, law, politics, peace and reconciliation. I wanted to find out if I could find someone to write about the content of religious materials in school texts, church property issues in Europe and elsewhere, megachurch developments inside and outside North America, new faces of Christianity in the Global South, new faces (even agnostics) in the North, religion and climate change debates, religious freedom and conflict, politics and religious voting. These were just a few of the topics that I thought merited inclusion here.

1.3 Organization of the Volume

In addition to this Introduction (Chap. 1), this volume has 207 chapters that are divided into 14 major parts, each which, in its own right could almost "stand alone" or constitute a separate volume. The major sections and the subjects of several chapters in each are:

Part I. Introduction

Part II. Nature, ethics and environmental change

- Planetary theologies
- Faith communities and climate change

- Responses to natural disasters
- Weber's Protestant ethic and ecological modernization

Part III. Sacred spaces and places

- Religion, ideology and globalization
- Sacred sites, heritage landscapes and place making
- Archaeological research into sacred places
- Cemeteries and visualizing the dead

Part IV. Pilgrimage landscapes and tourism

- Religion, nationalism and pilgrimage
- Landscapes and aesthetics of religious pilgrimages
- Tourism and spiritual journeys
- Postmodern museum pilgrimages

Part V. Education and changing worldviews

- Religion, science, education and the state
- Religion and language revitalization
- Missionaries, missions and mission trips
- Christian seminary library holdings

Part VI. Business, finance, economic development and law

- The faith/business nexus
- Islamic banking
- The economic logistics of the Church of Jesus Christ of Latter Day Saints
- Religion, moral exchange and investment

Part VII. Globalization, diversity and new faces in the Global North

- New diasporas and disaporic churches
- Immigration and religion issues
- Multifaith and secular cities
- California's new religious faces

Part VIII. Globalization, diversity and new faces in the Global South

- The global growth of Christianity and Islam
- Emerging missionary strategies
- Liberation theologies
- Global Pentecostal movements

Part IX. Secularization

- Urban secular spaces and forces
- Church property issues
- Geographies of atheism
- Marketing religion

Part X. Megachurches and architecture

- The global megachurch phenomenon
- Negotiating sacred spaces and the built environment
- Sacred space architecture
- Russian Orthodox churches in post-Soviet Russia

Part XI. Culture: museums, drama, fashion, food, music, sports and science fiction

- Marketing faith and fashion
- Landscapes and soundscapes
- Dietary laws and religion
- Religion's future through science fiction

Part XII. Organizations

- Multifaith responses to global environmental risks
- Evangelical geopolitics
- Religion's presence in the UN founding
- The global agenda of the World Council of Churches

Part XIII. Identity, gender and culture

- Religion, identity and gender
- Religion and issues of gay rights
- Islam and assisted reproduction
- Jewish urban intermarriage rates

Part XIV. Politics, reconciliation and advocacy

- Existential security and secularization
- Religious freedom indexes
- Faith communities and human trafficking
- Religion and peace building

Part XV. Virtual worlds and the visual media

- Virtual geographies and religion
- Religion, Facebook and social media
- Christian-atheist billboard wars
- Hollywood as secular religion

As expected, there is some overlap in the contents of chapters and sections. For example, secularization is a separate section, but also the processes are discussed in chapters on the New Faces in the Global North and South as well as the chapters on Nature, Ethics and Environmental Changes. The same is true with the overlap of chapters in Organizations, Politics, and Culture. The reader is encouraged to study the major themes discussed in individual chapters as well as accompanying illustrative materials: maps, photos, graphs and tables, all which are important ingredients in understanding the topics at hand.

All chapters in the volume are original. Some authors, understandably, have expanded on ideas from previous chapters, articles and books and elaborated on them here.

1.4 Authors, Countries of Origin and Disciplines

The effort to ensure this volume was international, interdisciplinary and interfaith paid off with many personal contacts and also with the assistance of many friends and friends of friends. The authors come from nearly 40 different countries: all continents and major world regions are represented. Religious topics and issues are discussed in more than 60 different countries. While the majority covers the Global North, especially Europe and the United States and Canada, I am very pleased with the number of chapters dealing with Africa, Asia and Latin America. The authors come from a wide variety of disciplines, all which have important perspectives, theories, models and approaches that are valuable in studying the world's changing religion maps. There are geographers, anthropologists, sociologists, psychologists, economists and political scientists as well as historians, philosophers, educators, architects and those who study music, theater, linguistics and education. There are a good number who teach in departments of religion and in divinity schools and others who work in museums and for religious organizations. And as one would expect in such a transdisciplinary and interdisciplinary volume, there are scholars who work in gender studies, environmental programs, tourism, planning and policy, linguistics, cross-cultural and global studies. Many of the authors have regional religious interests. Finally, I am pleased with the age mix and professional ranks of authors. While the majority are senior professors in major and minor universities, there are also those who are early in their professional training and career-development. A few are also undergraduate students. Almost half the authors are women.

1.5 Expectation and Uses

It is my hope that this volume will become *a* or *the* standard reference source for those studying religion and religion and society, culture, politics and the environment in the coming decade and beyond. The book, in its entirety or portions, could be used in classes on Religion and Society, World Religions, Religion and the Environment, Religion and the Media, Religion in the Global North and South, and also Religion and Culture. Chapters could be used in disciplinary courses about religion and history, sociology, geography, economics, gender studies, the media as well as the religion in the First and Third Worlds. They could be used in advanced undergraduate classes, in disciplinary and interdisciplinary seminars, in divinity schools and seminaries as well as readings for transdisciplinary and interfaith conferences and seminars. I also hope that many public and private university libraries will purchase copies for student and faculty use. The chapters are valuable, as are the chapter bibliographies, which contain recent articles, chapters and books, and also websites.

1.6 Where Do We Go from Here?

The curious or casual reader of the Table of Contents might ponder this question, especially since this book includes many topics not studied previously by scholars from various disciplines and many topics that are studied in some new theoretical framework. As someone who is constantly working at the peripheries of his own discipline and ferreting out common ground with those in related fields and subfields, I constantly ask myself, as well ask my own students, what are some legitimate topics that remain unstudied? The answer is simple. There are many topics that could and would legitimately merit study by undergraduate and graduate students, by doctoral and postdoctoral students and by junior and senior faculty from colleges, universities, seminaries and divinity schools around the world. Permit me to suggest only a few of many that I could list. Each is listed under a section in the book:

Part I. Introduction

Part II. Nature, ethics and environmental change

- Packaging nature for visual spiritual experiences
- Changing God/Nature debates: blessing, curses, and human error
- The westernization of eastern religions

Part III. Sacred spaces and places

- Secular/religious pilgrimages
- Virtual pilgrimages and religious tourism
- Archaeological religious pilgrimages

Part IV. Pilgrimage landscapes and tourism

- Religion, nationalism and pilgrimage
- Landscapes and aesthetics of religious pilgrimages
- Tourism and spiritual journeys
- Postmodern museum pilgrimages

Part V. Education and changing worldviews

- Religion and debates within evolutionary sciences
- Religious content in home school textual materials
- Translating religious texts
- Minority religions' role in national history

1 Changing World Religion Map: Status, Literature and Challenges

Part VI. Business, finance, economic development and law

- Religion/law/property issues in Europe, Middle East and elsewhere
- Marketing atheism and secularism
- The place/region origins of church/state constitutional cases
- Life and death place landscapes and the law

Part VII. Globalization, diversity and new faces in the Global North

- The De-Christianizing of America: regional variations
- Aperiodic worshippers: Christers (Christmas and Easter worshippers)
- The disestablishment of church/state religion in the West
- The spatial/temporal dynamics of church dropouts

Part VIII. Globalization, diversity and new faces in the Global South

- Christianity in China
- Cities in the Global South as crucibles for religious diversity
- Social media use by new diasporas in the urban South

Part IX. Secularization

- Marketing secularization in the Bible Belt
- The secularization of the urban South
- Secular issues the Catholic South
- Secular music threads in religious music

Part X. Megachurches and architecture

- Secular schools of architecture
- The built environments of new diasporic church communities
- The Islamic megamosque phenomenon
- Recycled church properties

Part XI. Culture: museums, drama, fashion, food, music, sports and science fiction

- Cartooning religious issues
- Revival of religious music traditions in Central Asia
- Church athletic leagues (women and men)
- Tours of religious musicians and museum exhibits

Part XII. Organizations

- The uses of GIS in defining religious communities
- Declining Southern Baptist membership in the U.S.
- Self-help religious community networks

Part XIII. Identity, gender and culture

- Religious communities' access to health services (abortion, genetic counseling, mental health, euthanasia)
- Interfaith and interracial marriage patterns
- Same-sex referenda
- Gender and Islam, Hinduism and Buddhism

Part XIV. Politics, reconciliation and advocacy

- Use of religion in the rhetoric of war
- Religious issues and the International Court of Justice
- The disappearing white evangelical vote in US presidential elections
- Religious political parties (old and new)

Part XV. Virtual worlds and the visual media

- The "electronic" missionary in a social media world
- The stamp issues of the Vatican, Saudi Arabia and Israel
- Use of social media in religious protests (in many locations)
- Religious (faith and interfaith) museum exhibitions in Russia, China, Turkey and elsewhere
- Social media, religious identity and belonging

1.7 A Map Constantly Changing

Maps might be considered as "frozen visual documents" in time, which is understandable when we often look at the contemporary patterns or processes. A more useful way to look at any current map, especially of a topic that is as fluid, dynamic and ever-changing as one with a religious content, is that it is one that constantly needs updating, reclassifying and relabeling of categories and place/region features, and also new and distinguishing features which previously were unknown or not considered salient. Many city, regional and national maps with a religious content have a palimpsest quality, that is, layers of unknown (and unmapped) place features.

Many of the subjects discussed in this book are accompanied by maps, some of which are straightforward and easy to understand and others which are more complex because of the processes underlying the pattern or complexities presented by the author. One should examine many of these maps carefully and ask a series of "what if" questions about the subject matter and about the processes that are behind the pattern. Even a casual inquiry into many of the map patterns will likely evoke many questions, not only cartographic, but also conceptual. Let me propose ten questions to challenge our thinking about what is on the forefront of future religion maps.

(a) Where are the Pentecostal groups in Latin America gaining at the expense of Catholicism?
(b) What new pilgrimage sites, routes and webpages are developed to promote religious tourism in Africa, South Asia or Europe?
(c) Where are the "new religious islands and archipelagos" of North African and Eastern European groups into west European cities? (The same question could be asked about the demise of relict religious communities.)
(d) How can one map an expanded "gendered religious landscape" in cities in the Global North and South?

(e) How has the "virtual religion" phenomenon affected church attendance, church belong and church shopping in North American cities?
(f) How can one map the increased Islamic presence (mosques, madrassas, religious shops, tourist sites, etc.) in expanding Muslim populated areas of Central Asia and Russia?
(g) How effective are secular marketing schemes (Hollywood, consumer products, and environmental ethics) in promoting interfaith and cross-cultural understanding?
(h) Where among the world's major religions are the epicenters of transfaith, interfaith, and ecumenical music and drama performances and educational initiatives at regional and global scales?
(i) What new faces of Christianity are emerging in Africa and Asia? Of Islam in Europe and the Global North? And Hinduism and Buddhism outside South Asia?
(j) What will be the "faces" of secularism that will emerge in the Global South and where will these faces first appear (cities and regions)? And what will be the key landscape features of places (in many regions on the planet) with growing numbers of atheists, agnostics, and those unaffiliated with any formal religion?

1.8 The Next Challenges

The challenges facing scholars and students of religion in the future are at least three; there are probably more that can be considered following a reading of this section. The first is to recognize that there are scholarly communities, especially disciplinary, that have yet to learn about the role of religion in society, culture, education and politics. As this volume illustrates and documents with many innovative and path-breaking initiatives, there remain many gaps in our study about religious groups and organizations, the origins and diffusions of minor and major faith belief systems, the role and influences or religion in people's daily work, worship, and professional lives, and the impacts of religious thinking, actions and policies on individuals, their communities and wider society. The gaps in our intellectual arenas exist at all levels, from individuals to families, to local communities, to national scenes and also at world scales. One way to think about these gaps is to consider a large matrix in which the rows are major world regions, subregions or even countries on the planet. One might think of 15 or 25 rows. The columns would be the many topics and issues discussed in this volume. There might be 25 or 40 columns. One could then fill in the cells of the matrix with what is known or how much (a little or much) about a given topic and region. Completion of this matrix would likely show there are many empty cells; the emptiness would not only be in countries and regions in the Global South, but also many in the Global North. Filling in these empty cells would be among the challenges of those studying religion in the coming decades. Those contributions will come, as in the case of this volume, from junior and senior scholars around the world, not only in the social sciences, humanities and fine arts and those in religion departments and divinity schools, but those in new and emerging environmental, health, gender, human sexuality, human behavior, linguistics, tourism, education, visual media, the fine arts, architecture and public policy programs.

A second challenge echoes a point made earlier in this chapter. That is, the subject of religion needs to be taken seriously by those in the universities around the world. Rather than being the topic of interest of only a few scholars or only a few disciplines, the subject is one that calls for major contributions by disciplinary, interdisciplinary and transdisciplinary scholars worldwide and of all major and minor faith belief systems. As more than one author has pointed out in her/his chapter, there remain significant gaps in existing knowledge about the role of religion in human behavior, our cultures, our economies, political institutions and governmental organizations as well as history and prehistory. From a geographical perspective the scholarly materials of published materials (books and articles) are grossly uneven; vast gaps exist in our prehistorical, historical and contemporary understandings of Central, West and Southern Africa, Central and East Asia, coastal and interior Latin America, and I would even submit, Mediterranean Europe-North Africa and the U.S. and Canada. The unevenness is further evident in our knowledge about early Christianity in the South, the spread of Islam into East Africa and Southeast Asia, the role of the Silk Road as a transfer route for religions of the East, West and South Asia, and the role of missionaries in new diasporic communities as producers (good and bad, including cartography, lithographs and photographs) of geographic knowledge about other places, peoples and their religions. The reluctance of scholars in the aforementioned disciplines and fields to study may be attributed to personal dislike or distaste for studying religion (often from unfortunate and even childhood experiences) or religion subjects or they did not think it was fashionable (help us if that becomes a guiding light in our research agendas!!) or simply their belief or that they did not consider it a meaningful enough subject for intellectual inquiry. Or they have had deep philosophical and ideological reasons for steering clear of studying religion. While these may be given as reasons, these are insufficient excuses for not studying a subject or topic that has been, is, and will be an important part of the vast majority of the planet's population and human behavior today and in the future. It could even be argued persuasively that even if one did not consider her/himself to be religious or have religious/spiritual feelings those sentiments are still not a sufficient reason to study the subject on purely intellectual grounds.

Third, the opportunities for the study of religion are, it seems to be, more propitious than at any time in recent memory. The subjects discussed in this volume and the authors' intellectual backgrounds attest to the growing importance of studying religion at all scales: individual, local, community, national and global. The opening up of the study of religion from not only those trained in a specific discipline, but who are engaged in interdisciplinary and interfaith programs are very positive signs. The study of religion and faith issues can and should be part and parcel of those who study issues about human behavior, whether physical or mental health, reproduction, genetic counseling, neurosciences, stem cell research, organ transplants or end-of-life decisions. The science/religion interfaces are important to more than those in the evolutionary sciences (anthropology, biology, physical geography, geology and astronomy), but also those who work in public/private partnerships, physical and psyche caring communities and in educating the planet's next generation. Related to this point is the growing importance of environmental programs which not only look at issues about environmental quality, natural disaster preparedness, loss of biodiversity, human risk, and climate change, but issues about ethics and values, all which are within the foundations of many world religions and are important in understanding the human condition. The print and visual media can also be singled out as playing increasing important roles in the dissemination of information about religion (also too often given short shrift by news producers and organizations), but also the production (especially visual) of religious news items. The social media, as several authors point out, offer opportunities for the study of religion and its impacts on individuals, identities, families, learning, religious and political organizations (ad hoc and permanent), and cultures. Many disciplines studying religion can, should and will make contributions to the impacts of information/communication technologies (ICTs), many which are inexpensive, portable, widely available and impacting individuals, gender roles, households, work places, schools as well as the public arenas. The widespread dispersal of new information about any subject, the speed with which information is communicated and the growing visual impact of information on public opinion and policy are especially important in understanding traditional institutions, such as religion, education and government, but also the new forms, shapes, and directions that are emerging and with consequences intended and not intended.

What is important in both the near and distant future in the study of religion and the changing world religion map is the realization that the final map has not been produced. If such a map existed, it would be out of date the moment it appeared in print, visual or some other textual form. A preferred way to look at future maps of world religion or the world religion map is one where there is ample fluidity in the topic addressed, the categories used and patterns mapped.

I welcome correspondence from authors, scholars and students of religion about missing topics that should appear in any subsequent volume. From my perspective there are clearly many more gaps or lacunae in the geographies of religions (both words plural) than what we know. Only through the work of disciplinary and transdisciplinary scholars and students and those from different faith traditions and living in different regions will we be able to fill in some of those gaps.

1.9 The Bibliography

The bibliography that follows illustrates the research contributions geographers, especially, and others have made to the study of religion. It is not meant to be an exhaustive list, but it is fairly complete, especially in English language geography sources about religion, of all relevant and published books, atlases, chapters and articles; only a few theses and dissertations are listed. The bibliography is intended as a guide to aid those interested in knowing the contributions of the human and environmental geography communities and to those pursuing the study of places, landscapes, pilgrimages, nature-society relations, and regional changes in denominational shifts and in faith belief systems in geography and related disciplines. The sources by nongeographers will aid those in the geography communities and the geography references should also aid in other social sciences and the humanities, including those studying theology and spiritual topics. One would also be well advised to consult disciplinary journals, many which have their contents available electronically, but also interdisciplinary journals. In the latter group I would especially recommend *Christian Scholar's Review; Conversations in Religion and Theology; Culture and Religion: An Interdisciplinary Journal, Ecology: Journal of Religion, Nature and the Environment; Fieldwork in Religion; International Journal of Public Theology; Journal for the Scientific Study of Religion; Journal of Church and State; Journal of the American Academy of Religion; Journal of Contemporary Religion; Journal of Public Theology; Journal of Religion; Journal of Religion, Conflict and Peace; Journal of Religion and Health; Material Religion: The Journal of Objects, Art and Belief; Journal of Religion and Society; Journal of Southern Religion; Reviews in Religion & Theology, R & T: Religion & Theology; Studies in World Christianity;* and *Teaching Theology and Religion.*

Stanley D. Brunn
February 2014

Bibliography

Geography Journal Issues Devoted to Religion

Annals of the Association of American Geographers, 96 (1), (2006), 165–202. Forum on "Theorizing and Studying Religion." Editor: James Proctor.
Forum on "Theorizing and Studying Religion," James Proctor, Introduction to the Forum, 165–168
Adrian Ivakiv, Toward a geography of "religion," mapping the distribution of an unstable signifier, 169–175
Michael P. Ferber, Critical realism and religion: Objectivity and the insider/outsider problem, 176–181
Julian Holloway, Enchanted spaces: The science, affect, and geographies of religion, 182–187
James Proctor, Religion as trust in authority: Theocracy and ecology in the United States, 188–196
Anne Buttimer, Afterward: Reflections on geography, religion, and belief systems, 197–202
Geopolitics, 11 (2), (2006), 183–347. Issue on "Religion and Geopolitics." Editor: John Agnew.
John Agnew, Religion and Geopolitics, 183–191
Dijkink, Gertjan, When geopolitics and religion fuse: a historical perspective, 197–208
Ian Wallace, Territory, typology, and theology: Geopolitics and the Christian scriptures, 209–230
Tristan Sturm, Prophetic eyes: The theatricality of Mark Hitchcock's premillennial geopolitics, 231–255

Ethan Yorgasen & Dale Robertson, Mormonism's raveling and unraveling of the geopolitical thread, 256–279

Jefferson West, II, Religion as dissent: Geopolitics, geopolitical discussions within the recent publications of Fethullah Gulen, 280–299

Catherine DeBusser, From exclusiveness to inclusiveness: The changing politico- territorial situation of Span and its reflection on the national offerings to the Apostle Saint James from the Second Half of the Twentieth Century, 300–316

Dmitri Sidorov Gurst. M/ & Arweck, E. (Eds.) (2012). Religion and knowledge: Sociological perspective. Earnhem, Surry and Burlington, VT: Ashgate.

Post-Imperial Third Romes: Reconstruction of a Russian Orthodox geopolitical metaphor, 317–347

Christian Scholar's Review. 31 (4), 2002, 351–467. Theme issue on "Geography in Christian perspective." Guest editor: Janel Curry.

Janel Curry, Introduction, 353–368

Tom Bulten, Community and propinquity in church members, 359–376

Iain Wallace, Globalization: Discourse of destiny or denial? 377–392

Mark B. Bjelland, Until justice and stewardship embrace; or, how a geographer thinks about brownfield sites, 393–412

Henk Aay, Geography's cultural landscape school: A Reformational reading, 413–434

Kathleen Braden, Exploring the notion of "good" in Sack's Geographic Theory of Morality, 435–447

Southeastern Geographer. 2000. Special issue on "Geographies of Religion in the Changing South," Guest editors: S. D. Brunn and G. W. Webster.

Stanley D. Brunn & Esther Long, The worldviews of southern seminaries: images, mission statements, and curricula, 1–24

Gerald R. Webster, Geographic patterns of denominational affiliation in Georgia: 1970–1990: Population change and growing urban diversity, 25–51

Ira Sheskin, The Dixie diaspora: the "loss" of the small southern Jewish community, 52–74

International Geographical Union. 1998. *Religion, Ideology and Geographical Thoughts*. Prace Instytutu Geografii WSP Nr 3. Kilce Studies in Geography 3. IGU Commission on the History of Geographical Thought. Ute Wardenga and Witold Wilczynski (Eds.).

Introduction, 7–12

Izhak Schnell, Transformations in Zionist geographic myth, 13–30

Boleslaw Domanski, The manifestation of ideology and power in the urban landscape and in geographical teaching in the case of Poland, 31–44

Witold J. Wilczynski, A well trod path to no-where: the ideological burden of geography in Poland, 45–57

Jozsef Benede, Ideological constructs, social connections, and geographical thought, 58–64

Giiuseppe Campione, Monotheism and conflict in the Mediterranean Basin, 65–72

Abel Albet-Mas and Maria-Dolors Garcia-Ramon, Two gods, two shores, one space: ideological construction and geographical justification of Spanish colonialism in Morocco, 73–82

Ana Maria Liberali, Geography and evolution of Argentinean state policy, 83–86

Keicchi Takeuchi. Geography and Buddhism in Tsunesburo Makiguchi's thought, 87–95

Hong-key Yoon. Geomentalities as reflected in the New Zealand Maori creation myth and the Chinese Yin-Yang concepts, 96–102

Hiam Goren, The chase after the Bible: Individuals and institutions – and the study of the Holy Land, 103–115

Ute Wardenga, Religion, ideology and geographical thought. Recent advances in the historiography of German geography, 116–123

Macief Jakubowski, Carl Ritter's work and the prospects for a renaissance of the theistic and teleological approach in geography, 124–131

Florian Pitt, Wayside crosses and statues of saints in Polish landscape, 132–136

Mateusz Wiercinski, World in progress: landscape change and cultural shift in India, 137–142

Perla B. Zusman, Civilization, progress and geography of Gran Chaco, 143–152

Francis Harvey, Geographic integration: from holism to systems, 153–162
Marion Hercock, Holism and its applications in land management: A case study of Garden Island, Western Australia, 163–177
The Pennsylvania Geographer. 36 (1), (1998), 1–168. "Geography and Religion." Editors: Carolyn Prorok and Sandra Mather.
Notes from the Editors: Carolyn Prorok and Sandra Mather, 1–4
Kari Forbes-Boyle, It is all sacred: Foothill Konkow perceptions of sacred places, 5–29
J. Mc Kim Mulville, The symbolic landscape of Vijayanagara, 30–54
Rana P. B. Singh, Sacred journey and faithscape: An experience of the Panchakroshi pilgrimage, Varanasi (India), 55–91
Simon R. Potter, A geographical interpretation of Noah's flood based on the literature and legends, 92–123
Dennis P. Tobin, Moving a monastery in modern times: the Abbey of St. Walburga, Colorado, 124–134
Oscar H. Horst, Building blocks of a legendary belief: The Black Christ of Esquipulas, 1595–1995, 135–147
Johnathan Bascom, The religions geography of Evangelical Christians in North America, 148–168
The Pennsylvania Geographer. 35 (2), (1997). "Geography and Religion." Editors: Carolyn Prorok.
Notes from the Editor: Carolyn Prorok, 1–4
Adrian Cooper, Sacred interpretations of the ecophysiology in tropical forest canopies, 5–31
John R. Mather & Sandra P. Mather, Deborah – Prophetess and possible applied climatologist, 32–44
Etan Diamond, Places of worship: The historical geography of religion in a Midwestern city, 1930–1960, 45–68
Zoriah Jack Tharan, The Jewish American environmental movement: Stewardship, renewal and the greening of diaspora politics, 69–97
Carolyn V. Prorok & Clarissa T. Kimber, The Hindu temple gardens of Trinidad: Cultural continuity and change in a Caribbean landscape, 98–135
G. Rebecca Dobbs, Interpreting the Navajo sacred geography as a landscape of healing, 136–150
Gerald R. Webster, Religion and politics in the American South, 151–172
Robert Kuhlken, Sacred landscapes and settlement mythology in the Fiji Islands, 173–202

Journal Articles

Aay, H. (2002). Geography's cultural landscape school: A reformational reading. *Christian Scholar's Review, 31*, 413–434.
Abedibu, B. (2011). Origin, migration, globalisation and the missionary encounter of Britain's black majority churches. *Studies in World Christianity, 19*, 93–113.
Afe, A. (2013). 'Insider" and 'outsider' in African Christianities. *Studies in World Christianity, 19*, 1–4 (editorial).
Ahlin, L., Borup, J., Fibiger, M. Q., Kűhle, V. M., & Pedersen, R. D. (2012). Religious diversity and pluralism: Empirical data and theoretical reflections from the Danish pluralism project. *Journal of Contemporary Religion, 27*, 403–418.
Alshamsi, M. J. (2012). Islam and political reform in Saudi Arabia: The question for political change and reform. *Reviews in Religion and Theology, 19*, 405–407.
Anderson, J. (2013). Cathedrals of the surf zone: Regulating access to a space of spirituality. *Social and Cultural Geography, 8*, 954–972.
Anderson, L. C., & Englehart, M. (1974). A geographic appraisal of the North American Hutterite Brethren. *Geographical Survey, 2*, 53–71.

Anderson, P. B., Gundelach, P., & Lüchan, P. (2013). A spiritual revolution in Denmark? *Journal of Contemporary Religion, 28*, 385–400.

Andersson, J., Vanderbeck, R. M., Valentine, G., Ward, K., & Sadgrove, J. (2010). New York encounters: Religion, sexuality and the city. *Environment and Planning A, 43*(3), 618–633.

Andren, A. (2013). The significance of places in the Christianization of Scandinavia from a spatial point of view. *World Archaeology, 45*, 27–45.

Andrews, A. (1990). Religious geography of Union County, Georgia. *Journal of Cultural Geography, 10*, 1–19.

Andrews, A. C., & Paul, T. R. (1996). Geographic patterns of religion in Virginia. *Virginia Geographer, 27*, 1–17.

Arya, R. (2011). The neglected place of religion in contemporary western art. *Fieldwork in Religion, 6*, 27–46.

Azaryahu, M. (1996). The spontaneous formation of memorial space: The case of *Kikar Rabin*, Tel Aviv. *Area, 28*, 501–513.

Azaryahu, M., & Golan, A. (2001). (Re)Naming the landscape: The formation of the Hebrew map of Israel 1949–1960. *Journal of Historical Geography, 27*, 178–195.

Bacchetta, P. (2000). Sacred space and conflict in India: The Babri Masjib affair. *Growth and Change, 31*, 255–284.

Báckström, A. (2014). Religion in the Nordic countries: Between private and public. *Journal of Contemporary Religion, 29*, 61–74.

Bahr, H. M. (1982). Shifts in the denominational demography of Middletown, 1924–1977. *Journal for the Scientific Study of Religion, 21*, 99–114.

Bahram, M. (2013). Habermas, religion and public life. *Journal of Contemporary Religion, 28*, 353–367.

Bailey, C. A., Harvey, D. C., & Brace, C. (2007). Disciplining youthful Methodist bodies in nineteenth century Cornwall. *Annals of the Association of American Geographers, 97*, 142–157.

Baillie Smith, M., & Jenkins, K. (2012). Existing at the interface: Indian NGO activists in strategic cosmopolitans. *Antipode, 44*, 640–662.

Baillie Smith, M., & Laurie, N. (2011). International volunteering and development: Global citizenship and neoliberal professionalization today. *Transactions of the Institute of British Geographers, 36*, 545–559.

Bainbridge, W. S. (1982). Shaker demographics 1840–1900: An example of the use of U.S. census enumeration schedules. *Journal for the Scientific Study of Religion, 21*, 352–365.

Baird, J. (2013). Multifaith education in American theological schools: Looking back, looking ahead. *Teaching Theology and Religion, 16*, 309–321.

Baje, V. (2007). Creating ritual through narrative, place and performance in evangelical protestant pilgrimage in the Holy Land. *Mobilities, 2*, 393–412.

Bakar, I. A. (2013). Woman, man and God in modern Islam. *Reviews in Religion and Theology, 20*, 447–450.

Baly, D. (1979). Jerusalem: City of our solemnities: Politics of the Holy City. *Geographical Perspectives, 42*, 8–15.

Baran, E. B. (2011). Jehovah's Witness and post-Soviet religious policy in Moldova and the Transnistrian Moldovan Republic. *Journal of Church and State, 53*, 421–441.

Barcus, H. R., & Brunn, S. D. (2004). Mapping changes in denominational membership in the Great Plains, 1952–2000. *Great Plains Research, 14*, 19–48.

Barro, R., Hwang, J., & McCleary, R. (2010). Religious conversion in 40 countries. *Journal for the Scientific Study of Religion, 49*, 15–36.

Bartkowski, J. P., & Swearingen, W. S. (1997). God meets Gaia in Austin, Texas: A case study of environmentalism as implicit religion. *Review of Religious Research, 38*, 308–324.

Bartolini, N., Cris, R., MacKian, S., & Pile, S. (2013). Psychics, crystals, candles and cauldrons: Alternative spiritualities and the question of their esoteric economics. *Social and Cultural Geography, 14*, 367–388.

Basil, J. D. (2009). Problems of state and church in the Russian Federation: Three points of view. *Journal of Church and State, 51*, 211–235.

Batuman, B., Jazeel, T., Shuttleworth, I., Barr, P. J., & Gould, M. (2013). Minarets without mosques: Limits to the urban politics of neo-liberal Islamism. *Urban Studies, 50*, 72–79.

Bauer, J. T. (2012). U.S. religious regions revisited. *The Professional Geographer, 64*, 521–539.

Baviskar, A. (1999). Vanishing forests, sacred trees: A Hindu perspective on eco-consciousness. *Asian Geographer, 18*, 21–32.

Beaumont, J. (2008a). Faith action in urban social issues. *Urban Studies, 45*, 2019–2034.

Beaumont, J. (2008b). Introduction: Faith-based organizations and social issues. *Urban Studies, 45*, 2011–2017.

Beaumont, J., & Dias, C. (2008). Faith-based organizations and urban social justice in the Netherlands. *Tijdschrift voor Economische en Sociale Geografie, 99*, 382–392.

Bechtel, R. B. (1994). Varieties of religious places: An analysis of William James classic. *National Geographical Journal of India, 40*, 111–113.

Bedford-Strohm, H. (2007). Tilling and caring for the earth: Public theology and ecology. *International Journal of Public Theology, 1*(2), 230–248.

Ben-Arieh, Y. (1975). The growth of Jerusalem in the nineteenth century. *Annals of the Association of American Geographers, 65*, 252–269.

Berdichevsky, N. (1980). New Orleans churches: An index of changing urban social geography. *Ecumene, 12*, 44–54.

Berg, E. (1998). The Sherpa pilgrimage to Uomi Tsho in the context of the worship of the protector deities: Ritual practices, local meanings and this-worldly request. *Himalayan Research Bulletin, 18*, 19–34.

Berger, P. L. (2001). Reflections on the sociology of religion today. *Sociology of Religion, 62*, 443–454.

Berghammer, C., & Fliegenschnee, K. (2012). Developing a concept of Muslim religiosity: An analysis of everyday lived religion among female migrants in Austria. *Journal of Contemporary Religion, 29*, 89–104.

Best, M. (2011). Reconciliation and the web. *Journal of Religion, Conflict and Peace.* 5 http://www.religionconflictpeace.org/volume-5-issue-1-and-2-fall-2012-spring-2013

Bhardwaj, S. M. (1987). Single religious shrines, multireligious pilgrimages. *National Geographical Journal of India, 33*, 457–468.

Bhardwaj, S. M., & Rao, M. N. (1998). The temple as a symbol of Hindu identity in America? *Journal of Cultural Geography, 17*, 125–143.

Bhatt, B. L. (1977). The religions geography of south Asia: Some reflections. *National Geographical Journal of India, 23*, 26–39.

Bhattacharya, A. N. (1961). Geography and Indian religion. *National Geographer (India), 4*, 12–17.

Bibby, R. W. (1997). Going, going, gone: The impact of geographical mobility on religious involvement. *Review of Religious Research, 38*, 289–307.

Biddington, T. (2013). Towards a theological reading of multifaith spaces. *International Journal of Public Theology, 7*, 315–335.

Bigelow, B. (1986). The Disciples of Christ in antebellum Indiana: Geographical indicator of the Border South. *Journal of Cultural Geography, 7*, 49–58.

Bilska-Wodecka, E. (2006). From multi-confessional to mono-confessional state. State-church relations in Poland following World War II. *GeoJournal, 67*, 341–355.

Biswas, L. (1984). Evolution of Hindu temples in Calcutta. *Journal of Cultural Geography, 4*, 73–85.

Biswas, L. (1985). Religious landscapes and Hindu temples of Calcutta. *Geographical Review of India, 47*, 64–73.

Bjorkland, E. M. (1964). Ideology and culture exemplified in southwestern Michigan. *Annals of the Association of American Geographers, 54*, 227–241.

Black, B., & Eerdmans, W. B. (2011). Christian ethics in a technological era. *Conversations in Religion and Theology, 9*, 35–47.

Blakeman, J. C. (2006). The religious geography of religions expression: Local governments, courts and the first amendments. *Journal of Church and State, 48*, 399–422.

Boal, F. W. (1969). Territoriality on the Shankill-Falls divide, Belfast. *Irish Geography, 6*, 30–50.
Boal, F. W. (2002). Belfast walls within. *Politics Geography, 21*, 687–694.
Boal, F. S., & Livingstone, D. (1984). The frontier in the city: Ethnonationalism in Belfast. *International Journal of Political Science Research, 5*, 161–179.
Boal, F. W., & Livingstone, D. N. (1986). Protestants in Belfast: A view from the inside. *Contemporary Review, 248*, 169–175.
Bondi, L. (2013). Between Christianity and secularity: Counseling and psychotherapy provision in Scotland. *Social and Cultural Geography, 8*, 668–688.
Bonine, M. E. (1987). Islam and commerce: Waqf and the bazaar of Yadz, Iran. *Erdkunde, 41*, 182–196.
Borchert, T. (2010). Abbot's New House: Thinking about how religion works among Buddhists and ethnic minorities in southwest China. *Journal of Church and State, 52*, 112–137.
Bowen, D. (2002). John Terpstra and the sacramental in urban geography. *Literature and Theology, 16*, 188–200.
Bowen, J. R. (2004). Beyond migration: Islam as a transnational public space. *Journal of Ethnic and Migration Studies, 30*, 879–894.
Brace, C., Bailey, A. R., & Harvey, D. C. (2006). Religion, place, and space: A framework for investigating historical geographies of religious identities and communities. *Progress in Human Geography, 30*, 28–43.
Braden, K. (1998). On saving the wilderness: Why Christian stewardship is not sufficient. *Christian Scholar's Review, 28*, 254–269.
Braden, K. (1999). Description of the earth in four spiritual maps. *Geography of Religions and Belief Systems, 21*, 1–3.
Braudy, D. (2013). Artist's statement: Sacred Los Angeles. *Journal of the American Academy of Religion, 81*, 1–5.
Brauer, J. C. (1985). Regionalism and religion in America. *Church History, 54*, 366–378.
Brennan, V. L. (2012). Truly we have a good heritage: Musical meditation in Yoruba Christian diaspora. *Journal of Religion in Africa, 42*, 3–25.
Brennan, V. L. (2013). 'Up from the rivers;' hymns and historical consciousness in the Cherubim and Seraphim churches in Nigeria. *Studies in World Christianity, 19*, 31–49.
Brenneman, W. L. (1994). Croagh Patrick and Lough Derg: Sacred and loris space in two Irish pilgrimage sites. *National Geographical Journal of India, 40*, 115–121.
Brown, M. S. (2000). Estimating the size and distribution of South Asian religious population in Britain: Is there an alternative to a religion question in the census? *International Journal of Population Geography, 6*, 87–109.
Brown, R. K. (2010). Religion, economic concerns and African American immigration attitudes. *Review of Religious Research, 52*, 146–158.
Brubaker, R. (2013). Categories of analysis and categories of practice: A note on the study of Muslims in European countries of immigration. *Ethnic and Racial Studies, 36*, 1–8.
Bruce, S. (2009). The importance of social science in the study of religion. *Fieldwork in Religion, 4*, 7–28.
Bruce, S. (2013). Post-secularity and religion in Britain: An empirical assessment. *Journal of Contemporary Religion, 28*, 369–384.
Brunn, S. D. (1987). A world of peace and military landscapes. *Journal of Geography, 86*, 255–262.
Brunn, S. D. (2002). The World Council of Churches as a global actor: Ecumenical space as geographical space. *Geographica Slovonica, 34*, 65–77.
Brunn, S. D. (2010). Cartooning and Googling God and natural disasters: Iceland's volcanic eruption and Haiti's earthquake. *Mitteilungen der Österreichischen Geographischen Gesellschaft, 152*, 251–275.
Brunn, S. D., & Appleton, T. (1999). Wet-dry referenda in Kentucky and the persistence of prohibition forces. *Southeastern Geographer, 39*, 172–189.
Brunn, S. D., & Barcus, H. R. (2004). New perspectives on the changing religious diversity in the Great Plains. *Great Plains Research, 14*, 49–76.

Brunn, S. D., & Campbell, C. (2004). Differential locational harmony: The Cristo Redentor statue in the Uspallata Pass. *Political Geography, 23*, 41–69.

Brunn, S. D., & Leppman, E. (2003). America's learning about foreign places through the eyes of missionaries. Writings in the *Friends' Missionary Advocate, 1885–1933. Quaker Studies, 7*, 165–192.

Brunn, S. D., & Long, E. (2000). The worldviews of southern seminaries: Images, mission statements and curricula. *Southeastern Geographer, 40*, 1–22.

Brunn, S. D., & Wheeler, J. O. (1966). Notes on the geography of religious town names in the U.S. *Names, Journal of the American Name Society, 14*, 197–202.

Brunn, S. D., Byrne, S., McNamara, L., & Egan, A. (2011a). Belfast landscapes: From religious schism to conflict tourism. *FOCUS: American Geographical Society, 53*, 81–91.

Brunn, S. D., Webster, G. R., & Archer, J. C. (2011b). The Bible Belt in a changing south: Shrinking, relocating and multiple buckles. *Southeastern Geographer, 51*, 513–549.

Brush, J. E. (1949). The distribution of religious communities in India. *Annals of the Association of American Geographers, 39*, 81–89.

Brykczynski, R. (2014). Religions modernism, nationalism and anti-Semitism in Polish Catholicism and Egyptian Islam. *Journal of Religion and Society*.

Budge, S. (1974). Perception of the boundaries of the Mormon culture region. *Great Plains- Rocky Mountain Geographical Journal, 3*, 1–9.

Bugg, L. B. (2012). Religion on the fringe. The representation of space and minority religious facilities in the rural-urban fringe of metropolitan Sydney, Australia. *Australian Geographer, 43*, 273–289.

Bugg, L. B. (2013). Citizenship and belonging in the rural fringe: A case study of a Hindu temple in Sydney Australia. *Antipode, 45*, 1148–1166.

Busto, R. V. (1996). The gospel according to the model minority? Hazarding an interpretation of Asian American evangelical college students. *Amerasia Journal, 22*, 133–147.

Buttimer, A. (1990). Geography, humanism and global concern. *Annals of the Association of American Geographers, 80*, 1–33.

Buttimer, A. (1999). Humanism and relevance in geography. *Scottish Geographical Journal, 115*, 103–116.

Büttner, M. (1974). Religion and geography: Impulses for a new dialogue between religionswissenschaftlern and geography. *Numen, 21*, 165–196.

Büttner, M. (1979). The significance of the Reformation for the re-orientation of geography in Lutheran Germany. *History of Science, 17*, 151–169.

Büttner, M. (1980a). On the history and philosophy of the geography of religion in Germany. *Religion, 10*, 86–119.

Büttner, M. (1980b). Survey article on the history and philosophy of the geography of religion in Germany. *Religion*, 86–119.

Büttner, M. (1987). Kasche and Kant on the physicotheological approach to the geography of religion. *National Geographical Journal of India, 33*, 218–228.

Campbell, H. (2005a). Making space for religion in Internet studies. *The Information Society, 21*, 309–315.

Campbell, M. (2005b). Sacred groves for forest conservation in Ghana's coastal savannas: Assessing ecological and social dimensions. *Singapore Journal of Tropical Geography, 26*, 151–169.

Campbell, H. (2007). Who's got the power? Religious authority and the Internet. *Journal of Computer-Mediated Communication, 12*, 1043–1062.

Campo, J. E. (1998). American pilgrimage landscapes. *Annals of the American Association of Political and Social Sciences, 558*, 40–56.

Cannon, D. F. (2009). Hacks or flacks? Roles played by religions communication in the United States. *Fieldwork in Religion, 4*, 101–207.

Carroll, B. E. (2002). Reflections on regionalism and United States religious history. *Church History, 71*, 120–131.

Carroll, J. W., & Marler, P. L. (1995). Culture Wars? Insights from two ethnographies in two Protestant seminaries. *Sociology of Religion, 56*, 1–20.

Carter, C. L., & Geores, M. E. (2006). Heaven on earth: The Shakers and their space. *Geographies of Religion and Belief Systems, 1*, 5–27.

Cesari, J. (2005a). Mosque conflicts in European cities: Introduction. *Journal of Ethnic and Migration Studies, 31*, 1015–1024.

Cesari, J. (2005b). Mosques in French cities: Towards the end of a conflict? *Journal of Ethnic and Migration Studies, 31*, 1025–1043.

Chalfant, H. P., & Heller, P. L. (1991). Rural/urban versus regional differences in religiosity. *Review of Religious Research, 33*, 76–86.

Chamberlain, P. (2001). Topomystica: Investigation into the concept of mystic place. *Journal of Cultural Geography, 19*, 97–123.

Chaves, M. (1989). Secularization and religious revival: Evidence from the U.S. church attendance rates, 1972–1986. *Journal for the Scientific Study of Religion, 28*, 464–477.

Chen, Z., Hood, R. W., Jr., Yang, L., & Watson, P. J. (2011). Mystical experience among Tibetan Buddhists: The common core thesis revisited. *Journal for the Scientific Study of Religion, 50*, 328–338.

Cheng, P. L. K. (2008). The brand marketing of Halal products: The way forward. *Journal of Brand Management, 5*(4), 37–50.

Chivallon, C. (2001). Religion as space for the expression of Caribbean identity in the United Kingdom. *Environment and Planning D: Society and Space, 19*, 461–483.

Christensen, J. (2013). 'We have our way of life:' Spiritual homelessness and the sociocultural dimensions of indigenous homelessness in the Northwest Territories of Canada. *Social and Cultural Geography, 8*, 804–828.

Cimino, R. (2005). "No God in Common:" American evangelical discourse on Islam after 9/11. *Review of Religious Research, 47*, 162–174.

Clark, M. (1991). Developments in human geography: Niches for a Christian contribution. *Area, 23*, 339–344.

Clarke, C. J. (1985a). Religion and regional culture: The changing pattern of religious affiliation in the Cajun region of southwest Louisiana. *Journal for the Scientific Study of Religion, 24*, 384–395.

Clarke, J. I. (1985b). Islamic population: Limited demographic transition. *Geography, 70*, 118–128.

Clarke, C., & Howard, D. (2005). Race and religious pluralism in Kingston, Jamaica. *Population, Space and Place, 11*, 119–136.

Clarke, G. (2008). Faith matters: Faith-based organizations, civil society and international development. *Journal of International Development, 18*, 835–848.

Cloke, P. (2013). Geographies of postsecular rapprochement in the city. *Progress in Human Geography, 37*, 27–51.

Cloke, P., Johnson, S., & May, J. (2005). Exploring ethos? Discourses of 'charity' in the provision of emergency services for homeless people. *Environment and Planning A, 37*, 385–402.

Cnaan, R. A., & Boddie, S. C. (2001). Philadelphia census of congregations and their involvement in social service delivery. *Social Service Review, 75*, 580–595.

Cobbold, E. (1935). Pilgrim to Mecca. *Geographical Magazine, 1*, 107–116.

Cohen, E. (1992). Pilgrimage centers: Concentric and excentric. *Annals of Tourism Research, 19*, 33–50.

Cohen, E. (2012). A Middle Eastern Muslim tourist enclave in Bangkok. *Tourism Geographies, 14*, 570–598.

Cohen, A., & Susser, B. (2010). Sabbatical year in Israeli politics: An intra-religious and religious-secular conflict from the nineteenth through the twenty-first centuries. *Journal of Church and State, 52*, 454–475.

Coleman, S. (2013a). Landscape, nation and globe: Theoretical nuances in the analysis of Asian Christianity. *Culture and Religion, 14*, 180–184.

Collins-Kreiner, H. (1997). Cartographic characteristics of current Christian pilgrimage maps of the Holy Land. *Cartographica, 34*, 45–54.

Conde, S. G. (1980). Moslem contributions to geography. *Philippine Geographical Journal, 24*(2), 91–93.

Condran, J. G., & Tamney, J. B. (1985). Religious "Nones:" 1957 to 1982. *Sociological Analysis, 46*, 415–423.

Connell, J. (2005). Hillsong: A megachurch in the Sydney suburbs. *Australian Geographer, 36*, 315–332.

Conradson, D. (2008). Expressions of charity and action towards justice: Faith-based welfare provision in urban New Zealand. *Urban Studies, 45*, 2117–2141.

Cook, A. S. (2004–2005). Encountering the other; evangelicalism and terrorism in a post 911 world. *Journal of Law and Religion, 20*, 1–30.

Cooper, A. (1991). Religio-geographical research and public policy. *Geography of Religions and Belief Systems, 13*, 3–6.

Cooper, A. (1992). New directions in the geography of religion. *Area, 24*, 123–129.

Cooper, A. (1993a). A need to reflect upon progress so far. *Geography of Religions and Belief Systems, 15*, 4–6.

Cooper, A. (1993b). Space and geography of religion – A rejoinder. *Area, 25*, 76–78.

Cooper, A. (1994). Negotiated dilemmas of landscape: Place and Christian community in a Suffolk parish. *Transactions. Institute of British Geographers, 19*(2), 202–212.

Corbett, J. M. (1993). Religion in the United States: Notes toward a new classification. *Religion and American Culture, 3*, 91–112.

Corrie, E. (2013). From civic engagement to circles of grace: Mid-range reflection on teaching for global citizenship. *Teaching Theology and Religion, 16*, 165–181.

Cowan, D. E. (2005). Online U-Topia: Cyberspace and the mythology of placelessness. *Journal for the Scientific Study of Religion, 44*, 257–263.

Cox, J. L. (2005). The land crisis in Zimbabwe: A case of religious intolerance. *Fieldwork in Religion, 1*, 35–48.

Crawford, T. W. (2005). Stability and change on the American religious landscape: A centrographic analysis of major U.S. religious groups. *Journal of Cultural Geography, 22*, 51–86.

Cross, J. A. (2004). Expansion of Amish dairy farming in Wisconsin. *Journal of Cultural Geography, 21*, 77–101.

Crowley, W. K. (1978). Old Order Amish settlement: Diffusion and growth. *Annals of the Association of American Geographers, 68*, 249–264.

Cunningham, H. (2013). The doctrine of vicarious punishment: Space, religion and the Belfast Troubles of 1920–22. *Journal of Historical Geography, 40*, 52–66.

Curanovic, A. C. (2010). Relations between the Orthodox Church and Islam in the Russian Federation. *Journal of Church and State, 52*, 503–539.

Curry, J. (2000). Community worldviews and rural systems: A study of five communities in Iowa. *Annals of the Association of American Geographers, 90*, 693–712.

Curry-Roper, J. M. (1990). Contemporary Christian eschatologies and their relation to environmental stewardship. *The Professional Geographer, 42*, 57–69.

Curtis, J. R. (1980). Miami's Little Havana: Yard shrines, cult religions and landscape. *Journal of Cultural Geography, 1*, 1–15.

Dadao, Y. (2013). Religious pluralism in the thinking of the Qur-amic scholars of the Indian subcontinent. *Religion and Theology, 20*, 129–152.

Dann, N. K. (1976). Spatial diffusion of a religious movement. *Journal of the Scientific Study of Religion, 15*, 351–360.

Darden, J. T. (1972). Factors in the location of Pittsburgh's cemeteries. *The Virginia Geographer, 7*, 3–8.

Dart, J. (2001, March 21–28). A "census" of congregations. *Christian Century*, pp. 8–9.

Davies, O. (2012). Songs of worship in life and in death. *International Journal of Public Theology, 7*, 458–475.

Dawson, A. H. (2000). Geography, religion and the state: A comment on an article by Michael Pacione. *Scottish Geographical Journal, 116*, 59–65.
Dawson, A. (2006). A phenomenological study of the Gnostic Church of Brazil. *Fieldwork in Religion, 2*, 27–48.
Day, A. (2009). Researching belief without asking religious questions. *Fieldwork in Religion, 4*, 86–104.
Day, A., & Lynch, G. (2013). Introduction: Belief as cultural performance. *Journal of Contemporary Religion, 28*, 199–206 (special issue; six articles).
Day, A., & Rogaly, B. (2014). Sacred communities: Constellations and connections. *Journal of Contemporary Religion, 29*, 75–88.
De Galembert, C. (2005). The city's 'nod of approval' for the Mantes-la-Jolie mosque project: Mistaken traces of recognition. *Journal of Ethnic and Migration Studies, 31*, 1141–1159.
De Jong, G. F., & Ford, T. R. (1965). Religious fundamentalism and denominational preference in the southern Appalachian region. *Journal for the Scientific Study of Religion, 5*, 24–33.
De Temple, J. (2013). Imaging development: Religious studies in the context of international economic development. *Journal of the American Academy of Religion, 81*, 107–129.
De Villers, E. (2011). Special Issue: Responsible South African public theologies in a global era. *International Journal of Public Theology, 5*, 1–4.
Deacon, G., & Lynch, G. (2013). Allowing Satan in? Moving toward a political economy of neo-Pentecostalism in Kenya. *Journal of Religion in Africa, 42*, 277–316.
Dean, K. C. (2011). Almost Christians: What the faith of our teenagers is telling the American church. *Conversations in Religions and Theology, 9*, 213–223.
Deffontaines, P. (1953). The place of believing. *Landscape, 3*, 22–38.
Delage, R. (2005). From Facebook to research theory in an Indian pilgrimage. *Fieldwork in Religion, 1*, 105–121.
Delvert, J. (1981). Aspects géographiques du système des castes en République Indienne. *L'Information Géographique, 45*, 5–13.
Denis, D. M., & Sharma, V. N. (1999). Assessment of social status of the Christian community in Gorakhpur District, Uttar Pradesh. *Geographical Review of India, 61*, 254–267.
Desai, R. (1993). The religious geography of an Indian town: Bharuch. *South Asia, 16*, 61–78.
Deshmukh, S. B., & Navale, A. M. (1997). Impact of pilgrimage tourism on host population of Pandharpur. *National Geographical Journal of India, 43*, 93–101.
DeVan, B. B. (2014). How Christians and Muslims can embrace religious diversity and each other: An evangelical perspective. *Journal of Religion and Society, 16*, 1–31.
Devine, P. (2013). Men, women, and religiosity in Northern Ireland: Testing the theories. *Journal of Contemporary Religion, 43*, 473–488.
Dianteill, E. (2007). Deterritorialization and reterritorialization of the Orisha religion in African and the new world (Nigeria, Cuba and the United States). *International Journal of Urban and Regional Research, 26*, 121–137.
Digance, J. (2003). Pilgrimage at contested sites. *Annals of Tourism Research, 30*, 143–159.
Dillon, L. Y. (2000). Integrating Canadian and U.S. historical census microdata: Canada (1871 and 1901) and the United States (1870 and 1900). *Historical Methods, 33*, 185–194.
Dod, O., Banda, R. G., & Dodo, G. (2014). African initiated churches, pivotal in peace-building: A case of the Johane Masowe Chishanu. *Journal of Religion and Society, 16*, 1–12.
Doeppers, D. (1973). The evolution of the geography of religious adherence in the Philippines before 1898. *Journal of Historical Geography, 2*, 95–110.
Doeppers, D. (1977). Changing patterns of Aglipayan adherence in the Philippines: 1918–1970. *Philippine Studies, 25*, 265–277.
Donkin, R. A. (1959). The site changes of medieval Cistercian monasteries. *Geography, 44*, 252–258.
Dubey, D. P. (1985). The sacred geography of Prayaga (Allahabad): The identification of holy spots. *National Geographical Journal of India, 31*, 319–340.
Dubey, D. P. (1987). Kumbha Mela: Origin and historicity of India's greatest pilgrimages. *National Geographical Journal of India, 33*, 469–502.

Dubey, D. P., & Singh, R. P. B. (1994). Chitrakut: The frame and network of the faithscape and sacred geometry of a Hindu tirtha. *National Geographical Journal of India, 40*, 307–332.

Dunbar, G. S. (1970). Ahimsa and shikar: Conflicting attitudes towards wildlife in India. *Landscape, 19*, 24–27.

Dutt, A. K. (1979). Religious pattern of India, with a factorial regionalization. *GeoJournal, 3*, 201–214.

Dutt, A. K., & Davgun, S. (1977a). Diffusion of Sikhism and recent migration patterns of Sikhs in India. *GeoJournal, 1*, 81–90.

Dutt, A. K., & Davgun, S. (1977b). Diffusion of Sikkism and recent migration patterns of Sikhs in India. *GeoJournal, 1*, 81–90.

Dutt, A. K., & Noble, A. G. (1977c). Diffusion of Hinduism in southeast Asia with special reference to Indo-China. *National Geographical Journal of India, 28*, 86–94.

Dutt, A. K., & Noble, A. G. (1985). Religious diversity patterns of Rajastan within an Indian framework. *Asian Geographers, 4*, 137–146.

Dutt, A. K., & Sen, A. (1992). Regional concentration of Muslims in India and their rural/urban bias. *National Geographical Journal of India, 38*, 219–232.

Dwyer, C. (1999a). Contradictions of community questions of identity for British Muslim women. *Environment and Planning A, 31*, 53–68.

Dwyer, C. (1999b). Veiled meanings: British Muslim women and the negotiation of differences. *Gender, Place and Culture, 6*, 5–26.

Dwyer, C. (2000). Negotiating diasporic identities: Young British South Asian Muslim women. *Women's Studies International Forum, 23*, 475–486.

Dwyer, C., Bindi, S., & Gurchathen, S. (2008). "From cricket lover to terror suspect". Challenging representations of young British Muslim men. *Gender, Place and Culture, 15*, 117–136.

Dwyer, C., David, G., & Shah, B. (2013a). Faith and suburbia: Secularism, modernity and the changing geographies of religion in London's suburbs. *Transactions of the Institute of British Geographers, 38*, 403–419.

Dwyer, C., Parutios, V., & $ Gale, R. (2013b). Faith in the system? State-funded faith schools in England and the contested parameters of community cohesion. *Transactions of the Institute of British Geographers, 38*, 267–284.

Eagle, D. E. (2011). Changing patterns of attendance at religious services in Canada, 1986–2009. *Journal for the Scientific Study of Religion, 50*, 187–200.

Ebaugh, H. R. (2004). Religion across borders: Transnational religious ties. *Asian Journal of Social Science, 32*, 216–231.

Ebaugh, H. R., O'Brien, J., & Chafetz, J. S. (2000). The social ecology of residential pattern and membership in immigrant churches. *Journal for the Scientific Study of Religion, 39*, 107–116.

Edmunds, J. (2012). The limits of post-national citizenship: European Muslims, human rights and the hijab. *Ethnic and Racial Studies, 35*, 1181–1199.

Efrat, E., & Noble, A. G. (1988). Planning Jerusalem. *Geographical Review, 78*, 387–404.

Emmett, C. (1996). The capital cities of Jerusalem. *Geographical Review, 86*, 233–258.

Emmett, C. (1997). The status quo solution for Jerusalem. *Journal of Palestine Studies, 26*, 16–28.

Endfield, G. H., & Nash, D. (2002a). Drought, desiccation and discourse: Missionary correspondence and nineteenth-century climate change in central southern Africa. *The Geographical Journal, 168*, 33–47.

Endfield, G., & Nash, D. J. (2002b). Missionaries and morals: Climatic discourse in nineteenth century Central Southern Africa. *Annals of the Association of American Geographers, 92*, 727–742.

Endfield, G., & Nash, D. (2005). 'Happy is the bride the rain falls on:' Climate, health and 'the woman question' in nineteenth century missionary documentation. *Transactions of the Institute of British Geographers, NS 30*, 368–386.

Ethridge, F. M. (1989). Under-reported churches in Middle Tennessee: A research note. *Journal for the Scientific Study of Religion, 28*, 518–529.

Eyre, L. A. (1985). Geography in the early Irish monastic schools: A brief review of Airbheartach MacCosse's geographical poems. *Geographical Viewpoint, 3*, 31–33.

Falah, G. W. (1989). Arabs versus Jews in Galilee: Competition for regional resources. *GeoJournal, 21*, 325–336.

Falah, G. W. (2005). Dynamics and patterns of the shrinking of Arab lands in Palestine. *Political Geography, 22*, 179–209.

Feige, M. (2001). Jewish settlement in Hebron: The place and the other. *GeoJournal, 53*, 323–333.

Fellmete, D. (2013). It's funny because it's true? "The Simpsons" satire and the significance of religious humour in popular culture. *Journal of the American Academy of Religion, 81*, 2–48.

Ferber, M., & Harris, T. (2013). Critical realism and emergence in a sacred geography of religion. *Journal of Critical Research, 12*, 183–201.

Fickeler, P. (1947). Grundfragen der Religionsgeographie. *Erdkunde, 1*, 121–144.

Finke, R. (1989). Demographics of religious participation: An ecological approach. *Journal for the Scientific Study of Religion, 28*, 45–58.

Finke, R., & Scheitle, C. P. (2005). Accounting for the uncounted: Computing correctives for the 2000 RCMS data. *Review of Religious Research, 47*, 5–22.

Finke, R., & Stark, R. (1986). Turning pews into people: Estimating 19th century church membership. *Journal for the Scientific Study of Religion, 25*, 180–192.

Fisher, G. (2010). Fieldwork on East Asian Buddhism: Toward a person-centered approach. *Fieldwork in Religion, 5*, 236–250.

Fisher-Nielsen, A. M. (2010). The making of representations of the religious adherent engaged in politics. *Fieldwork in Religion, 5*, 162–179.

Foley, R. (2013). Small health pilgrimages: Place and practice at the holy well. *Culture and Religion, 14*, 44–62.

Form, W., & Dubrow, J. (2005). Downtown metropolitan churches: Ecological situation and response. *Journal for the Scientific Study of Religion, 44*, 271–290.

Foroutan, Y. (2008). Gender, religion and work: Comparative analysis of South Asian migrants. *Fieldwork in Religion, 3*, 25–50.

Foster, R. H., Jr. (1981). Recycling rural churches in southern and central Minnesota. *Bulletin, Association of North Dakota Geographers, 21*, 1–10.

Foster, R. H., Jr. (1983). Changing uses of rural churches: Examples from Minnesota and Manitoba. *Yearbook, Association of Pacific Coast Geographers, 45*, 55–70.

Francaviglia, R. V. (1970). The Mormon landscape: Definition of an image in the American West. *Proceedings of the Association of American Geographers, 2*, 59–61.

Francaviglia, R. V. (1971). The cemetery as an evolving cultural landscape. *Annals of the Association of American Geographers, 61*, 501–509.

Francis, L. J., Williams, E., & Village, A. (2011). Research note: Multi-faith Britain and family life: Changing patterns of marriage, cohabitation and divorce among different faith groups 1983–2005. *Journal of Contemporary Religion, 26*, 33–41.

Frantzman, S. J., & Bar, D. (2013). Mapping Muslim sacred tombs in Palestine during the Mandate Period. *Levant, 14*, 96–111.

Freston, P. (2009). Researching the heartland of Pentecostalism: Latin Americans at home and abroad. *Fieldwork in Religion, 3*, 122–144.

Fridolfsson, C., & Elander, I. (2013). Faith and place: Constructing Muslim identity in a secular Lutheran society. *Cultural Geographers, 20*, 319–337.

Friedland, R. (2001). Religions nationalism and the problem of collective responsibility. *Annual Review of Sociology, 27*, 125–152.

Froese, P., & Bader, C. (2011). Americas four Gods: What we say about God – What that says about us. *Conversations in Religion & Theology, 9*, 95–102.

Fuson, R. H. (1969). The orientation of Mayan ceremonial centres. *Annals of the Association of American Geographers, 59*, 494–511.

Gale, R. (2007a). Representing the city: Mosques and the planning process in Birmingham. *Journal of Ethnic and Migration Studies, 31*, 1161–1179.

Gale, R. (2007b). The place of Islam in the geography of religion: Trends and intersections. *Geography Compass, 1*, 1015–1036.

Gale, R. (2008). Locating religion in urban planning: Beyond 'race' and 'ethnicity'. *Planning Practice and Research, 23*, 19–39.

Galindo, R. (1994). Amish newsletters in "The Budget:" A genre study of written communication. *Language in Society, 23*, 77–105.

Garbin, D. (2012). Marching for God in the global city: Public space, religion and diaspora identification in a transnational African church. *Culture and Religion, 13*, 425–447.

Garmany, J. (2013). Slums, space and spirituality: Religious diversity in contemporary Brazil. *Area, 45*, 47–55.

Gathogo, J. (2012). Reconciliation paradigm in postcolonial Africa: A critical analysis. *Religion & Theology, 19*, 74–91.

Geffen, J. P. (1998). Landscapes of the sacred: Avebury as a case study. *Geography of Religions and Belief Systems, 20*(1), 3–4.

Gerlach, L. P., & Hine, V. H. (1968). Five factors crucial to the growth and spread of a modern religious movement. *Journal for the Scientific Study of Religion, 7*, 23–40.

Gilbert, C. P. (1991). Religion, neighborhood environments, and partisan behaviour: A contextual analysis. *Political Geography Quarterly, 10*, 110–131.

Gill, R. (2010). Public theology and music. *International Journal of Public Theology, 4*, 410–425.

Glazier, S. D. (1988). Worldwide missions of Trinidad's spiritual Baptists. *National Geographical Journal of India, 34*, 75–78.

Glenn, N. D., & Simmons, J. L. (1967). Are regional cultural differences diminishing? *The Public Opinion Quarterly, 31*, 176–193.

Gökanksel, B. (2009). Beyond the official sacred: Religion, secularism and the body in the production of subjectivity. *Social and Cultural Geography, 10*, 657–674.

Gökariksel, B., & Secor, A. (2009). New transnational geographies of Islamism, capitalism and subjectivity: The veiling fashion industry in Turkey. *Area, 41*, 6–18.

Gökariksel, B., & Secor, A. (2010a). Between fashion and *tesettür*: Marketing and consuming veiling fashion. *Journal of Middle East Women's Studies, 6*, 118–148.

Gökariksel, B., & Secor, A. (2010b). Islamic-ness in the life of a commodity: Veiling-fashion in Turkey. *Transactions of the Institute of British Geographers, 35*, 313–333.

Gökariksel, B., & Secor, A. (2012). Even I was tempted: The moral ambivalence and ethical practice of veiling fashion in Turkey. *Annals of the Association of American Geographers, 102*, 847–862.

Golan, A. (2002). Israeli historical geography and the Holocaust: Reconsidering the research agenda. *Journal of Historical Geography, 28*, 554–565.

Goldstein, S., & Kosmin, B. (1992). Religious and ethnic self-identification in the United States 1989–1990: A case study of the Jewish population. *Ethnic Groups, 9*, 219–245.

González, A. L. (2011). Measuring religiosity in a majority Muslim context: Gender, religious salience and religions experience among Kuwaiti college students – A research note. *Journal for the Scientific Study of Religion, 50*, 339–350.

González, L. (2013). The Camino de Santiago and its contemporary renewal: Pilgrims, tourists and territorial identities. *Culture and Religion, 14*, 8–22.

Gorlizki, Y. (2000). Class and nation in the Jewish settlement of Palestine: The case of Merhavia, 1910–30. *Journal of Historical Geography, 26*, 572–588.

Graham, E. (2008). What makes a good city? Reflection on urban life and faith. *International Journal of Public Theology, 2*, 7–26.

Graham, E., & Scott, P. M. (2008). Public theology and the city: Urban theology as public theology. *International Journal of Public Theology, 2*, 1–6.

Granger, B. H. (1957). Early Mormon place names in Arizona. *Western Folklore, 16*, 43–47.

Gray, C. (1995). Compositional techniques in Roman Catholic Church music in Uganda. *British Journal of Ethnomusicology, 4*, 135–155.

Greeley, A. M., & Hout, M. (1988). Musical chairs: Patterns of denominational change. *Sociology and Social Research, 72*, 75–86.

Gregorius, B. (1982). Christianity, the Coptic religion and ethnic minorities in Egypt. *GeoJournal, 6*, 57–62.
Grugel, A. (2012). Culture, religion and economy in the American Southwest: Zuni Pueblo and Laguna Pueblo. *GeoJournal, 77*, 791–803.
Grundel, M., & Maliepaard, M. (2012). Knowing, understanding and practicing democratic citizenship: An investigation of the role of religion among Muslim, Christian and non-religious adolescents. *Ethnic and Racial Studies, 35*, 2075–2096.
Guelke, J. K. (1988). Concepts of nature in the Hebrew Bible. *Environmental Ethics, 10*, 309–327.
Guelke, J. K. (1989). Human dominion over nature in the Hebrew Bible. *Annals of the Association of American Geographers, 79*, 214–242.
Guelke, J. K. (2004). Looking for Jesus in Christian environmental ethics. *Environmental Ethics, 26*, 115–134.
Guelke, J. K., & Brown, C. J. (1985). Mormon beliefs about land and natural resources. *Journal of Historical Geography, 11*, 253–267.
Guenther, K. M. (2014). Bounded by disbelief: How atheists in the United States differentiate themselves from religious believers. *Journal of Contemporary Religion, 29*, 1–16.
Gurgel, K. D. (1976). Travel patterns of Canadian visitors to the Mormon culture hearth. *Canadian Geographer, 20*, 405–418.
Gutschow, N. (1994). Varanasi/Benares: The centre of Hinduism? *Erdkunde, 48*, 194–209.
Habashi, J. (2013). Children's religious agency: Conceptualizing Islamic idioms of resistance. *Area, 45*, 155–161.
Hackworth, J., Gullikson, E., Hasbullah, S., & Korf, B. (2013). Giving new meaning to religious conversion: Churches, redevelopment and secularization in Toronto. *The Canadian Geographer, 57*, 72–89.
Hadaway, C. K. (1980). Denominational switching and religiosity. *Review of Religious Research, 21*, 451–461.
Hadaway, C. K. (1981). The demographic environment and church membership change. *Journal for the Scientific Study of Religion, 20*, 77–89.
Hadaway, C. K. (1989a). Identifying American apostates: A cluster analysis. *Journal for the Scientific Study of Religion, 28*, 201–215.
Hadaway, C. K. (1989b). Will the real Southern Baptist please stand up? Methodological problems in surveying Southern Baptist congregations. *Review of Religious Research, 31*, 149–161.
Hadaway, C. K., & Marler, P. L. (1996). Response to Iannaccone: Is there a method to this madness? *Journal for the Scientific Study of Religion, 35*, 217–222.
Hadaway, C. K., & Marler, P. L. (1998). Did you really go to church this week? Behind the poll data. *Christian Century, 115*, 472–475.
Hadaway, C. K., & Marler, P. L. (2005). How many Americans attend worship each week? An alternative approach to measurement. *Journal for the Scientific Study of Religion, 44*, 307–322.
Hadaway, C. K., Marler, P. L., & Chaves, M. (1993a). What the polls don't show: A closer look at U. S. church attendance. *American Sociological Review, 58*, 741–752.
Hadaway, C., Marler, P. L., & Chaves, M. (1993b). Over-reporting church attendance in America: Evidence that demands the same verdict. *American Sociological Review, 63*, 122–130.
Hale, R. (2013). Universal salvation in a universal language. Trevor Steele's Kaj staros tre alte. *Religion and Theology, 20*, 19–35.
Hamerly, I., & Waltman, J. L. (2009). Congressional voting on religious issues. The case of the Religious Liberty Protection Act of 1969. *Journal of Church and State, 51*, 454–471.
Hancock, M. E. (2013). New mission paradigms and the encounter with Islam: Fusing volunteerism, tourism and evangelism in short-term missions in the USA. *Culture and Religion, 14*, 305–333.
Hansen, R. (2011). The two faces of liberalism: Islam in contemporary Europe. *Journal of Ethnic and Migration Studies, 37*, 881–897.

Hardwick, S. (1991). The impact of religion on ethnic survival: Russian old believers in Alaska. *The California Geographer, 31*, 19–36.
Hardwick, S. (1993). Origin and diffusion of Russian Baptists and Pentecostals in Russia and Ukraine. *The Pennsylvania Geographer, 31*, 2–13.
Hardwick, W., Claus, R., & Rothwell, D. C. (1971). Cemeteries and urban land uses. *The Professional Geographer, 23*, 19–21.
Harvey, T. (2006). Sacred spaces, common places: The cemetery in the contemporary American city. *Geographical Review, 96*, 295–312.
Harvey, G. (2012). Sacred places in the construction of indigenous environmentalism. *Journal of Religion, Nature and the Environment, 7*, 60–73.
Hasbullah, S., & Korf, B. (2013). Muslim geographies: Violence and the antimonies of community in eastern Sri Lanka. *Geographical Journal, 179*, 32–43.
Havlicek, T. (2006). Church-state relations in Czechia. *GeoJournal, 67*, 331–340.
Hayes, M. (2003). Vive la difference: Jewish women teachers' construction of ethnicity and identity and their experiences of anti-Semitism in secondary schools. *Race, Ethnicity and Education, 6*, 51–70.
Hayes, S. (2010). Orthodox diaspora and mission in South Africa. *Studies in World Christianity, 16*, 286–303.
Hayes, B. C., McAllister, I., & Dowds, L. (2013). Integrated schooling and religious tolerance in Northern Ireland. *Journal of Contemporary Religion, 28*, 67–78.
Haynes, J. (2001). Transnational religious actors and international politics. *Third World Quarterly, 22*, 143–158.
Heap, B., & Comin, F. (2009). Climate change and well-being. *International Journal of Public Theology, 3*, 42–62.
Heatwole, C. (1977). Exploring the geography of America's religious denominations: A Presbyterian example. *Journal of Geography, 76*, 99–104.
Heatwole, C. (1978). The Bible Belt: A problem in regional definition. *Journal of Geography, 77*, 50–55.
Heatwole, C. (1985). The unchurched in the Southeast. *Southeastern Geographer, 25*, 1–15.
Heatwole, C. (1986). A geography of the African Methodist Episcopal Zion Church. *Southeastern Geographer, 26*, 1–11.
Heatwole, C. (1989). Sectarian ideology and church architecture. *Geographical Review, 79*, 63–78.
Hemming, P. (2011). Educating for religious citizenship: Multiculturalism and national identity in an English multi-faith primary school. *Transactions of the Institute of British Geographers, 36*, 441–454.
Henderson, J. L. (1993). What is spiritual geography? *Geographical Review, 83*, 469–472.
Henkel, R. (1989). Christian missions in Africa. A social geographical study of their activities in Zambia. *Berlin: Geographica Religionum, 3*.
Henkel, R. (2005). Geography of religion- rediscovering a subdiscipline. *Hrvatski Geografski Glasnik, 67*, 5–25.
Henkel, R. (2011). Are geographers religiously unmusical? Positionalities in geographical research on religion. *Erdkunde, 63*, 389–399.
Henkel, R., & Sakaja, L. (2009). A sanctuary in post-conflict space: The Baptist Church as a "middle option" in Banovina, Croatia. *Geografiska Annaler B, 91*, 39–56.
Henry, C. (2011). Jehovah's Witness and post-Soviet religions policy in Moldova and the Transnistrian Moldovan republic. *Journal of Church and State, 53*, 401–420.
Heppen, J. (2002). The Green Party vote in California: An examination of race, religion, and wealth from 1996–2000. *Pennsylvania Geographer, 40*, 56–81.
Herschkowitz, S. (1987). Residential segregation by religion: A conceptual framework. *Tijdschrift voor economische en sociale geographie, 78*, 44–52.
Hervieu-Léger, D. (2002). Space and religion: New approaches to religious spatiality in modernity. *International Journal of Urban and Regional Research, 26*, 99–105.
Hill, S. S. (1985). Religion and region in America. *Annals of the American Academy of Political and Social Science, 480*, 133–141.

Hirschman, C. (2004). The role of religion in the origins and adaptation of immigrant groups in the United States. *International Migration Review, 38*, 1206–1233.
Hobbs, J. J. (1992). Sacred space and touristic development at Jebel Musa (Mt. Sinai), Egypt. *Journal of Cultural Geography, 12*, 99–113.
Holloway, J. (2000). Institutional geographies of the new age movement. *Geoforum, 31*, 553–565.
Holloway, J. (2003). Make-believe: Spiritual practice, embodiment and sacred space. *Environment and Planning A, 35*, 1961–1974.
Holloway, J. (2010). Legend-tripping in spooky spaces: Ghost tourism and infrastructures of enchantment. *Environment and Planning D: Society and Space, 28*, 618–637.
Holloway, J., & Valins, O. (2002). Editorial: Placing religion and spirituality in geography. *Social and Cultural Geography, 3*, 5–9.
Holt, J. B., Miller, J. C., Naimi, T. S., & Sui, D. (2006). Religious affiliation and alcohol consumption in the United States. *Geographical Review, 96*, 523–542.
Homer, L. B. (2011). Registration of Chinese Protestant house churches under China's 2005 Regulation on Religious Affairs: Resolving the implementation impasse. *Journal of Church and State, 52*, 50–73.
Hopkins, P. E. (2004). Young Muslim men in Scotland: Inclusions and exclusions. *Children's Geographies, 1*, 257–272.
Hopkins, P. E. (2006). Youthful Muslim masculinities: Gender and generational relations. *Transactions of the Institute of British Geographers, NS 31*, 337–352.
Hopkins, P. E. (2007). Young people, masculinities and emotion: New social geographies. *Progress in Human Geography, 31*, 163–177.
Hopkins, P. E. (2009). Men, women, positionalities and emotion: Doing feminist geographies of religion. *ACME: International Journal for Critical Geographers, 8*, 1–17.
Horner, A. (2010). Representing cultural divides in Ireland: Some nineteenth and early twentieth century mappings of variation in religion and language. *Irish Geography, 43*, 233–247.
Hout, M., & Greeley, A. (1987). The center doesn't hold: Church attendance in the United States, 1940–1984. *American Sociological Review, 52*, 325–345.
Hout, M., Greeley, A., & Wilde, M. J. (1998). What church officials' reports don't show: Another look at church attendance data. *American Sociological Review, 63*, 113–119.
Hout, M., Greeley, A., & Wilde, M. J. (2001). The demographic imperative in religious change in the United States. *American Journal of Sociology, 107*, 468–500.
Howell, J. D. (2013). 'Calling' and 'training:' Role innovation and religions de-differentiation in commercialized Indonesian realm. *Journal of Contemporary Religion, 28*, 401–419.
Hudman, L. E., & Jackson, R. H. (1992). Mormon pilgrimage and tourism. *Annals of Tourism Research, 19*, 107–121.
Huff, B., & Stallins, J. A. (2013). Beyond binaries: Conservative Catholic visions and real estate in Ave Maria, Florida. *Culture and Religion, 14*, 94–110.
Hunt, S. (2014). Christian lobbyist groups and the negotiation of sexual rights in the United Kingdom. *Journal of Contemporary Religion, 29*, 121–136.
Hunt, L. L., & Hunt, M. O. (1999). Regional patterns of African American church attendance: Revisiting the semi-involuntary thesis. *Social Forces, 78*, 779–791.
Huntsinger, L., & Fernandez-Gimenez, M. (2000). Spiritual pilgrims at Mount Shasta, California. *Geographical Review, 90*, 536–558.
Ibrahim, F. N. (1982). Social and economic geographical analysis of the Egyptian Copts. *GeoJournal, 6*, 63–67.
Ilesanmi, S. O. (2011). Introduction: Islam and just war tradition. Editorial. *Journal of Church and State, 53*, 1–3.
Imazato, S. (1995). Rural religious landscape constituents and social structures: A case study of a mountain village, Shiga Prefecture. *The Human Geography/Jimbun Chir, 47*, 42–64.
Ingalls, M. M. (2012). Singing praises in the streets: Performing Canadian Christianity through public worship in Toronto's Jesus in the City parade. *Culture and Religion, 13*, 337–359.
Isaac, E. (1957). A geographic interpretation of the diaspora. *Jewish Social Studies, 19*, 64–67.
Ismail, R. (2006). Ramadan and Bussorah Street: The spirit of place. *GeoJournal, 66*, 243–256.

Issac, E. (1959). The citron in the Mediterranean: A study in religious influences. *Economic Geography, 35*, 71–78.
Issac. E. (1959–1960). Religion, landscape, and space. *Landscape, 9*, 14–18.
Issac, E. (1961–1962). The act of the covenant: The impact of religion on the landscape. *Landscape, 11*, 12–17.
Issac, E. (1963). Myths, cults and livestock. *Diogenes, 41*, 70–93.
Issac, R. (1964). God's acre: Property in land, a sacred origin? *Landscape, 14*(2), 28–32.
Issac, R. (1973). The pilgrimage to Mecca. *Geographical Review, 63*, 406–409.
Jackowski, A. (1987). Geography of pilgrimage in Poland. *National Geographical Journal of India, 33*, 422–429.
Jackowski, A., & Smith, V. L. (1992). Polish pilgrim-tourists. *Annals of Tourism Research, 19*, 92–106.
Jackson, R. H. (1978). Mormon perception and settlement. *Annals of the Association of American Geographers, 68*, 317–334.
Jackson, R. H., & Henrie, R. (1983). Perceptions of sacred space. *Journal of Cultural Geography, 3*, 94–107.
Jackson, R. H., & Layton, L. R. (1976). The Mormon village: Analysis of a settlement type. *The Professional Geographer, 28*, 136–141.
James, A. B. (2013). Rehabilitating Willow Creek: Megachurches, De Carteau and the tactics of navigating consumer culture. *Christian Scholar's Review, 43*, 21–40.
Jarosz, L. (1992). Constructing the dark continent: Metaphor as geographic representation of Africa. *Geografiska Annaler, 74B*, 105–115.
Jeans, D. N., & Kofman, E. (1972). Religious adherence and population mobility in nineteenth century New South Wales. *Australian Geographical Studies, 10*, 193–202.
Jeffrey, A. (1929). Christians at Mecca. *Moslem World, 22*, 109–116.
Jelen, T. G., & Lockett, L. A. (2010). American clergy on evolution and creationism. *Review of Religious Research, 51*, 277–287.
Jeremiah, A. H. M. (2011). Dalit Christians in India: Reflections on the "Broken Middle". *Studies in World Christianity, 17*, 258–274.
Jespers, F. (2010). The paranormal market in the Netherlands: New Age and folk religion. *Fieldwork in Religion, 5*, 58–77.
Jett, S. C. (1992). An introduction to Navajo sacred places. *Journal of Cultural Geography, 13*, 29–39.
Joas, H. (2000). Social theory and the sacred: A response to John Milbank. *Ethical Perspectives, 7*, 233–243.
Johnson, H. B. (1967). The location of Christian missions in Africa. *Geographical Review, 57*, 68–202.
Johnson, N. B. (1993). Muso Kokushi and the cave in Zuisen Temple, Kamakura, Japan: Buddhist ethnics, environment, and behavior. *National Geographical Journal of India, 39*, 161–178.
Jones, P. N. (1976). Baptist chapels as an index of cultural transition in the South Wales before 1914. *Journal of Historical Geography, 2*, 347–369.
Jones, R. (2006). Sacred cows and thumping drums: Claiming territory as 'zones of tradition' in British India. *Area, 39*, 55–65.
Jonker, G. (2005). The Mevlana mosque in Berlin-Kreuzberg: An unsolved conflict. *Journal of Ethnic and Migration Studies, 31*, 1067–1081.
Jordan, T. G. (1976). Forest folk, prairie folk: Rural religious cultures in north Texas. *Southwestern Historical Quarterly, 80*, 135–162.
Jordan, L. M. (2007). Religious adherence and diversity in the United States: A geographic analysis. *Geographies of Religion and Belief Systems, 2*, 3–20.
Jung, D. (2010). Islam as a problem: Dutch religious parties in the East Indies. *Review of Religious Research, 51*, 288–301.
Kahn, N. (2012). Between spectacle and banality: Trajectories of Islamic radicalism in a Karachi neighborhood. *International Journal of Urban and Regional Research, 36*, 568–584.
Kamaara, F. (2010). Towards Christian national identity in Africa: A historical perspective to the challenge of ethnicity to the church in Kenya. *Studies in World Christianity, 16*, 126–144.

Kaneko, N. (1997). The spatial structure of mountain religion: The case of Mt. Iwaki. *The Human Geography/Jimbun Chiri, 49*, 1–20.

Kapaló, J., & Travagin, S. (2010). Religionizing fieldwork and fieldworking religion: Hermeneutics of the engagement between religion and research methodologies in the field. *Fieldwork in Religion, 5*, 133–143.

Kark, R. (1983). Milleniarism and agricultural settlement. *Journal of Historical Geography, 9*, 47–63.

Katz, Y., & Lehr, J. C. (1991). Jewish and Mormon agricultural settlement in western Canada: A comparative analysis. *The Canadian Geographer, 35*, 128–142.

Kelley, J., & de Graf, N. D. (1997). National context, parental socialization and religious belief: Results from 15 nations. *American Sociological Review, 62*, 639–659.

Kenny, E. (2007). Gifting Mecca: Importing spiritual capital to West Africa. *Mobilities, 2*, 363–381.

Khan, A. U. (1994). Islamic conversion of eastern Bengal. *Oriental Geographer, 38*, 59–81.

Khan, Z. (2000). Muslim presence in Europe: The British dimension – Identity, integration and community activism. *Current Sociology, 48*(4), 29–43.

Khian, A., Scourfield, J., Gillat-Ray, S., & Ott, S. (2012). Reflections on qualitative research with Muslim families. *Fieldwork in Religion, 7*, 49–69.

Kieh, G. K. (2011). Religious leaders, peacemaking and the First Liberian Civil War. *Journal of Religion, Conflict and Peace, 5*. http://www.religionconflictpeace.org/volume-5-issue-1-and-2-fall-2011-spring-2013

Kim, K. (2010). Christianity's role in the modernization and revitalization of Korean society in the twentieth century. *International Journal of Public Theology, 4*, 212–236.

King, R. (1972). The pilgrimage to Mecca – Some geographical and historical aspects. *Erdkunde, 26*, 61–73.

King, K., & Hemming, P. (2012). Exploring multiple religious identities through mixed methods. *Journal of Fieldwork in Religion, 7*, 29–47.

Kinney, N. T. (2012). The rise of a transnational religious network in the development of a weak state: The international link of the Episcopal Church of Sudan. Development in Practice. *Journal of Religion and Society, 22*, 749–762.

Kinney, N. T. (2013). The role of U.S. denominations in mobilizing international voluntary service (IVS). *Journal of Religion and Society, 15*, 1–19.

Kipnis, B., & Schnell, I. (1978). Changes in the distribution of Arabs in mixed Jewish-Arab cities in Israel. *Economic Geography, 54*, 168–180.

Kluegel, J. R. (1980). Denominational mobility: Current patterns and recent trends. *Journal for the Scientific Study of Religion, 19*, 26–39.

Kniffen, F. B. (1967). Necrogeography in the United States. *Geographical Review, 57*, 426–427.

Knippenberg, H. (2006a). The changing relationship between state and church/religion in the Netherlands. *GeoJournal, 67*, 317–330.

Knippenberg, H. (2006b). The political geography of religion in Europe. *GeoJournal, 67*((Special issue)).

Knott, K. (2009). From locality to location and back again: A spatial journal in the study of religion. *Religion, 39*, 154–160.

Knott, K., & Khoker, S. (1993). The relationship between religion and ethnicity in the experience of young Muslim women in Bradford. *New Community, 19*, 593–610.

Kocsis, K. (2006). Spatial and temporal changes in the relationship between church and state in Hungary. *GeoJournal, 67*, 357–371.

Koepping, E. (2013). Spousal violence among Christians: Taiwan, South Australia and Ghana. *Studies in World Christianity, 19*, 252–270.

Kokkonen, P. (1993). Religious and colonial realities: Cartography of the Finnish mission in Ovamboland, Namibia. *History in Africa, 20*, 155–171.

Kong, L. (1967). Ideological hegemony and political symbolism of religious buildings in Singapore. *Environment and Planning D: Space and Society, 11*, 23–46.

Kong, L. (1990). Geography and religion: Trends and prospects. *Progress in Human Geography, 14*, 355–371.
Kong, L. (1991a). Cemeteries and columbaria, memorials and mausoleums: Narrative and interpretation in the study of deathscapes in geography. *Australian Geographical Studies, 37*, 1–10.
Kong, L. (1991b). Mapping "new geographies" of religion: Politics and poetics in modernity. *Progress in Human Geography, 25*, 211–233.
Kong, L. (1992). The sacred and the secular: Exploring contemporary meanings and values for religious buildings in Singapore. *Southeast Asian Journal of Social Science, 20*, 18–42.
Kong, L. (1993a). Ideological hegemony and the political symbolism of religious buildings in Singapore. *Environment and Planning D: Society and Space, 11*, 23–45.
Kong, L. (1993b). Negotiating conceptions of sacred space: A case study of religious buildings in Singapore. *Transactions, Institute of British Geographers, New Series, 18*, 342–358.
Kong, L. (1996). The commercial face of God: Exploring the nexus between the religious and material. *Geographia Religionum, 10*, 123–141.
Kong, L. (2001a). Mapping "new" geographies of religion: Politics and poetics in modernity. *Progress in Human Geography, 25*, 211–233.
Kong, L. (2001b). Religion and technology: Refiguring place, space, identity, and community. *Area, 33*, 404–413.
Kong, L. (2002). In search of permanent homes: Singapore's house churches and the politics of space. *Urban Studies, 39*, 1573–1586.
Kong, L. (2005). Religious schools for spirit (f)or nation. *Environment and Planning D: Society and Space, 23*, 615–631.
Kong, L. (2006a). Music and moral geographies: Constructions of 'nation' and identity in Singapore. *GeoJournal, 65*, 103–111.
Kong, L. (2006b). Religion and spaces of technology: Constructing and contesting nation, transnation, and place. *Environment and Planning A, 38*, 903–918.
Kong, L. (2010). Global shifts, theoretical shifts: Changing geographies of religion. *Progress in Human Geography, 14*, 755–776.
Kong, L. (2013a). Balancing spirituality and secularism, globalism and nationalism: The geographies of identity, integration and citizenship in schools. *Journal of Cultural Geography, 30*, 276–307.
Kong, L., & Tan, R. (1997). Women in a Catholic world: A case study of Singapore. *Asian Profile, 25*, 473–489.
Korf, B. (2006). Geography and Benedict XVI. *Area, 38*, 326–329.
Kostelnick, J. C. (2001). The diffusion of the Catholic and Methodist churches across the Iowa frontier, 1833–1891. *The North American Geographer, 3*, 84–100.
Kourie, C. (2011). Crossing boundaries: The way of interspirituality. *Religion & Theology, 18*, 10–31.
Kreinath, J., & Solcott, W. (2013). Introduction: Politics of faith in Asia: Local and global perspectives of Christianity in Asia. *Culture and Religion, 14*, 180–184 (articles on Thailand, South Korea & India).
Kumar, N. (1987). The Mazars of Banaras: A new perspective on the city's sacred geography. *National Geographical Journal of India, 33*, 263–267.
Kumar, K. (2013). The sacred mountain: Confronting global capital at Niyamgiri. *Geoforum* (in press).
Kuroda, A. (1992). Religious landscape images from Hakusan Pilgrimage Mandala on Kunigami Shrine. *The Human Geography/Jimbun Chiri, 44*(6), 66–84.
Laatsch, W. G. (1971). Hutterite colonization in Alberta. *Journal of Geography, 70*, 347–359.
Laatsch, W. G., & Calkins, C. F. (1986). The Belgian roadside chapels in Wisconsin's Door Peninsula. *Journal of Cultural Geography, 7*, 117–128.
Laing, C. R. (2002). The Latter-Day Saint Diaspora in the United States and the South. *Southeastern Geographer, 42*, 228–247.
Lamme, A. (1971). From Boston in one hundred years. Christian Science 1970. *The Professional Geographer, 23*, 329–332.

Lancee, B., & Dronkers, J. (2011). Elite, religious and economic diversity in Dutch neighborhoods: Explaining quality of contact with neighbors, trust in the neighbourhood and inter-ethnic trust. *Journal of Ethnic and Migration Studies, 37*, 597–618.

Land, K. C., Dane, G., & Blau, J. R. (1991). Religious pluralism and church membership: A spatial diffusion model. *American Sociological Review, 56*, 237–249.

Landing, J. (1969a). Geographic models of old order Amish settlements. *The Professional Geographer, 21*, 328–343.

Landing, J. (1969b). Geographic models of the Old Order Amish settlements. *The Professional Geographer, 21*, 238–243.

Landing, J. (1972). The Amish, the automobile and social interaction. *Journal of Geography, 71*, 52–57.

Landing, J. (1982). A case study in the geography of religion: The Jehovah's Witnesses in Spain, 1921–1946. *Bulletin, Association of North Dakota Geographers, 32*, 42–47.

Lane, B. (2001). Giving voice to place: Three models for understanding American sacred space. *Religion and American Culture, 11*, 53–81.

Lang, G., & Ragvald, L. (2005). Grasping the revolution: Fieldwork on religion in China. *Fieldwork in Religion, 1*, 219–233.

Laurie, N. (2010). Finding yourself in the archives and doing geographies of religion. *Geoforum, 41*, 165–167.

Le Bruyns, C. (2012). The church, democracy and responsible citizenship. *Religion & Theology, 19*, 60–73.

Leavelle, T. N. (2004). Geographies of encounter: Religion and contested spaces in colonial North America. *American Quarterly, 56*, 913–943.

Lee, P. (2009). Towards a theology of communication rights. *Fieldwork in Religion, 4*, 191–207.

Leonard, J. M. (2006). Local geography and church attendance: Wayne County, West Virginia. *Geographies of Religions and Belief Systems, 1*, 28–50.

Leppman, E. J. (2005). Appalachian churchscapes: The case of Menifee County, Kentucky. *Southeastern Geographer, 45*, 83–103.

Levesque, P. J., & Cuillaume, A. M. (2010). Teachers, evolution and religion: No resolution in sight. *Review of Religious Research, 51*, 349–365.

Levine, G. J. (1986). On the geography of religion. *Transactions of the Institute of British Geographers, 11*, 428–440.

Levitt, P. (2008). Religion as a path to civic engagement. *Ethnic and Racial Studies, 31*, 766–791.

Ley, D. (1974). The city and good and evil: Reflections on Christian and Marxist interpretations. *Antipode, 6*, 66–74.

Ley, D. F. (2002). Mapping the metaphysical, plotting the pious: Assessing four new atlases of religion. *Church History, 71*, 143–151.

Ley, D., & Martin, R. B. (1993). Gentrification as secularization: The status of religious belief in the post-industrial city. *Social Compass, 40*, 217–231.

Lim, J., & Fanghanel, A. (2013). Hijabs, hoodies and otpants: Negotiating the "slut" in "slutwalk". *Geoforum, 48*, 207–215.

Lindenbaum, J. (2013). The neoliberalization of contemporary Christian music's new social gospel. *Geoforum, 44*, 112–119.

Liu, Y. (2010). From Christian aliens to Chinese citizens: The national identity of Chinese Christians in the twentieth century. *Studies in World Christianity, 16*, 145–168.

Livingstone, D. N. (1983). Environmental theology: Prospect in retrospect. *Progress in Human Geography, 7*, 133–140.

Livingstone, D. N. (1984). Natural theology and Neo-Lamarckism: The changing context of nineteenth century geography in the United States and Great Britain. *Annals of the Association of American Geographers, 74*, 9–28.

Livingstone, D. N. (1985). Evolution, science and society: Historical reflections on the geographical experiment. *Geoforum, 16*, 119–130.

Livingstone, D. N. (1986). Science, religion and ideology: The case of evangelicals and evolution. *Science and Faith, 6*, 5–15.

Livingstone, D. N. (1987). Preadamites: The history of an idea from heresy to orthodoxy. *Scottish Journal of Theology, 40*, 41–66.

Livingstone, D. N. (1988). Science, magic and religion: A contextual reassessment of geography in the sixteenth and seventeenth centuries. *History of Science, 26*, 269–294.

Livingstone, D. N. (1994). Science and religion: Foreword to the historical geography of an encounter. *Journal of Historical Geography, 20*, 367–383.

Livingstone, D. N., Keane, M., & Boal, F. W. (1998). Space for religion: A Belfast case study. *Political Geography, 17*, 145–170.

Locke, S. (2012). Spirit(ualities) of science in words and pictures: Syncretizing science and religion in the cosmologies of two comic books. *Journal of Contemporary Religion, 27*, 383–401.

Longkumer, A. (2009). Exploring the diversity of religion: The geo-political dimensions of fieldwork and identity in the North East of India. *Fieldwork in Religion, 4*, 46–66.

Louder, D. R. (1975). A simulation approach to the diffusion of the Mormon Church. *Proceedings of the Association of American Geographers, 7*, 126–130.

Louder, D. R. (1993). Canadian Mormons in their North American context: A portrait. *Social Compass, 40*, 271–290.

Ludueña, G. A. (2005). Asceticism, fieldwork and technologies of the self in Latin American Catholic monasticism. *Fieldwork in Religion, 1*, 145–164.

Luidens, D. A. (1996). Fighting "decline": Mainline churches and the tyranny of aggregate data. *Christian Century, 6*(Nov), 1075, 1077–1079.

Lunn, J. (2009). The role of religion, spirituality and faith in development: A critical theory approach. *Third World Quarterly, 30*, 937–951.

Maclellan, D. (1983). The ecumenical start of New France in Arcadia. *Canadian Geographic, 103*, 66–73.

Macourt, M. A. (1995). Using census data: Religion as a key variable in studies of Northern Ireland. *Environment & Planning A, 27*, 593–614.

MacRae, E. (2006). The religious uses of licit and illicit psychoactive substances in a branch of the Santo Daime religion. *Fieldwork in Religion, 2*, 393–314.

Maddrell, A. (2009). A place for grief and belief: The Witness Cairn, Isle of Whithorn, Galloway, Scotland. *Social and Cultural Geography, 10*, 675–693.

Maddrell, A. (2013). Moving and being moved: More-than-walking and talking on pilgrimage walks in the Manx landscape. *Culture and Religion, 14*, 63–77.

Maddrell, A., & Della Dora, V. (2013). Editorial: Spaces of renewal. *Culture and Religion, 14*, 1–7 (seven articles on renewal).

Madsen, M. N. (2006). The sanctification of Mormonism's historical geography. *Geographies of Religion and Belief Systems, 1*, 51–73.

Maffly-Kipp, L. F. (2002). If it's South Dakota you must be Episcopalian: Lies, truth-telling, and the mapping of U.S. religion. *Church History, 71*, 132–142.

Maier, E. (1975). Torah as mobile territory. *Annals of the Association of American Geographers, 65*, 18–23.

Malville, J. M. (1999). Complexity and self-organization in pilgrimage systems. In *Proceedings, International seminar: Pilgrimage and complexity*. New Delhi: India Gandhi National Centre for the Arts.

Marcum, J. P. (1999). Measuring church attendance: A further look. *Review of Religious Research, 41*, 122–130.

Markhum, I. (2010). Religion and the human future: An essay on theological humanism. *Conversations in Religion & Theology, 8*, 184–192.

Marks, L. (2000). Exploring regional diversity in patterns of religious participation: Canada in 1901. *Historical Methods, 33*, 247–254.

Marsh, C., & Zhong, Z. (2010). Registration of Chinese protestant house churches under China's 2005 regulation on religious affairs: Resolving the implementation impasse. *Journal of Church and State, 52*, 34–49.

Martikaien, M. (2006). Consuming a cathedral: Commodification of religious places in late modernity. *Fieldwork in Religion, 2*, 127–145.
Marty, M. E. (1982). Religious power in America: A contemporary map. *Criterion, 21*, 27–31.
Marty, M. E. (1997). Theology by the map. *Christian Century, 114*. http://www.guesta.com/magazine/161-19417014/theology-by-the-map
Marty, M. E. (2001). Book review of *The new historical atlas of religion in America. Annals of the Association of American Geographer, 91*, 754–757.
Mary, A. (2007). Pilgrimage to Imeko (Nigeria). An African church in the time of a "global village". *International Journal of Urban and Regional Research, 26*, 106–120.
Masev, V., Shirlow, P., & Joni, D. (2009). The geography of conflict and death in Belfast, Northern Ireland. *Annals of the Association of American Geographers, 99*, 893–903.
Mastagar, M. (2006). Icons and immigrant context. *Fieldwork in Religion, 2*, 146–159.
Matsui, K. (1993). Development of geography of religion in Japan. *The Human Geography/ Jimbun Chiri, 45*(5), 75–93.
Matthey, L., Felli, R., & Mager, C. (2013). We do have space in Lausanne, we have a large cemetery: The non-controversy of a non-existent Muslim burial ground. *Social and Cultural Geography, 30*, 245–265.
Maunder, C. (2013). Mapping the presence of Mary: Germany's modern apparition shrines. *Journal of Contemporary Religion, 28*, 79–93.
Mayberry, B. L. (2013). The land of allium: An exploration into the magic of place. *Journal of Cultural Geography, 30*, 245–265.
Mayhew, R. (1996). Landscape, religion and knowledge in eighteenth-century England. *Ecumene, 3*, 454–471.
Mayhew, R. T. (2000). Geography is twined with divinity. The Laudian geography of Peter Heylyn. *Geographical Review, 90*, 18–34.
McAlister, E. (2005). Globalization and the religious production of space. *Journal for the Scientific Study of Religion, 44*, 249–255.
McClymond, M. J. (2002). Making sense of the census, or, what 1,999,563,838 Christians might mean for the study of religion. *Journal of the American Academy of Religion, 70*, 875–890.
McConnell, J. (2009). "The fabric of our lives". Church perspectives on the Internet. *Fieldwork in Religion, 4*, 150–167.
McConnell, F. (2013). The geopolitics of Buddhist reincarnation: Contested futures of Tibetan leadership. *Area, 45*, 162–169.
McCullum, R. (2011). Micro public spheres and the sociology of religion: An evangelical illustration. *Journal of Contemporary Religion, 26*, 173–187.
McDonald, D. (1995). Changes along the Peyote road. *Geography of Religions and Belief Systems, 17*(2), 1–4.
McGinty, A. M., Sziarto, K., & Seymour-Jorn, C. (2013). Researching within and against Islamophobia: A collaboration project with Muslim communities. *Social and Cultural Geography, 14*, 1–22.
McGregor, J. (2012). Rethinking detention and deportability: Removal centres as spaces of religious revival. *Political Geography, 31*, 236–246.
McGuire, M. B. (1991). Religion and region: Sociological and historical perspective. *Journal for the Scientific Study of Religion, 30*, 544–547.
McIntosh, E. (2010). Special issue – Hearing the other: Feminist theology and others. *International Journal of Public Theology, 4*, 1–4.
McKinney, J. C., & Bourque, L. B. (1971). The changing South: National incorporation of a region. *American Sociological Review, 36*, 399–412.
McLoughlin, D. (1983). The Hejaz railroad. *Geographical Journal, 124*, 282–285.
McVicar, J. M. (2013). Take away the serpents from us: The sign of serpent handling and the development of southern Pentecostalism. *Journal of Southern Religion, 15*. http://jsr.fsu.edu/issues/vol15/mcvicar.html
Meezenbroek, E., & Garson, B. (2012). Measuring spirituality as a universal human experience: A review of spirituality questionnaires. *Journal of Religion and Health, 51*, 336–354.

Megoran, N. (2004). Faith and vocation: Christianity and political geography: On faith and geopolitical imagination. *The Brandywine Review of Faith and International Affairs, 2*, 40–46.
Megoran, N. (2006). God on our side: The Church of England and the geopolitics of mourning 9/11. *Geopolitics, 11*, 561–579.
Megoran, N. (2010). Towards a geography of peace: Pacific geopolitics and evangelical Christian Crusade apologies. *Transactions of the Institute of British Geographers, 35*, 383–398.
Megoran, N. (2013). Radical politics and the apocalypse: Activist readings of Revelation. *Area, 45*, 141–147.
Meinig, D. W. (1965). The Mormon culture region: Strategies and patterns in the geography of the American West, 1847–1964. *Annals of the Association of American Geographers, 55*, 191–220.
Merino, S. M., & Finke, R. (2008–2009). Stimulating research and discovery in the study of religion: The Association of Religious Data Archives (www.theARDA.com). *Geographies of Religions and Belief Systems, 3*, 3–17.
Messerschmidt, D. A. (1983–1984). Geography and sacred symbolism of Muktinath Shrine, Nepal. *The Himalayan Review, 15*, 39–57.
Meyer, J. W. (1975). Ethnicity, theology and immigrant church expansion. *Geographical Review, 65*, 180–197.
Meyer, J. F. (1994). The miracles of Wutaishan, China: The ambiguity of place in Buddhism. *National Geographical Journal of India, 40*, 141–148.
Meyer, K., Barker, E., Ebaugh, H. R., & Juergensmeyer, M. (2011). Religion in global perspective: Looking back, looking forward. *Journal for the Scientific Study of Religion, 50*, 240–251.
Michaels, A. (1993). God versus cars: On moveable and immovable gods at the Nepalese Pasupati Temple. *National Geographical Journal of India, 39*, 151–159.
Mikoski, G. (2013). Going places: Travel seminars as opportunities for interfaith education. *Teaching Theology and Religion, 16*, 352–361.
Miles-Watson, J. (2013). Pipe organs and *satsang*: Contemporary worship in Shimla's colonial churches. *Culture and Religion, 14*, 204–222.
Miller, A. S., & Nakamura, T. (1996). On the stability of church attendance patterns during a time of demographic change: 1965–1988. *Journal for the Scientific Study of Religion, 35*, 275–284.
Mills, S. (2011). Duty to God/my Dharma/Allah/Waheguru: Diverse youthful religiosities and the politics of performance of informal worship. *Social and Cultural Geography, 13*, 481–499.
Mitchell, R. D. (1966). The Presbyterian Church as an indicator of westward expansion in eighteenth-century America. *The Professional Geographer, 18*, 293–299.
Moats, F. I. (1928). The rise of Methodism in the Middle West. *Mississippi Valley Historical Review, 15*, 69–88.
Moberg, N. M. (2011). The concept of *Scene* and its applicability in empirically grounded research on the intersection of religion/spirituality and popular music. *Journal of Contemporary Religion, 26*, 403–417.
Moodley, T., Esterhuyse, K. G. E., & Beukes, R. B. I. (2012). Factor analyzing the spatial well-being questionnaire using a sample of South African adolescents. *Religion & Theology, 19*, 122–151.
Mori, M. (2001). Contemporary religious meaning of the pilgrimage route. *The Human Geography/Jimbun Chiri, 53*(2), 75–92.
Mori, M. (2002). Spatial formation and change in the Henro pilgrimage in modern Japan. *The Human Geography/Jimbun Chiri, 54*(2), 1–22.
Morin, K., & Guelke, J. K. (1998). Strategies of representation, relationship, and resistance: British women travelers and Mormon plural wives, ca 1870–1890. *Annals of the Association of American Geographers, 88*, 436–462.
Morse, S., & McNamara, N. (2008). Creating a partnership: Analyzing partnership in the Catholic church development chair. *Area, 40*, 65–78.
Moser, S. (2012). Circulating visions of 'High Islam:" The adoption of fantasy Middle Eastern architecture in constructing Malaysian national identity. *Urban Studies, 48*, 2913–2935.

Mudu, P. (2002). Repressive tolerance: The gay movement and the Vatican in Rome. *GeoJournal, 58*, 189–196.

Mueris, R. (2009). Scripture, reason and the contemporary Islam-West encounter: Studying the "other", understanding the "self". *Conversations in Religion & Theology, 7*, 156–165.

Mugambi, J. N. K. (2013). Missionary presence in interreligious encounters and relationships. *Studies in World Christianity, 19*, 162–186.

Mulder, M. T. (2008–2009). Mobility and the (in)significance of place in an evangelical church: A case study of the south side of Chicago. *Geographies of Religions and Belief Systems, 3*, 18–45.

Müller, L. F. (2010). Dancing golden stools: Indigenous religion as a strategy for identity construction in Ghana. *Fieldwork in Religion, 5*, 32–57.

Muller, R. (2013). Historiography and cross-cultural research into African indigenous Christianity (AIC). *Studies in World Christianity, 19*, 5–24.

Muse, E. A. (2011). New England through kingdom eyes: The quiet revival and recontextualizing the Chinese Christian Church. *Journal of Contemporary Religion, 26*, 73–90.

Muthieh, R. (2013). Teaching the seedbeds: Educational perspective on theological education in Asia. *Teaching Theology and Religion, 16*, 395–396.

Myers, G. C. (1962). Patterns of church distribution and movement. *Social Forces, 40*, 354–363.

Myers, S. M. (2000). The impact of religious involvement on migration. *Social Forces, 79*, 755–783.

Nagra, R. (1997). The meaning of Hindu communal organizations, places and identities in post colonial Dar es Salaam. *Environment and Planning D: Space and Society, 15*, 707–730.

Nagra, B. (2011). 'Our faith has been hijacked:' Reclaiming Muslim identity in Canada in a post-9/11 era. *Journal of Ethnic and Migration Studies, 37*, 425–441.

Naidoo, M. (2011). An empirical analysis in spiritual formation of Protestant theological training institutions in South Africa. *Religion & Theology, 18*, 118–146.

Nakagawa, T. (1983). The significance of religious factors in the formation of settlement characteristics: Two settlements on the eastern shore of Lake Kasumigaura. *The Human Geography/Jimbun Chiri, 35*(2), 1–19.

Nakagawa, T. (1990a). Cemetery landscape evolution of a Japanese rural community. *Annual Report, University of Tiukuba, Institute of Geoscience, 16*, 8–12.

Nakagawa, T. (1990b). Grave structures and decorations of Louisiana cemeteries. *Tiukuba Studies in Human Geography, 14*, 145–168 (in Japanese).

Nakagawa, T. (1990c). Louisiana cemeteries as cultural artifacts. *Geographical Review of Japan, Series B, 63*, 139–155.

Nardella, C. (2012). Religious symbols in Italian advertising: Symbolic appropriation and the management of consent. *Journal of Contemporary Religion, 27*, 217–240.

Nayak, A. (2012). Race, religion and British multiculturalism: The political responses of black and minority ethnic voluntary organizations to multicultural cohesion. *Political Geography, 31*, 454–463.

Naylor, S., & Ryan, J. R. (2002). The mosque in the suburbs: Negotiating religion and ethnicity in South London. *Social and Cultural Geography, 3*, 39–59.

Neitz, M. J. (2005). Reflections on religion and place: Rural churches and American religion. *Journal for the Scientific Study of Religion, 44*, 243–247.

Nelsen, H. M. (1988). Unchurched black Americans: Patterns of religiosity and affiliation. *Review of Religious Research, 29*, 398–412.

Newman, W. M. (1982). Updating the archives: Churches and Church membership in the U.S. *Review of Religious Research, 24*, 54–59.

Newman, W. M. (1984). Religion and regional culture: Patterns of concentration and change among American religious denominations, 1952–1980. *Journal for the Scientific Study of Religion, 23*, 304–315.

Newman, D. (1986). Culture, conflict and cemeteries: Lebensraum for the dead. *Journal of Cultural Geography, 7*, 99–115.

Newman, W. M. (1988). Counting America's Jews: Clarifying the procedures: A research note. *Review of Religious Research, 29*, 431–434.

Newman, W. M. (1990). An American diaspora: Patterns of Jewish population distribution and change, 1971–1980. *Review of Religious Research, 31*, 259–267.

Newman, W. M. (1993). The church membership studies: An assessment of four decades of institutional research. *Review of Religious Research, 35*, 55–61.

Newman, W. M., & Halvorson, P. L. (1979). American Jews: Patterns of geographic distribution and change, 1952–1971. *Journal for the Scientific Study of Religion, 18*, 183–193.

Newman, W. M., Halvorson, P. L., & Brown, J. (1977). Problems and potential uses of the 1952 and 1971 National Council of Churches "Churches and Church Membership in the United States" studies. *Review of Religious Research, 18*, 167–173.

Ng, P. T.-M. (2007). 'Glocalization; as a key to the interplay between Christianity and Asian culture: The vision of Francis Wei in early 20th Century China. *International Journal of Public Theology, 1*, 101–111.

Nice, V. (2013). Christianity and secularization in South Africa: Probing the possible link between modernization and secularisation. *Studies in World Christianity, 19*, 141–161.

Nitz, J. (1983). Church as colonist: The Benedictine Abbey of Lorsch and planned *Waldhafen* colonization in the Odenwald. *Journal of Historical Geography, 9*, 105–126.

Noble, A. G. (1986). Landscape of piety/landscape of provit: The Amish-Mennonite and derived landscapes of northeastern Ohio. *East Lakes Geographer, 21*, 34–48.

Nolan, M. L. (1983). Irish pilgrimage: The different tradition. *Annals of the Association of American Geographers, 73*, 421–438.

Nolan, M. L. (1986). Pilgrimage traditions and the nature mystique in western European culture. *Journal of Cultural Geography, 7*, 5–20.

Nolan, M. L. (1987a). A profile of Christian pilgrimage shrines in western Europe. *National Geographical Journal of India, 33*, 229–238.

Nolan, M. L. (1987b). Christian pilgrimage shrines in western Europe and India: A preliminary comparison. *National Geographical Journal of India, 33*, 370–378.

Nolan, M. L. (1992). Religious sites as tourism attractions in Europe. *Annals of Tourism Research, 19*, 68–78.

Norman, A., & Johnson, M. (2012). World Youth Day: The creation of a modern pilgrimage event for evangelical intent. *Journal of Contemporary Religion, 26*, 371–385.

Numrich, P. D. (1997). Recent immigrant religions in a restructuring metropolis: New religious landscapes in Chicago. *Journal of Cultural Geography, 17*, 55–76.

Obadare, E. (2012). A sacred duty to resist tyranny: Rethinking the role of Catholic Church in Nigeria's struggle for democracy. *Journal of Church and State, 55*, 92–112.

Olliver, A. (1989). Christian geographers' fellowship conference report. *Area, 21*, 106–108.

Olsen, E. (2013). Gender and geopolitics in 'secular time'. *Area, 45*, 148–154.

Olson, E. (1985). Muslim identity and secularisms in contemporary Turkey: "The headscarf dispute". *Anthropological Quarterly, 58*, 161–170.

Olson, D. (1998). Religious pluralism in U.S. counties. *American Sociological Review, 63*, 759–761.

Olson, D. (1999a). Religious pluralism and U.S. church membership: A reassessment. *Sociology of Religion, 60*, 149–174.

Olson, D. V. A. (1999b). Religious pluralism and U.S. church membership: A reassessment. *Sociology of Religion, 60*, 149–173.

Olson, E. (2008a). "What kind of Catholic are you?" Reflexivity, religion and activism in the Peruvian Andes. *Fieldwork in Religion, 3*, 103–121.

Olson, E. (2008b). Common belief, contested meanings: Development and faith-based organizational culture. *Tijdschrift voor Economische en Sociale Geografie, 99*, 393–405.

Olson, D., & Guelke, J. K. (2004). Spatial transgression and the BYU Jerusalem Center controversy. *The Professional Geographer, 56*, 503–515.

Olson, D. V. A., & Perl, R. (2011). A friend in creed: Does the religious composition of geographic areas affect the religious composition of a person's close friends. *Journal for the Scientific Study of Religion, 50*, 483–502.

Olson, E., & Silvey, R. (2006). Editorial: Transnational geographies: Rescaling development, migration and religion. *Environment and Planning A, 38*, 805–808.

Onizuka, K. (1995). Relationship between religious places and boundaries in ancient palace cities and provincial capitals. *The Human Geography/Jimbun Chiri, 47*, 1–20.
Openshaw, J. (2006). Home or ashram? The Vaishnavas of Bengal. *Fieldwork in Religion, 2*, 65–82.
Otterstrom, S. M. (2008). Divergent growth of the Church of Jesus Christ of Latter Day Saints in the United States, 1990–2004: Diaspora, gathering and the East-West divide. *Population, Space, and Place, 14*, 231–252.
Pacione, M. (1990). The ecclesiastical community of interest as a response to urban poverty and deprivation. *Transactions of the Institute of British Geographers, 34*, 1073–1089.
Pacione, M. (1991). The church urban fund: A religio-geographical perspective. *Area, 23*, 101–110.
Pacione, M. (1992). The church urban fund and inter-parish linkages. *Area, 24*, 171–173.
Pacione, M. (1999). The relevance of religion for a relevant human geography. *Scottish Geographical Journal, 115*, 117–131.
Pacione, M. (2001). Religion and relevance in human geography: Some further issues. *Scottish Geographical Journal, 116*, 67–70.
Pacione, M. (2005). The geography of religious affiliation in Scotland. *The Professional Geographer, 57*, 235–255.
Pacione, M. (2009). The geography of religious affiliation in Glasgow. *Journal of Cultural Geography, 26*, 369–391.
Packard, J., & Sanders, G. (2013). The Emerging Church as corporatization's line of flight. *Journal of Contemporary Religion, 28*, 437–455.
Paget, J. C. (2012). Albert Schweitzer and Africa. *Journal of Religion in Africa, 42*, 277–316.
Palmer, D. A., Shive, G., & Wickeri, P. (2012). Chinese religious life. *Reviews in Religion and Theology, 19*, 504–507.
Palmer-Boyes, A. (2010). The Latino Catholic parish as a specialist organization: Distinguishing characteristics. *Review of Religious Research, 51*, 302–323.
Palmisano, S. (2009). Reconstructors: Reinventing the spiritual path within Italian Catholicism. *Fieldwork in Religion, 4*, 29–45.
Pandey, L. M. (1987). Segregated pattern of religious communities: A case study of Almara town, Uttar Pradesh. *National Geographer, 22*, 15–20.
Pantoja, M. R., & de Conceição, O. S. (2006). The use of Ayahuasca among rubber tappers in the Upper Junuá. Translated by Robin Wright, revised by Matthew Meyer. *Fieldwork in Religion, 2*, 235–255.
Pattison, W. D. (1955). The cemeteries of Chicago: A phase of land utilization. *Annals of the Association of American Geographers, 45*, 245–257.
Payne, D. P. (2010). Spiritual security: The Russian Orthodox Church and the Russian Foreign Ministry: Collaboration or cooptation. *Journal of Church and State, 52*, 712–727.
Peach, C. (1992). Islam in Europe. *Geography Review, 5*, 2–6.
Peach, C. (2002). Social geography: New religions and ethnoburbs – Contrasts with cultural geography. *Progress in Human Geography, 26*, 252–260.
Peach, C. (2006a). Islam, ethnicity and South Asian religions in the London 2001 census. *Transactions, Institute of British Geographers, N.S., 31*, 353–370.
Peach, C. (2006b). Muslims in the 2001 census of England and Wales: Gender and economic disadvantage. *Ethnic and Racial Studies, 29*, 629–655.
Peach, C., & Gale, R. (2003). Muslims, Hindus and Sikhs in the new religious landscape of England. *Geographical Review, 83*, 469–490.
Peach, C., & Glebe, G. (1995). Muslim minorities in western Europe. *Ethnic and Racial Studies, 18*, 26–45.
Pearson, C. (2010). Editorial. Special issue on climate change and the common good. *International Journal of Public Theology, 4*, 269–270.
Pereira, C. J. (2013). Geografia da religião: Um alhar panorâmico. *Espaco Geografico em Aalise, 27*, 10–37.
Pérez, E. (2012). Staging transformations: Spiritist liturgies as theaters of conversion in Afro-Cuban religious practice. *Culture and Religion, 13*, 361–389.

Pessi, A. B. (2011). Religiosity and altruism: Exploring the link and its relation to happiness. *Journal of Contemporary Religion, 26*, 1–18.

Petersen, W. (1962). Religious statistics in the United States. *Journal for the Scientific Study of Religion, 1*, 165–178.

Petersen, J. H., Petersen, K., & Kolstrup, S. (2013). Autonomy, cooperation or colonization? Christian philanthropy and state welfare in Denmark. *Journal of Church and State, 55*.

Photiadis, J. D., & Schnabel, J. F. (1977). Religion: A persistent institution in a changing Appalachia. *Review of Religious Research, 19*, 32–42.

Pieper, J. (1989). The monastic settlements of the Yellow Church in Ladakh, central places in a nomadic habitat. *GeoJournal, 1*, 41–54.

Piggott, C. A. (1980). A geography of religion in Scotland. *Scottish Geographical Magazine, 86*, 130–140.

Pihlaja, S. (2011). "Are you religious or are you saved?" Defining membership categories in religious discussions on YouTube. *Fieldwork in Religion, 6*, 47–63.

Pillsbury, R. (1971). The religious geography of Pennsylvania: A factor analytic approach. *Proceedings of the Association of American Geographers, 3*, 130–134.

Pitman, J. (2011). Feminist public theologies in the Uniting Church in Australia. *International Journal of Public Theology, 5*, 143–164.

Poppora, D. V. (2006). Methodological atheism, methodological agnosticism and religious experience. *Journal for the Theory of Social Behavior, 36*, 57–75.

Porteous, J. D. (1991). Transcendental experience in wilderness sacred space. *National Geographical Journal of India, 37*, 99–107.

Porteous, J. D. (1993). Resurrecting environmental religion. *National Geographical Journal of India, 39*, 179–187.

Portugali, J. (1991). Jewish settlement in the occupied territories: Israel's settlement structure and the Palestinians. *Political Geography Quarterly, 10*, 26–53.

Possami, A., Bellamy, J., & Castle, K. (2006). The diffusion of new age practices and believes among Australian church attendees. *Fieldwork in Religion, 2*, 9–26.

Poudel, P. C., & Singh, R. P. B. (1994). Pilgrimage and tourism at Muktinath, Nepal: A study of sacrality and spatial structure. *National Geographical Journal of India, 40*, 249–268.

Power, M. (2011). Preparing evangelical Protestants for peace: The evangelical contribution on Northern Ireland (ECONI) and peace building 1987–2005. *Journal of Contemporary Religion, 26*, 57–72.

Presser, S., & Stinson, L. (1998). Data collection mode and social desirability bias in self-reported religious attendance. *American Sociological Review, 63*, 137–145.

Preston, L. W. (2002). Shrines and neighborhood in early nineteenth-century Pune. *Journal of Historical Geography, 28*, 203–215.

Price, L. (1966). Some results and implications of a cemetery study. *The Professional Geographer, 18*, 20–27.

Prorok, C. V. (1986). The Hare Krishna's transformation of space in West Virginia. *Journal of Cultural Geography, 7*, 129–140.

Prorok, C. V. (1987). The Canadian Presbyterian mission in Trinidad: John Morton's work among East Indians. *National Geographical Journal of India, 33*, 253–262.

Prorok, C. V. (1991). Evolution of the Hindu temple in Trinidad. *Caribbean Geography, 3*, 73–93.

Prorok, C. V. (1993). Creating the sacred from the ordinary. *Scholars, 4*(2), 4–11.

Prorok, C. V. (1994). Hindu temples in the western world: A study in social space and ethnic identity. *Geographia Religionum, 8*, 95–108.

Prorok, C. V. (1997). The significance of material culture in historical geography: A case study of the church as school in the diffusion of the Presbyterian mission to Trinidad. *Historical Reflections/Reflections Historiques, 23*, 371–388.

Prorok, C. V. (1998a). Dancing in the fire: Ritually constructing Hindu identity in a Malaysian landscape. *Journal of Cultural Geography, 17*, 89–114.

Prorok, C. V. (1998b). Patterns of pilgrimage behavior among Hindus in Trinidad. *Geographia Religionum, 4*, 189–199.

Prorok, C. V. (2000). Boundaries are made for crossing: The feminized spatiality of Puerto Rican Espiritismo in New York City. *Gender Place & Culture: A Journal of Feminist Geography, 7*, 57–80.

Prorok, C. V. (2003). Transplanting pilgrimage traditions in the Americas. *Geographical Review, 93*, 283–307.

Prorok, C. V., & Faiers, G. (1990). Pilgrimage to a "national" American shrine: Our Lady of Consolation in Carey, Ohio. *Geographia Religionum, 5*, 137–147.

Prorok, C. V., & Hemmasi, M. (1993). East Indian Muslims and their mosques in Trinidad: A geography of religious structures and the politics of ethnic identity. *Caribbean Geography, 4*, 28–48.

Pulido, L. (1998). The sacredness of "Mother Earth:" Spirituality, activism, and social justice. *Annals of the Association of American Geographers, 88*, 719–723.

Quack, J. (2012). Organised atheism in India: An overview. *Journal of Contemporary Religion, 27*, 67–85.

Raivo, P. J. (1997). Comparative religion and geography: Some remarks on the geography of religion and religious geography. *Temenos, 33*, 137–149.

Raj, D. S. (2000). 'Who the hell do you think you are?' Promoting religious identity among young Hindus in Britain. *Ethnic and Racial Studies, 23*, 535–558.

Ray, A. (1996). Allahabadkumble celebrations: A study in pilgrim geography. *Geographical Review of India, 58*, 68–76.

Razin, E., & Hazan, A. (2004). Municipal boundary conflicts between Jewish and Arab local authorities in Israel: Geography of administration or geopolitics? *Geografiska Annaler, 86B*, 79–94.

Rechlin, A. T. (1972). The Epsilon generalization. *Annals of the Association of American Geographers, 62*, 578–581.

Reiffenstein, T., & Selig, N. (2013). Shifting monument production chains and the implications for gravestone design on Prince Edward Island, 1920–2005. *Journal of Cultural Geography, 30*, 160–186.

Reinbold, J. (2013). Sacred institutions and secular law: The faltering voice of religion in the courtroom debate over same-sex marriage. *Journal of Church and State, 55*.

Ridanpää, J. (2009). Geopolitics of humor: The Muhammed cartoon crisis and the *Kaltio* comic strip episode in Finland. *Geopolitics, 14*, 729–749.

Rigby, C. L. (2011). Almost Christian. What the faith of our teenagers is tell us about the American church. *Conversations in Religion & Theology, 9*, 213–233.

Rinschede, G. (1986). The pilgrimage town of Lourdes. *Journal of Cultural Geography, 7*, 21–33.

Rinschede, G. (1987). The pilgrimage town of Lourdes. *National Geographical Journal of India, 33*, 379–421.

Rinschede, G. (1990). Religious tourism. *Geographische Rundschau, 42*, 14–20 (in German).

Rinschede, G. (1994). Catholic pilgrimage centers in Quebec, Canada. *National Geographical Journal of India, 40*, 287–305.

Rinschede, G., & Sievers, A. (1987). The pilgrimage phenomenon in socio-geographic research. *National Geographical Journal of India, 33*, 213–217.

Roeland, J., Klaver, M., Van der Meulen, M., Van Mulligen, R., Stoffels, H., & Versteeg, P. (2012). "Can we dance in this place?" Body practices and forms of embodiment in four decades of Dutch evangelical youth events. *Journal of Contemporary Religion, 12*, 241–256.

Roemer, J. E. (2010). Religion and subjective well-being in Japan. *Review of Religious Research, 51*, 402–410.

Rogers, A. P. (2013). Congregational hermeneutics: A tale of two churches. *Journal of Contemporary Religion, 28*, 489–506.

Roislien, H. E. (2011). Via Facebook in Jerusalem: Social media as a toolbox for the study of religion. *Fieldwork in Religion, 6*, 6–26.

Rose, G. S. (1986). Quakers, North Catalonians and blacks in Indiana's settlement patterns. *Journal of Cultural Geography, 7*, 35–48.

Rose, M. (2010). Pilgrims. An ethnography of sacredness. *Cultural Geographies, 17,* 507–524.
Rosenwaike, I. (1989). The geographical distribution of American's Jewish elderly. *Papers in Jewish Demography, 25,* 145–153.
Roucek, J. S. (1970). Census and estimated data on religion in the United States. *Annali di Sociologia, 7,* 15–25.
Rowley, J. (Ed.). (1979). The Muslim world. *People, 6*(4): 3–27.
Rowley, G. (1984). Divisions in a holy city. *Geographical Magazine, 56,* 196–202.
Rowley, G. (1989). The centrality of Islam: Space, form and process. *GeoJournal, 18,* 351–359.
Rowley, G. (1990). The Jewish colonization of the Nablus region – Perspectives and continuing developments. *GeoJournal, 21,* 349–362.
Rowley, R. J. (2012). Religion in sin city. *Geographical Review, 102,* 76–92.
Rowley, G., & El-Hamden, S. A. (1977). Once a year in Mecca. *Geographical Magazine, 49,* 753–759.
Rowley, G., & El-Hamden, S. A. (1978). The pilgrimage to Mecca: An explanatory and predictive model. *Environment and Planning A, 10,* 1053–1071.
Roy, B. K. (1987). Census count of religions of India, 1901–1981 and contemporary issues. *National Geographical Journal of India, 33,* 239–252.
Rubin, R. (2006). One city, different views: A comparative study of three pilgrimage maps of Jerusalem. *Journal of Historical Geography, 32,* 267–290.
Ruez, D. (2013). Partitioning the sensible at Park 51: Ranciére, Islamophobia and common politics. *Antipode, 45,* 1128–1147.
Rush-Midbar, M., & Zachman, N. (2013). 'Everything starts within;' New age values, images and language in Israeli advertising. *Journal of Contemporary Religion, 28,* 421–436.
Rutter, E. (1929). The Muslim pilgrimage. *Geographical Journal, 74,* 271–273.
Ryan, M. M., & McKenzie, F. H. (2003). A monastic touristic experience: The packaging of place. *Tourism Geographies, 5,* 54–70.
Sack, R. D. (1999a). A sketch of a geographic theory on morality. *Annals of the Association of American Geographers, 89,* 26–44.
Sack, R. D. (1999b). A sketch of a geographic theory of morality. *Annals of the Association of American Geographers, 89,* 26–44.
Sadgrove, J., Vanderbeck, R. M., Ward, K., Valentine, G., & Andersson, J. (2010). Constructing the boundaries of Anglican orthodoxy: An analysis of the Global Anglican Future Conference (GAFCON). *Religion, 40*(3), 193–206.
Sadgrove, J., Vanderbeck, R. M., Andersson, J., Valentine, G., & Ward, K. (2012). Morality plan and money matters: Toward a situated understanding of the politics of homosexuality in Uganda. *Journal of Modern African Studies, 50*(1), 103–129.
Saint-Blaneet, C. (2007). Islam in diaspora. Between reterritorialization and extraterritoriality. *International Journal of Urban and Regional Research, 26,* 138–151.
Saleh, H. A. K. (1990). Jewish settlement and its economic impact on the West Bank, 1967–1987. *GeoJournal, 21,* 337–348.
Samers, M. E. (2003). Diaspora unbound: Muslim identity and the erratic regulation of Islam in France. *International Journal of Population Geography, 9,* 351–364.
Sarkissian, A. (2009). Religious reestablishment in post-Communist politics. *Journal of Church and State, 51,* 472–501.
Saunders, R. A. (2013). Pagan places: Towards a religiogeography of neopaganism. *Progress in Human Geography, 33,* 786–810.
Schmelzkopf, K. (2002). Landscape, ideology, and religion: A geography of Ocean Grove, New Jersey. *Journal of Historical Geography, 28,* 589–608.
Schmitt, T. (2003). *Moscheen in Deustchland. Konflikte um ihre Errichtung und Nutzung.* Leipzig: Forschungen zur deutschen Landeskinde. 252.
Schnell, I. (1990). The Israeli Arabs: The dilemma of social integration in development. *Geographische Zeitschrift, 78,* 78–92.
Schnell, I. (1997). Nature and environment in Socialist-Zionist pioneers' perceptions. *Ecumene, 4,* 69–85.

Schoenecke, M. (1999). American sacred space. *Journal of American Culture, 22*, 111.
Schwartzberg, J. E. (1965). The distribution of selected castes in the Southern Indian Plain. *Geographical Review, 55*, 477–495.
Seavoy, R. (1983). Religious motivation for placer diamond mining. *Journal of Cultural Geography, 3*, 56–60.
Secor, A. (2001). Toward a feminist counter-geopolitics: Gender, space and Islamist politics in Istanbul. *Space and Polity, 5*, 199–219.
Secor, A. (2002). The veil and urban space in Istanbul: Women's dress, mobility, and Islamic knowledge. *Gender, Place and Culture, 9*(1), 5–22.
Secor, A., & Gökariksel, B. (2009). New transnational geographies of Islamism, capitalism, and subjectivity: The veiling fashion industry in Turkey. *Area, 41*, 6–18.
Sengupta, S. (1989). Marginality and identity: The Jewish community of Calcutta. *Geographical Review of India, 51*, 15–20.
Shair, I. M. (1981a). Frequency of pilgrimage to Makkahi and pilgrims' socio-economic attributes. *Journal, College of Arts, University of Riyadh, 8*, 13–22.
Shair, I. M. (1981b). Volume of Muslim pilgrims in recent years, 1975–1980. Source areas and ports of entry. *Journal, College of Arts, University of Riyadh, 9*, 293–320.
Shair, I. M., & Karan, P. P. (1979). Geography of the Islamic pilgrimage. *GeoJournal, 3*, 599–608.
Sharkansky, I., & Auerbach, G. (2000). Which Jerusalem: A consideration of concepts and borders. *Environment and Planning D: Society and Space, 18*, 395–409.
Sharma, S., & Guest, M. (2013). Navigating religion between university and home: Christian students; experiences in English universities. *Social and Cultural Geography, 14*, 59–79.
Sheleff, L. (2001). Jerusalem – Figment of the imagination. *GeoJournal, 53*, 297–309.
Shelton, T., Zook, M., & Graham, M. (2012). The technology of religion: Mapping religious cyberspaces. *The Professional Geographer, 64*, 602–617.
Sherkat, D. E. (2001). Tracking the restructuring of American religion: Religious affiliation and patterns of religious mobility, 1973–1998. *Social Forces, 79*, 1459–1493.
Sheskin, I. M. (1993). Jewish metropolitan homelands. *Journal of Cultural Geography, 13*, 119–132.
Shilav, Y. (1983). Principles for the location of synagogues: Symbolism and functionalism in a spatial context. *The Professional Geographer, 35*, 324–329.
Shilhav, Y. (2001). Religious factors in territorial disputes: An intro-Jewish view. *GeoJournal, 53*, 247–259.
Shilong, Z. (1993). Ashkenzai Jewish almshouses in Jerusalem. *Journal of Cultural Geography, 14*, 35–48.
Shirlow, P., & Murtagh, B. (2006). *Belfast. Segregation, violence and the city*. London: Pluto.
Shortridge, J. R. (1976). Patterns of religion in the United States. *Geographical Review, 66*, 420–434.
Shortridge, J. R. (1977). A new regionalization of American religion. *Journal for the Scientific Study of Religion, 16*, 143–153.
Shortridge, J. R. (1978). The pattern of American Catholicism, 1971. *Journal of Geography, 77*, 56–60.
Shuttleworth, I., Barr, P. J., & Gould, M. (2013). Does internal migration in Northern Ireland increase religious and social segregation? Perspectives from the Northern Ireland Longitudinal Study (NILS) 2001–2007. *Population, Space and Place, 19*, 72–86.
Sievers, A. (1987). The significance of pilgrimage tourism in Sri Lanka. *National Geographical Journal of India, 33*, 430–447.
Silvey, R. (2007). Mobilizing piety: Gendered morality and Indonesian-Saudi transnational migration. *Mobilities, 2*, 219–229.
Simoons, F. J. (1979). Questions in the sacred-cow controversy. *Current Anthropology, 50*, 467–493.
Sims, D. H. (1920). Religious education in Negro colleges and universities. *Journal of Negro History, 5*, 166–207.

Singh, R. P. B. (1987a). Emergence of the geography of belief systems (GBS) and a search for identity in India. *National Geographical Journal of India, 33*, 184–204.
Singh, R. P. B. (1987b). The pilgrimage mandala of Varanasi/Kasi: A study in sacred geography. *National Geographical Journal of India, 33*, 493–524.
Singh, R. P. B. (1988). The image of Varanasi: Sacrality and perceptual world. *National Geographical Journal of India, 34*, 1–32.
Singh, R. P. B. (1991a). Pancakrosi Yatra, Varanasi: Sacred journey, ecology of place and faithscape. *National Geographical Journal of India, 37*, 49–98.
Singh, R. P. B. (1991b). Rama's route after banishment: A geographic viewpoint. *Journal of Scientific Research, 41B*, 39–46.
Singh, R. P. B. (1992a). The geography of pilgrimages in India: Perspectives and prospects. *National Geographical Journal of India, 38*, 39–54.
Singh, R. P. B. (1992b). The geography of religion in India: Perspectives and prospects. *National Geographical Journal of India, 38*, 27–38.
Singh, R. P. B. (1993). Cosmos, theos, anthropos: An inner vision of sacred ecology in Hinduism. *National Geographical Journal of India, 39*, 113–130.
Singh, R. P. B. (1994a). The sacred geometry of India's holy city, Varanasi: Kashi as Cosmogram. *National Geographical Journal of India, 40*, 189–216.
Singh, R. P. B. (1994b). Time and Hindu rituals in Varanasi: A study of sacrility and cycles. *Geographica Religionum, 8*, 123–138.
Singh, R. P. B. (1997). Sacredscape, cosmic territory, and faithscape: Goddess territory of Vindhyachal, India. *National Geographical Journal of India, 43*, 237–263.
Singh, R. P. B., & Malville, J. M. (1995). Cosmic order and cityscape of Varanasi (Kashi): Sun images and cultural astronomy. *National Geographical Journal of India, 41*, 69–88.
Singh, R. P. B., & Shahi, R. P. (1989). The religious landscape in an urban renewal programme: A study in sacred geography of space and time. *National Geographical Journal of India, 35*, 163–190.
Sinha, A. (1993). Nature in Hindu mythology, art, and architecture. *National Geographical Journal of India, 39*, 131–140.
Sinha, A. (1994). Pilgrimage-journey to the sacred landscape of Braj. *National Geographical Journal of India, 40*, 239–248.
Sinha, A. (1998). Design of settlements in the Vaastu Shastras. *Journal of Cultural Geography, 17*, 27–41.
Sittler, J. (1976). Space and time in American religious experience. *Interpretation, 30*, 44–51.
Sitwell, O. F. G. (1994). Sacred space reconsidered. *National Geographical Journal of India, 40*, 101–110.
Skirbekk, V., Kaufmann, E., & Goujon, A. (2010). Secularism, fundamentalism or Catholicism? The religious composition of the United States in 2043. *Journal for the Scientific Study of Religion, 49*, 293–310.
Slater, T. R. (2004). Encountering God. Personal reflections on "geographer as pilgrim". *Area, 36*, 245–253.
Smidt, C. E., & Kellstedt, L. A. (1996, July–August). How to count the spirit-filled. *Books and Culture,* pp. 24–25.
Smith, D. M. (1988). How far should we dare? On the spatial scope of beneficence. *Progress in Human Geography, 22*, 14–38.
Smith, T. W. (1990). Classifying Protestant denominations. *Review of Religious Research, 31*, 225–245.
Smith, V. L. (1992). Introduction: The quest in guest. *Annals of Tourism Research, 19*, 1–17.
Smith, T. W. (1998). A review of church attendance measures. *American Sociological Review, 63*, 131–136.
Smith, T. W. (2002). Religious diversity in America: The emergence of Muslims, Buddhists, Hindus, and others. *Journal for the Scientific Study of Religion, 41*, 577–585.
Smith, G. (2005). Religious identities: Social networks and the power of information. *Fieldwork in Religion, 1*, 291–311.
Smith, J. I. (2013a). Islam in America. *Journal of Religion, 93*, 77–87.

Smith, C. (2013b). Anti-Islam sentiment and media framing during the 9/11 decade. *Journal of Religion and Society, 15*, 1–15.
Smith, J. (2013c). Creating a godless community: The collective identity at work in contemporary American atheists. *Journal of the Society for Studying Religion, 52*, 80–99.
Smith, C., & Faris, R. (2005). Socioeconomic inequality in the American religious system: An update and assessment. *Journal for the Scientific Study of Religion, 44*, 95–104.
Smith, M., & Marden, P. (2014). Capturing the religious spirit: A challenge for the secular state. *Journal of Church and State, 55*, 23–49.
Smith, C., Denton, M. L., Faris, R., & Regnerus, M. (2002). Mapping American adolescent religious participation. *Journal for the Scientific Study of Religion, 41*, 597–612.
Smith, M. B., Laurie, N., Hopkins, P., & Olson, E. (2012). International volunteering, faith and subjectivity: Negotiating cosmopolitanism, citizenship and development. *Geoforum, 45*, 126–135.
Smits, F., Ruiter, S., & Van Tubergen, F. (2010). Religious practices among Islamic immigrants: Moroccan and Turkish men in Belgium. *Journal for the Scientific Study of Religion, 49*, 247–263.
Smittle, W. R. (1976). Catholic colleges. *Journal of Higher Education, 7*, 87–92.
Sopher, D. E. (1964). Indigenous uses of turmeric in Asia and Oceania. *Anthropos, 59*, 93–127.
Sopher, D. E. (1968). Pilgrimage circulation in Gujarat. *Geographical Review, 58*, 392–425.
Sopher, D. E. (1981). Geography and religion. *Progress in Human Geography, 5*, 510–524.
Sopher, D. E. (1987). The message of place in Hindu pilgrimage. *National Geographical Journal of India, 33*, 353–369.
Southern, N. (2011). Strong religion and political viewpoints in a deeply divided society: An examination of the Gospel Hall Tradition in Northern Ireland. *Journal of Contemporary Religion, 26*, 433–449.
Stanislawski, D. (1975). Dionysus westward: Early religion and the economic geography of wine. *Geographical Review, 65*, 427–444.
Stanley, B. (2011). Edinburgh and world Christianity. *Studies in World Christianity, 17*, 72–91.
Stanley, B. (2013). Contested interpretations of Christian identity. *Studies in World Christianity, 19*, 205–207.
Stark, R. (1987). Correcting church membership rates: 1971 and 1980. *Review of Religious Research, 29*, 69–77.
Stark, R. (1992). The reliability of historical United States census data on religion. *Sociological Analysis, 53*, 91–95.
Stark, R., & Glock, C. Y. (1968). *Patterns of religious commitment.* Berkeley: University of California Press. Survey Research Center.
Starr, C. W. (2011). Faith with film is dull: C. S. Lewis connects evangelicals in art, movies and worldview analysis. *Christian Scholar's Review, 40*.
Starr, C. (2013). How theology is flourishing in China. Classroom Christianity. *Christian Century, 130*(3), 28–34.
Starrs, P. F., & Wright, J. B. (2005). Utopia, dystopia, and sublime apocalypse in Montana's Church Universal and Triumphant. *Geographical Review, 95*, 97–121.
Stausberg, M., & Tessermann, A. (2013). The appropriation of a religion: The case of Zoroastrianism in contemporary Russia. *Culture and Religion, 14*, 445–462.
Stoddard, R. H. (1968). Analysis of the distribution of Hindu holy sites. *National Geographical Journal of India, 14*, 148–155.
Stoddard, R. H. (1979–1980). Perceptions about the geography of religious sites in the Kathmandu Valley. *Contributions to Nepalese Studies, 7*, 97–118.
Stoddard, R. H. (1987). Pilgrimages along sacred paths. *National Geographical Journal of India, 33*, 448–456.
Stoddard, R. H. (1988). Characteristics of Buddhist pilgrimages in Sri Lanka. *Geographica Religionum, 3*, 45–61.

Stoddard, R. H. (1990). Some comments about the history of the geography of religion and belief systems. *Geography of Religions and Belief Systems, 12*, 1–3.

Stoddard, E. (2008). 'Yes. No. Cancel:' Clicking your way to a public theology of cyberdemocracy. *International Journal of Public Theology, 2/3*, 328–353.

Storrar, W. (2012). Special issue: Faith-based organizations in the USA. *International Journal of Public Theology, 6*, 377–381.

Strothmann, L. (2013). Giving comfort, dispelling fear: Social welfare at the shrine of Data Granj Bukhsh in Lahore, Pakistan. *Erdkunde, 67*, 49–61.

Stump, R. W. (1984a). Regional divergence in religious affiliation in the United States. *Sociological Analysis, 45*, 283–299.

Stump, R. W. (1984b). Regional migration and religious commitment in the United States. *Journal for the Scientific Study of Religion, 23*, 292–303.

Stump, R. W. (1985). Toward a geography of American civil religion. *Journal of Cultural Geography, 5*, 87–95.

Stump, R. W. (1986a). Patterns in the survival of Catholic national parishes, 1940–1980. *Journal of Cultural Geography, 7*, 77–97.

Stump, R. W. (1986b). Pluralism in the American place-name cover: Ethnic variations in Catholic Church names. *North American Culture, 2*, 126–140.

Stump, R. W. (1986c). Regional variations in the determinants of religious participation. *Review of Religious Research, 27*, 208–225.

Stump, R. W. (1986d). The geography of religion – Introduction. *Journal of Cultural Geography, 7*, 1–3.

Stump, R. W. (1987a). Regional contrasts within black Protestantism: A research note. *Social Forces, 66*, 143–151.

Stump, R. W. (1987b). Regional variations in denominational switching among white Protestants. *The Professional Geographer, 39*, 438–449.

Stump, R. W. (1998). The effects of geographical variability on Protestant church membership trends, 1980–1990. *Journal for the Scientific Study of Religion, 37*, 636–651.

Sturm, T. (2013). The future of religions geopolitics: Towards a reach and theory agenda. *Area, 45*, 134–169.

Stuvland, A. (2010). Emerging church and global civil society. Postmodern Christianity as a source for global values. *Journal of Church and State, 52*, 203–231.

Sutcliffe, S. J., & Cusack, C. (2013). Introduction: 'Making it (all?) up' – 'Invented religions' and the study of 'religion'. *Culture and Religion, 14*, 353–361.

Swan, J. A. (1991). Sacred places in nature and transpersonal experiences. *National Geographical Journal of India, 37*, 40–47.

Swan, J. A. (1994). Sacred places of the Bay Area. *National Geographical Journal of India, 40*, 123–130.

Swatos, W. H., & Kivisto, P. (1991). Max Weber as "Christian sociologist". *Journal for the Scientific Study of Religion, 30*, 347–362.

Takashasi, S. (2009). Church of museum – The role of state museums in conserving church buildings, 1965–1985. *Journal of Church and State, 51*, 502–517.

Talebi, M. K., & Desjardines, M. (2013). The immigration experience of Iranian Baha'is in Saskatchewan: The reconstitution of their existence, faith and religious experience. *Journal of Religion and Health, 51*, 293–309.

Tanaka, H. (1981). The evolution of pilgrimage as a spatial-symbol system. *The Canadian Geographer, 25*, 240–251.

Tanaka, H. (1984). Landscape expression of the evolution of Buddhism in Japan. *The Canadian Geographer, 28*, 240–257.

Tatum, C. E., & Sommers, L. M. (1975). The spread of the Black Christian Methodist Episcopal Church in the United States, 1870–1970. *Journal of Geography, 74*, 343–357.

Taylor, B. (2013). On departing empire: A mission of the churches in the Caribbean. *International Journal of Public Theology, 7*, 355–376.

Taylor, R. J., Chatters, L. M., Mattais, J. S., & Joe, S. (2010). Religious involvement among Caribbean blacks in the United States. *Review of Religious Research, 52*, 125–145.

Teather, E. K. (1999). High-rise homes for the ancestors: Cremation in Hong Kong. *Geographical Review, 89*, 409–430.

Temerius, S. (1996). The relationship between church attendance and selected socio-economic factors: A spatial analysis. *Transactions of Missouri Academy of Science, 30*, 135–136.

Tharan, Z. J. (1997). The Jewish American environmental movement: Stewardship, renewal and the greening of diaspora politics. *Pennsylvania Geographer, 35*, 69–97.

Thomas, O. H. W. (2013). Cricket in the Caribbean as theological practice. *International Journal of Public Theology, 7*, 398–408.

Thompson, T. J. (2011). Beyond empire: Post colonialism and missionaries in global context. *Studies in World Christianity, 17*, 93–94.

Tillman, B. F., & Emmett, C. F. (1999). Spatial succession of sacred space in Chicago. *Journal of Cultural Geography, 18*, 79–108.

Tilson, D. J. (2010). The 2006 winter Olympics and the Shroud of Turin: A confluence of town, vestment and media. *Fieldwork in Religion, 4*, 123–149.

Tinker, C., & Smart, A. (2012). Constructions of collective Muslim identity by advocates of Muslim schools in Britain. *Ethnic and Racial Studies, 35*, 643–663.

Tobin, G. A. (1989). Issues in the study of the urban and regional distribution of Jews in the United States. *Papers in Jewish Demography in 1985*, 66–75.

Togarasei, L. (2012). Mediating the gospel: Pentecostal Christianity and media technology in Botswana and Zimbabwe. *Journal of Contemporary Religion, 12*, 257–274.

Tomkinson, M. (1969). Seaside city for Mecca's pilgrims. *Geographical Magazine, 42*, 93–104.

Toney, M. B., & Stinner, C. M. (1983). Mormon and non-Mormon migration in and out of Utah. *Review of Religious Research, 25*, 114–126.

Tong, C. K., & Kong, L. (2000). Religion and modernity: Ritual transformations and the reconstruction of space and time. *Social and Cultural Geography, 1*, 29–44.

Torma, R., & Teussner, P. E. (2012). iReligion. *Studies in World Christianity, 17*, 137–155.

Toulouse, M. G. (2012). Christian churches and their peoples, 1840–1965: A social history of religion in Canada. *Journal of Religion, 92*, 554–556.

Trepanier, C. (1986). The Catholic Church in French Louisiana: An ethnic institution? *Journal of Cultural Geography, 7*, 59–75.

Trzebiatowska, M. (2010). When reflexivity is not enough: Doing research with Polish Catholics. *Fieldwork in Religion, 5*, 78–96.

Tse, J. K. H. (2013). Grounded theologies: 'Religion' and the 'secular' in human geography. *Progress in Human Geography*. doi:10.1177/0309132512475105.

Tuan, Y.-F. (1976). Humanistic geography. *Annals of the Association of American Geographers, 66*, 266–276.

Turner, B. (2001). Cosmopolitan virtue: On religion in a global age. *European Journal of Social Theory, 4*, 131–152.

Tweedie, S. W. (1978). Viewing the Bible Belt. *Journal of Popular Culture, 11*, 865–876.

Valentine, G., Vanderbeck, R. M., Ward, K., Sadgrove, J., & Andersson, J. (2010). Implacements: The event as a prism, for exploring intersectionality, a case study of Lambeth conference (2008). *Sociology, 44*(5), 925–943.

Valentine, G., Vanderbeck, R. M., Sadgrove, J., & Andersson, J. (2013a). Producing moral geographies: The dynamics of homophobias within a transnational religious network. *Geographical Journal, 179*, 165–176.

Valentine, G., Vanderbeck, R. M., Sadgrove, J., Andersson, J., & Ward, K. (2013b). Transnational religious networks: Sexuality and the changing power geometries of the Anglican Communion. *Transactions of the Institute of British Geographers, 38*, 50–64.

Valins, O. (2000). Institutionalized religion: Sacred texts and Jewish spatial practice. *Geoforum, 31*, 575–586.

Valins, O. (2003). Defending identities or segregation communities? Faith-based schooling and the UK Jewish community. *Geoforum, 34*, 235–247.

Van der Heyden, U. (1996). The archives of the Berlin Mission Society. *History in Africa, 23*, 411–427.
Van Tubergen, F., & Sindradottir, J. I. (2011). The religiosity of immigrants in Europe: A cross-national study. *Journal for the Scientific Study of Religion, 50*, 272–288.
Vanderbeck, R. M., Valentine, G., Ward, K., Sadgrove, J., & Andersson, J. (2010, May 20). The meanings of communion: Anglican identities, the sexuality debates and Christian relationality. *Sociological Research Online, 15*(2).
Vanderbeck, R. M., Andersson, J., Valentine, G., Sadgrove, J., & Ward, K. (2011). Sexuality, activism, and witness in the Anglican communion: The 2008 Lambeth conference of Anglican bishops. *Annals of the Association of American Geographers, 101*, 670–689.
Vann, B. (2007). The geotheolgoical imagings of a trans-Irish Sea Scottish community. *Geographies of Religions and Belief Systems, 2*, 21–39.
Verhoeven, T. (2013). The case for Sunday mails: Sabbath laws and the separation of state in Jacksonian America. *Journal of Church and State, 55*, 71–91.
Vertovec, S., & Peach, C. (1997). *Islam in Europe: The politics of religion and community*. Basingstoke: Macmillan.
Voas, D., & McAndrew, S. (2012). Three puzzles of non-religion in Britain. *Journal of Contemporary Religion, 12*, 29–48.
Voegt, J. W. (1977). Sacred space, architectural tradition, and the contemporary designer. *The Himalayan Review, 9*, 41–53.
Voeks, R. (1990). Sacred leaves of Brazilian candomble. *Geographical Review, 80*, 118–131.
Von Sinner, R., & Cavalcante, R. (2012). Special issue on theology in Brazil. *International Journal of Public Theology, 6*, 1–6.
Von Strukrad, K. (2013). Secular religion: A discourse-historical approach to religion in contemporary western Europe. *Journal of Contemporary Religion, 28*, 1–14.
Waitt, G. (2003). A place for Buddha in Wollongong, New South Wales? Territorial rules in the place-making of sacred spaces. *Australian Geographer, 34*, 233–238.
Wang, W. F., Zhou, S. Y., & Fan, C. C. (2002). Growth and decline of Muslim Hui enclaves in Beijing. *Eurasian Geography and Economics, 43*, 104–122.
Wanner, C. (2004). Missionaries of faith and culture. Evangelical encounters in Ukraine. *Slavic Review, 63*, 732–755.
Wanner, C. (2010). Southern challenges to Eastern Christianity: Pressures to reform the state-church model. *Journal of Church and State, 52*, 619–643.
Warburg, M., Schepelern, B. J., & Ostergaard, K. (2013). Counting *niqabs* and *burqas* in Denmark: Methodological aspects of quantifying rare and elusive religious sub-cultures. *Journal of Contemporary Religion, 28*, 33–48.
Ward, R. (1980). Migration, myth and magic in Papua New Guinea. *Australian Geographical Studies, 18*, 119–134.
Warf, B., & Vincent, P. (2007). Religious diversity across the globe: A geographic exploration. *Social and Cultural Geography, 8*, 597–613.
Warf, B., & Winsberg, M. (2008). The geography of religious diversity in the United States. *The Professional Geographer, 60*, 413–424.
Warhola, J. W., & Bezci, E. B. (2010). Religion and state in contemporary Turkey. Recent developments in Laiklik. *Journal of Church and State, 52*, 427–453.
Warkentin, J. (1989). Mennonite agricultural settlements in southern Manitoba. *Geographical Review, 49*, 342–368.
Waters, M. S., Heath, W. C., & Watson, J. K. (1995). A positive model of the determination of religious affiliation. *Social Science Quarterly, 76*, 105–123.
Weaver-Zercher, D. (1999). Putting the Amish to work: Mennonites and the Amish culture market, 1950–1975. *Church History, 68*, 87–117.
Webster, G. R., & Leib, J. I. (2002). Political culture, religion and the Confederate flag debate in Alabama. *Journal of Cultural Geography, 20*, 1–26.
Weightman, B. A. (1993). Changing religious landscapes in Los Angeles. *Journal of Cultural Geography, 14*, 1–20.

Weightman, B. A. (1996). Sacred landscapes and the phenomenon of light. *Geographical Review, 86*, 59–71.

Welch, C. (2013). Dead in the field: Utilizing fieldwork to explore the historical interpreting of death related activity and the emotional coping with death. *Fieldwork in Religion, 8*, 127–132 (8 articles on cemeteries, death and dying).

Welch, M. R., & Baltzell, J. (1984). Geographic mobility, social integration, and church attendance. *Journal for the Scientific Study of Religion, 23*, 75–91.

Wentz, R. E. (1981). Region and religion in America. *Foundations, 24*, 148–156.

Whelan, K. (1983). The Catholic parish, the Catholic chapel and village development in Ireland. *Irish Geography, 16*, 1–15.

Williams, R. H. (2005). Introduction to a forum on religion and place. *Journal for the Scientific Study of Religion, 44*, 239–242.

Williams, A. M. (2013). Surfacing therapeutic landscapes: Exploring cyberpilgrimage. *Culture and Religion, 14*, 78–93.

Williamson, R. (2013). Using Twitter to teach reader-oriented biblical interpretation: Teaching the Gospel of Mark. *Teaching Theology and Religion, 16*, 274–286.

Williamson, R. (2014). Teaching the Bible in a liberal arts classroom. *Teaching Theology and Religion, 17*, 94–95.

Wilson, D. (1993). Constructing social process and space in the geography of religion. *Area, 25*, 75–76.

Wilson, J., & Janoski, T. (1955). The contribution of religion to volunteer work. *Sociology of Religion, 56*, 137–152.

Winders, J. (2005). Changing politics of race and region: Latino migration to the U.S. South. *Progress in Human Geography, 29*, 683–699.

Wolfel, R. L. (2001). The diffusion of evangelical abolitionism. *The Pennsylvania Geographer, 39*, 38–62.

Woodberry, R. D. (1998). When surveys lie and people tell the truth: How surveys oversample church attenders. *American Sociological Review, 63*, 119–122.

Woodhead, L. (2009). Old, new and emerging paradigms in the sociological study of religion. *Nordic Journal of Religion and Society, 22*, 103–121.

Woods, A. (2013a). The place of religion in Chicago. *Social and Cultural Geography, 14*, 482–483.

Woods, O. (2013b). The spatial modalities of evangelical Christian growth in Sri Lanka: Evangelism, social ministry and the structural mosaic. *Transactions of the Institute of British Geographers, 38*, 652–664.

Wright, R. M. (2006). The Brazilian Ayahuasca religions. *Fieldwork in Religion, 2*, 171–186.

Wu, K. K. (1994). The road of Saint James-El Camino de Santiago: Power of place on a medieval pilgrimage route. *National Geographical Journal of India, 40*, 131–140.

Xie, X. (2010). Religion and modernity in China. Who is joining the Three-Self Church and why? *Journal of Church and State, 52*, 74–93.

Yang, F. (2010). Chinese views on church and state. *Journal of Church and State, 52*, 3–33.

Yeoh, B. S. A., & Hui, T. B. (1995). The politics of space: Changing discourses on Chinese burial grounds in post-war Singapore. *Journal of Historical Geography, 21*, 184–201.

Yiftrachel, O., & Zacobi, H. (2002). Discussion. Planning a bi-national capita: Should Jerusalem remain united? *Geoforum, 33*, 137–144.

Yip, A. K. T. (2005). Religion and the politics of spirituality/sexuality. *Fieldwork in Religion, 1*, 271–289.

Yip, A. K. T. (2010). Special feature; Sexuality and religion/spirituality. *Sexualities, 13*, 667–670.

Yoder, M. L. (1985). Findings from the 1982 Mennonite census. *Mennonite Quarterly Review, 59*, 307–349.

Yorgason, E., & Della Dora, V. (2009). Geography, religion and emerging paradigms. Problematizing the dialogue. *Social and Cultural Geography, 10*, 629–637.

Zelinsky, W. (1960). Geographical record: The religious composition of the American population. *Geographical Review, 50*, 272–273.

Zelinsky, W. (1961). An approach to the religious geography of the United States: Patterns of church membership in 1952. *Annals of the Association of American Geographers, 51*, 139–194.

Zelinsky, W. (1973). *The cultural geography of the United States* (pp. 94–100 specifically on religion). Englewood Cliffs: Prentice-Hall.

Zelinsky, W. (2001a). Book review of The New Historical Atlas of Religion in America. *Annals of the Association of American Geographer, 91*, 757–762.

Zelinsky, W. (2001b). The uniqueness of the American religious landscape. *Geographical Review, 91*, 565–585.

Zelinsky, W. (2007). The gravestone index: Tracking personal religiosity across nations, regions, and periods. *Geographical Review, 97*, 441–446.

Zhai, J. E., & Woodberry, R. D. (2012). Religion and educational ideals in contemporary Taiwan. *Journal for the Scientific Study of Religion, 50*, 307–327.

Zhange, N. (2011). Rewriting Jesus in Republican China: Religion, literature and cultural nationalism. *Journal of Religion, 91*, 223–252.

Books and Atlases (Mostly by Geographers)

Aapeis, S., & Houtman, D. (Eds.). (2010). *Religions of modernity: relocating the sacred to the self and the digital*. London: Brill.

Aay, H., & Griffioen, S. (Eds.). (1998). *Geography and worldview: A Christian reconnaissance*. Lanham: University Press of America.

Abruzzi, W. S. (1993). *Dam that river. Ecology and Mormon settlement in the Little Colorado River Basin*. Lanham: University Press of America.

Adogame, A. (Ed.). (2011). *Who is afraid of the Holy Ghost? Pentecostalism and globalization in African and beyond*. Trenton: Africa World Press.

Aitchison, C., Hopkins, P. E., & Kwan, M. P. (Eds.). (2007). *Geographies of Muslim identities: Diaspora, gender and belonging*. Aldershot: Ashgate.

Albera, D., & Couroucli, M. (Eds.). (2013). *Shared spaces in the Mediterranean: Christian, Muslims and Jews*. Bloomington: Indiana University Press.

Aldridge, I. I. R., & Sopher, E. D. (1974). *The changing shape of Protestantism in the South*. Macon: Mercer University Press.

Al-Faruqi, I., & Sopher, D. (1974). *Historical atlas of religions of the world*. New York: Macmillan.

Anderson, A., Bergrunder, M., & Droogers, A. F. (Eds.). (2010). *Studying global Pentecostalism: Theories and methods*. Berkeley: University of California Press.

Aquino, M. P., & Rosalo-Nunez, M. J. (Eds.). (2007). *Feminist intercultural theology. Latina explorations for a just world*. Maryknoll: Orbis.

Aquino, M. P., Machado, D., & Rodriguez, J. (Eds.). (2002). *A reader in Latina feminist theology: Religion and justice*. Austin: University of Texas Press.

Ballard, P. (2008). *The church at the centre of the city*. Peterborough: Epworth Press.

Baly, D. (1957, 1974). *The geography of the Bible*. New York/Harper/London: Butterworth Press.

Balzer, M. M. (Ed.). (2010). *Religion and politics in the Russian Federation*. Armonk: ME Sharpe.

Balzer, M. M. (2012). *Shamans, spirituality and cultural revitalization: Explorations in Siberia and beyond*. New York: Palgrave Macmillan.

Barbalet, J., Passami, A., & Turner, B. S. (Eds.). (2011). *Religion and the state: A comparative sociology*. London/New York: Anthem Press.

Bays, D. H. (2011). *A new history of Christianity in China*. New York: Wiley.

Beaumont, J., & Baker, C. (2011). *Postsecular cities: Space, theory and practice*. New York/London: Continuum.

Becci, I., Burchardt, M., & Casanova, J. (Eds.). (2013). *Topographies of faith religion in urban spaces*. Boston: Brill.

Berger, P. L. (1967). *The sacred canopy. Elements of sociological theory of religion*. New York: Doubleday.
Berger, P. L. (1999). *The desecularization of the world. Resurgent religion and world politics*. Grand Rapids: W. B. Eerdman.
Berger, P. L., Davie, G., & Fokas, G. (2008). *Religious America, secular Europe*. Aldershot: Ashgate.
Bhardwaj, S., Rinschade, G., & Sievers, A. (Eds.). (1990). *Pilgrimage in the old and new world*. Berlin: Dietrich Reimer Verlag.
Boal, F. W., Keane, M., & Livingstone, D. N. (1999). *Them and us? Attitudinal variation among Belfast churchgoers*. Belfast: Institute of Irish Studies.
Bossius, T., Häger, A., & Kahn-Harris, K. (2011). *Religion and popular music in Europe: Emergent expressions of sacred and secular identity*. London/New York: I. B. Tauris.
Bradley, M. B., Green, N. M., Jones, D. E., Lynn, M., & McNeil, L. (1992). *Churches and church membership in the United States 1990: An enumeration by region, state and county based on data reported for 133 Church Groupings*. Atlanta: Glenmary Research Center.
Brenneman, W. L., & Brenneman, M. G. (1995). *Crossing the circle at the holy wells of Ireland*. Charlottesville: University Press of Virginia.
Brockman, N. C. (1997). *Encyclopaedia of sacred places*. Santa Barbara: ABC-CLIO.
Bruce, S. (2002). *God is dead. Secularization in the west*. Oxford: Oxford University Press.
Buttimer, A. (1993). *Geography and the human spirit*. Baltimore: Johns Hopkins Press.
Campbell, C. (2004). *Images of New Jerusalem. Latter Day Saint faction interpretations of Independence, Missouri*. Knoxville: University of Tennessee Press.
Cao, N. (2010). *Constructing China's Jerusalem: Christians, power and place in contemporary Wenschou*. Palo Alto: Stanford University Press.
Carey, H. M. (2011). *God's empire: Religion and colonialism in the British world, c. 1801–1908*. Cambridge: Cambridge University Press.
Carmichael, D. L., Reeves, H. R., & Schanche, A. (Eds.). (1994). *Sacred sites, sacred places*. London: Routledge.
Carroll, B. E. (2000). *The Routledge historical atlas of religion in America*. New York: Routledge.
Carson, P. (2013). *The East India Company and religion: 1698–1858*. Woodbridge/Rochester: Boydell Press.
Casanova, J. (1994). *Public religions in the modern world*. Chicago: University of Chicago Press.
Cavanaugh, W. T. (2013). *Migrations of the holy: God, state and political meaning of the church*. Grand Rapids: Eerdmans.
Chidester, D., & Linenthal, E. T. (Eds.). (1995). *American sacred space*. Bloomington: Indiana University Press.
Chu, J. (2013). *Does Jesus really love me? A gay Christian's pilgrimage in search of God in America*. New York: Harper Collins.
Coleman, E. (Ed.). (2013). *Region, religion and English Renaissance literature*. Burlington: Ashgate.
Cooper, A. (1997). *Sacred mountains: Ancient wisdom and modern meaning*. Edinburgh: Floris Books.
Crumrine, N. R., & Morinis, E. A. (Eds.). (1991). *Pilgrimage in Latin America*. New York: Greenwood Press.
Cusack, C., & Norman, A. (Eds.). (2012). *Handbook of new religions and cultural production*. Boston/London: Brill.
Davidson, J., Bonde, L., & Smith, M. (Eds.). (2005). *Emotional geographies*. Aldershot: Ashgate.
Davie, G. (1994). *Religion in Britain since 1945. Believing without belonging*. Oxford: Oxford University Press.
Davie, G. (2002). *Europe the exceptional case. Parameters of faith in the modern world*. London: Darton, Longman & Todd.
De Planhol, X. (1959). *The world of Islam*. Ithaca: Cornell University Press.
Deffontaines, P. (1948). *Géographie et religions*. Paris: Gaillimard.

Derrida, J. (2001). *Acts of religion* (edited and with an introduction by G. Anidjar). London: Routledge.
Dowley, T. (Ed.). (1997). *Atlas of the Bible and Christianity*. Grand Rapids: Baker Books.
Eck, D. (2012). *India: A sacred geography*. New York: Harmony Books.
Eliade, M. (1959). *The sacred and the profane*. New York: Harcourt, Brace and Company.
Enders, K., & Lauser, A. (2011). *Engaging the spirit world: Popular beliefs and practices in modern Southeast Asia*. New York: Oxford University Press.
Falah, G.-W., & Nagel, C. (Eds.) (2005). *Geographies of Muslim women: Gender, religion and space*. New York: Guilford Press.
Fenster, T. (2004). *The global city and the holy city: Narratives on knowledge, planning and diversity*. Harlow: Pearson/Prentice-Hall.
Ferrari, S., & Pastorelli, S. (2012). *Religion in public space: A European perspective*. Earnhem/Burlington: Ashgate.
Fitzgerald, T. (2007). *Discourse on civility and barbarity: A critical history of religion and related categories*. Oxford/New York: Oxford University Press.
Foote, K. E. (1997). *Shadowed ground: America's landscapes of violence and tragedy*. Austin: University of Texas Press.
Francaviglia, R. (1978). *The Mormon landscape: Existence, creation, and perception of a unique image in the American West*. New York: AMS Press.
Francaviglia, R. (2003). *Believing in place: A spiritual geography of the Great Basin*. Reno: University of Nevada Press.
Gannett, H. (1898). *Statistical atlas of the United States, eleventh (1890) census*. Washington, DC: U.S. Government Printing Office.
Gaustad, E. S. (1962). *Historical atlas of religion in America*. New York: Oxford University Press.
Gaustad, E. S. (1976). *Historical atlas of religion in America* (Rev. ed.). New York: Harper and Row.
Gaustad, E. S., Barlow, P. L., & Dishno, R. W. (2001). *New historical atlas of religion in America*. New York: Oxford University Press.
Gay, J. D. (1971). *The geography of religion in England*. London: Gerald Duckworth.
Gelder, K., & Jacobs, J. M. (1998). *Uncanny Australia. Sacredness and identity in a postcolonial nation*. Melbourne: University of Melbourne Press.
Gesler, W. M. (2003). *Healing places*. Lanham: Rowman and Littlefield.
Glenmary Research Center. (2002). *Religious congregations and membership in the United States*. Atlanta: Glenmary Research Center.
Goldman, M. (2012). *The American soul rush: Esalen and the rise of spiritual privilege*. New York/London: New York University Press.
Gómez, L. (Ed.). (2012). *The sacred in the city*. London/New York: Continuum.
Good, C. M., Jr. (2004). *The steamer parish. The rise and fall of missionary medicine on the African frontier*. Chicago: University of Chicago Press.
Gornick, M. R. (2010). *World made global: Stories of African Christianity in New York City*. Grand Rapids: Eerdmans.
Gregory, I., Cunningham, N., Lloyd, C., Shuttleworth, I., & Eli, P. (2013). *Troubled geographies: A spatial history of religion and society in Ireland*. Bloomington: Indiana University Press.
Gruber, I. (2012). *Orthodox Russia in crisis: Church and nation in the time of troubles*. Dekalb: Northern Illinois University Press.
Guiley, R. E. (1995). *Atlas of the mysterious in North America*. New York: Facts On File.
Gurst, M., & Arweck, E. (Eds.). (2012). *Religion and knowledge: Sociological perspective*. Farnham/Burlington: Ashgate.
Habermas, J. (2002). *Religion and rationality: Essays on reason, God and modernity*. Cambridge: MIT Press.
Hackworth, J. (2012). *Faith based: Religious neoliberalism and the politics of welfare in the United States*. Athens: University of Georgia Press.
Halman, L., Sieben, I., & van Zundert, M. (2011). *Atlas of European values. Trends and traditions at the turn of the century*. Leiden/Boston: Brill.

Halvorson, P. L., & Newman, W. M. (1987). *Atlas of religious change in America, 1971–1980*. Atlanta: Glenmary Research Center.
Halvorson, P. L., & Newman, W. M. (1994). *Atlas of religious change in America, 1952–1990*. Atlanta: Glenmary Research Center.
Halvorson, P. L., Newman, W. M., & Nielsen, M. C. (1978). *Atlas of religious change in America, 1952–1971*. Washington, DC: Glenmary Research Center.
Hamilton, M. B. (2001). *The sociology of religion. Theoretical and comparative perspectives*. London: Macmillan.
Hardwick, S. (1993b). *Russian refuge: Religion, migration and settlement on the North American Pacific Rim*. Chicago: University of Chicago Press.
Harpur, J. (1994). *Atlas of sacred places: Meeting points of heaven and earth*. New York: Henry Holt.
Hart, J. (2006). *Sacramental commons: A Christian ecological ethics*. Lanham: Rowman and Littlefield.
Hedlund, R. E. (Ed.). (2010). *The Oxford encyclopedia of South Asian Christianity*. New Delhi: Oxford University Press.
Heelas, P., & Woodhead, L. (2005). *The spiritual revolution: Why religion is giving way to spirituality*. Oxford: Blackwell.
Hermkens, A.-K., Jensen, W., & Notermans, C. (2009). *Moved by Mary: The power of pilgrimage in the modern world*. Farnham: Ashgate.
Hewes, F. W., & Gannett, H. (1883). *Scribner's statistical atlas of the United States, showing by graphic methods their present condition and their political, social and industrial development*. New York: Charles Scribner's Sons.
Higgs, R. J. (1995). *God in the stadium. Sports and religion in America*. Lexington: University Press of Kentucky.
Hitchcock, S. T., & Esposito, J. L. (2004). *Geography of religion: Where God lives, where pilgrims walk*. Washington, DC: National Geographic Society.
Hoge, D. R., Johnson, B., & Luidens, D. A. (1994). *Vanishing boundaries: The religion of mainline Protestant baby boomers*. Louisville: Westminster/John Knox Press.
Hoover, S., & Emerich, M. (Eds.). (2011). *Media, spiritualities and social change*. London/New York: Continuum.
Hopkins, P., & Gale, R. (2009). *Muslims in Britain: Race, place and identities*. Edinburgh: Edinburgh University Press.
Hopkins, P., Kong, L., & Olson, E. (Eds.). (2013). *Religion and place. Landscape, politics and piety*. Dordrecht: Springer.
Hutchins, L., & Williams, H. S. (2012). *Sexuality, religion and the sacred: Bisexual, pansexual and polysexual perspectives*. London/New York: Routledge.
Jacquet, C. H. J. (Ed.). (1982). *Yearbook of American and Canadian Churches* (50th ed.). Nashville: Abingdon.
Johnson, D. W., Picard, P. R., & Quinn, B. (1974). *Churches and church membership in the United States: An enumeration by region, state and county, 1971*. Washington, DC: Glenmary Research Center.
Jones, D. E., Doty, S., Grammich, C., Horsch, J. E., Rouseal, R., Lynn, M., Marcum, J. P., Sanchagrin, K., & Taylor, R. H. (2002). *Religious congregations and membership in the United States 2000: An enumeration by region, state and county based on data reported for 149 religious bodies*. Nashville: Glenmary Research Center.
Kapferer, B., Telle, K., & Eriksen, A. (2012). *Contemporary religiosities: Emergent socialities and the post-nation state*. New York/Oxford: Berghahn Books.
Kedar, B. Z., & Werblowsky, R. J. Z. (Eds.). (1998). *Sacred space: Shrine, city, land*. New York: New York University Press.
Kirby, D. (2013). *Fantasy and belief: Alternative religions, popular narratives and digital cultures*. Sheffield/Bristol: Equinox.
Knott, K. (2005). *The location of religion. Toward a spatial analysis*. London: Equinox.

Kosmin, B. A., Mayer, E., & Keysar, A. (2001). *American religious identification survey 2001*. New York: The Graduate Center of the City University of New York.

Krause, A. E. (1952). *Mennonite settlement in the Paraguayan Chaco*. Chicago: Department of Geography, University of Chicago.

Krindatch, A. (1996). *Geography of religions in Russia*. Decatur: Glenmary Research Center.

Lau, S. (2012). *Popular music in evangelical youth culture*. New York/London: Routledge.

Lindner, E. W. (Ed.). (2002). *Yearbook of American and Canadian churches* (70th ed.). Nashville: Abingdon Press.

Livingstone, D. N. (1992). *The preadamite theory and the marriage of science and religion*. Philadelphia: American Philosophical Society.

Livingstone, D. N. (2008). *Adam's ancestors: Race, religion and politics of human origins*. Baltimore: Johns Hopkins University Press.

Livingstone, D. N., & Wells, R. A. (1999). *Ulster-American religion: Episodes in the history of a cultural connection*. Notre Dame: University of Notre Dame Press.

Livingstone, D. N., Hart, D. G., & Noll, M. A. (1990). *Evangelicals and science in historical perspective*. New York: Oxford University Press.

Luckmann, T. (1967). *The invisible religion*. New York: Macmillan.

McGrath, J. F. (2012). *Religion and science fiction*. Cambridge: Lutterworth Press.

McGuire, M. B. (2008). *Lived religion: Faith and practice in everyday life*. Oxford: Oxford University Press.

Mead, F. S., Hill, S. S., & Atwood, C. D. (2001). *Handbook of denominations in the United States* (11th ed.). Nashville: Abingdon Press.

Mitchell, C., & Ganiel, G. (2011). *Evangelical journeys: Choice and change in a Northern Irish religious structure*. Dublin: University College Dublin.

Moreton, B. (2009). *To serve God and Wal-Mart. The making of Christian free enterprise*. Cambridge: Harvard University Press.

Morin, K. M., & Guelke, J. K. (Eds.). (2007). *Women, religion and space: Global perspectives on gender and faith*. Syracuse: Syracuse University Press.

Müller, R. (2011). *African pilgrimage: Ritual travel in South Africa's Christianity of Zion*. Farnhamn: Ashgate.

Naquin, S., & Yu, C. F. (Eds.). (1992). *Pilgrims and sacred sites in China*. Berkeley: University of California Press.

National Council of Churches of Christ in the USA. (1956–1958). *Churches and church membership in the United States*. New York: National Council of Churches of Christ.

Newman, W. M. (1980). *Patterns in pluralism: A portrait of American religion*. Washington, DC: Glenmary Research Center.

Newman, D. (Ed.). (1985). *The impact of Gush Emunim: Politics and settlement in the West Bank*. London: Croom Helm.

Newman, W. M. (2000). *Atlas of American religion: The denominational era, 1776–1990*. Walnut Creek: AltaMira Press.

Ng, A. H. S. (2011). *Intimating the sacred. Religion in English language Malaysian fiction*. Hong Kong: Hong Kong University Press.

Ng, P. T.-M. (2012). *Chinese Christianity: An interplay between global and local perspectives*. London/Boston: Brill.

Nolan, M. L., & Nolan, S. (1989). *Christian pilgrimage in modern western Europe*. Chapel Hill: University of North Carolina Press.

Noll, M. A., & Livingstone, D. N. (1994). *What is Darwinism? And other writings on science and religion by Charles Hodge*. Grand Rapids: Baker.

Norris, K. (1993). *Dakota: A spiritual geography*. New York: Ticknor and Fields.

O'Brien, J., & Palmer. M. (Eds.). (1993). *The state of religion atlas*. New York: Simon and Schuster.

Packard, J. (2012). *The emerging church: Religion at the margins*. Boulder/London: First Forum Press/Lynne Rienner Publisher.

Park, C. C. (1994). *Sacred worlds. An introduction to geography and religion*. London: Routledge.

Paullin, C. O. (1932). *Atlas of the historical geography of the United States* (edited by J. K. Wright).Washington, DC/New York: Carnegie Institute/American Geographical Society.

Phan, P. C. (Ed.). (2011). *Christianities in Asia*. Chichester: Wiley-Blackwell.

Possamai, A. (Ed.). (2012). *Handbook of hyper-real religions*. London: Brill.

Putney, C., & Burlin, P. T. (Eds.). (2012). *The role of the American Board in the world: Bicentennial reflections in the organization's missionary work*. Eugene: Wipf and Stock.

Quinn, B., Anderson, H., Bradley, M. B., Goetting, P., & Shriver, P. (1982). *Churches and church membership in the United States, 1980: An enumeration by region, state, and county, based on data reported by 111 church bodies*. Atlanta: Glenmary Research Center.

Reader, I., & Walter, T. (Eds.). (1992). *Pilgrimage in popular culture*. Houndmills: Macmillan.

Richardson, M. (2003). *Being-in-Christ and putting death in its place. An anthropologist's account of Christian performance in Spanish American and the American South*. Baton Rouge: Louisiana State University Press.

Rinschede, G., & Bhardwaj, S. M. (Eds.). *Pilgrimage in the United States*. Berlin: Dietrich Reimer Verlag.

Scott, J., & Simpson-Housley, P. (Eds.). (1991). *Sacred places and profane spaces: Essays in the geographics of Judaism, Christianity, and Islam*. New York: Greenwood Press.

Shelton, J. P., & Emerson, M. O. (2012). *Blacks and whites in Christian America: How racial discrimination shapes religious convictions*. New York: New York University Press.

Shibley, M. A. (1996). *Resurgent evangelicalism in the United States: Mapping cultural change since 1970*. Columbia: University of South Carolina Press.

Shipps, J., & Silk, M. (Eds.). (2004). *Religion and public life in the Mountain West: Sacred landscapes in transition*. Walnut Creek: AltaMira Press.

Singh, R. P. B. (Ed.). (2013). *Hindu tradition of pilgrimage: Sacred space and system*. New Delhi: Dev Publishers.

Smart, N. (Ed.). (1999). *Atlas of the world's religions*. New York: Oxford University Press.

Smith, J. (1982). *Imaging religion: From Babylon to Jamestown*. Chicago: University of Chicago Press.

Smith, J. (1978). *Map is not territory: Studies in the history of religions*. Leiden: Brill.

Sopher, D. E. (1967). *Geography of religions*. Englewood Cliffs: Prentice-Hall.

Stoddard, R. H., & Morinis, A. E. (Eds.). (1997). *Sacred places, sacred spaces: The geography of pilgrimages*. Baton Rouge: Louisiana State University. Geoscience Publications.

Stump, R. W. (2000). *Boundaries of faith: Geographical perspectives on religious fundamentalism*. Lanham: Rowman and Littlefield.

Stump, R. W. (2008). *The geography of religion: Faith, place, and space*. Lanham: Rowman and Littlefield.

Taves, A. (2009). *Religious experience reconsidered: A building-block approach to the study of religion and other special things*. Princeton: Princeton University Press.

Thompson, T. J. (2012). *Light in darkness: Missionary photography of Africa in the late nineteenth century and early twentieth century*. Grand Rapids/Cambridge: Eerdmans.

Tuan, Y.-F. (2010). *Religion: From place to placelessness*. Chicago: University of Chicago Press.

Tweed, T. (1997). *Our Lady of the Exile: Diasporic religion at a Cuban Catholic shrine in Miami*. New York: Oxford University Press.

Wáhrishu-Oblau, C. (2013). *The missionary self: Perception of Pentecostal-Charismatic church leaders from the Global South in Europe: Bringing back the gospel*. Leiden: Brill.

Walker, F. A. (1874). *Statistical atlas of the United States based on the results of the ninth census, 1870*. New York: Julius Bien.

Werner, D., Esterhne, E., Kang, N., & Raja, J. (Eds.). (2010). *Handbook of theological education in world Christianity: Theological perspectives, ecumenical trends, regional surveys*. Eugene: Wipf and Stock.

Wheatley, P. (1971). *The pivot of the four quarters: A preliminary enquiry into the origin and character of the Chinese city*. Chicago: Aldine.

Whitman, L. B., & Trimble, G. W. (1956). *Churches and church membership in the United States: An enumeration and analysis by counties, states and regions.* New York: National Council of Churches.
Wilford, J. (2013). *Sacred subdivisions: The postsuburban transformation of American evangelicalism.* New York: New York University Press.
Williams, P. W. (2000). *Houses of God: Region, religion, and architecture in the United States.* Urbana: University of Illinois Press.
Wills, G. (2007). *Head and heart. American Christianities.* New York: Penguin Press.
Wilson, J. (2012). *Dixie Dharma: Inside a Buddhist temple in the American South.* Chapel Hill: University of North Carolina Press.
Winter, G. (1962). *The suburban captivity of the churches: An analysis of Protestant responsibility in the expanding metropolis.* New York: Macmillan.
Wunder, E. (2005). *Religion in der postkonfessionellen Gesellschaft. Ein Beitrag zur sozialwissenschaftlichen Theorieentwicklung in der Religionsgeographie.* Stuttgart Sozialgeographische Bibliotek 5.
Wuthnow, R. (2004). *Saving America? Faith-based services and the future of civil society.* Princeton: Princeton University Press.
Zelinsky, W. (1994). *Exploring the beloved country: Geographic forays into American society and culture.* Iowa City: University of Iowa.
Zelinsky, W., & Matthews, S. A. (2011). *The place of religion in Chicago.* Chicago: Center for American Places at Columbia College.

Chapters (Mostly by Geographers)

Aaftaab, N. G. (2005). (Re)Defining public spaces through developmental education for Afghan women. In G.-W. Falah & C. Nagel (Eds.), *Geographies of Muslim women: Gender, religion and space* (pp. 44–67). New York: Guilford Press.
Alkire, S. (2006). Religion and development. In D. A. Clark (Ed.), *The Elgar companion to development studies* (pp. 502–510). Cheltenham: Edward Elgar.
Ayari, L., & Brosseau, M. (2005). Writing place and gender in novels by Tunisian women writers. In G.-W. Falah & C. Nagel (Eds.), *Geographies of Muslim women: Gender, religion and space* (pp. 275–299). New York: Guilford Press.
Baker, C. R. (2008). Contemporary renewal in the centre of the city. In P. Ballard (Ed.), *The church at the centre of the city* (pp. 29–44). Peterborough: Epworth Press.
Balzer, M. M. (2011). Religions communities and rights in the Russian Federation. In T. Banchoff & R. Wuthnow (Eds.), *Religion and the global politics of human rights* (pp. 247–283). New York: Oxford University Press.
Balzer, M. M. (1999). Shamans emerging from repression in Siberia: Lightning rods of fear and hope. In D. Riboli & D. Torre (Eds.), *Shamanism and violence: Power, repression and suffering in indigenous religious contexts* (Series: Vitality of indigenous religions) (pp. 411–424). Burlington: Ashgate.
Bhardwaj, S. M. (1990). Hindu deities and pilgrimages in the United States. In G. Rinschede & S. M. Bhardwaj (Eds.), *Pilgrimage in the United States* (pp. 211–228). Berlin: Dietrich Reimer Verlag.
Bhardwaj, S. M. (1998). Geography and pilgrimage: A review. In R. H. Stoddard & A. Morinis (Eds.), *Sacred places, sacred spaces: The geography of pilgrimages* (pp. 1–23). Baton Rouge: Louisiana State University. Geoscience Publications.
Boal, F. W., & Livingstone, D. N. (1983). An international frontier in microcosm: The Shankill-Falls divide, Belfast. In S. Waterman & N. Kliot (Eds.), *People, territory and state: Pluralism in political geography* (pp. 138–158). London: Croom Helm.

Bompani, B. (2013). 'It is not a shelter, it is a church!' Religious organisations, and public sphere and xenophobia in South Africa. In P. Hopkins, L. Kong, & E. Olson (Eds.), *Religion and place: Landscape, politics and piety* (pp. 131–147). Dordrecht: Springer.

Buffetrelle, K. (1998). Reflections on pilgrimages to sacred mountains, lakes and caves. In B. Z. Kedar & R. J. Z. Werblowsky (Eds.), *Sacred space: Shrine, city, land* (pp. 18–34). New York: New York University Press.

Büttner, M. (1985). Zur Geschichte und Systematik der Religionsgeographie. In M. Büttner, K. Hocheisel, U. Kōpf, G. Rinschede, & A. Sievers (Eds.), *Grundfragen der Religionsgeographie* (pp. 11–121). Berlin: Geographia Religonum.

Cameron, C. (1990). Pilgrims and politics: Sikh Gurdwaras in California. In G. Rinschede & S. M. Bhardwaj (Eds.), *Pilgrimage in the United States* (pp. 193–209). Berlin: Dietrich Reimer Verlag.

Caplan, A. (1997). The role of pilgrimage priests in perpetuating spatial organization within Hinduism. In R. H. Stoddard & A. Morinis (Eds.), *Sacred places, sacred spaces: The geography of pilgrimages* (pp. 209–233). Baton Rouge: Louisiana State University. Geoscience Publications.

Chapman, G. (1990). Religious vs. regional determinism: India, Pakistan, and Bangladesh as inheritors of empire. In M. Chisholm & D. M. Smith (Eds.), *Shared space, divided space: Essays on conflict and territorial organization* (pp. 106–135). London: Unwin Hyman.

Ciechocinska, M. (1989). Geography of the Roman Catholic Church in Poland. *National Geographical Journal of India, 35*, 115–128.

Cloke, P. (2011). Geography and invisible powers: Philosophy, social action and prophetic potential. In C. Brace, A. Bailey, S. Carter, D. Harvey, & N. Thomas (Eds.), *Emerging geographies of belief* (pp. 9–29). Newcastle: Cambridge Scholars Publishing.

Conradson, D. (2013). Somewhere between religion and spirituality? Places of retreat in contemporary Britain. In P. Hopkins, L. Kong, & E. Olson (Eds.), *Religion and place. Landscape, politics and piety* (pp. 185–202). Dordrecht: Springer.

Curry, J.-R. (1998). Christian worldview and geography: Positivism, conventional relations and the importance of place. In H. Aay & S. Griffioen (Eds.), *Geography and worldview: A Christian reconnaissance* (pp. 49–60). Lanham: University Press of America.

Davie, G. (2007). Vicarious religion. A methodological challenge. In N. Ammerman (Ed.), *Everyday religion. Observing modern religious lives* (pp. 21–37). New York: Oxford University Press.

Davis, D. K. (2005). A space of her own: Women, work and desire in an Afghan nomad community. In G.-W. Falah & C. Nagel (Eds.), *Geographies of Muslim women: Gender, religion and space* (pp. 68–90). New York: Guilford Press.

Din, A. K., & Hadi, A. S. (1997). Muslim pilgrimage from Malaysia. In R. H. Stoddard & A. Morinis (Eds.), *Sacred places, sacred spaces: The geography of pilgrimages* (pp. 161–182). Baton Rouge: Louisiana State University. Geoscience Publications.

Doughty, R. W. (1994). Environmental theology: Trends in Christian thought. In K. E. Foote, P. J. Hugill, K. Mathewson, & J. Smith (Eds.), *Re-reading cultural geography* (pp. 313–322). Austin: University of Texas Press.

Dwyer, C. (1998). Contested identities: Challenging dominant representations of young British Muslim women. In T. Skelton & G. Valentine (Eds.), *Cool places: Geographies of youth cultures* (pp. 50–65). London: Routledge.

Dwyer, C. (1999). Negotiations of femininity and identity for young British Muslim women. In N. Laurie, C. Dwyer, S. Holloway, & F. Smith (Eds.), *Geographies of new femininities* (pp. 135–152). Harlow: Longman.

Dwyer, C., Uberoi, V., & Medwood, T. (2011). Feeling and being Muslim and British. In T. Modood & T. Salt (Eds.), *Global migration, ethnicity and Britishness* (pp. 205–224). London: Palgrave Macmillan.

Faiers, G. E., & Prorok, C. V. (1990). Pilgrimage to a "National" American shrine: "Our Lady of Consolation" in Carey, Ohio. In G. Rinschede & S. M. Bhardwaj (Eds.), *Pilgrimage in the United States* (pp. 137–147). Berlin: Dietrich Reimer Verlag.

Falah, G.-W. (2005). The visual representation of Moslem/Arab women in daily newspapers in North America: A geographical perspective. In G.-W. Falah & C. Nagel (Eds.), *Geographies of Muslim women: Gender, religion and space* (pp. 300–320). New York: Guilford Press.

Fickeler, P. (1962). Fundamental questions in the geography of religions. In P. L. Wagner & M. W. Mikesell (Eds.), *Readings in cultural geography* (pp. 94–117). Chicago: University of Chicago Press.

Fielder, J. (1995). Sacred sites and the city: Urban aboriginality ambivalence and modernity. In R. Wilson & A. Dirlik (Eds.), *Asia/Pacific as space of cultural production* (pp. 101–119). London: Duke University Press.

Freeman, A. (2005). Moral geographies and women's freedom: Rethinking freedom discourse in the Moroccan context. In G.-W. Falah & C. Nagel (Eds.), *Geographies of Muslim women: Gender, religion and space* (pp. 147–177). New York: Guilford Press.

Friedland, R., & Hecht, R. D. (1991). The politics of sacred space: Jerusalem's Temple Mount/al-Sharif. In J. Scott & P. Simpson-Housley (Eds.), *Sacred places and profane spaces: Essays in the Geographies of Judaism, Christianity, and Islam* (pp. 21–61). New York: Greenwood Press.

Giuriati, P., & Lanzi, G. (1994). Pilgrims to Fatima as compared to Lourdes and Medjugorje. In G. Rinschede & S. Bhardwaj (Eds.), *Pilgrimage in the old and new world* (pp. 57–79). Berlin: Dietrich Reimer Verlag.

Giuriati, P., Myers, P. M. G., & Donach, M. E. (1990). Pilgrims to "Our Lady of the Snows" Belleville, Illinois in the Marian Year: 1987–1988. In G. Rinschede & S. M. Bhardwaj (Eds.), *Pilgrimage in the United States* (pp. 149–192). Berlin: Dietrich Reimer Verlag.

Glass, J. W. (1979). Be ye separate, saith the Lord. In R. A. Cybriwsky (Ed.), *The Philadelphia region. Selected essays and field trip itineraries* (pp. 51–63). Washington, DC: Association of American Geographers.

Goh, R. (2003). Deus ex machina: Evangelical sites, urbanism and the construction of social identities. In R. Bishop, J. Phillips, & W.-W. Yen (Eds.), *Postcolonial urbanism: Southeast Asian cities and global processes* (pp. 305–322). London: Routledge.

Gökariksel, B., & Secor, A. (2013). "You can't know how they are inside:" The ambivalence of veiling and discourses of the other in Turkey. In P. Hopkins, L. Kong, & E. Olson (Eds.), *Religion and place: Landscape, politics and piety* (pp. 115–130). Dordrecht: Springer.

Grapard, A. B. (1998). Geotyping sacred space: The case of Mount Hiro in Japan. In B. Z. Kedar & R. J. Z. Werblowsky (Eds.), *Sacred space: Shrine, city, land* (pp. 215–258). New York: New York University Press.

Griffiths, G. (2001). Postcoloniality, religion, geography: Keeping our feet on the ground and our heads up. In J. Scott & P. Simpson-Housley (Eds.), *Mapping the sacred: Religion, geography and postcolonial literatures* (pp. 445–461). Amsterdam/Atlanta: Rodopi BV.

Guelke, J. K. (1995). Mormons and mountains. In W. K. Wyckoff & L. M. Dilsaver (Eds.), *The mountainous west: Explorations in historical geography* (pp. 368–396). Lincoln: University of Nebraska Press.

Guelke, J. K. (1997). Sweet surrender, but what's the gender? Nature and the body in the writings of nineteenth century Mormon women. In J. P. Jones III, H. J. Nast, & S. M. Roberts (Eds.), *Thresholds in feminist geography: Difference, methodology and representation* (pp. 361–382). Boulder: Rowman and Littlefield.

Guelke, J. K. (2003). Judaism, Israel, and natural resources: Models and practices. In H. Selin (Ed.), *Nature across cultures: Non-western views of nature and the environment* (pp. 433–456). Dordrecht: Kluwer.

Han, J. (2011). "If you don't work, you don't eat." Evangelizing development in Africa. In J. Song (Ed.), *New millennium South Korea: Neoliberal capital and transnational movements* (pp. 142–258). London: Routledge.

Handy, R. T. (1982). Protestant patterns in Canada and the United States: Similarities and differences. In J. D. Ban & P. R. Dekar (Eds.), *The great tradition: In honor of Winthrop S. Hudson essays on pluralism, voluntarism and revivalism* (pp. 33–51). Valley Forge: Judson Press.

Henkel, R. (1986). Die Verbreitung der Religionen und Konfessionen in Afrika südlich der Sahara und ihr Zusammenhang mit dem Entwicklungstand der Staaten. In M. Büttner, K. Hoheisel,

U. Köpf, G. Rinschede, & A. Sievers (Eds.), *Religion und Siedlungsraum* (pp. 225–243). Berlin: Geographia Religionum 2.

Henkel, R. (2006). Definition von Religion und Religonstheorien – was kann die Religonsgeographie aus der Diskussion in Religionssoziologie und Religionswissenschaft lernen? In *Acta Universitatis Carolinae* (pp. 77–90). Prague: Geographica 41.

Henkel, R. (2011). Identity, ethnicity and religion in the Western Balkans. In C. Brace, A. Bailey, S. Carter, D. Harvey & N. Thomas (Eds.), *Emerging geographies of belief* (pp. 91–111). Newcastle. Cambridge Scholarly Publishing.

Henkel, R., & Knippenberg, H. (2005). Secularization and the rise of religious pluralism. Main features in the changing religious landscape of Europe. In H. Knippenberg (Ed.), *The changing religious landscape of Europe* (pp. 1–13). Amsterdam: Het Spinhuis.

Hermansen, M. K. (2003). How to put the genie back in the bottle: "Identity Islam" and Muslim youth cultures in America. In O. Safi (Ed.), *Progressive Muslims: On pluralism, gender and justice* (pp. 303–319). Oxford: Oneworld Publications.

Holloway, J. (2003). Spiritual embodiment and sacred rural landscapes. In P. Cloke (Ed.), *Country visions* (pp. 158–175). Harlow: Pearson Education.

Holloway, J. (2011). Spiritual life. In V. Del Casino, M. E. Thomas, P. Cloke, & R. Panelli (Eds.), *A companion in social geography* (pp. 385–401). Chichestser: Wiley-Blackwell.

Holloway, J. (2013). The space that faith makes: Towards a (hopeful) ethos of engagement. In P. Hopkins, L. Kong, & E. Olson (Eds.), *Religion and place. Landscape, politics and piety* (pp. 203–218). Dordrecht: Springer.

Hopkins, P. E. (2008). Politics, race and national: The difference that Scotland makes. In C. Dwyer & C. Bressey (Eds.), *New geographies of race and racism* (pp. 113–124). Aldershot: Ashgate.

Horsley, A. D. (1978). The development and spatial diffusion of gospel quartet music in the states. In G. O. Carney (Ed.), *The sounds of people and places: Readings in the geography of music* (pp. 173–195). Washington, DC: University Press of America.

Jackowski, A. (1990). Development of pilgrimages in Poland: Geographical – Historical study. In G. Lallanji & D. P. Dubey (Eds.), *Pilgrimage studies: Text and context* (pp. 241–250). Allahabad: The Society of Pilgrimage Studies.

Jackson, R. H., Rinschede, G., & Knapp, J. (1990). Pilgrimage in the Mormon Church. In G. Rinschede & S. M. Bhardwaj (Eds.), *Pilgrimage in the United States* (pp. 27–61). Berlin: Dietrich Reimer Verlag.

Jordon, T. G. (1980). A religious geography of the hill country Germans of Texas. In F. C. Leubke (Ed.), *Ethnicity on the Great Plains* (pp. 109–128). Lincoln: University of Nebraska Press.

Karan, P. P. (1997). Patterns of pilgrimage to the Sikh shrine of Guru Gobind Singh at Patna. In R. H. Stoddard & A. Morinis (Eds.), *Sacred places, sacred spaces: The geography of pilgrimages* (pp. 257–268). Baton Rouge: Louisiana State University. Geoscience Publications.

Kaschewsky, R. (1994). Muktnah-A pilgrimage place in the Himalayas. In S. Bhardwaj & G. Rinschade (Eds.), *Pilgrimage in the old and new world* (pp. 139–168). Berlin: Dietrich Reimer.

Katz, Y. (1991). The Jewish religion spatial and communal organization: The implementation of Jewish religious law in the building of urban neighborhoods and Jewish agricultural settlements in Palestine at the close of the nineteenth century. In J. Scott & P. Simpson-Housley (Eds.), *Sacred places and profane spaces: Essays in the geographies of Judaism, Christianity, and Islam* (pp. 3–19). New York: Greenwood Press.

Kokkonen, P. (1999). Early missionary literature and the construction of the popular image of Africa in Finland. In A. Buttimer, S. D. Brunn, & U. Wardenga (Eds.), *Text and image: Social construction of regional knowledges* (pp. 205–214). Leipzig: Institute für Länderkunde.

Kong, L. (2008). Religious processions: Urban politics and poetics. In A. E. Lai (Ed.), *Religious diversity in Singapore* (pp. 295–314). Singapore: ISEAS.

Kong, L. (2013). Christian evangelizing across national boundaries: Technology, cultural capital and the intellectualization of religion. In P. Hopkins, L. Kong & E. Olson (Eds.), *Religion and place: Landscape, politics and piety* (pp. 21–38). Dordrecht: Springer.

Ley, D. (2000). Geography of religion. In R. J. Johnston, D. Gregory, G. Pratt, & M. Watts (Eds.), *The dictionary of human geography* (pp. 697–699). Oxford: Blackwell.

Ley, D., & Tse, J. (2013). *Homo religiosus*? Religion and immigrant subjectivities. In P. Hopkins, K. Long, & E. Olson (Eds.), *Religion and place: Landscape, politics and piety* (pp. 149–165). Dordrecht: Springer.

Livingstone, D. N. (1988). Farewell to arms: Reflections on the encounter between science and faith. In M. A. Noll & D. F. Wells (Eds.), *Christian faith and practice in the modern world* (pp. 239–262). Grand Rapids: Eerdmans.

Livingstone, D. N. (1998). Geography and the natural theology imperative. In H. Aay & S. Griffioen (Eds.), *Geography and worldview* (pp. 1–17). Lanham: University Press of America.

Livingstone, D. N. (1999). Situating evangelical responses to Darwin. In D. N. Livingstone, D. G. Hart, & M. A. Noll (Eds.), *Evangelicals and science in historical perspective* (pp. 193–219). New York: Oxford University Press.

Livingstone, D. N. (2008). Re-placing Darwinism and Christianity. In D. C. Lindberg & R. L. Numbers (Eds.), *When science and Christianity meet* (pp. 183–202). Chicago: University of Chicago Press.

Livingstone, D. N. (2009). Evolution and religion. In M. Rose & J. Travis (Eds.), *Evolution: The first four billion years* (pp. 348–369). Cambridge: Harvard University Press.

Livingstone, D. N. (2011). Which science? Which religion? In J. H. Broke & R. L. Numbers (Eds.), *Science and religion around the world* (pp. 278–296). New York: Oxford University Press.

Luz, N. (2013). Metaphors to live by: Identity formation and resistance among minority Muslims in Israel. In P. Hopkins, L. Kong, & E. Olson (Eds.), *Religion and place: Landscape, politics and piety* (pp. 57–74). Dordrecht: Springer.

McCormick, T. (1997). The Jaina Ascetic as manifestation of the sacred. In R. H. Stoddard & A. Morinis (Eds.), *Sacred places, sacred spaces: The geography of pilgrimages* (pp. 235–256). Baton Rouge: Louisiana State University. Geoscience Publications.

Moser, S. (2013). New cities in the Muslim world: The cultural politics of planning an 'Islamic' city. In P. Hopkins, K. Long, & E. Olson (Eds.), *Religion and place: Landscapes, politics and piety* (pp. 39–55). Dordrecht: Springer.

Noble, A. G., Dutt, A. K., & Vishnukumari, P. (1987). Daily and diurnal fluctuations in the attendance patterns of the Meenakshi Temple, Maduri, India. In V. S. Datye (Ed.), *Explorations in the tropics* (pp. 290–294). Pune: University of Poona Press.

Nolan, M. L. (1997). Regional variations in Europe's Roman Catholic pilgrimage traditions. In R. H. Stoddard & A. Morinis (Eds.), *Sacred places, sacred spaces: The geography of pilgrimages* (pp. 61–93). Baton Rouge: Louisiana State University. Geoscience Publications.

Olsen, D., & Guelke, J. K. (2004). "Nourishing the Soul": Geography and matters of meaning. In D. J. Janelle, K. Hansen, & B. Warf (Eds.), *WorldMinds: Geographic perspectives on 100 problems* (pp. 595–599). Dordrecht: Kluwer.

Olson, E. (2013a). Introduction – Religion and place: Landscape, politics and piety. In P. Hopkins, L. Kong, & E. Olson (Eds.), *Religion and place: Landscape, politics and piety* (pp. 1–26). Dordrecht: Springer.

Olson, E. (2013b). Myth, *miramiento,* and the making of religious landscapes. In P. Hopkins, L. Kong, & E. Olson (Eds.), *Religion and place: Landscape, politics and piety* (pp. 75–93). Dordrecht: Springer.

Osterrieth, A. (1997). Pilgrimage, travel and existential quest. In R. H. Stoddard & A. Morinis (Eds.), *Sacred places, sacred spaces: The geography of pilgrimages* (pp. 35–39). Baton Rouge: Louisiana State University. Geoscience Publications.

Peach, C. (2005). Britain's Muslim population: An overview. In T. Abbas (Ed.), *Muslim Britain: Communities under pressure* (pp. 18–30). London: Zed Books.

Price, N. (1994). Tourism and the Bighorn Medicine Wheel: How multiple use does not work for sacred land sites. In D. L. Carmichael, H. R. Reeves, & A. Schanche (Eds.), *Sacred places, sacred sites* (pp. 259–264). New York: Routledge.

Prorok, C. V. (1992). Geography of religions and belief systems: A learning activity. In T. Martinson & S. Brooker-Gross (Eds.), *Revising the new world: Teaching and learning the geography of the Americas* (pp. 204–207). Washington, DC: National Council for Geographic Education.

Prorok, C. V. (1997). Becoming a place of pilgrimage: An Eliadean interpretation of the miracle at Ambridge, Pennsylvania. In R. H. Stoddard & A. Morinis (Eds.), *Sacred places, sacred spaces: The geography of pilgrimages* (pp. 117–139). Baton Rouge: Louisiana State University. Geoscience Publications.

Rinschede, G. (1990). Catholic pilgrimage places in the United States. In G. Rinschede & S. M. Bhardwaj (Eds.), *Pilgrimage in the United States* (pp. 63–135). Berlin: Dietrich Reimer Verlag.

Rinschede, G. (1997). Pilgrimage studies at different levels. In R. H. Stoddard & A. Morinis (Eds.), *Sacred places, sacred spaces: The geography of pilgrimages* (pp. 95–115). Baton Rouge: Louisiana State University. Geoscience Publications.

Roof, W. C. (1988). Religious change in the American South: The case of the unchurched. In S. S. Hill (Ed.), *Varieties of southern religious experience* (pp. 192–210). Baton Rouge: Louisiana State University Press.

Rovisco, M. (2009). Religion and the challenges of cosmopolitanism: Young Portuguese volunteers in Africa. In M. Nowicka & M. Rovisco (Eds.), *Cosmopolitanism in practice* (pp. 181–199). Aldershot: Ashgate.

Rowley, G. (1985). The land of Israel: A reconstructionist approach. In D. Newman (Ed.), *The impact of Gush Emunim: Politics and settlement in the West Bank* (pp. 125–136). London: Croom Helm.

Rowley, G. (1997). The pilgrimage to Mecca and the centrality of Islam. In R. H. Stoddard & A. Morinis (Eds.), *Sacred places, sacred spaces: The geography of pilgrimages* (pp. 141–159). Baton Rouge: Louisiana State University. Geoscience Publications.

Secor, A. (2005). Islamicism, democracy and the political production of the headscarf issue in Turkey. In G.-W. Fallah & C. Nagel (Eds.), *Geographies of Muslim women: Gender, religion and space* (pp. 203–225). New York: Guilford Press.

Secor, A. (2007). Afterword. In K. M. Moran & J. K. Guelke (Eds.), *Women, religion, and space* (pp. 148–158). Syracuse: Syracuse University Press.

Sellers, R. W., & Walters, T. (1993). From Custer to Kent State: Heroes, martyrs and the evolution of popular shrines in the USA. In L. Reader & T. Walters (Eds.), *Pilgrimage in popular culture* (pp. 179–200). Houndmills: Macmillan.

Sherrill, R. A. (1995). American sacred space and the contest of history. In D. Chidester & E. T. Linenthal (Eds.), *American sacred space* (pp. 313–340). Bloomington: Indiana University Press.

Sheskin, I. M. (1991). The Jews of South Florida. In T. D. Boswell (Ed.), *Florida. Winds of change* (pp. 163–180). Washington, DC: Association of American Geographers.

Sheskin, I. M. (1998). The changing spatial distribution of American Jews. In H. Brodsky (Ed.), *Land and community: Geography in Jewish studies* (pp. 287–295). Bethesda: University Press of Maryland.

Shimazuki, H. T. (1997). The Shikoku pilgrimage: Essential characteristics of a Japanese Buddhist pilgrimage complex. In R. H. Stoddard & A. Morinis (Eds.), *Sacred places, sacred spaces: The geography of pilgrimages* (pp. 269–297). Baton Rouge: Louisiana State University. Geoscience Publications.

Shortridge, J. R. (1982). Religion. In W. Zelinsky, J. Rooney Jr., & D. Louder (Eds.), *This remarkable continent: An atlas of the United States and Canadian society and culture* (pp. 177–203). College Station: Texas A&M University Press.

Shortridge, J. R. (1984). The geography of southern religion. In S. S. Hill (Ed.), *The encyclopedia of religion in the South* (pp. 284–288). Macon: Mercer University Press (with Roger Stump).

Shortridge, J. R. (1989). Religious regions. In W. Ferris & C. Wilson (Eds.), *The encyclopedia of southern culture* (pp. 557–558). Chapel Hill: University of North Carolina Press (with Roger Stump).

Shortridge, J. R. (2004). Distribution (of religion). In *Encyclopedia of the Great Plains* (pp. 741–742). Lincoln: University of Nebraska Press.

Simpson-Housley, P., Freeman, D., & Scott, J. (1990). The highest holy ground: Mountains as metaphors and symbols in western religious and secular literature. *National Geographical Journal of India, 36*, 65–72.

Singh, R. P. B. (1996). The Ganga River and the spirit of sustainability in Hinduism. In J. Swans & R. Swans (Eds.), *Dialogues with the living earth* (pp. 86–107). Wheaton: Theosophical Publishing House.

Singh, R. P. B. (1997). Sacred space and pilgrimage in Hindu society: The case of Varanasi. In R. H. Stoddard & A. Morinis (Eds.), *Sacred places, sacred spaces: The geography of pilgrimages* (pp. 191–207). Baton Rouge: Louisiana State University. Geoscience Publications.

Singh, R. P. B. (1999). Nature and cosmic integrity: A search in Hindu geographic thought. In A. Buttimer & L. Wallin (Eds.), *Nature and identity in cross-cultural perspective* (pp. 69–86). Dordrecht: Kluwer Academic Publishers (Reprinted from *GeoJournal*, 1992, 26(2): 139–147).

Singh, R. P. B. (2005). The geography of Hindu pilgrimage in India: From trend to perspective. In B. Domanski & S. Skiba (Eds.), *Geografia I Sacrum: Festschrift to Prof. Antoni Jackowski* (pp. 417–429). Krakow: Institute of Geography and Spatial Management of the Jagiellonian University.

Singh, R. P. B. (2006). Pilgrimage in Hinduism: Historical context and perspectives. In T. J. Dallen & D. H. Olsen (Eds.), *Tourism, religion and spiritual journeys* (pp. 220–236). London/New York: Routledge.

Singh, R. P. B. (2008). Kashi as Cosmogram: The Panchakroshi route and complex structures of Varanasi. In J. M. Malville & B. N. Saraswati (Eds.), *The sacred and complex landscapes of pilgrimage* (pp. 97–109). Delhi: DK Printworld.

Singh, R. L., & Singh, R. P. B. (Eds.). (1987). *Trends in the geography of belief systems: Festschrift to Angelika Sievers*. Varanasi: National Geographical Society of India.

Singh, R. P. B., Malville, J. M., & Marshall, A. L. (2008). Death and transformation at Gaya: Pilgrimage, ancestors, and the sun. In J. M. Malville & B. N. Sarawati (Eds.), *The sacred and complex landscapes of pilgrimage* (pp. 110–121). Delhi: DK Printworld/IGNCA.

Sopher, D. E. (1997). The goal of Indian pilgrimage: Geographical considerations. In R. H. Stoddard & A. Morinis (Eds.), *Sacred places, sacred spaces: The geography of pilgrimages* (pp. 183–190). Baton Rouge: Louisiana State University. Geoscience Publications.

Staeheli, L., & Nagel, C. (2013). Different democracy? Arab immigrants, religion and democratic citizenship. In P. Hopkins, L. Kong, & E. Olson (Eds.), *Religion and place: Landscape, politics and piety* (pp. 115–130). Dordrecht: Springer.

Stoddard, R. H. & Morris, A. (1994). Major pilgrimage places of the world. In S. M. Bhardwaj, S. M. Rinschede, & G. Sievers (Eds.), *Pilgrimage in the old and new world* (pp. ix–xi). Berlin: Dietrich Reimer Verlag.

Stoddard, R. H. (1997a). Defining and classifying pilgrimages. In R. H. Stoddard & A. E. Morinis (Eds.), *Sacred places, sacred spaces: The geography of pilgrimages* (pp. 41–60). Baton Rouge: Louisiana State University. Geoscience Publications.

Stoddard, R. H., & Morris, A. (1997b). Introduction: The geographic contributions to studies of pilgrimage. In R. H. Stoddard & A. E. Morinis (Eds.), *Sacred places, sacred spaces: The geography of pilgrimages* (pp. ix–xi). Baton Rouge: Louisiana State University. Geoscience Publications.

Stoddard, R. H. (2010). The geography of Buddhist pilgrimage in Asia. In A. Proser (Ed.), *Pilgrimage and Buddhist art* (pp. 2–4). New Haven: Yale University Press/Asian Society Museum.

Stoddard, R. H., & Prorok, C. V. (2003). Geography of religion and belief systems. In G. I. Gaile & C. J. Wilmott (Eds.), *Geography in America at the dawn of the 21st century* (pp. 761–769). New York: Oxford University Press.

Stump, R. W. (1991a). Spatial implications of religious broadcasting: Stability and change in patterns of belief. In S. D. Brunn & T. R. Leinbach (Eds.), *Collapsing space and time: Geographical aspects of communication and information* (pp. 354–375). London: Harper Collins Academic.

Stump, R. W. (1991b). Spatial patterns of growth and decline among the Disciples of Christ, 1890–1980. In D. N. Williams (Ed.), *Case study of mainstream Protestantism* (pp. 445–468). Grand Rapids: Eerdmans.

Stump, R. W. (2005). Religion and the geographies of war. In C. Flint (Ed.), *The geography of war and peace* (pp. 149–173). Oxford: Oxford University Press.

Thumma, S., & Leppman, E. J. (2011). Creating a new heaven and a new earth: Megachuahes and the re-engineering of America's spiritual soul. In S. D. Brunn (Ed.), *Engineering earth: The impacts of mega engineering projects* (pp. 903–932). Dordrecht: Springer.

Troll, C. (1975). Religonsgeographie als Teilspekt der Kultur- und Sozialgeographie. In M. Schwind (Ed.), *Religonsgeographie* (pp. 250–253). Darmstadt: Wege der Forschung.

Tuan, Y.-F. (1978). Sacred space: Explorations of an idea. In K. Butzer (Ed.), *Dimensions of cultural geography* (pp. 84–89). Chicago: University of Chicago Press.

Ulack, R., Raitz, K. B., & Pauer, G. (Eds.). (1998). Cultural: Religious denominations. In *Atlas of Kentucky* (pp. 73–75). Lexington: University Press of Kentucky.

Van Spengen, W. (1998). On the geographical and material contextuality of Tibetan pilgrimage. In A. Mckay (Ed.), *Pilgrimage in Tibet* (pp. 35–51). Richmond: Curzon Press.

Vincent, G. (2013). 'There's just no space for me there:' Christian feminists in the UK and the performance of space and religion. In P. Hopkins, L. Kong, & E. Olson (Eds.), *Religion and place: Landscape, politics and piety* (pp. 167–184). Dordrecht: Springer.

Wagner, P. L. (1997). Pilgrimage: Culture and geography. In R. Stoddard & A. Morinis (Eds.), *Sacred places, sacred spaces: The geography of pilgrimages* (pp. 299–322). Baton Rouge: Louisiana State University. Geoscience Publications.

Warf, B. (2012). Religion. In S. D. Brunn, G. R. Webster, R. Morrill, F. Shelley, J. Clark Archer, & S. Lavin (Eds.), *Atlas of the 2008 elections* (pp. 229–240). Lanham: Rowman and Littlefield (individual maps of major denominations and 2008 presidential vote).

Webster, G. R., & Leib, J. I. (2012). Race, religion and the southern debate over the Confederate flag. In N. Wadsworth & R. Jacobson (Eds.), *Faith and race in American political life* (pp. 103–124). Charlottesville: University of Virginia Press.

Zelinsky, W. (1973). Unearthly delights: Cemetery names and the map of the changing afterworld. In D. Lowenthal & M. Bowden (Eds.), *Geographies of the mind* (pp. 171–196). New York: Oxford University Press.

Zelinsky, W. (1990). Nationalistic pilgrimages in the United States. In G. Rinschede & S. M. Bhardwaj (Eds.), *Pilgrimage in the United States* (pp. 253–267). Berlin: Dietrich Reimer Verlag.

Other Works

Brunn, S. D. (2009). *Changing map of world religions: Twelve emerging trends*. Presentation at interdisciplinary conference on Alternative Spiritualities: The New Age and New Religious Movements in Ireland. Unpublished paper. Maynooth: National University of Ireland.

Issac, E. (1965). Religious geography and the geography of religion. In *Man and the Earth* (Series in earth sciences no. 3). Boulder: University of Colorado Press.

Jackson, R. H., & Jackson, M. W. (2003). *Geography, culture, and change in the Mormon West, 1946–2003* (Pathways in geography series, no. 27). Jacksonville: National Council for Geographic Education.

Kostelnick, J. C. (2006). *An interactive mapping system for exploring the geography of American religion*. Ph.D. dissertation. University of Kansas, Department of Geography, Lawrence.

Licate, J. (1967). *The geographic study of religion: A review of the literature*. Unpublished MA thesis. University of Chicago, Chicago.

Rainey, L. M. (2001). *Community by design: Spaces for community in American megachurches*. MA thesis. University of Kentucky, Department of Geography.

Smith, T. W. (1996). Measuring church attendance. *GSS methodological report* 12. Chicago: NORC/University of Chicago.

Stoddard, R. H. (1966). *Hindu holy sites in India*. Ph.D. dissertation, Department of Geography, University of Iowa.

Stoddard, R. H. (2003). History of GORABS. *GORABS Newsletter, 25*(2), 1, 3–4. www.gorabs.org

Stump, R. W. (1981). *Changing regional patterns of White Protestantism in the United States, 1906–1971*. Ph.D. dissertation, Department of Geography-Meteorology, University of Kansas.

Part II
Nature, Ethics and Environmental Change

Chapter 2
Nature, Culture and the Quest of the Sacred

Anne Buttimer

2.1 Introduction

Few concepts so commonplace in everyday vocabulary are as elusive to grasp as "nature," "space" and "the sacred." Like mirrors, these notions also reflect the eye of the beholder. And beholding eyes wear the lenses of distinctive cultures, disciplines and biophysical environments. The themes, therefore, invite engagement in dialogue across the divides of language, religious denominations and lived experience, with hopes for mutual understanding and the re-discovery of shared values with respect to humanity and planet earth. A shepherding of the Earth has often been cited as a central concern for theology, but it has always been a central concern for geography, the field devoted to caring for Gaia, that marvelous orchestra of life where humanity fashions its home.

Geographic approaches to research on religion generally involved a mapping of global patterns of major world denominations, their spatial distributions, diffusion and imprints on cultural landscapes. Now and then attention was paid to cultural ecology, that is, ways in which religion dictated certain taboos with respect to livelihood and food. But the dominant methods bore the stamp of the "spatial tradition" which had by mid-century assumed pride of place with American geographers. Nature and the sacred were not unfamiliar with European colleagues, as in the classical texts of Sorre, Deffontaines and de Planhol (Buttimer 1971; Claval 2008). In recent years, too, there has been a renewed interest in religion and belief systems on both sides of the Atlantic and a growing recognition of ways in which belief systems shape cultural stereotypes and perceptions of natural resources (Bergman and Eaton 2011; Bergman 2006; Keller and Kearns 2007; UNEP).

A. Buttimer (✉)
Department of Geography, University College Dublin, Dublin, Ireland
e-mail: anne.buttimer@ucd.ie

2.2 Symbols, Experience and the Sacred

People everywhere use symbols: texts, maps, equations, graphics – symbols to unravel and analyze the parts, symbols to put them back together again. The most characteristically human activity of all is to transform direct experience into symbols, whether these be via sound, taste, literature or numbers, in cave drawings, town plans, architecture or mosaics on Cathedral walls; symbolic transformations are the stuff of human creativity on the earth (Cassirer 1953; Langer 1957). But the interpretation of signs and symbols, so clearly obvious for the "insider," may be a matter of scandal, shock, puzzlement, for the "outsider" It is also a universally human trait to use one's categories to interpret the symbols of the other. Consider the dragon, sacred symbol of palace and temple in China, symbol of evil to European eyes (Huxley 1979). The painting of *Saint George and the Dragon*, which for English eyes represented the triumph of virtue over vice, was once displayed in Beijing and was not surprisingly construed as a symbol of European aggression.

If one of humanity's traits is to make symbols, so it is quintessentially human to create and live by myths (Cassirer 1944). The mytho-poetic way of knowing is the necessary complement to the rational way – one without the other is like a "broken-winged bird afraid to fly". This is particularly true in the study of religion which appeals to the intuitive, the aesthetic and emotional aspects of humanness. Religion also motivates one to reach out beyond one's own milieu with concern for others. *Religio* means "bond," its opposite *negligio* or neglect. To gain holistic understanding of nature, then, one needs to explore not only the scientific "explanations" of natural dynamics, but also the religious context in which perceptions, uses and abuses of various resources are formed. Water symbolism in diverse milieux offers a dramatic example.

2.3 Water Symbols of Experience, Expertise and Myth

At the basic level of sensory experience, water appeals to the whole: one can see, feel, smell, touch and taste it. What would the day be like without the morning shower, the cool drink of water after a hike, the refreshment of a swim, or the beauty of falling snow? Socially speaking, the beach, the oasis, the river or stream, has been the meeting place for humans and animals throughout history. Water functions as magnet and shrine, in whose presence all kinds of communication barriers seem to dissolve. The establishment of international boundaries, one of the most important in humankind's symbolic interaction, has nearly always used watersheds, rivers straits and sounds. The United Nations still seeks international agreement on offshore limits or provisions for deep sea mining. The future survival of humanity on the earth may depend upon whether agreements can be reached about access to and use of water (Robinson et al. 2007).

At the scientific level, enquiry into the nature and dynamics of water is an ideal starting point for cross-disciplinary communication. As a recent historian of ideas put it:

> Life first assembled itself in the primeval sea. We are born from a water-filled womb. Our bodies are 90 % water (sic); 70 % of the world's surface is covered with water.... Water regulates our body's temperature and moderates the world's climate. We listen to the songs of whales; we wish we could speak with dolphins. Water has tempted the eye of the artist, the sinews of the engineer, the intelligence of the scientist, and its mysteries are not yet fathomed. (Judson 1980: 12)

Ninety-seven percent of all the world's water lies in oceans and seas; only 3 % on land. Of this 3 %, 77 % is locked up in icecaps and glaciers, 22.5 % is underground, and only a tiny 1.5 % is available for plants, animals, and humans. The West has long sought explanation for the disposition of land and water. The theory of continental drift, readily dismissed as "myth" by hard-nosed scientists in the early twentieth century, is again validated thanks to the discovery of tectonic plates. Intellectual historians are today less cavalier about dismissing the role of myth in scientific discovery.

Thales, one of the earliest Western philosophers, once hypothesized that "all is water." In this one finds perhaps the best symbolic prototype of the perennial quest for simple propositions and unifying principles of the "whole" which has characterized the Western intellectual heritage. What a contrast this is to the Oriental approach where imagination and intuition play a far greater role than the cerebral. While the West stubbornly pursued the route of empirical and hypothetico-deductive reasoning in its investigations of nature, the Orient appealed to art, music and poetry, to explore its world. The twelfth century Chinese philosopher Han Cho declared that the fundamental aim of landscape painting, for example, was to *display* the "principle of organization connecting all things." A fundamental distinction needs to be made immediately, of course, between the philosophy of Tao, which advocated sensitivity to nature's own "nature," and the Confucian stance which endeavored to master it. For both, however, the aesthetic sense was important for understanding the whole (Sokyo Ono 1962).

On Buddhist temples in Kyoto one can observe carvings that depict water turbulence, a puzzle which in the West became a central one for physics. A seventeenth century Japanese artist Kano Motobunu depicted the four moods of water: a mountain stream plunging into a waterfall which bursts into anger and then rolls away in a braided torrent. Liquid motion can be seen in a sixteenth century Japanese silk painting which depicts rolling waves against the background of open sea (Judson 1980). Via art, then, and an appeal to the human experience of particular milieux, the Oriental mind probed toward understanding the nature of water. But there was another lesson which Nature taught. Floods, earthquakes and other disasters, were regarded as signs of Divine displeasure with political regime: wise government was something which should be learned from Nature itself (Nakamura 1980).

Orient and Occident also differ in their understandings of myth. Ever since Xenophanes in the sixth century B.C. chided Homer and Hesiod for their "mythological" expressions, the main thrust of Hellenic thought has been to empty *mythos* of all possible religious or metaphysical value (Kirk and Raven 1960; Glacken

1967). Myth has come to connote all that is false. Proper knowledge ("truth") required *logos* or at least *historia*. In Western traditions, both Christian and Socratic, myth became suspect, the very opposite of truth. Historians of science have acknowledged how myth and metaphor have inspired research models and theories. But this cognitive role by no means exhausts the full significance of myth in human experience (Jung 1964). In societies where myth is alive, Mircea Eliade claims, there is a commonly held interpretation of how Creation happened, *in illo tempore*, and how it continues to unfold (Eliade 1961). The big difference between such societies and so-called "modern" ones is that the latter are trapped in history; the former can live in a sacred history where the events of primordial time may be re-enacted periodically (Eliade 1963). Christians who experience Easter, sing the *Exsultet,* and celebrate the Resurrection, can appreciate what this means.

And water plays such a central role in this sacred story. In the Judaeo-Christian account of Creation, it is the Spirit which breathes over the waters. When men disobeyed, a great flood came to cleanse the world. Water symbolizes the Holy Spirit who comes to dwell within the believer upon Baptism. To the Samaritan woman at the well, Christ said: "Whoever drinks the water that I will give him will never be thirsty again. The water that I will give him will become in him a spring which will provide him with living water, and give him eternal life" (John 4:13).

Water plays a cardinal role in most Creation myths, frequently associated with the female element, in reciprocal relationship to the male elements of Sky and Earth. For the Navajo, living in a semiarid environment, for example, the process of creation is seen to emerge through the conjunction of Mother Earth and Father Sky, the basic ingredients being cornmeal, pollen and powdered plants or flowers. Japanese cosmology shows a heterogeneous array of *Kami* (Gods) and levels of being and the sacred lotus (floating on the ocean) assumes a central role (Sokyo Ono 1962; Nakamura 1980). Water symbols, thus, show an attunement to both the biophysical and cultural context. Indeed, in many cosmologies, water symbols have been used to reach understanding of the sacred in nature, appealing as they do to not only the visual sense, but also to the moral and emotional aspects of lived experience.

But Western theology has departed somewhat from such holistic conceptions of nature and the sacred. The dominant emphasis on "Fall and Redemption" and the vital role assigned to "original sin" have exercised a powerful influence on believers, often causing fear and guilt with respect to natural emotion and passion and an insensitivity to the sacredness in natural processes. Matthew Fox (1983) has offered a brilliant critique of this heritage and argues for a rediscovery of a "creation theology" which would view all elements of nature and humanity in terms of "original blessing," all in a perpetual state of becoming, co-evolution.

2.3.1 Water Symbolism, Religion and Milieu

Common sense would suggest that natives of Oceania would have a different conception of sacredness than those of Gobi desert natives; that Egyptian peasants would have a different conception of water than those of the Ukraine. The Hopi Indian might pray for rain to come, while natives of the Amazon might pray that it stop. In milieux where extreme conditions (heat or cold, drought or excessive rain) exist, one often finds the attitude of fear and insecurity vis-à-vis the whole. Myth counsels passivity with respect to nature; social power can easily be vested in the control of natural resources. There is little motivation to interfere with the natural course of events, and often there is an elaborate liturgy surrounding, for example, the coming of the rain.

2.3.2 *Nature and Desire in Monsoon Lands*

At Jaipur in India, the yearly drama of the monsoon has enormous social significance. During the burning months of May and June, a desiccated Nature (landscapes, humans, animals and trees) pants for rain. Dark clouds gather over the Ganges in whose sacred waters people stand in prayer. When the rains come there is great rejoicing at the prospect of new energy and creative life (Narottam Das). As soon as the rains stop all kinds of new life appears; gardens are alive with insects and at night the air is filled with the deafening noise of crickets and frogs. Villagers sit on porches telling "monsoon stories."

In a milieu such as the Ganges, it is no surprise to find in the Hindu Scriptures, and especially in the Upanishads, a reference to water and desire as the source of Creation itself (Zaehner 1966). "Water is greater than food," the Chandogya Upanishad (VII, x) sings, "Truly, earth and atmosphere and sky are nothing but water transmuted into different forms ... who so reveres water as Brahman, obtains all his desires and will be well satisfied. He gains freedom of movement in the whole sphere of water – who so reveres water as Brahman" (Zaehner 1966; De Nicolas 1978).

Hindu belief and practice acknowledges the sacredness of water. The notion of Ganganization, perceiving the Ganges as goddess and naming various water channels as Ganga, continues as unique feature of monsoon lands (Singh 1999). Literary opinions vary on the claim that India is either the origin or crossroads of all human civilization. There should be no problem in tracing India's influence westward through Plato, Islam, and later Western mythology, and eastward through Southeast Asia and the Pacific.

2.3.3 *Cargo Cults of Oceania*

If rain and flood are the perennial experience of people on the Ganges, then Ocean is the ubiquitous horizon for Polynesians (Handy 1927). In the beginning, it is believed, there were only the Waters and Darkness:

> IO, the Supreme God, separated the waters by the power of thought and of his words, and created the Sky and the Earth. He said: "Let the Waters be separated, let the Heavens be formed, let the Earth be." Men still utter those same powerful words when there is something important to be done or created like making a sterile womb fecund, cheering a gloomy heart,

helping the feeble-aged and decrepit; for shedding light into secret places and matters, for inspiration in composing songs and in times of despair or war. (Eliade 1963: 30–31)

But the islands of the Pacific have been colonized by alien civilizations. How do natives "resolve" this peculiar happening?

The cargo of cults of Polynesia announce the imminence of a fabulous age of plenty and happiness. Natives will again be masters of their islands and will no longer have to work, for the dead will return in magnificent ships laden with goods like the giant cargoes that the whites receive in their ports… all these actions and beliefs are explained by the myth of the destruction of the world, followed by a new Creation and the establishment of the Golden Age (Eliade 1963: 2–3).

When the Congo became independent in 1960, some villagers tore the roofs off their huts to give passage to the gold coins that their ancestors were to *rain down*. Elsewhere everything was allowed to decay except the roads to the cemetery by which the ancestors would make their way to the village (Eliade 1963). Myth supplies the story which can make sense of a perplexing social history, and the metaphors of salvation bear the stamp of milieu. But similar milieux have been colonized by different peoples; political power can redefine both myth and reality. All over the world, cultural interpretations of nature and resources were imposed upon diverse bio-ecological milieux. Consider how much Christian liturgy is imbued with the water symbolism of a nomadic people from arid lands; water as symbol of life and re-birth may have less sense for natives of rainy places like Ireland or the Philippines. The succession of protracted theological debates over liturgy and calendar, for example, indicates the fundamental challenge of symbolism and milieu between the Mediterranean and Northern Europe.

2.4 The Middle East: Cradle of Water Symbolism

For virtually all European nations, the Mediterranean has been the source of symbols, many of which emanated from the Asian heartland as well as the Fertile Crescent. In the religions of the Near East, water is central, either as ally or foe of human life. In Mesopotamia, the New Year festival celebrated the triumph of Marduk (God) over Tiamat (the dragon symbolizing the primordial ocean). Each New Year shared something of the first day of creation when the cycle of seasons started. The *beginning* was organically linked with an *end* which preceded it, as the chaos preceding creation. For Egyptians too the New Year symbolized Creation. The Jewish New Year (Yom Kippur) brought the enthronement of Yahweh as king of the world, the symbolic victory over his enemies, both the forces of chaos (ocean) and the historical enemies of Israel.

From the Middle East hearth comes not only Yahweh, but also Socrates as progenitor of Western myths about truth and being. An important distinction, of course, needs to be made between the dialectical and fluid world of the pre-Socratics, and the fixed cosmologies of the Socratics. Compare Empedocles, for instance:

> He makes the material elements four in number: fire, air, water, earth, all eternal, but changing in bulk and scarcity through mixture and separation; but his real first principles, which impart motion to these, are love and strife. The elements are continually subject to an alternate change, at one time mixed together by Love, at another separated by Strife, so that the first principles are, by his account, six in number. (Simplicius in Kirk and Raven 1960; cited in Glacken 1967: 10)

With Plato:

> Now of the four elements the construction of the cosmos had taken up the whole of every one. For its Constructor had constructed it of all the fire and water and air and earth that existed, leaving over, outside it, no single particle or potency of any one of these elements. (Plato 1952; cited in Glacken 1967: 44)

The dynamic and constantly changing world of the pre-Socratics becomes a *unified Cosmos*, with a fixed geometry and physics, for the Socratics. Moving westward, one finds two distinct streams of cosmological development diffusing outward from that same cultural heart: one toward the desert (Islamic world) and the other toward the temperate, well-watered, European extension.

2.4.1 The Qu'ran, Oasis and Desert

"We made from water every living thing" (Qu'ran 21:30) and "It is he (God) who sends down rain from the sky from it ye drink and out of which grows the vegetation on which ye feed your cattle" (Qu'ran 16:10). Water, in Islamic cosmology, is seen as the source of life, coming down from heavens to earth (Hanafi 1999). But a clear distinction is made between water as flowing, pouring, strong, abundant, incorruptible and blessed on the one hand, and water as stagnant, lost, obscure, salty, bitter, boiling and despised on the other (Qu'ran 22:45). One of the worst penances for a sinner, for example, is to be thrown to the bottom of the well (Qu'ran 12:10, 12:5). The Qu'ran approves of mankind using his own ingenuity to harness and purify water. Paradise is full of springs from which believers drink (Hanafi 1999). The ocean (sea) in the Qu'ran is a symbol of the omniscience and omnipotence of God. The sea is also, of course, a field of action for humans who can be guided by the stars. Humans need vessels, ports, and stars to cross the sea. Man eats from the sea, even drinks, if he can desalinize the water, drowns or is safe. The sea also carries precious resources like pearls and coral. The word Fult (vessel) is mentioned 23 times to symbolize safety, transport, and commerce.

The study of Nature was considered as an essential part of education: Nature was regarded as the handiwork of God, a purposeful domain in which the power and wisdom of the Creator is manifested. In Islamic cosmology elements of both Hindu cosmology and Aristotelian metaphysics can easily be discerned: the world in becoming and its relation to being. The Universe at all levels emanates from Pure Being and ultimately returns to it. But there was another element: knowledge is operative (e.g., in Ibn Sina's cosmology), it is a process by which the being of the knower is transformed. This is also a Sufi principle: knowledge of the cosmos is reached through an effective journey through it. Nature becomes a background for the Gnostic's journey, the symbolic pilgrimage from pure matter (Occident) to pure form (Orient) and the means of deliverance (Hasr 1964).

Islamic ways of knowing point to the interrelatedness among all parts of the cosmos. The aim of knowledge is not just the discovery of an unknown domain outside the being of the seeker, but a return to the Origin of all things which lie in the heart of man as well as within "every Atom of the Universe" (Hasr 1964). Thus there is a belief in the ultimate return of all things to their origin, and the integration of multiplicity into Unity. Creation, in fact, is the bringing into being of multiplicity from Unity, while gnosis is the complimentary phase of the integration of the particular into the Universal.

2.4.2 Cosmology and Metaphor in Western Conceptions of Nature

Permeating Western myths and symbols one can discern key images of Nature, "root metaphors" of world, which have provided a common thread among artists and scientists in successive periods (Pepper 1942; Glacken 1967; Mills 1982). From Greece indeed the Western tradition has inherited that conviction that intellect was queen among human faculties, and that human reason (logos) should provide the ultimate criteria for assessing truthfulness. From the Hellenic stream of the Judaeo-Christian tradition, too, comes a suspicion of the emotional, biophysical/corporeal, sensory and intuitive features of our humanness. The distinction, and eventual separation, of intellectual and moral virtues, as epitomized in the controversy between Bernard and Abelard in Paris, is one which has had a special significance for university life down the ages. Yet in certain enduring metaphors of world one can find lurking assumptions the nature of being, that is, the true, the good, the sacred and the secular (Buttimer 1993). Four of these assumptions illustrate this point:

World as Text: A strongly held Medieval view was that landscapes could be read as evidence that Divine Providence had created a fit abode for humankind. "Some peoples, in order to discover God, read books," Augustine wrote, "but there is a great book: the very appearance of created things…. God, whom you want to discover, never wrote that book with ink; instead He set before your eyes the things He had made. Can you ask for a louder voice than that?" (Cited in Glacken 1967: 204)

World as Organic Whole: A classical view, rediscovered in Renaissance times, that the human body could be seen as microcosm of creation as a whole. "In every man," Bruno taught, "a world, a universe, regards itself" (cited in Koyré 1957: 42). This view is reflected in the "Creation theology" of Matthew Fox (1983).

World as Mechanical System: a view strongly defended in Greek and Roman science, gained fresh appeal in Enlightenment times. "The world is like a rare clock," Robert Boyle wrote, "where all things are so skillfully contrived, that the engine being once set a moving, all things proceed according to the artificer's first design…" (Mills 1982: 245). So God could be seen as a "retired engineer" (Nicolson 1960; Lenoble 1969).

World as an Arena of Spontaneous Events: a more modern contextual view. Strongly defended by French authors such as Bergson (1911) and Merleau-Ponty (1962, 1963), this approach was especially welcomed by American pragmatists such as William James (1955). Contextual approaches ushered in a radical transformation in conceptions of knowledge and belief, moving scientists from stances of "observation" to "participation."

In each of these visions of the whole there are claims to truth, each with implications for art, theology, philosophy and science; each has implicit or explicit guidelines for human behavior vis-à-vis nature, culture and the sacred (Buttimer 1993).

2.5 Concluding Queries

Water symbolism has yielded a rich harvest in the West through metaphor. Notions such as "organism," "mechanism," "cycle" and "flow" have helped provide a cognitive grasp of the whole or its internal dynamic. With water symbols, time and space have been mapped and measured, natural and social processes controlled. This primarily cognitive trust stands in marked contrast to the Indian use of metaphor to generate a sense of participating in creation, along with other living beings, spiritually and emotionally. It also contrasts with the East Asian way of intuition and aesthetic appreciation of the whole and the analogues between human sociality and cosmic reality. Western ways of exploiting water symbolism have laid emphasis on the visual over other senses: art and literature (to be seen or read) take precedence over sound and touch (to be heard and felt). Finally, the West has shown a peculiar attraction for symbols of how things work – from the balancing of moistures to the hydrological cycle – the West has fulfilled the Baconian formula "knowledge is power" without heeding his other counsel, *Natura nisi parendo vincitur* (Nature can only be overcome by obeying it). Still today, water symbolism invites ecumenical perspectives. For natural scientists and engineers, water still exemplifies best the principle of continuity from oceanic to micro-cosmic levels. For theologians, grace, so often described in water symbolism, is the principle of holiness, gratuitousness, healing and recreation throughout Buddhist, Hindu, and Judaeo-Christian traditions. And for geography the hydrological cycle is perhaps the one key metaphor which enables physical and human geographers to share a common concern wherever life exists on planet Earth (Hutton 1795; Tuan 1968).

But there are realms of human experience common to all civilizations which have enduring concern: health, creativity, and the reciprocity of male and female. Health, to the medieval anatomist, meant a balance of the humors in the body. In Oriental water symbolism one finds, as a cardinal principle of creation, as well as in health, the reciprocity of Yin and Yang: life, its maintenance and re-creation, demands the mutually-respecting union of male and female. In Western thought and practice could one not claim that the forces of Yang, boosted by Promethean and Faustian myths, have threatened to stifle the forces of Yin? (Buttimer and Hägerstrand 1980). And what of human creativity? Even Western scholars, when asked about their moments of creative insight, do they not spontaneously resort to water symbolism: "flow of imagination," "rush of insight," "stream of consciousness," "channeling of energy?" Isaac Newton, shortly before his death described himself as "a little boy playing on the sea shore and diverting myself in now and then finding a smoother pebble or a prettier shell than ordinary, whilst the *great ocean of truth* lay all undiscovered before me."

Symbols of health, creativity, identity and all such experientially grounded probings toward the "whole" are thus understandable in the context of particular cosmologies, and need to be interpreted with rational as well as mytho-poetic lenses. This enormous hermeneutical challenge is confounded by the ever present shadow of Narcissus. During the era when Nature was regarded as the handiwork of God, I presume theology had its heyday. When the secrets of civilization and climate were to be studied in terms of the human body, geographers could be regarded as "the anatomists of the great world." In the "machine" era I presume the technologist has felt at home. But the rejection and succession of basic metaphors was never a function of epistemology alone; such changes were born and steered from aesthetic, moral, and emotional judgments.

Finally, how well do scholars today encourage mutual understanding across cultural and linguistic divides? Do recent revolutions in information technology which open up potential interactions among people all over the world, actually encourage a sharing of ideas? Or does a concomitant fear of homogenization trigger a retreat into fundamentalist attachments to one's "own" world? In this context, what might the prospects be for a shared conception of evolving creation as "original blessing"? Could scholars contribute in this regard? Some would argue that our role is simply to provide credible descriptions from our own respective disciplinary cells – as though a "cloister" still existed for the cultivation of intellectual and moral virtues Yet few could deny that even the most esoteric speculation within the world of thought has ramifications for society's welfare. Does the fragmentation and specialization of our expertise not reflect, and also find itself reflected in the landscapes and life forms within which we pursue our everyday agenda? One cardinal myth of our present academic situation, the *ceteris paribus* assumption – blithely accepted in current scientific practice, is one of those habitual taken-for-granted patterns of thought and practice inherited from an era when science made its reformation and cast off the tyrannies of theologians and kings. Socialist *Techne*, too, assigned experts to applied sectors within national administrative machines. Viewed metaphorically, both scientists and planners occupy separate cells within a Babel Tower.

Yet a look at water symbolism and its cardinal role in world cosmologies screams out: *ceteris* are never *paribus*. Water symbolism beckons us all beyond our academic niches and encrusted routines toward alternative ways of perceiving ourselves and our world. Water permeates the whole of life, inviting all to ongoing creation. So might one dare to dream of a theology where Eros rejoins Logos, where spirituality, emotion, and worship rejoin intellect as equally valuable sources of insight into truth and goodness? Might one dream of a technology where Prometheus is reconciled with Epimetheus? That is, a geography ready to be tamed of its managerial *hybris* and listen to geosophy, a reading of the earth's surface in terms of the accumulated wisdom of civilizations, and thereby led to a sense of Creation as a whole? Teilhard de Chardin's poetic vision of the universe has something to offer us all: a vision of humanity finally become conscious of itself and aware of a Unity, founded on Infinite Love, which supports diversity and the integrity of each (Teilhard de Chardin 1961). I can think of few metaphors which could be more helpful in the journey toward understanding nature, culture and the sacred.

References

Bergman, S. (2006). Editorial for ecotheology. *The Journal of Religion, Nature and the Environment, 11*(3), 261–267.
Bergman, S. & Eaton, H. (2011). *Exploring religion, ethics and aesthetics*. Berlin/London: LIT Verlag. Studien zur Religion und Umwelt, Bd. 3.
Bergson, H. L. (1911). *Creative evolution*. New York: Holt.
Buttimer, A. (1971). *Society and Milieu in the French geographic tradition* (AAG Monograph, No. 6). Chicago: Rand McNally and Co.
Buttimer, A. (1993). *Geography and the human spirit*. Baltimore: The Johns Hopkins University Press.
Buttimer, A., & Hägerstrand, T. (1980). *Invitation to dialogue: A progress report* (Dialogue Paper, No. 1). Lund: University of Lund.
Cassirer, E. (1944). *The philosophy of symbolic forms*. New Haven: Yale University Press.
Cassirer, E. (1953). *Language and myth*. (trans. S. Langer). New York: Dover.
Claval, P. (2008). *Religion et Idéologie. Perspectives Géographiques*. Paris: Presses de l'Université Paris-Sorbonne.
De Nicolas, A. T. (1978). *Meditations through the Rg Veda. Four dimensional man*. Boulder/London: Shambala Press.
Eliade, M. (1961). *The sacred and the profane*. New York: Harper.
Eliade, M. (1963). *Myth and reality*. New York: Harper.
Fox, M. (1983). *Original blessing*. Santa Fe: Bear & Company.
Glacken, C. G. (1967). *Traces on the Rhodian Shore: Nature and culture in western thought from ancient times to the end of the 18th century*. Berkeley: University of California Press.
Hanafi, H. (1999). World views of Arab geographers. In A. Buttimer & L. Wallin (Eds.), *Nature and identity in cross-cultural perspective* (pp. 87–94). Dordrecht: Kluwer.
Handy, E. S. C. (1927). *Polynesian religion*. Honolulu: The Museum.
Hasr, S. H. (1964). *An introduction to Islamic cosmological doctrines. Conceptions of nature and methods used for its study by the Ikhwan al Safa, al Biruni and Ibn Sina*. Cambridge: Harvard University Belknap Press.
Hutton, J. (1795). *Theory of the earth with proofs and illustrations*. Edinburgh: Cadell, Junion and Davis.
Huxley, F. (1979). *The dragon: Nature of spirit, spirit of nature*. London: Thames and Hudson.
James, W. (1955). *Pragmatism and four essays from "The meaning of truth"*. New York: New American Library.
Judson, H. F. (1980). *The search for solutions*. New York: Holt, Rinehart and Winston.
Jung, C. G. (1964). *Man and his symbols*. New York: Doubleday and Co.

Keller, C., & Kearns, L. (Eds.). (2007). *Eco-spirit, religions and philosophies for the earth*. New York: Fordham University Press.
Kirk, G. S., & Raven, J. E. (1960). *The presocratic philosophers*. London: Cambridge University Press.
Koyré, A. (1957). *From the closed world to the infinite universe*. Baltimore: The Johns Hopkins University Press.
Langer, S. K. (1957). *Philosophy in a new key. A study in the symbolism of reason, and art*. Cambridge: Harvard University Press.
Lenoble, R. (1969). *Esquisse d'une Histoire de l'idee de Nature*. Paris: Editions Albin Michel.
Merleau-Ponty, H. J. (1962). *Phenomenology of perception* (trans: Smith, C.). New York: Humanities Press.
Merleau-Ponty, H. J. (1963). *The structure of behavior* (trans: Fisk, A. L.). Boston: Beacon Press.
Mills, W. J. (1982). Metaphorical vision: Changes in Western attitudes to the environment. *Annals of the Association of American Geographers, 72*(2), 237–253.
Nakamura, H. (1980). *The idea of nature, East and West* (p. 284). London: Encyclopedia Britannica.
Nicolson, M. (1960). *The breaking of the circle: Studies in the effect of the "New Science" upon seventeenth century poetry* (Rev. ed). London: Oxford University Press.
Pepper, S. (1942). *World hypotheses*. Berkeley: University of California Press.
Plato. (1952). *Timaeus, Critias, Cleitophon, Menexenus Epistles* (trans: Bury, R. G.). Cambridge: Harvard University Press (Reprinted and revised from the edition of Loeb Classical Library).
Robinson, P. J., Jones, T., & Woo, M.-k. (Eds.). (2007). *Managing water resources in a changing physical and social environment*. Rome: IGU Home of Geography.
Singh, R. P. B. (1999). Nature and cosmic integrity: A search in Hindu geographical thought. In A. Buttimer & L. Wallin (Eds.), *Nature and identity in cross-cultural perspectives* (pp. 69–86). Dordrecht: Kluwer.
Sokyo Ono. (1962). *Shinto. The Kami way*. Tokyo: Charles E. Tuttle.
Teillard de Chardin, P. (1961). *L'avenir de l'homme*. Paris: Editions du Seuil (trans. (1964) by W. Collins & Co. as *The future of man*). London/New York: W. Collins/Harper & Row.
Tuan, Y.-F. (1968). *The hydrological cycle and the wisdom of god: A theme in geoteleology*. Toronto: Department of Geography, University of Toronto.
UNEP Programme on World religions and ecology. http://environment.harvard.edu/religion/publications/massmedia/index.html
Zaehner, R. C. (Ed.). (1966). *Hindu scriptures*. London: J. M. Dent and Co.

Chapter 3
Church, Politics, Faceless Men and the Face of God in Early Twenty-First Century Australia

Mary C. Tehan

3.1 Introduction

In the two realms of religion and politics, Australians are currently experiencing both a holy and unholy mess. Although Professor Des Cahill, Professor of Intercultural Studies at RMIT University, asserts that, on balance globally, "religion adds to personal wellbeing and national social wealth, including in Australia," he also argues that fundamentalism and pathologies have emerged (Transcript, Victorian Family and Community Development Committee (2012) leading an Inquiry into the handling of child abuse by religious and other organizations October, 2012, pp. 2; 4). This perspective assumes that the foundational structures of, and cultures in, our current Australian religious and political institutions are still embedded and function in, and on, solid ground. But does this perspective offer a fullness of the current state-of-play and what, if any, glimpses of the face of God that might be emerging in the midst of a global civilization shift from the current technozoic era to the birthing of the ecozoic era? This chapter will seek to address this question and highlight some of the promising, perilous (at times) and emerging manifestations of the Holy Spirit, its fruits bringing light through the cracks of this fragmented, broken world; being expressed specifically through Australia's politics and Catholics.

In Genesis 32: 24–31 Jacob wrestles with an angel at Peniel; he asks to be blessed, and his life prevails, for:

> I have seen God face to face, and yet my life is preserved.

This wrestling, this awesome struggle to seek a blessing and preserve life is a wonder-filled image that more fully represents the global civilization transitus currently underway. At its most primal level there have been, and are, tectonic shifts at work – at physical, geographical, financial, cultural, social, spiritual and metaphorical

---

M.C. Tehan (✉)
Stirling College, University of Divinity, Melbourne, Australia
e-mail: mctehan@hotmail.com

levels. Aging populations, tsunamis and other earth-speaking disasters, addictive consumerism, increased global migration and mobility, depression, suicide, violence, trauma and earth-rape are manifesting in waves around the world, and a deep, anguished wrestling, sighing (and pausing) are taking place within many individuals, communities and societies regardless of their origins (Tehan, unpublished component of Masters in Public Health project, La Trobe University 2007).

3.2 Global Religious Context

Although twenty-first century Christian issues are not, in total, the same as the issues for Christians in the first century, there are common threads worth noting in relation to key tensions experienced in both eras. Specifically, in the first century, these tensions (not in any particular order of occurrence or importance) included the power relationship between church and state, the unity of Christian believers and how to maintain it (interfaith conversations), the rule of Rome over provincial and general councils (Eurocentric Church), the validity of already baptized lapsed catholics (defective clergy) and their re-integration into the One Holy Catholic Apostolic Church (hierarchy of holiness issue), and the authority of laity in relation to the clergy and bishops (Ellingsen 1999; Guy 2004; MacCulloch 2009). Additional, distinctive societal issues in the early twenty-first century include the speed of global communication and increasing insecurity of basic needs (for example, food, shelter, clean water, right livelihood) in socially, culturally and spiritually fragmented first-third world societies; no nation is immune from the impacts of these factors. Moreover, through a Royal Commission, Australia is poised to commence its first-ever public accountability process for dealing with child sex-abuse in *both* church *and* state institutions.

With the Australian bishops' approval, structures are being established, such as the establishment of a the Truth, Justice and Healing Council, a national coordinating body comprising men and women with professional and broader expertise (both laity and non-Christian) to both facilitate and strengthen ongoing relationships with the abused people, and provide expertise, wisdom and guidance over the course of the Royal Commission. This Church-State-laity and professionals-approved response is also distinctive from prior Church-State relations, which were one subordinate to the other and vice-versa throughout Church history; for example the Donatist controversy where the Donatists, Catholic bishops and the Empire became embroiled in resolving whether the tradidores or lapsed repentants could be forgiven, and who, ultimately, had legitimate authority to forgive (MacCulloch 2009: 211). This profound shift in church-government-professionals relationships to address child sex-abuse is new to church history in Australia and is yet to unfold. The current decision has occurred within the context of widespread public and clerical condemnation of Cardinal George Pell's public response on November eighteenth 2012 to the Federal Government decision to establish a Royal commission, where in the *National Catholic Reporter*, according to journalist Peter Crittendon, His Eminence stated that:

> We are not interested in denying the extent of misdoing in the Catholic church. We object to it being exaggerated. We object to being described as the only cab on the rank … that pressure

for the royal commission had been provoked by a one-sided media "smear" campaign against the Catholic church ... and that He (Cardinal Pell) welcomed the royal commission as "an opportunity to clear the air, to separate fact from fiction". (cited in Crittendon 2012)

Cahill argues that it is the celibate clerical caste system that needs restructuring as part of a preventative child sex-abuse strategy in the Church (2012: 3). Fowler (1996: 195) suggests that, just as in the covenant with Noah (Genesis 9: 8–17), God's praxis can be seen, biblically, "within the holding power of covenant" and three great comprehensive patterns; that are:

1. God creating
2. God governing; and
3. God liberating and redeeming

Specifically, God's generative and nurturing of an evolving universe; God's aim at "right-relatedness" and flourishing of creation; and God's costly praxis of absorbing suffering, restorative justice and shaping the fulfillment of divine hopes (1996: 195). Cardinal Pell's statement about media pressure for the Royal Commission highlights the challenge of engaging with a 24 h media cycle campaign that has the capacity to reveal and persist with issues at an unprecedented intensity and rate of revelation. It has fundamentally changed the extent to which media can influence societal, political and religious issues and responses, and the way in which church-state relationships need to be maintained.

Alongside this new emerging Australian church-state-nation power relationship towards "right-relatedness" in relation to child sex-abuse, the Catholic Religious of Australia (2012) have been pondering on the role they need to play at this point, and in the future, in history and from a local and global perspective. St. Christine Burke IBVM, in reflecting on themes from the 2011 seminar held in Rome on the *Theology of Religious Life: Identity and Significance of Consecrated Apostolic Life*, notes that:

> Today, consecrated religious life feels and deeply experiences issues on identity, credibility, and visibility ... [and ponders on] ... (1) questions around the title 'consecrated life;' (2) implications of resituating the apostolic focus within a cosmological and global context; (3) interconnecting any theology of religious life to emerging issues in Christology and ecclesiology. (Burke 2011: 1)

Burke noted that a terminology was needed that did not imply it was a better, more committed or pure way of following Christ in the world today (Burke 2011: 2). The notion of a "higher consecrated" vs. "lower lay" hierarchy of holiness is under question again, of which first century Donatists fought to maintain. Burke also reflected on Fr. Antonio Pernia, SVD General Superior, from the Philippines' contribution to resituating the apostolic focus to the global, ecozoic era, and his insight that:

> They ... [some church leaders] ... prefer religious as a workforce rather than a questioning alternative, which most founders pioneered. The 'irrelevance' of religious life within wider society is a result of the need for some new expression of religion, that takes account of the cyber age and its many and massive changes: this demands what he called a 'post-religious religious life.' The implication in this is that we are part of shaping this post (traditional) religious community by the way we seek God through our collaboration with others in this fragmented world. (Pernia, cited in Burke 2011: 2)

Importantly, Pernia states that the "questioning alternative" needs to include questioning "Religious dis-Orders" (rather than "Religious Orders") as well as contemporary societal issues and needs (cited in Burke 2011: 2). Pernia asserts that this period of chaos is due, in part, to the emergence of a global church, with its "corresponding reality of multi-cultural membership and multi-directional mission;" and that three opportunities exist for religious life into the future, that is: (1) religious members who work to create truly intercultural communities; (2) inter-congregational collaboration in ministries; and (3) partnership with the laity (cited in Burke 2011: 2). All three options invite an interdependent and shared-responsibility approach to contemporary religious-laity membership and mission; the Donatists would not have approved.

Sr. Ilia Delio (cited in Burke 2011: 2–3), a Franciscan from Jesuit Woodstock Centre in Georgetown University, also informed Burke's reflection by asking three questions related to the cosmos and environment and prioritizing them for future membership and mission:

1. What influence does environmental change and the interrelatedness of all creation have on what we discuss?
2. Is "communio" at the heart of how we live? and
3. What does Christogenesis mean for us today when the Spirit is aching to express Godself through us? Delio stated that "We must make our way to heaven *through* earth."

Burke reflected on Mary Maher, superior general of the SSND's insight about the need to "risk in hope … as part of congregation formation … and a post-modern path to holiness" as a means of shifting the focus from "church" to the "world" within a "communion dynamic" (cited in Burke, 2011: 3). Burke's key insight (2011: 4) is that consecrated, life-long commitment within religious life is a sign that is:

> an existential statement on the value of God. If we can nurture and deepen our inner experience of God, this in itself is a witness empowering others to see God in themselves and in the world.

The above religious context highlights the dynamic, transition and tectonic global membership and missionary shift currently underway today in the Church, including in Australia.

3.3 The Australian Religious Context

In 2010, Catholic Religious Australia (2012) and the Australian Catholic Bishops Conference launched a Report on the 2009 Survey of Catholic Religious Institutes of Australia. This Report's findings include the following data:

- Australia is a very urbanized nation: nearly three-quarters of Australia's Catholics (72 %) live in the major cities, but this figure rises to 89 % for Catholics born in non-English speaking countries;

- the majority of Catholics now grow up and live their lives without ever meeting a member of any religious congregation;
- the religious are fast becoming invisible in the Australian Church. (2010: 10; 20)

The 2010 Report also included individual reflections by religious on their insights from the 2009 data. Francis Moloney SDB, Provincial of the Australia-Pacific Province of the Salesians of Don Bosco, challenged conventional wisdom by stating that:

> the drop-off of vocations to the religious life might also have something to do with the fact that we no longer reflect the primacy of God. We are not called to be better, but to be living icons of the vocation of all humanity – 'to be conformed to the image of his Son' (Rom 8:29). Our priorities of spirituality, community and mission … [foundational to the consecrated life] … must be understood in this light. (2010: 33)

Moloney also highlights that "experiences" are the drivers for *both* the secular world of illusion, superficiality, and addictive, consumerist behaviours, *and* the religious world of "what it means to love God and to live, love and serve as Jesus did" (2010: 35). In addition, these experiences are either grounded in coming "from the head" or "coming from the heart" (2010: 35). Moloney invites the possibility that the religious life and its three priorities need to be articulated as "coming from the heart," where, according to Maloney, traditionally for many consecrated religious, these priorities were, in the past, learned and inculcated through "coming from the head" (2010: 35).

In reading online the history of different female religious founders, for example Mother Catherine MacAuley, Foundress of the Sisters of Mercy (1831); Mercy International Association (2012); Cabrini Mission Foundation (2012); Mother Cabrini, Foundress of the Missionaries of the Sacred Heart of Jesus (MSC) (1880); Sister Maude O'Connell, Foundress of The Company of Our Lady of the Blessed Sacrament (The Grey Sisters) (1930), there is an underlying congruence in their stories. These stories express personal experiences of heartache, of life-ache and of multiple losses, coupled with enormous resistance by established Orders and conventional wisdom when their new ministries became visible (refer to these Religious Order websites for specific examples). Patty Fawkner SGS, a member of the Council of the Sisters of the Good Samaritans of the Order of St Benedict, in her reflection of the 2010 Report on the Catholic Religious Institutes of Australia (2012), offers a unique insight about gender and consecrated religious in the future:

> A hidden statistic in this report is that women's means of support is precarious. Women have to find 'something' to earn money. Thank God for the pension. Priests can 'earn' through saying Mass. Women can't become chaplains on pilgrimages or cruise ships, because they can't say Mass. As women continue to be excluded from ordination and decision-making, is it saying something about survival of what seems to be the fittest in our church? Will a changing profile of religious contribute to an even more clerical church? (2010: 36)

The sighing and wrestling of peoples, therefore, is not just within the laity, or between laity and consecrated religious, but is also between the genders within the consecrated religious communities. A contemporary challenge then, is for the Catholic clergy and Bishops, to find a way to empower women in the Church that is congruent with the notion of shared-responsibility and interdependence. Is it any wonder then, in such a fragmented world, that partnerships between and within both the secular and religious realms, are so challenging to embrace, foster, strengthen, uphold, and articulate to both self and others? Cahill, Irarrazaval, and Wainwright acknowledge that the Councils of Nicea (325 CE) and Chalcedon (451 CE) recognized Jesus' humanity as essential to salvation and that Jesus of Nazareth was a man (2012: 7). According to Cahill, Irarrazaval, and Wainwright, this acknowledgement does not suggest that "it's Christ's maleness (rather than his humanity) that unites humans with God's saving grace" (2012: 7). As a way forward, might whispers of the presence of God, the God of surprise, be found in robust, functioning blended families; in the ordinariness and mess of their everyday life? Might a glimpse of God's face also be found in the midst of single parenting (of heartache, life-ache and multiple losses)? For Australia, perhaps the face of God is in the enduring generosity and forbearance of the Aboriginal peoples and wisdom learned from the experiences of all of Australia's oppressed peoples, for example: refugees, immigrants, addicts (cyber, drug, alcohol, gaming), unpaid careers, un- and under-employed, homeless, prisoners, ageing populations, ill people and war veterans?

Moloney suggests in the 2010 Report that:

> the time has come to recover the specific nature of the experience of religious life. Only in this way will it be credible, effective and significant. We have now been led by the Spirit to a place where the vows, the constitutions, the habit, and not even the mission, can be regarded as the identity of the religious life. (p. 33). It all depends upon a visible experience of a relationship with Christ. It is time for us to spell out again what a religious is: that something unique that gives religious their role and place within the mystery of the Church. Religious life must appear less as an organiser of works geared to education, health care and human development. We must work to become a living sign of the experience of the tender presence of God in the service of women and men in need. Our evangelisation must be directed more explicitly to them, and more openly in the name of the Gospel. (2010: 33)

Van Matre argues that "we need a new kind of love" … away from an egocentric love towards an ecocentric love (1983: 125). New rituals and parables are needed whereby humans can humbly relate to mother earth and express gratitude for all it offers to self and others (Van Matre and Weiler 1983: 125–126). Christ-centered expressions of tenderness and humility are central to broader society becoming aware of the unique nature of religious life and what it has to offer it members and the world today. Overall, these two expressions of the Holy Spirit appear missing in current authentic Church and State institutional responses to child-sex abuse. Although policies, financial compensation and legal proceedings are important, perhaps God's grace will be found more within experiences of shared-vulnerability than in public statements of justification and defense.

Awad, in God Without a Face? explores the subordination of the Holy Spirit in Trinitarian theology and proposes that "alongsideness" is the key missing notion for a balanced Trinitarian theology; and that reciprocity also plays a key role (2011: 133–134). Shared-vulnerability implies "alongsideness" as a contemporary manifestation of Christ-centered ministry. Fowler observes that Western society "shows greater willingness to grant the status of "reality" to the expressions of human violence and hard-heartedness than to relationships of tenderness, fidelity and empathetic understanding" (1996: 223). New forms of spirituality and religious entities must therefore have the above traits to ensure authentic relay of Christ's message that manifests the "image of the Son of God," regardless of charism.

Although, according to Cardinal Pell (2012), Christians are the largest single community-group in Australia, Protestant, Unitarian, Buddhist and other faith groups are also responding in various degrees to contemporary Australian societal issues as noted in the Catholic reports (Pell (2012), cited in Benson 2011). In 2009, the World Parliament of Religions met in Melbourne, Australia and identified seven religious leadership research themes for the twenty-first century: (1) theories and theologies of inter-religious relations; (2) social inclusion and countering radicalisation; (3) religion and poverty; (4) youth engagement; (5) reconciliation; (6) family, gender and sexuality; and (7) religion and ecology (Tehan, cited in Creative Ministries Network Board meeting minutes, December 2009). These themes proffer a sociological focus that do not attend directly to the issues needing to be addressed in Catholic religious communities. Arbuckle reveals the current common thread across faith traditions: the spirituality of grieving. Each faith tradition needs to grieve deeply for its 'emptying out', its 'kenosis', and is struggling to do so (cited in 2010: 40). Bouma (2006) argues that "interspirituality and bricolage, the piecing together of cultural (and religious) elements drawn from various sources" are on the rise in Australia, just as "the book of Genesis borrowed pre-existing creation stories and adapted them to its purpose" (2006: 211). A new group, the Progressive Christian Network of Victoria Inc., has even defined seven characteristics of progressive Christians, found on their website (2012). New attempts to address injustices are also emerging in the broader public sphere. For example, all local government agencies in Victoria are now mandated to embed human rights policies and practices in organizations and service delivery (Victorian Equal Opportunity and Human Rights Commission website, 2012). Ecological and social isolation issues are also being addressed via local government policies and activities, albeit on a voluntary basis.

3.4 The Australian Political Context

Currently, Australia governs federally through a minority Labor government, with three Independent MPs and The Greens political party holding the balance of power. With another Federal Election due in 2013, this term of Labor governance has also proved to be an "holy and unholy mess." While the Catholic Church in Australia is experiencing issues related to identity, credibility and visibility, the Australian Federal government is also encountering a similar crisis, as the Liberal-National Coalition voices a consistent cry of "No" to every debate, regardless of the specific issue or policy. Although the 24 h media cycle ensures "visibility," as determined by the media, the extent to which "behind the scenes" politics have influenced policy decisions and political appointments have left Australian citizens concerned about the integrity of their democracy.

For example, in 2010, Kevin Rudd, the elected Australian Prime Minister, was ousted from his role by four "faceless men." Since then, Paul Howes (2010), one of these "faceless men," documents the campaign to remove the Prime Minister in an overnight coup, in his book In Confessions of a Faceless Man. Howes (2010) is the Australian Workers Union national secretary, Vice President of the Australian Council of Trade Unions and member of the National Executive of the Australian Labor Party. Although the media have disclosed the names of the four "faceless men," this image has remained in the Australian psyche and is now broadly associated with the experience of Kevin Rudd's demise as Prime Minister of Australia. Moloney, in the 2010 Report, declared that "*religious* (author's emphasis) are the face of Christ's compassion in our world" (cited in Norton 2010: 37).

As scientific compassion and human rights research policies and practices gain traction and toxic power abuses are publicly revealed in contemporary political, business, workers' union, government, and church entities, Moloney's assertion needs to be challenged. It is not only the "religious" who are Christ's messengers of compassion; but also informed laity and enlightened educational institutions. Examples of this re-newing traction include Stanford University Centre for Compassion and Altruism Research and Education; Hearts in Healthcare, New Zealand; Compassionate Communities Network, Australia; Swann's (2001, 2002) Compassionate Leadership model in schools; and Tehan's *Compassionate Workplace* model (partially published 2007 Masters in Public Health project, Latrobe University). These secular manifestations suggest a "lack" of authentic compassion and human rights in some existing institutions. This renewal and unfolding, as Moloney has indicated, might be distinguished at its source by either "coming from the head" or "coming from the heart." The Foundresses of Orders, as outlined above, offered solid ground for their charisms to flourish, as their Orders charisms manifested from the heart … the heart of the Founder's experience of God's perceived abandonment and rejection of humanity prior to commencing their ministries.

For example, Kane, in the History of the Grey Sisters, notes that "suffering is the progenitor of endurance" (1980: 149). Maude O'Connell, Foundress of the Grey Sisters, was only 11 years old when her mother died; she was "old enough to realize her loss and have her heart marked by it" 1980: 5). Maude O'Connell herself wrote:

> at this time the working conditions for women were not good and in order to attempt to improve these conditions I went into a factory and worked alongside other women. Later I became a union official and in this capacity strove to improve the conditions I had experienced. (cited in Kane 1980: 5–6)

The widespread protest against "State Aid" to registered Catholic Schools and a lock-out of female workers at the factory they worked at (without money they could not pay for accommodation) evoked a strong sense of injustice in Maude and fueled her deep desire to improve women's equality both at work and at home (cited in Kane 1980: 6–9). In the late 1920s, Maude O'Connell met Dr. William Collins, priest at St. Francis Church in the Melbourne central business district, who would help her found the Grey sisters. His memory of her was vivid: "her criticism of Catholic social work, or rather the defects of it, was devastating" (Collins, cited in Kane 1980: 12). Each Foundress had her own personal experience that shaped their particular spiritual charism and Religious Order.

Similarly, in the 1980s, in Victoria, Australia, the historical and political landscape also irrefutably changed. According to Professor Graham Hodges, an expert in Public/Private Partnerships, during Jeff Kennett's period in government as Liberal premier of Victoria, the State became more privatized than the whole of the UK under Thatcher's rule of government (cited in Coghill 1997: 2). During the Kennett

(Liberal government) era, community-based palliative care services were tendered out by Government, resulting in a profound loss of trust and goodwill in the sector/s involved in the Tendering process. The fracturing and fragmenting of relationships within and between health care, community health and local government services, severely impacted on the capacity of local communities to look after their own dying and bereaved community members. In 1996, under the Howard Liberal government, The Australian Workplace Relations Act (Attorney-General's Department 2007), was enacted; it was known as "WorkChoices." Some of its provisions upon first being enacted included (1999: 125, 203–204, 252–262, 441, 465–466, 625–639):

- the introduction of Australian Workplace Agreements, a form of individual contract which can override collective agreements;
- expansion of the use of enterprise bargaining agreements;
- a reduction of the allowable matters in federal awards to 20;
- restrictions on union activity; and
- outlawing closed shops

Tehan's *Compassionate Workplace model* (unpublished Masters in Public Health project, Latrobe University, 2007) was a public health/grief and pastoral response to Workchoices and a disruptive (and very destructive), neo-liberal, privatization agenda. In particular, it was a contemporary response to an increasing number of women carers (including bereaved carers) working in low-paid employment while trying to manage and transition both aging parents and semi-independent adult youth into the next phases of their lives. Again, it was gender-specific and drew on personal experience including multiple prior losses by its project worker and author, as a consequence of the prevailing government policy of the day. It was also shaped by, and manifested as, a criticism of the existing institutions' inability or unwillingness to address these issues for women of faith and no faith, even with government-funded financial safety nets (although inadequate) and democratic voting rights.

One of Australia's key democratic strengths is its mandatory electoral voting system for Federal, State and local government elections. "Faceless men" and unaccountability are held in tension with mandatory policies that seek to maintain the Australian democratic political system. Australian citizens have the right to choose spiritualities of "faith" or "no faith," yet do not have that choice to "vote" or "not to vote." This bias is paradoxically strengthening the political system's voice/s in determining policy positions without a commensurate capacity and strength to respond to any harm caused by these same policies, by existing social justice arms of the different churches and religious communities. An "emptiness" is emerging at both religious and political levels that has the potential to profoundly impact on contemporary Australian culture. The question to reflect, and ultimately act, on is found in Matthew 32: 24–32; that is, whether in Peniel, the place Jacob named meaning "the face of God," this kenosis is metaphorically poised, or leaning, towards the compass of life or compass pin of death. Is life to be preserved at any cost or does exodus or death need to be encountered and wrestled with first?

3.5 The Australian Cultural Context

In 2012, Tim Storrier, an Australian artist, was awarded the annual Australian Archibald Art prize for his painting, a self-portrait, of "The Histrionic Wayfarer (after Bosch)." Storrier's painting refers to both a c.1510 Hieronymus Bosch painting titled '*The Wayfarer*' and the American Poet Robert Frost's (1920) poem titled "*The Road Not Taken.*" Ironically, Storrier's 2012 painting is a richly complex image of a "faceless man" that has been acknowledged and honored by an Australian portrait prize! It portrays a faceless man journeying across a desolate terrain with all the paraphernalia he needs to survive in life attached to his faceless body. It seems to epitomize the current Australian cultural landscape within which religions are also trying to find their way. The poem, *The Road Not Taken,* speaks of sorrow, lament and apology, doubt, sighing, differentiation, individuation, invitation, wonder and of an unfolding and evolving.

The Road Not Taken, by Robert Frost (1920)
Two roads diverged in a yellow wood,
And sorry I could not travel both
And be one traveler, long I stood
And looked down one as far as I could
To where it bent in the undergrowth;
Then took the other, as just as fair,
And having perhaps the better claim,
Because it was grassy and wanted wear;
Though as for that the passing there
Had worn them really about the same,
And both that morning equally lay
In leaves no step had trodden black.
Oh, I kept the first for another day!
Yet knowing how way leads on to way,
I doubted if I should ever come back.
I shall be telling this with a sigh
Somewhere ages and ages hence:
Two roads diverged in a wood, and I--
I took the one less traveled by,
And that has made all the difference.
(**Poem reprinted with permission from The Road Not Taken and Other Poems, edited by Stanley Appelbaum, Dover Thrift Edition: Dover Publications, 1993.**)

In Table 3.1, Stark argues that during the early Christian era, Christian unity was driven less by moral degeneration than by the combination of two major epidemics of smallpox and the need to revitalize the capacity of a culture to deal with its problems in relation to death (1996: 74, 78).

The smallpox plagues offered an opportunity for Christians to demonstrate their distinctive Christian identity, ministry and community in ways that drew pagan skeptics and lapsed Catholics into the Christian fold. As Stark highlights, the pagans and Christians had similar beliefs in the supernatural, in making behavioral demands on humans through sacrifice and worship, and in the supernatural responding to offerings through gods exchanging services for sacrifices (1996: 86). What was new and distinctly Christian, however, was the belief that God loves those who love him; God loves humanity; Christians please God when they love one another (Stark

Table 3.1 Different beliefs and responses to epidemics and the question, "Why?"

Pagan beliefs	Christian beliefs
Declared ignorance as to why the gods bring misery. Philosophers couldn't offer meaning to the experience	Life is meaningful, even amid death. A belief and a vision of a heavenly existence for relatives and friends who were "sent before" and who "led the way" provided hope and solidarity
No meaning/meaningfulness; therefore an individual's life was trivial if death by plague was only due to 'luck'	Placing oneself 'at risk' by remaining amidst the plague (to care for others) was a sign of martyrdom, and therefore pleasing to God
No coherence	Seek martyrdom while learning not to fear death
No explanation	
Beyond human control (claimed the cause was senility)	
Pagan belief encouraged action for the good of oneself	Christian doctrine was a prescription for action for the good of others
Pagan responses	Christian responses
Priests, highest civil authority, and wealthiest families fled the plagued city resulting in increased chaos and suffering	The well care for the sick
	Relatives love their kinsmen (kith and kin)
	Masters show compassion for ailing slaves
	Physicians do not desert the afflicted
	Epidemics were seen as schooling and testing (of and in Christian faith)

Source: The Rise of Christianity: A Sociologist Reconsiders History, by Rodney Stark 1996, pp. 79–81 © 1996 Princeton University Press, Reprinted by permission of Princeton University Press

1996: 86). Kent suggests that the challenge is to offer this belief in ways that serve humanity and not through seduction (1987: 12).

3.6 A Way Less Traveled

The 2010 Report, *See I am Doing a New Thing,* on the 2009 survey of Catholic Religious Institutes of Australia (2012), provides evidence of a resistance to kenosis so that new life can manifest, be nurtured and embraced; that is, there are 36 congregations with fewer than 10 members, and 40 congregations are not contemplating change.

Noel Connolly SSC declared that deep trust is required to be able to give up control with hope that God is doing a "new thing" (2010 Report: 39). Just as with the Royal Commission on child sex-abuse, the undertow of compassion and lament draw all the above threads (religious, political and cultural) together. The first century Donatists upheld a world-denying stance in order to be able to receive God's saving grace. In the 2010 report, Connolly noted that Vatican Two encouraged religious to seek out and affirm the emerging presence of the Holy Spirit in the world (2010: 39). Arbuckle, an anthropologist, priest and co-founder of the Refounding and Pastoral Development Unit, Sydney, in the same 2010 document calls on the religious to engage fully, individually and communally, in the spirituality of lamentation because this grieving and keening is the path to renewed energy and openness to new life (2010: 40). Arbuckle reflected on Jesus' acute awareness of his imminent passion and death, the meaning of his mission and what this will cost him personally, and the need for consecrated individuals and communities to also encounter deeply this painful path (2010: 40).

Prophetic ministry, according to Walter Brueggemann, consists of "offering an alternative perception of reality." Grieving, he says, "is a pre-condition … Only that kind of anguished disengagement permits fruitful yearning and only the public embrace of deathliness permits newness to come" (cited in Arbuckle 2010: 41). The Foundresses aforementioned knew the importance of grieving and lamentation and both experienced and expressed it in abundance through-out their lives. Arbuckle (2010: 41) suggests Moses as an example of ritual leadership in grieving and offers four insights into this approach:

1. Permit people to experience and name the chaos;
2. Acknowledge one's own limitations and pray into the chaos to God for help;
3. Keep ties with the collective roots of his people back in Egypt (Ex 13: 19); and
4. Undertake rituals to help let go of the past and express hope in the vision of the future.

Expressions of gratitude also need to be part of these rituals; although Euro-centric, the "Spiritual exercises of St Ignatius of Loyola" is one such example of the above process that also includes gratitude (Fleming 1978). These insights are also similar to Haberecht and Prior's suggestions for attending to grief, with an additional element – to be "presence-centred" in ministry (2002: 6).

3.7 Implications for Christian Identity, Ministry and Distinctive Community

As a large, desert continent, Australia is well-positioned to engage in, and find a way through the societal and church-faith transitions currently underway, especially in partnership with the wisdom and guidance of Australian Aboriginal people, the oldest, surviving indigenous people in the world. Lamentation, grief and human rights are familiar and constant companions in Aboriginal peoples' search for justice, reconciliation and healing against oppression and the rape of their souls and songlines through their experience of the "stolen generations" (Reid et al. 2012; Read 1999: vii–viii). Aboriginal people have long carried the wounds and scars of bearing "unacceptable faces," and of their loss of identity, credibility and visibility, while at the same time, also carrying sustainable, life-giving knowledge about interdependence and shared – responsibility as individuals and within their own communities; in particular their spirituality in relation to land, "country" and the environment. For Aboriginal people, their life-affirming spirituality, skin names (individuals are identified through their skin mother's name) and kinship codes underpin their very survival in the Australian desert and cities (Greene et al. 1987: 38–43).

A Eurocentric understanding of the Face of God does not necessarily connect with the current pluralism within contemporary Australian multicultural communities and spiritualities. Distinctive Australian Christian communities may need to find religious ritual and expression of lament, loss and hope in light of Australia's landscape and indigenous heritage, and alongside apostolic relay, in order to successfully transition along the less travelled road. As Arbuckle suggests, the Face of God may be present, at this time in history, through signs of God's consolation (2010: 40). Prophets are aware that gossamer threads of the Face of God may manifest where least expected, such as in the thin spaces of committed interdependent relationships, and the minutiae of daily life. In the end, to encounter the Face of God is one aspect of religious experience; to be still and know that I am God (Ps 46:10) is quite another.

References

Arbuckle, G. (2010). 'Final reflection on the survey'. In Catholic Religious Australia and Australian Catholic Bishops Conference. *See I am doing a new thing: The changing face of religious life and the challenges.* Mulgrave: John Garrett Publishing.

Attorney-General's Department. (2007). *Workplace Relations Act 1996.* Canberra: Office of Legislative Drafting and Publishing.

Awad, N. (2011). *God without a face?* Tubingen: Mohn Siebeck.

Benson, R. (2011, April 6). *Rendering to Caesar: What church leaders told the PM.* ABC online religion and ethics. www.abc.net.au/religion/articles/2011/04/06/3184075.htm. Accessed 22 Nov 2012.

Bouma, G. (2006). *Australian soul: Religion and spirituality in the twenty-first century.* Sydney: Cambridge University Press.

Burke, C. (2011). *Religious life – An international perspective. Catholic religious Australia 2011 National Assembly 'See I am doing a new thing: the changing face of religious life and the challenges.'* www.catholicreligiousaustralia.org/index.php/what-we-do/national-assembly/84-website/content/what-we-do/national-assembly/95. Accessed 13 Dec 2012.

Cabrini Mission Foundation. (2012). *Who is mother Cabrini?.* http://cabrinifoundation.org/2011/06/15/who-is-mother-cabrini/. Accessed 13 Dec 2012.

Cahill, D. (2012, October 22). Hearing Transcript. *Inquiry into the handling of Child Abuse by religious and other organizations.* Melbourne: Family and Community Development Committee. http://www.parliament.vic.gov.au/fcdc/article/1786

Catholic Religious Australia and Australian Catholic Bishops Conference. *See I am doing a new thing: The changing face of religious life and the challenges.* Mulgrave: John Garrett Publishing. www.catholicreligiousaustralia.org/index.php/what-we-do/national-assembly/84-website/content/what-we-do/national-assembly/95. Accessed 13 Dec 2012.

Coghill, K. (Ed.). (1997). *Globalisation and local democracy.* Melbourne: Monash University. Graduate School of Government Montech Pty. Ltd.

Connolly, N. (2010). 'A final reflection on the survey and the discussions at Hobart'. In Catholic Religious Australia and Australian Catholic Bishops Conference. *See I am doing a new thing: The changing face of religious life and the challenges.* Mulgrave: John Garrett Publishing.

Crittendon, S. National Catholic Reporter. (2012). http://ncronline.org/news/accountability/australian-commission-investigate-abuse-sydney-cardinal-provokes-backlash-citing. Accessed 11 December.

Ellingsen, M. (1999). *Reclaiming our roots: An inclusive introduction to church history* (Vol. 1). Harrisburg: TPI.

Family and Community Development Committee. (2012). *Transcript of inquiry into the handling of child abuse by religious and other organizations.* Cahill, D. (witness). www.parliament.vic.gov.au/images/stories/committees/fcdc/inquiries/57th/Child_Abuse_Inquiry/Transcripts/Professor_Des_Cahill_22-Oct-12.pdf. Accessed 13 December.

Fleming, D. L. (1978). *The spiritual exercises of St. Ignatius: A literal translation and a contemporary reading.* St. Louis: Institute of Jesuit Sources.

Fowler, J. W. (1996). *Faithful change: The personal and public challenge of postmodern life.* Nashville: Abington Press.

Frost, R. (1920). The road not taken. In *Mountain interval.* Henry Holt. New York: Henry Holt.

Greene, G., Tramacchi, J., & Gill, L. (1987). *Tjarany roughtail*. Broome: Magabala Books.

Guy, L. D. (2004). *Introducing early Christianity: A topical survey of its life, belief and practices*. Downers Grove: IVP.

Haberecht, J., & Prior, D. (2002, March). *Grief as spiritual chaos*. Presented at the NALAG National Conference 1995. National Association for Loss and Grief (Vic) Inc. Newsletter, Melbourne, Autumn.

Howes, P. (2010). *Confessions of a faceless man: Inside campaign 2010*. Melbourne: Melbourne University Press.

Kane, K. D. (1980). *The history of The Grey Sisters 1930–1980*. Melbourne/Canterbury: The Grey Sisters.

Kent, J. (1987). *The unacceptable face: The modern church in the eyes of the historian*. London: SCM Press Ltd.

MacCulloch, D. (2009). *A history of Christianity: The first Three thousand years*. London: Allen Lane.

Mercy International Association. (2012). *Foundress: Introducing Catherine McAuley*. www.mercyworld.org/foundress/story.cfm?loadref=49. Accessed 12 Dec 2012.

Moloney, F. (2010). 'A theological reflection'. In Catholic Religious Australia and Australian Catholic Bishops Conference. *See I am doing a new thing: The changing face of religious life and the challenges*. Mulgrave: John Garrett Publishing.

Pell, G. (2012, November 18). *Royal Commission. Catholic Archdiocese of Sydney*. www.sydney-catholic.org/people/archbishop/stc/2012//20121118_98.shtml. Accessed 12 December.

Progressive Christian Network. (2012). www.pcnvictoria.org.au/. Accessed 22 December.

Read, P. (1999). *A rape of the soul so profound: The return of the stolen generations*. St Leonards: Allen and Unwin.

Reid, S., Dixon, R., & Connolly, N. (2012). *A report on the 2009 survey of catholic religious Institutes in Australia*. Catholic Religious Australia and Australian Catholic Bishops conference 2012. Australian Catholic University, Fitzroy. www.catholicreligiousaustralia.org/images/docs/publications/see_i_am_doing_a_new_thing.pdf

Stark, R. (1996). *The rise of Christianity: A sociologist reconsiders history*. Princeton: Princeton University Press.

Storrier, T. (2012). *The histrionic wayfarer. The annual Australian Archibald art prize winner, 2012*. www.artgallery.nsw.gov.au/prizes/archibald/2012/29250/. Accessed 13 December.

Swann, R. (2001). Ecocentrism, leader compassion and community. *Leading and Managing, 7*(2), 163–180.

Swann, R. (2002, October). Compassion in school leadership. Australian Council for Educational Leaders, Monograph 31.

Tehan, M. (2007). *Leading the way: Compassion in the workplace*. Unpublished component of Masters in Public Health, Latrobe University. Melbourne.

Tehan, M. (2009). *World Parliament of religions 2009 report to Creative Ministries Network Board*. In minutes of meeting held on December 10th.

Van Matre, S., & Weiler, B. (1983). *The earth speaks: An acclimatization journal*. Warrenville: The Institute for Earth Education.

Victorian Equal Opportunity and Human Rights Commission. (2012). www.humanrightscommission.vic.gov.au/. Accessed 22 December.

Chapter 4
The Island Mystic/que: Seeking Spiritual Connection in a Postmodern World

Laurie Brinklow

4.1 Introduction

Since ancient times islands have been equated with sacred space. From Ulysses' Greek Islands and Islands of the Blessed to Saint Brendan's Island and the Scottish island of Iona, islands have been seen by Pagans and Christians alike as "place(s) of power and revelation: a place to cross from this world to the other" (Weale 2007: 10). "Indeed," writes social and cultural historian John Gillis, "islands have long been a symbolic presence in western culture, closely associated with the sacred in its multiple incarnations as holy isles and earthly paradises" (2001: 56). Further, the Celts called many of these islands "thin places," where the veil between heaven and earth is thinnest – thresholds where humans feel connected to the cosmos.

In the twenty-first century, islands continue to be regarded as special places. Not only do we recognize the significant role islands have played in Old World explorers' discovery of the New World, evolutionarily we regard islands and island ecosystems as laboratories for all life. And islands continue to attract humans to their shores. There is something in the human imagination – an island "drive," if you will – that finds islands mysterious, safe, exotic, otherworldly – even if they may not be so in reality. And it is a "tenacious hold," writes geographer Yi-Fu Tuan. "Its importance lies in the imaginative realm. Many of the world's cosmogonies, we have seen, begin with the watery chaos: land, when it appears, is necessarily an island" (1974: 118). The glossy magazine, called, simply, *PRIVATE ISLANDS*, is devoted to the buying and selling of islands. The tourism industry thrives on islands and the cruise ship industry depends on islands. Writers write about islands, singers

L. Brinklow (✉)
School of Geography and Environmental Studies, University of Tasmania,
Private Bag 78, Hobart, TAS 7001, Australia

Interim Co-ordinator, Master of Arts in Island Studies Program,
University of Prince Edward Island, Charlottetown, PE, Canada
e-mail: brinklow@upei.ca

sing about islands, artists make art about islands, and island studies scholars study islands. Indeed, *island* has become a central (Gillis 2004: 3) – and some would say *the* central – metaphor for Western civilization. The allure is such that islands – through the millennia – have achieved a kind of celebrity status, up there with tracking the stars – off earth and on.

Through the lens of phenomenology and citing in-depth interviews, this paper explores how, for some island-dwellers, island living is another kind of pantheism – one where living in tune with the rhythms of the island brings meaning, often expressed through art and writing. Based on research on island artists who are poles apart – on the islands of Newfoundland (population 515,000) and Prince Edward Island (140,000), Canada, and Tasmania (508,000), Australia – this chapter looks at the role islands play in artistic, and, more often than not, spiritual expression.

4.2 Islands as Pilgrimages

What is so special about being surrounded by water, where land meets sea, sea meets sky, and sky meets universe? If, as Tuan says, geography is "a mirror for man [sic]" (1971: 181) – and further described by Relph as "reflecting and revealing human nature and seeking order and meaning in the experiences that we have of the world" (1976: 4) – then islands are a mirror with an emphatic frame, with boundedness serving to accentuate the experience because of the intensity and perceived unity of an island. Travel to an island becomes a pilgrimage, with the outer journey mirroring the inner journey. Writes Prince Edward Island historian and author David Weale (2007: 167),

> Where better to discover a doorway to some neglected corner of the soul than on an island, a place so clearly in the beyond… for a high percentage of visitors to small islands the journey is a spiritual pilgrimage; a quest for the "blessed isle" of self-awareness.

Weale's reference to the "blessed Isle" is a nod to one of those ancient islands imbued with mythic qualities; the "Blessed Isles" stem from Greek and Celtic mythology (or Islands of the Blessed or Blest, and also called the Fortunate Isles, thought to be in the Canary Islands or the Azores) and were the site of the Elysian Fields, in which heroes, the virtuous, and relatives of the gods lived out their afterlife (Atsma 2000). Other mystical islands include the island of Avalon, or island of apples, originating in the legends of King Arthur. It is said that Arthur's sword Excalibur was forged on Avalon, and he was taken there to heal from his battle wounds (Driver 1999). Saint Brendan's Island was a mythical island visited anywhere between 565 and 573 by priest Brendan of Clonfert while on his voyage across the Atlantic in a curragh or leather boat; described as "a fantastic land: more beautiful and golden than he could have imagined" (Gill 2001), the island was never found by subsequent explorers. Even philosopher Thomas More's fictional *Utopia*, written in 1516, was posited as an imaginary island, since "an ideal society, and consequently happiness, was possible only on an island, a delimited, finite space" (Vigneau 2009: v).

The sacred island of Iona, off the coast of Scotland, is known for both Christian and Pagan traditions. Before Iona, it was called *Innis nan Druidhneach,* which translates as the Island of the Druids (Dunford 2009). The island is now regarded as the symbolic center of Scottish Christianity and is closely associated with Saint Columba and the monastery he established in 563 AD, "seeking seclusion among the 'desert' of the Atlantic Ocean" BBC Scotland (2012). A site of pilgrimage, for peasants and kings alike, Iona is where the Book of Kells was written circa 800 AD Trinity College Dublin (2012).

But Iona is perhaps best known for being a "thin place:"

> Something knocking on the heart which speaks of mystery and holiness, of dreams and truths which have outlived time… There is an indescribable atmosphere in Iona as if a "Presence" dwells in the hallowed soil of the tiny island which has been washed by the waters of prayer down through the ages. Hallowed and blessed by St. Columba and countless Christians for about 1,400 years, as well as by those who were there long before St. Columba, is it any wonder that an aura of spiritual peace surrounds the island. (Sandwith and Sandwith 1959, quoted in Dunford)

Both Pagans and Christians viewed islands as places where land and sea are at their closest:

> mesmerizing places like the wind-swept isle of Iona… or the rocky peaks of Croagh Patrick. Heaven and earth, the Celtic saying goes, are only three feet apart, but in thin places that distance is even shorter. (Weiner 2012)

Called "sanctuaries of creation," "Where the unpredictable becomes the means of discovery" (Maddox 1999), islands help us "to anchor our longing in the ancient longing of Nature" (O'Donoghue 1999: 15). Burgoyne (2001) writes,

> They probe to the core of the human heart and open the pathway that leads to satisfying the familiar hungers and yearnings common to all people on earth, the hunger to be connected, to be a part of something greater, to be loved, to find peace.

With the land laid bare, and the sky unveiled, in a "thin place" you could catch a glimpse of the glory of God, have a direct line to the cosmos. Or, as Eric Weiner goes on to say, "Maybe thin places offer glimpses not of heaven but of earth as it really is, unencumbered. Unmasked" (Weiner 2012).

4.3 Islands: A Kind of Secular Spirituality

In today's increasingly secular world, islands have become a place to "unmask:" as vacation destinations to get away from it all, as refuges from mainstream society, or as a place to strip down to essentials and get in touch with nature and the self. For many, these spiritual pilgrimages become permanent relocation, as people seek to find "place" in an increasingly placeless world. To seek connection, or belong, is the goal for many, and an island – particularly a small one – set apart from the mainland by an encircling sea, is the ideal place to get away from the lumbering weight of progress, reassess one's priorities, strip down to the essentials and become one with nature and the elements, and to feel like maybe you belong to something bigger than yourself that isn't necessarily God, or organized religion, but is, rather, a secular spirituality. David Weale tends to think of this secular spirituality as a kind of neo-paganism:

> The pagans were so much more aware of the interconnectedness of everything. There were, of course, the Christian mystics with the same sensibility, but they were marginalized and suspect. The best of them were ex-communicated or burned at the stake. It's always been there, but it's been so suppressed in our culture. But now it's coming out full-force and I think it's great. If there's anything that will halt the kind of depredation of the earth, it's that. You can fight a lot of fights, environmentally and politically and economically, but the most powerful force is mysticism, because it's the consciousness that recognizes that we are not separate from the earth. To me it's the most significant shift in my lifetime – the shift away from a sense of the transcendence of the sacred – the transcendent spirit in the sky – to an imminence, to a sense of being that yourself. (Weale 2012)

And small islands, for Weale, invite an intimate encounter with nature, which allows for a kind of connectedness.

> Whether it's a deep relationship with an island or a forest or a plain or a mountain, that relationship tends to take the person outside the narrow confines of head consciousness or ego consciousness, which is the enemy of creativity. For me, because I live on a small island, the quintessential experience in that regard is the experience of the ocean and the shore. (Weale 2012)

While the ocean has been arguably the most powerful and enduring metaphor for the eternal, the island is an equally powerful metaphor for the self.

> The self is surrounded by mystery, the way an island is surrounded by the vastness of the ocean, a domain that connects you with every other part of the earth. It can, of course, be experienced as a separation, but the mystic doesn't experience nature or the ocean as something that separates you, it's a reminder of our connection with the universe. And island consciousness does that. (Weale 2012)

But, says Weale, the shore, where land and sea meet, is an especially powerful liminal space. There, he writes, "the eternal makes love to the temporal" (2007: 10). He calls it a "threshold location," a

> thin place… where the thick veil of ordinary consciousness, that limits our vision, becomes less opaque; a place of liberation from the constraints of ego consciousness, and of deep and joyous connectedness with others, and with all that is, or ever has been. It is, in a word, an awakening to our own depth. (Weale 2011)

In 2007, on Prince Edward Island, off Canada's east coast, Weale helped establish a group called Shorewalkers, a secular group that meets bi-weekly – and not necessarily on the shore – to

> support and encourage one another in the movement away from the constructed boundaries of everyday consciousness into a wider field of awareness, and to discuss topics ranging from the mysticism of Julian of Norwich, William Blake, or Krishnamurti, to grief, fear, judgment, and ego. (Weale 2012)

The name has become a metaphor for the insights that can happen while walking the shore, where the everyday is momentarily transcended by proximity to the ocean.

4 The Island Mystic/que: Seeking Spiritual Connection in a Postmodern World

> The experience of walking on the shore, and taking in the sound of the ocean, the smell of the ocean, the sight of the ocean, can promote a kind of wonderful mindlessness that is liberating. I often come away feeling rinsed. What it liberates you from are all the cares and worries and anxieties and compulsions that bedevil us. I can go there in a knot, and when I leave, that knot has disappeared – or at least has greatly loosened – just from being there. (Weale 2012)

Like Weale, Zita Cobb of Fogo Island (population just under 3,000), off the northeast coast of the island of Newfoundland in the North Atlantic Ocean, expresses her attachment to her island place in similarly spiritual terms. A devout islander, Cobb has returned to Fogo Island after making her fortune in the fiber-optics industry, and is committed to revitalizing the island's once-vibrant fishing economy by using culture and the arts as an economic driver. In 2006, she and her brothers Anthony and Alan created the Shorefast Foundation, which aims for Fogo Island to become a center for sustainable fishery, arts, discovery, and innovation by making Fogo Island into a world-renowned destination for artistic, cultural, ecological, and culinary pursuits at "the edge of the earth" (shorefast.org). The fervor with which Zita Cobb speaks about her island shows a deep personal connection to her place that comes from being in tune with the land and the ocean and its people. She says,

> Nature doesn't have any artifice. It's geology. Newfoundland, Fogo Island, especially, is really naked, you can see, it reveals itself to you all the time. Trees on the other side, but trees and grass, they cover up the rocks, they get in the way of knowing the place. Because I think Newfoundland is so naked, it draws that out in the people... Here you feel your own vulnerability so much more, and when you feel vulnerable, then you're more open, you're more you. Here, your character has more currency than your personality. (Cobb 2012a)

Although she spent her working life in urban areas, during the summer she used to come home to Fogo Island to "reset her compass." Now she's home to stay. She says,

> There's something in the way that people here were so open, spirited – something in the constancy of the place, where – I suddenly felt given back to myself, and I felt I could just go back to being who I'd been all along, and what a great relief that was. (Cobb 2012b) (Fig. 4.1)

Fig. 4.1 An island off an island off an island. The view from Fogo Island, Newfoundland and Labrador (Photo by Laurie Brinklow)

Of course, many have experienced the negative sides of being in close proximity to the ocean. The vastness of the sea can call up emotions that reveal an existential questioning, the smallness of self in relation to God and all existence. Many become mute in the face of the ocean, such as one of Wayne Johnston's characters in his novel, *The Colony of Unrequited Dreams*, set on the island of Newfoundland. Joey Smallwood watched his father rail against all the injustices of the world when he looked inward toward the land, but when he stood to face the sea, he was silent, stifled, paralyzed with fear and a sense of emptiness and being inconsequential. Says Smallwood,

> For though I had an islander's scorn of the mainland, I could not stand the sea…. It was not just drowning in it I was afraid of, but the sight of that vast, endless, life-excluding stretch of water. It reminded me of God … Melville's God, inscrutable, featureless, indifferent, as unimaginable as an eternity of time or an infinity of space, in comparison with which I was nothing. (Johnston 1998: 131)

Devlin Stead, another of Johnston's characters in his novel, *The Navigator of New York*, says, "Nothing so reminds you like the sea that the enemy of life is not death but loneliness" (Johnston 2002: 472). Although Johnston now lives and writes in Toronto, "feeding off a homesickness that I need and that I hope is benign and will never go away" (Johnston 1999: 236), he grew up in Newfoundland, calling it his "circumscribed geography of home" (Johnston 1999: 89). He is always conscious of his "islandness," believing that "The other side of the gulf was remoter than the moon" (Johnston 1999: 94).

4.4 Islands and Art-Making

Writing and other ways of making art on and about islands goes back to ancient times and is pervasive as the winds and tides that define islands. From Homer's *The Odyssey* and Plato's *Atlantis* to Shakespeare's *The Tempest* and Defoe's *Robinson Crusoe,* island narratives are an integral part of the world's canon of great literature. Artists find islands particularly attractive for their inward journeys to creativity – free from the distractions of the everyday world, where living in tune with the rhythms of the island brings meaning, or as Canadian poet John Steffler writes, islands are a "blunt place… where excuses stop" (Steffler 1985: 9). On islands artists connect meaningfully with a place geographically, psychically, spiritually, and emotionally, surrounded by the ocean – which itself has become a powerful symbol for the unconscious, a metaphor for limitless possibility and inspiration.

Artists from both Newfoundland and Tasmania seem to find their inspiration in the geology of their islands: Don McKay and John Cameron both express through their writing a spiritual connection to the island that spans deep time. Indeed, Newfoundland poet McKay uses the term "geopoetry," coined by geologist Harry Hess in the 1960s, to describe "the place where materialism and mysticism, those ancient enemies, finally come together, have a conversation in which each harkens to the other, then go out for a drink" (McKay 2011: 11)

4 The Island Mystic/que: Seeking Spiritual Connection in a Postmodern World

McKay's Griffin-Poetry-Prize-winning volume of poetry, *Strike/Slip*, is inspired by his feelings of awe and wonder at the uniqueness of Newfoundland's geologic history, where the earth's tectonic plates have collided to form the island. The island was "formerly part of a micro-continent called Avalonia," writes McKay. "During much of the Paleozoic era Avalonia existed as a separate island in the middle of the Iapetus Ocean (the Atlantic's predecessor)" (McKay 2012: 81). On just why it inspires his writing, he says,

> Such scientific reflections may serve to extend the condition of wonder from its peak epiphany into everyday existence. We might find it spreading from exceptional instances, like a trip to Mistaken Point [on Newfoundland's Avalon Peninsula and site of some the planet's oldest fossils], to the nondescript rock in my backyard, which turns out to have travelled here from its birthplace in a volcano on the continent that became today's Africa. (McKay 2011: 19)

John Cameron, an essayist and former professor of social ecology who has retired to Bruny Island off the southeast coast of Tasmania (population 600), also finds meaning in the geology of his island – in particular, the ancient rock formations on his shore that are revealed by constant wave action – through the perspective of Goethean science (Fig. 4.2). Instead of the Western science model of observation, experimentation, and prediction, he believes that by opening oneself up to the phenomenon and using imagination and feeling as well as the senses and observation, it is possible to grasp something about the essential nature of the phenomenon, and then, in the final stage, having been given the gift of the phenomenon's presence, determining how to be of service to it (Cameron 2005).

Fig. 4.2 The art of nature. Bruny Island, Tasmania (Photo by Laurie Brinklow)

Writ[ing] about the way I'm responding to the place and what it's giving to me [is] a fairly essential meaning-making process. The lovely thing about Goethean science, it's not like you leave your logical mind at the door and simply just open up poetically, because I think there's great richness in bringing the best of western mainstream science and the best of the poetic imagination together. But there is a wonderful presence about rocks that I can't call them inanimate anymore. (Cameron 2011)

Cameron has found that living on an island to be an essentializing experience. All the things he'd managed to keep more or less hidden from the world, and himself, have become more evident on the island. "Here you can't hide from yourself. You see your warts and habits and things that get in the way, reflected back from the mirror that is the place – and particularly the water" (Cameron 2011). After nearly a decade of working to restore his 55, over-farmed, "clapped-out" acres to its original state, he's become attuned to the rhythms of the land, and to what the land needs from him (Fig. 4.3).

I've had a sense of entering into participation with the land in this process. By which I mean that I feel like I am being watched as much as I am watching the land. But it's reciprocal because the more I pay attention, the more I'm aware in paying attention that the creatures of the land are paying attention to me. I've written about a particular encounter I had with an eagle. It became very clear that the eagle was keeping its eye on me. On another level, I've been quite aware that the land has called us here. I don't want to ascribe a magical separate consciousness to the land. It's more a growing sense of a huge whole, a unity that includes us and all the creatures here, and all the people who have been here, and the Aboriginal people who have been here, and who's here now. That involves a level of consciousness and awareness that we're part of, but only a part of. (Cameron 2011)

Cameron (2011) echoes Weale's commitment to changing our approach to place when he says,

If we can imagine ourselves in different relationship with place, and with the island, that's an important step, as well as thinking through logically all the things we can do to reduce our ecological footprint.

Fig. 4.3 Blackstone. Bruny Island, Tasmania (Photo by Laurie Brinklow)

4 The Island Mystic/que: Seeking Spiritual Connection in a Postmodern World

Peter Adams of Tasmania was also attracted to the ancientness of Tasmania, to something so primal and present in the land that he felt compelled to "attach" himself to it.

> I'm still learning from this landscape… seascape… windscape. I'm still apprenticed to this place. It's also for me, as an artist educator, where I gather my material. I'm both learning from here and taking from this particular place raw materials, physical raw materials, and the kind of more intellectual, philosophic understandings that come with it to give my art work and my teaching more meaning, more juice, more authenticity. But being here, you can just go into yourself and you can start feeling the different clocks, there's the geologic clock that is really slow, and then there's the wind clock which is a hell of a lot faster. (Adams 2011a)

Adams finds much of his inspiration in his surroundings, but goes on to meld it with Greek mythology, creating wood sculptures from native Tasmanian wood and integrating them with materials found on his land at Roaring Beach, on the southwest coast of the island. His piece, *Ovum d'Aphrodite*, is comprised of a stone, representing Aphrodite, the goddess of fertility, love, and beauty, as a fertilized egg, nestled in a scallop shell carved out of huon pine. The story goes that Cronus, the youngest son of the Earth goddess Gaia, castrated his father Uranus and tossed the genitals into the ocean. Aphrodite was born of the sea foam that resulted. Adams's intention was "to bring an awareness of eros, of love and beauty, feeling and intuition, mystery and passion back into our overly masculine perception of the world" (Adams 2011a).

> Ultimately, the goal – if there is a goal – is to provide a means, to allow people, including myself, to find a way to love this earth. What I'm ultimately after is that we don't destroy this earth. These old stories, although much forgotten, are still a part of us. They need to be remembered, made alive once again through the artist's eyes, hands and heart and given a new life; a new birth, so to speak. (Adams 2011b) (Fig. 4.4)

Visual artist Victoria King, also of Bruny Island, Tasmania, finds herself in a storied relationship with her land and gives voice to it through her paintings and sculptures.

> There are often uncanny presences within the pieces of wood and rusted metal I find on Blackstone's shore and on our land. My sculptures are all made of driftwood and found objects that have traces of human presence, some from previous misguided attempts at agriculture. I'm very interested in the interface between driftwood and previously worked wood, especially those with layers of paint peeling off that speak to me of the vulnerability and futility of human existence. (King 2011) (Fig. 4.5)

Fig. 4.4 Ovum d'Aphrodite (Photo by Peter Adams, used with permission)

Fig. 4.5 Nature morte (Photo by Victoria King, used with permission)

She says that after living and making art in the US, England, and mainland Australia, the art she makes on Bruny Island is stripped down to the bones.

> The main reason is that this is the first time in my life I've really felt that my art practice has come together with my life. It's very integrated; the content is not separate from myself. Within each of my bird sculptures and bird-women figures there is a story, where I'm not the center, but I'm part of it. (King 2011) (Fig. 4.6)

Her studio is filled with hundreds of these small birds and tall, totem-like figures made with hand tools from driftwood and found objects; a handful of feathers spurting from a rusted water faucet; barbed wire twisted to make a boat or hanging like skeletal fingers from an old chain; a handcrafted book with paper made with thistles.

After she and husband John Cameron bought their property called Blackstone, they discovered that the adjacent land had special historic significance: the remains of an earthen "sod hut," from which George Augustus Robinson set out on his "Friendly Mission" to round up the Aboriginal population in the 1830s[1] (Fig. 4.7).

---

[1] George Augustus Robinson's "Friendly Mission" was an attempt by Governor George Arthur of Van Diemen's Land (now Tasmania) to relocate, between 1832 and 1835, the Aborigines to Flinders Island peacefully and under Robinson's protection. But, as time would tell, this was anything but "friendly"; writes historian James Boyce, "The colonial government from 1832 to 1838 ethnically cleansed the western half of Van Diemen's Land and then callously left the exiled people to their fate. The black hole of Tasmanian history is not the violence between white settlers and the Aborigines – a well-recorded and much-discussed aspect of the British conquest – but the government-sponsored ethnic clearances which followed it" (Boyce 2010: 296).

4 The Island Mystic/que: Seeking Spiritual Connection in a Postmodern World 107

Fig. 4.6 Two birds (Photo by Victoria King, used with permission)

Fig. 4.7 Vessel for grief (Photo by Victoria King, used with permission)

Fig. 4.8 Vessel of consequences (Photo by Victoria King, used with permission)

Through a series of coincidences that can only be called serendipity,[2] Vicki and John were able to purchase that property, too, and set about ensuring it would be given special designation by Heritage Tasmania.

> Something is being articulated through this place. My artwork is very particular to Blackstone, and particular to my being here through my own personal history. I feel like I can be a channel for the pain and betrayal felt by the indigenous Nuenone people who were once custodians of this land (King 2011) (Fig. 4.8).

Like many island artists, she is attempting through her art to confront the injustices humans have perpetrated on the land, as well as each other, and giving voice to

---

[2] To read more about John Cameron and Victoria King's serendipitous arrival to Blackstone on Bruny Island, Tasmania, and their subsequent purchase of the adjacent land, see John Cameron's "Letter from Far South" in Environmental & Architectural Phenomenology Newsletter, David Seamon, Ed., www.arch.ksu.edu/seamon/Cameron_letter_south.htm and "Third Letter from Far South: Inhabiting Intercultural History" www.arch.ksu.edu/seamon/Cameron_letter_3.htm

Fig. 4.9 Natural resources (Photo by Victoria King, used with permission)

the land – or, in the language of Goethean science, being of service to it to bring the stories to the fore, so that maybe history won't be repeated (Fig. 4.9).

Then there is the practicality of making art in a small community that is an island: Tasmanian painter Richard Wastell says,

> Tasmania is unique in a way. The artistic community is small enough here that I've always thought there were these really amazing opportunities to come together and do something here – not as usual in big cities where people are locked into their own fields.

At the same time, Wastell appreciates the fact that he's making art on an island.

> I think it's a real human want to just know where your boundaries are – where one thing starts and another ends… a human necessity really. I don't feel I could work anywhere else. I could, but I'd be making work of this place, from memory – I have enough to work for the rest of my life. I could work from my house, and never go out… I'd be insane, but I could do it. (Wastell 2011)

A native of Tasmania, Wastell says he grew up with an intimate contact with the land, swimming and surfing, hiking and fishing. His art is testament to that intimacy: his extraordinarily detailed work creates patterns that make up the bigger whole.

> I'm always struck by its wildness… these hills that are tangled, messy scrub, unruly, a raw power. I always worked with the landscape here, though at Art School it wasn't cool. Everyone was looking overseas for their models. I was a little guilty of that myself – only really when I met [writers] Richard Flanagan and Pete Hay and [photographer] Matt Newton that I had an epiphany and realized that it was almost the only thing that I could do. Because I saw that I was them. (Wastell 2011)

Through his art, Wastell has taken on some large subjects that are part of the "pain and the beauty" of Tasmania, including the devastation that has been wrought through the clear-felling of Tasmanian forests. Writes Richard Flanagan in his introduction to Wastell's exhibition, "We are making a new world,"

> They are intensely spiritual paintings by a painter whose close technique becomes ever more capable of conveying an enormous emotion… These paintings represent something new in Australian painting: they mark the point where we finally acknowledged our connection with the land in the most profound way possible: by acknowledging the spiritual cost of its destruction. (in Wastell 2011; Flanagan (2006): 5–6)

All this is not to say that organized religion does not play a role in island artists' creativity. Indeed, the art of Newfoundland painter, stained-glass artist, and sculptor Gerald Squires, a native of Change Islands off the northeast coast of Newfoundland, is at once nature-based *and* deeply spiritual. Indeed, critic Caroline Stones says that his book, *Where Genesis Begins,* "would not be out of place behind an altar." Writes Stan Dragland in the book's "Afterword,"

> Squires' response to the component of darkness that has its source in religion has not been to turn away, but toward it. More than one of his critics has noticed that his relationship with nature is religious in spirit. (Dragland 2009: 117)

Squires himself agrees:

> Art is prayer. It is communion with nature, with God. That is what you are aiming at in your lifetime. If you can tap that source, which is the greatest source in the world – the communion, the unknown – tap it and bring it through onto canvas then you can make art… (Winter 1991: 29)

But he also says that it is the "dark side" of religion that inspires his art:

> There was a fear that the church had created in me as a child, and it's something I've had to overcome as a man. Most of my work dealt with that and still does today. (Bruce 1987: 24)

And that work continues to this day to be perceived as being "pushed in an ecumenical direction" (Dragland 2009: 117).

4.5 Discussion: The Gift of Islands

What, then, can we take away from the deep spiritual connection people have with their islands, and islands have with their inhabitants, as witnessed through the voices of their artists? That the intimacy of an island craves connection, that the dramatic landscapes, seascapes, and windscapes have agency, that the very fact of separation from a mainland creates a special bond with an island's inhabitants that transcends ego-consciousness, resulting in a creativity that is rooted in a deep spiritual communion with place-that-is-an-island? As John Gillis writes,

> Today islands are more likely to be associated with salubrity than with sacrifice, but if we probe beneath the surfaces we find what Eliade would call a "'mythical geography' – the only geography man could never do without," displaying many of the same features of sacredness as the earlier holy and utopian isles. (2001: 57)

David Weale tends to believe

> …that any kind of earth or nature-based spirituality always discovers that "the holy land" is right there in the back yard, whereas Christianity, with its emphasis on heaven, de-sacralizes the whole planet. So what we are seeing today in the spirituality of many is a return to paganism. My bias is showing, but for humans I don't think there is any more powerful metaphor for the Eternal than the ocean. That makes island spirituality especially powerful. (Weale 2012)

Weale and other writers and island artists like him continue to make art on the edge, where ancient echoes reverberate through the deep time of geology, where increasingly volatile world weather patterns evoke fear – and at the same time pride in having survived yet another weather bomb – and where close-knit community stemming from these shared experiences thrives – in a place that is separate from the mainstream – all feed a deep spiritual connection to the special places that are islands – "thin places," if you will. And it is on islands such as these that they find inspiration for art that is dynamic, edgy, and confronting. Indeed, as Tasmanian writer Pete Hay has written,

> If island art embodies a psychological distinctiveness, if it is concerned with a politics of identity, and constructed in reaction to the particular stresses of a hard-edged, bounded existence, it follows that island art should be confrontational, abrasive and often concerned with the negative aspects of existence. (Hay 2002: 81)

Newfoundland's Stan Dragland in his Afterword to *Where Genesis Begins* writes,

> Ferment creates creativity… determination, anger, the rebellious attitude toward political and cultural centers… Newfoundland has become a crucible for jangling cultural impulses. (Dragland 2009: 111)

And it is particularly on islands that you can learn about walking lightly on the earth. Or in the words of Prince Edward Island poet Milton Acorn, in his poem "I, Milton Acorn":

> To be born on an island's to be sure
>   You are native with a habitat.
>   Growing up on one's good training
>   For living in a country, on a planet. (Acorn 2002: 92)

4.6 And What of Other Island Cultures?

Whether islands have this same symbolic presence in non-Western culture is something to be explored further. For instance, the Aborigines of Australia's Gulf of Carpentaria do not regard islands as separate entities; rather they see them as one vast entity, linked by songlines beneath the ocean: says researcher John Bradley, "There is no separation between land and sea: it's all Country" (Brinklow 2010). What of island mythology in Buddhist, Muslim, and Hindu cultures? Is the replica of the world, "World Islands," being constructed off the coast of Dubai symbolic of that culture's longstanding attachment to islands, or is the massive real estate project a testament to human ingenuity and greed – and hubris? Because, after all, "Unlike continents, islands, real material ones, can be rented, bought and sold; and in that way prostrate themselves to human imagination" (Baldacchino 2009: XIV). As our knowledge of our "world of islands" expands and grows, so, too, will our understanding of the scope of island spirituality. Meanwhile, people seeking something deeper will continue to seek out islands to fill a void in their lives, where, as researcher Bob Kull, who spent a year in solitude on a tiny Patagonian island, says, "the only way out was further in" (2004: 10).

Acknowledgements I would like to thank David Weale, John Cameron, Victoria King, Richard Wastell, Peter Adams, and Zita Cobb for their generosity in sharing their time, words, and wisdom. Artwork is used with permission of the artists.

I would also like to thank my PhD supervisor, Dr. Pete Hay, and the School of Geography and Environmental Studies at University of Tasmania – along with the Endeavour International Postgraduate Research Scholarship – for their tremendous support.

References

Acorn, M. (2002). *The edge of home: Milton Acorn from the Island*. Charlottetown: Island Studies Press.
Adams, P. (2011a, May 6). *Interview*, Roaring Beach, Tasmania.
Adams, P. (2011b). *Ovum Aphrodite*. Retrieved July 10, 2012, from www.windgrove.com
Atsma, A. J. (2000–2011). *The Thoi project: Greek mythology*. Retrieved July 10, 2012, from www.theoi.com/Kosmos/Elysion.html
Baldacchino, G. (2009). The island as figurae utopia. In J.-Y. Vigneau (Ed.), *Utopiae insulae figura* (pp. XI–XIV). Saint-Jean-sur-Richelieu: Action Art Actuel.
BBC Scotland. (2012). *Scotland's history: Iona*. Retrieved July 3, 2012, from www.bbc.co.uk/scotland/history/articles/iona/
Boyce, J. (2010). *Van Diemen's land*. Melbourne: Griffin Press.
Brinklow. (2010, October 25). *Songlines and storylines*, blog entry. www.tasmania-bound.blogspot.ca/2010/10/songlines-and-storylines.html
Bruce, M. (1987). Gerald Squires' last supper. *Atlantic Insight, 9*(6), 24–26.
Burgoyne, M. (2001). *Walking through thin places*. Retrieved July 3, 2012, from www.thinplaces.net/openingarticle.htm
Cameron, J. (2005). Place, Goethe and phenomenology: A theoretic journey. In B. Bywater & C. Holdrege (Eds.), *Janus head* (pp. 174–198). Amherst: Trivium Publications.
Cameron, J. (2011, April 10). *Interview*, Bruny Island, Tasmania.
Cobb, Z. (2012a, February 15). *Interview*, Fogo Island, NL.
Cobb, Z. (2012b). *TEDxFortTownshend-Zita Cobb-The Way Forward*: Fogo Island Shorefast Foundation, March 11, 2012. Retrieved June 19, 2012, from www.youtube.com/watch?v=1dVDpvVwGWw&feature=share
Dragland, S. (2009). Afterword. In T. Dawe & G. Squires (Eds.), *Where genesis begins* (pp. 91–135). St. John's: Breakwater Books.

Driver, M. (1999). *The legends of King Arthur*. Retrieved July 3, 2012, from http://csis.pace.edu/grendel/projs993a/arthurian
Dunford, B. (2009). *Iona: Sacred isle of the west*. Retrieved July 3, 2012, from www.sacredconnections.co.uk/holyland/iona.htm
Flanagan, R. (2006). Love walks naked. In R. Wastell (Ed.), *We are making a new world*. Hobart: Bett Gallery.
Gill, B. (2001). *Who is Brendan?* Retrieved July 9, 2012, from www.brendans-island.com/brendan.htm
Gillis, J. (2001). Places remote and islanded. *Michigan Quarterly Review, 40*(1), 39–58.
Gillis, J. (2004). *Islands of the mind*. New York: Palgrave Macmillan.
Hay, P. (2002). *Vandiemonian essays*. Hobart: Walleah Press.
Johnston, W. (1998). *The colony of unrequited dreams*. Toronto: Vintage Canada.
Johnston, W. (1999). *Baltimore's mansion*. Toronto: Knopf Canada.
Johnston, W. (2002). *The navigator of New York*. Toronto: Knopf.
King, V. (2011, April 11). *Interview*. Bruny Island, Tasmania.
Kull, B. (2004, May/June 2–12). *My year alone in the wilderness*. Canadian Geographic.
Maddox, S. (1999–2006). *Where can I touch the edge of heaven?* Retrieved July 3, 2012, from www.explorefaith.org/mystery/mysteryThinPlaces.html
McKay, D. (2011). *The shell of the tortoise*. Wolfville: Gaspereau Press.
McKay, D. (2012). *Paradoxides*. Toronto: McClelland & Stewart.
O'Donohue, J. (1999). *Eternal echoes*. New York: Harper Collins.
Relph, E. (1976). *Place and placelessness*. London: Pion Limited.
Sandwith, G., & Sandwith, H. (1959). *The miracle hunters*. Quoted in Dunford. Retrieved July 3, 2010, from www.sacredconnections.co.uk/holyland/iona/htm
Steffler, J. (1985). *The grey islands*. Toronto: McClelland and Stewart.
Trinity College Dublin. (2012). *The book of Kells*. Retrieved July 3, 2012, from www.tcd.ie/Library/bookofkells/book-of-kells/
Tuan, Y.-F. (1971). Geography, phenomenology, and the study of human nature. *The Canadian Geographer, 15*(3), 181–92.
Tuan, Y.-F. (1974). *Topophilia: A study of environmental perception, attitude and values*. Englewood Cliffs: Prentice-Hall.
Vigneau, J.-Y. (2009). *Utopiae insulae figura*. Saint-Jean-sur-Richelieu: Action Art Actuel.
Wastell, R. (2011, April 21). *Interview*, Hobart, Tasmania.
Weale, D. (2007). *Chasing the shore*. Charlottetown: Tangle Lane.
Weale, D. (2011). *Shore walkers: A company of friends*. Retrieved July 11, 2012, from www.shorewalkers.ca/
Weale, D. (2012, April 11). *Interview*, Charlottetown, PEI.
Weiner, E. (2012, March 11). Where heaven and earth come closer. *New York Times*. Retrieved July 3, 2012, from http://travel.nytimes.com/2012/03/11/travel/thin-places-where-we-are-jolted-out-of-old-ways-of-seeing-the-world.html?pagewanted=all
Winter, K. (1991, March 3). *Communion with the silent centre: A visit to Gerald Squires' studio*. St. John's, Sunday Express.

Chapter 5
The Spatial Turn in Planetary Theologies: Ambiguity, Hope and Ethical Imposters

Whitney A. Bauman

5.1 Introduction

Charting the territory of theology is no easy task. There are many ways one can begin to map the situation. However, there are at least three maps I want to draw on for the purposes of this chapter. Mind you, these are not the only possible maps, but rather three that I think will help map out the turn toward horizontal spatial relations in religion, which we might refer to as "spiritual cartographies." (Tweed 2006: 112) First is the map that marks the move in the Enlightenment from God and Revelation to Nature and Natural Laws. This narrative map, among other things, traces the ways in which there is a shift from viewing the world as a Great chain of being that literally removes humans and human meaning-making practices from the world to a world that becomes instrumental for human ends. Though perhaps no longer out of this world, humans are still in an objective space above the world from which we can turn the world into "standing reserve" (Heidegger 1977). The Death of God and the Death of Nature mark this shift. Though humans are perhaps no longer tied vertically to an Imago Dei, humans become the top/center of the world towards which the rest of the world bends and in this way are still located in a vertical position "above." The second narrative map that marks a turn toward the horizontal in religious thought is the attempt at retrieving the "sacred" in the world. This turn is marked by the pantheism of Spinoza, early emergentists such as Henri Bergson, the panentheism of Whitehead, and the Romantic responses to the Industrial Revolution. It is squarely in this second narrative that much of what is known as "religion and ecology" or "religion and nature" finds its conceptual ancestry. Here, the removal of humans from the horizontal plane of nature is seen as problematic and the hope is to "return" to the realm of nature. As such, humans are still seen as "apart from" and

W.A. Bauman (✉)
Department of Religious Studies, Florida International University, Miami, FL 33199, USA
e-mail: whitneyabauman@mac.com

in some sense "above" nature in that it is our actions upon nature that will either lead to doom or a glorious return. The third and final narrative map that I will discuss here has to do with a series of "deaths" that result from the processes of globalization: the death of identity, the death of certainty, and the death of truths. In a word, this third narrative map marks a turn toward radical immanence and embodied thinking about what it means to be human vis-à-vis a world where there are only proliferations of becoming life. It is from this third narrative that thinking and meaning-making practices become, finally, planetary; that is, radically of and for the becoming bodies of this planet. From such a horizontally located place, we can think of meaning-making as always and already an imaginative part of the rest of the natural world or even as "lines of flight" (Deleuze and Guattari 1987).

Lest this sound like yet another adventure in western triumphalism, I should note that many aboriginal and indigenous peoples have, as far as recorded history can tell, always been more horizontally oriented. Further, these turns toward the spatial do not happen the same in western cultures as they do in say, India, China, or Japan. Salvation in Heaven is not the same as liberation from the world of suffering that we find in both Hinduism and Buddhism. However, I would argue that the turns toward the horizontal in meaning-making practices are not "ready made" in any extant tradition of philosophical or religious thought. Transcendence is still the name of the game whether we are talking about salvation in the afterlife or liberation from the cycle of suffering. Further, and though much can be learned from indigenous cartographies, no indigenous cartography was created to deal with the hybrid identities we experience in a globalized world nor with a post-Industrial revolutionary world. It is the very process of thought (and bodies) having exhausted the "frontiers" and "locals" of specific places and the very knowledge that we are through and through planetary creatures that leads to this turn toward the horizontal in meaning-making practices. Finally, and along with other "post" epistemologies, we must always remember that map is not territory (Smith 1978). None of these maps map directly onto the reality that is becoming, bubbling up and being constantly reinterpreted by present seafarers. Rather, all maps help to shape the ways in which bodies become. It is my hope that these maps will shed some light on how meaning-making matters in and to earth-bodies today in order that we might take some responsibility for these meaningful materializations. In other words, as Butler notes, the "matter of bodies will be indissociable from the regulatory norms that govern their materialization and the signification of those material effects" (Butler 1993: 2). The following maps help to draw out some of these regulatory norms that have been dependent on transcendent or vertical understandings of Religion and Nature and that reveal possible different regulations in the turn toward horizontal meaning-making practices.

5.1.1 Map 1: From God's Revelation to Natural Law

As noted above, I am focusing this map on a specific trajectory in history that finds itself caught up in western forms of thinking and the globalization of that thought. This is not to say that these are the only maps that exist. On the contrary, the processes that I describe here happen differently in different places. However, processes of globalization have led to similar shifts worldwide. We could take as our starting point the movement of Vedic traditions in India to the more ascetic practices of Buddhism and Jainism that emerge out of those traditions. These practices, to some degree, chart a move toward transcending the everyday world of experience. Or alternatively, we might start with the history of symbols in cave paintings and discuss how the move toward symbol use in homo sapiens led to a certain amount of abstraction that took us "out of this world" (see Van Huyssteen 2006; Abram 1997). However, I want to start here with the narratives that lead into globalization because it is this phenomena that marks the context of the first generation of planetary identities. In other words, for the first time in history, the past century has seen a connection of all places on the globe. Through advances in transportation and communication, our identities have literally been shaped by identities on a global scale, and consciously so. Though this globalization has roots in colonization and still causes much violence in social and ecological terms, it is nonetheless that which marks the contemporary millieu. What marks this is hybridity, but we should never imagine that such hybridity is always creative. As Zygmunt Bauman (1998: 100) notes,

> The cultural hybridization of the globals may be a creative, emancipating experience, but cultural disempowerment of the locals seldom is; it is an understandable, yet unfortunate inclination of the first to confuse the two and so to present their own variety of 'false consciousness' as a proof of the mental impairment of the second.

Though I might disagree with Bauman's strict categories of "global" and "local," this process of globalization and the hybridity that marks it is quite destructive even as it creates new possibilities for becoming. The task here is to recognize the process as something for which we can take responsibility. It is, I would argue, this very process of globalization and the multiperspectivalism that comes out of it that begins to move us towards the very idea that there are options for becoming, rather than becoming being a result of what is Natural or God-given.

One initial step toward the process of globalization as we know it today is the breakdown of the "great chain of being" and the emergence of a secular space. The great chain of being secured the *Ancien Regime* and its way of life, right up until the Reformation and the Scientific Revolution in Western Europe. Such transcendent thinking suggested a rule by divine right that created an ordered hierarchy from God to the Church and Kings, Nobles, and their subjects. This form of transcendent thought and its corresponding technologies of economics (feudalism) and politics (monarchy) coded the human as existing apart from the rest of the natural world and the rest of the natural world as something which should either be feared (forests, desires of the flesh, wilderness) or made useful (agriculture, wood, coal, and other "resources") (See Merchant 2004). Though the "natural world" was both feared and seen as the source of life, it was controlled by a transcendent creator and it was toward this creator that one wanted to move throughout life. In other words, meaning-making only included the non-human world in a very limited sense. Such transcendent or vertical meaning-making practices were not unique to the Christian west, but could also be found in Hinduism and Buddhism. We can think of the caste systems of India or the royal dynasties that persisted in China and Japan, for instance.

During the Protestant Reformation in Western Europe, the authority that continuously affirmed this state of affairs was challenged. Subsequently, the scientific revolution, and eventually the Enlightenment and social revolutions that overthrew rule by divine right in favor of social contracts began challenging the great chain of being as well. At least three major shifts in thought characterize this period.

The first such shift is the shift from the founding of life, knowledge, and truth in God to the founding of life, knowledge and truth in Nature. It is no mere coincidence that the European Age of Exploration, the birth of Cartesian skepticism and the Reformation all happen at roughly the same time. The Reformation and the era of European colonization provide fodder for deep questioning of the God preached by the Catholic Church. Such skepticism topples the God that founds the great chain of being and begins looking toward the horizon, toward Nature for answers. Priests of the church such as Bruno, Copernicus, Galileo and Newton begin the process of re-thinking theology based upon observations in Nature. Rather than a theology that dictates nature, they all played their own part in developing a natural theology.

Furthermore the Reformation and Counter-reformation are to a great extent focused on bodily practices and habits rather than theology *per se*. This is an important turn to immanence: viz., the control of bodies. As Foucault notes, the sacrament of confession is one form of controlling bodies and such confessional practices will eventually be taken over by therapists and scientists from the priests. (Foucault 1978) Whereas this shift was an "ontological" shift, human beings are thought more and more "down to earth" in terms of nature rather than in terms of God, it brought about an epistemological shift from Revelation to Natural law.

Like in many other religious traditions, nature, understood as creation in Christianity, was not thought of as an End in itself. Instead, nature was one of the "two books" by which one could come to knowledge of god.[1] God was the end of all inquiry and thus it was God's Revelation through creation that was the source of knowing. "In the Beginning was the Word" and this word is precisely what one should come to know through nature. Though a complete shift to atheism was beyond most people in this period, many people shifted to a sort of deism. Whatever and wherever God was, was not as important as the fact that the world was created according to Natural Laws that could be understood through Reason. Hence, again, epistemology that was once "vertical" (focused on God) now becomes "horizontal" (focused on Nature). Both shifts, toward nature and toward reason and natural laws, also implied shifts in the ways that human were governed. Thus, the third shift discussed under this mapping is political.

Once the great chain of being is ontologically and epistemologically challenged, the social system that gives rule by divine right to kings and some degree of authority to the Church in governing over social and political matters, gives way. Authority and power shift from the vertical, top-down model to a horizontal rule by the people. I am thinking here of the Glorious Revolution in Great Britain, the American Revolution, and the French Revolution, the rise of national powers, and

---

[1] In many eastern traditions, I should note, one finds a similar symbolic understanding of nature, viz., nature is not what is really real but there is something beyond that constitutes the really real.

in general turns toward rule by the people. Religion responds to such shifts and in many ways aids such shifts: the Reformation leads to the protestant idea that one can have an individual relationship with God (without the mediation of the church) and one can also be his/her own interpreter of scripture (which helps spread literacy). Rather than religious justifications for monarchy, thinkers such as John Locke begin to find religious justifications for democracy. Thus, the turn in politics is also toward horizontal forms of government.

Far from being seamless, these shifts did not occur without exception nor did they occur overnight. Furthermore, many vertical ways of thinking persisted. For instance, one hangover from this era that still persists today is human exceptionalism. (Peterson 2001) If there is no longer a transcendent source for human beings but rather nature is the source of all that exits, why does the idea that humans are somehow above nature persist? Various versions of "religion and ecology" and "religion and nature" challenge the ideas of anthropocentrism and human exceptionalism. Often, these challenges fall into the same sort of pitfalls that a turn towards the horizontal does. For instance rather than seeking one's true nature in relationship to God or Ideas or Forms, one seeks to find one's Natural identity or an authentic relationship to Nature. Regardless, the events that lead up to the Scientific Revolution and the Enlightenment are crucial for the turn towards horizontal thinking in religion. Whereas Nature was once the Revelation of a Divine Creator, now that God is dead, and with it the sacrality of nature. As such, as Carolyn Merchant suggests, we can also speak of this transition as the beginning of "The Death of Nature." (Merchant 1980) In response to the Death of God and Nature, a second mapping of the horizontal turn is drawn which has much more to do with pantheism and the Romantic movement.

5.1.2 Map 2: Pantheism, Panentheism, and Romantic "Returns"

As the technologies of science and capitalism led more and more to a regime of truth that is captured by the phrase, the "death of nature," and more and more to an understanding of the rest of the natural world as "standing reserve" for human ends, two major religious responses began mapping out a resacralization of the natural world. One such map that was more theological was the turn toward pantheism and panentheism in the face of the death of transcendence. The second came from a newly emergent spirituality that saw nature itself as the source for the spiritual and divine. Though these two turns toward the horizontal understanding of religious thought overlap, I will discuss them separately here as one is a response, more directly, to the philosophical and theological "Death of God," (pantheism/panentheism) and the other is a response to the death of Nature most visible in the processes of Industrialization (the Romantics/early Environmentalists). While the former can be seen as a source for re-thinking more traditional understandings of religion in an immanent way, the latter is the source for thinking about newly emerging knowledge of nature, for example, the sciences, as generating their own earth-based spirituality's.

5.2 Panetheism, Panentheism, and the Recovery of "Traditional" Religions

The declaration of "the death of God," comes at a time in the nineteenth Century when the world is becoming a much smaller place. Through colonization, industrialization, and the increases in transportation and communication that would feed into the processes of globalization, many different religious, cultural, and philosophical beliefs began to butt up against one another. The recognition of these multiple beliefs and multiple ways of knowing in the West can be understood as the philosophical declaration that transcendent Objectivity is dead: there is no transcendent male God in the sky that secures any one way of knowing. For many religious people, this death of God has still not taken place, but other religious leaders, philosophers and academics took this death of God and began to re-think religion in a more immanent way.

One of the earlier and most well-known attempts at such a rethinking comes from Spinoza, who equates God with Nature. (Spinoza 1996) As such, Nature and God signify the same sort of thing, and accordingly, religious thought must become more concerned with geography, place, and other sciences that tell us about nature. Henri Bergson's *elan vital* would be another source for thinking religion in an immanent manner. For Bergson, this vital impulse is inherent to all life and is that which forces all life to change, in other words, to live. The vital source is not superimposed upon a mechanical nature, but internal to nature. Hence, an attempt to overcome vertical, dualistic thought in which energy/spirit is imposed upon matter from the outside. (Bergson 1911) Another option for re-thinking religion from an emergent perspective and one based upon science was the evolutionary theology of Teilhard de Chardin. For Teilhard, evolution was all leading toward an Omega/Christic point, which was the goal of life. However, this final goal was based upon the movement of life toward higher and higher levels of consciousness rather than a conscious imposing that end from the outside. (de Chardin 1955) Still another model for such thinking, no doubt influenced by Spinoza and Bergson is the philosophy/theology of Alfred North Whitehead. Whitehead developed a very complex, and immanent system of thinking about the relationship between the world and God, based upon his mathematical understanding of the newly emerging understanding of quantum physics. This "process thought," Whitehead and subsequent Whiteheadeans describe as panentheistic. (Whitehead 1978) Whereas with the pantheism of Spinoza or the emergence theory of Henri Bergson whatever we might call God and Nature are the same, with panentheism, the world is understood as existing within God: thus God is in the world, but the world of nature does not exhaust the reality of God.

Often drawing from these philosophical and theological strands that begin to think God and spirituality down to earth, mid-twentieth century theologians began reforming religious traditions from within to become more "earth friendly," in a movement that is now widely regarded as "religion and ecology," or "the greening of world religions." Granted, the emergence of "religion and ecology" had also to do with the emerging problems of environmental degradation in a post-Hiroshima, Silent Spring, and post- "little blue ball" image of the planet world. However, it is these earlier immanent traditions of thinking religion/god that scholars began to draw upon. For instance, Thomas Berry finds conceptual help from both Teilhard and Whitehead in rethinking the universe as a "communion of subjects" (Berry 1988). Sallie McFague, finds help from Whitehead in particular in thinking of the universe as "the body of God" (McFague 1993). The point is that this theoretical groundwork for re-thinking religion in an immanent way paved a path for religious studies scholars and theologians to begin doing the work of re-thinking religious traditions for the earth community. The sources for thinking "religion and ecology" are different in other traditions, of course. Hindu and Buddhist scholars began to draw upon the non-substantial and interrelated metaphysics within those traditions to green their religions from within. Still, such greening did not occur until there was a need for a response to the Industrial Revolution.

Such scholarship has now produced peer-reviewed journals and volumes on every world religion, the most standard being the Religions of the World and Ecology series coming out of Harvard University Press. (Tucker and Grim 1997–2003) Further there is a long-standing Group within the American Academy of Religion devoted to Religion and Ecology (and a newer one devoted to Religion and Animals), and the Forum on Religion and Ecology sponsors publications, events, and hosts a major web-site, all devoted to thinking religions down to earth. There is also, I would argue, a second movement that finds its roots in the turn to horizontal thinking about religious thought, spirituality, and value, viz., religion and nature. However, as Bron Taylor has aptly noted in his book, *Dark Green Religion*, this spatialization of religious thought draws much more from the Romantics and sciences in order to develop new forms of religiosity rather than reform extant world religions (Taylor 2010).

5.3 The Romantic Return: Indigenous Thought and Religious Naturalisms

Whereas Pantheism, Panentheism, and emergent understandings of religious attitudes in the West were responding to the philosophical/theological "death of God," the romantic return was responding to the "death of nature" brought about by the increase of technologies that turned all of nature into standing reserve. Though not often couched in religious terms, the romantic return is an attempt to make the rest of the natural world matter beyond human utility once again, and in this sense it can be thought of as a way of making nature valuable or sacred. The difference between the "romantic return" and the movements of immanent theologies and philosophies that I mentioned earlier has to do with their sources. Whereas the above-mentioned movements and scholarship draw from within the history of major world philosophies and religious traditions to re-think them within contemporary scientific contexts and circumstances of environmental decline, the romantic return develops new "nature religions" out of the sciences and also out of many idealized understandings of the pre-industrial relationship between native peoples and their lands.

One such strand of religious thought was the emergence of an environmentalism in the early twentieth Century that held a great reverence for the rest of the natural world. Muir, Thoroeu, Emerson, and other Romantic writers and activists went well beyond thinking of nature as resources to be preserved or conserved, but actually felt a sense of divinity in their experiences with nature. As Bron Taylor (2010) so aptly describes, these early Romantic writers would become the spiritual gurus for environmental activists. This field of "religion and nature" also has its own society and peer-reviewed journal, The Society for the *Study of Religion, Nature and Culture*. Furthermore, "religious naturalisms" have been under investigation by several scholars within the field of "religion and science" over the past fifty or so years in organizations (and their publications) such as the Zygon Center for Religion and Science, the Institute for Religion in an Age of Science, and the Center for Theology and the Natural Sciences.

For better and worse, there is a romantic understanding of nature that still underwrites much environmentalism today and so many emerging "nature religions." This romantic return is problematic for at least two reasons. First, it rewrites the human being (and human culture and technology) as somehow different from the rest of the natural world (which is something that humans inhabit, but is not something we are necessarily a part of). American Landscape paintings in the early 20th reveal that nature is best and most pure when humans are not a part of it. Second, this romantic return often idealizes native peoples, drawing from their alleged harmonious relationships with nature, while at the same time becomes the justification for removing them from areas that become national parks or in order to preserve endangered species and rainforests. Whereas many of the religions we think of as world religions began to be studied as "religions" in the eighteenth and nineteenth centuries, it was not really until the twentieth century that indigenous studies emerged (Cox 2007). As such, the study of indigenous people by academicians has always been tied not only to the history of colonization, which often placed indigenous peoples in close relationship to the land, but also to the history of environmentalism, which often does so in a romantic way. Despite these problems, the religious sensibility of environmentalism has indeed shaped our twenty first century concern for other animals on the planet, the stability of eco-systems, waters, and atmosphere.

Environmentalism is not the only type of spirituality informed by the romantic return and the emerging earth sciences. Gaian traditions inspired by James Lovelock's work and ancient Earth Goddess worship has also inspired a spirituality that is both female-positive (unlike many male centered religious traditions) and sees the earth as a living organism. (Lovelock 1979) There are also contemporary religious naturalisms that understand culture and religion as an emergent part of the process of evolution Goodenough (2000). Brian Swimme and Thomas Berry (1994) have now famously created a similar spirituality based upon contemporary big-bang cosmology. Within this "universe story," human beings become the universe in conscious form (Swimme and Berry 1994). This universe story has been the inspiration for a documentary film and a whole movement of "green nuns," among other things. (McFarland-Taylor 2009) Still others understand ecological practices such as restoration ecology as a spiritual practice (Jordan 2003).

This second mapping, then, is really the story of how contemporary religious traditions have expanded their circle of moral concern to include the rest of the natural world. As Mary Evelyn Tucker notes, it is the process of religions "entering their ecological phase" (Tucker 2003). Furthermore, it is also the story of how "natural sciences" begin opening up to spiritual interpretations, sometimes referred to as religions of nature or religious naturalisms. Both of these trends in religious thought begin focusing on the spatial relations among humans and between humans and the rest of the natural world. There is still, however a third map that we can identify, that deals more directly with the spatial concerns of embodiment, and it is to this map that I now turn.

5.3.1 Map 3: The Emergence of Planetary Identities

It may seem a bit strange at first to begin describing the death of identities, certainties, and truths with a discussion of the turn towards embodiment, but it is precisely this turn that leads into both the emergence of multiple perspectives based upon embodiment (identity politics) and the breakdown of the very categories of identity, certainty, and truth thought necessary for political or liberation movements (such as is found in deconstruction, post-structural, post-colonial, and queer analyses). In other words, just as the very conditions that spread the Modern worldview (the era of colonization and globalization) contained within them the conditions for postmodernity, so here the very conditions for identity politics contain the seeds of the destruction of identity politics. What comes "after" identity politics is anyone's guess, but I argue here for yet another horizontal metaphor of identity that goes beyond the human and political and rather regards humans as first and foremost planetary creatures. I take here the metaphor of planetarity from Gayatri Spivak and it suggests that far from being "global" creatures that have access to a global or objective vision, we are contextual, of the planet, always viewing planetary others from within our own context y*et al*ways already involved with a whole host of planetary others. It is through our differences that we can begin to link up and construct a planetary culture, not in writing over these differences with a single way of being becoming (as in globalization of a free-market capitalism). (Spivak 2003) Thus, planetary identities resist attempts at writing over bodies with sameness, whether the sameness of nation, gender, sex, sexuality, etc., while at the same time recognizing the textures of different bodies that make up the becoming planetary community (human and non). However, this move toward planetary embodiment is only made possible through previous embodied moves toward liberation and identity politics.

5.4 Liberation Theologies, Identity Politics and the Death of Identities

Just as the Industrial Revolution (and the spread of Industrial technologies around the planet) led to the emergence of environmentalism and religious concern with the more than human world, so liberation theologies begin as a result of the class, environmental, and sexist effects of economic globalization. The very emergence of Latin American liberation theology can be seen as a result of applying post "death of God" theologies to the conditions of people in Latin American countries. In other words, if religion is no more about salvation in the afterlife, then salvation becomes focused on this world and the conditions of this world that prevent salvation. For Latin American liberation theologians of the latter twentieth century, this meant identifying with "the poor" of the world that were kept poor by Bretton Woods style economics (see Gutierrez 1973). A similar approach can be seen in feminist theologies, black theologies, *mujerista* theologies, *Dalit* theologies, and even the ways in which Gandhi, Martin Luther King Jr., and Thich Nhat Hanh in various ways began to apply their religious principles to political movements for social and ecological justice.

Such movements formed on the basis that there was more than one History; rather, there are multiple perspectives on any given event based upon the embodiment of the person interpreting that event. Furthermore, due largely to a patriarchal and European hegemony, embodiment came to be understood as more or less difficult based upon sex, gender, sexual orientation, class, race, etc. These factors came to be seen as essential in politics of liberation that were more often than not religiously justified. From these multiple-embodied positions, there emerged also the need for breaking open essential identity constructions.

In the United States, "feminist" theologians soon began to realize that there was no single "female," but multiple embodiments based upon sexuality, class, race, etc. Similarly, "black" theologians and Latin American "liberation" theologians were critiqued for what was often their inherent sexism and heterosexism. Even now as more and more Gay and Lesbian scholars begin to find theological justifications for multiple versions of sexuality, bisexual and transgendered voices are often very minimalized if not excluded. Eventually these multiple identities even within a certain category began to critique the very idea of stable identities: hence the emergence of queer theory. It is not necessarily that identities are "constructed" by culture alone, but rather that there are multiple expressions of sexuality, race, gender, sex, etc. that can never be captured by a single umbrella that seeks to essentialize these identities. Rather, like nations, these identity communities are "imagined" (Anderson 1983) and "performed" (Butler 1993) in such a way that we are always talking about multiplicities when we speak of identities. It is this "death" of identity that I refer to. Critiques of liberal humanism for being egocentric, androcentric, anthropocentric, and even Eurocentric and heterosexist have broken down the very boundaries of self/other and thus have questioned the very concepts of agency on which identity politics have been built. There has been both recognition that religions have historically played a huge role in constructing these identity boundaries, and recognition, as queer theory becomes more and more a part of religion, that they can also help to deconstruct these identity boundaries and open us onto the "interstitial" nature of our open and evolving identity construction with multiple "others." (Bhabha 1994) It is these very issues of embodiment leading to the breakdown of identity essentialisms, and the politics that go with it, that also begin to erode our concepts of truth and certainty as well.

5.5 The Death of Certainty and Truth

As mentioned throughout this chapter, the death of God, the death of Nature, and the death of Identity are all a part of the process of colonization and globalization whereby multiple identities make multiple and often conflicting truth claims upon reality. In many western cultures, we like to think of multiple religious truth claims as existing within a politically, legally, and scientific "secular" space. Another way of putting it is that there may be many personal beliefs, but the secular space of nature, politics, the legal system, science, and economics is neutral and "for all." Secular studies have challenged the idea of a neutral, secular space. The death of certainty and truth, then, extends not just to "religious" meaning-making practices, but also to "secular" meaning-making practices. As such, some proponents of the New Atheism such as Richard Dawkins (2008) and the late Christopher Hitchens (2007) are just as much guilty of "fundamentalism" as the religious fundamentalists against which they argue.

The death of certainty and truth in our meaning-making practices means that there are no smooth or objective spaces left from which we can dictate reality (again, no "global" vision or gaze). It also means that our identities and our knowledge are always in constant flux toward an open and evolving future. This opens us onto the textures of the planet, onto the textures of planetary others in our meaning-making practices, and highlights the co-construction of the very meanings to which we give allegiance. Such recognition allows us to pay close attention to how our meaning-making practices affect planetary bodies and our relationships to the rest of the natural world. In other words, and as I have discussed elsewhere, (Bauman 2011) such attention opens us onto the way our meanings materialize in the world.

Post-Death theologies such as those of Mark Taylor, Thomas Tweed, and Richard Kearny are beginning to help us think of our meaning-making practices from within these planetary contexts (Kearney 2011; Taylor 2007; Tweed 2006). Furthermore, more and more religious studies scholars and theologians are engaged in how to make meaning from a multiperspectival context and are even suggesting that such polydoxy has always and already been the case for meaning-making throughout human histories (Keller and Schneider 2010.) Finally philosophers, theologians, religious studies scholars, literary critics, and scholars in the newly emerging field of animal studies are beginning to break open our outdated concepts based upon species boundaries (speciesism) and rather suggesting that we ought to think of ourselves as assemblages (Deleuze and Guattari 1987) or in terms of collectives (Latour 2004), Companion Species (Haraway 2007) or in constant negotiation with the earth others that we find ourselves in spatial relationships with.

Such planetary negotiations mark a transition from meaning-making that was based on transcendent certainty toward a horizontal, evolving, ambiguity. Meaning-making becomes more involved in imaginative "lines of flight" toward ever new ways of relating with planetary others in a way that brings about the flourishing of the planetary community. Again, such meaning-making practices will never reach a transcendent utopic space, but rather will always be involved in renegotiating at every step of the way. The turn toward horizontal space, toward context, in religious thought over the past 500 or so years means not that religion is superseded by science, or revelation by reason, or god by nature, but that we begin taking responsibility for how meaning-making practices and constructions affect other planetary bodies. Imagine if appeals to what is "Natural" or "God-Given" ceased to be an acceptable rhetorical trick and we had to continuously argue for certain ways of being and becoming? Would not our meaning-making practices become more responsive to becoming others? Would not we begin to recognize that we are planetary creatures and that spaces of removal that provide transcendent certainty have caused enough damage? That is one hope of the turn toward the horizontal, planetary context in meaning-making practices.

References

Abram, D. (1997). *The spell of the sensuous: Perception and language in a more than human world*. New York: Vintage.
Anderson, B. (1983). *Imagined communities: Reflections on the origin and spread of nationalism*. London: Verso.
Bauman, Z. (1998). *Globalization: The human consequences*. Cambridge: Polity.
Bauman, W. (2011). Meaning-making practices and environmental history: Toward an ecotonal theology. In J. Haag, G. Peterson, & M. Spezio (Eds.), *Routledge companion for religion and science* (pp. 368–378). New York: Routledge.
Bergson, H. (1911). *Creative evolution*. Mineola: Dover.
Berry, T. T. (1988). *The dream of the earth*. San Francisco: Sierra Club Books.
Bhabha, H. (1994). *The location of culture*. London: Routledge.
Butler, J. (1993). *Bodies that matter: On the discursive limits of sex*. New York: Routledge.
Cox, J. (2007). *From primitive to indigenous* (Vitality of indigenous religions series). Aldershot: Ashgate.
Dawkins, R. (2008). *The God delusion*. New York: Mariner Books.
De Chardin, T. (1955). *The phenomenon of man*. New York: Harper and Row.
Deleuze, G., & Guattari, F. (1987). *A thousand plateaus: Capitalism and schizophrenia*. Minneapolis: University of Minnesota Press.
de Spinoza, B. (Ed.). (1996). *Ethics*. New York: Penguin.
Foucault, M. (1978). *The history of sexuality* (Vol. 1). New York: Random House.
Goodenough, U. (2000). *The sacred depths of nature*. New York: Oxford University Press.
Gutierrez, G. (1973). *A theology of liberation: History, politics and salvation*. Maryknoll: Orbis.
Haraway, D. (2007). *When species meet: Posthumanities*. Minneapolis: University of Minnesota Press.

Heidegger, M. (1977). *The question concerning technology, and other essays.* New York: Harper and Row.
Hitchens, C. (2007). *God is not great: How religion poisons everything.* New York: Hachette Book.
Jordan, W. (2003). *The sunflower forest: Ecological restoration and the new communion with nature.* Berkeley: University of California Press.
Kearney, R. (2011). *Anatheism: Returning to God after God.* New York: Columbia University Press.
Keller, C., & Schneider, L. (Eds.). (2010). *Polydoxy: Theology of multiplicity and relation.* New York: Routledge.
Latour, B. (2004). *Politics of nature: How to bring the sciences into democracy.* Cambridge: Harvard University Press.
Lovelock, J. (1979). *Gaia: A new look at life on earth.* Oxford: Oxford University Press.
McFague, S. (1993). *The body of God: An ecological theology.* Minneapolis: Fortress Press.
McFarland-Taylor, S. (2009). *Green sisters: A spiritual ecology.* Cambridge: Harvard University Press.
Merchant, C. (1980). *The death of nature: Women, ecology and the scientific revolution.* New York: Harper and Row.
Merchant, C. (2004). *Reinventing Eden: The fate of nature in western culture.* New York: Routledge.
Peterson, A. (2001). *Being human: Ethics, environment, and our place in the world.* Berkeley: University of California Press.
Smith, J. Z. (1978). *Map is not territory: Studies in the history of religions.* Chicago: University of Chicago Press.
Spivak, G. (2003). *Death of a discipline.* New York: Columbia University Press.
Swimme, B., & Berry, T. (1994). *The universe story: From the primordial flaring forth to the ecozoic era.* New York: Harper Collins.
Taylor, M. (2007). *After God: Religion and postmodernism.* Chicago: University of Chicago Press.
Taylor, B. (2010). *Dark green religion: Nature spirituality and the planetary future.* Berkeley: University of California Press.
Tucker, M. E. (2003). *Worldly wonder: Religions enter their ecological phase.* Peru: Open Court.
Tucker, M. E., & Grim, J. (1997–2003). *Religions of the world and ecology* (Vol. 9). Cambridge: Harvard University Press.
Tweed, T. (2006). *Crossing and dwelling: A theology of religion.* Cambridge: Harvard University Press.
Van Huyssteen, J. W. (2006). *Alone in the world?* Grand Rapids: Eerdmans.
Whitehead, A. N. (1978). *Process and reality.* New York: The Free Press.

Chapter 6
The Age of the World Motion Picture: Cosmic Visions in the Post-*Earthrise* Era

Adrian Ivakhiv

6.1 Introduction

Philosopher Martin Heidegger once characterized the modern world as the "age of the world picture," an era when the world itself became conquered by humanity as a picture or representation set fully and clearly before our gaze. In the 1960s, the first images of the Earth from space delivered a glimpse of a world picture that was global and ecological, but that also suggested humanity's domination both of the earth (today) and of outer space (tomorrow) (Chaikin 2009; Cosgrove 1994; Garb 1985; Helmreich 2011; Jasanoff 2001; Lazier 2011). Fifty years later, we have not colonized other planets, but we might speak instead of an age of the world *motion* picture, an era when our colonization extends to imaginary planets – like the Pandora of James Cameron's blockbuster film *Avatar* (2009) – and where we see our own world and our very selves in turbulent and uncontrollable motion – on screens around the globe.

The moving image in all its variations – from the first short films through the eras of talkies, technicolor, IMAX and 3-D, alongside television, videos, virtual reality games and the rest – has been with us a little more than a century, but over that time it seems the world itself has come to move faster and faster all around us. For philosopher and cineaste Gilles Deleuze, it was cinema that provided the greatest resource for reviving our lost "belief in this world" (Deleuze 2005: 166). This chapter will ask how cinema is faring today, on the cusp of a digital era that heightens the speed of life in every direction, in supporting "belief in this world" and in the universe that sustains it. It will provide glimpses from the last five decades of cinema by focusing on images of the Earth as seen from space. Drawing on the semiotic phenomenology of Charles Sanders Peirce, I will shed light on some of the ways in

A. Ivakhiv (✉)
Environmental Program, University of Vermont, Burlington, VT 05405, USA
e-mail: aivakhiv@uvm.edu

which such images have moved their viewers, affectively and cognitively, and how their use in movies, from *2001, A Space Odyssey* (1968) onward, has facilitated new forms of global identity construction built on new emotional and spiritual geographies.

6.2 From the World Picture to the World *Motion* Picture

To say that we are living in the age of the world motion picture is to rephrase and update Martin Heidegger's claim that we are living in the "age of the world picture" (1977). This is an age in which the world has become for us a picture, something we can grasp frontally, measure and assess, evaluate and transform to our whim. That world, according to Heidegger, stands before us and apart from us, its secrets unveiled, to be analyzed, taken apart, and put together in new ways. Humanity has become separate from nature: its judge, assessor, manipulator, and master.

For Heidegger, the world grasped as picture is only one among many possible worlds or, rather, many ways of "worlding" – many ways of bringing forth a world into sense and meaning. Beneath and behind them is a generative openness that he called "earth," which subsists, subtends, and renders possible any and all worlding. When this earth itself is brought into our picture of the world, when we think we have encompassed it without remainder, we risk losing what is most essential: an awareness of how our world itself emerges from an openness that is mysterious and unknowable in its essence.

There is a paradox in the way our world of the motion picture has arisen. We understand ourselves as sharing a planet together, a blue-green biosphere, in part because we, or some among us, have tried to remove ourselves from its gravitational pull. We are torn between the rocket-fueled movement away from the planet and the motion of looking back onto it. It is this double movement that has given us our sense of what our world looks like from space, and, therefore, what it *really* looks like to an imaginary, but objective, observer. Recalling this double movement can help ensure that our moving images do not displace us, because we *ourselves* are – and have always been – moving images. We, like all things, are in motion, always becoming *other* than who we just were. At least that is the claim of a loose tradition of (what I will call) process-relational philosophers, from C. S. Peirce, Henri Bergson, and A. N. Whitehead to Gilles Deleuze and Felix Guattari, all of whose insights I wish to apply in what follows (cf. Ivakhiv 2013). The question underlying this chapter might be phrased as follows: Is there a cinema that can remind us of this fundamentally ungraspable mobility of all things, human and unhuman, and, therefore, of the ways we can *become other,* with each other and with more-than-human others, so as to create a new collectivity on a shared earth?

Any history of the "world as picture" would necessarily note the innovations of the Renaissance and of early modernity in picturing the view of a world as it is seen by a stationary observer. The development of linear-perspectival representation gave Europe's maritime powers a powerful tool for mapping, navigation, and ultimately for colonizing new lands and conquering space. By the nineteenth century, the modern impulse to see, know, order and map the world was supplemented by technologies of display and exhibition, which laid the world out as something already known and ordered – given to us in museums, zoos, dioramas and panoramas, exhibitions, arcades, and tourist spectacles. Yet behind those spectacles was a world increasingly in motion. By harnessing energies (of coal and oil) that had been locked up and stored underground for hundreds of thousands of years, society was launched into turbocharged motion: the world became a world to be produced, transformed, and ultimately consumed as an endless parade of ostensible novelties. We live today in a visual world, a world dominated by technologies that have given us the clearest, starkest, and most seemingly objective picture of the universe ever held. At the same time, our inundation by images has made it difficult to make sense of the "imageworlds" that "cover the planet like a sheath" (Burnett 2005: xxii).

If paintings, pictures, and photographic images have always quietly *moved* their viewers, a world of moving images – from silent and sound films to YouTube videos replicating at near light-speed across computer terminals around the planet – moves us further, projecting our imagination more extensively across the territory of the world. Those images draw us into their motion, engage us in the movement of their storyline, the actions and reactions unfolding in and through and around the places and characters portrayed. They immerse us in the flow of sounds and words, places and landscapes, bodily movements and performative gestures. The moving pictures that surround us are pictures in motion: they take us on cognitive and emotional journeys, and today they take us further and deeper than ever as they become more immersive, more powerful, more animated, and less constrained by the physical realities they depict. They have also become ever more global, and in this sense more unifying, as the same movies (such as *Avatar*) are seen by millions, if not billions, around the world. What is it that happens when a global world comes to share the same powerful, moving images?

6.3 Images of an Earth in Semiotic Motion

American philosopher, logician, and polymath Charles Sanders Peirce developed useful tools for understanding how images (or what he preferred to call "signs") are intimately related, in one direction, to the meanings that inform and shape our lives and, in the other, to the material reality that underpins them. For Peirce, the world is "perfused with signs" through and through. It is not only humans who signify; signification, or semiosis, occurs whenever there is a "taking account" of anything by something else. Peirce's definition of a sign is "something which stands to somebody for something in some respect or capacity" (Peirce 1934: par. 228). To be a sign there must be three interrelated elements: a *sign-vehicle*, which is the form or medium that carries meaning (as, for instance, smoke, or a word); an *object* (such as fire), which is the absent or inaccessible referent that is being pointed to by the sign; and an *interpretant*, which is the sense or meaning made of the sign by an interpreter ("danger, fire!"). If there is no meaning being produced *about* something *for* someone at a given moment, there is no signification occurring. Semiosis is an active event of meaning-making.

The best known way of classifying signs arising from Peirce's framework is the one that distinguishes between icons, indexes, and symbols. These refer to the relationship between the sign-vehicle and the object. They are in turn grounded in Peirce's phenomenology, or what he called phaneroscopy. While Peirce ultimately distinguished at least 66 classes of signs, all the distinctions emerge from three, which are the categories he found applying to everything in the universe, and which he labeled Firstness, Secondness, and Thirdness. For Peirce, there are things in themselves, *firsts*, which are what they are: pure qualities. There are things in relation with other things, or *seconds*, which are existential realities, events of interaction. The universe measured by the physical sciences is a universe of Secondness, actual encounters between substances. And there are *thirds*, things that mediate relations between other things. Relations of Secondness, when mediated by thirds – by observations, measurements, interpretations, judgments, syntheses – become Thirdness, which means pattern, habit, meaning, regularity, and law.

Images are each of these. An image can resemble something, that is, it can be related to another thing simply by virtue of it being what it is, as a first; this kind of image is an *icon*. It can be directly and existentially related to something, by virtue of being caused by the other thing, such as a footprint is directly caused by the step of a foot, or a weathervane by the wind that directs it, that is by Secondness; this is an *index*. And it can be related to something by interpretive convention or habit, that is, by Thirdness, making it what Peirce called a *symbol*.

Photographs of the Earth from space are each of these. The two best known and most widely distributed of such photos – and among the most widely reproduced images of all time – are those known as *Earthrise,* taken aboard Apollo 8 (1968; Fig. 6.1), and *Blue Marble,* from Apollo 17 (1972; Fig. 6.2). In its resemblance to the Earth, *Blue Marble* is an icon. It resembles what we recognize as the Earth seen from space *because* we had seen those images before – before they were photographically produced. They have a prehistory in globe-shaped maps dating back at least to the seventeenth century, and in the places where such maps and globes have been reproduced: as emblems and logos for universal exhibitions and world's fairs, for missionary organizations, newspaper and media organizations, airlines and art museums, on covers of science-fiction books and magazines, and in movie studio logos and political commentary. This iconic resemblance of the photos to what we now know of as the Earth from Space is one that these images helped to create, since we had not seen the Earth quite like that until then. Their iconicity is part of their history; they are, in this sense, always already *indices* of that history. They are causally related to the processes that led up to them.

Two of these processes were the development of photographic technology and the Space Race. The space programs of the U.S. and the USSR functioned, to some extent, like the Public Relations arms of technological and military development in both countries, employed to garner public support for things that would have been deemed less easily supportable otherwise. But these were not merely political and technological projects. They were extensions of human-biospheric activities outward and away from the planet's gravitational pull. Both Earthrise and Blue Marble were taken as afterthoughts by Apollo astronauts with more important things on their minds. To get an image like Blue Marble, one has to travel at least 20,000

6 The Age of the World Motion Picture: Cosmic Visions in the Post-*Earthrise* Era

Fig. 6.1 Earthrise (Photo taken aboard Apollo 8, Dec. 1968, by William Anders)

Fig. 6.2 Blue Marble (Photo taken aboard Apollo 17, Dec. 1972)

miles from the Earth's surface – which is something that only 24 humans have ever done, all between 1968 and 1972, in that 4-year period separating these two photographs. Only the last three Apollo missions saw a full disk of the Earth in sunlight (Reinert 2011). All the images made after 1972 that show the entire surface of (one side of) the Earth are not single-exposure photographs, but composite, computer-aided digital images. Google Earth is a kind of composite and in-progress descendant of these images. It, like the others but even more so, is an image in motion because it is always being added to, transformed, and deployed for new purposes.

Finally, as a multivalent sign that elicits varying responses among viewers, these images are Peircian symbols. This means that they elicit further meanings, further interpretations, in an ongoing semiotic process that never ends. They have come to represent and embody a sense of awe, wonder, sadness, and beauty; of wholeness and globality, the unity of a world without political or cultural borders and divisions, and of a common destiny shared by all organisms; of planetary vitality, but also of the vulnerability and fragility of life on Earth, and of fear and trepidation for its, and our, future; and, finally, of disappointment, boredom (because we've seen it used so often for mundane ends), or even disgust – for what we have done to the Earth, or for even being "up there" where, as one of my students put it, "we don't belong," so that it signifies not home, but as Heidegger put it, a "vertiginous unmooring" (Helmreich 2011: 1215; cf. Cosgrove 1994; Garb 1985; Jacobs 2011; Lazier 2011).

Their meanings return some viewers back to a collective moment of radio and television contact between a handful of humanoid bodies and voices on a distant moon, an earthly base, Ground Control in Houston, Texas, and some 600 million people – one in five humans at the time – watching or listening (Chaikin 2009: 55). To younger viewers, the walk on the moon that memorialized these images has in turn been reframed and rememorialized by performers like David Bowie ("Space Oddity"), Elton John ("Rocket Man"), and Michael Jackson, with his famous "moonwalk." These images evolve and change their meanings, and as they cycle through from Peircian Firstness to Secondness to Thirdness, so they change the conditions for human inter-involvement with the world around us. As Benjamin Lazier (2011: 627) writes,

> The sedimentation of Whole Earth iconography into the mental architecture of the West means that for the foreseeable future, environment will be inflected by planet, cityscape by globe, and skyline by space – not the "space of experience" but the void. The lived experiences of earthliness and worldliness, at least as Heidegger and Arendt imagined them, are available, if they are available, only against the background of this new dispensation.

6.4 Five Visions: Toward a New Earth and a People to Come

If our experience of earthliness and worldliness is only thinkable against the background of the "new dispensation" made possible by the view from space, can we create a viable future within this dispensation? This is the question Gilles Deleuze and Felix Guattari indirectly pose through their claim, in their last collaborative work, that we must create a "new earth" and a "people to come" (Deleuze and Guattari 1994: 109). They suggest that we ought to do this not by remaking ourselves

6 The Age of the World Motion Picture: Cosmic Visions in the Post-*Earthrise* Era   135

in separation from the nonhuman, but in remaking our relations with nonhuman others – animal and machine others – so as to transform the possibilities for new forms of political collectivity.

In what remains, I will focus on five films: two from the 5-year period in which *Earthrise* and *The Blue Marble* were produced, two recent films that show the deep imprint of the earlier pair, and one film that appeared in between. I will suggest that these films represent points on a continuum of a kind of cinematic-geophilosophical thinking about humanity and the Earth, a thinking that was made possible by the view from space, but that has, or can, reframe that view for us in interesting ways. These present a range of visions by which a cinematic humanity is reshaping its understanding of our spiritual and cosmic environment.

Cinematic images of the future range widely: from the space-opera optimism of *Star Trek* and the epic battles of *Star Wars* and the *Aliens* and *Terminator* franchises, to the more bleak and pessimistic earth-based future-fiction that grew popular in the 1970s (with films like *Soylent Green* and *Silent Running*), to the equally dark visions of later films such as *Blade Runner*, *Total Recall*, *12 Monkeys*, *Gattaca*, *AI: Artificial Intelligence*, *Minority Report*, *The Matrix*, and *Children of Men*. Each of the following five films depicts a planet in its cosmic environment, a planet that either is intended to represent ours, or that has come to be taken as a double for ours. Planetary images figure centrally in all of these films, and together they provide different ways of imagining both the human relationship to life in its biospheric totality and the relationship between earthly life and the cosmos.

6.4.1 *2001, A Space Odyssey*, or, an Earth and a People Who May or May Not Be to Come

This movie (1968) was the first big-budget film to realistically depict the whole Earth as it might be seen from outer space. It featured partial images of the whole Earth as these were available to director Stanley Kubrick at the time, several months before Apollo 8 made those images available to a mass audience. In its mixture of science-based futuristic realism, metaphysical speculation, and psychological suspense, the film became a groundbreaking critical and commercial success.

It had been planned by MGM to be a blockbuster on several levels. According to Peter Krämer, "the very same factors that made the space programme controversial – the gigantic expenditure of taxpayers' money, the possibility of national humiliation should the Soviets win the race, the enormous risk for astronauts," made *2001*, or *Journey Beyond the Stars* as it was originally called, look "even more attractive as a virtual alternative: only MGM's money would be spent, the Soviet film industry could never rival Hollywood and no one would die." This film

> …was going to be a roadshow attraction for the whole family; a big-budget historical epic and futuristic Cinerama travelogue which promised to take audiences on the most spectacular journey of their lifetime while also dealing with crucial developments in human history;

a spiritual film which raised questions about humankind's relationship to higher powers in the universe; a speculative, yet scientifically based and thus educational docudrama which extrapolated from the current state of knowledge and exploited the intense public interest in the space race; a science-fition film featuring futuristic hardware and exotic, yet humanoid, aliences; the latest work of a young director who was increasingly perceived as one of the great masters of Hollywood cinema. (Kramer 2010: 39–40)

The film evolved somewhat differently than planned, becoming along the way a visually evocative but narratively ambiguous feature. And this ambiguity arguably made it all the more resonant with audiences. The film still makes it onto many critics' "best films of all time" lists and its influence has been cited by many directors.

Two things in particular are worth our singling out about the film's imagery. First is what we might call the musical vertigo of the outer space it depicts. *2001* intends to relate (and, in turn, to shape) the distinct form of embodiment that goes along with leaving the Earth's atmosphere. That experience, as portrayed in the film, is disorienting as well as releasing: it releases us from the verticality of gravity, but this release can be accompanied by disorientation and by fear just as it can be accompanied by beauty. In the film, the circular and spiraling movements of space stations in orbit around the Earth, accompanied by Johann Strauss' well-known "Blue Danube" waltz, become a kind of space ballet. Elsewhere, the music of Richard Strauss and Gyorgi Ligeti make outer space seem liberating yet more disturbing in its openness.

The second feature worth noting is the metaphysics and mystery in the film's depiction of outer space. The space age as portrayed here recontextualizes the historical emergence of our species, humanity's ontogeny, as birthed by otherness and as lured forward by mystery, to return, in the film's final section, to an utterly alien otherness. This alienness is represented in the film by mysterious rock monoliths that appear both at the "dawn" of hominid evolution to tool-makers (and, not coincidentally, as killers), and again at the arrival of modern humanity on the Moon. These monoliths are signs – we do not know of what, but at the very least signifying that the universe is itself *significant*, though its significance may be beyond our current comprehension.

Other movies were to take on this ambiguous otherness by interpreting it either in beneficent terms (as in *ET* or *Close Encounters of the Third Kind*) or in maleficent ones (as in Ridley Scott's *Alien* and, if more ambiguously, in his recent *Prometheus*). In *2001*, this mystery constituted both a psychic threat – particularly the threat of a human creation, the computer HAL 2000, who is aware of something that the astronauts themselves are not aware of and who acts in accordance with this awareness – and the inner psychological threat and possibility represented by the trip "Beyond the Infinite" taken by the film's lead character. In its depiction of humanity's relationship to earthly life, then, *2001* is anthropocentric: humans are portrayed as the leading edge of earthly evolution – a premise that goes unquestioned, as would have been the case for most people in 1968 (though not 4 or 5 years later) – and it is our destiny to reach the next level, to trigger the set of events that might fulfill our cosmic potential. But in the cosmic environment, things are much more mysterious and elusive. The universe transcends us in ways we cannot even fathom. It is so advanced,

Fig. 6.3 Star child (Image from "2001, A Space Odyssey," 1968)

so different, so alien to us that – as scriptwriter Arthur C. Clarke presented it in his earlier novel *Childhood's End,* by which *2001* was partly inspired – it might require the extinction of humanity as we know it for us to get there (Fig. 6.3).

6.4.2 *Solaris:* Or, an Earth and a People Who Have Never Left Us

Solaris (1973) is considered by some to be a kind of Soviet art cinema response to *2001*, a psychological and metaphysical thriller that portrays a relationship between humans and an alien planet, the planet Solaris, but in ways intended to help us rethink our relationship to *this* planet. Where *2001* was based on a novel by one of the most celebrated American science-fiction writers, futurists, and space travel advocates (Arthur C. Clarke), *Solaris* was based on a novel by one of the Eastern Bloc's most famous science-fiction writers, Polish author Stanislaw Lem. Like Stanley Kubrick, director Andrei Tarkovsky remade the story to make it his own. Specifically, he made it more metaphysical and psychological, a kind of speculative metaphysical thriller that toyed with ideas that were becoming popular in the West under the guise of James Lovelock's and Lynn Margulis's "Gaia hypothesis," but which had been developed in an earlier incarnation by Ukrainian geochemist Volodymyr Vernadsky, who first popularized the idea of a *biosfera* and *noosfera* in the Russian-speaking world.

Tarkovsky took this to be a kind of planetary intelligence that challenges humans morally and spiritually by reading its visitors' minds and materializing their fears and desires. These phantoms from their past in turn challenged them to what Russian linguist Mikhail Bakhtin called an "answerability" for their actions. Called to investigate strange goings-on at a space station orbiting the newly discovered planet Solaris, psychologist Kris Kelvin repeatedly encounters his wife, who in reality had committed suicide some years earlier on Earth, but who returns incessantly to haunt him on the space station. The surface of the planet Solaris is a watery surface that seems to take recognizable forms only to dissipate them, as if it were a giant screen full of fluid static that sometimes showed the contents of the mind that was looking into them. In the end, what seems to happen – though the ending is as ambiguous as Kubrick's – is a merging of the main character with this other planet, the planet that represents our hauntedness by our own psychological, moral and spiritual relatedness with others here on our planet.

Solaris is a kind of double to the Earth, one that we need to go *out* to only in order to discover what is *within*. The film features no celebration of human technological prowess; it is, in this sense, quite the opposite of Kubrick's (and Clarke's) anthropocentrism. Instead, *Solaris* combines a kind of Heideggerian nostalgia for one's home locale and roots with the hint of a transcendent force that remains mysterious, but tied to the expression of material reality, and whose point seems ultimately be to ask us: how we are relating to our others? Tarkovsky's Earth (as in all of his films) is a haunted earth, and in Solaris we find the planetary double that brings its haunted remainders – our haunted others – to visibility (cf. Žižek 2000).

6.4.3 *Contact, or, an Earth and a People Just Like Us (Americans)*

Robert Zemeckis's *Contact* (1997) is indicative of a more mainstream and optimistic American sentiment about space: this is both a more domesticated portrayal of the anthropocentric humanism that we saw in *2001*, and a transcendentalist optimism concerning the presence of others *like us* in the universe. The film's imagery epitomizes what could be called the "post-*Earthrise* cosmic sublime." Yes, despite the anthropocentric superiority of humans as they are portrayed on *this* planet, the effect of this cosmic sublime, for many viewers, is to re-enchant the planet in ways very much like the original images of Earth from space.

Two scenes in particular aim for this effect, and for many viewers successfully achieve it. The first is the 3-min opening sequence, in which the camera (using CGI) apparently zooms back gradually from the Earth, revealing the vastness of space, passing planets and asteroids, star clusters and galaxies. This movement backward in space is accompanied by a movement backward in time on the soundtrack, as clips from well-known radio transmissions are heard that include Richard Nixon's admission of guilt in the Watergate scandal, Neil Armstrong's "One small step for [a] man…," Martin Luther King's famous "I have a dream" oration, and others. The second scene is the "wormhole" sequence, in which astronomer Eleanor Arroway (Jodie Foster), in a time travel machine built from instructions supposedly encoded in a radio transmission from the star system Vega, meets her father, or perhaps it is an extraterrestrial being disguised as her father. The ambiguity of the scene is left unresolved, as even Arroway herself later admits that she may have hallucinated the entire event. But this is followed by an insistence that concurs with a very American preoccupation with the value of personal experience, a preoccupation that arguably runs from transcendentalist and pragmatist philosophers, like Emerson and James, and nineteenth century metaphysical spiritualists through to the New Age spirituality of the late twentieth and twenty-first centuries:

> I ... had an experience. I can't prove it, I can't even explain it, but everything that I know as a human being, everything that I am tells me that it was real. I was given something wonderful, something that changed me forever. A vision of the universe, that tells us undeniably, how tiny, and insignificant and how ... rare, and precious we all are! A vision that tells us that we belong to something that is greater than ourselves, that we are not, that none of us are alone.

In this experientialist insistence on a metaphysic that is "at home in the universe," the film finds its compromise between the theological currents so resonant in American culture and the humanist atheism of scientists like the film's co-scriptwriter, Carl Sagan. (The dialogue between these two religious-philosophical currents is a central theme within the film.) In this, the film presents a tamed American "cosmism" for a post-Cold War world.

6.4.4 *The Tree of Life*, or, an Earth and a People Moving (and Mediating) Ever Forward

Terrence Malick's *The Tree of Life* (2011) takes up a similar transcendentalist impulse, but complexifies it substantially. *Tree of Life* is "a family chamber drama" staged "on a cosmic scale" (Vishnevetsky 2011). It is an exquisitely realized period piece about growing up in suburban Waco, Texas, in the 1950s, that comes wrapped in a thickly painted overlay of cosmic reference points, narrative voiceovers, extra-diegetic classical music, and poetic nature imagery. Its early scenes of family life, punctuated by images of sunflowers, waterfalls, figures swinging from trees, and the sound of Tavener's "Funeral Canticle," are followed by a cosmic "creation" or "evolution" sequence in which colorful swirling gases and stellar nebulae, dinosaurs, planets, and an asteroid on a collision course with earth, are accompanied by classical requiems and by Jessica Chastain's voice-over questions. The latter are seemingly addressed alternately to a creator and to a child: "Where were you? ... Who are we to you? ... Answer me," and later, "Life of my life ... I search for you ... My hope ... My child."

The film's scenes of growing up – boys playing with hoses and sprinklers, climbing trees, sneakily throwing grasshoppers down shirts, lobbing balls up on the roof, creeping into a female neighbor's vacant home and stealing her negligee from a bedroom drawer, then embarrassedly sending it floating down the river – are exuberant in their movement, both of the camera and of bodies, and in a general unpredictable fluidity of perspectives, words, and visual and bodily expressions depicted. As many critics have noted, *Tree of Life* is a film about growing up, but this is a growing up with an indeterminate end-point in a complex universe.

Many have pointed out the Heideggerian foundations of Malick's film-philosophy; Malick was in fact a Heideggerian philosopher before he took up the camera. I would like to point, instead, to C. S. Peirce to make sense of this film. There could not be two more different philosophers in their style than Heidegger and Peirce, yet they end up arriving at a very similar place, which is the nature of world-making, in Heidegger's terms, or of semiosis, in Peirce's. Both world-making and semiosis constitute a kind of poetic and interpretive act by which the givenness (or Firstness) of things proceeds to be encountered (in Secondness) and interpreted (in Thirdness). For Peirce, this is the movement of all things in the universe: Firstness is generated out of the random creativity of things; Secondness arises as they enter into relations with other things; and Thirdness denotes the emergence of meaning, pattern, and regularity. This is no contemplative world where nature is a background to human activities. It is a world with an active, dynamic, and ceaselessly conflictual core at its heart, a core of Peircian Secondness, that is, of the existential encounter of one thing and another, one force and another.

The Tree of Life is a vortex that circles a traumatic kernel. It is not entirely clear what that kernel is: perhaps the death of the youngest brother, the fair-haired, talented brother who learns the guitar (Malick's own youngest brother had studied with Andres Segovia, but later committed suicide); or the middle brother (Malick's apparently was severely disfigured in a car accident); or the tension that simmers and grows between Jack's mother and father; or, perhaps most obviously, the tension between the father and Jack, which accounts for much of the film's narrative dynamic; or all of these tensions that hold together, yet pull apart, the nuclear family core. But if it is a vortex, it is one that continually spirals outward in Jack's, or the film's, quest to wrest meaning, or "grace" (in the film's dialectic between "nature" and "grace"), out of the signs that configure this tension. The film is in this sense about the flow from Firstness to Secondness to Thirdness, and the openings made available by this flow. It is about the flow: of images, fragments, glimpses, memories, feelings, and dreams; of emotionality in its thickness and tensility; of thought, in its many voices and the questions that punctuate this quest for sense (especially Jack O'Brien's/Sean Penn's, but also that of the other characters); of connections, felt and probed but never rendered exact, between past and present, cause and complex effect, moments of loss and the haunting abysses they leave behind; and, finally, the flow of cinematic light and sound, of music (several requiems among them), and of the camera-eye, under Emannuel Lubezki's direction, almost ceaseless in its elliptical motion in and around, toward and away from, the people and things that populate this unsettled world. *The Tree of Life* is in this sense a film about the movement away from the Earth which never manages to *get* away, but which always comes upon something new or not-quite-the-same. It is about the eternal recurrence of the different: the continuing probing at the boundaries of the human and the divine, without ever attaining its goal except in the movement itself. It is, in this sense, about the self-transcendence of the immanent, the creative drive of evolution itself.

6.4.5 *Melancholia, or, an Earth and a People Becoming Sweetly Extinct*

Finally, Lars von Trier's *Melancholia* (2011), which premiered at the Cannes film festival two days after the premiere of *The Tree of Life*, takes a very different, yet equally "geophilosophical" approach to the human-universe relationship. In *Melancholia*, a mysterious sister planet to the Earth, called Melancholia, appears in the sky and comes to take on the embodiment of the depressive psychological condition of its main character, Justine (Kirsten Dunst). The film's premise – that a planet has "hidden behind the sun" all this time, but now comes hurtling out toward the Earth – is scientifically unrealistic. Yet as a psychological metaphor, it could hardly have been given a more aesthetically potent treatment. The blue planet Melancholia is a mesmerizing metaphor for a kind of double to the known world – a hidden, deadly intruder that is destined to come and destroy with the direct impact of irresistible doom. Melancholia is the herald and the medium of utter destruction, the sign of extinction and nonexistence (Fig. 6.4).

The final scene, which is presaged in an 8-min long introductory sequence that plays like a Wagnerian overture, depicts the main characters as the planet comes on its supposed "fly-by," which we know by now will be a direct hit of planet Earth. The lead character, the melancholic Justine, comforts her young nephew, Leo, by building a protective "magic cave" out of wooden sticks to shelter him, herself, and her sister Claire, from impending doom. This futile gesture is probably all that can be done to bring a little comfort to the frightened child. It is, in a sense, von Trier's joke at humanity's helplessness in the face of a hostile universe, and yet it also signifies that we can still act, existentially, through art and imagination, despite the inefficacy of our action.

As in *Solaris*, the planetary double is an instrument of seduction and destruction, as well as of realization: it tears down anthropocentric illusions (or any illusions), renders them impotent, and swallows us all in its embrace. This blue sister planet

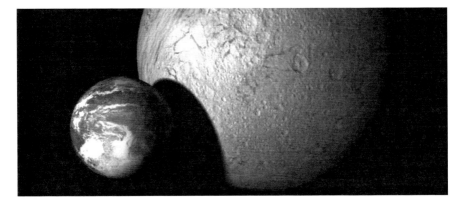

Fig. 6.4 The planet Melancholia striking the Earth, in "Melancholia" (Image from "Melancholia," Lars Von Trier, 2011)

takes out everything that has ever been known, experienced, and celebrated by humanity. To recontextualize astronomer Carl Sagan's famous "Pale Blue Dot" speech, given at Cornell University in 1994 and accompanied by an image of the Earth taken from Voyager 1 as it sped out of our solar system, what is taken out by von Trier's fictional intruder is the planet on which

> …everyone you love, everyone you know, everyone you ever heard of, every human being who ever was, lived out their lives. The aggregate of our joy and suffering, thousands of confident religions, ideologies, and economic doctrines, every hunter and forager, every hero and coward, every creator and destroyer of civilization, every king and peasant, every young couple in love, every mother and father, hopeful child, inventor and explorer, every teacher of morals, every corrupt politician, every "superstar," every "supreme leader," every saint and sinner in the history of our species lived there – on a mote of dust suspended in a sunbeam.

If we, the viewers, are at all moved by *Melancholia's* image of a massive blue planet closing in on the Earth and finally decimating it (and humanity), this is to say that we have in some sense *moved with* this image. We have made the decimation of the earth a cognitive and affective virtuality for ourselves (as Deleuze might say), something we can feel for and about because we have practiced this feeling while, and after, watching it. It has become thinkable and "feelable." This does not mean that such a planetary collision has become any more likely than it ever was. But it does mean that we may be psychically changed by the experience of having imagined, in the graphic way that cinema makes possible, the end of all that we know, love, hope for, and believe to be true.

6.5 Conclusion

If each of these films can be considered a composite and complex set of moving images – images of movement in a moving universe – then each provides a means for us to reimagine our own movement with that universe. With Stanley Kubrick's *2001*, we move *out* into an open, unearthly universe with trepidation, knowing we are poised at the edge of an evolution that may or may not lead to some unimaginable place. Kubrick's is a view outward toward the open space of the heavens, and simultaneously inward toward a soul-searching, and maybe soul-crunching, destiny that both attracts and repels us. With Tarkovsky's *Solaris*, we move *back and inward* to face our own ghosts, our own relations, memories, fears, and traumas, in the knowledge that until we have dealt with them, the way out for us is blocked, or in any case irrelevant. With Zemeckis's *Contact*, by contrast, we move out with confidence that our rightful place is in the intergalactic conversation, and that everything we are doing to get there – such as the Search for Extraterrestrial Intelligence program that both Carl Sagan and the fictional Eleanor Arroway championed – is warranted. With Malick's *Tree of Life*, we tumble into a spiraling (and sprawling) movement forward, an arching of the head upward to the light that is visible through the branches of a tree, yet always elusive behind those branches. This is a movement forward in a struggle that connects us with all things moving, all things struggling, all things lit in the graceful sunlight that ever eludes us while it lures us forward. And with Von Trier's *Melancholia*, we move in an erratic dance inward in the knowledge that all of life is finally destined to extinction. In the process we defy all human pomp, building our magic cave as an existential act in the face of a meaningless universe.

Each of these films provides a flow of images that renders real the relationship between transient subjectivity – or what we might call *anthropomorphosis*, the becoming-human of us who call ourselves human – and the larger ecologies of the Earth and universe. All follow in a line of descent that goes back to *Earthrise, Blue Marble,* and other images of the Earth from space. As I have tried to suggest here, all are images in motion, variations on a motion that evolves from the Firstness of its sheer quality (the image as it first appears) to the Secondness of its impact (upon its viewers, and as a result of the technological and political forces that shaped it) to the Thirdness of its many, changing meanings, and back again. These are five visions by which we – always somewhat homeless – humans make a home for ourselves in a universe that is mysterious, elusive, yet there for us in its brute facticity. To the extent that we *believe* in the world captured by these images, they contribute to the remaking of our spiritual and cosmic bearings.

Our belief, however, is what is in question. In a world where everything has seemingly been turned into an image, and in which digitally coded images are woven together into a streaming spectacle, accessible to us 24–7 on television screens, desktop computers, and mobile and hand-held devices, can we still *believe* in images, or in the world they ostensibly represent? My implicit argument has been that the only thing we *can* believe in is the movement by which images emerge, encounter each other, and deliver up meanings, though these meanings change and mutate just as did the surface of Tarkovsky's alien planet Solaris. Images like *Earthrise* and *Blue Marble* reshaped our ability to visualize a collective "we" and our cosmic environment; and filmmakers like Kubrick, Tarkovsky, Zemeckis, Malick, and Von Trier have worked within the parameters of the space they opened up to wrest some meaning out of its relational possibilities. Both the original images from space and these five films elicited a range of religious responses: for instance, much of the commentary surrounding *Contact* concerned its theological themes, while *The Tree of Life,* despite its unabashed evolutionism, also found itself embraced by many Christian commentators (and critiqued by others who found the apparent Christian themes too cloying).

Visualizations of our cosmic environment in the post-*Earthrise* era will continue to challenge, provoke, and resonate with religious themes; and religions, to remain relevant, will continue to respond to the "new dispensation" these images have set out. To the extent that future religion is unthinkable without this imagistic dispensation, the map of world religion will be shaped by the maps that world religions make of the cosmos, aided in part by images like these. If artists constitute a creative edge of humanity's imaginative capacities, the spectrum of images presented by these films indicates some of the diversity by which the religious imagination will unfold in years to come.

References

2001, A Space Odyssey. (1968). Stanley Kubrick, dir. U.S.
Burnett, R. (2005). *How images think.* Cambridge: MIT Press.
Chaikin, A. (2009). Live from the moon: The social impact of Apollo. In S. J. Dick & R. D. Launius (Eds.), *Societal impact of spaceflight* (pp. 53–66). Washington, DC: NASA Office of External Relations, History Division.
Contact. (1997). Robert Zemeckis, dir. U.S.
Cosgrove, D. (1994). Contested global visions: One-world, whole-earth, and the Apollo space photographs. *Annals of the Association of American Geographers, 84*(2), 270–294.
Deleuze, G. (2005). *Cinema 2: The time-image* (trans: Tomlinson, H. & Galeta, R.). London: Continuum.
Deleuze, G., & Guattari, F. (1994). *What is philosophy?* (trans: Tomlinson, H. & Burchell, G.). New York: Columbia University Press.
Garb, Y. J. (1985, March). The use and misuse of the whole earth image. *Whole Earth Review,* pp. 18–25.
Heidegger, M. (1977). The age of the world picture. In W. Lovitt (Ed.), *The question concerning technology and other essays* (pp. 115–154). New York: Harper & Row.
Helmreich, S. (2011). From spaceship earth to Google ocean: Planetary icons, indexes, and infrastructures. *Social Research, 78*(4), 1211–1242.
Ivakhiv, A. J. (2013). *Ecologies of the moving image: Cinema, affect, nature.* Waterloo: Wilfrid Laurier University Press.
Jacobs, R. (2011). Whole earth or no earth: The origin of the whole earth icon in the ashes of Hiroshima and Nagasaki. *The Asia-Pacific Journal,* 9. 13. 5, March 28, 2011. www.japanfocus.org/-Robert-Jacobs/3505. Accessed 5 Sept 2012.
Jasanoff, S. (2001). Image and imagination: The formation of global environmental consciousness. In C. A. Miller & P. Edwards (Eds.), *Changing the atmosphere: Expert knowledge and environmental governance* (pp. 309–337). Cambridge: MIT Press.
Krämer, P. (2010). 2001. London: British Film Institute.
Lazier, B. (2011). Earthrise; or, the globalization of the world picture. *The American Historical Review, 116*(3), 602–630.
Melancholia. (2011). Lars Von Trier, dir. Denmark/Sweden/France/Germany.
Peirce, C. S. (1934). *Collected papers* (Vol. 2). Cambridge: Harvard University Press.
Reinert, A. (2011, April 12). The blue marble shot: Our first complete photograph of earth. *Atlantic,* pp. 24–25. www.theatlantic.com/technology/archive/2011/04/the-blue-marble-shot-our-first-complete-photograph-of-earth/237167/. Accessed 21 Aug 2012.
Solaris. (1973). Andrei Tarkovsky, dir. USSR.
The Tree of Life. (2011). Terrence Malick, dir. U.S.
Vishnevetsky, I. (2011, May 26). 'The tree of life': A Malickiad. *Notebook.* http://mubi.com/notebook/posts/the-tree-of-life-a-malickiad. Accessed 15 Oct 2011.
Žižek, S. (2000). The ting from inner space. In R. Salecl (Ed.), *Sexuation* (pp. 248–250). Durham/London: Duke University Press.

Chapter 7
Weber's Protestant Ethic Thesis and Ecological Modernization: The Continuing Influence of Calvin's Doctrine on Twenty-First Century Debates over Capitalism, Nature and Sustainability

Ernest J. Yanarella

7.1 Introduction

Max Weber's thesis about the elective affinity between work ethic and the spirit of capitalism has engendered continued comment and controversy since his book, *The Protestant Ethic and the Spirit of Capitalism*, was published in 1905. It has also spawned powerful and sweeping musings upon and interpretations of Weber's historical sociological study of Calvin's doctrine and its role in the proliferation of capitalism throughout the world by many of the most illustrious philosophers and social thinkers, including Maurice Merleau-Ponty (1973), Raymond Aron (1967), Anthony Giddens (1971), Talcott Parsons (1967), Alasdair MacIntyre (1962), Frederic Jameson (1988), Jürgen Habermas (1987), and Wang Hui (2011). These works have been supplemented by probing historical and biographical accounts by other noted twentieth century intellectual historians and psycho-historians like H. Stuart Hughes (1961) and Arthur Mitzman (1970). Latter-day Calvinist theologians and theological scholars (e.g., Little 1981; Dommen 2007; Stückelberger 2007) too have entered the fray, exploring the historical and theological Calvin in relation to contemporary ecological problems brought on by capitalism and modernity. There is even a small literature (Breiner 2004; Scaff 2005/2006; Gerhardt 2007) exploring controversies over Talcott Parsons' translation into English of Weber's classic.

This chapter begins with an examination of the subtle and nuanced historical and cultural hermeneutic that underpins Weber's analysis of the Calvinist roots of the capitalist spirit. Working from the historical sociological writings of Benjamin Nelson, we seek to show how Nelson's historical sociological work provides persuasive evidence of the richness and durability of Calvin's religious revolution

E.J. Yanarella (✉)
Department of Political Science, University of Kentucky, Lexington, KY 40506, USA
e-mail: ejyana@email.uky.edu

and its subsequently radiating influence in Europe, America, and the Asia, in particular. We then place this religious revolution within a broad framework of two other revolutions contributing to the emergence and shaping of modernity. In the process, we develop an operational definition of what Weber called a uniquely Western "way of life" influencing all facets of human life and what Nelson termed a central intellectual (or cultural) paradigm. The next section places Weber's studies of Calvinism within his more far-reaching studies of economic ethics and world religions (Confucianism and Taoism, Hinduism and Judaism) in order to test the singularity of his original thesis.

The next part of this chapter begins by exploring what Jürgen Habermas has called the project of modernity and the impact around the world of the rationalization processes that Weber so assiduously investigated and somberly pondered. Then in the face of the spread of societal modernization through developmental strategies across the globe and the mounting threats to the world's Ecosystem from top-down globalization, Western-style economic development, and climate change, the succeeding section speculates on the shape of a new ecologically-grounded paradigm or worldview going beyond the flawed project of modernity driven by the dialectic of Enlightenment. Informed by Weber's cautionary words on capitalism and modernity at the close of his classic work, the chapter concludes by identifying potential resources found in contemporary debates about capitalism, modernity and sustainability that point to alternative modernities and a new sustainability paradigm transcending the Enlightenment legacy.

7.2 The Protestant Ethic, the Spirit of Capitalism, and the Making of Modernity

Our scholarly guide or leitmotif in interpreting this classic work is Benjamin Nelson, one of Weber's most careful and perceptive students. Nelson is best known for two major works: a book, *The Idea of Usury: From Tribal Brotherhood to Universal Otherhood* (Nelson 1969a), and his lengthy article, "The Making of Early Modern Cultures and Early Modern Minds: The Protestant Ethic Beyond Max Weber" (1969b). Many of his other writings deepen his critical analysis of the Weberian problematic (1973b) and expand its wanderings and future to other cultures and civilizations (1973a, 1976). His general concern is to trace the background and evolution of Weber's thesis, that is, that the origins of the spirit of capitalism lay in its religious roots in the Calvinist work ethic, beyond the Western world. While he is respectful of Weber's towering achievement as a pioneering historical sociologist and student of the sociology of religion, the Calvinist revolution for him is seen as only one of three revolutions in Western thought and practice that brought into existence a new worldview—the modern world outlook (Fig. 7.1).

For our purposes, Nelson's other key contribution is the way he ties these revolutions to the making of early modern minds and culture. Like Weber, he seeks to

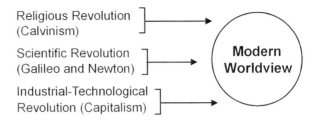

Fig. 7.1 Three revolutions (Source: Ernest J. Yanarella)

transcend the stale debate over the priority of ideas and material forces by underlining the crucial role that the concatenation of factors that shape human beings and invest their way of life with meaning and purpose plays. In one of his earliest statements on Weber's study of Calvinism and capitalism, he issues this penetrating dictum that goes to the heart of Weber's method: "to understand the makings of early modern culture we need to understand the makings of early modern minds and, therefore, need to have a proper sense of the change in the central [cultural or intellectual] paradigm as well as the restructuring of axial institutions in society" (1969a: 240). If we probe into the meaning and significance of the phrase, "early modern minds," we can derive the following basic sense or meaning: unless a person's deepest beliefs are altered, unless he or she sees the world and his/her relations with its various spheres in a fundamentally different way, no real change is possible—even if the basic social (i.e., axial) institutions are forcibly altered (See Nelson's extension of Weber beyond Weber in Nelson 1969a: 239–245). The recent experiences of the Velvet revolution in Czechoslovakia and the overturning of the other Communist regimes in East Europe point to how the imposition of Soviet economic and institutions in the late 1940s could not ultimately hold back the native political cultures of these countries before Communization—experiences where the so-called cultural superstructure eventually repulsed the foreign economic structure (the Stalinist command economy) imposed upon them.

What key concepts and relations constitute a central intellectual or cultural paradigm? Others (Nelson 1969a, b; Taylor 1995) have endeavored to devise a framework for conceptualizing such a cultural paradigm or worldview. Nelson's strategy is to emphasize the shift from "universal brotherhood" to "universal otherhood" accompanying the religious and scientific revolutions and to explain how the search for subjective certitude in religion and objective certainty in knowledge prompted leading figures in the Reformation and the modern scientific revolution to challenge the medieval system of conscience, moral casuistry, and the cure of souls (Nelson 1969a, b; Nielsen 1998). In defending a cultural theory of modernity, Taylor embraces a vista on a cultural paradigm or worldview that highlights the changing interpretations of person, time, nature, society, cosmos, the good, and God (1995: 27, 30 and *passim*). My reading of Weber's comparative work on the sociology of religion, as well as my speculation on a potential worldview beyond capitalism and modernity, leads me to offer this heuristic, but non-arbitrary, depiction of a worldview (Fig. 7.2).

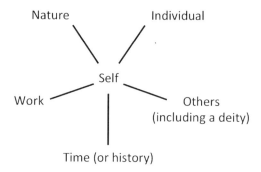

Fig. 7.2 Intellectual (cultural) paradigm or worldview (Source: Ernest J. Yanarella)

7.2.1 The Place of Calvinism in the Making of Early Modern Culture

Weber's study is generally regarded as an idealistic counter-explanation of the rise of capitalism that stands in opposition to Marx's materialistic interpretation. It is said that Weber offers a one-sided spiritualistic causal explanation of the emergence of the capitalistic system while Marx put forth a materialistic or economic one, which Weber flatly denied (1958a: 183). Weber's view is more subtle, just as Marx's is, too. Equally, Weber's thesis does not claim that the Protestant ethic is the only factor that contributed to its metamorphosis into the spirit of capitalism. Further it is grounded in a methodology drawn from historical sociology and critical hermeneutics that is largely immune to facile positivist studies that have sought to refute it (for example, Schaltegger and Torgler 2009).

What then is Weber concerned with explaining? Basically, it is that which he believes is unique to Western capitalism, viz., its accent on the rational organization of free labor. For him, the central problem in need of explanation is not capitalist activity itself, for capitalism and capitalist institutions existed precariously (for example, Florence in the fourteenth and fifteenth centuries) prior to its modern form in Europe elsewhere and earlier (Weber 1958a: 74–75). Rather "it is…the sober bourgeois capitalism with its rational organization of free labour" (Weber 1958a: 24). As he cogently puts it:

> It is hence our first concern to work out and to explain genetically the special peculiarity of the modern Occidental form. Every such attempt at explanation must, recognizing the fundamental importance of the economic factor, above all take account of the economic conditions. But at the same time the opposite correction must not be left out of consideration. For though the development of economic rationalism is partly dependent on rational technique and law, it is at the same time determined by *the ability and predisposition of men to adopt certain types of practical rational conduct*. (Weber 1958a: 26, emphasis added)

This position leads Weber to what he terms the historical question: why business leaders and owners of capital, higher skilled labor, and upper echelons of technically and commercially trained personnel in Germany in his time were predominantly Protestant in religious identification and, more tellingly, why those regions in the sixteenth century with the highest levels of economic development were simultaneously most disposed to the religious revolutionary upheaval in the Church (Weber 1958a: 35, 36).

As Weber discovers, the overthrow of economic traditionalism, far from leading to the relaxation of religion's control over Protestant renegades in these districts, triggered the tightening of controls in the religious and every other sphere of everyday life. So the capitalist organization of free labor, tied as well to the separation of business and household and rational bookkeeping, becomes the defining feature of Occidental (western) capitalism. Such rationalization, in his religio-hermeneutic analysis, leads to the crucial role of a calling as an aspect of the rationalization of work. As Tawney (1958) notes in his foreword to the work (Weber 1958a), the Calvinist notion of calling differs significantly from medieval and even Lutheran versions:

> For Luther, as for most mediæval theologians, it had normally meant the state of life in which the individual had been set by Heaven, and against which it was impious to rebel. To the Calvinist,…the calling is not a condition in which the individual is born, but a strenuous and exacting enterprise chosen by himself, and to be pursued with a sense of religious responsibility. (Weber 1958a: 2)

With it, labor as an economic means is supplanted by its transformation into a spiritual end (Weber 1958a: 3). Without it, human works are casual, fleeting, and episodic. Or as Calvinist theologian Richard Baxter states it starkly, "A man without a calling thus lacks the systematic, methodical character which is…demanded by worldly asceticism" (quoted in Weber 1958a: 161). Such innerworldly asceticism is required to assure that the focus of labor and its fruits is directed to the glorification of God (see below).

No less critical to Weber's interpretation of the linkage between the Protestant work ethic and the spirit of capitalism is the central tenet, the characteristic and essential dogma, of Calvinism: the doctrine of predestination. Working from Calvin's enunciation of this doctrine in the Westminster Confession of 1647, he spells out its icy, awful, and inexorable logic, viz., that God through God's secret wisdom and humanly inscrutable determination before the beginning of the world and time has chosen some to be the elect and appointed others to be damned without godly foresight of their future faith or good deeds or steadfastness in either (Weber 1958a: 99–101). Flowing from this doctrine then is an attitude and orientation to the world characterized by the regimen of worldly rationalization and the ethic of innerworldly self-denial of the fruits of one's rational labor.

The deep and profoundly troubling problem of this doctrine for the first-generation faithful Calvinist is this: the necessity of working in a calling and Calvin's command to believe in one's election, despite the absence of any positive signs and clues. In the face of its outward uncertainty and inner imperatives, two varieties of pastoral advice were issued:

> On the one hand it is held to be an absolute duty to consider oneself chosen, and to combat all doubts as temptations of the devil, since lack of confidence is the result of insufficient faith, hence of imperfect grace….On the other hand, in order to attain that self-confidence intense worldly activity is recommended as the most suitable means. It and it alone disperses religious doubts and gives the certainty of grace. (Weber 1958a: 111–112)

To this advice, Calvin added another. Speaking *ex cathedra* (that is, outside the logic and bounds of the doctrine of predestination), he held that *extra ecclesium nulla salus* (no salvation outside the Church); still, even the damned must be compelled to membership in the existential Church in order to fulfill God's commandments and to participate in worldly transformation for the good and glory of God, just as the elect do (Weber 1958a: 104). Surrendering to the magnificent [psychological] consistency of Calvin's doctrine, dogma, and teachings in a time of social upheaval, religious tumult, and historical dislocation, the Calvinist believer held firm to its preachings and strapped him/herself to the economic and social dicta and imperatives that flowed from them.

7.2.2 Fundamental Changes in the World Paradigm from the Medieval to Modern Epoch

We are now in a position to illuminate what the religious revolution shaped by Calvinism had wrought in playing out its restructuring upon the central cultural paradigms of the West. In *The Protestant Ethic and the Spirit of Capitalism* (1958a), Weber's historical sociological analysis of the rise of the spirit of capitalism into a "way of life" discloses the contribution of this religious revolution to the transition from the medieval worldview to the modern worldview.

Working from the operationalized framing of intellectual paradigm or worldview offered earlier (see Fig. 7.2), one can discover in Weber's text how these culturally variable concepts and relationships constituting a cultural paradigm changed with the movement from medieval to modern worldviews.

7.2.3 Individual

In the spirit of the Nelson's theme of the shift from Tribal Brotherhood to Universal Otherhood later uncovered by him in his investigation of the usury, Weber sees the life of the Calvinist faithful increasingly experiencing a sense of "unprecedented loneliness of the single individual" (Weber 1958a: 104) Despite the profound importance of the Protestant soul-interest (that is, discerning whether the faithful was among the elect or the damned), the Calvinist saint had to face a solitary fate that had been decreed before all eternity by God. Shedding the communal bonds so prominent under the hegemony of the Catholic church during the height of the medieval epoch, the Calvinist believer found him/herself alone to face the world alienated from priests, the sacraments (including confession and forgiveness), and the church itself. Thus, the "Calvinist was doomed by an inexorable fate, admitting of no mitigation … for such friendly and human comforts [of the medieval Catholic community and its sacraments of atonement, grace and forgiveness] did not exist"

(Weber 1958a, b: 117). In the chilling words of Sebastian Franck, the Reformation meant that "now every Christian had to be a monk all of his life" (quoted by Weber 1958a: 121).

7.2.4 Others, Including a Deity

As for others, communal bonds of neighborliness, friendship, and sociality were dissolved as the Calvinist believer was admonished to put trust exclusively in God and to withdraw "any trust in the aid of friendship of men" (Weber 1958a: 106). Trading the mediated relationship between God in exchange for an individual, direct, unmediated one, the faithful were counseled by Calvin's theological lieutenants like Richard Baxter and Robert Barclay to practice "deep distrust of even one's closest friend" and strongly urged to "trust no one and to say nothing compromising to anyone" (Weber 1958a: 106). Likewise, they were exhorted to depend upon no friend, neighbor or loved one for worldly assistance or spiritual aid, since "God helps those who help themselves" and the Calvinist "creates his own salvation, or as would be more correct, the conviction of it" (Weber 1958a: 115). The demand by Calvinism for the psychological experience of certitude of election "was accompanied by an attitude of hatred toward the sin of one's neighbor, not of sympathetic understanding based on consciousness of one's own weaknesses, but of hatred and contempt for him as an enemy of God bearing the signs of eternal damnation" (Weber 1958a: 122). As for God, in spite of the substitution of intermediate churchly agents like priest, bishop, and cardinal for unmediated relations with God, the Calvinist saint experienced the estrangement of even a deity, since "even Christ died only for the elect, for whose benefit God had decreed His martyrdom from eternity" (Weber 1958a: 104).

7.2.5 Time

In *Technics and Civilization* (1934: 14–15), Lewis Mumford took note of the revolutionary impact the clock had in the regimentation of everyday life and the mechanization of work with the rise of modernity. While Calvinism predated the full effects of mechanical clock-time, the shift from the medieval to modern world was certainly abetted by the emergent experience of time that supplanted the hegemonic role of the Catholic Church calendar with its numerous feasts and saints' days and the annual rhythms of nature associated with farming in a predominantly agrarian economy (Sennett 1977). With Calvinism, time too becomes quantified, fungible, resourceful, and intimately tied to money, credit, and investment. Here, he turns to Benjamin Franklin, whom he sees as embodying Calvinism's teachings, where "time is money … credit is money … [and] money is of the prolific, generating nature" (Weber 1958a: 48–49). Connecting the Protestant work ethic to the experience of temporality, he goes on to note that "waste of time is thus the first and in principle the deadliest of sins. The span of human life is infinitely short and precious to make sure of one's own election" (Weber 1958a: 157). Since it is the task of every Calvinist to express his/her certitude of election, "every hour lost is lost to labour for the glory of God" (Weber 1958a: 158). Socializing, unproductive chatter, luxurious living, even sleeping more than necessary for one's health—all of these things merit the most stern moral rebuke!

7.2.6 Work

Clearly, with the rise and proliferation of Calvinism during the Protestant Reformation, a new time consciousness became imbued in Christian attitudes toward work. As we have seen, the Calvinist idea of a calling required converts to work in a self-chosen precise and strenuous life activity, not a godly vocation established in heaven. The link between work and this new temporality can be seen in Weber's negative example of how the introduction of piece rates as a means of increasing worker productivity in agrarian societies and economies often was met with the reduction of worker productivity because the "opportunity of earning more was less attractive than that of working less" (Weber 1958a: 59–60). Calvinism adds to work as its overriding goal the glorification of God through the products of one's labor and through the organization of one's entire life around that objective.

7.2.7 Nature

If the conception of nature that dominated the medieval epoch was invested with organic and female characteristics and virtues (Merchant 1990) such that the Church imposed sanctions against mining and other practices that might disrupt or violate an organic, nurturing Nature, the imposition of a fully mechanistic image of nature borrowing upon the Great Machine metaphor had to await the cultural ramifications of early modern science in its Newtonian-Galilean revolution. The religious revolution inaugurated by Calvinism, however, did its part undercutting the medieval concept of nature by calling upon the faithful Calvinist to use the resources or raw materials of humankind to make over the world as built environment for the greater good and glory of God. "The world exists," according to Calvinism, "to serve the glorification of God and for that purpose alone" (Weber 1958a: 108). As work is rationalized and social life gives way to rational organization, rational calculation, and instrumental rationality, the fruits of such labor and organization derive from the increasing domination and exploitation of nature and its conversion into economic products and a built environment reflecting its higher purpose. Giving way to rationality, science, and industry, nature loses its mystery and magic and leads to the disenchantment of both nature and the world.

In his now classic essay, Fredric Jameson (1973, 1988) has brilliantly interpreted the metamorphosis of the Protestant ethic into the spirit of capitalism and by implication the transformation of the religious way of life of reverent and ascetic Protestants driven by the motives of psychological relief from unrelievable spiritual uncertainty into the economic world of capitalism with its materialistic acquisition and accumulation, mass consumption, the will to dominate nature and humans, creative destruction, and ceaseless change without change (fashion). He calls the Calvinist revolution a "vanishing mediator" (Jameson 1973: 78, 1988: 25).

Indeed, as Weber shows, the capitalist economic system ultimately grows into a prodigious cosmos "now bound to the technical and economic conditions of machine production which to-day determine the lives of all the individuals who are born into this mechanism, not only those directly concerned with economic acquisition, with irresistible force" (Weber 1958a: 181). In the process, the "care of external goods" hitherto only draped lightly like a cloak or shawl, begins to weigh heavy on the Calvinist believer and the religious ethos—the spirit of religious asceticism--is secreted, leaving the shell of bondage or "iron cage," which no longer requires its supporting beliefs and assumptions (Weber 1958a: 181).

For Jameson, the tenets and doctrines of Calvinism that elicit the Protestant ethic and rationalize worldly activity and generates a planful organization of major spheres of individual and social life act as "catalytic agents" of social change. Collectively, they then become a vanishing mediator in the sense that "once Protestantism has accomplished the task of allowing the rationalization of inner-worldly life to take place, it has no further reason for being and disappears from the historical scene" (Jameson 1973: 78).

7.3 The Import of Weber's Comparative Studies of World Religions

Already, in his study of the Protestant work ethic and the capitalist spirit, Weber strongly suspected that the rise of capitalism from its Protestant religious moorings was unique and the other world religions lacked different characteristics and qualities distinctive to the West (Weber 1952: 25). What is originally a bold, but unverified, claim or assertion triggered years more study into his last years to try to demonstrate. While Weber's studies of other world religions and supporting evidence have been criticized by more recent religious scholars for its incompleteness, biases, and errors, his sociology of world religions for its time exhibited remarkable scholarship and insight. Such critiques aside, what drove Weber to undertake this lengthy exploration were two questions: why did the conditions for the spirit of capitalism only emerge in Protestantism, not Confucianism or Taoism in China, not Hinduism or Buddhism in India, not Judaism in the Middle East, not Islam in west Asia? What factors or circumstances were lacking in these institutionalized religions elsewhere? The relevance of answers to these queries, for Weber, helped to clarify those other factors in association with Calvinism's direct contribution that made the spirit of capitalism unique to the Euro-American experience.

It must be remembered that Weber believed that the rationalization process he uncovered in the West for him had "*universal* significance and value" and was/is a universal development (Weber 1958a: 13, Weber's italics). His expansion of world religious studies in his later years, as Habermas observes, involved the effort to address a "problem of universal history"—viz., why, outside of Europe, "the scientific, the artistic, the political or the economic development … did not enter the path of rationalization which is peculiar to the Occident?" (Habermas 1987: 1). While we will take up the import and meaning of Habermas's reading of Weber for what he calls the incomplete "project of modernity," in this section we briefly take up the preceding two questions by examining Weber's studies of China, India, and ancient Judaism.

7.3.1 Ancient Judaism

The last and perhaps most neglected of Weber's works on world religions, his interpretation of ancient Judaism, has suffered this fate in large part because, in contrast with his examinations of China and India, it seems to have the least bearing on the overall structure and purposes of his wider project. Weber's stated aim in taking up his study of ancient Judaism is to explain the "pariah" status of Jewish settlements during the Diaspora (Fahey 1982: 63). Working from the early history of the early Hebrews, he identifies the distinguishing feature of the political organization of the Israelite confederacy in its unique religious foundations in the Covenant. Its singularity stems from this confederacy's binding force between community and deity residing in the idea of Yahweh as an "active *partner* in the social contract as opposed to the more widespread notion of God as a witness and guarantor of the pledges and agreements men make among themselves" (Fahey 1982: 65, emphasis added). This element in early Israeli history marks its distinctive cultural significance for later history.

From here, Weber seeks to carry his historical-sociological analysis to the monotheism of the classical prophetic period, providing a complex and textured reading of this era. A peasant confederacy is slowly supplanted by a centralized monarchy and urban organization. This rich, often contradictory and tension-filled social evolution toward Yahwist maturation, in Weber's view, is regarded as a critical connector in the emergence of Judaic ethical rationality, abetted by the key role and success of the Levitical Torah teachers.

If Weber's stated purpose in exploring the early development in Judaism, China, and India of world religions is to unearth the economic ethic of each of those world religions, his volume on ancient Judaism radically departs from that major focus. Rather, for Weber, the world-historical significance of his socio-historical study of ancient Judaism lay in how key characteristics of Judaism succeeded in maintaining intact its solidarity and religious foundations in spite of periods of diaspora and most particularly how, beyond the economic role the Jews played in the history of medieval and modern Europe, Judaism bequeathed to Christianity at least three features critical to the unfolding of modernity: (1) the sacred nature of the Old Testament, which played such a crucial function in the rise of Protestantism; (2) the development of an ethic emphasizing communal and political conduct over individual behavior; and (3) a theology of salvation grounded in this-worldly, rather an other-worldly, promise. Weber also suggests in text and the many, elaborate, and often illuminating footnotes in the *Protestant Ethic* (1958a: 164–166 and footnote 58; also 270–272) that, on the one hand, the English Puritans saw Jewish capitalism as a "pariah-capitalism" quite different from its sober, rationalizing capitalism (1958a: 271, footnote 58) while, on the other hand, certain affinities and possible contributions can be found between Puritanism and Judaism, especially the self-righteousness and formal legality derived from Old Testament morality (Weber 1952: 165).

7.3.2 China

In the *Religion of China* (1951) and the *Religion of India* (1958b), Weber focuses more directly on the economic ethic in the world religions emergent and institutionalized in these two countries. Further, he highlights those features in them that were absent and inhibited or that blocked the kind of capitalism facilitated by Protestantism in Europe and America. These investigations might be considered as a number of "thought-experiments" in comparative historical sociology of these world religions (Nelson 1976: 118). In his explorations of Chinese religions (Confucianism and Taoism), Weber dwelled on the singular features of the Occidental city that failed to appear in Chinese cities. As Mumford has illuminated in *City in History* (1968) and elsewhere, the Occidental city has its origins as fortress cities that were centers of commerce, trade, and craft production, were frequently tied to the agricultural production in the surrounding rural lands, found parts of it under the control of guilds, and was based upon a money economy, which usually generated and flowed from rents as well as trade. As a result, the Western city was largely politically autonomous, generally self-governing, and characterized by a separate and independent militia, a central marketplace, and a functioning court system (Nelson 1976: 119–120). As the quintessential model of the Occidental city, the medieval city was distinguished by a sense of brotherhood (Weber called it "fraternization") and a high degree of universalization, which derived from ancient Greco-Roman, Hebraic, and Christian traditions (Molloy 1980).

By contrast, early Chinese cities failed to acquire the political autonomy more characteristic of Western cities and lacking it, remained bound to the shackles of the kinship group and the practice of ancestor worship (Weber 1958b). As a result, the Chinese city's comparative weakness and kinship "militated against the citizens' political solidarity," even undercutting the potential solidarity of Chinese merchant and craft guilds, since they fragmented into competitive, protective associations, shielding individual members from economic hardship and seeking imperial aid and favors from the Emperor (Bendix 1962: 100).

Patrimonialism was another factor that militated against the appearance of urban autarky and autonomy in China. Weber's cites the priority given to the countryside and its protection over the city by rising centralized imperial government that eroded urban autonomy and led to a political evolution of a comparatively large-scale imperial state with centralized administrative offices and a centrally-controlled military that curbed tendencies toward characteristics observed in the Occidental city. On the other hand, Weber notes that patrimonialism and the absence of urban political autonomy led to a sort of urban-imperial balance such that while local groups and organizations built upon kinship groupings were relatively strong, they were too fractured and competitive to unite in opposition to central government. Also abetting imperial power was the fact that political contestation between city and state turned on the allocation of offices instead of the distribution of land, a political game that gave distinct advantage to the emperor and his officials (Weber 1958b: 58). Patrimonial government organization also worked against the emergence of bureaucratic administration along rational lines fostered in the West (Weber 1951: 151).

Particularly noteworthy was yet another factor militating against China duplicating the Western Protestant feat. This had to do with Chinese religious organization and, in particular, its lack of a prophetic tradition and the absence of a strong and powerful priestly hierarchy. Instead, Chinese religious cults became tools of secular leaders, who used them as means to control and manipulate the mass citizenry. As Bendix notes, "this unity of cult and state and the separation of state cult and popular religions contrasted with the place of religion in medieval Europe, where the Church curbed the power of secular rules for many centuries and where the same faith was professed by rules and people alike" (Bendix 1962: 104; Weber 1951: 142–1944 and passim). Thus, the emperor was recognized as a symbol of cultural unity. As "Son of Heaven," he was, in Weber's words, "the old rainmaker of magical religion translated into ethics … and the Lord approved by Heaven insofar as the people fared well unto him" (Weber 1951: 31).

Nelson's interpretation of Weber's socio-historical analysis of China's two world religions underscores the failure of Chinese rationalism to open itself to a belief in a transmundane horizon or reality that might catalyze prophetic energies that could break the hold from its traditionalism. Instead, it "reinforced traditionalist particularism, patrimonialism, and praxis of ritual etiquette and sacromagical observance" (Nelson 1976: 122). Such a cultural orientation leads to relative stasis expressed in an overweening embrace of order, classical learning, and loyalty to family or kinship group. It also inhibits building upon Chinese inventions like the compass, printing, paper money, and the like, that might have affected a breakthrough to universalizing rational science. Moreover, the Chinese did not succeed in establishing a universalistic basis for instituting natural law and rationalizing judicial processes and procedures, which Weber believed were crucial elements observed in Europe for advancing capitalism and modern rationalization. In sum, *contra* the West, the traditionalism and particularism shaped in large measure by Confucianism and

Taoism proved such a cultural drag on Chinese society, culture, and economy that China remained virtually "untouched by the great transformations associated with scientific rationalism of the Greeks, the juridical rationalism and universality of the Roman, the prophetic demagicization of the Hebrews, and the universal fraternization associated with Christianity" (Nelson 1976: 122).

7.3.3 India

If kinship groups and bureaucracy were the dominant realities of China, Weber sought to demonstrate that the caste and otherworldly asceticism were prime hegemonic forces shaping Hinduism and Buddhism (Bendix 1962: 142; Nelson 1976: 124). In the *Religion of India* (1958b), he puts great stock in the "world-historical difference" between caste and guild, saying:

> The uniqueness of the development of India lay in the fact that these beginnings of guild organization in the cities led neither to the city autonomy of the Occidental type nor, after the development of the great patrimonial states, to a social and economic organization of the territories corresponding to the 'territorial economy' of the Occident. (Weber 1958b: 33; cited in Nelson 1976: 125)

Hinduism is also associated with the transmigration of the souls from one life to another through a series of rebirths—the idea that one's actions within his/her caste are "imprinted" upon the soul in one lifetime, and, through adherence to the principle of *dharma*, as one remains faithful to the duties and traditions of one's caste, one may be elevated to a higher caste in a succeeding reincarnation. The absence of an otherworldly creator-god in Hinduism and the existence of many gods lead to an otherworldly conception of salvation grounded in the notion of "a permanent escape from the worldly existence by mysticism or asceticism" (Eisenstadt 1973: 134). The implications of caste, the transmigration of souls, and *dharma* in Hinduism for preparing the ground for capitalism are profound: the search for salvation through traditional ritual acts and the depreciation of thisworldly existence have the effect of obstructing key facilitating conditions necessary for the capitalist spirit to flower.

Buddhism, if anything, carries tendencies toward contemplation and otherworldliness even further than Hinduism. As Weber put it, Buddhism

> ..is a specifically unpolitical and anti-political status religion, more precisely, a religious 'technology' of wandering and of intellectually-schooled mendicant monks. Like all Indian philosophy and theology it is a 'salvation religion,' if one is to use the name 'religion' for an ethical movement without a deity and without a cult. More correctly, it is an ethic with absolute indifference to the question of whether there are 'gods' and how they exist. Indeed, in terms of the 'how,' 'from what,' 'to what end' of salvation, Buddhism represents the most radical form of salvation-striving conceivable. Its salvation is a solely personal act of the single individual. There is no recourse to a deity or savior. From Buddha himself we know no prayer. There is no religious grace. There is, moreover, no predestination either. (Weber 1958b: 206–207)

Buddhism in addition lacks an idea of sin that must be overcome to achieve salvation, a sense of passion for a deity or a concept of brotherly love or fraternization as evidenced in the West. Thus, it becomes a negligible resource for challenging existing social arrangements, giving birth to rational science or promoting sweeping social and cultural change through collective means (see also Gellner 1982).

7.4 The Project of Modernity and World Rationalization in an Era of Capitalist Modernization

At the beginning of the foregoing section, I argued that Weber's studies of world religions in China and India were motivated by the question of why these nations from their traditional settings did not proceed in a direction of European modernity and capitalism. He did not ask: how might they yet follow a path toward modernity beyond their traditional conditions or how might they develop differently from the species of capitalism and modernity emergent in the West? In the decades since Weber wrote the volumes comprising his sociology of world religions, these questions have occupied revolutionaries and scholars alike with growing interest and passion.

Perhaps the most influential theoretical work of the early twentieth century following in Weber's footsteps was Max Horkheimer and Theodor Adorno's *Dialectic of Enlightenment* (1972). In this work, the two founding figures of the Frankfurt School offered a critique of domination grounded in their critical appreciation of Weber's interrogation of the world-historical process of rationalization, which in many respects supersedes Weber's founding work. (Part of this critical reading is drawn from Reid and Yanarella 1974.) Providing perhaps the most succinct summary of their critical analysis, Irving Wohlfarth took note of their depth historical investigation of the

> …inextricability of reason and violence, philosophy and domination … the central theme of *Dialektik der Aufklärung*, which argues that Western reason has never liberated itself from myth (itself already a form of rationality) and has in recent history rapidly, but immanently, reverted to it; the dialectic is a largely Freudian one and it is, as always, with a vengeance that the repressed returns. It is because, on this argument, the constitution of human (or at least Western identity), the control of nature and the domination of man by man have hitherto interacted as the inseparable moments of one long fatal syndrome, that such central categories of the Frankfurt School as 'identity' and "non-identity,' and the domination of the latter by the former, refer simultaneously to the relation between society and individual, man and nature, ego and id, male and female, concepts and their objects. If this in turn sounds reductive, it is the reductiveness of history itself, which philosophy, the logic of domination, has helped to codify. (Wohlfarth 1967: 65)

With regard to modern science, the ultimate truth of the world guiding the cumulative development of scientific knowledge is to be captured and made totally explicit in mathematical formalism. Thought is confused with calculative thinking, the task of which is identified as the search for the conclusive formula to the mathematical schema deemed the essential structure of existence. Over against this objectivistic structure stands the subjectivistic self that would control it through the knowledge bearing, in Bacon's world, "the sovereignty of the Man."

Drawing out the social and economic implications of this new regime prefigured in Weber's closing words in the *Protestant Ethic and the Spirit of Capitalism* (1958a), Horkheimer and Adorno argue that:

> Being is apprehended under the aspect of manufacture and administration. Everything— even the human individual, not to speak of the animal—is converted into the repeatable, replaceable process, into a mere example for the conceptual models of the system. ...The conceptual apparatus determines the senses, even before perception occurs; a prior, the citizen sees the world as the matter from which he himself manufactures it. Intuitively, Kant foretold what Hollywood consciously put into practice: in the very process of production, images are pre-censored according to the norm of the understanding which will later govern their apprehension. (1972: 84)

So the intent of the Enlightenment and the rationalizing processes conspire to convert reason and liberation into their opposite: myth and domination.

The socio-historical inquiries on capitalism and modernity of Weber and the critical theoretical ones on enlightenment and domination of Horkheimer and Adorno were again taken up in Jürgen Habermas' theorizations on what he called the project of modernity. On the occasion of the city of Frankfurt, Germany's conferral to him of the Theodor W. Adorno prize in 1981, Habermas opened up a critical debate over the idea of modernity as an incomplete, but redeemable, project (Habermas 1983, 1997). This work was complemented the same year by the publication of the *Theory of Communicative Action* in two volumes (1985a, b) and then in 1985 by *The Philosophical Discourse of Modernity: Twelve Lectures* (1987). Habermas' speech involved an excursus on the long history of modernity and the modern in Western culture that he saw taking expression in two forms, the cultural or aesthetic modernity associated with Baudelaire and his theory of art and the societal modernity and the project of modernity stemming from the French Enlightenment and Weber's dark critique of the Enlightenment faith expounded in his work on the Protestant ethic.

For Habermas, one of Weber's key contributions to our understanding of modernity involves the carving up of substantive reason, formerly grounded in religion and metaphysics, into three value spheres: science, morality and law, and art. As these value realms are institutionalized, they lead to "autonomous structures intrinsic to the cognitive-instrumental, the moral-practical and the aesthetic-expressive knowledge complexes" (1997: 45), which are increasing absorbed into professional communities of experts who become greatly distanced from the general public.

As Habermas argued, the project of modernity, as seen by these eighteenth century philosophers, was animated by a limitless faith in the emancipatory possibilities of human reason. They anticipated that the advance of reason through scientific inquiry would promote moral and material progress, foster the mastery of nature through technology, and create efficient administrative and planning techniques leading to the good society. As Gaonkar (1999) has observed, another achievement borne of this project was that "the rationalization of cultural and social life resulting from the spread of scientific knowledge and attitude would lead, among other things, to the progressive eradication of traditional superstitions, prejudices, and errors and to the gradual establishment of a republican form of government" (Gaonkar 1999: 8).

In "Modernity: An Incomplete Project" and subsequently in the *Theory of Communicative Action*, Habermas reads Weber's critique of the Enlightenment project carefully and critically. He notes how Weber's understanding of reason or rationality that drives his conception of modernity is that of an instrumental-purposive rationality, one that involves the loss of meaning and unity formerly given to existence by traditional religious worldviews. Ultimately, such rationalization associated with the transformation of the Protestant ethic into the spirit of capitalism and the emergence of the "prodigious cosmos" of capitalism freed from its religious moorings leads, not to the utopian imaginings of Enlightenment thinkers but to Weber's, and our, iron cage," ushering in alienation, anomie, and disenchantment.

Habermas demurs from this Weberian vision of "mechanized petrification," spiritless specialists, and heartless sensualists inhabiting the projected endpoint of modernity and seeks instead to rehabilitate faith in reason and the completion of the Enlightenment project. While conceding some points about the impact of rationalization process upon the socio-cultural world, he flatly refuses to accept Weber's definition of reason as purposive or instrumental rationality, just as he eschews the ramifications of Horkheimer and Adorno's critique of Enlightenment modernity. Similarly, he rebuffs Lukacs' equation of rationalization with reification or Foucault's entwinement of knowledge and power. For Habermas, such readings of rationality mistake one aspect of reason *under capitalist modernization* for reason itself and further attributes a truncated or reductionist mode of reason (instrumental rationality) with a subject-centered reason linked to the tradition of philosophizing based on consciousness. Instead, he opens up reason to two types of rationality: cognitive-instrumental rationality and communicative rationality (Habermas 1985a: 10–11).

Communicative rationality stands at the core of Habermas' alternative rendering of the project of modernity, a post-capitalist social existence and a democratic politics. In his *Theory of Communicative Action*, Habermas summarizes its meaning:

> To sum up, we can say that actions regulated by norms, expressive self-presentations, and also evaluative expressions, supplement constative speech acts in constituting a communicative practice which, against the background of the lifeworld, is oriented to achieving, sustaining, and renewing consensus—and indeed in a consensus that rests on intersubjective recognition of criticizable validity claims. The rationality inherent in this practice is seen in the fact that a communicatively achieved agreement must be based in the end on reasons. And the rationality of those who participate in this communicative practice is determined by whether they could, if necessary, under suitable circumstances, provide reasons for their expressions. Thus the rationality proper to the communicative practice of everyday life points to the practice of argumentation as a court of appeals that makes it possible to continue communicative action with other means when disagreements can no longer be repaired with everyday routines and yet are not to be settled by the direct or strategic use of force. For this reason I believe that the concept of communicative rationality, which refers to an unclarified systematic interconnection of validity claims, can be adequately explicated on in terms of a theory of argumentation. (Habermas 1985a: 17)

This extended definition is predicated on the idea that communicative rationality is not built upon subjective or subjectivistic foundations but inheres in language itself, unfettered speech acts, and rational argumentation.

Since its articulation, the nuanced and seemingly magisterial architecture of Habermas' theorization of community rationality and project of modernity has drawn extensive commentary and critique. Some like Thomas McCarthy (1985) and Seyla Benhabib (1992) have become among the most ardent philosophical defenders of Habermas' project, though in a more toned down, qualified rendering. Others, like Fred Dallmayr (1988) have challenged Habermas' conception of communicative rationality as unduly narrow and have questioned his consignment of philosophical inquiry to the role of "usher" and "mediating interpreter," offering instead an embodied understanding of reason based upon the interpenetration of thought and world that surpasses Habermas' reversion to dualistic thinking. Still others like Charles Taylor question Habermas' communicative rationality theory because its "proceduralist ethics" fails to acknowledge the world-disclosing facet of language (Taylor 1991: 30–31). Recently, Andrew Feenberg (1995: 48 and *passim*) has faulted Habermas for his embrace of Arnold Gehlen's theory of technological development as the progressive objectification and supplementation of the human body and mind unconnected to class or social values, which implicitly has led to Habermas' complete neglect of technology in his two-volume treatise on communicative rationality and action.

In what follows, I would like to carry critique into the terrain of alternative modernities and to the possible recuperation of premodern, anti-modern, and some strains of post-metaphysical strains that Habermas' corpus has sought to bar as inimical to a redeemed modernity. Central to this effort will be the conjoining of alternative forms of modernity with the rising concerns for sustainability as a focal theme for a post-modern (as opposed to a post-modernist) response to globalization and ecological catastrophe. These musings and perspectives should be regarded less as sustained and definitive treatments of these two landscapes of the future than openings for further reflection and research.

7.5 Resources for a Sustainable Future: A Research Prolegomenon

Weber is ambivalent, even contradictory, in evaluating his Protestant ethic thesis and especially with respect to his speculations on the triumph of capitalism in his day. On the one hand, the spirit of capitalism, he says, has congealed into an iron cage and has deployed *Zweckrationalität* (purposive-instrumental rationality) in the economic practices over Europe, America and increasingly around the world. In the process, this tremendous cosmos "is now bound to the technical and economic requirements of machine production which today determine the lives of all the individuals who are born into this mechanism, not only those directly concerned with economic acquisition, with irresistible force" (Weber 1958a: 181). "Perhaps," he goes on in an ironic and prescient speculation, "it will so determine them until the last ton of fossilized [Kentucky? Wyoming? Chinese?] coal is burnt" (Weber 1958a: 181). Yet, on the other hand, he pulls back from the reified and deterministic cast of this fatalistic imagination of the future. For, as he goes on to acknowledge, "no one knows who will live in this cage in the future, or whether at the end of this tremendous development entirely new prophets will arise, or there will be a great rebirth of old ideas and ideas, or, if neither, mechanized petrification, embellished with a sort of convulsive self-importance" (Weber 1958a: 182).

In our day, Weber's and Horkheimer and Adorno's gloomy depth-historical excavations and theoretical diagnoses of the present and near future seem closer to the mark than Habermas' optimism grounded in the promise of communicative rationality and action, as societal modernization has gone global, largely breaking free from its nation-state anchorage as markets, financing, production are extended by the global reach of multi-national and increasingly transnational corporations. Meanwhile, economic and technological assaults on the earth, while felt locally and regionally and often contradictorily, are posing global challenges to the Ecosystem and humankind in the form of, among other things, global warming, soil erosion, acid precipitation, genetically modified crops and animals, and disappearing phytoplankton in the oceans.

One avenue forward in thinking about modernity *qua* societal modernization has been presented by Charles Taylor (1991, 1995, 1999) in his reflection on two theories of modernity and Dilip Gaonkar (1999, 2001) in his critical appropriation of Taylor's essay. Tacitly criticizing Weber's baleful iron cage rendition and perhaps Habermas' project of modernity, Taylor characterizes the acultural theory that "conceives of modernity as the growth of reason, defined in various ways: as the growth of scientific consciousness, or the development of a secular outlook, or the rise of instrumental rationality, or an ever-clearer distinction between fact-finding and evaluation" (Taylor 1995: 24). On this account, the shift from traditional society to modernity takes place through a series of culturally neutral processes that any and perhaps every society will undergo voluntarily or by force. As against this theory, the cultural theory of modernity is based on the image of "a plurality of cultures, each of which has a language and a set of practices that define specific understandings of personhood, social relations, state of mind/soul, goods and bads, virtues and vices" (Taylor 1995: 24). As Taylor argues, "modernity is not that form of life toward which all cultures converge as they discard beliefs that held our forefathers back." Instead, he states, "it is a movement from one constellation of background understandings to another, which repositions the self in relations to others [personhood, nature, God] and the good" (Taylor 1995: 24). In other words, a new cultural paradigm or worldview.

On this reading of modernity, modernity is not a singular, but a plural, noun that justifies our speaking of alternative modernities. As a result, as Gaonkar argues, this cultural theory "holds that modernity always unfolds within a specific cultural or civilizational context and that different starting points for the transition to modernity lead to different outcomes" (Gaonkar 1999: 15). Out of this unavoidable "dialectic of convergence and divergence" emerges the distinctive "lived experience and cultural expressions of modernity" and the "site-specific 'creative adaptations'" informing these alternative modernities (Gaonkar 1999: 16).

Whether this plurality of modernities unfolding in the world today will lead to what Gaonkar (1999: 18) calls an "ethic of the global modern" is an open question (see Dirlik 2002). But one of the global imaginaries that has surfaced in the face of

Table 7.1 Alternative ecological paradigm (Source: Ernest J. Yanarella, derived from Marcuse 1972 and others)

Humans	Identical and non-identical with nature: a post-humanist, decentered conception of the embedded, social self-entwined in a more encompassing framework of natural processes
Nature	Humankind's organic body guided by the non-exploitative appropriation of nature and old and newly-discovered sustainability practices
Work	Non-alienated work combining physical labor and intellectual play and creativity
Time	Subordination of mechanical clock-time to efflorescence of multiple temporalities, including leisure- and play-time and forms of embodied temporality
Environment	Biosphere/ecosystem—the life-sustaining natural processes of which humankind is a part
Science	An ecologically-grounded science purged of mere scientific instrumentalism and the technological impulse to dominate
Technology	Alternative, appropriate technology grounded in sustainability and non-exploitative social relations
Good society	A democratically planned, decentralized, but regionally coordinated society promoting discursivity and active political participation, ecological sustainability, meaningful work, and elegant frugality

capitalist modernization, globalization and ecological assault is sustainability. Though an essentially contested concept (Yanarella and Levine 2011), it is also and more fundamentally a powerful cultural force shaping the global debate about economic growth and development and possibly transforming the modern cultural paradigm and its constellation of understandings of the concepts comprising it.

In some respects, we might conceive of this epoch as a transition period where, as Gramsci once noted, "the old is dying and the new cannot be borne; [and] in the interregnum a whole assortment of morbid symptoms appear[s]" (Gramsci 1971: 276). Provisionally, such a new ecological or genuinely postmodern paradigm or worldview might be schematized in something like the following way (Table 7.1).

7.5.1 Habermas' Modernist Modernity Solution to the Ravages of Capitalist Modernization

In offering this ecological or sustainability worldview, one must acknowledge that Habermas' theory of communicative rationality and action has much to contribute to its development and institutionalization. Though largely silent on the ramifications of his theory for addressing capitalist modernization run amok and the promise of sustainability, some of his students and followers have sketched outlines of possible contributions of Habermas' modernist modernity to the sustainability debate using his model of the free speech situation and better argument strategy, as well as his media theory of power and money, into policy debate and assessment (Brulle 2002;

Sköllerhorn 1998; Elling 2008, 2010; see also Hendricks 2010; Ganis 2010). Utilizing Habermas' modernist modernity framework, much is to be gained by bringing stakeholders into a concrete sustainability/economic development context of free-wheeling discourse and argumentation grounded in the best available evidence with irresponsible power and money marginalized. (Levine and Yanarella have attempted to draw upon the spirit of Habermas' theory in the context of a game-based discourse focuses on a soft landing for coal—see Levine and Yanarella 2014.)

My concern is with the limits that this approach grounded in Habermas' assumptions underpinning his communication action theory imposes. That is, his implicit endorsement of the modernists' belief that "modernity can and will no longer borrow the criteria by which it takes its orientation from the models supplied by another epoch: *it has to create its normativity out of itself*" (Habermas 1987: 7, my italics). In addition, his quasi-positivistic assumptions about modern science and his bodily extension theory of technology prompt me to worry that the "rules of the game" will incline modernist solutions toward measures derived from ecological modernization theory and practice. That is, much of the appeal of ecological modernization stems from the fact that changes in institutional adaptations and greater technical efficiencies avoid changes in social, economic, and political relations that are arguably the underlying sources of ecological problems (see, among others, York and Rosa 2003; Yanarella and Levine 2008; Foster et al. 2011). In the face of its many flaws, ecological modernization as technological modernity hardly seems to be the solution to the shortcomings of the Enlightenment project of modernity and capitalist modernization. Rather, it is more likely to perpetuate them in newer and more destructive forms on the model of Freud's return of the repressed.

7.5.2 Alternative Modernities and the Sustainability Paradigm: Alternatives to Habermas' Modernist Modernity

Against the stream of Habermas' counterarguments and cautions and his characterization of Bloch as a "Marxist romantic" (Habermas 1969/1970), I wish instead to limn a set of apparently anti-modern alternatives building upon Ernst Bloch's search for the cultural surplus of the pre-modern and anti-modern and his notion of non-contemporaneous contradictions (Bloch 1977). As an exercise in the radical political task of critical inheritance of past traditions, the theoretical thrust of this apparently heretical philosophical move draws heavily upon Ernst Bloch's work, especially his writings on the philosophy of hope (1986) and the spirit of utopia (2000). Beginning with his definition of human beings as creatures who hope, Bloch explores the many expressions of human hope in the classical literature, the professions, the arts, mass advertising and popular entertainment. His intent is to offer a positive hermeneutic of figures of hope and anticipatory consciousness in popular culture and everyday life. Regarding religious believers, farmers, and small business proprietors as continuing repositories of older ideas of land, community, work, and home subverted by the advancing of modern industrializing forces and secularizing trends, Bloch stressed the need of change agents to shoulder the task of active inheritance as a form of hegemonic rearticulation of the cultural surplus of these ideas (Schroyer 1982). That is, Bloch recognized that these premodern and even antimodern elements of earlier phases of economic development act as noncontemporaneous contradictions in present-day politics often expressed in conservative and even reactionary articulations. But he believes that their continuing articulation of antiquated cultural ideals and values points to their failure to become fully sedimented in social relations in the past and to their spawning of a cultural surplus of these traditions that can reactivate these old dreams and seed continuing political mobilization on their behalf in the living present and hopeful future. For Bloch, their political potential as part of a forward coalition of forces promoting radical change requires Marxists to pay the debts of the past in order to receive the present by articulating the futurity contained in every value and ideal expressed by existing remnants of older economic being and political consciousness—a futurity that could only be truly realized in an open, democratic, post-capitalist society yet to be made.

7.5.3 Pre-modern

One example of the refunctioning of the pre-modern into vistas on sustainability is the theoretical, architectural, and design work at the University of Kentucky Center for Sustainable Cities and the CSC Design Studio undertaken over the past 25 years. This work has involved drawing upon the genius of the design of the medieval Italian hill-town—the city on a hill—and recentering its social, political, and urban architectural values into the design of the City as a Hill. This refunctioning process has also drawn heavily upon the design principles of aesthetic modernism and cultural modernity in a manner that yields a synthesis of ecological design values and modern urbanity into a model of the built environment ensconced in many features of the inchoate ecological/postmodern paradigm. (For an overview and compilation of some of this work, see: Yanarella and Levine 2011).

7.5.4 Traditional Religion—Confucianism

Another resource for the cultural construction of a sustainability paradigm may also come from renewed interest in the traditional, non-Western religions, among them, Confucianism. With the rise of China and India and the pursuit of economic development in these emerging players on the global scene, a renewed interest in Calvinism and the Weber Protestant ethic thesis has erupted in China, India, and other nations in South and East Asia. This attention is reflected in Web pages with content commenting on Calvinist religion and Weber's thesis (Stückelberger 2007: 122–123) and primary and secondary publications of Calvin's work and especially

Weber's studies of world religions (2007: 123). Most notably, native scholars in south and east Asia have begun to reassess Weber's studies of Confucianism, Taoism, Hinduism, and Buddhism, drawing upon indigenous language and cultural traditions, as well as the wealth of materials old and new that Weber's work could not have accessed. This inquiry has led many of these researchers to question both Weber's and Habermas' putative Eurocentric biases with respect to interpreting those Eastern religions studied by Weber and Habermas' historical-theoretical treatment of the public sphere and his communicative action theory (see Gunaratne 2006, for some of this literature).

Recently, Kassiola (2008) issued a call for Confucianizing modernity and modernizing Confucianism. While suffering from a weak historical or theoretical grounding in the literature and animated by a characteristically moralistic impulse, it does point to the spaciousness of tradition highlighted by Kolb (1992: 81–84) that counters Habermas' resistance to the role of tradition in constituting modernity as evidenced in the Habermas-Gadamer debate (Ricoeur 1973; Mendelson 1979). A better guide in imagining the productive role that Confucianism may take with respect to the alternative form and substance that the future of Chinese modernity may take is Wang Hui's recent ruminations on the subject (2011). Also indispensable is Hui's earlier investigation of the many Chinese intellectual schools of thought shaping the native and expatriate Chinese intellectual debate on the question of Chinese modernity (Hui 1998), where he also outlines the Chinese communist regime's strategy of societal modernization in the context of global capitalism. There, he makes several salient points: that some New Enlightenment scholars have resorted to traditional values, particularly Confucianism, to question whether the Western model of development had any particular applicability to Chinese society and culture (Hui 1998: 21); and that the idea animating the move above—Confucian capitalism—"enables exponents to embrace the capitalist mode of production and the global capitalist system … while adding a layer of cultural nationalism on top" (1998: 22). In any case, Hui observes that the Chinese Communist party elite has so far succeeded in co-opting and rendering harmless Confucianism as a cultural resource beyond state socialism and economic marketization. On the other hand, if the discourse of modernity becomes a truly intercultural and global one that breaks the grip of global capitalism (Nelson 1973a), Confucian values may yet have a formative influence upon the articulation of sustainability.

7.5.5 Traditional Religion—Calvinism (!)

Recognizing the small role that Presbyterianism as a mainline religion plays in politics and social change nationally and globally, it may be worthwhile to underline that Calvin's theology may yet have a voice in the debate about alternative modernities and a new ecological cultural paradigm. As Weber continually stressed, his investigation of ethic was less centered on an evaluation of the spiritual teachings of Calvin than the "historical significance of the dogma" (Nelson 1973a: 101) and its unintended economic and other consequences. Recent Protestant theologians have performed close readings of Calvin's words and demonstrated that many of Calvin's specific teachings run quite contrary to the fate of the earth being shaped under the regime of capitalism (Little 1981; Parker 1995; Elwood 2002; Dommen 2007; Stückelberger 2007; Partee 2008; Straub 2009). Much of their efforts to rediscover the original or "authentic" Calvin buried beneath Weber's sociohistorical interpretation of Calvinism is devoted to the treatment of God's creation by his creatures. Among other things, Calvin, like Thomas Aquinas, underlines the importance of common possession and common benefit as qualifications on the use of private property. Although God gave human beings dominion over the world, human use of God's creation—nature—could be rightly regulated by the state to assure that fallible and sometimes venal human beings do not exercise that dominion in a malicious and self-interested fashion. Far from countenancing a selfish and even rapacious orientation to nature, Calvin counseled stewardship of the land, seeing property as being held in trust for the common good and for the benefit of all human beings alive and yet to be born.

Surpassing a stewardship ethic toward the earth and the environment, Calvin goes so far to say:

> There is a general rule which we must note well: each time we are inclined to ravage or pollute nature in some way, we should remind ourselves that our Lord has placed us to live in this world and he has given us things which he knew would be beneficial to our life...Am I worthy that the earth should sustain me when I wish to obliterate God's grace for my neighbors as well as for myself? Do I deserve to be sustained by nature, if I do not let nature follow its own course, obey it own rule? If I do not, am I not then a monster? To remind ourselves of the true purpose of nature ought to hold us back when we are malicious and defiled enough to despoil trees, houses, and similar objects. [If we do these things we are waging war] against him who shows us a mirror of his goodness, not toward one man only, but toward all men, including us. (Biéler 1961: 434–435; cited in Little 1981: 214–215)

As Little perceptively notes, this passage, especially its reference to nature as a mirror of God's goodness, points to Calvin's belief in the "relative autonomy of nature [that] calls into question the typical characterization of Calvin as breeding nothing but a crudely instrumentalist attitude toward nature" (Little 1981: 215). This line of doctrinal analysis and exegesis points to realms of possibility that a reawakened spiritual attitude toward God's creation might take in seeding sites of critique and affirmation with religious tradition that goes well beyond Habermas' recent concessions and limited *rapprochement* with religion (Dillon 1999; Habermas 2006, 2010).

7.5.6 *Heideggerian Postmetaphysical Eco-philosophy*

Without trying in any sense to complete this political order, I would like to point to the relevance of Heidegger's difficult, but rewarding, comments on building, dwelling, and home to this portentous theoretical and practical exercise.

In his essay, "Building Dwelling Thinking," Martin Heidegger (1971: 143–161) offers a series of simply stated and allusive arguments dovetailing with the manifest ecological feelings and aspirations and the latent political concerns of eco-philosophers and political ecologists from Heinrych Skolimowski and Raymond Dassman to Arne Naess and Murray Bookchin (see Zimmerman 1997; Yanarella 2001). His central concept is the notion of dwelling and his concern is how humankind dwells authentically. On his view, whenever human beings build, whether it be a house or a bridge or a city, they are already dwelling. Ideally, dwelling is the way in which we are on the earth. Moreover, true dwelling implies a sparing and preserving attitude-- tolerance of a thing in its essence, a willingness to let something be. In his words, "to dwell, to be set at peace, means to remain at peace within the free, the preserve, the free sphere that safeguards each thing in its nature" (Heidegger 1971: 149). Respectful sparing then stands in contrast to the modernist impulse to dominate things, to seize things and to thoughtlessly refashion them according to our subjectivistic designs.

Authentic dwelling also incorporates the fourfold—earth, sky, gods, mortals: "to dwell is to spare the earth, receive the sky, expect the gods, and have a capacity for death" (Vycinas 1969: 15). Sparing the earth and receiving the sky mean respectfully allowing the things of the earth and in the sky to be how they are in their essence and not try to subjugate or strap them to our subjectivistic values or whims. Awaiting the gods means adopting openness to the signs of the divine in our intimate surroundings or acknowledging their absence. Accepting our mortality and the inevitability of death involves understanding human life in its completeness and in its ultimate ties to the rhythms of nature. On the other hand, this sparing and preserving attitude does not mean that human beings reside in the world in a passive way. Human beings must act in and upon the earth, but when they do so authentically, they build in the company of the fourfold. Tools, constructions, artifices are rightly made and used by us (who must appropriate from nature in order to live) within the attitude of sparing. Thus, a product of human labor (a bridge or a house or a city) assembles the fourfold. And, by assembling the fourfold, a building "receives from the foursome the references or standards for organization or mutual interrelation of places" (Vycinas 1969: 16).

Natural and humanly-made things can, according to Heidegger, be perverted by not being permitted to be the way they are. For example, home was once an emotion-laden place and experience that subsumed us as creatures and oriented our way of life even if we left our abode and journeyed far away. Today, its exchangeability and our impulse to rule things make us moderns homeless, even if we have a place to live. So, to make a thing, to build a home (in the widest and narrowest senses) means to be disposed to the higher realities of the fourfold and to stand under their direction.

The opening Heidegger provides in contributing to a sustainability ethic based upon Heidegger's notion of world disclosure, explored most recently by Kompridis (2005, 2011) in his effort to rethink critical theory in a post-Habermasian phase, remains foreign to Habermas' philosophy of communicative rationality. Still, it arguably points to an alternative basis for participating in the long task of rehabilitating critical theory and charting new pathways for overcoming modernity's limitations and paradoxes.

7.6 Conclusion

I would like to close with some partings comments drawn from Merleau-Ponty (1973) and his interrogation of the "crisis of understanding" in Weber—a dialogue that Arnason (1993) still considers unfinished. The achievement of Weber in his study of the Protestant ethic and the spirit of capitalism, Merleau-Ponty says, is that if "Weber succeeds in understanding the basic structure of the facts, it is because he has discovered an objective meaning in them, has pierced the appearances in which reason is enclosed, and has gone beyond provisional and partial perspectives by restoring the anonymous intention, the dialectic of the whole" (Merleau-Ponty 1973: 13–14). While history has a meaning, he goes on to say, "there is no pure development of ideas" (Merleau-Ponty 1973: 16). Seeing the emergence of capitalism from its Protestant roots, Weber begins his historical hermeneutic by looking for signs and tokens scattered here and there, just as the emerging system of capitalism at the beginning "is not an all-powerful idea; it is [instead] a sort of historical imagination which sows here and there elements capable one day of being integrated" (Merleau-Ponty 1973: 17). Later, historians and sociologists "can talk of "rationalization" or 'capitalism' when the affinity of these products of the historical imagination becomes clear. But history does not work according to a model; it is, in fact, the advent of meaning" (Merleau-Ponty 1973: 17).

These words summarizing the great achievement of Weber are humbling and cautionary to scholars and activists as they ponder, assay, and act upon the future and reflect upon the past in community with others seeking to give good weight to helping build a future that will surpass us and judge us by our actions and inactions, by values we lived by and sought to embed in a collective future. If Weber has offered us a guide to investigate how the strains of Calvinism seeded a spirit that grew into a prodigious cosmos, perhaps we can as knowers and actors take responsibility for working beyond global capitalism and its varying forms of societal modernization toward an ecological paradigm that may yet be recognized in the future as essential but today is acknowledged as in no sense inevitable. Perhaps in the cross-cultural conversations and movements for social justice, peace, and ecology being spawned here and there across the world, the resources for that new paradigm may yet be uncovered and disclosed, may yet be woven together through political and cultural processes still eluding us today.

References

Arnason, J. P. (1993). Merleau-Ponty and Max Weber: An unfinished dialogue. *Thesis Eleven, 36*, 82–98.
Aron, R. (1967). *Main currents of sociological thought: Durkheim, Pareto, Weber* (Vol. II). New York: Basic Books. Chapters IV and V.
Bendix, R. (1962). *Max Weber: An intellectual portrait*. Berkeley: University of California Press.
Benhabib, S. (1992). *Situating the self: Gender, community and postmodernism in contemporary ethics*. New York: Routledge.

Biéler, A. (1961). *La pensée économique et sociale de Calvin*. Genève: Georg.
Bloch, E. (1977). Nonsynchronism and the obligation to its dialectics. *New German Critique, 11*(Spring), 22–38.
Bloch, E. (1986). *Principle of Hope* (trans: Neville Plaice et al.) (3 Vols.). Cambridge: The MIT Press.
Bloch, E. (2000). *Spirit of Utopia* (trans: Nassar, A.). Stanford: Stanford University Press.
Breiner, P. (2004). Translating Max Weber: Exile attempts to forge a new political science. *European Journal of Political Theory, 3*(2), 133–149.
Brulle, R. J. (2002). Habermas and green political thought: Two roads converging. *Environmental Politics, 11*(4), 1–20.
Dallmayr, F. (1988). Habermas and rationality. *Political Theory, 16*(4), 553–579.
Dillon, M. (1999). The authority of the holy: Habermas, religion and emancipatory possibilities. *Sociol Theory, 17*(3), 290–306.
Dirlik, A. (2002). Modernity as history: Post-revolutionary China, globalization and the question of modernity. *Social History, 27*(Jan), 16–39.
Dommen, E. (2007). Calvin and the environment: Calvin's views examined through the prism of present-day concern especially of sustainable development. In E. Dommen & J. D. Bratt (Eds.), *John Calvin rediscovered: The impact of his social and economic thought* (pp. 53–66). Louisville: Westminster/John Knox Press.
Eisenstadt, S. N. (1973). The implications of Weber's sociology of religion for understanding process of change in contemporary non-European societies and civilizations. In C. Y. Glock & P. E. Hammond (Eds.), *Beyond the classics?* (pp. 131–155). Berkeley: University of California Press.
Elling, B. (2008). *Rationality and the environment: Decision making in environmental policy and assessment*. New York: Earthscan.
Elling, B. (2010). A record on modernity, rationality and sustainability. In K. A. Nielsen, B. Elling, M. Figueroa, & E. Jelsoe (Eds.), *A new agenda for sustainability* (pp. 31–42). Burlington: Ashgate Publishing Company.
Elwood, C. (2002). *Calvin for armchair theologians*. Louisville: Westminster/John Knox Press.
Fahey, T. (1982). Max Weber's ancient Judaism. *American Journal of Sociology, 88*(1), 62–87.
Feenberg, A. (1995). *Alternative modernity: The technical turn in philosophy and social theory*. Berkeley: University of California Press.
Feenberg, A. (1996). Marcuse or Habermas: Two critiques of technology. *Inquiry, 39*, 45–70.
Foster, J. B., Clark, B., & York, R. (2011). *Ecological rift: Capitalism's war on the earth*. New York: Monthly Review Press.
Ganis, R. (2010). *Politics of care in Habermas and Derrida: Between measurability and immeasurability*. New York: Lexington Books.
Gaonkar, D. G. (1999). On alternative modernities. *Public Culture, 11*(1), 1–18 (Republished in Gaonkar, 2001, 1–23).
Gaonkar, D. G. (Ed.). (2001). *Alternative modernities*. Durham: Duke University Press.
Gellner, D. (1982). Max Weber: Capitalism and the religion of India. *Sociology, 16*(4), 526–543.
Gerhardt, U. (2007). Much more than a mere translation—Talcott Parsons's translation into English of Max Weber's Die protestantische Ethik und der Geist des Kapitalismus: An essay in intellectual history. *Canadian Journal of Sociology, 32*(1), 41–62.
Giddens, A. (1971). *Capitalism and modern social theory: An analysis of the writings of Marx, Durkheim and Max Weber*. New York: Cambridge University Press.
Gramsci, A. (1971). *Selections from the prison notebooks*. New York: International Publishers.
Gunaratne, S. A. (2006). Public sphere and communicative rationality: Interrogating Habermas's Europecentrism. *Sage Journalism and Communications Monographs, 8*(2), 93–156.
Habermas, J. (1969/1970). Ernst Bloch—Marxist romantic. *Salmagundi*, No 10/11 (Fall–Winter): 311–325.
Habermas, J. (1983). Modernity – An incomplete project. In H. Foster (Ed.), *The anti-aesthetic* (pp. 3–15). Port Townsend: Bay Press.
Habermas, J. (1985a). *Theory of communicative action* (Reason and rationalization of society, Vol. 1). Boston: Beacon.

Habermas, J. (1985b). *Theory of communicative action* (Lifeworld and system: A critique of functionalist reason, Vol. 2). Boston: Beacon.
Habermas, J. (1987). *The philosophical discourse of modernity: Twelve lectures* (trans: Lawrence, F. G.). Cambridge: The MIT Press.
Habermas, J. (1997). Modernity: An unfinished project. In M. P. d'Entréves & S. Benhabib (Eds.), *Habermas and the unfinished project of modernity* (pp. 38–55). Cambridge: The MIT Press.
Habermas, J. (2006). Religion and the public sphere. *European Journal of Philosophy, 14*(1), 1–25.
Habermas, J. (2010). *An awareness of what is missing: Faith and reason in a post-secular age*. New York: Polity.
Heidegger, M. (1971). Building dwelling thinking. In *Poetry, language, thought* (pp. 143–161). New York: Harper & Row.
Hendricks, C. M. (2010). Inclusive governance for sustainability. In V. A. Brown, J. A. Harris, & J. Y. Russell (Eds.), *Tackling wicked problems through transdisciplinary imagination* (pp. 150–160). New York: Routledge.
Horkheimer, M., & Adorno, T. W. (1972). *Dialectic of enlightenment*. New York: Continuum Books.
Hughes, H. S. (1961). *Consciousness and society: The reorientation of European social thought, 1890–1930*. New York: Vintage Books. Chapter 8.
Hui, W. (1998). Contemporary Chinese thought and the question of modernity. *Social Text, 16*(2), 9–44.
Hui, W. (2011). Weber and the question of Chinese modernity. In *Politics of imagining Asia* (pp. 264–306). Cambridge: Harvard University Press.
Jameson, F. (1973). The vanishing mediator: Narrative structure in Max Weber. *New German Critique, 1*(Winter), 52–89.
Jameson, F. (1988). *The vanishing mediator; or, Max Weber as storyteller. The ideologies of theory: Essays, 1971–1986* (Syntax of history, Vol. 2, pp. 3–34). Minneapolis: University of Minnesota Press.
Kassiola, J. J. (2008). Confucianizing modernity and 'modernizing' Confucianism: Environmentalism and the need for a Confucian positive argument for social change. In J. J. Kassiola & S. Guo (Eds.), *China's environmental crisis: Domestic and global political impacts and responses* (pp. 195–218). New York: Palgrave Macmillan.
Kolb, D. (1992). *Postmodern sophistications: Philosophy, architecture, and tradition*. Chicago: University of Chicago Press.
Kompridis, N. (2005). Disclosing possibility: The past and future of critical theory. *International Journal of Philosophical Studies, 13*(3), 325–351.
Kompridis, N. (2011). *Critique and disclosure: Critical theory between past and future*. Cambridge: The MIT Press.
Levine, R. S., & Yanarella, E. J. (2014). A soft landing for coal (Manuscript under review).
Little, D. (1981). Land use and 'The Common Good': Religious backgrounds. *Environmental Education and Information, 1*(3), 209–223.
MacIntyre, A. (1962). A mistake in causality in social science. In P. Laslett & W. G. Runciman (Eds.), *Philosophy, politics and society* (Vol. 2, pp. 48–70). Oxford: Basil Blackwell.
Marcuse, H. (1972). *Counterrevolution and revolt*. Boston: Beacon.
McCarthy, T. (1985). Translator's introduction. Jürgen Habermas. *Theory of communicative action* (Reason and rationalization of society, Vol. 1, pp. v–xli). Boston: Beacon.
Mendelson, J. (1979). The Habermas-Gadamer debate. *New German Critique, 1*(Autumn), 44–73.
Merchant, C. (1990). *Death of nature: Women, ecology, and the scientific revolution*. San Francisco: HarperOne.
Merleau-Ponty, M. (1973). The crisis of understanding. *Adventures of the dialectic* (pp. 9–29) (trans: Bien, J.). Evanston: Northwestern University Press.
Mitzman, A. (1970). *The iron cage: An historical interpretation of Max Weber*. New York: Knopf.
Molloy, S. (1980). Max Weber and the religions of China: Any way out of the maze? *British Journal of Sociology, 31*(3), 377–400.
Mumford, L. (1934). *Technics and civilization*. New York: Harcourt, Brace & Company, Inc.
Mumford, L. (1968). *City in history*. New York: Harcourt, Brace & World.

Nelson, B. (1969a). *The idea of usury: From tribal brotherhood to universal otherhood.* Chicago: University of Chicago Press.
Nelson, B. (1969b). Conscience and the making of early modern cultures: The protestant ethic beyond Mas Weber. *Social Research, 36*(1), 5–21.
Nelson, B. (1973a). Civilizational complexes and intercivilizational encounters. *Sociological Analysis, 34*(2), 79–105.
Nelson, B. (1973b). Weber's protestant ethic: Its origins, wanderings, and foreseeable futures. In C. Y. Glock & P. E. Hammond (Eds.), *Beyond the classics?* (pp. 113–130). Berkeley: University of California Press.
Nelson, B. (1976). On orient and occident in Max Weber. *Social Research, 43*(1), 114–129.
Nielsen, D. A. (1998). Benjamin Nelson. In W. H. Swatos & P. Kivisto (Eds.), *Encyclopedia of religion and society* (pp. 327–328). Walnut Grove: AltaMira Press.
Parker, T. H. L. (1995). *Calvin: An introduction to his thought.* Louisville: Westminster/John Knox Press.
Parsons, T. (1967). *Structure of social action* (Vol. 2). New York: Free Press.
Partee, C. (2008). *Theology of John Calvin.* Louisville: Westminster John Knox Press.
Reid, H. G., & Yanarella, E. J. (1974). Toward a post-modern theory of American political science and culture: Perspectives from critical Marxism and phenomenology. *Cultural Hermeneutics, 2*, 91–166.
Ricoeur, P. (1973). Ethics and culture: Gadamer and Habermas in dialogue. *Philosophy Today, 17*(Summer), 153–165.
Scaff, L. A. (2005/2006). The creation of the sacred text: Talcott Parsons translates 'The Protestant Ethic and the Spirit of Capitalism.' *Max Weber Studies, 5 & 6*: 195–219.
Schaltegger, C. A., & Torgler, B. (2009). Was Weber wrong? A human capital theory of protestant economic history: A comment on Beck and Woessmann. *Working paper 2009–06.* Basel: CREMA Research.
Schroyer, T. (1982). Cultural surplus in America. *New German Critique, 26*(January), 703–711.
Sennett, R. (1977). *Fall of public man.* New York: Vintage Books.
Sköllerhorn, E. (1998). Habermas and nature: The theory of communicative action for studying environmental policy. *Journal of Environmental Planning and Management, 41*(5), 555–573.
Straub, C. B. (2009). *Honorable harvest: Shakers and the natural world.* New Gloucester: United Society of Shakers.
Stückelberger, C. (2007). Calvin, Calvinism, and capitalism: Challenges of the new interest in Asia. In E. Dommen & J. D. Bratt (Eds.), *John Calvin rediscovered: The impact of his social and economic thought* (pp. 121–131). Louisville: Westminster/John Knox Press.
Tawney, R. H. (1958). Foreword. In *Protestant Ethic and the Spirit of Capitalism* (trans: Parsons, T.). New York: Charles Scribner's Son.
Taylor, C. (1991). Language and society. In A. Honneth & H. Joas (Eds.), *Communicative action: Essays on Jürgen Habermas's theory of communicative action* (pp. 23–25). Cambridge: The MIT Press.
Taylor, C. (1995). Two theories of modernity. *Hastings Center Report, 25*(2), 24–33.
Taylor, C. (1999). Two theories of modernity. *Public Culture, 11*(1), 153–174.
Vycinas, V. (1969). *Earth and gods: An introduction to the philosophy of Martin Heidegger.* The Hague: Martinus Nijhoff.
Weber, M. (1951). *Religion of China* (trans. and ed. Gerth, H.). New York: Macmillan.
Weber, M. (1952). *Ancient Judaism* (trans. and ed. Gerth, H., & Martindale, D.). Glencoe: Free Press.
Weber, M. (1958a). *Protestant Ethic and the Spirit of Capitalism* (trans: Parsons, T.). New York: Charles Scribner's Sons.
Weber, M. (1958b). *Religion of India: Hinduism and Buddhism* (trans. and ed. Gerth, H., & Martindale, D.). Glencoe: Free Press.
Wohlfarth, I. (1967). Presentation of Adorna. *New Left Review, I/46* (November-December), pp. 63–66.

Yanarella, E. J. (2001). *The cross, the plow and the skyline: Contemporary science fiction and the ecological imagination*. Leland: Brown-Walker Press.

Yanarella, E. J., & Levine, R. S. (2008). Research and solutions: Don't pick the low-hanging fruit! Counterintuitive policy advice for achieving sustainability. *Sustainability: The Journal of Record, 1*(4), 256–261.

Yanarella, E. J., & Levine, R. S. (2011). *The city as fulcrum of global sustainability*. New York: Anthem Press.

York, R., & Rosa, E. (2003). Key challenges to ecological modernization: Institutional efficacy, case study evidence, units of analysis and pace of eco-efficiency. *Organization & Environment, 16*(3), 273–288.

Zimmerman, M. (1997). *Contesting earth's future: Radical ecology and postmodernity*. Berkeley: University of California Press.

Chapter 8
Exploring the Green Dimensions of Islam

Mohammad Aslam Parvaiz

8.1 Introduction

Our planet distinguishes itself from the rest in a number of ways, the foremost being its ability to harbor and nurture life – the wonderful manifestation of Allah's creativity. During its early phase, the earth was as hostile to life as most of the planets we find now. However, as it evolved, it changed its contours and ultimately became the cradle for life. The sum total of all conditions and influences that affect the development and life of organisms is called Environment. The primitive young earth neither had free oxygen in its atmosphere nor pure clean water. The air was full of toxic gases and water was a dark thick solution of chemicals, almost like diesel in appearance. Earth's environment also evolved from a hostile one to a friendly one. That this process was indeed an evolution is proved by the fact that it was a change for the better, with a purpose and towards a goal. As per Allah's scheme earth was destined to become the abode of life, hence conditions became favorable. The Quran says:

> And the earth, He has set it for living creatures. (55: 10)

Once conditions became favorable, life appeared and an important dimension, in the form of living creatures, was added to the environment.

M.A. Parvaiz (✉)
Director, Islamic Foundation for Science and Environment (IFSE),
New Delhi 110 025, India
e-mail: maparvaiz@gmail.com

8.2 Humans – A Unique Creation

Everything that Allah has created, submits to His will and acts according to His commands:

> And whatever creature that is in the skies and that is in the earth and the angels (too) bow down (submit) to Allah and they do not consider themselves great (to disobey). They fear their sustainer above them and do what they are commanded. (16: 49–50)

Since everything in the universe behaves in accordance with laws enacted by Allah, the whole universe is therefore *Muslim*,[1] surrendering to the will of Allah. However, humans are the only exception to this universal law, for they are the unique beings endowed with the free choice of obeying or disobeying the command of Allah. They may follow their own desires, whims and fancies and disorient and dissociate themselves from the rest of the universe (which is Muslim) or follow the divine guidance and enjoy oneness with rest of the universe:

> Do they seek for other than the Deen (Law, guidance, submission) of Allah? While all creatures in the skies (heavens) and on earth have, willingly or unwillingly, bowed to His will and to Him shall they all be brought back. (3: 83)

All cohabitants of our planet work under, and according to the Natural Laws or Divine Guidance, which one often refers to as their natural instinct. Thus, under favorable conditions and if allowed to function naturally, a root would always absorb water and a green leaf would always photosynthesize, a clay particle would always hold water strongly and wind currents would always move towards a low pressure area. They have no choice but to act according to Natural Laws, that is, they have no freedom to act otherwise. Since Islam is a system which has been revealed to ensure peace and harmony, it guides the humanity also to a set of Laws and Directives which preserve and maintain the Divine balance and order on our planet. It defines the position as well as responsibilities of human beings on this temporary abode.

Another uniqueness conferred on human beings by Allah is their position on this planet. Allah defines the humankind as "*Khalifa*" on this earth which literally means the guardian or vicegerent who inherits the planet from its forerunners:

> And He it is Who has made you successors (vicegerents) in the land … (6: 165)

According to Islam man[2] is not the conqueror or master of nature, he is its guardian and hence protector who ensures continuity and availability of all its bounties. The earth is given in the custody of the vicegerent with duties of its maintenance and where possible, improvement for the benefit of subsequent generations. In

---

[1] The term Muslim comes from the word Islam, and the latter is derived from the root s.l.m., which means "to be safe," "to be whole and integral," "not to be disintegrated." The basic idea is that by accepting the Allah's guidance and "surrendering" to it one avoids disintegration.

[2] In this chapter the word "man" is used generically only. Similarly "Allah" is referred to as "He" and "His." This usage is not intended to refer to a specific gender.

this role humankind is responsible for the survival and good condition of various communities and commodities it has inherited, be it plants, animals and fellow human beings or soil, water and air.

8.3 The Dimorphism of Aya – The Knowledge for the Seeker

The word *Aya* (sign) which finds its mention, in singular or plural form, 280 times in the Quran, refers to anything or phenomenon which reflects our attention towards Allah – the Creator. Every sign (*Aya*) gives some news of Allah. The word of Allah or the verses of Quran are called *Aya* because they bring the divine guidance to its reader. However, when Quran invites people to believe in Allah, it bases its claim on sound arguments. It does not invite people to believe in Allah who is incomprehensible. The Quran begins its invitation by inviting people to ponder over their environment. According to Quran, the universe and everything in it, is a sign or *Aya* pointing to something beyond themselves. In Nasr's words, nature is "the theatre wherein are manifested His signs" (Nasr 1989: 3).

There are various verses in the Quran which invite people to read everything around them as well as within themselves – as signs of Allah:

> We will soon show them Our signs in the Universe and in their own souls, until it will become quite clear to them that it is the truth. (41: 53)
>
> And in the earth there are signs for those who are sure, and in your own souls (too); will you not then see? (51: 20–21)

This means that the signs of the natural world and the Quranic verses are two forms of the same guidelines, confirming each other. The entire universe is an embodiment of Allah's guidance to humanity. It is our duty to use our intellectual faculties to understand the guiding signs provided in both these modes and logically, both deserve equal consideration. Interestingly, while Allah has assured about the protection of the Quranic verses, He has left the natural signs (*Aya*) at our disposal, making us their custodian and hence protector.

8.4 Natural Bounties – How to Handle

Nature provides necessities of life to all creatures through its bounties. Some of these resources are available naturally in large quantities, while others need to be tapped, reared or cultivated. At the dawn of civilization, human population was thin and scattered. Humans traveled through land and used all natural resources freely and without any restriction. However, the situation changed with the establishment of kingdoms and empires and emergence of class systems based on riches. Money started multiplying through spreading trade, interest and banking systems, till it established its own clan. Since then world remains divided between haves and have not's.

8.5 Equitable Distribution

Since justice, equity and balance are the main planks of Islamic system, Islam asks believers to treat all natural resources with the same spirit. Quran declares it without any ambiguity that whatever is in earth, He has created it for everyone:

> He it is Who created for you (all), all that is in the earth … (2: 29)

This seemingly small verse has very deep meaning and revolutionary message for humanity. Apparently, and particularly to a "material conscious" (as opposed to "God conscious") world, this concept of just distribution may seem very difficult and too theoretical to be put into practice. However, when one turns to Allah's signs operating in nature, one gets ample manifestations of this just distribution. In nature, everything flows or moves from a region of its high concentration to a region of low concentration, provided there are no barriers or restrictions. This is called the Law of Diffusion. It operates everywhere in nature and in fact the very survival of life and its sustenance depends on this law. In soil, water and nutrients move according to diffusion gradients and hence maintain a balance within the soil body. Within our body it is this phenomenon which ensures that each and every cell receives due amount of oxygen and gets rid of carbon dioxide. So is the case in plants. The entire gas exchange operates on this principle. In the atmosphere, all around us, the movement of air, moisture and even heat follows the same directive set by Allah. So if one has the vision and a submitting (*Muslim*) mind, he or she would immediately conclude that when this Law of Diffusion (read Law of Just Distribution) is operative everywhere in nature, how can we afford to ignore it (Parvaiz 2003). These *Ayat* of nature strengthen one's belief and faith and clearly demonstrate that equitable distribution is not only possible, but in fact, it is the key for the survival of any system. But for those who do not believe in divine guidance, none such signs, whether in words or in natural phenomena, make any difference:

> Say: Consider (observe) whatever there is in the heavens and on earth! But of what avail could all the signs (Aya) and all the warnings be to people who will not believe? (10: 101)

According to the Islamic system, the wealth, produce, or any other resource which anyone gets, earns or inherits does not belong to him or her alone. It must be shared with all the needy, starting from his own close relatives to neighbors, travelers, displaced and dispossessed, to anyone in need:

> By no means shall you attain to righteousness until you spend (benevolently) out of what you love; and whatever thing you spend, Allah surely knows it. (3: 92)

> Say to My servants who believe that they should keep up prayer and spend out of what We have given them secretly and openly before the coming of the day in which there shall be no bartering nor mutual befriending. (14: 31)

> And give to the near of kin his due and (to) the needy and the wayfarer, and do not squander wastefully. (17: 26)

Allah thus judges the resourceful and elevated ones, clarifying the concept of guardianship:

> And He it is Who has made you successors in the land and raised some of you above others by (various) grades, that He might try you by what He has given you … (6: 165)

Allah warns those who accumulate wealth of a crushing disaster:

> Who amasses wealth and considers it a provision (against mishap); he thinks that his wealth will make him immortal. Nay! he shall most certainly be hurled into the crushing disaster. (104: 2–4)

The concept of resource sharing and taking care of needy has been elaborated very clearly in the Quran:

> Have you ever considered (the kind of man) who gives the lie to Deen (the day of judgment, moral law). Behold it is this (kind of man) who thrusts the orphan away and does not urge (others) to feed the poor. So woe to the praying ones, who are unmindful of their prayers, who do (good) to be seen, and withhold the necessities of life. (107: 1–7)

These commands of Allah (about resource sharing) are also found operating in nature. The process of photosynthesis may be taken here as a relevant *Aya* (sign). Just like any other creation, a tree is also a Muslim and let us assume it as a society where green leaves are that part of the society which receives Allah's *Fazl* (bounties) in the form of sugar. These leaves have the capacity to synthesize sugar by using carbon dioxide and water as raw material and light as source of energy. The so formed sugar is their produce which is needed by every part of the plant or member of this Muslim society. Roots which are situated deep down in the soil, are not capable of synthesizing sugar because they do not have access to light, so are other non-green (nonproductive) parts of this society. However, it must be noted that although these nonproductive parts do not synthesize sugar, but they perform some other equally important functions for the well-being of their society. They are neither idle, nor irresponsible. The productive section of this society, that is, leaves neither shows any hesitations, nor any inhibition in transferring their own produce, that is, sugar to non-green parts of their body. They do so according to the same Law of Just Distribution (diffusion) whereby every component of the system, takes just as much food as it needs, and lets the rest move to other needy parts. This continues throughout the growing phase of the plant. Once it stops growing, the sugar which was consumed for growth, becomes surplus. Now this is the surplus of the society which is deposited in *Bait-ul-Maal* or storehouses situated at every appropriate part of this society. These are the fruits or seeds or other edible parts of this plant, where this surplus food is stored for anybody who needs it, without any restriction or prejudice. No sane person would doubt the efficacy of this system which provides us food and almost everything organic from our fellow green cohabitants of this planet.

To facilitate the distribution and ensure the availability of resources, Islam has made it mandatory on all rich people to establish the institution of "*Zakat*," which in essence means a Development Fund for the needy (2:43,83,110,177,277; 4:162; 5:12; 7:156; 9:5,11,18,71; 21:73; 22:41,78; 24:37,56; 27:3; 30:39; 31:4; 33:33; 41:7). "*Zakat*" prevents hoarding of money and causes the wealth to grow for the welfare of needy so that people can earn their living instead of depending on charity. The Islamic system does not encourage the provision of sustenance to poor and needy merely through charity, it can be a short term remedy, but on a long term basis, the surplus with the rich should be invested to generate gainful employment for the needy, distressed and displaced.

Islam encourages individual Muslims to participate in the conservation and proper development of the resources by creating endowments or "*Awqaf*," which constitute the major avenue for private contribution to the public welfare. In India, and elsewhere as well, there are thousands of such "*Awqaf*" taking care of, and maintaining mosques, schools, hospitals and other welfare activities.

8.6 Judicious Use

Islam does not approve of a lavish or unjust consumption of resources, wasteful attitude and extravagance:

> … and do not act extravagantly; surely He does not love the extravagant. (6: 141)

> … and eat and drink and be not extravagant; surely He does not love the extravagant. (7: 31)

> And give to the near of kin his due and (to) the needy and the wayfarer, and do not squander wastefully. (17: 26)

The permissible provisions of modern development in Islam can include all those articles which enhance efficiency in terms of time, space and material utilization, provided they do not disturb the socio-economic equilibrium at any particular place or situation. Islam links "*Israf*" or extravagance to "*Fasad*," that is, chaos, disorder and mischief in society and declares extravagant as corruptors of society and spoilers of social order and harmony. It forbids people to follow such people or systems:

> And do not obey the bidding of the extravagant, who make mischief in the land and do not act aright. (26: 151–152)

It considers extravagance of one as economic deprivation of other because extravagance by the former is certainly an encroachment upon the accessibility right of the latter. Instead of wasting resources in demonstrative and extravagant life styles, it asks the Believers to spend, whatever surplus they have, on needy people:

> … And they ask you as to what they should spend. Say: What you can spare. Thus does Allah make clear to you the communications, that you may ponder. (2: 219)

8.7 The Significance of Balance

The concept of measure, or balance (*al-mizan*), among the various components of our environment dawned on us when we noticed some very disturbing phenomena in nature. One such phenomenon is the so-called greenhouse effect. The Swedish scientist Ahrrenius discovered in 1898 the causal link between carbon dioxide and global warming, but it was only in the 1970s that this began to be recognized as a serious environmental concern.

The temperature of the earth's atmosphere is maintained by a process in which the amount of energy the earth absorbs from the sun, mainly as high-energy ultraviolet radiation, is balanced by the amount radiated back into space as lower-energy infrared radiation. Playing a key role in regulating this temperature are the greenhouse gases, mainly carbon dioxide, water vapor, nitrous oxide and methane. They are called "greenhouse gases" because, like a pane of glass in a green house, they let in visible light from the sun, but prevent some of the resulting infrared radiations from escaping and re-radiate them back to the earth's surface. The buildup of heat that results from this re-radiation raises the temperature of the earth's lower atmosphere, a process commonly known as the greenhouse effect. Over the past few decades, human activity, especially the burning of fossil fuels like coal and petroleum and the use of chlorofluorocarbons (CFCs), has increasingly overloaded the earth's natural greenhouse system, slowing down the escape of heat into space and increasing the average temperature of the earth's atmosphere (Parvaiz 2003). In 1988, the United Nations Environment Program (UNEP) and the World Meteorological Organization established the Intergovernmental Panel on Climate Change (IPCC). In 1996 the IPCC published its second assessment report, written and reviewed by two thousand scientists and experts. It has established that:

> based on current emission levels, it is believed that the global temperature will rise by between 1 °C and 3.5 °C between now and the year 2100; that even after emission levels are stabilized, climate change will continue to occur for hundreds of years. (UN Environmental Program 1996)

The mean sea level will rise by 15–95 cm (5.9–37.4 in.) between now and the year 2100, causing floods and threatening the existence of some island countries. All this disturbance in the climate is occurring because the proportion of certain gases has increased slightly in the atmosphere, thereby disturbing the natural balance that exists among all the creations of Allah. The Quran says:

> And the earth we have spread out (like a carpet), set there on mountains firm and immoveable; And produced therein all kinds of things in due balance. (15: 19)

Global warming is also producing another catastrophic phenomenon, known as El Niño, which is not only disturbing the climate the world over, but is also spreading epidemics and reviving many fatal diseases in the affected regions. While global warming and El Niño are caused by disturbing the natural balance among the components of the environment, the ozone layer depletion has occurred because of the introduction of synthetic industrial chemicals – specifically CFCs – into the atmosphere.

These examples very clearly establish the importance of balance in nature. Since the onset of the industrial revolution we have created disturbances in physical or nonliving components of our environment. The consequences have been horrible. We are breathing poisonous air, water is polluted, loaded with toxic chemicals, the earth is parched, its soil is barren and contaminated by pesticides. The scenario needs no elaboration.

After doing what we should not have done to the physical components of our environment, we have now fixed our greedy eyes on the living components of this beautiful planet. Armed with selfish and utilitarian objectives, we are trying to modify different living organisms according to our needs through genetic engineering. Using various techniques of genetic engineering, we are trying to incorporate "useful" genes in those organisms where they are not found naturally. This is being done with utter disregard to the natural balance that exists among all the genes of a gene pool. It is totally illogical and unscientific to assume that the balance, which occurs among different nonliving components of this planet, does not exist among life-forms or within their gene pools. In fact, Allah has created everything with due balance and He forbids us from disturbing this balance:

> And the sky he has uplifted; and He has set the balance (measure). That you exceed not the balance. But observe the balance strictly, nor fall short thereof. (55: 7–9)

However, genetic engineers have brought out many innovations in Allah's creations. We have genetically engineered fish that grow much faster than wild and traditional aquaculture varieties. We have crops bioengineered for pest resistance, crops engineered to produce oil-derived chemicals, and recombinant bovine growth hormone (rBGH) that enables cows to produce more milk. To elaborate upon the risks, which are associated with our efforts to change Allah's creations, one may point out the case of transgenic fish. They are produced by transferring a fish growth hormone gene from one species to another. One such work was done by Robert H. Devlin and others (Devlin et al. 1994). He modified the growth hormone gene in Coho salmon. The transgenic Coho grew on average eleven times faster than unmodified fish and the largest fish grew 37 times faster.

Transgenic fish are wild types, or nearly so, often created from eggs hatched from gametes collected in the wild, so they are fully capable of mating with wild fish. Consequently, one important risk associated with transgenic fish is that, if they escape to fresh water or to the ocean and mate with wild fish, they could destroy the diversity of the wild population gene pool. Such an event occurred in Norway, with farmed salmon. Seals occasionally broke the net cages where the salmon were being raised and some of the salmon escaped and mated with Norway's wild salmon. Because the numbers of wild salmon had already been depleted as a result of acid rain on fresh water spawning grounds, the wild salmon were easily overwhelmed by aquaculture salmon. As a result, says Anne R. Kapuscinski, professor of fisheries and sea grant extension specialist at the University of Minnesota, the genes of the wild salmon were homogenized and Norway lost one of its most important resources – a tremendous amount of genetic diversity in its wild salmon – and the associated commercial and sport fishing industries (Kapuscinski and Hellerman 1991).

Besides, transgenic fish could also eliminate whole aquatic ecological systems by preying on and out-competing native species, as many introduced exotic (nonindigenous) fish have done.

As for transgenic crop plants, some experts warn that herbicide resistance, or insect-resistance genes, could spread from transgenic crops to their wild relatives and create new weeds that are especially difficult to control, or that the crop itself could become a weed. Similarly, despite its advantageous yields, the virus-resistant squash is a subject of scientific controversy. Some experts worry that when the squash is infected with other viruses, re-combinational events could occur that would generate new viral strains. Plant biologists Anne Green and Richard Allison of Michigan State University found that re-combinational events occurred in transgenic cow pea plants modified with a virus coat protein. When the plants were inoculated with a different virus, viral RNA or DNA recombined with genetic material from the invading virus to form a new, more virulent strain (Green and Allison 1994).

These few examples clearly indicate that it is very dangerous to modify or change Allah's creations. Allah says in the Holy Quran that His creations are perfect and flawless. They have been created in harmony and balance with the environment. As the Holy Quran states:

> (Hallowed be) He who has created seven heavens in full harmony with one another: no fault will you see in the creation of the Most Gracious. And turn your vision (upon it) once more: can you see any flaw? Then turn your vision (upon it) again and yet again: (and every time) your vision will fall back upon you, dazzled and truly defeated. (67: 3–4)

One can very well imagine what will happen when an attempt in made to change the most sacred of the sacreds, that is, the gene pool of an organism. These changes are deep seated and run through generations. There is no return pathway and no way out. Once a new gene or a new genetic combination has been introduced into a life-form, it is there to stay and take its own course. Interestingly, most of this manipulation work is being done with microorganisms, and this is the group about which very little is known. In fact, we do not even have full details of their diversity. Very few species are known to us and much remains to be characterized. How one would justify, even on scientific grounds, our attempts to alter such life-forms just to suit our needs, when we do not have the full knowledge of the components of the system, the interrelationships that exit, or the balance among different species? These piecemeal efforts are bound to create unknown, unforeseen disturbances on the planet. When even a slight change in the gaseous balance of our atmosphere has created such problems for us, it follows logically that changes in balance among various life-forms will produce catastrophic effects of at least similar magnitude, though perhaps not for 100 years.

8.8 Pollution – Consequence of Imbalance

Whenever any imbalance occurs in nature, it results in chaos or disorder. Our current problems of pollution and ozone depletion are very relevant examples. We have polluted our rivers, seas and air. We are generating huge amount of hazardous wastes every year. The soil suffers from misuse and chemical overload, as does the air we breathe, while above our heads the ozone layer fades away, exposing us to cancer-causing rays.

Carbon dioxide and other "greenhouse gases" that come from increasing numbers of people and places burning gas, coal, oil and wood are more than 25 %

higher than at any time in the last 160,000 years. They are heating the planet in a way that many scientists believe is likely to disrupt our weather and food production (Bassett 2000: 37).

The Quran terms the mischief which results in chaos or disorder as "*Fasad*" and forbids it:

> And do not spread corruption on earth after it has been so well ordered. And call unto Him with fear and longing: surely the mercy of Allah is near to those who do good (to others). (7: 56)

Elaborating upon different types of mischief or disorder. Quran declares the destruction of tilth. And the stock as mischief:

> And when he turns back, he runs along in the land that he may cause mischief in it and destroy the crops and the stock, and Allah does not love mischief making. (2: 205)

Similarly the incomplete measurements insufficient payment for someone's labor, economic disparities and encroaching upon other's right are also termed *Fasad*:

> And to Madyan (We sent) their brother Shu'aib. He said: O my people! serve Allah, you have no god other than Him; clear proof indeed has come to you from your Lord, therefore give full measure and weight and do not diminish to men their things, and do not make mischief in the land after its reform; this is better for you if you are believers. (7: 85)

Disrupting a just system and commitment of crimes is also termed as mischief:

> Said she: Verily, whenever kings enter a country they corrupt it ... (27: 34)
> They said: By Allah! You know for certain that we have not come to make mischief in the land, and we are not thieves. (12: 73)

All this mischief by people of evil intent results in an all-around chaos and corruption:

> Corruption has appeared in the land and in the seas on account of what the hands of men have wrought, that He may make them taste a part of that which they have done, so that they may return. (30: 41)

Quran uses the term "*Musleheen*" (rectifiers, correctors or reformers) in contrast to *Mufsideen* (corruptors, spoilers, mischief makers) who have spoiled the natural balance to serve their own ends:

> And when it is said to them, Do not make mischief in the land, they say; We are but peace-makers. (2: 11)

Allah characterizes believers as *Musleheen* of the society. They have been commanded to do "*Aml-e-Saleh*" (acts of correction and reformation which would undo the damage done by the spoilers or corruptors). Thus it is the duty of all believers to take up corrective measures for improving the society and to ameliorate the condition of people suffering because of inequalities, imbalances and disorders in the society (The Quran: 2: 82; 95: 6; 103: 1–3). As the social problems and maladies vary with time and space, these "good deeds" to be performed by believers would also be different and according to the challenges faced by the society at any particular time and place.

8.9 Conclusion

Human being's role on this planet is that of a guardian or vicegerent. Hence, the relation to nature and its resources should be one of stewardship and not mastery. All the creations of Allah are a Divine work of Art. They all have been called as "*Aya*" or signs to man, indicative of the greatness, the goodness, the subtlety, the richness and so on of the Creator. To deface, defile or destroy Nature would be an impious or even blasphemous act.

Though man is accorded the right to use the natural resources, he is not permitted to abuse it with impunity. Besides, this earth is a temporary abode for man and according to his deeds done on earth, Allah rewards or punishes him here in this world as well s hereafter. Therefore, those who act against Allah by damaging defacing or destroying His creations will certainly be punished.

Secondly, human beings are an integral and important component of nature. Hence, they are a natural entity, subjected fully to the laws of nature like any other entity, participating as an integral component in the overall environmental balance (*Mizan*) that exists in the entire universe. It means that to damage or destroy this natural balance is to destroy oneself. If a system is affected, all its components face the consequence. Quran precisely draws our attention to this doctrine:

> … and whoever goes beyond the limit of Allah, he indeed does injustice to his own soul … (65: 1)

In fact, whenever a person deviates from Allah's path, Allah calls it a *Zulm* (wrong doing, injustice) he committed on himself:

> They said: Our Lord! We have been unjust to ourselves, and if You forgive us not, and have (not) mercy on us, we shall certainly be of the losers. (7: 23)

> The same doctrine has been elaborated elsewhere also in the Quran. (11: 101; 26: 16; 34: 19; 43: 76)

To sum it up it may be concluded that the destruction of one part of the environment will have its repercussions on its each and every component, including humans. This is almost a self-destruction which is strictly prohibited in Islam. Allah has sent His guidance and if we choose to ignore it and refuse to practice His model, operating throughout the universe and wish to follow our own desires, we are the losers:

> … and who is more erring than he who follows his low desires without any guidance from Allah? Surely Allah does not guide the unjust people. (28: 50)

References

Bassett, L. (Ed.). (2000). *Earth and faith – A book of reflection for action* (Interfaith partnership for the environment). New York: United Nations Environment Programme (UNEP). 37.

Devlin, R. H., Yesaki, T. Y., Biagi, C. A., Donaldson, E. M., Swanson, P., & Chan, W. K. (1994). Extraordinary salmon growth. *Nature, 371*(6497), 209.

Green, A., & Allison, R. (1994). Recombination between viral RNA and transgenic plant transcripts. *Science, 263*(5152), 1423.

Kapuscinski, A. R., & Hallerman, E. M. (1991). Implications of introduction of transgenic fish into natural ecosystems. *Aquatic Sciences, 48*, 99–107.

Nasr, S. H. (1989). *Man and nature: Beyond current alternatives.* Paper delivered at the International Seminar on Islamic Philosophy and Science, Kuala Lumpur, 30 May–2 June 1989, p. 3.

Parvaiz, M. A. (2003). Scientific innovation and al-Mizan. In R. C. Foltz, F. M. Denney, & A. Baharuddin (Eds.), *Islam and ecology – A bestowed trust* (pp. 393–401). Cambridge: Harvard University Press.

United Nations Environment Program, Second assessment report No. FCCC/SBSTA/1996/7/Add.1 (Nairobi: UNEP, 22 Feb 1996), pp. 7–10.

Chapter 9
Making Oneself at Home in Climate Change: Religion as a Skill of Creative Adaptation

Sigurd Bergmann

9.1 Introduction

Religions offer substantial cultural skills for the "making-oneself-at-home" of humans (German: *Beheimatung*) as they locate believers in a world and at a place which is inhabited by the Divine.[1] Religious practices "reflect the natural environments and ways of life in which they emerged" (Buttimer 2006: 200).

"Religion" is in such a view neither approached in an essentialist way, as belief in supernatural powers, nor in a purely functionalist way, as a (somehow) specific cultural practice, but in a synthetic way that does justice both to the believer's deep embeddedness in the "factuality" (cf. Geertz 1973: 90f) of the Divine/Sacred/Spiritual, and the deep feeling of interconnectedness with and "dependence"[2] on it, *and* to the manifold sociocultural, historical and ecological functions. My circumscription of religion as skill stretches, however, beyond a simple combination of essential and functional aspects and seeks an understanding of religion as a cultural capacity that is nurtured by place and an all-embracing space. And as mobilities are among those characteristics of our contemporary world, which are mapped so colorfully in this book, place and space should be understood not as static but dynamic, or better: in a dialectic reciprocity of static and dynamic. This tension between the fast and fluent works similarly as the well known tension between the global and the local.

---

[1] cf. Yi-Fu Tuan (Tuan and Strawn 2009) who emphasizes the difference as well as the common for geography and religion, and regards, 15, religions in general as practices with its central meaning "to make a home for humans." Religion, for him, 70, is "the core idea of which is that humans are most deluded when they believe that they can feel, even in the best of times, at ease and at home on Earth."

[2] cf. Friedrich Schleiermacher's influential understanding of Religion as "das Gefühl schlechthinniger Abhängigkeit," the feeling of being absolutely (utterly) dependent (of God).

S. Bergmann (✉)
Department of Philosophy and Religious Studies, Norwegian University
of Science and Technology, Trondheim 7491, Norway
e-mail: sigurd.bergmann@ntnu.no

Religion appears in such a view as a trans-phenomenon rather that an -ism. It is not the-ism, buddh-ism, anim-ism or islam-ism, that is an upset of cognitive convictions. Rather it works as a fluent quality that transcends borders of historical, geographical and cultural kinds. Religion can and should, therefore, not easily be identified as a differentiated social sub-system but as an internal deep driving force with both unique and common cultural qualities. It is characterized by transcultural, transcontextual and translocal abilities as well as it allows for the roots of identity, community and tradition in resistance and/or adaptation to the turmoil of change. Normatively, it can produce destructive as well as constructive qualities, pathologies as well as drivers of liberation. Religion, shortly, can appear both as accelerator and brakeman with regard to glo-c-al change, sometimes as "either-or" and sometimes as "and" integrating them both.[3]

Does it make sense to approach religion as a skill[4] of *Beheimatung*, that is, roots and inhabitation as well as a skill of movement, transformation and creative adaptation in a world "in turmoil" to use Rilke's striking expression? Might religion not just include a spatial dimension but serve as *Raum* itself? Using another of Rilke's images: religion as "Weltinnenraum" im "Wohnen im Gewoge" (Religion as the inner space of the world dwelling in turmoil) (Rilke 1955: 145; cf. Bergmann 2014: 28)? The spatial dimension of reality would in such a view move from the periphery to the center and it might accelerate fruitfully the spatial turn of theology and religious studies in our lenses even more (Bergmann 2007).

Geographers of religion would scarcely be alien to such a view of the entanglement of *Raum* (space/place), mobility and religion through the ages and contexts. Landscapes in Aboriginal Australia, for example, are shaped by spiritual powers in the dreamtime. The movement of mythological snakes shape the bed of the river which again impacts on the formation and nature of the landscape. The journeys of ancestors create a map which is spiritual and physical at the same time. Narrations take place as stories and places emerge as stories. Stories and lands again can be visualized in paintings, songs and rituals.[5] Tracking melts together places, stories and events (cf. Gill 1998). The social dwells in the spatial. Identifying religion analytically as a specific delimited part of the culture does not make any sense in such a context as belief grows out of the spiritual-physical landscape.

How the skill of religious imagination impacts on the construction of the social and built geography can furthermore clearly be explored in classical Mayan contexts. Cities in Mayan Yucatan are placed on the edge of subterranean water streams and the stars, both dwelling places for ancestors and gods. Mayan cities are knots of a large complex weaving of lines, between subterranean, earthly and heavenly,

---

[3] cf. Kandinsky (1973) who declared the end of the age of either-or and replaced it with the age of the and.

[4] I am inspired by Ingold's understanding of skill, although this is developed in the context of transforming our (poor) view of technology. Skills, for him, are "not … techniques of the body, but the capabilities of action and perception of the whole organic being (indissolubly mind and body) situated in a richly structured environment" Ingold (2000: 5).

[5] On the synergy and synaesthetic of Aboriginal Australian art where the "inside" of the land represents some kind of a fourth dimension, see: Morphy (2000: 130); cf. Bergmann (2008).

Fig. 9.1 The City of Mayapan with the Cenote close to the temple, Yucatan (Photo © Sigurd Bergmann, November 2005)

between gods, ancestors and humans, and between the natural and the built environments. Subterranean water streams seem to form the most significant layer of this universe (Fig. 9.1) (cf. Brady and Prufer 2005; Prufer and Brady 2005).

In a city like Mayapan, for example, the holy cave and dwelling, the "cenote," forms the center of the city, as well as being a manifestation of the sacred geography of Yucatan. Its limestone ground can encourage a perception of this place as land floating on a holy stream of subterranean water, although it is more appropriate to claim that the cenote has determined the location of the city by providing itself for the land and its people. The cenote offers a knot of encountering lines to the city, connecting what is under the land and what carries, sustains and nourishes its ecological environment. The city knits the divine life-giving powers of the subterranean world to the formation, development and history of the people between water, earth and heaven. It seems obvious that the ancient Mayan city, which was built on a territorialized hierarchy, ordered by reference to religion, geography and gender, integrated natural, human, ecological, historical, cultural, political and aesthetic dimensions into a whole. This general cosmic order regulated and preserved the city as one common spiritually ritualized space, even if it often led the inhabitants into destructive conflicts with other cities. The complexity of such synekism in a small urban space offers us, in miniature, an illustration of the challenge to perceive and conceive urban space as a deeply religious manifestation. In a city like Mayapan lived religion works as a cultural skill to perceive the spiritual in the physical and to design, build and develop the lived space as a knot of lines within natural and built environments.[6]

Also Christianity illuminates the embedding of belief in spatiality, the integration of "lived religion" in "lived space." Even if Jesus moved around in a quite homeless way, places and movements are loaded with meanings. Especially the gospel of John locates the whole narration about the incarnated Creator's salvation of all in a complex geography, where the Creator and God's creatures, places and

---

[6] More in S. Bergmann (2014). Cities on the stream of Gods: Wandering in Mayan sacred geography. In Bergmann (2014, Chap. 4); German version in Bergmann (2010, Chap. 3).

events, the Divine, and the human are encountering on one common stage with many scenographies (cf. Kieffer 1984). And even if Christianity transformed the code of a split between profane and holy places by deifying the whole creation, Christian churches have also shaped a specific sacred geography. Places for the graves and commemoration of the martyrs, churches and graveyards, pilgrimages, and more have shaped a rich net of sites for skilled practices of adoration, remembrance and solidarity in a deep synthesis of ethics and aesthetics. Due to Christianity's root in the mystery of God's incarnation in the earthly and historical Christ, the cosmic and universal dimension on the one hand and the local and bodily individual/communal dimension on the other hand have kept their balance. The Creator of all between heaven and earth could be recognized in the most insignificant and earthly small. Macro- and microscales of belief remained to be interwoven in the mystery of the theandric (the god-human) synergy. Belief in the Trinity has, as we know, changed the mapping of reality, world and power in a sustainable way that continues to impact on our mapping (Bergmann 2005). Knowledge about what religion can do to our perception of and action with environments seem to be centrally necessary if one wants to understand the world surrounding us and us within it.

Even if premodern codes of the spiritual within the spatial still are at work, they have also undergone change in the processes of colonialization, modernization and globalization. Not only the shift from modes of local belief to the utopization of belief which reduces the significance of the spatial and superordinates time over place, diagnosed by Jonathan Z. Smith, must be taken into consideration. Also the countervailing forces which produce new modes of re-localization and new modes of *Beheimatung* must be included in the interpretation of contexts of accelerating global change. Furthermore, current technically and economically driven mobilities seem to catalyze an increasing homelessness in the ongoing globalization (Pallasmaa 2008), which challenges and changes also religious modes of making-oneself-at-home. The mobilization of traditional indigenous spirituality as well as the emergence of the "global Sacred" (Szerszynski 2005: 159ff) can be analyzed in such a horizon. Crucial in methodology for such analysis is not to think of the inhabited world as "composed of mutually exclusive hemispheres of sky and earth, separated by the ground" but attend "to the fluxes of wind and weather" (Ingold 2011: 115). Climate would then appear not only in the sense of empirical climate impact science, but as a human experience of being alive in a weathered environment, which in its diversity flows through and in between human bodies and minds. Humans as well as other living beings are impacted by wind and weather at the same time as they contribute to nature's "ever-evolving weave" (Ingold 2008: 158).

9.2 Climate Change, Homelessness and Creative Adaptation

In the following section I will focus on a specific context where it would be promising to mine deeper how religion as a cultural microcosm and skill of perceiving and acting is at work, viz., the context of climatic change on global, regional and local scales. As such research still is in its infancy, I will content myself with formulating a couple of arguments for why such an objective makes sense and conclude with three examples from Indonesia, Siberia and Tanzania in order to illuminate the skill of religion to serve as a tool for creative adaptation in the lived space of climatic change. One of several hypotheses in this field is that ongoing dangerous climatic change accelerates homelessness and causes increasing waves of forced migration and, therefore, challenges the spiritual skills of religion as a mode of (anew) making-oneself-at-home.

On a general level climate change creates a painful spiritual dilemma. For so called theistic religions, this stance can be summarized in a conflict of belief in the good Creator and Creation on the one hand and the human capacity of structural sin and long-term global destruction on the other. How can one feel oneself at home in a good creation when this is destroyed by humans themselves? How can God be the Creator, Sustainer and Liberator, and how can human beings be understood as being in the image of God, if they destroy the gift of life? Do climate change and the environmental catastrophe indicate a punishment for human sin? Why are Earth and God angry (Machila 2008)? Is God absent or present with the suffering? What are the spiritual sources for climate justice (cf. Deane-Drummond 2009)? Also other religions can identify such a conflict in the mirror of dramatic environmental change, where the environment not any longer appears in harmony with the human community and where balance anew must be achieved through social changes and specific interaction of the believers with the powers (see Religions of the World and Ecology 1997–2004).

On a local level, vulnerable populations need to experiment with new modes of adaptation to a changing environment with increasing turbulences and disasters. Such creative climate adaptation impacts on physical as well as on the sociocultural and spiritual dimensions of life. Not only human ecologies but also belief systems are threatened in the process of accelerating climatic change. Climate change changes also religion and one can, therefore, summarize the analytic question that arises: how can religion bring about a change?

And why is it meaningful to explore religion *in* climatic change? (cf. Gerten and Bergmann 2012)

The first reason is simply demographic. The majority of the world's inhabitants are practicing religious believers. A sociocultural analysis of practices, values and worldviews with regard to climate change must, therefore, necessarily include the religious dimension with all its ethical, aesthetic and political ambiguities. As the majority of the planet's inhabitants are increasingly affected by climate change and as a majority identifies themselves as believers, the question arises what religion means for climate change mitigation and adaptation.

As Michael Northcott (2007) has made us aware of the fact that local populations are usually regarded and treated as objects rather than subjects of power. It is not only a central demand to negotiate about the even more increasing geopolitical injustice but also to enhance and empower local people to become the central agents of change, even if this must take place in alliances with others (cf. Northcott 2007: 15). It is contradictory to impose systemic solutions over large parts of local populations without involving them as central agents of adaptation and mitigation. Climate change and the ecologic and social injustice that is accelerated by it represent a radical global threat to democracy and, thereby, to all existing power constellations. An analysis of the complex interplay of religion and the environment in this context produces necessary insights in the sociocultural dynamics of global and local change.

Faith communities have in recent years developed a rapidly increasing and intense activity with regard to climate change. Evangelical Christians in the U.S., for example, are divided into two camps, the one arguing for the urgent moral challenge of climatic change while the other is accusing the fear of climate change to promote a new pagan religion similar to medieval fear of hell (an argument that practically encourages to go on with "business as usual").[7] Churches and other religious bodies intervene in the public spheres with declarations about the spiritual dimension of a civilization crisis. Religious contributions emphasize especially the ethical dimension and the geopolitical and ecological injustice, which becomes even more inescapable in the ongoing climate change and its asymmetrical distribution of risks and sufferings between the rich and the poor. Some examples can be mentioned. The Interfaith Climate Summit in Uppsala gathered 2008 a large group of religious leaders from many traditions all over the planet.[8] Muslims have created a global network of exchange (see IFEES). Many local church leaders and synods, such as the German Protestant Church (EKD), have agreed on declarations about the radical challenge of climate change.[9] International ecumenical institutions such as the World Council of Churches have contributed to the discourse and mobilized believers, mainly in the poor world regions (see WCC 2000). Since 2001 a working group for climate issues works intensively in the European Churches Environmental Network (ECEN), and focuses especially on climate justice (www.ecen.com). The Lutheran World Federation (LWF) has recently started a process to enhance local churches on all continents to develop a pastoral theology for adaptation (Bloomquist and Machila 2009).[10]

Another discourse where "religion" is used in an interesting denunciatory mode provides further evidence for the underlying hypothesis that religion is at work in climate change, viz., namely the criticism of climate scientists by self-described skeptics, active in and outside the academy, as nothing more than religious missionaries threatening the identity of true objectivist science. Examples such as these show how climate change alters religious belief systems and provokes them to

---

[7] For a map of Christian responses to climate change in the USA see Kearns (2012) and Roberts (2012).
[8] www.svenskakyrkan.se/default.aspx?di=143415, where one also can download the Uppsala Interfaith Climate Manifesto in six languages.
[9] The EKD published 1995 its study Gefährdetes Klima, www.ekd.de/EKD-Texte/44652.html, and its former chair, bishop Wolfgang Huber, has addressed decision makers in: Es ist nicht zu spät für eine Antwort auf den Klimawandel, Mai 2007, www.ekd.de/EKD-Texte/20070530_appell_klimawandel.html. As representatives for many others one can refer to A Southern Baptist Declaration on the Environment and Climate Change, 2008, THE OTIN TAAI DECLARATION – The Pacific Churches Statement on Climate Change, 2004, and the Statement by His All Holiness Ecumenical Patriarch Bartholomew I. for the WCC working group on climate change, 12 August 2005.
[10] Further materials, incl. Climate Change – Facing Our Vulnerability, at: www.lutheranworld.org/News/LWI/EN/2300.EN.html.

respond in new ways, and further illustrates how the notion of religion, in a well-known modernist code, is used as a concept in opposition to rationality and science. Nevertheless, such a problematic one-eyed, rationalist dogmatic view appears in late (pluralist) modernity to an increasing degree to be outpaced and replaced by an intense interest for integration of different dimensions of sociocultural and subjective aspects of being-human-in-and-with-surroundings. Furthermore it does not do justice to the state of art in religious and cultural studies or the increasing insight into the so-called return of religion, which in fact is not simply a return but an operative religious dynamics "in between worldview and ethos" (C. Geertz 1957), in traditional as well as in modern social systems. We might therefore posit: Religion and what is regarded as "Sacred" affects perceptions, actions and thinking about climate change. If images of the Sacred change images of nature are also in flux, and if environmental change accelerates, images of the Sacred will also be affected. And if the images of the Sacred again will change, these will also influence climate change anew through human behaviors.

If the climate change discourse aims at more than simple "managing" the world through economic and technological strategies for mitigation to be imposed on populations by elites, it seems necessary to regard climate change itself as a religious process that is embedded in the sociocultural process of producing the problems, constructing the discourse and striving for change. A strong reason to explore religion with regard to climate change is to increase the manifold of voices (cf. Leduc 2007) and diverse perspectives about weather, climate, nature, and the "common future." Without such a plurality of approaches the "one-dimensional man" (H. Marcuse 1964) can scarcely find his way out of human-made self-going systems.

For the reader of this book it might be of interest to reflect further on the spatiality of religion itself, which is challenged radically in the process of environmental and climatic change. Space and place have been marginalized or are even absent in religious studies and theology for long periods of its history. This mirrors a strong characteristic of the so-called Western culture and its history which has valued time over space. The marginalization of spatiality in Western understandings of nature and reality has a long history, linked to the continuing dominance of Leibniz's well known preference for time over space in Western worldviews. Even if Einstein and others have voided the dominance of "empty space" in theory, ordinary culture and the social sciences still employ the Newtonian worldview (Schroer 2006). Physical space is thus separated from mental space, where the form of space is regarded as identical to the expanding form of time, where only the eye and not all senses can "see space," and where places are experienced as mere containers for contents. In the twentieth century's last years, however, space/place reflections have moved from the margins into several sectors of the sciences and the humanities in what has been characterized as "the spatial turn." Spatial metaphors and the central image of the Earth as "our home" are also at the heart of environmentalism. Theology has also mobilized its forces slowly but safely and promoted a spatial turn in its own fields. Religious studies and theology therefore need to develop a strong emphasis on identifying the religious dimension of topographies, architecture, urban studies and technical mobility systems, which for religious studies and theology represents an unconventional path that, nevertheless, has been very positively received by others. As far as climatic change impacts on all kinds of physical, imagined and lived space, so also religion is affected on all kinds of its dimensions.

9.3 In-sites and Insights

Three insights into sites and regions can illuminate the religious skill of making oneself at home anew in a lived space of climatic change (Fig. 9.2).

Environmental anthropologists Undine Frömming and Christian Reichel (2012) have investigated the situation of indigenous groups in vulnerable coastal regions of Indonesia with regard to the role of local knowledge, environmental perception and cosmologies in the context of climate change related disasters. Their research shows in detail how local knowledge includes numerous strategies concerning the prevention and management of climate-relevant natural dangers and catastrophes as well as sustainable resource management. Religious restrictions, like taboo areas or calendar systems, can hereby play an ambiguous role as some seem inefficient while others "bear a striking resemblance to modern strategies which seem innovative because they draw upon the latest research" (Frömming and Reichel 2012: 229) The category of local knowledge, which includes religious worldviews, moralities and practices, is in this approach a valuable tool if regarded in the frame of social-ecological resilience; it describes how societies understand uncertainties, disturbances and surprises as a trigger of a recursive learning process making them sustainable for the long term and capable of surviving (Fig. 9.3). Religion serves in this context

Fig. 9.2 The Ngada clan's Reba ceremony, dance for calling the ancestors and natural spirits, Flores, Indonesia (Photo © Undine Frömming, December 1997, used with permission)

Fig. 9.3 Increasing water on the lands in Northern Siberia, interfering directly with Viliu Sakha's ability to harvest hay for their herds through winter (Photo © Susan A. Crate, used with permission)

as a cultural skill to respond to environmental change and danger as it enables the population "to provide diverse sustainability strategies combined with cultural techniques … in order to protect against hazards or extreme events in the long term" (Frömming and Reichel 2012: 231).

A similar insight arises among the Viliui Sakha communities of northeastern Siberia, which anthropologist Susan Crate (2012) has followed for a long time (Crate 2012; cf. Crate and Nuttall 2009). Her research makes clear how global climate change is not only affecting their physical, but also their cultural and cosmological worlds and animistic belief systems. Crate shows how belief and cosmology shapes a local people's perceptions of and responses to climate change, especially with regard to water abundance and the dramatically changing pattern of frost and

Fig. 9.4 Mount Kilimanjaro, view from Moshi (Photo © Sigurd Bergmann, December 2010)

seasonal cold in winter times. She, with many other anthropologists, pleas rightly for methodologies that capture the cultural, cosmological and religious elements in order to inform interdisciplinary projects and policy efforts, an argument that is relevant for all kinds of climate policies I would say.

Another region where the significance of global warming and climate change, especially with regard to the water flows,[11] and its impact on culture and religion can be studied is the Kilimanjaro region in northeast Tanzania (Fig. 9.4). As a research project on religion in climatic change in this region, conducted by the author and Undine Frömming, is still in the beginning I can only offer preliminary insights. The slopes of Kilimanjaro and its melting glaciers provide highly interesting insights into the religious-cultural dimension of the consequences of a melting glacier and its impact on the local availability of water, in combination with the increasingly serious problems of water supply aggravated by increasing temperatures, droughts and forest fires. Furthermore, also the long colonial and postcolonial history of intercultural encounters between traditional and modern belief systems and human ecologies still have impacts on contemporary ideologies and practices. Melting glaciers offer important representations and metaphors of global environmental change

---

[11] Or should one rather than of flows talk about water cycles and investigate how these relate to what Tuan calls "the wisdom of God?" (Tuan 1968).

Fig. 9.5 Irrigation in the Kilimanjaro region, in a traditional mode at the higher slopes nearby Marangu (Photo © Sigurd Bergmann, December 2010)

and the symbolic values of Kilimanjaro cannot be underestimated for inhabitants as well as for visitors and tourists. The slopes of Mt. Kilimanjaro have an extensive farming history, as demonstrated, for example, by the probable introduction of bananas in the late first millennium. Intensive agricultural systems developed, and during the last centuries the region developed the largest cluster of indigenous irrigation in SubSaharan Africa, with more than 2,000 irrigation schemes, based on a technology called "hill furrow irrigation" (Fig. 9.5).

The current practice represents a continued development of technologies, institutions and cultures of water that have evolved over several centuries among several Bantu- and Nilotic-speaking ethnic groups in Eastern Africa. Indigenous models remain relevant to resource management, despite changes relating to colonialism, socialism and formal religion, but are currently subject to multiple stresses from globalization, environmental and socioeconomic change. Over the past decade, Mt. Kilimanjaro has emerged as the best known and most visible icon of climate change in Africa, with attention focused on the retreat of its high altitude glaciers followed by studies documenting a decline in precipitation from twentieth century records. The geography of Northern Tanzania is dominated by the contrast between semiarid plains and the mountainous "water towers." These mountain areas are important as the source areas of water for the lowlands and the Pangani River, because they capture orographic precipitation, and because they are home to considerable population concentrations and economic activities. Communities of smallholders managing their own irrigation schemes may find that their access to water becomes insecure through the double exposure of climate change and globalization.

Globalization changes the access to water, not only by causing increased demand for water for sugar, starches and vegetable oils and a decline in the coffee-based economy, but through environmental governance policies, management and entitlements (Tagseth 2009). In this context, the project will investigate how farming communities interpret, and respond to, environmental and socioeconomic change and uncertainty. Preliminary results suggest that local actors in the region draw on local and Bantu religious-cultural narratives to explain these changes through discourses of aridity and marginalization, and recollect how ritual was used to bring about conditions of peace, fertility and rain in the past.[12] In more practical terms, vulnerability depends on the agronomic and economic strategies available to farmers to reduce risks from climate variability. And there are also the related issues of indigenous technical/agricultural knowledge, innovations and technical advice. Farming systems in the region tend to vary with agroecological zones with differences in flexibility and drought risk between main agricultural uses of land. Religion appears in such a frame as a tool itself as well as a kit that interconnects different other historical, technical, social and political "tools" to respond to environmental change.

9.4 Lived Religion in Lived Space?

To sum up. In this chapter religion has been regarded as a skill of orientation, perception and action in a complex environment. Especially its ability of *Beheimatung* (making oneself at home) has served as a heuristic tool to become aware of the role of belief systems in a globalizing world, where the "dwelling in turmoil" and increasing homelessness, either as migrants or as mental nomads, have turned into existential challenges for most of the world inhabitants. Ongoing climatic change is accelerating this development even more and challenges religious belief systems to respond to dangerous environmental change which changes environments as well as cultures and religions. Climatic change seems in such a view to appear as a change of culture and religion as much as it represents a change of the natural earth system. An approach, such as envisioned in this chapter, provokes the theory of religion to take the spatiality and mobility of religion much more serious than before. It leads us to the question of whether or not religious studies in general should be re-envisioned and developed as a contribution to third space study, that is, the study of lived religion in the lived, physical-and-imagined, space of both natural and built environments.

---

[12] The significance of ritual for the theme of this book as well as for this chapter should be investigated much deeper in the future. While ritual theory has developed in a differentiated way in its sociocultural dimensions, it mostly still dwells in first space reflections about the spatiality of rituals and ritual place. Van Herck, for example, rightly points towards the advantages of the concept of a place-based "ritual unity" over "consensual unity," 21f., and Bergmann (2009, 2012) pleas, for a fruitful exchange between ritual theory and urban studies. cf. also the deep linkages between images of the Earth, wisdom, and practices of spatial design in classical Asian and modern global geomancy.

In a similar way as geographers are challenged to study the religious and the sacred "as ways of distributing particular kinds of significance across geographic spaces" (Ivakhiv 2006: 169) scholars of religions and theologians should explore deeper the spatial as ways of embracing, carrying and nurturing contextual expressions of belief across diverse, fluid and multivalent (cf. Kong 2001: 212), religious "lands." The here envisioned focus on *Beheimatung* might take us a step closer to an interconnection of both tasks. If sacred places are places which are treated as such over time, and if the stories about them "that are told to make ourselves "at home" in the world already include nonhuman others" (Ivakhiv 2001: 234) such a focus might assist contemporary humanity that "needs to remember more harmonious ways of dwelling on Planet Earth" (Buttimer 2006: 201) for human as well as others being alive in "our common future" and our common home.

References

Bergmann, S. (2005). *Creation set free: The Spirit as liberator of nature* (Sacra doctrina: Christian theology for a postmodern age 4). Grand Rapids: Eerdmans.
Bergmann, S. (2007). Theology in its spatial turn: Space, place and built environments challenging and changing the images of God. *Religion Compass, 1*(3), 353–379.
Bergmann, S. (2008). "It can't be locked in" – Decolonising processes in the arts and religion of Sápmi and aboriginal Australia. In S. J. Stålsett (Ed.), *Religion in a globalised age: Transfers and transformations, integration and resistance* (pp. 81–101). Oslo: Novus.
Bergmann, S. (2009). Ecological geomancy: Earth energy and the wisdom of spatial design. In S. Bergmann & Y.-B. Kim (Eds.), *Religion, ecology & gender: East-west perspectives* (Studies in religion and the environment 1, pp. 147–174). Berlin/Münster/Wien/Zürich/London: LIT.
Bergmann, S. (2010). *Raum und Geist: Zur Erdung und Beheimatung der Religion – eine theologische Ästh/Ethik des Raums* (Research in contemporary religion 7). Göttingen: Vandenhoeck & Ruprecht.
Bergmann, S. (2012). Religion in the built environment: Aesth/ethics, ritual and memory in lived urban space. In L. Gómez & W. Van Herck (Eds.), *The sacred in the city* (pp. 73–95). London/New York: Continuum.
Bergmann, S. (2014). *Religion, space & the environment*. New Brunswick (USA) and London (UK): Transaction Publishers.
Bloomquist, K. L. (Ed.). (2009). *God, creation and climate change* (Theology in the life of the Church, Vol. 5). Geneva: Lutheran World Federation.
Bloomquist, K. L., & Machila, R. (Eds.). (2009). *God, creation and climate change: A resource for reflection and discussion*. Geneva: Lutheran World Federation.
Brady, J. E., & Prufer, K. M. (Eds.). (2005). *In the maw of the earth monster: Mesoamerican ritual cave use*. Austin: University of Texas Press.
Buttimer, A. (2006). Afterword: Reflections on geography, religion, and belief systems. *Annals of the Association of American Geographers, 96*(1), 197–202.
Crate, S. A. (2012). Climate and cosmology: Exploring Sakha belief and the local effects of unprecedented change in North-Eastern Siberia, Russia. In Gerten & Bergmann (Eds.), (pp. 175–199).
Crate, S. A., & Nuttall, M. (Eds.). (2009). *Anthropology & climate change: From encounters to actions*. Walnut Creek: Left Coast Press.
Deane-Drummond, C. E. (2009). *Seeds of hope: Facing the challenge of climate justice*. London: Cafod.
Frömming, U. U., & Reichel, C. (2012). Vulnerable coastal regions: Indigenous people under climate change in Indonesia. In Gerten & Bergmann (Eds.), (pp. 213–235).

Geertz, C. (1957). Ethos, world-view and the analysis of sacred symbols. *The Antioch Review, 17*(4), 421–437.

Geertz, C. (1973). Religion as a cultural system. In *The interpretation of cultures* (pp. 87–125). New York: Basic Books.

Gerten, D., & Bergmann, S. (Eds.). (2012). *Religion in environmental and climate change: Suffering, values, lifestyles*. London/New York: Continuum.

Gill, S. D. (1998). *Storytracking: Texts, stories, and histories in Central Australia*. Oxford/New York: Oxford University Press.

IFEES (Islamic Foundation for Ecology and Environmental Sciences). http://ifees.org.uk

Ingold, T. (2000). *The perception of the environment: Essays in livelihood, dwelling and skill*. London/New York: Routledge.

Ingold, T. (2008). The wedge and the knot: Hammering and stitching the face of nature. In S. Bergmann, P. Scott, M. Jansdotter Samuelsson, & H. Bedford-Strohm (Eds.), *Nature, space and the Sacred: Transdisciplinary perspectives* (pp. 147–161). Aldershot: Ashgate.

Ingold, T. (2011). *Being alive: Essays on movement, knowledge and description*. London/New York: Routledge.

Ivakhiv, A. (2001). *Claiming sacred ground: Pilgrims and politics at Glastonbury and Sedona*. Bloomington: Indiana University Press.

Ivakhiv, A. (2006). Toward a geography of 'religion': Mapping the distribution of an unstable signifier. *Annals of the Association of American Geographers, 96*(1), 169–175.

Kandinsky, W. (1973). "und." In M. Bill (Ed.), *Essays über Kunst und Künstler* (pp. 97–108). Bern: Benteli 3. ed., (1955).

Kearns, L. (2012). Religious climate activism in the United States. In Gerten & Bergmann (Eds.), (pp. 132–151).

Kieffer, R. (1984). Rum och tid i Johannesevangeliet. In *Svensk Exegetisk Årsbok 49* (pp. 109–125). Uppsala: Svenska Exegetiska Sällskapet.

Kong, L. (2001). Mapping 'new' geographies of religion: politics and poetics in modernity. *Progress in Human Geography, 25*(2), 211–233.

Leduc, T. B. (2007). Sila dialogues on climate change: Inuit wisdom for a cross-cultural inter-disciplinarity. *Climatic Change*. doi:10.1007/s10584-006-9187-2.

Machila, R. (2008). Why are earth and God angry? In *Issue 20 in the LWF "Thinking It Over" series*. Available at: www.lutheranworld.org/What_We_Do/Dts/DTS-Welcome.html

Morphy, H. (2000). Inner landscapes: The fourth dimension. In S. Kleinert & M. Neale (Eds.), *The Oxford companion to aboriginal art and culture* (pp. 129–136). Oxford: Oxford University Press.

Northcott, M. S. (2007). *A moral climate: The ethics of global warming*. London: Darton, Longman and Todd.

Pallasmaa, J. (2008). Existential homelessness – Placelessness and Nostalgia in the age of mobility. In S. Bergmann & T. Sager (Eds.), *The ethics of mobilities: Rethinking place, exclusion, freedom and environment* (pp. 143–156). Aldershot: Ashgate.

Prufer, K. M., & Brady, J. E. (Eds.). (2005). *Stone houses and earth lords: Maya religion in the cave context*. Boulder: University of Colorado Press.

Religions of the World and Ecology. (1997–2004). Cambridge, MA: Harvard University Press. http://fore.research.yale.edu/publications/books/book_series/cswr/index.html

Rilke, R. M. (1955). *Das ist die Sehnsucht. In Sämtliche Werke, Band 1: Gedichte • Erster Teil* (p. 145). Frankfurt am Main: Insel.

Roberts, M. (2012). Evangelicals and climate change. In Gerten & Bergmann (Eds.), (pp. 107–131).

Schroer, M. (2006). *Räume, Orte, Grenzen: Auf dem Wege zu einer Soziologie des Raums*. Frankfurt am Main: Suhrkamp.

Szerszynski, B. (2005). *Nature, technology and the sacred*. Oxford: Blackwell.

Tagseth, M. (2009). Social and cultural dimensions of irrigation management in Kilimanjaro. In T. A. R. Clack (Ed.), *Culture, history and identity: Landscapes of inhabitation in the Mount Kilimanjaro area, Tanzania* (pp. 89–105). Oxford: Archaeopress.

Tuan, Y.-F. (1968). *The hydrological cycle and the wisdom of God: A theme in geoteleology.* Toronto: University of Toronto Press.

Tuan, Y.-F. (text) & Strawn, M. A. (photographs and essays). (2009). *Religion: From place to placelessness.* Chicago: The Center for American Places at Columbia College Chicago.

World Council of Churches (WCC). (2000). The atmosphere as global commons: Responsible caring and equitable sharing. www.wcc-coe.org/wcc/what/jpc/cop6-e.html

Chapter 10
Scale-Jumping and Climate Change in the Geography of Religion

Michael P. Ferber and Randolph Haluza-DeLay

10.1 Introduction

Global climate change is a vector of environmental change that is challenging the global community on multiple fronts and to which the world's religions are quickly coming to pay attention. The cross-scalar aspect of climate change is noted by many observers as climate change has consequences that vary with global, regional and local dimensions (Henrikson 2009). Given the resounding scientific consensus from a variety of evidentiary streams that the global climate is rapidly changing, religious traditions are confronted on practical and cosmological terms. These challenges can be captured by an ontological query: since when did humans take on the role of the divine and become so powerful as to alter the very stability of creation? In response to the "explosion in the awareness and attention" given to environmental degradation, Kong (2010: 767) asks, "What role can religion play in alleviating environmental crisis?"

The scientific consensus for over a decade is that humans are causing the climate to change (Good et al. 2011; Solomon et al. 2007; Stott et al. 2010). In the past century global surface temperature has already increased approximately one degree centigrade (Meehl et al. 2007). Carbon dioxide and the atmospheric concentration of other greenhouse-producing gases appear to have increased due to anthropogenic causes, leading to projected transformations in global climate and a cascade of effects including coastal impacts, ocean acidification, ecosystems and biodiversity, water resources and desertification, agriculture and food security, and human health (Gosling et al. 2011). Climate transformation will have far-reaching consequences

M.P. Ferber (✉)
Department of Geography, The King's University, Edmonton T6B 2H3, AB, Canada
e-mail: michael.ferber@kingsu.ca

R. Haluza-DeLay
Department of Sociology, The King's University, Edmonton T6B 2H3, AB, Canada
e-mail: randy.haluza-delay@kingsu.ca

for human populations. Estimations in the United States are that by 2030 up to 20 million people will be affected by sea level rise leading to large scale migration or expensive adaptation efforts (Curtis and Scneider 2011). More extensive effects will be felt in subtropical and equatorial zones (Pelling and Dill 2010).

Yet, although there is substantial scientific evidence indicating anthropogenic global warming is advancing, responses have been paradoxical. On one hand, there have been concerted efforts across scales as municipal, regional, and federal jurisdictions launch major initiatives to mitigate and reduce GHG (greenhouse gas) emissions (Benson 2010). But on the other hand, recalcitrance on the parts of many individuals, entities and governments to adapt have been pronounced, and even outright opposition has appeared, with numerous explanations (Hulme 2009). The paradoxical response extends especially to religion. In some cases nonaction and disbelief or denial are associated with religious belief, as in the considerable research on evangelical Christian attitudes in the United States (summarized by Wilkinson 2012). Some of these studies suggest a strong negative correlation between conservative theology and environmental concern, and weak, occasional correlations between denominational affiliation and environmental concern (Guth and Green 1995). Biblical literalism may also be negatively correlated with environmental concern (Eckberg and Blocker 1989), although this is controversial (Guth and Green 1995). Yet religious institutions have also issued many statements and engaged in ecumenical and interfaith efforts to address climate change (Ferber 2010).

In this chapter we explore the complexity of reactions of the religious to climate change. We use the intersection of religion – as values and beliefs, faith practices, collective agents and institutions – and climate change to articulate a sociospatial notion of scale as a framework for an explanation of the general inertia in addressing what may be the most profound anthropogenic alteration of the planet in human history. We concentrate especially on Christian responses due to our own areas of expertise and because focusing on one tradition allows illustration of the cross-scalar complexity of religious practice *vis-a-vis* the cross-scalar impacts of global climate change.

Recent scholarship in geography has added considerably to the notion of scale. A decade long scale debate has permeated many of geography's key journals and has moved definitions of scale from representation (Johnston et al. 2000) to being socially produced (Marston 2000), to serving as a category of practice rather than an analytical tool (Moore 2008), to scale as a polymorphic, multidimensional and sociospatial relationality constituted by the intertwining of territories, places, scales and networks (Jessop et al. 2008). The debates are too rich and detailed to summarize here, but there are significant implications for rethinking both religion and climate change using these new conceptualizations of scale. Climate change is almost always conceptualized as a global phenomenon, yet it produces localized environmental effects. Religious adherents are embedded in scaled religious milieu of a personal spirituality as well as a local congregation or assembly associated with a larger faith tradition. We recognize a fluidity in how individual and institutional actors approach scaled lifeworlds. The interaction of religion and the phenomena collectively understood as climate change illustrates this fluidity well.

Religions are geographically scale-crossing networks, providing material resources, communication, skills, commitment and workers. They cross social boundaries such as ethnicity or class. Faith traditions – like the worldwide Anglican Communion or Greek Orthodox Church – are transnational entities. Religious groups such as the World Council of Churches and the World Parliament of Religions have been at the leading edge of institutions advocating for climate justice. Religious communities have a broad impact, one that extends beyond their own adherents. They influence attitudes, public policy, education and human rights – all of which are part of the climate change conundrum.

Major religious institutions address climate change through official statements aligned with their religious tradition, and even actions – such as the Vatican's carbon-neutral commitment. Regional bodies, such as dioceses, often pass their own more nuanced statements and action recommendations, with localized facets of the issues perceived as more important than the planet-changing scale of climate change. At the even more nuanced scales of individuals and congregations there are extensive variations. Much of the scholarship on the intersection of religion and climate change has been as theological or normative expressions. Social scientists have been slow to empirically assess the ways that religion and climate change intersect, although this may be changing (Veldman et al. 2012; see also, Gerten and Bergman 2012; Haluza-DeLay et al. 2013). This chapter contributes to this discussion by responding favourably to Kong's (2010) challenge for geographers to investigate the macro-level and "tackle larger scale issues so that we can intervene in public debate," while not jettisoning "the microscale of analysis while addressing the bigger questions" (p. 763). In the chapter we do not enter the larger debates on theorizing scale, but the essay has a scalar structure and below we demonstrate how well documented challenges of crossing or jumping scale contribute to the difficulty religious adherents have accepting and reacting to global climate change.

10.2 Macro Scale

It is difficult to gauge religious reactions to climate change at the national scale without analyzing institutional responses from denominations and associations. These bodies frequently assert their institutionally representative views on major issues through theological or policy statements, often joining together in interfaith initiatives through ecumenical agreements. Care for the environment is a growing area of emphasis within these kinds of documents. In the United States, the website of the National Religious Partnership for the Environment (NRPE), an association of independent faith groups including the Evangelical Environmental Network (EEN), the U.S. Conference of Catholic Bishops, the National Council of Churches (NCC) and the Coalition on the Environment and Jewish Life, hosts a robust collection of statements from many religious denominations and organizations. Such texts voice denominational goals to carry out certain programs or adopt policy positions concerning ecological issues. Numerous researchers have also documented environmental developments within evangelical Christianity that have occurred over the last two decades (Ackerman 2007; Harrington 2009; Kearns 2007, 2012; McCammack 2007; Nagle 2008; Prelli and Winters 2009; Roberts 2012; Simmons 2009; Skeel 2008; Wilkinson 2009, 2010, 2012).

Samples of these statements and declarations include:

- Evangelical Declaration of Creation Care (1993): As followers of Jesus Christ, committed to the full authority of the Scriptures, and aware of the ways we have degraded creation, we believe that biblical faith is essential to the solution of our ecological problems. Because we worship and honor the Creator, we seek to cherish and care for the creation.
- National Council of Churches (2006): The National Council of Churches in Christ calls on all Christians, people of faith and people of good will the world over to lead by example and seek active means whereby they may, individually and in community, quickly reduce their emissions of green house gas emissions and speak out for engagement by their elected officials on matters of global warming.
- American Baptists Resolution on Global Warming (2007): Global warming affects hunger, access to clean water, environmental stewardship, health and peace. Addressing global warming will make it more possible for all to live the life of possibility that God intends. Therefore, based on our faith in the Creator God who makes us a part of a unified creation, the General Board of the American Baptist Churches USA, calls on national boards, regions, American Baptist institutions, congregations and individuals to join in ways to build a culture that can live in harmony with God's creation.

Despite many resolutions, contestation regarding climate change remains even at this macro scale. An alternative to calls for GHG reductions called the *Cornwall Declaration* (2000), sponsored by the Cornwall Alliance, critiques the pro-climate change stance for viewing humans principally as consumers and polluters rather than producers and stewards. Their primary contention revolves around the science of climate change itself. Following the resignation of climate change activist Richard Cizik as Vice-President of the National Association of Evangelicals (NAE), the Cornwall Alliance celebrated a statement by NAE President Leith Anderson asserting that the NAE has no formal position on climate change. Other Evangelicals reject the creation care movement altogether primarily due to theological reasons, with more extreme positions arguing that based on Genesis 1: 28 God gave humans complete dominion over the earth carte blanche.

In spite of this contestation, many new denominational and parachurch organizations have reported support for creation care and an acknowledgment of the realities of climate change. For instance, *Renewal: Students Caring for Creation* is a growing movement Christian college students dedicated to mobilizing and equipping their campuses to better steward the environment. Their website states, "Our Creator took chaos and transformed it into indescribable beauty, form and creative order. What's more, God breathed life into humankind and commanded us to 'tend and keep' His blessed creation. In the past humans have neglected this charge, instead participating in environmental harm that degrades ecosystems, as well as human lives. We have made a mess of God's creation. But with His strength and grace, Christian students across North America are uniting to work for its renewal." (Renewal 2012) Renewal recently released the *Green Awakenings Campus Report* chronicling environmental initiatives on 52 Christian college campuses across the United States and Canada (Lowe 2010).

Other non Evangelical Christian examples of macro scalar climate advocacy include groups such as the Coalition on Environment and Jewish Life and the Catholic Climate Covenant, which have embarked upon numerous campaigns to mobilize members of their respective faiths. The World Council of Churches has increasingly focused on climate change as a paramount issue (Reuter 2011) and the Alliance of Religions and Conservation (ARC), a secular body that helps the major religions of the world to develop their own environmental programs based on their own core teachings, beliefs and practices, has partnered with the United Nations Development Program (UNDP) to help religious leaders of 11 major faiths develop 7 year plans of action to address climate change. In 2000 this coalition developed a Climate Change Partnership Initiative that resulted in 15 countries agreeing to make it official policy to work with religions on climate change.

While statements, agreements and initiatives like these are legion at the macro scale, there is an evident disconnect between what national and international religious leaders and bodies say and what the faithful do in the context of their communities and families. Hence, understanding what is happening at the micro scale within faith communities is critical for appreciating how and why some adherents and congregations deny climate change and resist initiatives within their movements to lower emissions.

10.3 Micro Scale

While the national scale is helpful for gauging broad policy initiatives among denominations, micro scales such as congregations and individual adherents provide a more fluid medium for investigating the growing religious conflicts surrounding climate change. Stump (2008) suggests that the primary scale for understanding religious territoriality in general is the local community of religious adherents, and that from this pivotal scale "narrower scales" such as the body and family and "wider scales" such as denominations are illuminated. Stump places greater importance at the scale of the local community because it represents the nexus of life-world interactions as various scale levels interpenetrate each other. Curry (2008) also suggests that the formation of environmental attitudes toward nature in general and climate change in particular are captured at the community scale. This is consonant with the expectations of practice theory, which understands competent behavior expected by social others as enabled by sociocultural fields (Shove and Walker 2010). The actions of the religious are enabled, constrained and contextualised by the social, cultural and technonatural (White and Wilbert 2009) systems in which people are embedded.

To illustrate the complexities of climate change in local communities of faith we point to focus groups conducted among Alberta church congregations exploring Christian reactions to the movie *Avatar* and its mix of spirituality and environmental messaging (Haluza-DeLay et al. 2013). These kinds of studies are important because, according to Kong (2010: 767), "there has perhaps been more analysis of theological texts and what they say about the environment than there has been about actual human behavior and how that may have religious influences." Participants indicated knowledge of the problems of oil dependence as a significant emissions contributor to global climate change. At the same time, the greater effects of a structure that superceded their potential efforts was also keenly believed [sic], as represented in a comment by one member: "To some degree we're stuck.... We're stuck in the system, and we like the system, even though we hate what it does to the earth." This was accompanied by belief that the system could not be changed. Our analysis indicates that when lived practices rather than beliefs are the unit of analysis, the local category of scale (that is, place: Escobar 2001; Massey 2004) becomes activated (Moore 2008). The territoriality of the local believer and social context of the congregation (what others say and do in Alberta) are particularly salient, as Stump suggests. Believers are also imbricated in local networks which have considerable effect on the application of faith principles to practical action. Thus, as Jessop, Brenner and Jones (2008) argue scale, place, territory and networks are entwined, and affect the sociospatial application of religious characteristics.

DeLashmutt (2011) conducted similar focus groups in the Church of England and concluded that, "despite the effort of the Church to engender a pro-environmental culture, there is very little evidence to suggest that individual churchgoers can effectively connect their religious faith and their environmental dispositions." The interviews by Carr et al. (2012) show that most Christians were unaware of the rich information available in the Christian community regarding climate change (or the more general topic of environmental stewardship). In both studies, it is clear that climate change is handled differently across the scale spectrum. Different religious processes and properties are at work at different scales of church structure (below we will refer to this as emergence). DeLashmutt believes this is due to a disconnect of scale, though he does not use these terms. He states, "the grassroots church, though informed by valuable local wisdom and pastoral contexts, has failed to successfully identify with the bigger context of environmental action." We argue this practical disconnect has name: ecological fallacy.

10.4 Climate Change, Ecological Fallacy and Emergence

Classic *ecological fallacy* occurs when variables observed at one level of aggregation are applied to explain a finer or broader scale (Martin 1999), and when the characteristics of a population are attributed to individuals within that population or the characteristics of an individual are applied to an aggregated population (Johnston et al. 2000). The term "ecological" stems from its early use in ecology to establish that the aggregated characteristics of a forest could not be attributed to an individual tree, nor could the characteristics of a particular tree be attributed to the aggregate of a forest. Robinson (1950) first identified the issue of ecological fallacy in a sociological context when he was able to demonstrate that the association between the percentage of black population and the percentage of illiterate populations were not transferable to differing scales of analysis.

According to Harris (2006), issues of ecological fallacy and scale underpin both the observed patterns and our interpretations of hidden processes of geography. Harris asks, "Can the operation of a forest ecosystem be determined if the problem were studied at the scale of a leaf? Equally, can one understand the physiology of a leaf if the problem were examined at the scale of the forest?" (Harris 2006: 47). The dilemma of global climate change demonstrates a similar disconnect, only in this case from individual human lifestyle to planetary crisis.

While global climate change does have local manifestations in some places, such as forced migration due to the rise of sea levels, demonstrating clear links to human lifestyles can be challenging. Despite the virtual consensus among scientists, many people will not allow themselves to jump scale from their relatively minor personal impact to the combined effect of the world's populace.

A conceptual tool some social scientists use to understand how mechanisms and processes at one scale influence and are influenced by mechanisms and processes at another is *emergence* (Ferber 2006). An emergent approach to scale recognizes that the world is stratified hierarchically (Danermark et al. 1997). Each stratum is formed from the powers and mechanisms of the underlying strata. Emergence occurs "when the properties of underlying strata have been combined," and "qualitatively new objects have come into existence, each with its own specific structures, forces, powers and mechanisms" (Danermark et al. 1997: 60). Inkpen (2005) suggests that "each stratum is composed of entities that are not reducible to entities found at lower strata and that each stratum may have unique entities which interact according to relations and processes appropriate to the strata while also interacting with relations and processes found in lower strata."

Ferber and Harris (2011) utilize emergence to demonstrate the relationships between individual adherents, congregations and denominations. While a congregation is comprised of individual adherents, it has causal powers not present at the individual scale. Likewise, denominations have causal powers and generative properties not present at the scales of the individual or congregation. One example is the capacity of a church leader to impact geo-political issues at or beyond the scale of the nation. The statements on climate change described above (for example, denominational statements) are examples of the emergent powers present at the denominational scale.

Emergence and ecological fallacy have many implications in regard to religion and climate change, the simplest being the implication that many smaller individual impacts scale up to a global impact, and that the emergence of global effects is very difficult to fathom at the individual scale and are thus disassociated from practice in the personal lifeworld. There are also political implications. In advocating for sustainability, the collective actor has more impact on policy than an individual actor: 20 people together have a collective voice louder than 20 disconnected solo voices. This is multiplied in the emergence in power of national and multinational organizations to influence and impact global climate change awareness (McCright and Dunlap 2011; Brulle et al. 2012). Ecological fallacy is demonstrated in the challenges associated with moving between these scales, especially between the individual adherent and the national or global denomination. At the individual scale adherents re-constitute social order in the face of threats by following norms of emotion and conversation (Norgaard 2011). At the meso level congregations and regional assemblies of congregations practice, shape and re-inscribe what is talked about, felt and done. And at the macro level, the industrial system continues spewing GHGs seemingly independent of moral admonition or steering mechanisms aimed at a different direction.

10.5 Bridging the Gap

Are there solutions to bridging the massive jump in scale between the individual and the planet? Bergmann (2009: 109) argues from a theological perspective that "Christianity needs to accelerate its spatial turn." He asserts that theologians have addressed God in terms of time and history, but that "the context of climatic change clearly turns our focus to the spatiality of Creation. The all embracing space of our life represents for all religions one common gift of life." Trying to facilitate a scale-jump by association with the familiar lifeworld, Bergmann refers to the Earth Charter statement "the Earth is our home" and comments, "the statement sounds simple but it summarizes a deep wisdom that has been guarded by religions for many ages," (109).

A second way that Bergmann tries to facilitate awareness of the fluidity of scale is by reference to spatial justice. Faith traditions can be renewed, he says, by appreciating how the good gift of spatiality has been become "a global space where risks and damages are socioeconomically shared in a violent and unjust way."(109) He concludes with a few riveting questions about spatial justice that could assist Christians in bridging the ecological chasm between individual humans and the planet. First, "Will God's good all-embracing space turn into a catastrophic space where some are victimized for the survival of others?" Second, "How does God's love to the poor relate to the situation where the most vulnerable become the most victimized?" Finally, "What does climate justice imply and why is it a central Christian virtue today?" These critical questions are particularly important in light of recent efforts to expand the purview of justice beyond anthropocentric concerns to include the entire circle of creation (Bender 2003).

Another theological approach involves transcending scale through the sacrament of the Eucharist, described in some Christian circles as Communion. Galbraith (2009) asserts that this Christian sacrament provides opportune time to reflect materially upon the environment. He draws upon Sallie McFague's (1993) metaphorical understanding of the universe as God's body, making the notion of body fundamental to every aspect of our existence. Hence, sin and salvation are bodily matters that correspond to planetary equivalencies of ecological degradation and global warming mitigation. Galbraith (2009: 290) states, "how one attends to (or neglects) the bodies around her or him is a matter of sin and redemption. To act contrary to humankind's place within the world as revealed by the common creation story – which McFague says is provided to us by science – is to not give other bodies their proper due, to miss the mark, to sin." In the sacrament of the Eucharist Christians are offered forgiveness of sins through the broken body and shed blood of Jesus Christ.

The solutions of Bergmann and Galbraith represent only two of countless ways Christians can jump scale, if only theologically, from individual lifestyle choices to collective global impacts. Suggestions like these help believers transcend scale in ways akin to what Tweed describes as moving across four chronotypes of body, home, homeland and cosmos. According to Tweed (2006: 111), religions "orient devotees temporally and spatially by creating cosmogonies and teleographies" in a process of religious homemaking that "maps social space." Tweed asserts that since the homeland is an imagined territory inhabited by an imagined community, a space and group continually figured and refigured in contact with others, its borders shift over time and across cultures." (110) Hence, Christians can, indeed, transcend ecological fallacy at body and home scales to make a leap of faith by collectively embracing the earth (cosmos) as home (homeland in Tweed's schematic).

10.6 Conclusion

The climate change story, as presented by numerous natural science disciplines, now presents humans as powerful enough to change our planet, a power once reserved for God or the creative ontology of the universe. We have entered the Anthropocene, the geological era wherein human activities have become the pre-eminent geophysical force (Steffen et al. 2011). It is not just those from historic faith traditions who are struggling to understand the heretofore implausibility of this position, with its capacity to negatively impact the planet. Along with other human social institutions, religious traditions are adapting to the conditions of new times, but institutions and practice change slowly.

Science can be seen as creating new cosmological stories and portraying a newly determined role for humans in the cosmic order. Some faith groups resist, others accept, while for others the interaction of faith, reason and science cause adaptation. Any cosmology purports to present the world as it is, not merely as one choice among possible choices. This is the "sacred canopy" that Berger (1967) names religion as fulfilling – providing an overarching "plausibility structure" of shared assumptions among co-religionists that cohere together to become orderly and meaning-full. These assumptions include views of science, economics, religious and secular authority, political rights, and so on. More importantly for sustainability concerns, regularized practices operate within the plausibility structure provided by local communities, not just denominational scales or the historical geographical imagination of faith traditions. The changing map of religion includes cosmological impacts on the "sacred canopy" of creation itself, to which all humanity is struggling to comprehend and adapt.

References

Ackerman, T. (2007). Global warming: Scientific basis and Christian responses. *Perspectives on Science and Christian Faith, 59*(4), 250–264.

American Baptists. (2007). *American baptists resolution on global warming*. Retrieved from www.abc-usa.org/LinkClick.aspx?fileticket=0JpsQst6Agw%3d&tabid=199

Bender, F. L. (2003). *The culture of extinction: Toward a philosophy of deep ecology*. Amherst: Humanity Books.

Benson, M. (2010). Regional initiatives: Scaling the climate response and responding to the conceptions of scale. *Annals of the Association of American Geographers, 100*(4), 1025–1035.

Berger, P. L. (1967). *The sacred canopy: Elements of a sociological theory of religion*. Garden City: Anchor Books.

Bergmann, S. (2009). Climate changes religion: Space, spirit, ritual, technology – through a theological lens. *Studia Theologica, 63*, 98–118.

Brulle, R. J., Carmichael, J., & Jenkins, J. C. (2012). Shifting public opinion on climate change: An empirical assessment of factors influencing concern over climate change in the U.S., 2002–2010. *Climatic Change, 114*, 169–188.

Carr, W., Patterson, M., Yung, L., & Spencer, D. (2012). The faithful skeptics: Evangelical religious beliefs and perceptions of climate change. *Journal for the Study of Religion Nature & Culture, 6*(3), 276–299.

Curry, J. (2008). Christians and climate change: A social framework of analysis. *Perspectives on Science and Christian Faith, 60*(3), 156–164.

Curtis, A., & Scneider, K. (2011). Understanding the demographic implications of climate change: Estimates of localized population predictions under future scenarios of sea-level rise. *Population Environment, 33*, 28–54.

Danermark, B., Ekstrom, M., Jakobsen, L., & Karlsson, J. (1997). *Explaining society: Critical realism in the social sciences*. New York: Routledge.

DeLashmutt, M. (2011). Church and climate change: An examination of the attitudes and practices of Cornish Anglican churches regarding the environment. *Journal for the Study of Religion, Nature and Culture, 5*(1), 61–81.

Eckberg, D. L., & Blocker, T. J. (1989). Varieties of religious involvement and environmental concerns: Testing the Lynn White thesis. *Journal for the Scientific Study of Religion, 28*(4), 509–517.

Escobar, A. (2001). Culture sits in places: Reflections on globalism and subaltern strategies of localization. *Political Geography, 20*, 139–174.

Evangelical Environmental Network. (1993). *Evangelical declaration of creation care*. Retrieved from http://creationcare.org/blank.php?id=39

Ferber, M. (2006). Critical realism and religion: Objectivity and the insider/outsider problem. *Annals of the Association of American Geographers, 96*(1), 176–181.

Ferber, M. (2010). American pastors creation care covenant. In J. Newman & P. Robbins (Eds.), *Green education: An A-to-Z guide* (pp. 20–21). Thousand Oaks: SAGE Publications.

Ferber, M., & Harris, T. (2011). Resurrecting scale in an emergent geography of religion. *International Journal of Humanities and Social Sciences, 1*(12), 1–8.

Galbraith, K. (2009). Broken bodies of God: The Christian Eucharist as locus for ecological reflection. *Worldviews, 13*, 283–304.

Gerten, D., & Bergmann, S. (2012). *Religion in environmental and climate change: Suffering, values, lifestyles*. New York: Continuum.

Good, P., Caesar, J., Bernie, D., Lowe, J., van der Linden, P., Gosling, S., Warren, R., Arnell, N., Smith, S., Bamber, J., Payne, T., Laxon, S., Srokosz, M., Sitch, S., Gedney, N., Harris, G., Hewitt, H., Jackson, L., Jones, C., O'Connor, F., Ridley, J., Vellinga, M., Halloran, P., & McNeall, D. (2011). A review of recent developments in climate change science, Part 1: Understanding of future change in the large scale climate system. *Progress in Physical Geography, 35*(3), 281–296.

Gosling, S., Warren, R., Arnell, N., Good, P., Caesar, J., Bernie, D., Lowe, J., van der Linden, P., O'Hanley, J., & Smith, S. (2011). A review of recent developments in climate change science. Part 2: The global-scale impacts of climate change. *Progress in Physical Geography, 35*(4), 443–464.

Guth, J. L., & Green, J. C. (1995). Faith and the environment: Religious beliefs and attitudes on environmental policy. *American Journal of Political Science, 39*(2), 364.

Haluza-DeLay, R., Ferber, M., & Wiebe-Neufeld, T. (2013). Watching Avatar from 'AvaTar Sands' land. In B. Taylor (Ed.), *Avatar and nature spirituality* (pp. 123–140). Waterloo: Wilfrid Laurier University Press.

Harrington, J. (2009). Evangelicalism, environmental activism and climate change in the United States. *Journal of Religion and Society, 11*, 1–24.

Harris, T. (2006). Scale as artifact: GIS, ecological fallacy, and archaeological analysis. In G. Lock & R. Molyneaux (Eds.), *Confronting scale in archaeology* (pp. 39–53). New York: Springer.

Henrikson, J. (2009). Theology between nature and culture. *Studia Theologica, 63*, 95–97.

Hulme, M. (2009). *Why we disagree about climate change: Understanding controversy, inaction and opportunity*. Cambridge: Cambridge University Press.

Inkpen, R. (2005). *Science, philosophy and physical geography*. New York: Routledge.

Jessop, B., Brenner, N., & Jones, M. (2008). Theorizing sociospatial relations. *Environment and Planning D: Society and Space, 26*(3), 389–401.

Johnston, R. J., Gregory, D., Pratt, G., & Watts, M. (Eds.). (2000). *The dictionary of human geography*. Oxford: Blackwell.

Kearns, L. (2007). Cooking the truth: Faith, science, the market and global warming. In L. Kearns & C. Keller (Eds.), *Ecospirit* (pp. 97–124). New York: Fordham University Press.

Kearns, L. (2012). Religious climate activism in the United States. In D. Gerten & S. Bergmann (Eds.), *Religion in environmental and climate change: Suffering, values, lifestyles* (pp. 132–151). London/New York: Continuum.

Kong, L. (2010). Global shifts, theoretical shifts: Changing geographies of religion. *Progress in Human Geography, 34*(6), 755–776.

Lowe, B. (Ed.). (2010). *Green awakenings: Students caring for creation*. www.renewingcreation.org

Marston, S. (2000). The social construction of scale. *Progress in Human Geography, 24*(2), 219–242.

Martin, D. J. (1999). *Geographical information systems and their socioeconomic applications* (2nd ed.). London: Routledge.
Massey, D. (2004). Geographies of responsibility. *Geografiska Annaler, 86B*(1), 5–18.
McCammack, B. (2007). Hot damned America: Evangelicalism and the climate change policy debate. *American Quarterly, 59*(3), 645–668.
McCright, A. M., & Dunlap, R. E. (2011). The politicization of climate change and polarization in the American public's views of global warming, 2001–2010. *The Sociological Quarterly, 52*(2), 155–194.
McFague, S. (1993). *The body of God: An ecological theology*. Minneapolis: Fortress.
Meehl, G. A., Stocker, T. F., Collins, W. D., Friedlingstein, P., Gaye, A. T., Gregory, J. M., Kitoh, A., Knutti, R., Murphy, J. M., Noda, A., Raper, S. C. B., Watterson, I. G., Weaver, A. J., & Zhao, Z.-C. (2007). Global climate projections. Climate Change 2007: The physical science basis. In S. Solomon, D. Qin, M. Manning, Z. Chen, M. Marquis, K. B. Averty, M. Tignor, & H. L. Miller (Eds.), *Contribution of working group I to the fourth assessment report of the intergovernmental panel on climate change* (pp. 747–845). Cambridge: Cambridge University Press.
Moore, A. (2008). Rethinking scale as a geographical category: From analysis to practice. *Progress in Human Geography, 32*(2), 203–225.
Nagle, J. (2008). The evangelical debate over climate change. *University of St. Thomas Law Journal, 5*(1), 52–86.
National Council of Churches. (2006). *Resolution on global warming*. Retrieved from www.ncccusa.org/NCCpolicies/globalwarming.htm
Norgaard, M. (2011). *Living in denial: Climate change, emotions and everyday life*. Cambridge: MIT University Press.
Pelling, M., & Dill, K. (2010). Disaster politics: Tipping points for change in the adaptation of socio-political regimes. *Progress in Human Geography, 34*(1), 21–37.
Prelli, L., & Winters, T. (2009). Rhetorical features of green evangelicalism. *Environmental Communication, 3*(2), 224–243.
Renewal. (2012). *Our spirit*. www.renewingcreation.org/about-renewal/our-spirit. Last accessed July 21, 2012.
Reuter, T. (2011). Faith in the future: Climate change at the World Parliament of Religions, Melbourne 2009. *Australian Journal of Anthropology, 22*(2), 260–265.
Roberts, M. (2012). Evangelicals and climate change. In D. Gerten & S. Bergmann (Eds.), *Religion in environmental and climate change: Suffering, values, lifestyles* (pp. 107–131). London/New York: Continuum.
Robinson, A. H. (1950). Ecological correlation and the behaviour of individuals. *American Sociological Review, 15*, 351–357.
Shove, E., & Walker, G. (2010). Governing transitions in the sustainability of everyday life. *Research Policy, 39*(4), 471–476.
Simmons, J. (2009). Evangelical environmentalism: Oxymoron or opportunity? *Worldviews, 13*, 40–71.
Skeel, D. (2008). Evangelicals, climate change and consumption. *Environmental Law Reporter, 38*, 10868–10872.
Solomon, S., Qin, D., Manning, M., et al. (2007). *Climate change 2007: Contribution of working group I to the fourth assessment report of the intergovernmental panel on climate change*. Cambridge: Cambridge University Press.
Steffen, W., Grinevald, J., Crutzen, P., & McNeill, J. (2011). The Anthropocene: Conceptual and historical perspectives. *Philosophical Transactions of The Royal Society of Mathematical, Physical and Engineering Sciences, 369*, 842–867.
Stott, P. A., Gillett, N., Hegerl, G., et al. (2010). Detection and attribution of climate change: A regional perspective. *Wiley Interdisciplinary Reviews: Climate Change, 1*, 192–211.
Stump, R. (2008). *The geography of religion: Faith, place and space*. New York: Rowman and Littlefield.

Tweed, T. (2006). *Crossings and dwellings: A theory of religion*. Cambridge: Harvard University Press.
Veldman, R. G., Szasz, A., & Haluza-DeLay, R. (2012). Climate change and religion: Research review and gaps (Special issue on Global religions and climate change.) *Journal for the Study of Religion, Nature and Culture, 6*(3), 255–275.
White, D. F., & Wilbert, C. (Eds.). (2009). *Technonatures: Environments, technologies, spaces and places in the twenty-first century*. Waterloo: Wilfrid Laurier University Press.
Wilkinson, K. K. (2009). *Caring for creation's climate: Climate change discourse, advocacy and engagement among American evangelicals, geography & the environment*. Oxford: University of Oxford.
Wilkinson, K. K. (2010). Climate's salvation: Why and how American evangelicals are engaging with climate change. *Environment, 52*(2), 47–57.
Wilkinson, K. K. (2012). *Between God and green: How evangelicals are cultivating a middle ground on climate change*. Oxford: Oxford University Press.

Chapter 11
All My Holy Mountain: Imaginations of Appalachia in Christian Responses to Mountaintop Removal Mining

Andrew R.H. Thompson

11.1 Introduction

Mountaintop removal (MTR) coal mining is unarguably changing the face of Appalachia in a variety of ways. This dramatic method of surface mining, which uses explosives to demolish the top levels of a mountain to access the seams of coal beneath them, has had a wide range of environmental and social impacts, including a variety of alleged health issues in nearby communities, negative environmental changes resulting from frequently poor reclamation, and divisive and even violent debate throughout the region and around the nation.

At the foundation of the debate over this practice, however, is a struggle over the meaning of a geographic region. Assumptions about what Appalachia means, who is or is not truly Appalachian, and the relationship of these Southern Mountains to the rest of the nation figure prominently, either explicitly or implicitly, in arguments about MTR and the future of the region. These assumptions are part of a long history of the social construction of the region around political interests (Batteau 1990: 33). Thus at the same time that it transforms the physical landscape of Appalachia, the practice of MTR negotiates and manipulates its complicated ideological terrain as well.

The work of Christian ethicist H. Richard Niebuhr seeks to address precisely this kind of cultural complexity. Niebuhr argues that the fundamental character of human moral action is that of response to actions and events that affect us, and that the way we interpret those events – our imaginations of the world in which we respond – shapes our actions (Niebuhr 1963: 55–68, 2006: 49–72). For Niebuhr these imaginations can be either self-centered, interpreting the world in ways that reflect and advance human interests and purposes, or God-centered (or theocentric), viewing all actions upon us as expressing the actions of God, and orienting our own

---

A.R.H. Thompson (✉)
The School of Theology, The University of the South, 335 Tennessee Ave.
Sewanee, TN 37383, USA
e-mail: andrew.thompson@sewanee.edu

actions to conform with what we understand to be God's purposes. He believes that human centered imaginations inevitably lead to alienation and destruction; theocentric imaginations alone provide a foundation for integrated and unifying moral action (Niebuhr 2006: 53–69). The task for the believing community, then, is to scrutinize carefully its imaginations – the discourses and images believers use to shape their understandings of and responses to events – and to displace selfish imaginations with theocentric ones.

This is, obviously, an abbreviated description of Niebuhr's ethical approach; it is, I believe, nonetheless sufficient to suggest the usefulness of his work with respect to the culturally freighted debate around MTR. In this essay, therefore, I will use Niebuhr's approach as the framework for a Christian ethical response to MTR, arguing that theocentric imaginations of MTR can transform the inadequate discourses that have represented and manipulated Appalachian identity. Specifically, I will address imaginations of power and powerlessness, of insiders and outsiders, and of destruction and reclamation as they apply to MTR. With each pair of terms, after briefly discussing its history and its influence on religious responses to MTR (based on a variety of statements and interviews), I will describe what I take to be a basic theocentric perspective, based principally on Niebuhr's theology, and indicate its implications for this debate. While Christians have naturally been influenced by the typical views, there are also many extant examples of theocentric imaginations being enacted around this issue; I will describe some of these as well.

A key insight of Niebuhr's work is that the Church can finally only speak from within the context of its own history and beliefs, and that theology falls into its greatest error when it seeks to justify itself to others on some worldly basis or another (Niebuhr 2006). In this sense my goal is not apologetic: I do not wish to commend a theocentric perspective universally, based on something like practical efficacy. Rather, the intention is to argue that, from the particular standpoint of the Christian community of faith, the theocentric approach most consistently and completely addresses the intricacies of this culturally freighted issue.

11.2 "King Coal": Power and Powerlessness

11.2.1 The Story of Power in Appalachia

There is a standard narrative about power in Appalachia, the nature of it and who wields it over whom (for example, Caudill 1963; Gaventa 1980). Before the Civil War, this story goes, the people of the Southern Mountains, mainly of German and Scotch-Irish descent, were independent and isolated. In the second half of the nineteenth century, with the discovery of vast reserves of coal and the development of the infrastructure (in the form of railroads) to export it, capitalists from the rest of the nation and beyond came to the region, either to "develop" and "uplift" it, or to exploit it (depending on one's perspective). These outsiders took advantage of the ignorance, need, or simple guilelessness of the mountaineers and acquired the vast majority of their land through deceit, force, or legal manipulations. In addition to building the facilities necessary to extract, treat, and transport the coal, the coal companies also built housing for workers, as well as stores, banks, and saloons, creating the notorious company town. In these towns, the company controlled every aspect of life, down to printing its own currency, "scrip," good only in company-owned establishments. Based on the region's economic dependence on coal, the companies were also able to establish significant, and in some cases, near-absolute, political influence. Thus the early history of the coal-mining regions of Appalachia is portrayed as "the establishment of the hegemony of industrial economic interests over a particularly independent, roughly equal, and relatively content enclave society" (Gaventa 1980: 80).

One of the most insightful and influential inquiries into the dynamics of power in the Appalachian coal regions, *Power and Powerlessness* by John Gaventa, both retells and complexifies this story. Gaventa sets out to understand why the powerless often seem to offer little resistance to their oppression. He considers two standard views of the mechanisms of power: First, it is maintained by being exercised over actual decision-making processes. Those with political or economic power use it directly to affect outcomes to serve their own interests. Second, power may be maintained by foreclosing the options of the powerless, through force, threats, or a biased system, such that they are compelled to concede power to their oppressors. They are simply "left out" of decision-making processes. To these conceptions, Gaventa adds a "third dimension" of power, whereby power is maintained by shaping the wants and perceptions of a community through the manipulations of symbols, myths, and language, to legitimate the status quo and foster quiescence in the powerless (Gaventa 1980: 14–20).

Gaventa traces the workings of this third dimension of power in the development of one coal town, Middlesboro, Kentucky. As the town developed, the coal company brought not only industrial and commercial institutions, but also a particular worldview that celebrated a modern consumptive lifestyle and denigrated more traditional mountain ways (Gaventa 1980: 63–68). Thus while the mountaineers had a choice, it was really no choice at all – either a backward, archaic, inferior way of life, or the modern, enlightened, consumptive world offered by industry. The dependence that this created was perpetuated by churches in the company towns, which, under the control of the company, preached an "otherworldly" religion that de-emphasized issues of justice in this world.[1] As a result of this ubiquitous ideological pressure, the mountaineers internalized an image of themselves as dependent and incapable of action, such that resistance simply ceased to exist as a live option (Gaventa 1980: 93).

Gaventa's examination of power in the coalfields adds nuance and insight to the typical story outlined above. Rather than attribute the powerlessness of miners and their communities to their own deficiencies, or to the structures of power themselves, he shows how quiescence is meticulously perpetuated to serve the interests of those

---

[1] This role of religion, and particularly of mountain churches, such as Pentecostal and Holiness Churches, in encouraging quiescence and dependence, continues to be invoked in discussions of the role of religion in Appalachia and in the debate over MTR. Of course, such a straightforward view of the Church's complicity, or even the complicity of these particular churches, is overly simplistic (Cf. Almquist 2009; Billings and Samson 2012; Corbin 1981).

in power. Accordingly, rather than advocating a change in the attitudes of the mountaineers, or a change in political institutions, Gaventa argues that conscientization and education are required to counter the ideological colonization mobilized by the industry. This is an important corrective to more traditional conceptions of power and powerlessness. Nonetheless, Gaventa leaves the main contours of the standard narrative in place; for all its insight, his understanding of power remains inadequate for this reason. Power is unidirectional, exercised by the coal industry upon the "powerless" for economic benefit. The legacy of vastly unequal power relations and carefully cultivated quiescence is seen as the biggest obstacle to overcoming the current injustices in the region (Gaventa 1980: 252–261).

11.2.2 *Power and Religious Responses to MTR*

This story, whether understood in a straightforward way or as nuanced by scholars like Gaventa, remains central in discussions of Appalachia and MTR, both among academics and in popular discourse (Burns 2007; Morrone and Buckley 2011; Howard 2012; House 2011). Power is seen as the possession of the "coal interests," the "ruling elite," generally understood to be outsiders. Appalachia is described as a "sacrifice zone," a peripheral region exploited for the interest of the rest of the nation. And all of this is part of the legacy of those first capitalists and their deceitful broad form deeds.

Naturally, this narrative has influenced the Church's thinking about Appalachia and MTR as well. Indeed, in discussions of MTR among Christians, as among others, it is difficult to escape the notion that the coal industry maintains absolute control over the region. Allen Johnson, one of the lead Christian activists against MTR, reflects this conception when he asserts, "Coal is beyond an industry. It has a supernatural hold over the entire region" (Almquist 2009: 152). An article on MTR in *Sojourners* magazine refers to the decades of domination by the coal industry, comparing it to the infamous and exploitative system of sharecropping, and the editor's introduction to the issue begins by saying that "King Coal has long ruled in West Virginia" (Alston 2010). A piece in the evangelical magazine *Prism* characterizes the coal industry as "ubiquitous [and] powerful" (MacIvor-Andersen 2007).

Among Christian perspectives this view of power finds its fullest expression in two pastorals by the Catholic Bishops of Appalachia. In "This Land is Home to Me," published in 1975, the bishops state, "[The saying that coal is king is] not exactly right. The kings are those who control big coal, and the profit and power which come with it. Many of these kings don't live in the region." They then proceed to outline the history of coal in the region in terms very similar to those of the story above (Catholic Bishops of Appalachia 2007: 14). The drama is heightened in the second letter, "At Home in the Web of Life," where the story is retold in such a way that comparison with the Biblical narrative of Paradise and Fall is inescapable. Before the coming of industry, first the Native Americans, then settlers and escaped slaves, lived a simple life of spiritual harmony with nature and with one another. In the nineteenth century, however, "giant corporations" came with "outside workers," introducing social division "in rejection of God's teaching." In spite of the workers' noble efforts to unite in the face of the industry's power, their power was no match for the control of outside corporations (Catholic Bishops of Appalachia 2007: 59–60). In both these letters the image of a legacy of the industry's near-absolute power that remains an obstacle to justice is central.

Other denominational statements draw on a similar conception, either explicitly or implicitly. A resolution by the Unitarian Universalist Association[2] asserts that the cheap energy that Appalachian coal provides is exploitative, enriching the rest of the nation at the expense of Appalachia (Unitarian Universalist Association 2006). More than this, though, by the means the churches choose, they implicitly express a certain view of power. By issuing national resolutions that focus on legislation and policy and call for action by national and state agencies or by the companies themselves, the national churches show a "top-down" understanding of agency and power. In Gaventa's terms, this is a two-dimensional view, since the implicit belief is that the obstacles to justice are institutional and structural (Gaventa 1980: 14–15). If power is concentrated in the hands of the coal industry, mainly outside of Appalachia, the only apparent recourse is to appeal to either the companies themselves or to other equally external and institutional powers. While they express a great deal of concern for communities in the region, nowhere do the resolutions by mainline denominations or the West Virginia Council of Churches appeal to the agency or knowledge of those communities (West Virginia Council of Churches 2007). Allen Johnson, the leader quoted above, notes that many in these communities may feel "angry and snubbed" when presented with these hierarchical statements (Almquist 2009: 99).

Certainly these resolutions, and the growing national awareness that they illustrate, ought to be seen as positive steps. And it seems appropriate that national churches speak out on MTR, since, as the Unitarian Universalist statement quoted above suggests, the entire nation is complicit in this issue (though in this regard it is worth noting that only the statement by the Presbyterian Church, USA, makes any reference to its own need to "[turn] away from sin" [Presbyterian Church, USA 2006; Reprinted with the permission of the Office of the General Assembly, Presbyterian Church, U.S.A.]). Nonetheless, these statements reveal a certain incomplete conception of the mechanisms of power in Appalachia. A more thorough understanding would attend to the ways power operates on a variety of levels and in a variety of relationships. Such a view, present (if underemphasized) in the pastorals by the Catholic Bishops of Appalachia, is also in evidence in the reflections of several other Christian groups and activists, though it remains a minority viewpoint.

---

[2] I acknowledge that the Unitarian Universalist Association is not, strictly speaking, a Christian organization; I include the UUA statement here because it most clearly expresses a view of power that seems more or less operative in the other statements, and because the similarities between this statement and the others are enough that I believe it may be taken as representative.

11.2.3 A Higher Power and a Lower Power: A Theocentric View

Even as the familiar narrative of power in Appalachia has influenced Christian responses to MTR, there is an alternative view of power extant in many of these responses. This perspective challenges the conventional view by calling attention to the power that is both above and in a sense below the power of the coal industry. The conventional view, in its various formations, tends to portray power as resting completely in the hands of the coal industry, which exercises near-absolute control over all aspects of life in the coalfield. The current struggle against MTR, in this view, confronts the same monolithic power that created company towns. From a theocentric perspective, on the other hand, this represents an inadequate imagination, in that it places human ideas and interests at its center, viewing power as a human possession and in worldly terms of political and economic power. The theocentric view, in contrast, would argue that power ultimately lies with God, and that this divine power is exercised throughout creation. This means that all exercises of power in creation are expressions of that unified power behind creation; conversely, it means that divine power is enacted in an infinite number of finite relationships and actions (Niebuhr 1963: 125–126). Power in the world is not monolithic, a possession of one group to be imposed upon another. It is multidirectional, operating in the innumerable actions of life. Moreover, as the radical understanding of power expressed in Christian revelation makes clear, if one wishes to see clearly the exercise of divine power, it is to those who are apparently weak – the "powerless" – that one must turn (Niebuhr 1963: 126, 164–166, 2006: 97–98).

Among Christians engaged in the debate over MTR, there are examples of this view of power and its implications. Even as they describe oppression and power in somewhat monolithic terms, the pastorals from the Catholic Bishops of Appalachia acknowledge that "despite the theme of powerlessness, we know that Appalachia is already rich here in the cooperative power of its own people" (Catholic Bishops of Appalachia 2007: 33). The Bishops point to the creativity present in local communities and churches and confirm the possibility of transformation, "for it is the weak things of the world which seem like folly that the Spirit takes up and makes its own" (Catholic Bishops of Appalachia 2007: 98). Other activists share a similar confidence in God's power working in unlikely places. Lon Oliver, executive director of the Appalachian Ministries Educational Resource Center (AMERC), perhaps puts it most clearly when he argues that the Church's role in a debate like MTR is to seek really to listen to those involved in order to find "what God is up to" in a given situation. This listening, he believes, can lead to "life-giving" discussions that can call attention both to shared complicity and to truly transformative action (Oliver 2010b).

AMERC believes that this kind of careful listening can help overcome the assumption that the people of Appalachia are powerless and in need of rescuing, instead finding "stories of hope" and positive, transformative visions that can then lead to action (Oliver 2010b). AMERC's approach expresses humility in that it seeks first to discern God's action and purpose in a particular context, but also a hopeful confidence that such action is present and already transforming the situation. In one example not directly related to MTR, AMERC director Lon Oliver points to an ABC news report in 2009 that called attention to the crime and drug problems of Cumberland, KY (ABC News 2009). He argues that while this awareness was salutary, it completely ignored the successful efforts by local churches and community groups to improve law enforcement and strengthen the relationship between the community and police (Oliver 2010b). This is the kind of transformative power that is overlooked by typical discourses, but emphasized by a theocentric view of power. A response to MTR that incorporates such a view would pay special attention to movements like this one as they relate to that particular issue.

Christians for the Mountains (CFTM), a group that has challenged MTR in more direct ways than AMERC, also sees power in unexpected places, and sees the hope that results as one of the main contributions of the Church to this issue. Allen Johnson, director of CFTM, says that hope and joy are essential in confronting such an overwhelming issue, one that has driven many activists to despair. Without resorting to fatalism or what he calls "Polyanna" optimism, Johnson sees Christian activists enacting a faith that God loves humans and the world, and is the one power greater than evil (Johnson 2010). CFTM sees one of its main goals as nurturing this hopeful confidence in the benevolence of divine power in creation. Practically speaking, for CFTM this has involved providing pastoral support for those struggling against MTR to prevent the burnout and frustration that too often comes from opposing, often in isolation, such an overwhelming issue. Like AMERC, CFTM exemplifies a hopeful awareness that divine power is at work in unexpected and transformative ways. These approaches, shared by others in the region, illustrate an awareness of power that is, as I have put it, both above and below the power of the coal industry: a divine creative power that finds expression in unexpected, apparently weak or "powerless" places.

11.3 Appalachian Identity: Insiders and Outsiders

11.3.1 The "Real" Appalachia

The view of power described above, and the narrative of Appalachia's fall from paradisiacal harmony with nature to national sacrifice zone, relies heavily on the distinction between insiders and outsiders, between "local communities" and manipulative elites from abroad, between those who are genuinely Appalachian and those who wish only to oppress and exploit them. The claim is sometimes made explicitly: "hired managers became agents of the absentee owners whose corporate offices were far removed from the coal towns themselves. This domination of the coalfields by large, often multinational, corporations … persists to this day" (Burns 2007: 5). Opponents of MTR argue that what meager benefits it offers go to outsiders, and the precise identification of a reviled coal executive as local or outsider – based not only on where, but also how, he lives – can be a matter for serious scrutiny (Shnayerson 2008: 19–20). More frequently, though, this attention to distinguishing Appalachians from outsiders is implicit in arguments about the exploitation and manipulation of the region, where words like "local" and "residents of Appalachia" invariably carry a clear moral weight.

Of course, it is not only opponents of MTR or purveyors of the "national sacrifice zone" narrative that orient their claims around this distinction between insiders and outsiders. Those who support the practice regularly characterize their adversaries as environmentalists from elsewhere (Reece 2006; Loeb 2007; Scott 2010). A statement released by a Massey Energy company spokesperson in response to protests against MTR illustrates clearly the power of the insider/outsider dichotomy around this issue. The spokesperson dismisses the protests, attributing them to "residents of states other than West Virginia," seeking "hype and media attention for their out of state funders" (Ward 2009).

There is a more profound discourse operating within this dichotomy, a discourse about who and what can be considered authentically Appalachian, and who may speak for Appalachia. Claims of legitimacy as insiders are not primarily about geographic boundaries – where one (or one's kin) was born or raised – but rather about identity. When they mobilize the dualism of insiders versus outsiders, participants in contemporary debates trade on the notion – long established in discourse about Appalachia – that there is a coherent class of "Appalachians," and that the members of this class share certain characteristics and some common interest; at the same time they challenge past portrayals of this class as, in some way, inaccurate. Sociologist Rebecca Scott contends that the debate about MTR and about the future of coal mining more generally is at bottom a struggle between different conceptions of this Appalachian identity as it relates to an imagined American identity that has continually denied its validity (Scott 2010).

The people of Appalachia have long posed an enigma to the larger culture: like the mountains themselves, they have been understood to be both a part of the nation and irremediably set apart. This paradoxical otherness is, of course, a political creation, constructed to serve the interests of one group over against another (Batteau 1990: 33). As historian Allen Batteau argues, in the case of Appalachia this construction of otherness first took place in a nation that was seeking to solidify its identity based on the image of the frontier and the conquest of nature, and where, at the same time, the conflict between North and South was becoming increasingly tense (Batteau 1990: 29, 33). In this context, the image of a region that was a kind of internal frontier, where nature was still and continually being conquered by heroic means, could be mobilized for political purposes. As outside of and more pure (because closer to nature) than the dominant American culture, Appalachia was used to criticize the Northern establishment; simultaneously, as a part of the South yet innocent of that region's special vices; it was also a critique of Southern slaveholders. Thus in the period leading up to the Civil War the basic characteristics of an image of Appalachia were already in place: a region whose close relationship to nature and place represented the ideals of the nation, yet whose primitive culture and traditions inevitably set it apart as backward (Batteau 1990: 37).

This politically constructed notion of Appalachian difference is what makes possible the economic exploitation of Appalachia as a national sacrifice zone, according to Rebecca Scott (2010: 31). Such exploitation is only conceivable because Appalachia and its people are understood in the national consciousness as different, simultaneously ideal and deviant, viewed with romantic nostalgia and bemused derision, but always decidedly other. "These epistemologies of disgust and social distance," she argues, "help create the conditions of the possibility for some of the most dangerous environmental exploitation in the United States and the designation of Appalachia as a sacrifice zone" (Scott 2010: 63). At the same time that this social distance makes MTR possible, the construction of mining as normatively masculine (in a traditional sense, emphasizing provider status, toughness, and progress) and white creates a context wherein MTR makes ideological sense by appealing to racial and gender biases (even as, she argues, it cannot be said to make economic or environmental sense) (Scott 2010: 113).

11.3.2 *Appalachian Identity and Religious Responses to MTR*

As with narratives of power and powerlessness, the Church's response to MTR has, to a certain extent, engaged this dichotomy between insiders and outsiders, and the attendant images of Appalachian identity. The statement by the Unitarian Universalist Association cited above, for example, trades on the moral weight of the division between locals and outsiders (Unitarian Universalist Association 2006). The statement by the United Methodist Church more subtly invokes the idea that Appalachians have a special connection to their land and to the past, describing the mountains as home places where families have been born, lived, married, and died (United Methodist Church 2000). The Presbyterian Church (USA) makes reference to several established images of Appalachian identity: a special relationship to the land and nostalgic association with a bucolic past (seen in the use of terms like "roots," "family home place," and a "lifetime of memories," from which people have been exiled) and Appalachia as a sacrifice zone (mining "destroys the beauty and productive capacity of the land … eliminating future or alternative economic opportunities" and creates "a cycle of poverty") (Presbyterian Church 2006).

The "Fall narrative" recounted in the Roman Catholic Pastorals (discussed above) exemplifies these assumptions about Appalachian identity. Familiar images of Appalachian identity – spiritual harmony with nature, a nostalgic pre-modern folk culture, a combination of fierce independence with close community and kinship ties – are prominent throughout the letters (Catholic Bishops of Appalachia 2007: 21–22, 48, 55–59). The bishops place commendable emphasis on listening to "the voice of the region;" yet the pastorals minimize whatever heterogeneity those voices present, relying instead on established conventions. Threats to "Appalachia's old traditions" are seen as coming from an alien culture, the "modern consumer society" (Catholic Bishops of Appalachia 2007: 61–63).

Unsurprisingly, the attitudes of individual religious activists in the mountains show a greater awareness of this ambivalence than the statements of the larger church bodies. Some of the conventional images are invoked, especially by those who have been involved in or influenced by the Roman Catholic pastorals. Carol Warren, a Roman Catholic activist with the Ohio Valley Environmental Coalition, refers to the close connection to place and the loss of home-places as one of the most serious effects of MTR (Warren 2010). Fr. John Rausch, director of the Catholic Committee of Appalachia, refers to the region as a "mineral colony," indexing the image of Appalachia as a "sacrifice zone" (Rausch 2010). At the same time, these activists and others are very aware of the heterogeneity of the people living in Appalachia, and of the need to avoid relying on easy dichotomies and polarizing characterizations and thereby foreclosing real understanding. Parish ministers, in particular, are cognizant of the variety of cultures and interests that make up the region, and of the need to balance prophetic witness with respect for this diversity (Holmes 2010). As with understandings of power, the approaches of many Christians to this issue reflect, even if secondarily, a more complex, theocentric conception of identity.

11.3.3 *Identity and Inclusivity: A Theocentric View*

For a theocentric understanding of identity, I turn to Niebuhr's book *The Responsible Self*. In Niebuhr's anthropology, the central characteristic of personhood is response to a multiplicity of actions. The person-as-answerer is defined by his or her responses to other persons, as well as to "that to which [the others] respond," to the events and ideas of the world around her (Niebuhr 1963: 71–84). The question this relational anthropology raises for Niebuhr is where unity – where identity – is to be found in this network of relationships (Niebuhr 1963: 121–125). He contends that unity in the midst of this plurality is only found in the response of trust in "the radical action by which I am." What unity one has as a self cannot be found in any one of the conflicting and changing relationships in which one participates, nor in the finite causes to which one may devote oneself, but rather in the fundamental relationship, the existential action by which we interpret and respond to all other actions.

In contrast to this fundamental identity, the finite conceptions of identity enacted in a multiplicity of roles and responses are no more monolithic or unified than those roles are. Moreover, Niebuhr argues that our own narrow images of ourselves (that is, our identities, both individual and collective) are at the center of the destructive imaginations that divide us from one another. When these imaginations are reconceived with God, who values all being, at the center, we understand ourselves in continuity and unity with all of humanity and creation (Niebuhr 2006: 64–65). This unity is, from a theocentric perspective, the only absolute identity; other enactments of identity are partial and shifting, characteristic of the self who responds in a wide variety of interconnected relationships. These incomplete notions of identity are thus relativized – though not denied – by the revelation of God the universal valuer.

In light of this relativization of all finite notions of identity, the construction of Appalachian identity and demarcation of boundaries between insiders and outsiders, between who may and who may not represent Appalachia, is seen to reify one finite conception of identity and establish value judgments based on that human notion, rather than on the inclusive intentions of the universal valuer. In Niebuhr's terms, these claims absolutize what is properly relative; he calls this "the great source evil in life (Niebuhr 2006: xxxiv)." Moreover, notions of Appalachian identity are invariably constructed around the same issues: coal, poverty, nature, or history, for example. Niebuhr's anthropology reminds us, in contrast, that just as selves are constantly responding in relationship to other selves, they are also always responding in relationship to other ideas and causes. The identities of Appalachians are forged in relation to a wide range of issues beyond coal or poverty; narratives of identity that limit themselves to these issues are incomplete.

Rebecca Scott believes that such a conception of identity as shifting and fluid can be mobilized strategically to build coalitions for change. Rather than being concerned with the accuracy of claims of Appalachian authenticity, those involved in issues like MTR can draw on a variety of identities and understandings in their struggle against "the divisive strategies of the industry, which support its unsustainable practices" (Scott 2010: 212–214). Her argument indicates, in a general way, the main implication of a theocentric view of Appalachian identity as dynamic and relative in the response to MTR: a move toward greater inclusivity. If the radically inclusive center of being relativizes human boundaries and narrow conceptions of identity, and if revelation of this inclusive center means "that all our values are transvaluated by the activity of the universal valuer," the appropriate response is to seek to conform conceptions of identity to this radical inclusivity (Niebuhr 2006: 80).

Christians active on both sides of the debate are challenging reified conceptions of Appalachian identity and making efforts toward greater inclusivity. Again, the contextual educational approach of AMERC is illustrative. By encouraging Church leaders living outside the region to explore theological questions in local communities, AMERC shows students how the identities of Appalachia and of Appalachians are enacted in relationships with other people and in relationship with issues other than those typically associated with the region. In AMERC's classes, Appalachia is not understood only in relation to poverty or mining, nor poverty only in relation to Appalachia; rather, these issues are explored in relation to, for example, prophecy or eschatology (Oliver, 2010, July 28, Personal communication). AMERC focuses many of its efforts on connecting these students with mountain churches such as Mountain Pentecostals and Old Regular Baptists (while most environmental activism remains centered around mainline Protestant denominations). Thus the complex networks of relationships that shape fluid identities are engaged. Moreover, this is done with the belief that "the Gospel is borne on local culture," that the purposes and values of God are more inclusive than the human boundaries that separate one group from another. AMERC's approach, then, is seeking Scott's more inclusive coalitions, but also seeking to include a greater diversity of theological views and to encourage a more inclusive conversation about the breadth of issues surrounding MTR. Director Lon Oliver has seen numerous instances of the kind of "life-giving discussion" that this openness can lead to, such as a staunch environmentalist who admitted, in the spirit of real dialogue, that a reclaimed mine site was literally indistinguishable from the untouched mountains around it. Oliver also cites similar admissions from MTR supporters that some mining and reclamation has been truly irresponsible. Real changes in attitude like these are the fruit of AMERC's inclusive approach.

Andrew Jordon, the owner of a small (by MTR standards) surface mining operation, regularly invites members of his church community to his MTR site, even holding Bible studies there. He seeks to expand the conversation to include energy policy and the future of the region, even as he also narrows the focus to this particular mine and its impact on the community and individual workers' lives (Jordon 2010). Clearly efforts toward greater inclusion are not limited to MTR opponents. And leaders like Allen Johnson, of CFTM, point to hopeful stories where conversation between activists and miners has led to genuine understanding on both sides. At one prayer service on an active mine site in 2008, participants were able to engage in real, positive dialogue with miners who had gathered to disrupt the service. The miners were able to articulate their own conflicted position of both concern for the effects of mining and the need to support their families in a place with few other options (Johnson 2010). From the perspective of a theocentric view of identity as fluid and as relativized by God the universal valuer, inclusive efforts like these are to be celebrated, and the divisive claims and imaginations that foreclose on them challenged and reevaluated.

11.4 Changing Places: Destruction and Reclamation

11.4.1 Destruction as a Political Claim

I have argued that the conceptions of power and powerlessness and of insiders and outsiders, as they are mobilized in the debate around MTR, have complicated and ambivalent histories. While both of these point to important realities and may be, therefore, both inevitable and useful, to invoke them uncritically obscures the dynamic, multifaceted social and political processes behind them and the rich heterogeneity of the mountains. A theocentric perspective, on the other hand, is able to incorporate and comprehend this multiplicity. In this section, I will advance a similar argument about the concepts of destruction and reclamation. Rather than being straightforward descriptions of processes involved in MTR, these terms involve a variety of interrelated and politically driven claims. Their uncritical application (or in the case of reclamation, uncritical omission) is question-begging, in that it takes for granted precisely what needs to be examined from a moral standpoint: what kinds of changes does MTR effect on a place, and how can or should those changes be mitigated?

At first glance, this third pair of terms seems far less problematic than the first two. Compared to questions of power or questions of identity, the notion of destruction seems to refer to a relatively concrete and straightforward reality. It is likely that anyone looking at an active MTR site would agree with this assessment: the destruction of the mountaintop seems complete and undeniable. And indeed, for most opponents of MTR, the fact that it destroys mountains is self-evident. One important work on MTR opens with large-format photographs of active mine sites under the words "destruction," "devastation," and "desecration" in enormous font (Butler and Wuerthner 2009). The same book features a chapter entitled: "Mountaintop Removal: The Destruction of Appalachia" (Butler and Wuerthner 2009: 59). Even Rebecca Scott's careful analysis of the meaning-making dynamics at work around MTR asserts plainly that "MTR is simultaneously destroying forests and ecosystems [and] flattening the beautiful Appalachian Mountains," and is "annihilating the place" (Scott 2010: 96, 174). These are just a few examples; the claim that MTR is incontrovertibly and absolutely destructive is axiomatic among anti-MTR activists.

Among the religious opposition to MTR this claim virtually becomes an article of faith. A common theme is that MTR destroys God's creation, and is, therefore, blasphemous. One prominent activist, Larry Gibson, cites Revelation 11:18 ("The nations raged, but your wrath has come, and the time for judging the dead and for destroying those who destroy the earth") to illustrate God's wrath against those responsible for MTR (Almquist 2009: 162). Kathy Selvage, vice president of the Virginia nonprofit Southern Appalachian Mountain Stewards, states plainly, "If you believe there is a God, and you believe that he created this earth, then every time you blow up a mountain, it's like slapping his face" (Almquist 2009: 172). Activist and Episcopal deacon Denise Giardina proclaims that mine owners and operators are "bent on destruction. Destruction is their life's work." Consequently, "they are cut off from God" (Giardina 2005). For Christians involved in the fight against MTR, the complete destruction it causes is not only undeniable, it is blasphemous.

Far from being self-evident, however, destruction is an ambiguous and politically constructed notion that comprises a variety of questions and claims, as political ecologists remind us. Even a straightforward definition of the term – for example, "the substantial decrease in either or both of an area's biological productivity or usefulness due to human interference" (Robbins 2004: 91) – leaves a variety of questions unanswered: decrease compared to what? What sort of productivity or usefulness? How is human interference assessed? Is the decrease necessarily permanent? Responding to these questions involves choosing among criteria based on particular interests (Robbins 2004).

Even where parameters exist for determining some of these conditions (for example, measuring a decrease by comparing an area's productivity with that of a similar area, or with the same area in past years), certain judgments remain necessary. Frequently a decrease in one sort of productivity is accompanied by an increase in another sort. What counts as an appropriate sort of "usefulness" in a particular region is an explicitly political decision (Robbins 2004: 94). In the case of MTR, this can be seen in the claim by many supporters of the practice that the level grasslands that are frequently the result of reclamation are, in fact, more useful than the mountaintop was. Proponents of MTR also point out that similarly dramatic processes are used to build roads and other human edifices in the mountains; yet these are not as vehemently decried as "destruction," presumably because their usefulness is more accepted. Besides this question of usefulness, human interference is also not always easily determined, since natural systems are highly dynamic. Measuring the effects of human action against such a variable standard can be a challenge (Robbins 2004: 97).

Certainly, in any particular case, some of these questions are more compelling than others. With MTR, while usefulness or productivity remain open questions for some, the question human interference brooks little debate. Whatever natural dynamism exists, the impact of human actions on the environment is readily measurable, whether in tons of overburden displaced or miles of streams buried. Yet even given this clear human interference, and even after answering the question of productivity (for example, advancing what many would see as the uncontroversial argument that a forested mountaintop and free-flowing streams are more productive in a variety of ways than the "moonscape" of an active mine site), the applicability of the term "destruction" is still not incontrovertible. For some degree of permanence is also implied in the meaning of destruction (Robbins 2004: 90). To argue that an ecosystem has been destroyed suggests that recovery is impossible, or at least requires such a long span of time as to be virtually impossible. Thus in the case of MTR, examination of the notion of destruction necessarily involves consideration of reclamation.

11.4.2 Reclamation as a Political Claim

The claim that MTR is destructive entails the belief that what is lost – the productivity or usefulness of forest, streams, or soil – can never be regained. And if opponents of MTR see its destructiveness as self-evident, the belief that reclamation is impossible is equally axiomatic. The statement by the Presbyterian Church (USA) (the only of the denominational statements to refer to reclamation at all) is illustrative. It asserts, "Streams, mountains, and forests damaged by mountaintop removal coal mining *can never be restored* to support the community of life that God created" (Presbyterian Church USA 2006; emphasis added). A wide variety of claims are made about the possibility of anything approaching real reclamation, the restoration of a functioning mountaintop ecosystem. Publications from the West Virginia Coal Association and the University of Kentucky show lush forest landscapes that were once surface mines (WV Coal Association 2011; University of Kentucky College of Agriculture 2008). Yet opponents of MTR see this as deceptive propaganda, and emphasize that most so-called reclamation involves replacing hardwood forests and mountain peaks with desolate grassland (Burns 2007: 118–140; Scott 2010: 95).

Again, for Christians the arrogance and deception of reclamation is elevated to blasphemy. When supporters argue that MTR improves on nature by creating more useful land than what had existed prior (frequently exemplified by developments such as Twisted Gun Golf Course, a regional airport, or the Earl Ray Tomblin Industrial Park [Scott 2010: 176–179]), many Christians interpret this as an affront to the mountains' Creator. At least one mountaintop miner agrees with this interpretation, although he denies that this is what coal operators are claiming: "It would be blasphemy if they said they were going to make this land better than what God made it. All they are saying is that they are making it more useful" (Scott 2010: 83). Perhaps no statement better expresses Christian resistance to the notion of reclamation than the sign posted by one activist: "God was wrong: support mountaintop removal" (Reece 2006: 142).

Of course, what constitutes reclamation is as selective and politically constructed as the notion of destruction. Even with minimal or no human intervention toward reclamation, ecosystems recover from human impacts in a variety of ways (Robbins 2004: 101–103). Some systems may be relatively resilient, where others may be more fragile, and still others may show resilience to a low level of human activity, but may be unable to recover from higher levels of human impact. In some ecosystems, recovery may naturally lead to a new state, rather than simply reverting to the original state; thus it may be that an impact is not "reversible," yet recovery in some form is still possible (Robbins 2004: 101–103). Again, these variations in the recovery of natural systems occur even without human interventions toward reclamation.

These observations about environmental recovery in general are relevant to the specific case of MTR. Surface mine sites that predate the Surface Mining Control and Reclamation Act of 1977, which required that certain criteria be met for post-mining reclamation, are home to mature hardwood trees and show significant recovery (University of Kentucky College of Agriculture 2008). In other words, these sites had a minimum of human-led reclamation, yet have recovered well, suggesting that the mountain ecosystem is capable of a degree of resilience. On the other hand, these sites are now more than 30 years old; what time-scale for recovery ought to be considered permissible?

As with defining destruction, the original state of the land that is the goal of reclamation is not easily established. Environmental assessments of mining and reclamation plans tend to view the land as wilderness prior to mining, and the goal of reclamation as a return to wilderness in some form (Hufford 2002: 113). People on both sides of the MTR debate echo this assumption in different ways: proponents argue that the land was "just sitting there," useless, before mining and reclamation, and some opponents want land preserved as pristine wilderness (Scott 2010: 117–121, 211). Yet people living in the mountains have been using this land for recreation and sustenance since well before European settlers arrived, and current inhabitants continue to do so, using the land as a local commons (Scott 2010: 123). Images of reclamation as returning the land to some constructed image of wild nature, without taking into account the usage patterns of local residents, are illusory, whether they are employed by advocates or critics of mining and reclamation.

Defining reclamation involves these and various other judgments and decisions. On a technical level, of course, it requires careful assessment of the various factors necessary for a particular environment to function, from plant and animal species to water conductivity. But like the idea of destruction, reclamation also involves similar judgments about what kinds of use or productivity are preferred, about what makes a particular ecosystem valuable. Regardless of the apparent absurdity of a golf course on a leveled mountaintop, or wine grown on a reclaimed mine site (Scott 2010: 131–132, 178), decisions about what constitutes appropriate use are not foregone conclusions. The question of reclamation also necessitates an awareness of the dynamism of natural systems. It is not enough to argue that reclamation cannot return a site to "the way it was before," if this "before" has always been a moving target (Robbins 2004: 102). Most importantly, perhaps, this question requires the recognition that humans unavoidably interact with and change their environments, and that careful consideration of and debate about which kinds of changes are acceptable and which are not is more productive than the insistence that one particular kind of change is irrevocable.

11.4.3 Reclamation for God: A Theocentric View

A theocentric view of destruction and reclamation sees them, like all that happens, as reflecting God's action in the world, in which humans participate (Niebuhr 1963: 126). As part of human participation in God's purposive action in the world, they are inevitably ambiguous. Without a clear knowledge of God's telos, we are forced to admit that there is no clear justification for any particular moral stance in this case (Gustafson 1994: 66–73). There are interrelated values, benefits and consequences to be considered, and these in light of the various choices already described: choices about the original state of the land, how human impact is measured, and what the appropriate use of a place might mean.

Thus the theocentric view, with its focus on the ambiguity of our actions, emphasizes that we do best to think of terms like destruction and reclamation as ways of talking about how humans change places. If this ambiguity of human interventions lead us to recognize that "God is in the details" (Gustafson 1994: 13), or that the morality of our actions lies in the specific consequences of those interventions, then our discussion of how we change places will need, at times, to be more precise than terms like destruction and reclamation allow.

Accordingly, the assumptions frequently implicit in Christians' use of these terms can be seen as representing, from a Niebuhrian perspective, inadequate imaginations. When MTR is approached with the *a priori* belief that it destroys something that God created and values, this presumes to know what God values and how, and, further, that God's valuing is static and singular, as opposed to dynamic and universal. That God values mountains can, from a theocentric view, be certain, since Niebuhr affirms that God values being universally; how God values mountains, or a particular mountain, or a particular part of a particular mountain, is a question that must be approached with much greater humility if we are to avoid absolutizing the relative. Likewise, the *a priori* assertion that because only God creates mountains, reclamation is therefore blasphemous, denies God's ongoing participation in creation through human actions, and forecloses the possibility of discerning how to make reclamation best express the divine purpose.

The theocentric perspective, therefore, commends greater attention to the details – the values and disvalues of human interaction with the environment, and the choices and relationships in which those values are negotiated – with the belief that it is in these details that the purposes of God are to be discerned, so far as is possible. There are better and worse ways of interacting with nature; interventions can be more or less conducive to God's intentions. As with notions of identity, it may be that discussion of destruction and reclamation can be strategically mobilized in effective and non-absolutizing ways; yet it is essential that the choices and assumptions behind them be brought to light. Even those who are firmly opposed to MTR may see the importance of substantive discussion of what kinds of extraction and what kinds of reclamation are being practiced on mines that already exist. When Andrew Jordon of Pritchard mining takes over abandoned and unreclaimed surface mines for the purpose of mining and then carefully reclaiming them, perhaps even opponents of the practice might see something of God's sustaining actions in Jordon's work (Jordon 2010).

This, then, suggests the direction a theocentric perspective on reclamation might lead: toward more substantive, precise discussion of the kinds of changes to the environment we believe to be warranted or unwarranted, in light of our limited knowledge of divine purposes. Jordon provides one example of this kind of engagement. There is bound to be disagreement with his claim to be practicing good environmental stewardship by carefully attending to the reclamation of his mines, yet this disagreement may then lead to further consideration on both sides of what constitutes effective reclamation. Even if agreement is impossible, a more complete understanding may not be. AMERC, again, provides another compelling example. Even as director Oliver argues, referring to MTR, that "some things need to be stopped," he believes it is crucial that Christian activists, particularly from mainline churches, pay more attention to the real work that is being done toward effective reclamation, as this reflects part of "what God is up to" in this context (Oliver 2010b).

There is another related approach to this issue that I believe can be derived from the Niebuhrian view. If human interventions are interpreted as part of God's action in the world, and the goal of the theocentric ethic is to direct those interventions, so far as is possible, toward what we believe to be God's purposes, I propose that the notion of reclamation might be broadened. We might think of it in terms of reclamation on God's behalf; that is, of reclaiming mountains, and the human relationships with them, for God's purposes. This would encompass the above discussion, framing conversation about values and disvalues in the reclamation process as one way of participating in God's sustenance of creation. Whatever one's feeling on the practice of MTR, there are sites waiting to be restored, and responsible reclamation like Jordon's can, I believe, be understood as (always ambiguous) participation in God's action. Even in the context of dynamic and multiple value-relationships, few people would disagree that a responsibly reclaimed mine is more valuable than an unreclaimed one.

My idea of reclamation for God, however, can be broadened even further, beyond the practices typically thought of as reclamation. It can include the practices Christians employ symbolically to claim a mountain or part of a mountain as belonging to God, a part of God's good creation, prior to (and I would argue, throughout and in spite of) mining activity. Jordon's mine site Bible studies are one example, expressing the belief that God may be encountered as fully in the midst of an active mine site as elsewhere. Another powerful example of reclaiming the mountains for God are the liturgies practiced by groups like CFTM and the Catholic Committee of Appalachia. In these services, Christians pray for the health and renewal of the mountains and mountain communities around them, and mourn the greed and thoughtlessness that is destroying that environment for the sake of cheap energy. In one such service, worshippers carefully planted wildflower seeds on a mined mountaintop, powerfully symbolizing hope and faith in God's creative and sustaining power in the most damaged of places (Rausch 2003). A third, very intriguing example is the partially successful effort by the Ohio Valley Environmental Coalition (OVEC) and the West Virginia Council of Churches (WVCC) in collaboration with the West Virginia Coal Association – normally an opponent – to strengthen legal protection of cemeteries and families' access to them. The recognition of the sanctity of these small plots literally provided common ground for both sides to recognize and defend the limits of human intervention in God's creation (Warren 2010; McGlynn 2012).

These practices, though very different from one another, all powerfully illustrate the divine creative action that continually upholds and embraces creation, both through and outside of human activity. They remind all of the ultimate, original source of the mountain's value, beyond all finite and changing ways of valuing it. They symbolically reclaim the mountains for God.

11.5 "All My Holy Mountain": Conclusions

In this essay, I have applied one particular Christian ethical approached, based on the theocentric ethic of H. Richard Niebuhr, to the complex and culturally freighted debate around mountaintop removal mining. I have argued that the typical imaginations employed with respect to this issue, specifically about power, identity, and reclamation, are incomplete and inadequate, and have indicated ways that these imaginations might be reconceived theocentrically. In all of this, the goal has been to show that, from the perspective of the Christian believing community, the theocentric imagination is capable of dealing more completely with the difficulties of this issue. Whether this approach is compelling from the standpoints of other communities and other faiths (since, as Niebuhr avers, every standpoint belongs in some community and begins from faith in something), is not mine to argue.

In any case, the suitability of this approach may be challenged by Christians and non-Christians alike, on the grounds that its emphasis on divine action leads inevitably to resignation and quiescence. Niebuhr believes that this, too, represents an inadequate imagination. The symbols of fatalism, he argues, are simply not the right symbols: the deistic "Determiner of Destiny" is not the loving and transforming God of Christianity. Niebuhr's approach calls for nothing less than replacing the selfish and finite interests at the center of human moral imaginations with divine purposes; consistently applied, this theocentrism has radical, not complacent, implications.

References

ABC News. (2009). A hidden America: Children of the mountains. Retrieved October 4, 2012, at http://abcnews.go.com/2020/story?id=6845770&page=1
Almquist, C. (2009). *I have been to the mountaintop, but it wasn't there: Christian responses to mountaintop removal coal mining in Appalachia.* Middlebury: Middlebury College. Unpublished senior thesis (used by permission of author).
Alston, O. (2010). Destroying West Virginia, one mountain at a time: Christians battle king coal to save Appalachia. *Sojourners, 39*(6), 18–20.
Batteau, A. (1990). *The invention of Appalachia. The anthropology of form and meaning.* Tucson: University of Arizona Press.
Billings, D. B., & Samson, W. (2012). Evangelical Christians and the environment: 'Christians for the Mountains' and the appalachian movement against mountaintop removal coal mining. *Worldviews: Global Religions, Culture & Ecology, 16*(1), 1–29.
Burns, S. S. (2007). *Bringing gown the mountains: The impact of mountaintop removal on southern West Virginia communities.* Morgantown: West Virginia University Press.
Butler, T., & Wuerthner, G. (Eds.). (2009). *Plundering Appalachia: The tragedy of mountaintop = removal coal mining.* San Rafael: Earth Aware.
Catholic Bishops of Appalachia. (2007). *This land is home to me (1975) and at home in the web of life (1995): Appalachian pastoral letters.* Martin: Catholic Committee of Appalachia.
Caudill, H. M. (1963). *Night comes to the Cumberlands: A biography of a depressed area.* Boston: Little, Brown.
Corbin, D. (1981). *Life, work, and rebellion in the coal fields: The southern West Virginia Miners, 1880–1922.* Urbana: University of Illinois Press.
Gaventa, J. (1980). *Power and powerlessness: Quiescence and rebellion in an Appalachian valley.* Urbana: University of Illinois Press.
Giardina, D. (2005). Keynote address: Christians for the Mountains Conference, November 2005. Retrieved October 4, 2012, at http://christiansforthemountains.org/site/Topics/About/deniseGiardina.html
Gustafson, J. M. (1994). *A sense of the divine: The natural environment from a theocentric perspective.* Cleveland: Pilgrim Press.
Holmes, S. (2010). Interviewed August 9, Charleston.
House, S. (2011, February 19). My polluted Kentucky home. *The New York Times*, sec. opinion. Retrieved October 4, 2012, at www.nytimes.com/2011/02/20/opinion/20House.html?_r=1&scp=1&sq=silas%20house&st=Search
Howard, J. (2012, July 8). Appalachia turns on itself. *The New York Times*, sec. opinion. Retrieved October 4, 2012, at www.nytimes.com/2012/07/09/opinion/appalachia-turns-on-itself.html
Hufford, M. (2002). Reclaiming the commons: Narratives of progress, preservation, and ginseng. In B. J. Howell (Ed.), *Culture, environment, and conservation in the Appalachian South* (pp. 100–120). Chicago: University of Illinois Press.
Johnson, A. (2010). Interviewed August 6, Marlinton.
Jordon, A. (2010). Interviewed August 9, Kanawha County.

Loeb, P. (2007). *Moving mountains: How one woman and her community won justice from big coal*. Lexington: University Press of Kentucky.

MacIvor-Andersen, J. (2007). A brief history of coal. *Prism: America's Alternative Evangelical Voice, 14.6*, 14.

McGlynn, D. (2012). Move not those bones. *Sierra, 97*(2), 28–33.

Morrone, M., & Buckley, G. L. (Eds.). (2011). *Mountains of injustice: Social and environmental justice in Appalachia*. Athens: Ohio University Press.

Niebuhr, H. R. (1963). *The responsible self: An essay in Christian moral philosophy*. New York: Harper & Row.

Niebuhr, H. R. (2006). *The meaning of revelation*. Louisville: Westminster John Knox Press.

Oliver, L. (2010). Interviewed August 10, Berea.

Presbyterian Church (USA) 217th General Assembly. (2006). Commissioners' resolution. Retrieved October 4, 2012, at http://ilovemountains.org/resolutions

Rausch, J. S. (2003, January 31). Sowing my community back. *The Steubenville Register*. Steubenville.

Rausch, J. S. (2010). Interviewed August 10, Staunton.

Reece, E. (2006). *Lost Mountain: A year in the vanishing wilderness: Radical strip mining and the devastation of Appalachia*. New York: Riverhead Books.

Robbins, P. (2004). *Political ecology: A critical introduction* (Critical introductions to geography). Malden: Blackwell.

Scott, R. R. (2010). *Removing mountains: Extracting nature and identity in the Appalachian coalfields*. Minneapolis: University of Minnesota Press.

Shnayerson, M. (2008). *Coal river*. New York: Farrar Straus and Giroux.

United Methodist Church. (2000). Cease mountaintop removal coal mining. Retrieved October 4, 2012, at http://ilovemountains.org/resolutions

University of Kentucky College of Agriculture/Kentucky State University. (2008). *Reclaiming the future: Reforestation in Appalachia*. Lexington: University of Kentucky College of Agriculture/Kentucky State University.

Ward, K. (2009). Massey protest update: 14 arrested; Accusations fly: Coal tattoo. Retrieved October 4, 2012, at http://blogs.wvgazette.com/coaltattoo/2009/06/18/massey-protest-update-14-arrested-accusations-fly/

Warren, C. (2010). Interviewed August 9, Charleston.

West Virginia Coal Association. (2011). *West Virginia coal: Fueling an American renaissance* (Coal facts 2011). Charleston: West Virginia Coal Association.

Chapter 12
God, Nature and Society: Views of the Tragedies of Hurricane Katrina and the Asian Tsunami

Janel Curry

12.1 Introduction

Humans have always been fascinated by extreme natural events. A relatively new field of study, natural hazards research, focuses on natural disasters in the context of human society. This group of scholars builds on the foundational work of geographer Gilbert White, who stated in his 1942 doctoral dissertation, "Floods are 'acts of God' but flood losses are largely acts of man" (Sullivan 2006: B4). Such scholars are interested in how the results of extreme events are partially mitigated or enhanced by social structures and societal and individual perceptions of risk (Tobin and Montz 1997; Cutter 1994; Resources for the Future 2009). Natural hazard studies see disasters as not so much natural events as they are natural events that interact with human decisions and responses to create disasters. For example, if societies chose to limit development along coastlines, then the threat of hurricanes on human structures would be lessened. And an ability to do such a task is intertwined with societal property rights structure. Or in the case of Hurricane Katrina, human actions have destroyed the wetland barrier south of New Orleans, reducing the natural protection that was at one time present. And on the psychological level, individuals and societies either confront risk or deny its existence.

My own research addresses an understudied aspect of the study of natural hazard, viz., the role of religious worldviews in shaping interpretations, responses and explanations for such events. Religious commitments, as part of an overall cultural system, express themselves in a form of a worldview. A worldview is a community's picture of the way things are, that is, their concept of nature, of self, and of society. Worldviews create social and cultural "maps." These maps influence the direction of the development of institutional and societal structures (Aay and Griffioen 1998: xii; Wallace 1998:

J. Curry (✉)
Provost, Gordon College, Wenham, MA 01984, USA
e-mail: janel.curry@gordon.edu

46; Griffioen 1998: 126). Extreme natural events bring to the fore the basic structures of these belief systems, beliefs about what are identified as the most basic human problems and their solutions, concepts of good and evil and the trajectory of history (Leege and Kellstedt 1993: 231). Several works that specifically explore the relationship between religion and natural disasters include Brunn (2010) who analyzed the portrayals of religion and natural disasters in cartoons found through the use of a Google search engine. His research primarily explored interpretations of the religious community's responses to the events of the Haitian earthquake and the Icelandic volcanic eruption through the eyes of the cartoonist. Lang and Wee speculated more broadly on the different religious responses to the tsunami in Southeast Asia, describing a broad range of religious traditions and their responses to such disasters (Lang and Wee 2005).

The interactions of worldviews and extreme events are illustrated in the example of a famous study reported in *Science* in 1972. Researchers attempted to explain the disproportionately higher frequency of tornado deaths in the southern United States compared to the North (Sims and Bauman 1972). The hypotheses they tested included the following: the South had more tornados or higher population density, the South had a greater number of tornados at night, the strength of tornados was greater in the South, poorer quality of housing in the South led to more deaths, and the South had a poor quality of warning system.

None of these factors could explain the difference in death rates. Researchers turned to human perception and responses to the tornado threat in an attempt to find the explanation. What did they find? Southerners placed more weight than northerners on forces external to themselves. They exhibited a greater sense of being powerless in responding to outside threats. Southerners were less confident in themselves as causal agents and less convinced of their ability to engage in effective action. This view meant they did not use the weather service and media to assess risk, but rather went outside and looked at the sky, and they were less proactive after the event, that is, their first response was to feel bad rather than act. The southern worldview exemplifies an external locus of control, characterized by a perceived inability to shape one's future. And this worldview increased the risks associated with tornadoes because it resulted in a lack of effective action.

Previous historical work has been done on the impacts of major natural events on various societies, including religious interpretations of the tragedies that have resulted (de Boer and Sanders 2005). More recent work on the relationship between religion and natural hazards has come from a variety of disciplines. For example, R. S. Sugirtharajah (2007) explored the theological reactions of different faith communities to the Asian Tsunami which tended to either blame God for such events or attribute them to human misbehavior. Paradise (2005), in his study of earthquake risk perception in Morocco, found that less educated people tended to attribute earthquakes to divine action and retribution, and equate prediction with fortune telling and correlated higher risk to less devotion to Allah. Other researchers have explored more general beliefs and structures of trust in relation to perceptions of risk (Sjoberg 1979, 2008). These beliefs and structures of trust are studied as part of culturally constructed orientations or worldviews that are held by subgroups in a society (Boholm 1998: 151). Studies that focus on religion are in the minority, however. The vast majority of natural hazards researchers have primarily focused on socioeconomic factors affecting risk assessment (Ali 2007; Armas 2006, 2008; Gustafson 1998).

My research attempted to extend our understanding of the role of more clearly defined religious traditions in hazard events. The goal was to contribute to our conceptual understanding of ethical frameworks of religious communities that shape their perceptions and actions (Buranakul et al. 2005: 246–247). Such beliefs may be more essential to understanding responses than once thought. As Sjoberg stated, "The properties of belief structures which account for risk judgments may yet be of great interest … semantically distant contents may be connected by means of beliefs about causal relationships" (Sjoberg 1997: 127).

12.2 Asian Tsunami and Hurricane Katrina

The Asian Tsunami (December 2004) and Hurricane Katrina (August 2005) occurred close together in time. Both were devastating disasters and were widely covered in a variety of media. Similarly, the responses to both events that were reported in the press ranged from the perceptions of these events being "acts of God" to "forces of evil" to totally natural events. The range of responses reflected and illuminated the shape of the religious commitments that underlay them (Freeberg 2005; Martel 2006; Hart 2005; Sugirtharajah 2007: 125–126). Their occurrence offered an opportunity to study the effects of religious worldviews on responses to natural disasters.

I started this study with some specific assumptions. The primary assumption was that such intense experiences illuminate key relational aspects of religious worldviews. First, they elicit views on the relation between God and the natural world. Secondly, they illuminate the perceived relationship between humans and the natural world, central to understanding and interpreting natural hazards. Finally, the exploration of responses to such events allow for the study of various religious impulses in terms of their response to, development of, and understanding of societal structures that shape the impacts of natural hazards on a larger scale. For example, how much internal resilience is seen in a community? How effective are the responses of the various governmental units? Are there strong nonprofit organizations that fill in the gaps, collaborate amongst themselves and with government? And finally, how do religious worldviews affect how cultures understand and respond to risk. These moments of great stress thus illuminate religious worldviews, but they also demonstrate how worldviews shape the strength and vitality of the social structures within which we live and work.

12.3 Methodology

This research is part of a larger interdisciplinary project that explored the relationship between Christian worship and civic life. Initially, ten churches in a transect in a Midwestern city that extended from the center of the city outward were studied over a year's time. Urban and suburban churches were paired from four Christian traditions representing a range of religious perspectives (mainline—Methodist, fundamentalist—Baptist, Catholic, and Calvinist—Reformed). This particular urban transect was chosen because of its great ethnic and denominational mix; the variables taken into account in the choice of the ten churches in the transect included an urban-suburban mix, ethnicity/race, and a range in economic status. Members of a larger research team completed ethnographic research of the congregations, including in-depth personal interviews and analysis of church histories, examining how members defined and understood notions of race, class, gender, and community (Mulder 2009, 2011, 2012; Heffner 2007; Isom 2012).

From these ten churches, audio-taped services from the second Sunday of each month were collected. The worship services were transcribed and analyzed using both close textual analysis and content analysis to identify the themes, language, and rhetorical appeals to individuals and communities. Worship, it is worth mentioning, remains an understudied but vital feature of American social life. Pastors' sermons act as meaning-makers for congregations, shaping and framing beliefs and actions. It is estimated that every week over 100 million North Americans attend worship services. Prominent American theologian Stanley Hauerwas (2008) has noted that the practice of worship is transformative for the participants through its communal nature, its sustained presence, and its focus on the formation of its participants. In short, because of its socially formative role and sustained pervasiveness, worship continues to have a profound effect on the social life of many Americans.

Initially my study intended to focus on the relationships expressed among God, nature and humankind that were articulated among the churches in the study transect. Unfortunately, analysis of the 120 worship services showed virtually no mention of the relationships among God, nature, and humans! A significant finding in itself, the study was reshaped in order to find contexts in which events would prompt comments and analysis related to the relationships of the study. Consequently, the transect church sermons were supplemented by the analysis of sermons from Reformed, Methodist, Baptist, and Catholic churches following both the Asian tsunami (January 2, 9 and 16, 2005) and Hurricane Katrina (September 4, 11 and 18, 2005), from both the transect churches, when available, and from the web. Both the Asian Tsunami and Hurricane Katrina events were used because it was hypothesized that closer geographic proximity to the events would lead to greater nuance in interpretation and much greater understanding of the role of societal structures in the creation of natural hazards (Sjoberg 2000: 2).

A total of 52 sermons across four Christian traditions were included in the study (Table 12.1). The Catholic and Baptist samples included several African American congregations and also included several sermons from outside the United States, viz., the United Kingdom, Australia, and Canada. The analysis showed little influence from either ethnicity or nationality and no difference based on geographic proximity to the events.

Initial categories for content analysis were established prior to the reading of the sermons and were informed by previous research on relationships among God,

Table 12.1 Categorization of sermons analyzed in the study

	Asian Tsunami	Hurricane Katrina
Reformed	8	6
Methodist	3	4
Catholic	7	5
Baptist	11	8

Source: Janel Curry

nature, and society of different Christian denominations in North America (Curry 2000; Curry and Groenendyk 2006; Guth et al. 1993, 1995; Hand and Van Lierre 1984; Schultz et al. 2000). These categories included eschatology, communal versus individualistic views of society, and locus of control.

The full group of 52 sermons was read by two independent researchers using these initial categories. The two researchers then met and revised and expanded the categories based on this first reading, and established agreed-upon subcategories and definitions. A more fine-grained analysis of the sermons was then carried out based on these categories. In addition, information was collected on each congregation from either the larger study (in the case of the city transect) or from the websites of the churches. These characteristics included location, denomination, general size, and racial composition of the congregation. Sermon summaries and scriptural references to the sermon were also included in the analysis. A tabulation of quotes and perspectives illustrating each subcategory per subject was compiled. Based on this database, computations were done by religious tradition. The findings and interpretations were further subjected to assessment from the larger research team and to several groups who represented a broad range of Christian denominational backgrounds and were involved in planning worship services within their congregations.

The final categories of analysis that arose in the analysis included: (1) eschatology (the study of "end times")—where is history going?, (2) integration of God, nature, and humans (conceptual framework)—how do sermons theologically construct the relationship among humans, nature and God?, (3) responsibility and responses to events (agency)—who, or what is responsible for responding to events?, and (4) otherness (identification with)—how are those who have suffered from these events depicted?

12.4 Findings

Eschatology—where is history going? The subcategories of analysis under this worldview element ranged from the total absence of comments related to eschatology to those that illustrated (a) the relationship between these events and the "end times," (b) an emphasis on the present only, (c) a general reference to the second coming of Christ and the future perfection of the earth in terms of ultimate hope, and (d) a very explicit emphasis on these events being "signs of the times" that predict the coming judgment.

The majority of Reformed sermons made no mention of eschatological themes (71 %). References were made to the second coming of Christ and future perfection of the world (21 %). Some emphasis was placed on the events leading to repentance.

The Methodists likewise made few references to eschatological themes (71 %). They put an emphasis on the present—the redeemed have been changed (present tense) and need not fear death. The events were primarily used to call out for compassion.

The Catholic sermons were similar to the Methodists in the lack of eschatological themes (75 %), and their emphasis on present reality of the redeemed being changed, and need not fear death (17 %). But an added uniqueness to the Catholic sermons was the allowance of mystery, along with models of religious figures. For example, we do not know "why" but we know that Jesus has gone before us; we have the model of Mary pondering the mystery of the incarnation and the Holy Family, who were also refugees.

The Baptist sermons put a much greater emphasis on eschatological connections to the events. Forty-seven percent referred to the events as "signs of the times" predicting the return of Christ and 11 % called for repentance in light of the coming judgment. Thus the events were used as warnings of coming judgment and used as lessons.

Integration of God, nature, and humans asks this basic question: how do sermons theologically construct the relationship among humans, nature and God? The conceptual integration of God, nature and humans was the second category of analysis. This category of analysis was by far the most complex. The view of the relationship between nature and God ranged from nature being totally apart from God to God being active in sustaining nature, to God using nature directly to teach lessons to humans. Perspectives also ranged in terms of the moral autonomy of nature. For example, some sermons depicted nature as a neutral force governed by natural laws. Others viewed nature as good and beautiful because it was created by God. On the other extreme, sermons also depicted nature as off-kilter and broken due to original sin. In this latter view, sermons connected the fallenness nature of humanity with an imperfect natural world. This view of nature's brokenness did not refer to environmental pollution, but to something more inherently wrong with nature that included the presence of hurricanes and earthquakes. Finally, this category of analysis looked at the relationship between humans and nature. Perspectives represented in the sermons ranged from seeing humans and nature co-equal to seeing nature as a terrifying, brute force to which humans were subject.

The Reformed sermons emphasized the role of God as an on-going sustainer of nature (43 %) over all other groups. Reformed sermons emphasized God's sovereignty over nature rather than nature as a subjugating force on humanity. Responses were evenly distributed among views that saw nature as off kilter, neutral, or good.

The Methodist sermons were characterized by the absence of reflection on the relationship among God, nature, and humanity (71 %). Nature itself was largely absent among the Methodists. Other categories that were minor but present included humans and nature as co-equal (14 %) and human sin as having warped nature (14 %).

Catholic sermons also showed a general lack of integrative thought (77 %). Where nature was mentioned, the Catholic sermons were the only ones to emphasize that nature was good (17 %). This finding is confirmed by previous research on Catholics and their views of nature (Greeley 1993). Other evidence of this emphasis on the goodness of nature was reflected in comments on the need for humans to harmonize with nature. Catholic sermons were unique in their emphasis on humans being subject to nature, which fit the need to harmonize. This unique emphasis on harmonization with and inherent goodness of the natural world may reflect the tradition's emphasis on natural law, an ethical framework that has a starting point of seeing humans participating in eternal law by, as rational beings, discerning and participating in it (The Natural Law Tradition in Ethics 2008). Influenced by this body of thought, sociologist Andrew Greeley has found Catholics to be characterized by the belief that God is present in the world and is disclosed in and through creation. But such goodness is not limited to the natural world. Society is also seen as natural and good in its expression of a set of ordered relationships governed by both justice and love (Greeley 1989: 486).

The Baptist sermons showed a consistent use of nature in the abstract, that is, God used nature to teach lessons or give warnings to humans (33 %). They were the only group to use nature in this way. Baptists pointed to human sin as the source of imperfection in nature (19 %) and their sermons emphasized God's direct control of nature, and nature as being off-kilter.

This category included an analysis of understanding of social structures, that is, to what extent are human social structures understood to contribute to these tragedies? All groups showed a general lack of understanding of natural hazards and events like Katrina and the Asian Tsunami in relation to human structures. Only 14 % of the Reformed sermons showed at least minimal understanding of human structures and the events. The Catholics were similar (18 %), but several sermons constructed an analysis of the issues of global warming and human culpability. Methodists showed the greatest extent of structural analysis. The relationships between poverty and ability to evacuate were clearly stated in several cases. This emphasis may reflect the Methodist tradition of taking creeds and confessions as boundaries to work within. Instead of focusing only on "what they believe," a Methodist oriented identity emphasizes the implications of beliefs on what one does. John Wesley, the founder of Methodism, taught that the highest purpose of Christian doctrine was to provide practical guidance for Christian life in the world (Maddox 1997: 126). Methodist theology is thus shaped by what is called the Quadrilateral made up of scripture, tradition, experience, and reason (Gunter et al. 1997: 10). The Baptists showed the lowest level of references to structural understanding to the events. Only one sermon showed even the most minimal understanding (6 %) and several verged on blaming the victims. In fact, one sermon blamed the Asian countries for having no warning system, glossing over the public discussion related to the distribution of wealth in the world and poor countries' ability to maintain such systems.

Sermons called for a variety of responses to events (agency). Again, the sermons ranged greatly in the responses they regard from their congregants. In some cases, no response was requested. In others, congregants were asked to discover the meaning of the suffering. In general the responses could be categorized between individualized responses and communal or institutional responses. Individualized responses included everything from asking for individual acts of service or prayers to asking individuals to put their faith in Jesus and witness to others. Communal/institutional responses ranged from teaming up with others in joint relief efforts and corporate giving to changing the policies that contribute to these tragedies.

The Reformed sermons showed a balance between individual (47 %) and communal responses. Sermons that called for individual compassion and putting faith in Jesus dominated (72 %) of the individualized responses. Twenty-seven percent of the communal responses focused on being community/agents of hope and supporting relief efforts or corporate giving. No references were made to structural change. Methodists also showed a balance between individualized (55 %) and communal responses. What dominated in the sermons were calls for individual compassion, for putting faith in Jesus, and for prayers. Among the communal responses (45 %) the sermons emphasized being community/agents of hope and also supporting relief efforts. In comparison to other groups, the Methodist sermons showed an emphasis on changing policies (18 %). But in general, congregants were asked to be willing hands individually and collectively, to pray and to grieve.

Catholics showed a tendency toward more communal responses (67 %). While compassion and prayers were requested at the individual level, calls for communal responses included a strong emphasis on structural change (13 %), support for relief work and being agents of hope. The Catholics were the only group to talk about valuing the poor, hungry, and sorrowing and seeing them as through God's eyes (13 %). One phrase that showed up several times spoke of the need to seek peace and justice, dignity for all humans, and live out lives of concern and love for the poor and suffering.

The Baptists in the study, unsurprisingly, had a much greater level of individualization than the other three groups. Ninety percent of the comments were individualized, and among those, 24 % focused on compassion, service, and generosity, 43 % called for reflection and repentance and 14 % called for people to witness to others.

This pattern reflects previous studies that divide religious traditions between those that are believing (mental) and belonging (social). Catholics fall into the "belonging," ritualistic and sacramental category (Wald and Smidt 1993: 32, 34). This category is characterized by centralized religious authority and institutional mediation between humanity and God (Kellstedt and Green 1993: 57). Pietists, at the other extreme emphasize unmediated contact between believers and God and individual religious experience (Kellstedt and Green 1993: 57). Mainline denominations, such as Methodists, have a mixed pattern, reflecting both tendencies (Kellstedt and Green 1993: 61).

The variable of "otherness" related to identification with those affected by the events. How were those who have suffered from these events depicted? What is the relationship between sin and suffering, if any? What is the distance placed between "us" and those who experienced these events? Reformed sermons, more than any others, purposefully said we are sinful like those who experienced these natural disasters (36 %). Fifty percent of the sermons purposefully led to some form of self-identification of the congregation with those who were suffering. The Reformed sermons purposefully went out of their way to reject any relationship between sin and suffering. Suffering was seen as the result of the general fallenness of the world and was used by God to bring people back to him. No direct connection between an individual's state and the events was implied. This approach, unique in Reformed sermons, is similar to the findings of Karen Halttunen (2000), in her book, *Murder Most Foul*. She shows how the Puritans regarded even those convicted of crimes as a moral representative of all sinful humanity, not somehow unique in their "sinfulness." The Puritan preaching emphasized the cliché: "There but for the grace of God, go I" (Halttunen 2000).

Methodists emphasized the universality of those who suffer, naming and personalizing those affected. Thus they showed the strongest measure of self-identification (8 %), with many references to these "neighbors." Methodists drew no connection between sin and suffering. Rather, the problem of evil was seen as a mystery, and one sermon turned the table to ask congregants to answer the questions: Who suffers? Who is left behind and why? Randy Maddox argues that the Methodist tradition emphasizes both scripture along with experience, which results in a call to be sensitive and willing to learn from those who have suffered (Maddox 1997: 110).

Catholic sermons were especially strong in their self-identification (39 %) with the suffering. Illustrations or self-identification with the suffering of the Holy Family intensified this link. In addition, comments were made on the need to help the poor and the universality of those who suffered from these events; one sermon went yet the next step to be critical of those who separate themselves from the suffering.

Baptists were the only group to show strong boundaries between themselves and those who experienced these events. This lack of identification (42 %) with those who suffered in these events tended toward blaming the victim. When identification was expressed, it was from the mutual common fear of the coming judgment. An example of the psychological distance of the Asian tsunami victims from Baptist congregants was seen in one sermon, where the only reference to the tsunami in the sermon was the following comment: "Who among us would have predicted that the Red Sox would win the World Series after 100 years or that the year would have ended with a lifetime tsunami disaster?" Baptists were the only group to argue for a connection between sin and suffering. The events were used as spiritual lessons illustrating the failure of people to heed the warning signs and evacuate. The African American Baptist church was an exception to this trend. This sermon argued that the event could have happened at their own front door, bringing some self-identification with those who experienced the disasters.

12.5 Summary and Further Research

Differences amongst the groups were significant. While the Baptists identify themselves as biblical "literalists" (Carpenter 1997), their sermons largely used natural disasters as very abstract lessons, distancing themselves from the events and those affected.

Catholics differentiated themselves both in their communally strong identification with those affected by these events, intensified through identification with religious figures who had also experienced suffering, but also in their placement of people as being subject to nature, which is good.

Methodism reflected its mainline denominational heritage in its emphasis on social justice, but also reflected its own individual heritage, which emphasizes personal spiritual growth (the "method" in Methodism). This appeared in the absence of any reflection on the relationships among God, nature, and humans.

Reformed theology was seen in both the emphasis of God as the on-going sustainer of the natural world as well as in the expressed belief that this actual earth will be made new when Christ returns.

Overall, it appears that across the groups, they exhibit little understanding of the role of social structures such as race, incomes, etc. on the distribution of the effects of natural disasters. Furthermore, little understanding was expressed concerning the relationship between human choices and nature, which in fact create natural disasters. Unexpectedly, this relationship was equally true with Hurricane Katrina and the Asian Tsunami, in spite of greater press coverage related to structural inequality and human impacts contributing to the destruction in New Orleans.

What can be drawn from this analysis of worldviews? The findings illuminate a range of worldview elements among Christian traditions that shape interpretations of natural hazards. These structures involve a variety of elements of vertical and horizontal orientations with vertical orientations focusing on the relationship between humans and God while a horizontal orientation emphasizes the outworking and understanding of religious faith in terms of relationships with other humans and nature. Davidson has argued for a similar categorization of religious worldviews (Leege and Kellstedt 1993: 232).

The first vertical and horizontal orientation element is related to nature. John Calvin embraced the created world. He saw the created world as something to be studied, understood and from which God and humans could take pleasure. For Calvin, God's concern and thus our human concerns should incorporate the physical earth into its understanding of the horizontal reality. Calvin argued that knowledge of nature was a gift of God that God lavished on believers and unbelievers alike (Young 2007: 8) and that the downgrading, neglect, or willful ignorance of science and learning were reprehensible (Young 2007: 10). Yet, he rejected a mechanistic view of nature. He saw all of the creation as ultimately owing its existence and character to the sovereign will of God, being sustained and subject to the will of God (Young 2007: 58). He saw God as continuously providing the various entities the energy and ability to function in such a way that they behave and operate in a consistent regular manner (Young 2007: 203).

Many sermons, especially amongst the Reformed, attempted to incorporate the physical earth in their understanding of these events, but ended up with a focus on the meaning of the fall when it came to nature and these events. In these cases the very existence of hurricanes and earthquakes was used as evidence of the fallenness of the creation. For example, one sermon depicted nature as fallen and off-kilter, while using the "15° inclination of the earth" as evidence of this. The reality is that the tilt of the earth creates the seasons, the wind and ocean circulation patterns, and the diversity of our ecosystems. Hurricanes are part of the system of distribution of energy across the globe as are earthquakes.

If the Reformed struggled with how to depict these physical events, among the Methodists, nature was absent, and among Baptists, it was merely a tool used by God to either punish, teach us lessons, or used to give warnings of the second coming of Christ. Catholics uniquely emphasized the goodness of the natural world.

This struggle with the place of the physical earth and the meaning in the Christian religious tradition is addressed by N.T. Wright in his newest book, *Surprised by Hope: Rethinking Heaven, the Resurrection, and the Mission of the Church*. While the Catholic tradition has a long history of theological thought on the natural world, the Protestants in this study struggled at the integration. This reasoning is perhaps why Wright's recent book has received a great deal of attention within that community (Wright 2008). The meaning of the fall for the natural world remains a challenge for the Protestants.

The second aspect of vertical and horizontal orientation involved the human community. Are events such as the Asian Tsunami and Hurricane Katrina used to call people to focus on their own relationship with God (act vertically), or used to call people to horizontal action, or both? The dominance of a vertical relational response to these events comes out of several roots. It can reflect a static and hierarchical view of the world where structures are accepted as given. It can also reflect a very individualized view of society, where individual concern is focused on the state of their individual soul, or the exploration of their own heart (Goudzwaard 1979: 11). The dominance of a horizontal perspective reflects a different view of the individual person, where the human is capable of changing circumstances, and their own fortunes (Goudzwaard 1979: 21).

One aspect of this vertical and horizontal orientation is seen in the tension between the individual or collective response (Kuyper 1931). Reformed and Methodists expressed a balance in response to these events—both individual reflection, but also action as a community. John Calvin's theology does emphasize the person in community and the tie of mutual love and regard (Graham 1987: 60). From different roots, Methodists were at the forefront in the twentieth century in uniting worship and social action. This arises out of their emphasis on connecting theology with real-life situations and experiences (White 1993: 475).

The Catholic response was even more communal, reflecting the nature of the Catholic Church, its parish model, and the communitarian nature of its theology and structures (Leege and Trozzolo 1985; Welch and Leege 1988). Not surprisingly, Baptists were the most individualized and vertical in their orientation. John Stott argues that the values and spirit that created and sustained free institutions throughout the English-speaking world were initially grounded in evangelical revivals, maintaining both a vertical and horizontal orientation (Stott 1990: 2). The great reversal of this relationship came in the early twentieth century and resulted from fights against theological liberalism, the social gospel, widespread pessimism following WWI, and the growth of premillennialism eschatology (Stott 1990: 6–8). The results of this worldview are evident in this study. Very little horizontal expressions of faith were evident among fundamentalist Baptists.

The strength of a vertical versus a horizontal orientation is also evident in attitudes toward social structures. The dominance of a vertical orientation, combined with individualism leaves an absence of theorizing in terms of the role of the state and human institutions in both contributing to these crises, but also their role in responding to such crises. Protestants and particularly Pietists remain focused on personal ethics whereas Catholics are more likely to see human nature and human structures as good (Greeley 1989: 493).

A Reformed perspective sees governmental structures as needed due to sin, on the negative side, but on the positive side, as an outgrowth of humanity's social nature, viz., the horizontal and relational aspect of our being made in God's image. Abraham Kuyper (1931: 79–80) argued that even without sin there still would have been political life. We are horizontal beings. This strong valuing of the development of both critique and involvement in shaping human institutions leads to a strong internal locus of control. Historian Fred Graham pointed out that John Calvin moved politics out of its previous state where the common person was a passive nonparticipant. The valuing of public life and institutions led to the common person learning habits of managing affairs that allowed them to engage society rather than withdraw (Graham 1987: 172).

In spite of the differences amongst groups toward government, with the Baptists being the group that was most suspicious of any secular organizational responses, no sermons from any of the groups exhibited much understanding of the role of social structures in the creation of these two natural disasters. The sermons remained very flat in comparison to the reality of the complexity and richness of society. The dominance of a vertical orientation, combined with individualism, left an absence of theorizing in terms of the role of the state, in this case in responding and preventing disasters. Any response from the state was viewed with suspicion because the dominant relationship is between the individual and God.

Finally, vertical and horizontal orientation is seen in expressed connections between the present and the future. In this study, the Baptists were shown to be future oriented—focused almost entirely on vertical eschatological hope. Consistent with previous studies, a premillenial eschatology remains one of the most powerful influences (Guth et al. 1995; Curry-Roper 1990).

Catholics and Methodists emphasized the horizontal present, more focused on the realism of the pain and suffering in the events. Reformed sermons attempted to hold the two in tension. The connection between sin and suffering can also be considered a temporal factor related to the vertical and horizontal. How do past sins relate to present suffering? How do present sins relate to future suffering and judgment? Graham argues that Weber was wrong and that John Calvin argued that poverty and misfortune are not evidence of God's disfavor, nor is prosperity a sign of God's blessing (Graham 1987: 66). Calvin saw both wealth and poverty as sacramental, that is, channels of grace from God, and making it clear that most often the rich failed the test of using it well (Graham 1987: 67). Calvin did not see the personal character of the needy, nor their gratitude or lack of it, as in any way affecting their position as God's ambassador to receive what the rich owed to God (Graham 1987: 69).

12.6 Implications

Religious belief systems shape responses to natural disasters. However, it appears that the groups in this study exhibited little understanding of the role of social structures in the creation of these two natural disasters. Furthermore, little understanding was expressed concerning the relationship between human choices and nature, which in fact create natural disasters. This was true with Hurricane Katrina as well as the Asian Tsunami, in spite of all the press coverage related to structural inequality and human impacts contributing to the destruction in New Orleans.

The sermons in this study also showed very minimal understanding of the natural environment. The absence of any mention of nature in the original 120 sermons of the study transect illustrate the absence of religious teaching, in its most prominent and popular form, on nature, on the relationship between humans and nature, and on the religious meaning and place of the physical earth within Christian doctrine.

Several works in theology and in the field of natural hazards research have added to our interpretation of the events surrounding Hurricane Katrina and the Asian Tsunami since the delivery of the sermons that were used in this study. Theologian Terence Fretheim (2010) in his book, *Creation Untamed: The Bible, God, and Natural Disasters*, argues that creation is not finished but is a dynamic process open to a number of possibilities and that natural disasters are part of God's design. He goes on to link these natural events to the human choices reflected in our built environment and social structures when he states that "specific natural events may be made more severe by human sin" which leads the effects of it to be connected to divine judgment (Fretheim 2010:150–152). He reflects on the range of views found within the Christian Bible and also evident in the sermons in this study concluding that:

> …the way in which God is related to these events is not as crystal clear as many Bible readers would like to think. Some texts are thought to focus on issues of sin and judgment (e.g., the flood), and others on issues of the nature of the created order (Job). And how one might bring such understandings together is not altogether evident.
>
> The way in which God is thought to be involved in natural disasters in our own time is heavily influenced by the way in which one reads these disaster texts in the Bible. (Fretheim 2010: 149)

While theological reflection such as Fretheim has proceeded, bringing a more complex and nuanced understanding of the relationship between religion and natural disasters, geographers who study natural hazards have continued to bring deeper analysis to the specific disaster rout by Hurricane Katrina. For example, Watkins and Hagelman (2011) found that race was the most salient variable related to the impacts of Hurricane Katrina. The impacts of the hurricane intersected with social-spatial changes resulting from a historical pattern of white-flight and increasing racial segregation (Watkins and Hagelman 2011: 124). The extensive body of literature produced by Craig Colten has furthered our understanding of the history of human choices made over the centuries that ultimately led to the disaster (Colten and Sumpter 2009; Colten 2006). As Colten and Sumpter conclude, in the end, natural events are not experienced equally and,

> All too often, the record focuses on the geophysical and not the social dimensions of past events…. The social-construction of disaster perspective considers long term processes and acknowledges that human changes to the landscape contribute to calamities. (Colten and Sumpter 2009: 355–356)

Social scientists and religious scholars need to come together to bring a more nuanced and deeper understanding of the role of religion in natural disasters, both prior to and after such events. The horizontal and vertical orientations developed out of this research may provide a conceptual analytical tool for both scholars and those from religious communities. Analysis of the strength of the orientations amongst groups may provide yet one more way to think about how religious worldviews are shaping responses to natural disasters.

Natural disasters are only going to increase from the combination of increasing populations, particularly concentrated along coastal margins, and climate change. Faith-based groups provide a great deal of social capital in times of such disasters. The great range of perspectives found in this study among a group of faith-based entities illuminate some of the theological reflection that is going to be needed in order to strengthen the engagement of the Christian community with others in addressing these challenges.

Acknowledgement I want to thank the Calvin Center for Christian Worship, the Lilly Foundation, and the research team made up of Gail Heffner, Denise Isom, Kathi Groenenyk, and Mark Mulder for their support and contributions to this study.

References

Aay, H., & Griffioen, S. (1998). Introduction. In H. Aay & S. Griffioen (Eds.), *Geography and worldview* (pp. xi–xiv). Lanham: University Press of America.

Ali, A. M. S. (2007). September 2004 flood event in southwestern Bangladesh: A study of its nature, causes, and human perception and adjustments to a new hazard. *Natural Hazards, 40*, 89–111.

Armas, I. (2006). Earthquake risk perception in Bucharest Romania. *Risk Analysis, 26*, 1223–1234.

Armas, I., & Avram, E. (2008). Patterns and trends in the perception of seismic risk. Case study: Bucharest Municipality/Romania. *Natural Hazards, 44*, 147–161.

Boholm, A. (1998). Comparative studies of risk perception. A review of twenty years of research. *Journal of Risk Research, 1*, 135–163.
Brunn, S. D. (2010). Cartooning and Googling God and natural disasters: Iceland's volcanic eruption and Haiti's earthquake. *Mitteilungen Der Osterreichischen Geographischen Gesellschaft, 152*, 251–275.
Buranakul, S., Grundy-Warr, C., Horton, B., Law, L., Rigg, J., & Tan-Mullins, M. (2005). Intervention: The Asian Tsunami, academics and academic research. *Singapore Journal of Tropical Geography, 26*, 244–248.
Carpenter, J. (1997). *Revive us again: The reawakening of American fundamentalism* (p. 47). New York: Oxford University Press.
Colten, C. E. (2006). Vulnerability and place: Flat land and uneven risk in New Orleans. *American Anthropologist, 108*, 731–734.
Colten, C. E., & Sumpter, A. R. (2009). Social memory and resilience in New Orleans. *Natural Hazards, 48*, 355–364.
Curry, J. M. (2000). Community worldview and rural systems: A study of five communities in Iowa. *Annals of the Association of American Geographers, 90*, 693–712.
Curry, J. M., & Groenendyk, K. (2006). Place and nature seen through the eyes of faith: Understandings among male and female seminarians. *Worldviews: Environment, Culture, Religion, 10*, 326–354.
Curry-Roper, J. M. (1990). Contemporary Christian eschatologies and their relation to environmental stewardship. *The Professional Geographer, 42*, 157–169.
Cutter, S. L. (Ed.). (1994). *Environmental risks and hazards*. Englewood Cliffs: Prentice-Hall.
de Boer, J. Z., & Sanders, D. T. (2005). *Earthquakes in human history: The far-reaching effects of seismic disruptions*. Princeton: Princeton University Press.
Freeberg, D. L. (2005). Free to hate: An American church rejoices in Swedish tsunami victims. *Perspectives: A Journal of Reformed Thought, 20*, 18–28.
Fretheim, T. E. (2010). *Creation untamed: The Bible, God, and natural disasters*. Grand Rapids: Baker Academic.
Goudzwaard, B. (1979). *Capitalism & progress: A diagnosis of western society*. Grand Rapids: William B. Eerdmans Publishing Company.
Graham, W. F. (1987). *The constructive revolutionary John Calvin & his socio-economic impact*. Lansing: Michigan State University Press. Originally published by John Knox Press, Atlanta in 1971 and reprinted in 1978.
Greeley, A. (1989). Protestant and Catholic: Is the analogical imagination extinct? *American Sociological Review, 54*, 485–502.
Greeley, A. (1993). Religion and attitudes towards the environment. *Journal for the Scientific Study of Religion, 32*, 19–28.
Griffioen, S. (1998). Perspectives, worldviews, structures. In H. Aay & S. Griffioen (Eds.), *Geography and worldview* (pp. 125–143, p. 126). Lanham: University Press of America.
Gunter, W. S., Jones, S. J., Campbell, T. A., Miles, R. L., & Maddox, R. L. (1997). *Wesley and the quadrilateral: Renewing the conversation*. Nashville: Abingdon.
Gustafson, P. E. (1998). Gender differences in risk perception: Theoretical and methodological perspectives. *Risk Analysis, 18*, 805–811.
Guth, J. L., Kellstedt, L. A., Smidt, C. E., & Green, J. C. (1993). Theological perspectives and environmentalism among religious activists. *Journal for the Scientific Study of Religion, 32*, 373–382.
Guth, J. L., Green, J. C., Kellstedt, L. A., & Smidt, C. E. (1995). Faith and the environment: Religious beliefs and attitudes on environmental policy. *American Journal of Political Science, 39*, 364–382.
Halttunen, K. (2000). *Murder most foul: The killer and the American Gothic imagination*. Cambridge: Harvard University Press.
Hand, C. M., & Van Liere, K. D. (1984). Religion, mastery-over-nature and environmental concern. *Social Forces, 63*, 555–570.

Hart, D. B. (2005). *The doors of the sea. Where was god in the tsunami?* (pp. 205–214). Grand Rapids: William B. Eerdmans Publishing Company.

Hauerwas, S. M. (2008). Worship, evangelism ethics: On eliminating the 'and'. In P. W. Chilcote and L. C. Warner (Eds.), *The study of evangelism: Exploring a missional practice of the church* (pp. 205–214). Grand Rapids: William B Eerdmans Publishing Company.

Heffner, G. G. (2007). *Congregations and neighborhoods: Race and connections to place*. Seattle: Urban Affairs Association.

Isom, D. (2012). Fluid and shifting: Racialized gender and sexual identity in African American children. *International Journey of Interdisciplinary Social Science, 6*, 127–137.

Kellstedt, L. A., & Green, J. C. (1993). Knowing God's many people: Denominational preference and political behavior. In D. C. Leege & L. A. Kellstedt (Eds.), *Rediscovering the religious factor in American politics* (pp. 53–69). Armock: ME Sharpe.

Kuyper, A. (1931). Calvinism and politics. In *Lectures on Calvinism* (pp. 78–109). Grand Rapids: William B. Eerdmans Publishing Company. Reprinted, 1994.

Lang, G., & Wee, W. (2005). Religions and the tsunami disaster in Southeast Asia. ASA: Sociology of Religion: Newsletter of the Sociology of Religion Section of the American Sociological Association 11, 3 and 8.

Leege, D. C., & Kellstedt, L. A. (1993). Religious worldviews and political philosophies: Capturing theory in the grand manner through empirical data. In D. C. Leege & L. A. Kellstedt (Eds.), *Rediscovering the religious factor in American politics* (pp. 216–231). Armock: ME Sharpe.

Leege, D. C., & Trozzolo, T. R. (1985). Religious values and parish participation: The paradox of individual needs in a communitarian church. *Notre Dame Study of Catholic Parish Life, 4*, 1–8.

Maddox, R. L. (1997). The enriching role of experience. In S. J. Gunter, S. J. Jones, T. A. Campbell, R. L. Miles, & R. L. Maddox (Eds.), *Wesley and the quadrilateral: Renewing the conversation* (pp. 107–127). Nashville: Abingdon.

Martel, B. (2006, January 17). Mayor Nagin says 'God is mad.' *The Grand Rapids Press*, p. A3.

Mulder, M. (2009, October 24). Churches and neighborhoods: The link (?) between worship and place. *Society for the Scientific Study of Religion, Annual Meeting*, Denver.

Mulder, M. (2011, March 18). Maintaining internal homogeneity by avoiding the local: A case study of urban and suburban congregations. *Urban Affairs Association Annual Meeting*, New Orleans.

Mulder, M. (2012). Worshipping to stay the same: Avoiding the local to maintain solidarity. In R. Hawkins & P. Sintiere (Eds.), *Christians and the color line*. New York: Oxford University Press.

Paradise, T. R. (2005). Perception of earthquake risk in Agadir, Morocco: A case study from a Muslim community. *Environmental Hazards, 6*, 167–180.

Resources for the Future. (2009). Are catastrophes insurable? *172* (Summer), 19–23.

Schultz, P. W., Zelezny, L., & Dalrymple, N. J. (2000). A multinational perspective on the relation between Judeo-Christian religious beliefs and attitudes of environmental concern. *Environment and Behavior, 32*, 576–591.

Sims, J. H., & Bauman, D. D. (1972). The tornado threat: Coping styles of the North and South. *Science, 176*, 1386–1392.

Sjoberg, L. (1979). Strength of belief and risk. *Policy Sciences, 11*, 39–57.

Sjoberg, L. (1997). Explaining risk perception: An empirical example of cultural theory. *Risk Decision and Policy, 2*, 113–130.

Sjoberg, L. (2000). Factors in risk perception. *Risk Analysis, 20*, 1–11.

Sjoberg, L. (2008). Antagonism, trust and perceived risk. *Risk Management, 10*, 32–55.

Stott, J. (1990). *Decisive issues facing Christians today*. Old Tappan: Revell.

Sugirtharajah, R. S. (2007). Tsunami, text and trauma: Hermeneutics after the Asian tsunami. *Biblical Interpretation, 15*, 117–134.

Sullivan, P. (2006, October 9). Gibert F. White: Altered flood-plain management. *Washington Post*, p. B4.

The Natural Law Tradition in Ethics. (2008). The Stanford encyclopedia of philosophy. http://plato.stanford.edu/entries/natural-law-ethics/. Accessed 12 Oct 2009.

Tobin, G. A., & Montz, B. E. (1997). *Natural hazards: Explanation and integration*. New York: Guilford Press.

Wald, K. D., & Smidt, C. E. (1993). Measurement strategies in the study of religious factor in American politics. In D. C. Leege & L. A. Kellstedt (Eds.), *Redisovering the religious factor in American politics* (pp. 26–49). Armock: ME Sharpe.

Wallace, I. (1998). A Christian reading of the global economy. In S. Aay & S. Groffopem (Eds.), *Geography and worldview* (pp. 37–48, p. 46). Lanham: University Press of America.

Watkins, C., & Hagelman, R. R. (2011). Hurricane Katrina as a lens for assessing socio-spatial change in New Orleans. *Southeastern Geographer, 51*, 110–132.

Welch, M. R., & Leege, D. C. (1988). Religious predictors of Catholic parishioners' sociopolitical attitude, devotional style, closeness to God, imagery, and agentic communal religious identity. *Journal for the Scientific Study of Religion, 27*, 536–552.

White, J. F. (1993). Methodist worship. In R. E. Richey, K. E. Rowe, & J. M. Schmidt (Eds.), *Perspectives on American Methodism: Interpretive essays* (pp. 460–479). Nashville: Kingswood Books.

Wright, N. T. (2008). *Surprised by hope: Rethinking heaven, the resurrection, and the mission of the church*. New York: Harper Collins.

Young, D. (2007). *John Calvin and the natural world*. Lanham: University Press of America.

Chapter 13
Japanese Buddhism and Its Responses to Natural Disasters: Past and Present

Yukio Yotsumoto

13.1 Introduction: Natural Disaster-Prone Japan

Japanese people were surprised at the scenes of the *tsunami* on television on March 11, 2011. Many houses were destroyed and washed away and cars and trucks are floating on black water. People who were able to reach high ground were watching the water and cried in pain. Television also broadcasted the images of cities that were burning fiercely. For many Japanese it was unbelievable that it was happening in Japan; many of them thought it was like an incident happening in other countries. The 2011 Tohoku Earthquake was the most powerful earthquake ever recorded in Japan, registering at a magnitude of 9.0. It caused 15,829 deaths, 3,724 missing, 5,943 injured and 118,816 houses totally destroyed (National Police Agency 2011). In terms of damage, this was the largest in Japan after World War II.

Due to the location of the Japanese Archipelago, the country has two major natural disasters, typhoons and earthquakes. Typhoons develop in low altitudes in spring and move westward to the Philippines. In summer, the genesis of typhoons occurs at higher altitudes and many typhoons go northward to Japan. On an average, there were 23 typhoons developed per year between 2001 and 2010. Among those, on an average, 11.3 typhoons approached to Japan and 2.8 typhoons actually landed on Japan each year. During the Showa period (1926–1989), 6 typhoons induced more than 1,000 casualties that included death and missing (Japan Meteorological Agency 2011). The Isewan Typhoon (Typhoon Vera) in 1959 was the deadliest typhoon in the Showa era. It resulted in 4,697 deaths, 401 missing, 38,921 injured, and 40,838 houses totally damaged. In the Heisei period (1989-present), there were 5 typhoons that caused more than 40 deaths and missing. Among those, The Heisei 16 Typhoon 23 was the worst one causing 95 deaths, 3 missing, 721 injured, and 907 houses

Y. Yotsumoto (✉)
College of Asia Pacific Studies, Ritsumeikan Asia Pacific University, Beppu, Oita, Japan
e-mail: yotsumot@apu.ac.jp

totally damaged. Typhoons have also caused damage to agricultural products because they hit Japan just before harvest season. In the pre-modern era a typhoon was a life-threatening incident as their life was based on a self-sufficient economy.

Japan is also threatened by earthquake. Although Japanese land area is only 0.25 % of the area of the world, almost 10 % of earthquakes happen in Japan. The Japanese Archipelago is on the Pacific Ring of Fire; the archipelago and waters are where four tectonic plates converge. West Japan is on the Eurasian plate and Philippine plate pushes it from the southeast. East Japan is on the North American plate and the Pacific plate thrusts it. Every year, Japan has 500–1,000 perceptible earthquakes. When we look at the Japanese history, between 1605 and 1995, there were about 35 large earthquakes of magnitude greater than 7.0 hit Japan causing devastation (Karan 2005: 34). In a very recent earthquake, the 2008 Iwate Earthquake there were 17 deaths, 6 missing, 426 injured, and 30 houses totally destroyed; the 2007 Chuetsu Offshore Earthquake in Niigata prefecture resulted in 15 deaths, 2,346 injured and 1,331 houses totally destroyed; and The Hanshin-Awaji Earthquake in 1995 had 6,434 deaths, 3 missing, 43,792 injured, and 104,906 houses totally destroyed (Japan Meteorological Agency 2011).

The two violent natural disasters of typhoons and earthquakes are still major threats to Japan. In the past, Japan also faced droughts and long-continued rains that gradually impoverished people and caused famine. These incidents were more frequent than earthquakes and thus were pressing issues in the past. However, in contemporary Japan, these are not very problematic anymore as extensive irrigation systems and an effective distribution system of goods have been developed.

As shown, Japan is a natural disaster-prone country. Throughout history, Japanese people tried to overcome or manage the natural disasters. Buddhism as the main Japanese religion has faced the problem and tried to challenge them. In this manuscript, I explore two ways Japanese Buddhism has taken to face them; prayers and volunteer activities. The first one is prayers that tried to suppress natural threat by the supernatural power of Buddhism; prayers were offered to stop droughts, long-continued rains, typhoons and even earthquakes (Kobayashi 2008: 14). This was the practice carried out by monks often asked by power holders in the pre-modern era. I discuss prayers to calm the nature in three phases in Japanese history. However, as Buddhism was secularized and Japanese society experienced modernization, people began to lose interest or faith in mystic power of religion. Also, Buddhism became a formality often derogatorily referred to as funeral Buddhism. Traditional Japanese Buddhism is not socially as active as new religions (for example, Risho Kosei Kai, Soka Gakkai and Shinnyo-en) which had been established since the late Tokugawa Period. However, after the 2011 Tohoku Earthquake, some monks of traditional Buddhism actively participate in reconstruction assistance especially by participating in volunteer activities. Through describing these active monks, I would like to point out a possibility of Japanese Buddhism's new social role in contemporary Japan.

13.2 Japanese Buddhism's Response to Natural Disasters in the Past: Prayers

The importance of prayers to overcome natural disasters is observed most vividly between the late sixth century and late thirteenth century. During this period, the role of Buddhism had three stages in relation to social strata. From the sixth century to the late eighth century (the late Yamato Period and Nara Period), Buddhism had a role to protect the nation by supporting the Imperial system. Buddhism in this period is often called National Buddhism as it was used to solidify the centralized government based on the Imperial system. In Heian Period (794–1185), as the Imperial system was firmly established and the aristocracy emerged, Buddhism turned inward and mostly served the nobility who longed for Paradise and developed culture based on its image. Thus, Buddhism in this period is labeled as Aristocratic Buddhism. In Kamakura Period (1185–1333), many distinguished monks appeared and transformed the inward-oriented Buddhism. It became a dynamic social institution that served for the mass of the people. Therefore, Buddhism in this period is termed as Buddhism for the Masses (Hongo 2002: 181). It is helpful to look at the prayers to overcome natural disasters along these three stages since the nature of Buddhism to society is different in each phase.

13.2.1 Chingo Kotsuka (The Tranquil Nation by Prayers)

Chingo Kotsuka means "to protect the nation or make it tranquil by the reciting of Buddhist prayers such as the Lotus Sutra, the Golden Light Sutra and the Humane King Sutra and other Buddhist ceremonies" (Shinmura 2008: 1847). Although the idea of *Chingo Kotsuka* had existed during the three stages of Buddhism mentioned above, it was the idea most powerfully used in the period of National Buddhism.

The rulers of Yamato consolidated Japan into a single nation in the early fourth century and successive emperors fortified the system of government by importing continental knowledge from social system to culture (Nippon Steel Human Resource Development 2002: 45). As part of the efforts, Buddhism was introduced to Japan from Korea probably in 538. Initially, it was practiced among immigrants from China and Korea, and Soga clan, one of the most powerful clans at that time. It spread rapidly after receiving patronage from Soga clan who eventually controlled the Imperial Court. Asuka (a part of Nara) where the Imperial Court was located became the first center of Buddhist culture in Japan. In the seventh century, the Chinese (Tang) *ritsuryo* legal system was introduced. This created the constitutional form of government that put the land and people under the direct control of the government instead of powerful clans. Continental knowledge was the foundation of the nation building.

When we look at religion at this time, three religions (Taoism, *Jingi* worship[1] and Buddhism) coexisted although Taoism was the minor one without a religious order. In the consolidation process of the Imperial system, the position of these religions were contested but gradually put in order. In the course of learning of continental knowledge from China (Tang), the components of Taoism were also brought into Japan. In the Tang Dynasty, Taoism was regarded as an indigenous religion and Buddhism as a foreign religion; Taoism was placed as the first national religion and Buddhism as the second national religion. As religion, Taoism formed a religious order in Tang and Emperor Xuanzong of Tang encouraged Japanese missionaries to import it to Japan. However, the Japanese emperors intentionally excluded Taoism as a religions order from import items. The idea of Taoism was planted in Buddhism in China and Buddhism's introduction to Japan was meant to bring the idea of Taoism, but not as a religious order. In China, Buddhism had brought in some aspects of Taoism when the Buddhist scriptures were compiled in China. Thus, magical rites performed under Buddhism during the seventh and eighth centuries were originated in Taoism (Kanegae 2008: 228).

Looking at *Jingi* worship, large scale ceremonies of *Jingi* worship as the Imperial court rituals were established and the ceremonies conducted by emperors became important matter of the nation. Various objects of worship in different regions of Japan were placed under the system of Japanese mythology that legitimized the Imperial system (Kanegae 2008: 226–228). Emperors were the highest religious figures who performed rituals and its position was the basis of sovereignty. Emperors were regarded as the authority to perform rites to overcome the threat of nature especially they had a role of rainmaker during droughts (Tsukushi 1964).

Buddhism was believed to be a part of magic to pray for the happiness of the deceased and make statues of Buddha for cure of disease (Inoue et al. 1984: 36). From the beginning stage of Japanese Buddhism, people conceived Buddhism as having miraculous power. As discussed, Buddhism brought in this characteristic from Taoism. As emperors' legitimacy originated from their performance of *Jingi* worship, prayers based on *Jingi* worship had priority over prayers based on Buddhism to overcome the threat of nature, especially droughts and long-continued rains. Buddhism had a supplementary role to bring about magical effects that would protect the nation (Hongo 2004: 199). However, the position of these two religions changed over time as the *ritsuryo* legal system was gradually established and Buddhism which had a sophisticated system of teaching and written documents was thought to be more universal than *Jingi* worship. Empress Komyo (701–760) tried to save her husband, Emperor Shomu (reign 724–749) by asking him to rely on Buddhism instead of *Jingi* worship. Shomu was doubtful about his position as the holder of worship because he was not able to overcome natural disasters that took place year after year by means of *Jingi* worship (Hongo 2002: 187). Emperor Shomu and Empress Komyo were avid followers of Buddhism and established

---

[1] Jingi worship became Shinto after the Middle Ages when it was systematized. At this time, the name, Shinto, did not exist.

kokubunji (provincial temples) and *kokubunniji* (provincial convents) throughout Japan and giant Buddha in Todai-ji temple.

The Lotus Sutra, the Golden Light Sutra and the Humane King Sutra (*Ninno-kyo*) have been regarded as the three sutras of protecting the nation. *Ninno-e* started during Empress Saimei's time (reign 655–661). It is a Buddhist service held at the Imperial court to pray for protection from natural disasters and epidemics by reciting the Humane King Sutra. In *kokubunji*, the Golden Light Sutra was used as its main teaching and in *kokubunniji*, it was the Lotus Sutra. The Humane King Sutra states that "This sutra has immeasurable benefits. It is called the benefit of the protection of national land (Mizuno 2009: 262)." It simply expounds that by believing and reciting the Humane King Sutra, the king and people are protected. It also states that:

> When the land is in disorder, the devils definitely get out of control first. As devils are out of control, people will be out of control. Foreign bandits invade the country and peasants will die. The king, his subjects, crown princes, hundreds officials will quarrel. The heaven and earth are not in a normal state of affairs. Twenty eight constellations, sun and moon will be out of ordinary. Many insurgencies will take place. Great King! If there are calamities caused by fire, water and wind and all the difficulties, you should read and explicate this sutra. (Mizuno 2009: 262)

The sutra explains calamity in detail and teaches that the nation is protected by reciting this sutra.

13.2.2 Japanese Esoteric Buddhism

In Heian Period, Japanese esoteric Buddhism flourished by the establishment of two sects, the Tendai sect of Buddhism and Shingon Buddhism. Tendai sect of Buddhism was formed by Saicho (767–822). He conducted ascetic practices in mountains and forests, and established Ichijyo Shikan-in temple in Mount Hiei in 788, which became Enryaku-ji temple in 823, the center of Japanese Buddhism until the blossoming of Kamakura Buddhism. He participated in a Japanese mission to Tang-dynasty China in 804 and stayed in China for 8 months and learned the Lotus Sutra from Tiantai School. Through the envoy, Saicho convinced that the Lotus Sutra was the best teaching of Buddhism, but he also brought some aspects of esoteric Buddhism. In Tendai School, Ennin and Enchin, Saicho's disciples, adopted esoteric Buddhism and it became the major doctrine of the school.

Shingon sect of Buddhism was formed by Kukai (774–835). He also did ascetic practices in mountains and forests[2] and he was a member of a Japanese mission to Tang-dynasty China with Saicho in 804. He stayed in China for 3 years and learned esoteric Buddhism from Hui-quo at Qinglong temple. He established two seminaries,

---

[2] Ascetic practices in mountains and forests have a strong affinity with esoteric Buddhism (Uejima 2002: 249). Thus, Saicho and Kukai had an inclination to select esoteric Buddhism when they studied in Tang-dynasty China.

one in Koyasan and the other in Kyoto. The latter is To-ji temple which was given to Kukai by Emperor Saga in 823 and became an important place for Japanese esoteric Buddhism. Kukai's close relationship with Emperor Saga advanced his scheme to make Shingon esoteric Buddhism as an independent Buddhist sect. Due to the relationship, Shingon sect had a role of protecting the nation especially emperors (Kitayama 1965: 179).

Esoteric Buddhism is different from exoteric Buddhism because the power of Buddhism can be felt by secret incantations which are only transmitted by generation-to-generation instruction from master to disciple. Monks chose the types of Buddha depending on the characteristics of problems (such as sickness, a drought and warding off evil fortune) which were wished to be solved. Then they prayed according to the methods of secret incantations to overcome the problems. Esoteric Buddhism of Shingon and Tendai sects was embraced by the Imperial Family and aristocrats as Buddhism that aimed at gaining spiritual and material benefits by performing incantations (Inoue 1984: 62). In esoteric Buddhism, *Sokusaiho* is one of four secret incantations that extinguish one's bad karma and avoid natural disasters, sickness, the turmoil of war and sudden deaths; it was widely practiced to overcome these problems.

During Emperor Saga's reign (809–823), rainmaking by Buddhist prayers reappeared and it continued throughout the Heian period without interruption. Buddhist prayers to overcome drought were conducted by Soga Family in Yamato and Nara periods, but it was a supporting role to rainmaking by *Jingi* worship that had disappeared for 70 years until 809 when prayers for rain was performed in the Yoshino Mountains. Dynamism of Buddhism stimulated by Saicho and Kukai is one of the factors that Buddhist prayers resurfaced in rainmaking (Yabu 1992: 63).

According to Yabu (1992), there were four types of Buddhist rainmaking in the Heian period. The first one was rainmaking conducted at various large temples within the vicinity of the capital or outside it. This was the first Buddhist rainmaking among them. However, this type had not been popular; it was practiced sporadically for only 25 years out of 397 years of the Heian period. The Imperial Court asked the temples to perform it and monks recited sutras. The second type is rainmaking by reciting sutras at *Daigokuden* (an imperial audience hall). It was performed for 49 years in the Heian period. At *Daigokuden*, 100 monks chanted the Wisdom Sutra for 3 days to bring about rain. The third type was rainmaking at Todai-ji temple. It started in 839, but it was only practiced for 17 years in the Heian period. Usually, more than 100 monks from 7 large temples of Nara got together and recited the Wisdom Sutra or the Humane King Sutra. These three types of Buddhist rainmaking were conducted jointly by monks from different sects; reciting sutras was its method to bring about rain. The fourth type of Buddhist rainmaking was conducted exclusively by esoteric Buddhism of Shingon sect. As esoteric Buddhism, its method is secret incantations. This type became very popular after 908 and continued even after the end of Heian period (1191), while the other types disappeared by 1089. In the Heian period, it was practiced for 69 years. During cloistered rule (1186–1192), only rainmaking by esoteric Buddhism of the Shingon sect was performed except for 2 years when rainmaking at *Daigokuden* and Todai-ji temple was performed.

The popularity of rainmaking by the Shingon sect during this time was due to the further development of esoteric Buddhism.

13.2.3 Nichiren Buddhism

Kamakura period (1185–1333) is the time when Buddhism became a religion for all; monks such as Honen, Shinran, Dogen and Nichiren established new sects that were for a wider social strata including the warrior, farmers, merchant classes not just for the Imperial Family and aristocrats. Among those, Nichiren (1222–1282) was the notable figure who squarely confronted the misery of the people caused by natural disasters; his serious challenge to overcome natural disasters was different from Honen and Shinran, monks who established Nembutsu sects in which suffering was just a divine test for the Buddhist paradise, and Dogen, a monk who established the Soto school of Zen Buddhism in which the individual level of enlightenment was emphasized. Nichiren wanted to remove people's sufferings that stem from social as well as individual problems of the present existence of life. Thus, it is fair to say that, among new sects of Buddhism, only Nichiren squarely confronted the misery of the present existence brought about by natural disasters. As one of the new sects emerged in Kamakura, his idea was also different from Nara and Heian Buddhism. In Nara Buddhism, monks prayed to overcome natural disasters for emperors and in the Heian period; these were prayers for the Imperial Family and the aristocrats.[3] For Nichiren, his main concern was the peace of the land and its people (Sato 2003: 117).

Nichiren was born in Awa (Chiba) in 1222 as a child of fisherman and at the age of 12, went to study at Seicho-ji temple in Awa, a temple that belonged to the Tendai sect of Buddhism. At that age, he had an ambition that he would become the wisest man in Japan and prayed for it to *Kokuzo Bosatsu* (a bodhisattva). He became a monk at the age of 16 and studied in various temples in Kamakura, Enryaku-ji temple, Koyasan and other influential temples in Japan and at the age of 32 (1253), he established his own sect at Seicho-ji temple. His teaching places the Lotus Sutra the highest status among Buddhist teachings. Reciting the Lotus Sutra and chanting Nam-myoho-renge-kyo were the practices. He moved to Kamakura, the capital city of the Kamakura shogunate government, to propagate his Buddhism around 1256. During this time, natural disasters drove Japanese land and its people into miserable conditions. Nichiren began to refer to the incidents repeatedly.

When Nichiren was in Kamakura, many natural disasters took place. Let's look at occurrences of natural disasters for 5 years between 1256 and 1260, a period when Nichiren moved to Kamakura and penned three major writings related to calamities called "The Cause of Misfortunes," "Treatise on the Elimination of Calamities," and "Treatise on Spreading Peace throughout the Country by

---

[3] The majority of higher ranking monks in large temples were from the Imperial Family and the upper Court nobles.

Table 13.1 Natural disasters documented between the years 1256 and 1260 (Based on a Chronicle table of disasters in Nihon Tensai Chihen Shi, edited by Tokyo-fu Gakumu-bu Shakai-ka, 1938)

	1256	1257	1258	1259	1260
Earthquake		February	April 22	March 12	March 25
		May	December 16		August 5
		August			
		September			
		November (all large scale)			
Windstorm, heavy rain and flood damage	February		June (long-continued rain)		June 1 (heavy flood damage)
	August				
	(Heavy rain, windstorm, flood damage)		August 1 (Kyoto and Kanto, windstorm and heavy rain)		August 5 (heavy rain and windstorm)
			October 16 (rain storm)		August 14 (Kyoto and Kinki, windstorm)
Drought, cold-weather damage	(temperatures like February or March)	July 1 (drought)	June 14 (cold like winter)		
Epidemic	September 1 (Shogun caught measles; spread to all ranks of people)	Epidemic	Epidemic	Epidemic	Epidemic
Famine	June (Famine throughout the nation)	(Many died from hunger)	(Famine) (Countless people died from hunger)	(Famine since January)	(Famine throughout the nation)

Establishing the True Dharma." Table 13.1 shows major natural disasters (earthquakes, windstorm, heavy rain, floods, drought, and cold-weather), epidemic and famine, calamities that affected people of all ranks. Every year, there were natural disasters.

The major natural disaster was Shoka Earthquake on August 23, 1257. It was a large scale earthquake with a magnitude of 7.0 which caused huge damage to temples and shrines in Kamakura as its epicenter was nearby, the east of Boso Peninsula (Kobayashi 2010: 61). Large earthquakes in February, May, September and November on that same year were recorded in The *Azuma Kagami*, an official historical text of the Kamakura shogunate. There were probably also aftershocks of the Shoka Earthquake. The *Azuma Kagami* describes the earthquake as follows:

> Around 7–9 pm, there was a large earthquake with a roar. There were no shrines and temples that escaped from damaging. Mountains crumbled, houses were destroyed, landfill collapsed and everywhere land had cracks and water gushed out. Near Nakagebabashi, the land was violently torn and blue flame was blowing off. Aftershocks continued afterwards. (Cited by Sato 2003: 85)

Although this was a huge calamity, more people suffered from a large scale famine that followed.

Nichiren described the conditions of this period in "Treatise on Spreading Peace throughout the Country by Establishing the True Dharma" that was submitted in 1260 to Hojo Tokiyori, the former regent of the Kamakura shogunate and the most powerful figure at that time. He wrote:

> In recent years, strange phenomena in the sky, natural calamities on earth, famines, and epidemics have occurred and spread over all the land of Japan. Oxen and horses lie dead at crossroads and the streets are filled with skeletons. A majority of the population has perished and everyone has been touched by grief. (Nichiren Shu Overseas Propagation Promotion Association 2003: 107)

Nichiren who believed in the power of the Lotus Sutra identified the cause of people's suffering induced by calamities in wrong teachings that spread throughout Japan; Honen's Nembutsu sect of Buddhism, especially, was pinpointed as the cause of calamities and misfortunes by referring to various sutras as evidence. Because people embraced Nembutsu by abandoning the true teaching of Buddha (the Lotus Sutra), deities who protect people and land left this country. In "Treatise on the Elimination of Calamities," he stated that:

> As protective deities can no longer savor the taste of dharmas, they lose their divine power and abandon this country. The sages called the Four Reliances whom Buddhists can rely on for guidance also leave this country and do not return. This is exactly what is meant by sutras such as the Sutra of the Golden Splendor (The Golden Light Sutra) and the Sutra of the Benevolent King (The Humane King Sutra) when they preach, "When sages all leave the country, the seven calamities will not fail to befall it;" and "We, the Four Heavenly Kings, … all will abandon this kingdom…. When we all abandon this kingdom, various disasters will befall it. (Nichiren Shu Overseas Propagation Promotion Association 2003: 99)

Nichiren's reasoning was based on sutras which he studied extensively since his childhood. He preached to embrace the Lotus Sutra and also wrote what kind of society it will be when people believe in the true teaching and chant Nam-myoho-renge-kyo. In "On Practicing the Buddha's Teachings," he explained that:

> When the people all chant Nam-myoho-renge-kyo, the wind will no longer buffet the branches, and the rain will no longer break the clods of soil. The world will become as it was in the ages of Fu His and Shen Nung.[4] In their present existence the people will be freed from misfortune and disasters and learn the art of living long. Realize that the time will come when the truth will be revealed that both the person and the Law are unaging and eternal. There cannot be the slightest doubt about the sutra's promise of "peace and security in their present existence." (Nichiren Daishonin and Gosho Translation Committee 1991: 392)

---

[4] Fu His and Shen Nung were legendary kings who reigned over ideal societies in ancient China (Nichiren Daishonin and Gosho Translation Committee 1991: 397).

Nichiren believed in the power of a prayer to true Buddhism that can overcome natural disasters. There is an episode of a contest to pray for rain between Nichiren and Ninsho Ryokan, a Shingon Ritsu Shu sect monk; it was held in 1271 when Japan faced a major drought; it resulted in Nichiren's victory.

As discussed in this section, Buddhism in the past responded to natural disasters squarely and tried to overcome them by prayers, although the targeted beneficiary was different in each stage. In the next section, I illustrate the process of secularization of Japanese Buddhism which weakened people's belief in the power of prayers.

13.3 The Secularization of Japanese Buddhism

The secularization of Japanese Buddhism affected the way Buddhism tackles natural disasters. Secularization, a process of declining significance of religion in society, is probably the most influential and debated theme in sociology of religion. Wilson (2002: 169) lists eight notable meanings of secularization: (1) political power seizes assets and facilities of religious organizations; (2) various activities and functions of society that were the responsibility of religion succumb to secular control; (3) people's time, energy and resources devoted to metempirical interests decrease; (4) various aspects of a religious system decline; (5) religious injunctions that regulated people's action were replaced by instructions based on strictly technical standards; (6) unique religious consciousness (from a spell and rituals, magic and prayers to spiritually inspired ethical interests) is gradually replaced by empirical, rational and specific orientation; (7) mythological, poetic and aesthetic interpretations of nature and society were abandoned in favor of a description of facts; and (8) strict distinctions were made between evaluative emotional orientation and cognitive and positive orientation. The secularization hypothesis is universal in scope, thus, scholars found contradictory evidence in their empirical studies such as the sacralization phenomena.

By integrating critics' arguments, Demerath (2007) constructed *four types of secularization* using two distinctions: internal versus external sources and directed and non-directed scenarios. The types are: emergent secularization, coercive secularization, imperialist secularization and diffuse secularization. This is a typology that accommodates the dialectical nature of secularization and sacralization of social change by widening sacredness into the cultural realm. In order to explain this typology, he describes two instances of Japanese religious experiences as examples, one for imperialist secularization and the other for coercive secularization. The former is the secularization of State Shinto caused by the separation of State Shinto from the government apparatus and the demotion of Emperor's status to mere symbolic one directed by the allied forces after World War II.

The latter is a formation of State Shinto as the ideological base for the Meiji government in the 1860s that forcefully subdued Buddhism nationwide, resulting in secularizing Japanese Buddhism. When the Meiji government issued *Shinbutsu hanzenrei* (an order to separate the Kami and Buddhas) in 1868, *Haibutsu kishaku* (a movement to abolish Buddhism and destroy Shakamuni) took place in various part of the country (Makihara 2008: 92). Although the attempt eventually failed as it was based on a small number of urban intellectuals (Ienaga 1965: 26), so called the New Buddhism emerged in response to *haibutsu kishaku*. The New Buddhism aimed at Buddhism as modern religion that is rational and empirical and in accordance with a modernization policy of the government (Sharf 1993: 3–5). As a part of disestablishing Buddhism (Jaffe 1998: 25), in 1872, the government promulgated a law that allows monks to eat meat and marry. If the New Buddhism contributed to secularization at the realm of philosophy, this change was secularization at a behavioral sphere of monks. Although there was resistance to clerical marriage in the beginning, it became the norm in Japanese Buddhism.

I want to add one instance of coercive secularization of Japanese Buddhism which was not mentioned in Demerath's article. It is the enforcement of the "temple registration" by the Tokugawa shogunate in the seventeenth century and the subsequent development of the *danka* (households that support a temple) system. This time was when Japanese Buddhism was secularized the most while spreading to all ranks of Japanese people, thus, universalizing its religion (Nishimura 2004).

In the process of solidifying power, one of the threats Tokugawa government faced was the spread of Christianity. In 1637 the Shimabara Rebellion broke out in Kyushu region of Japan. It was a peasant revolt consisting of mostly Catholics. As it was also a time when the Tokugawa shogunate was consolidating its power, the uprising was ruthlessly suppressed. In 1638 the Tokugawa government ordered all the Japanese to obtain certificates of temple registration, not just for *korobi Kirishitan* (Japanese Catholics who abandon their Christian faith). A practice that started around 1614 as a verification of one's anti-Christian standing (Hur 2007: 49) became a means of social control for all Japanese people; each person had to belong to a particular temple. The temple registration was the basis for the formation of the *danka* system which was firmly established by 1700. By this system, Buddhist temples were able to secure a stable membership and financial base. *Danka* were required to participate in funeral and memorial rites and to contribute to construction of temples and various anniversaries of the sects (Tamamuro 2001: 261). In this process, many monks adopted vulgarized lifestyles in which fame, money and sexual behaviors of the laity were pursued (Tamamuro 1999: 229). Buddhism which had no tradition of funeral rites originally (Abe 2006) became the institution that perform burial and memorial ceremonies and later began to be called funeral Buddhism disdainfully (Ama 2007: 193).

Currently, there is a tendency that Japanese people to regard religion as not very important (Matsutani 2007: 39). A 2005 international comparative survey indicates that 32 % of the Japanese think religion is not very important in life and 40 % think it is not important at all. Also, 58 % of the Japanese do not have any religion at all. Among those who have religion, Buddhism is the most popular religion with 31 % followed by other religions (4 %) and Christianity (2 %) (Yoshida 2010: 95–97). Through the secularization of Japanese Buddhism, not only was its vitality to engage in social problems lost but also the belief in the power of prayers to calm violent nature was forgotten.

13.4 Japanese Buddhism's Response to Natural Disasters at the Present Time: Volunteer Activities

In present-day Japan, the role of Buddhism against natural disasters is not preventive one through prayers like in the past but reactive one. Japanese Buddhism sees volunteer works as their main role faced with natural disasters but their degree of participation is different between the traditional Japanese Buddhist community and new religions[5] with the former less active and the latter very active. Currently, Japanese Buddhism is divided into two groups; the traditional Japanese Buddhist community that consists of 13 sects (Hosso-shu, Kegon-shu, Ritsu-shu, Tendai-shu, Shingon-shu, Yuzu Nembutsu-shu, Jodo-shu, Rinzai-shu, Jodo Shin-shu, Soto-shu, Nichiren-shu, Ji-shu and Obaku-shu), all of which were established before the Kamakura period except for Obaku-shu, which was established in 1661 and new religions (Soka Gakkai, Risho Kosei Kai, Shinnyo-en and others) that were established since 1800s, and some which have adopted Japanese Buddhist doctrines.

Traditional Japanese Buddhism has been ridiculed as funeral Buddhism because of its minor role in society. Social contribution as a religious role in society has not been fully embraced by the traditional Japanese Buddhist community. In contrast, since their establishment, new religious organizations have actively participated in solving social problems. The traditional Japanese Buddhism tended to distance itself from social engagement as adherents, they have a perception that social contribution should be taken care of by new religions. Tatsuya Yumiyama, a professor at Taisho University which was established by Tendai sect of Buddhism and other sects, comments that the community "has a perception that it is OK to leave social contribution activities to Risho Kosei Kai and Soka Gakkai. We are different from them" (Fujio 2011: 34). Social contribution activities have been carried out eagerly by new religions and Christian churches. For example, Shinnyo-en has four areas of social contribution: humanitarian assistance in conflict-ridden and disaster areas, educational supports (scholarship provision to students in the developing countries and managing a free nursing school in Sri Lanka), cultural preservation in Angkor Wat, and environmental protection activities such as reforestation project in China. The Catholic and Protestant Churches have been visible in support of the homeless by distributing food.

During the Hanshin-Awaji Earthquake in 1995, these religious organizations mobilized their members to support the victims. Soka Gakkai was very active in relief works; they formed motorcycle teams that were able to transport emergency supplies while trucks and trains were incapable to do so as the infrastructure was damaged. They also provided their buildings as shelters not only to their members, but also for anyone who needed help while many traditional Buddhist temples closed their doors. In response to the disaster, Shinnyo-en established a permanent relief work section within the organization called Shinnyo-en Relief Volunteers

---

[5] New religions have mainly two different types, Buddhism-based and Shinto-based. In this paper, I focus on Buddhism-based new religions.

(SeRV) which mobilized 11,330 volunteers for 7 months. So far, it was deployed to 70 places in three foreign countries and various areas of Japan where earthquakes, volcanic eruptions and typhoons caused damage.

In response to the 2011 Tohoku Earthquake, new religions such as Soka Gakkai and Risho Koseikai that have accumulated social contribution experiences, and which provide a relief work, mobilized their organizations quickly. They opened up their facilities to the victims, donated land for temporary housing, held counseling sessions, distributed food, collected contributions to the relief of the victims and cleaned up debris (Chugai Nippo 2011). Many of these tasks were carried out by members of these organizations as volunteers.

For the traditional Japanese Buddhist community that has been not very active in social contributions including relief works, this natural disaster presents a challenge for its existence in society. Traditional Japanese Buddhist monks whose life have been secured by the *danka* system now have to reflect on themselves asking questions like "How monks should live? What are the roles of monks? How do monks relate to society?" and "Is our existence needed for the victims of the earthquake?" (Takahashi 2011: 113). Like a new religion, some monks participated in volunteer works for people in Tohoku, ranging from offering a shelter, providing foods, sorting out emergency supplies and removing debris to collecting contributions for the relief of earthquake victims. Yet, there are still conservative voices within the traditional Japanese Buddhist community such as "Why do we participate in volunteer works? Our duty is to pray (Isomura 2011: 14)." Therefore, some monks are not able to take a step forward despite of their interest in social contribution.

Among volunteer works carried out by traditional Japanese Buddhist monks, a café operation and sutra-chanting are unique support that are different from regular volunteer works and can be a clue for the future direction of the traditional Japanese Buddhist community in relief works. A touring café, "Café de Monk (A Café by Monks)" began to move around shelters and temporary housings along the coastal area of Miyagi Prefecture since the mid-May 2011. This is a café where people air their grievances. The café provides free cake, confectionery, coffee, tea and Japanese tea. The tables are decorated with flowers and jazz is being played in the background. This is operated by Taio Kaneda, the chief priest of Tudaiji-temple of Soto-sect in Kurihara City, Miyagi Prefecture. He operates this café for two purposes. First, he wants to provide choices of foods, in this case, sweets because people have been eating mainly rations and emergency foods such as rice balls, canned foods, bread and instant noodles in a cup. So, he asked a cake maker of his friend to make varieties of cake. Second, he wants to provide a breathing space to the victims because they live in a stark landscape which reminds them of the earthquake and tsunami. They enjoy eating sweets and drink coffee and tea, and then they start talking to him about hardships they are bearing. By confiding their sorrow and worries to him, people can relieve their mental and emotional pains. Since he thinks Buddhism cannot solve all the mental problems, he receives a support from "Mental Advice Center" in Sendai City. It is an interfaith network of Buddhism and Christianity in which medical doctors and researchers participate in building a support system for the victims in medical, mental and life aspects. Now, his café is supported by young volunteer monks who heard about it and came from different parts of Japan. Due to the support, the second Café de Monk opened at the end of July 2011 and there is a plan to open the third café (Isa 2011).

Sutra-chanting is another volunteer work conducted by Buddhist monks and is closer to a traditional task of Buddhists than the café operation. Sutra-chanting practiced in the disaster area is different from a memorial service which is the main task of traditional Japanese Buddhism because in the latter, it is a ritualized service and an income generating activity of the monks. Sutra-chanting in the disaster area grew out of the monks' genuine feeling for the deceased and the bereaved. A good example is a sutra-chanting voluntary service in Ishinomaki City, Miyagi Prefecture. Kitamura Akihide, the vice chef of Hozan-ji temple (Soto-sect) and secretary-general of Ishinomaki Buddhist Association, describes how the voluntary service started in the city. Because of the large number of victims in the city, cremation was not able to keep up with the demand. Thus, the city decided to have an emergency measure in which a temporary inhumation was conducted with the help of Self-Defense Force. In inhumation, Self-Defense Force personnel sent off the victims with a salute. It was a respectable scene, but monks were not satisfied with it from the point of view of religion. Thus, Ishinomaki Buddhist Association decided to send 7–8 monks for sutra-chanting during inhumation. In total, they mourned 630 people (Fujio 2011). Similarly, in Sendai City, Miyagi Prefecture, sutra-chanting was conducted in two crematories. Some dead bodies were unidentified and thus it was for reposing the soul of the deceased people (Fujio 2011). Although it consoles the surviving families, this sutra-chanting mainly grew out of monks' desire to mourn for the dead.

Another example of sutra chanting is an offering of a funeral service by Jifuku-ji temple in Yamagata Prefecture that is 3–4 h away by car from Ishinomaki City. Unlike the previous examples, this service was established mainly for the surviving members of the deceased. When inhumation started in disaster areas, Zenkyo Uno, the chief priest of Jifuku-ji temple decided to hold a funeral service to the tsunami victims. He felt it is necessary to cremate the remains as soon as possible for the relief of the survivors. His participation in a spirit-consoling service for the dead in the 1993 southwest earthquake off Hokkaido that killed 230 people made him realize that it takes long time for the survivors to accept death of the family members that is suddenly forced upon them. Thus, he believes that monks' duty is not just to read sutras but to listen to the survivors' stories about their deceased family and ease their sufferings. Mr. Uno held 26 funerals at Jifuku-ji temple and 44 funeral services at a hall and crematory in which he listened to the stories of the bereaved family (Yamakawa 2011).

These examples show that sutra-chanting volunteer services by traditional Japanese Buddhism emerged in response to the earthquake are really prayers for the victims (dead and survived) not prayers for pacifying nature practiced in the past.

13.5 Conclusion

The 2011 Tohoku Earthquake will be a turning point for the Japanese Buddhist religious community in terms of how to respond to natural disasters and sudden deaths caused by them. In the past, this community believed in the power of prayers. During the consolidation period of the Imperial system, prayers based on three major sutras were believed to have power for protecting the nation to overcome natural disasters. During the Heian Period, esoteric Buddhism became influential mainly serving for the aristocracy and secret incantations were practiced for tranquil nature and rainmaking which creates a favorable condition for rice cultivation. In Kamakura Period, many passionate Buddhist monks appeared and preached Buddhism for the masses. Their teachings tried to remove the sufferings of people. Among those monks, Nichiren squarely confronted the sufferings caused by natural disasters by teaching the power of the Lotus Sutra and chanting Nam-myoho-renge-kyo. Although the target beneficiary was different in each period, the Japanese Buddhist community before Edo Period believed in the power of Buddhism; monks recited sutras or conducted secret incantations to pacify nature.

However, in highly secularized contemporary Japan, people do not believe that natural disasters can be overcome by prayers. Instead, they participate in voluntary relief works such as offering shelters and foods. New religions which adopted some Buddhist doctrines have been very active in social contributions. During the Hanshin-Awaji Earthquake in 1995, their voluntary relief works were significant. In the 2011 Tohoku Earthquake, they were able to mobilize their organizations quickly because of their experiences in day-to-day social contribution activity. On the other hand, the traditional Japanese Buddhist community has not been active in social contribution, partly due to the *danka* system that created an inward nature of the community. However, ignorance to this large scale disaster is not possible as they were in the Hanshin-Awaji Earthquake. Thus, some monks have begun to participate in voluntary relief works. Among them, sutra chanting is a unique practice that is carried out for the survivors and the deceased. This voluntary work of sutra chanting may clarify the role of the traditional Japanese Buddhism in contemporary Japanese society.

References

Abe, D. (2006). Syohyo: Nakajima Takanobu cho 'Oterano Keizaigaku' (Book review: Economics on temples by Takanobu Nakajima). *Ryukoku Daigaku Keizaigaku Ronshu* (Ryukoku University Economics Bulletin), *46*(1), 47–56.

Ama, T. (2007). *Bukkyo to Nihonjin* (Buddhism and the Japanese). Tokyo: Chikuma Shobo.

Chugai Nippo. (2011, September 17). Higashi Nihon Daishinsai Kyodan Anketo: Kaku Kyodan wa Ikani Taiou shitaka – Shinto, Shin Syukyo, Kirisuto Kyo nado (A questionnaire about the 2011 Tohoku earthquake to religious organizations: How did each organization respond to it? – Shinto, New Religion, Christianity and others).

Daishonin, N., & Gosho Translation Committee. (1991). *The writings of Nichiren Daishonin*. Tokyo: Soka Gakkai.
Demerath, N. J., III. (2007). Secularization and sacralization deconstructed and reconstructed. In J. A. Beckford & N. J. Demerath III (Eds.), *The SAGE handbook of the sociology of religion* (pp. 57–80). London: Sage Publications Ltd.
Fujio, A. (2011). Hisaichi ga Kitaeru Bukkyo no Michi: Shinko wa Chikarato nariuruka, Higashi Nihon Dai Shinsai (A way of Buddhism that is trained in disaster areas: Faith can be a power? The 2011 Tohoku Earthquake), AERA. April 23 issue.
Hongo, M. (2002). Nara Bukkyo to Minshu (Nara Buddhism and the masses). In *Nihon no Jidai Shi 4: Ritsuryo Kokka to Tenpyo Bunka* (Japanese period history 4: Ritsuryo nation and Tenpyo culture). Tokyo: Yoshikawa Kobun Sha.
Hongo, M. (2004). Nara, Heian Jidai no Syukyo to Bunka (Religion and culture in Nara and Heian periods). In Rekishigakukenkyukai & Nihonshikenkyukai (Eds.), *Nihonshi Kouza 2: Ritsuryo Kokka no Tenkai* (Studies in Japanese history 2: Imperial rule under the Ritsuryo penal and administrative codes) (pp. 191–222). Tokyo: University of Tokyo Press.
Hur, N. (2007). *Death and social order in Tokugawa Japan: Buddhism, anti-Christianity, and the Danka system*. Cambridge: The Harvard University Asia Center.
Ienaga, S. (1965). Japan's modernization and Buddhism. *Contemporary Religions in Japan, Nanzan University, 6*(1), 1–41.
Inoue, M., Kasahara, K., & Kodama, K. (1984). *Shosetsu Nihonshi* (Detailed explanation of Japanese history). Tokyo: Yamakawa Publishing Company.
Isa, K. (2011, June 27). Toinaosu Syukyosha no Arikata: Souryo, Muryo Kissa de Hisaisha no Koe wo Kiku (Reexamine the state of religious figures: Monks listen to the voices of the victims in a free cafe). *Asahi Shimbun*.
Isomura, K. (2011, December 17). Opinion: Hinkon Mondai, Syokuryo Shien, Jiin no Chikara ni Kitai (Opinion: Poverty issue, food assistance, expectation for the power off Buddhist temples). *Asahi Shimbun*.
Jaffe, R. (1998). Meiji religious policy. Soto Zen, and the clerical marriage problem. *Japanese Journal of Religious Studies, 25*(1/2), 45–85.
Japan Meteorological Agency. (2011). Taifu niyoru Saigaino Rei (Examples of disasters by typhoon). www.jma.go.jp/jma/kishou/know/typhoon/6-1.html#shobohakusho
Kanegae, H. (2008). *Nohon no Rekishi, Asuka, Nara Jidai: Ritsuryo Kotsuka to Manyo no Hitobito* (Japanese history, Asuka and Nara periods: Ritsuryo nation and people in Manyo). Tokyo: Shogakkan Publishing Company.
Karan, P. P. (2005). Japan in the 21st century: Environment, economy and society. The University Press of Kentucky.
Kitayama, S. (1965). *Nihon no Rekishi 4: Heiankyo* (Japanese history 4: Heian period). Tokyo: Cyuokoronsha.
Kobayashi, T. (2008). Saigai no Hatusei to sore eno Hitobito no Taisho ni kansuru Bunkashi: Kodai Niigata Keniki ni okeru Jirei no Kenshutsu to Hitobito no Saigai kan (Cultural history concerning people's reactions to facing natural disasters: Detected cases in ancient Niigata prefectural region and people's views on disasters). *Niigata Sangyo Daigaku Keizaigakubu Kiyo* (Bulletin of Niigata Sangyo University Faculty of Letters) No. 19.
Kobayashi, T. (2010). Nihon no Cyuse Zenhanki ni Okeru Saigai Taisho no Bunkashi: Niigata Keniki ni okeru Jirei no Kenshutsu to Hitobito no Saigaikan wo Cyushin toshite (A cultural history of dealing with disasters in Japan from the early period in the Middle Ages: Centering on the detection of the case in the Niigata prefecture region and people's conception of disasters). *Niigata Sangyo Daigaku Keizaigakubu Kiyo* (Bulletin of Niigata Sangyo University Faculty of Letters) No. 21.
Makihara, N. (2008). *Nohon no Rekishi, Bakumatsu kara Meiji Jidai Zenki* (Japanese history from the end of Tokugawa Bakufu to the early Meiji era, toward a civilized nation). Tokyo: Shogakkan Publishing Company.
Matsutani, M. (2007). Syukyo Kenkyu no Houho: Seron Chosa (Methods of religious study: A public opinion poll). In S. Yoshihide & M. Hideki (Eds.), *Yokuwakaru Syukyo Shakaigaku* (Easy to understand sociology of religion) (pp. 38–39). Kyoto: Minerva Publishing Company.

Mizuno, S. (2009). Cyugoku Bukkyo ni okeru Gokoku Shiso no Jyuyo Katei ni tsuite (The reception process of the idea of protecting a nation in Chinese Buddhism). *Indogaku Bukkyogaku Kenkyu, 58*(1), 261–266. The Japanese Association of Indian and Buddhist Studies.
National Police Agency. (2011). Heisei 23 Nen Tohoku Chiho Taiheiyo Oki Jishin no Higai Jyokyo to Keisatsu Syochi (The damage of the 2011 Tohoku earthquake and police action). www.npa.go.jp/archive/keibi/biki/higaijokyo.pdf
Nichiren Shu Overseas Propagation Promotion Association. (2003). Writings of Nichiren Shonin: Doctrine 1. Jay Sakashita (Ed.) (trans: Kyotsu Hori), Tokyo.
Nippon Steel Human Resource Development. (2002). *Nippon: The land and its people*. Tokyo: Gakuseisha Publishing Co. Ltd.
Nishimura, R. (2004). *Nihon Kinse Bukkyo Shiso no Kenkyu – Gakuso Fujyaku wo meguru Syomondai* (A study on Buddhist thought in Japanese early-modern times: Various issues in connection with a learned monk, Fujyaku). Ph.D. dissertation. Department of Arts & Letters, Tohoku University, Sendai
Sato, H. (2003). *Mineruva Nihon Hyoden Sen: Nichiren – Ware Nihon no Hashira to Naramu* (Mineruva critical biography of Japanese: Nichiren – I will be the pillar of Japan). Kyoto: Mineruva Publishing Company.
Sharf, R. H. (1993). The Zen of Japanese nationalism. *History of Religions, 33*(1), 1–43.
Shinmura, I. (Ed.). (2008). *Kojien* (6th ed.). Tokyo: Iwanami Publishing Company.
Takahashi, T. (2011). Otsunami ga Nomikonda Mono: Shinsai to Dento Bukkyo (Things that were engulfed in a large tsunami). In K. Uchihashi (Ed.), *Daishinsai no Nakade: Watashitachi wa Nani wo Subeki ka* (In a large earthquake: What should we do?) (pp. 107–113). Tokyo: Iwanami Publishing Company.
Tamamuro, F. (1999). *Rekishi Bunka Raiburari 70: Soshiki to Danka* (Historical and cultural library 70: Funeral and the Danka). Tokyo: Yoshikawa Kobun Kan.
Tamamuro, F. (2001). Local society and the temple-parishioner relationship within the Bakufu's governance structure. *Japanese Journal of Religious Studies, 28*(3–4), 262–292.
Tokyo-fu Gakumu-bu Shakai-ka. (1938, 2004). *Nihon Tensai Chihen Shi* (The history of natural disasters and terrestrial upheavals in Japan). Tokyo: Kaiji Shoin (Originally published in 1938)
Tsukushi, N. (1964). *Nihon no Shinwa* (Japanese mythology). Tokyo: Kawade Shobo.
Uejima, S. (2002). Heian Bukkyo: Kukai, Saicho no Jidai (Heian Buddhism: Age of Kukai and Saicho). In S. Yoshikawa (Ed.), *Nippon no Jidaishi 5: Heiankyo* (History of Japanese periods 5: Heian period) (pp. 229–269). Tokyo: Yoshikawa Kobun Kan.
Wilson, B. (2002). *Syukyo no Syakaigaku: Toyo to Seiyo wo Hikaku shite* (Religion in sociological perspective) (trans: Tsuyoshi Nakano & Toshie Kurihara).Tokyo: Hosei University Press.
Yabu, M. (1992). *Kodaini okeru Kiu nitsuite* (On rainmaking in ancient times). Master's thesis. Faculty of School Education, Hyogo University of Teacher Education, Kato.
Yamakawa, T. (2011). *Izokuno Kokoro wo Uketomeru: Yamagata, Jifukuji no Sougi Shien* (Take the emotions of the family of the deceased seriously: Funeral support by Jifuku-ji Temple, Yamagata Prefecture). Purejidento (President), August 16 issue.
Yoshida, S. (2010). Kachikan to Seikatsu Ishiki ni kansuru Teiryo Bunseki: Syukyoishiki wo meguru Kosatsu (Study guide for large scale international comparative surveys and quantitative analysis. Some findings about values and religiosity). *The Bulletin of Toyama University Faculty of Arts and Culture, 4*, 86–104.

Chapter 14
Reshaping the Worldview: Case Studies of Faith Groups' Approaches to a New Australian Land Ethic

Justin Lawson, Kelly Miller, and Geoff Wescott

14.1 Introduction

This chapter will provide a brief background on the development of religious organizations focusing on environmental issues in Australia and the importance placed on environmental sustainability by religion. A discussion on maintaining pro-environmental behaviors through values and worldviews with links to religious beliefs is provided. Methods of the research are outlined followed by a discussion of the results, focusing on survey and interview comments. Case studies are provided, with a brief site description and a discussion demonstrating how differing worldviews are integrated into effective environmental sustainability strategies. The implications of these case studies are given collectively, followed by conclusions on the status of faith groups engaging on environmental sustainability issues.

14.1.1 Background

Inasmuch as Australia is traditionally a Protestant country, an increase of population (partly due to immigration), a change in disclosing religious belief in census reports, failures in serving the needs of various social classes, and conversions to different faiths and traditions has seen Australia become a country of many faiths (Porter

---

J. Lawson (✉)
Health, Nature and Sustainability Research Group, School of Health and Social Development, Deakin University, Burwood, VIC 3125, Australia
e-mail: justin.lawson@deakin.edu.au

K. Miller • G. Wescott
Centre for Integrative Ecology, School of Life and Environmental Sciences, Deakin University, Burwood, VIC 3125, Australia
e-mail: kelly.miller@deakin.edu.au; geoffrey.wescott@deakin.edu.au

1990; Carey 1996; Tacey 2000; Bouma 2006; Jupp 2009a). All of these faiths are at various stages of increase and decline (Jupp 2009c).

Despite these conflicting patterns, religion in Australia has played a significant role in its recent settlement history with preserving and defending morality, maintaining cultural practices, preserving tradition, developing intellect, and providing education, welfare, healthcare and counseling (Jupp 2009b). Concern for the environment has only been a recent affair, reflecting similarly held views with the wider community (Black 1997) and yet there is a reluctance in some denominations to enact on those concerns (Douglas 2009). Changes are afoot, however, as media, political parties and industries have increased their attention to such issues as climate change and carbon emissions, such that new organizations and alliances are beginning to form, *e.g.* the Australian Religious Response to Climate Change (ARRCC), Faith and Ecology Network (FEN), GreenFaith Australia, Jewish Ecological Coalition (JECO), Al-Ghazzali Centre, Catholic Earthcare, Australian Anglican Environment Network (AAEN), and the Uniting Church in Australia's (UCA) Green Church and Earthweb, amongst various environmental committees within other denominations. These organizations are attempting to integrate profound spiritual values into positive environmental behaviors; a process which is difficult and with considerable hurdles.

The United Nations has also recognized the significance of including cultural and spiritual values into effective environmental management. The *Cultural and Spiritual Values of Biodiversity* (Posey 1999) and the *Earth Charter* (ECICaS 2010) are two examples of this recognition, serving to guide and inform national and local governments as well as non-government organizations (NGOs), institutions and businesses. To properly address the current environmental crisis;

> Values and ethics, religion and spirituality are important factors in transforming human consciousness and behavior for a sustainable future [and as such] we are called to a new intergenerational consciousness and conscience. (Tucker and Grim 2007: 4)

Gardner (2002, 2006) asserts that the religious community is in an enviable position of having five solid assets that can assist in the integration of attitudes and behaviors with respect to sustainability issues:

1. The capacity to shape cosmologies;
2. Moral authority;
3. A large base of adherents;
4. Significant material resources, and;
5. Community-building capacity.

With these five assets combined, religion as a whole potentially can have immense influence on the attitudes toward nature, the use of natural resources and their distribution, as well as shaping behaviors and policies that affect the environment (Gardner 2002). These influences can be either positive or negative. Numerous theories have been proposed that are addressing the interaction of values on nature (Manfredo 2008). Kellert's typology of wildlife attitudes (Kellert 1984) has shown that people hold a diversity of attitudes and beliefs about the environment and wildlife in particular. Shwartz's theory of social values (Schwartz 1992) demonstrates that values can be 'stable motivational constructs' and religion can play a part (Manfredo 2008: 148).

Gardner is hopeful that the five assets can be leveraged; these range from assessing spiritual teachings, using the pulpit as a platform for addressing the congregation on issues of environmental sustainability, encouraging members to join boycotts and protests, using facilities as showcases of sustainable activities and increasing bonds of trust and communication within the community amongst other suggestions (Gardner 2002). These, combined with the skills and perspectives of environmentalists, can help embark on a 'quest for a new cosmology, a new worldview for our time' (Gardner 2002: 51). Gardner refers to the cultural historian Thomas Berry who calls this 'emerging perspective a New Story – the story of a people in an intimate and caring relationship with their planet, with their cosmos, and with each other' (Gardner 2002: 51). It is the 'Great Work' (Berry 1999); an 'ecology of care' (Fuller 1992) that requires 'a feeling of identification with what one's doing' (Pirsig 1974 quoted by Fuller 1992: 100). It is a call that that has been made previously by Griffin (1988), Rasmussen (1992), Ruether (1992), McFague (1993) and Haught (1993). Yet, those who are prepared to care, however, are extremely challenged; for 'the religious solutions to environmental concerns are visionary and […] there is a gap between theory and practice' (de Silva 2002: 159). This gap will be discussed further in light of the results below.

14.1.2 *Values and Worldviews: The Social Amplifiers*

Cary et al. (2002) refer to Stern et al.'s (1995) framework of environmental concern and adjust it slightly by including the causal arrows that Stern et al. made a point of omitting (Fig. 14.1). The framework outlines that individuals are located within a social culture, which then influences the development of values, beliefs attitudes and eventually behaviors. Inasmuch as the arrows help in demonstrating the general flow of influence of variables from the top to the bottom, it is nevertheless a convoluted flow, with loops within loops. The variables immediately adjacent to each other exert the strongest influence, yet the nonadjacent ones can also have an effect. Thus;

> …values and worldview act as filters for new information and ideas. Information congruent with an individual's values and worldview will be more likely to influence beliefs and attitudes. (Stern et al. 1995: 726)

As the diagram shows, there is a loop of attitudes and behaviors. There is a vast wealth of research dedicated to the causal or otherwise relationships between attitudes and behaviors (Thurstone 1928; LaPiere 1934; Doob 1947; Blumer 1955; Wicker 1969; Triandis 1971; Figà-Talamanca 1972; Ajzen and Fishbein 1977; Liska 1984; Davis 1985; Ajzen 1991; Madden et al. 1992; Holland et al. 2002; Armitage

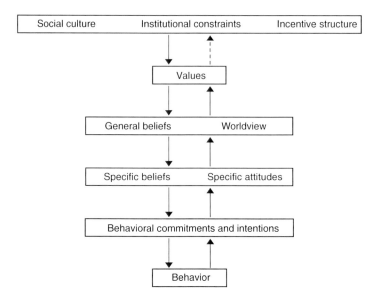

Fig. 14.1 A framework of environmental concern (Revised and adapted from Cary et al. 2002: 36; derived from Stern et al. 1995: 727)

and Christian 2003; Chaiklin 2011).[1] These authors amongst a plethora of others have provided scales of measurement, discussed or reviewed the processes involved in changing attitudes and behaviors but little has been improved to predict behavior from attitude and it is still 'murky' (Chaiklin 2011: 48). Yet the framework provided by Stern et al. (1995) shows entry points where change will eventually occur further along the loop(s). Chaiklin (2011) is adamant that changing attitudes is not necessary to change behavior as this inevitably threatens freedom, especially if civil liberties are curtailed in the pursuit of affecting attitudinal change. Nevertheless, values and worldviews are significant as these operate as 'social amplifiers,' such that 'a strong value orientation may lead someone to seek information selectively or to attend selectively to information about the consequences of an environmental condition for particular valued objects, and therefore develop beliefs about those consequences that will guide action' (Stern and Dietz 1994: 68). Certainly, the 'fundamental lesson [is that] worldviews and beliefs do matter' (Berkes 2008: 252). Herring (2009) also sees the importance placed on values and worldviews, by drawing a hierarchical structure of motives, consequently linking behaviors (*e.g.*, using energy efficient lighting) to worldviews (*e.g.*, the meaning of life) (Fig. 14.2).

Is it too much, however, to draw a line from the meaning of life to changing a light bulb? Stern and Dietz (1994) reflect on the 'ideological struggle' that was

---

[1] The list of authors is not exhaustive; Chaiklin (2011) refers to Schneider (2004) who provides some 15,000 references relating to one aspect of attitudinal research.

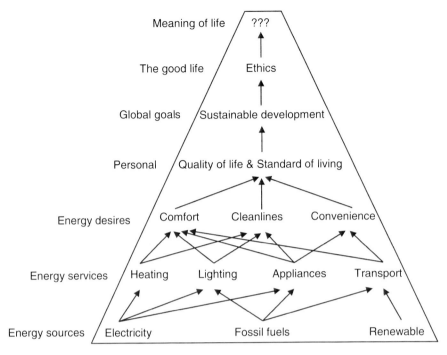

Fig. 14.2 A hierarchy of motives: from energy sources to the meaning of life (Source: Energy Efficiency and Sustainable Consumption, edited by Horace Herring and Steve Sorrell, 2008, Palgrave Macmillan Publishers, p. 227, used with permission)

underway in the 1990s that is still occurring now, where 'proponents of biospheric values have not yet succeeded in generating a clear distinction in general public consciousness between valuing nature in itself and valuing nature because of the human benefit it provides' (Stern and Dietz 1994: 78). Therein lies the challenge; in light of increased awareness of the plight of the earth, to reset the plots and narratives of the stories pertinent to the meaning of life. It calls for a re-imagining of our place in nature, what Pinn (2003: 38) calls 'restor(y)ing a sense of place, self and community.' It is the 'dream of the earth,' where 'our human destiny is integral with the destiny of the earth' (Berry 1988: xiv). What is required, Berry continues, is a 'deeper meaning of the relationship between the human community and the earth process' (Berry 1988: 10). The process calls for a 're-enchantment,' where sacredness is 'the key to environmental integrity' (Tacey 2000: 162).

This re-enchantment, however, is very tricky to implement; even Tacey admits that it is difficult and covers 'sensitive ground' (Tacey 2000: 242). The 'new Australian spirituality' is caught between two difficult dichotomies, where on one hand there is the dominant presence of a secular, postmodern yet patriarchal and Euro-centric culture (with a USA influence as well), while on the other is a spirituality sourced from indigenous and Eastern traditions, also postmodern yet inclusive of marginalized perspectives (gender, race, sexuality) and certainly multi-cultural.

How this affects positive developments in environmental sustainability is in some ways subtle yet very complex. The following sections will thus reveal whether or not 're-enchantment' is taking place and what the implications are for environmental sustainability.

14.2 Methods

During 2009 and 2011 a mixed methods study (surveys, interviews, site visits) of faith groups in Australia was undertaken. A total of 2,907 surveys were delivered to 1,303 locations throughout Australia, targeting 56 different faith groups. Of these, 423 were completed by individual representatives from 36 different faith communities. Semi-structured interviews were also conducted with 34 individual representatives, increasing the total number of faith groups represented to 40 and consequently all six major faith groups present in Australia were represented, *i.e.* Buddhism, Christianity, Hinduism, Islam, Judaism, and other religious groups (including Bahá'í, nature religions, Sikhism, spiritualism, scientology, Eckankar and new age).[2] Site observations were undertaken at 35 sites of worship throughout Australia (excluding Tasmania and Northern Territory). For the purpose of this discussion focusing on integrating a spiritual worldview in environmentally sustainable practice, comments will be drawn from surveys and interviews with representatives from ten different faiths/denominations supplemented with site observations at three Christian communities to compare and contrast points of convergence and divergence.

14.3 Results and Discussion: Re-enchantment and Restor(y)ing – Perspectives from the Field

14.3.1 Gaps

Understandably, not everyone is in agreement as to where and how faith groups can engage effectively on environmental issues. Lawson and Miller (2011) have shown that within the Abrahamic communities, differing positions on stewardship, eco-justice, creation spirituality, and safeguarding creation are held by different denominations. Including Eastern traditions in this study revealed similar and other positions (and hence gaps), *e.g.* a Zen Buddhist did not identify with the 'stewardship' label but rather 'an awareness that if we don't look after it [the earth], we're not looking after ourselves' (Zen Buddhist, male, 60–69, rural

---

[2] These categories are used in accordance with those used by the Australian Bureau of Statistics (2005).

Queensland). This is in keeping with the 'non duality' worldview maintained in Buddhism and has links to deep ecology or an eco-centric view (Nisker and Gates 2000; Khisty 2006). A Hindu devotee, however, supported a 'safeguarding creation' ethic where he recognized that the existence of god is in 'everything animate or inanimate that exists in this world [and] therefore Hindus feel it is right to protect any sacred cultural ikons [sic] and control any excesses in use of monetary or other resources' (Hindu, male, 70–79, urban Adelaide). Interestingly, a progressive Christian has undertaken an understanding of non-separate realities, yet identifies with 'stewardship':

> In my branch of the Christian faith there is more emphasis [on] the Creator God rather than the God of salvation – hence our place in creation is one of stewardship with a creation the whole of which has value. The emphasis is also on creating the kingdom of God here and now by modeling one's life on Christ's life rather than the rewards of a future arrival based on salvation. Hence problems on earth and in the material world are not separate from the Spiritual world. There is no escape from here and now so here and now is where action must be taken. (UCA, female, 40–49, rural Northern Territory)

Extending the matrix of environmental positions of Abrahamic faith groups to include other faiths is difficult and quite likely impossible.[3] The conflict of redefining the spiritual worldview also extended into the prime directive of some of the faiths, in particular Christianity and its role of evangelizing:

> We have encountered a great deal of resistance amongst some conservative evangelical congregations who think we should focus on the main game, *i.e.* conversion. The project is gaining momentum in other parts of the Diocese. The plan is to have 80 % of parishes on board the project in 4 years' time. The resistance we have encountered suggests this faction in the church […] see[s] no value in it. (Anglican, male, 50–59, Perth)

> You ought to understand that Religion – and Cathecism in particular – has as its objective the Worship of God. That is what we do! Secondly, there is the matter of evangelisation [sic] […] and this involves encouraging people to live rightly. It includes, but is not restricted to, care for the earth. And finally, Christianity is "apocalyptic" – that means that we do expect that the world as we know it will come to an end, and there will be a "new heaven and a new earth." All this is part of Christian belief, although we do not expect that it will happen soon. But the essence of Christian belief is the "waiting." So to sum up, we live, expecting to die and to rise from the dead. But while we live on the earth we must treat it with care so that we can hand it on to those who will live after us! (Catholic, male, 60–69, Melbourne)

> I deliberately reach out to people in Jesus' name for their present and eternal well being […] God's priority is salvation for people, not the planet. Our rebellion and sinfulness against God and one another is far more destructive than greenhouse gas and pollution. Jesus came to make a heaven-of-a-difference on earth and so must the church while we have opportunity and time, until Earth and Heaven have a time limit on them (Revelation 21:1). We […] look for new heavens and a new earth, 'all landscaped with righteousness' (The Message Bible) (UCA, male, 40–49, northern coastal New South Wales).

---

[3] Buddhism and Hinduism are notoriously difficult to categorize, due to some adherents proclaiming that Buddhism is not a religion, and Hinduism incorporating many conflicting belief systems.

The importance placed on environmental issues, such as greenhouse gases and climate change were consequently seen as secondary concerns. Thus the reshaping of worldviews is very difficult to reconcile, however, a response from the Bahá'í tradition provided a clue:

> I think for the fostering of things such as humility and unity and respect and those kind of qualities are definitely very proactively fostered with the idea that they then become applied in a whole range of situations, so that informs the way that then Bahá'ís would interact with other human beings, with animals, with the environment, and informs a whole lot of choices. (Bahá'í, female, 30–39, Darwin)

The view expressed here fits well with Herring's hierarchical framework of motives, where humility, unity and respect could be aspects of the 'good life.' These values can be seen to feed into two directions; one into the 'meaning of life' (and hence worldview), the other into choices (and hence into behaviors). Humility and respect also keeps in check greed, an issue that was raised by many across significantly divergent faiths:

> …the real problem is people are not satisfied in spiritual life, they are not personally satisfied and therefore they're looking externally for happiness and they're fighting amongst each other and being greedy. (Hare Krishna, male, 50–59, rural New South Wales)

> …there are [five] things that we try to remove from our lives, or qualities – lust, anger, greed, attachment and egoism […] if you do look at those negative things you can often translate […] that those are the things that cause humanity to lose its sense of sustainability and environmentalism […] And just about everything on environmental issues can in one way or another be ultimately brought back to 'well we're doing this for the greater good rather than for our own selfish ends. (New Age, male, 60–69, rural New South Wales)

> …in our community that we believe that the deserts basically were caused by man's greed, way back […] I can understand that. (Exclusive Brethren, male, 50–59, rural New South Wales)

Seeing the impact of greed is not a new revelation; it has been recognized long before (*q.v. Bhagavad-gītā*, 16.22 (Trans. Prabhupāda 2001 [1983]: 763)). Indeed, the Buddha saw greed, or craving (*taṇha*), as being the fundamental root of all suffering, placing it as the second Noble Truth (*Ariya sacca*). Addressing greed, however, is inherently difficult. For Buddhists, adhering to the Eight-fold Noble Path (*Aṭṭhangika magga*) effectively extinguishes this craving. Other traditions use mantras, prayer or contemplations on scriptures as the path away from greed. Yet, in a pluralistic, postmodern society that we find ourselves in, we are bound to needs that are inextricably linked to desires. Unfortunately, this is the dilemma; in 1930, economist John Maynard Keynes made this observation, which is still disturbingly current:

> I see us free, therefore, to return to some of the most sure and certain principles of religion and traditional virtue – that avarice is a vice, that the extraction of usury is a misdemeanor, and the love of money is detestable, that those who walk most truly in the paths of virtue and sane wisdom [are those] who take least thought for the morrow […] But beware! The time for all this is not yet. For at least another hundred years we must pretend to ourselves and every one that fair is foul and foul is fair; for foul itself is useful and fair is not. Avarice and usury and precaution must be our gods for a little longer still. For only they can lead us out of the tunnel of economic necessity into daylight. (Keynes 1963 [1930]: 371–372)

Thus we are confronted with the convoluted problem that in order to stem consumption and resource depletion we need to perpetuate the very same (consider, for example, the materials, energy and advertising resources required to manufacture, distribute and sell solar panels, wind turbines, hybrid vehicles *etc.*). It is no wonder then, that reshaping the worldview is such a big challenge. It is a 'human predicament' (DeWitt 2002: 46) that has in part been addressed by various members of the faith community.

In summary, we see that there is an ongoing 'crisis' of sorts in identifying with the prevalent environmental issues and reconciling them within a spiritual/religious perspective. Given that greed was seen as a core malady, the responses of humility and respect were viewed as being positive influences in addressing this problem.

14.3.2 Closing the Gaps: Case Studies

To demonstrate viable attempts of environmentally sustainable living within a spiritual context are the following three rural, Christian case studies.

14.3.2.1 Case Study 1: Monastery of Blessed Virgin Mary for the Sisters of the Community of St Clare and the Hermitage of St Bernadine of Siena for the Brothers of the Society of St Francis, Stroud, New South Wales

Site Description/Background Saint Francis of Assisi's *Canticle of the Sun* (also known as *Laudes Creaturarum* (*Praise of the Creatures*)) is a keystone document of deep awareness of the interconnectedness of nature. The song, composed in 1224, has guided the Franciscans through a way of life filled with a deep sense of humility and respect. In 1979 when St Francis' was made patron saint of ecology, construction of a monastery and a hermitage began in rural New South Wales that typified St Francis' worldview, where simple adobe buildings were constructed amongst a natural setting (Figs. 14.3 and 14.4).

The area surrounding the monastery is extensively cleared for pastoral purposes, however, the 8 ha spiritual retreat has retained much of the native vegetation and has also had considerable augmentation with exotic, non-local indigenous and fruit-bearing floral species. Consequently, there is an abundance of fauna, notably avian, with kangaroos and koalas as well as reptiles and amphibians also present throughout the property. The potential to be an 'ecostery,' a monastery supporting ecological principles (LaFreniere 1985) is readily apparent.

The very existence of the monastery and hermitage is an interesting example of the convergence of mainstream and alternative; the Franciscan Sisters and Brothers are part of an order that is within the Anglican tradition, one of the largest mainstream Christian traditions in Australia. The buildings, however, are a monument of

Fig. 14.3 Satellite image of monastery and hermitage in Stroud, New South Wales (Source: Google Earth, image © 2012 GeoEye, © Whereis® Sensis Pty Ltd)

Fig. 14.4 Monastery of the Blessed Virgin Mary for the Sisters of the community of St Clare (Photo by J. Lawson)

sorts to part of a movement that was inspired by the counterculture movement of the late 1960s. While not new, mud-brick buildings, received considerable attention as a viable alternative to the mainstream weather-board or brick-veneer suburban house, and featured in alternative magazines such as *Grass Roots*, and *Earth Garden*, amongst others. Inspired by functioning examples in the Melbourne urban fringe, the Sisters and Brothers were assisted by scores of volunteers throughout the state to build the monastery and hermitage in a project that took about 2 years from 1979

Fig. 14.5 Mudbrick chapel of the Hermitage of St Bernadine, nestled among a variety of indigenous and non-indigenous flora (Photo by J. Lawson)

to 1980. Another building, called *Gunya Chiara* (House of Healing of St Clare), and maintaining the established style of the others, was constructed in 1997.

In keeping with St Francis' sense of austerity, there is nothing flamboyant or 'baroque' about the place; rooms are furnished simply and decorations are minimal. Of note is the Brothers' chapel (Fig. 14.5), the altar of which has been adorned with a *fresco* painting; a method of painting first used over 3,000 years ago, and the image of Christ is in a symbolic Byzantine style with Coptic inscriptions in keeping with the traditional technique.[4]

Thus, the traditional, mainstream and alternative conjoin in a natural setting that appears to evoke a sense of harmony and unity. The reality, however, is slightly different. Despite the steady growth of the monastery in the 1970s, facilitating the need for the buildings, and the relatively recent construction of the *Gunya Chiara*, by 2000 the monastery was no longer functioning as such and the hermitage currently is occupied by fewer than four friars. In order to maintain the property, the monastery has had to be opened to the public, becoming a center where people from all traditions can come to focus on their spiritual practice. Tibetan and Zen Buddhist monks have made use of the facilities and yoga retreats have occurred but the place is generally under-used. Monasteries, hermitages and retreats usually are meant to be away from the 'hustle and bustle' of ordinary life but the isolation can inadvertently undermine the inhabitants' efforts in supporting the development of worldviews

---

[4] Frescoes by Giotto, world-renowned painter of the fourteenth century, feature at the chapel of St Francis in Assisi.

that enrich empathy, humility and respect, where no-one outside of the tradition gets the experiential understanding of such teachings. The brother-in-charge noted the difficulties in travelling to the hermitage with public transport and that they resided in a 'different world.'

Case Study 1 Discussion Nevertheless, the order of friars has persisted in maintaining links to relevant issues, and recently facilitated a forum on global warming, attended by locals and parishioners, where attendees were given the opportunity to voice their concerns and information was distributed. The friars are also challenged by differing ideas of the 'sacralising' of Nature; St Francis simply addressed all forms as either a Brother or a Sister, thereby establishing a sense of kindred spirit with all beings, not unlike the Native American traditions or the Buddhist reverence for all sentient beings (Sessions 1995). However, as the Franciscan Brother explained, there were difficulties in encompassing a deep spiritual respect for all creatures that did not become trivialized into owning St Francis bird bath statues and hosting blessings for pet animals:

> … it's not about blessing pets […] although at the same time it is an avenue, in a sense, of reaching out to people saying, 'well, yes, you enjoy your pet animals, [but] what about the rest? What about the whole environment? (Anglican, male, 50–59, rural New South Wales)

What the Brother alludes to is a quality that was addressed by Berry (1999); that of authenticity:

> The universe carries in itself the norm of authenticity of every spiritual as well as every physical activity within it. The spiritual and the physical are two dimensions of the single reality that is the universe itself. There is an ultimate wildness in all this, for the universe, as existence itself, is a terrifying as well as a benign mode of being…we have at times thought that we could domesticate the world, for it sometimes appears possible… (Berry 1999: 49–50)

Berry continues further that integration is key to maintaining the authenticity, where 'each individual being is supported by every other being in the Earth community [and] in turn, each being contributes to the well-being of every other being in the community' (Berry 1999: 61). Integration, integral and integrity are all etymologically linked to *integer*, that which is untouched, whole. Of note then, is the term 'integrity of creation,' recognized as one of the Anglican tradition's 'Five Marks of Mission':

1. To proclaim the Good News of the Kingdom
2. To teach, baptise and nurture new believers
3. To respond to human need by loving service
4. To seek to transform unjust structures of society
5. To strive to safeguard the integrity of creation and sustain and renew the life of the earth.
   (Anglican Consultative Council 2011)

The St Franciscan Order is thus attempting to maintain the five marks but is struggling due to low numbers of members, and lack of supporting infrastructure from the wider community. Yet, we can see a causal chain from worldviews and beliefs to the behaviors of choosing to live within one means; a working example of Herring's hierarchy of motives outlined earlier. In subsequent discussions with other interviewees about viable alternatives, mention was made of another spiritual community working along similar ideals but from a slightly different tradition, a type of Jewish-Christian hybrid, called the Twelve Tribes. This community shall be the subject of the next case study.

14.3.2.2 Case Study 2: The Twelve Tribes, Picton, New South Wales

Site Description/Background The Twelve Tribes arose out of the counterculture movement of the 1960s in the United States, recruiting mainly the youth who were disillusioned with the direction of the mainstream culture of capitalism as well as the hedonistic and spiritually diffracted lifestyle of 'hippies' (The Twelve Tribes, NDa). These young and intelligent 'drop outs' saw the 'free love' in a very different light to their contemporaries and wanted to form a solid foundation of love, respect and simple living based on fundamental teachings and not on drug-induced insights or eastern mysticism that were becoming increasingly influential at the time (The Twelve Tribes, NDb). They found the support they needed in Christ's teachings and set about creating alternative communities throughout the US and around the world. Their presence in Australia is very small, notably in Katoomba and Picton, New South Wales, where a 10 ha property is being reworked (Fig. 14.6). The property belonged to the family of one of the current 'elders,' who decided to donate it to the movement after his 'conversion.' There are numerous buildings on the site, most notable being the original farmer's cottage from the 1850s which is still used as a

Fig. 14.6 Satellite image of Twelve Tribes' property near Picton, New South Wales (Source: Google Earth, image © 2012 GeoEye)

Fig. 14.7 Original mid-nineteenth century farmer's cottage with new two-story residence under construction in background to the right (Photo by J. Lawson)

residence for visitors and new male members (Fig. 14.7). Segregation is strictly adhered to, however, couples and young families live in shared accommodation and the intention is for all to live in one building (currently being built next to the original cottage).

In contrast to the previous example, there is very little native vegetation on the property; it was extensively cleared for grazing, however, a small agricultural plot has been established to supplement the food requirements of the residents. This is maintained by some members of the community; all adults have an occupation that is directed towards the functioning of the community, therefore most are multi-skilled out of necessity. Businesses have been established to generate income and resources so that the community can continue to thrive. One of these is in demolition, while the others are in baking and catering. The demolition business is a very astute operation, where it generates the most income and is also able to provide the materials for the community's building needs. The majority of material that is removed from the demolished sites is kept and sorted at the property; very little is taken to the tip. An industrial-sized wood-fired oven has been built entirely out of recycled materials at a second site a kilometer away, currently leased by the community. This site, the former Razorback Hotel (built in 1849), is planned to be the center of the baking operations, where various organic breads and cakes will be made and packaged for delivery throughout the state. Part of the hotel will also serve as a café, a 'court of the gentiles' as one member called it, in reference to the second temple of Jerusalem built by Herod.

Of particular note, there is no church or chapel; the members meet in one of the main rooms of the buildings twice daily for scriptural study and discussion. There is no minister; all are allowed to make a contribution to the discussion, children included. They are serious, yet joyous occasions; the discussions broach profound issues on the members' relationship with each other and God and culminate in lively singing and dancing that has a particular form, similar to a structured folk dance. Thus, a very strong sense of community is established with regular meetings, shared work and responsibilities as well as a collective goal being held. Again, there is a strong correlation to Herring's hierarchy of motives, with the beliefs filtering into the choices being made. Producing organic food, recycling materials, preserving cultural property, and living simply all point towards an engaged ethic; demonstrating a comprehensive integration of a spiritual awareness with living in a practical world.

Case Study 2 Discussion The integration, however, is the issue; members of the Twelve Tribes have consciously chosen to remove themselves from the usual activities of mainstream society. Despite the presence of businesses interfacing with secular community members, there are general misgivings of the wider community because of the different values that the wider community holds. The theory of evolution was rebuked in favor of creationism by one member, although this was not held in consensus. The lack of integration and affiliation with the public, the structured roles and type of dress within the community, and the presence of a 'charismatic leader' in the US (Elbert Eugene Spriggs, from Tennessee) has led to their label as a 'cult' (Kohn 2006 broadcast 12 February 2006). Former members have had a difficult time removing themselves from the community and it is here that we see problems with how differing worldviews can co-exist. There is an unfortunate irony; where, on one hand there is a sector of an established mainstream tradition attempting a sustainable lifestyle and struggling to maintain it, while on the other are growing communities demonstrating commendable qualities of sustainable living and yet are labeled as cults (and, therefore implied as being dangerous).

There are no simple solutions; the two case studies so far outlined demonstrate the various ways individuals and groups have attempted to integrate their belief system into a viable environmentally sustainable lifestyle but they are not by any means the directions all of us need to follow. For some, it is likely that the reshaping of a worldview is a deeply personal affair and all that is needed is to simply provide a space for the 're-enchantment' that Tacey (2000) advises; a 'nature sanctuary,' where 'wild Nature is honored, and human activities limited to contemplative, aesthetic and ceremonial ones' (TEFNA 2004).

One of the reasons for the establishment of national parks, championed by the likes of John Muir, was to provide a space for the admiration of the beauty and spirituality of nature (Nash 1990). The spaces for reflection and connection in national parks, however, are varied and diffuse; the shrine and altar is everywhere and nowhere. A towering Mountain ash (*Eucalyptus regnans*) or a thousand year old yew (*Taxus baccata*) could be a shrine in a forest, but so could a monolithic rock or a waterfall. There needs to be some kind of demarcation that informs the visitor or 'pilgrim' that they are have entered a sacred space; for some, the only way of knowing is by invitation and instruction as is the case with numerous native traditions, for others, modification of the landscape has sufficed (*e.g.* painting or carving rocks, clearing vegetation to highlight other types of vegetation or natural features, or placing stones in certain patterns *etc.*). Granted, this appears as a basic, animistic or paganistic approach to redefining or rediscovering a worldview and for some this is unacceptable. Nonetheless, there are attempts at maintaining this approach, and a final example follows, where there is no community immediately present but a space is provided to establish a presence with nature and with spirit.

14.3.2.3 Case Study 3: Green Cathedral, Tiona, New South Wales

Site Description/Background There are numerous pagan sites under or besides the foundations of Christian churches; the existence of ancient yew trees in or near Christian gravesites is evidence of this, as these were revered in pre-Christian times throughout Europe (Hageneder 2005). There are, however, churches which have been built without having deconsecrating another tradition, and the Green Cathedral near Forster, NSW is one example (Fig. 14.8). Very little, however, is actually constructed; a simple set of pews, an altar, a cross and a lectern mark the area (Fig. 14.9). There is no roof, wall or window of any sort; it is completely open to nature, indeed there are no gates obstructing the path to the church. The Cabbage Palms (*Livistona australis*) are the cathedral columns, and the view across Wallis Lake to the mountains

Fig. 14.8 Location of the Green Cathedral, Tiona, New South Wales (Source: Google Earth, © 2012 Europa Technologies, image © 2012 GeoEye, Data SIO, NOAA, U.S. Navy, NGA, GEBCO, © 2012 CNES/Spot image)

Fig. 14.9 Entrance to Cathedral, with Wallis Lake and Wallingat National Park in background (Photo by J. Lawson)

of the Wallingat National Park is the substitute stained glass window behind the altar. The church is completely immersed in nature (namely, Booti Booti National Park); a space has been provided for the visitor or 'pilgrim' to know that he/she has entered a place for spiritual contemplation.

On the path leading to the Cathedral, a wooden sign with instructions for conduct and a dedication serve as a reminder of what the purpose of the place is to be. The dedication, from a poem entitled *A Forest Hymn* (1824) by William Cullen Bryant, is significantly potent as the connections of the sacredness of nature to personal behavior are explicitly made.

> **Transcript of *A Forest Hymn* by William Cullen Bryant, inscribed on wooden panel at the entrance of the Green Cathedral, Tiona NSW:**
> The groves were God's first temples. Ere man learned
> To hew the shaft, and lay the architrave,
> And spread the roof above them […]
> In the darkling wood amid the cool and silence he knelt down,
> And offered to the Mightiest solemn thanks
> And supplication […]
> Father, thy hand
> Hath reared these venerable columns, thou
> Didst weave this verdant roof. Thou didst look down
> Upon the naked earth, and, forthwith, rose
> All these fair ranks of trees. They, in thy sun,
> Budded, and shook their green leaves in thy breeze,
> And shot towards heaven […] thou art here—thou fill'st
> The solitude. Thou art in the soft winds
> That run along the summit of these trees

> In music; [...] the barky trunks, the ground,
> The fresh moist ground, are all instinct with thee.
> Here is continual worship;—Nature, here,
> In the tranquillity that thou dost love,
> Enjoys thy presence [...] Thou hast not left
> Thyself without a witness, in these shades,
> Of thy perfections. Grandeur, strength, and grace,
> Are here to speak of thee [...]
> [Let us] often to these solitudes
> Retire, and in thy presence reassure
> [our] feeble virtue [...]
> Be it ours to meditate,
> In these calm shades, thy [gracious] majesty,
> And to the beautiful order of thy works
> Learn to conform the order of our lives.

Thus the invitation is made to be in stillness and awe of nature, to come at any time without interference. Certainly, services are held here regularly as it is supported by the Community of Christ based in Forster but the ministry has graciously allowed for the church to be open for 'private devotion' since 1940, when the church was officially dedicated. That a church like this should exist from 70 years ago is quite remarkable; while the act of worshipping in groves is indeed ancient, the idea that it can be maintained within a mainstream tradition and also provide an insight as to how we should conduct our lives in following the ways of nature is certainly very progressive. As a result, the process of integration is the sole responsibility of the visitor or pilgrim.

Case Study 3 Discussion The presence of the Green Cathedral and the process of integration are difficult to reconcile partly because of the connotations associated with sacred groves and organized religion in general. For Frazer (1993 [1922]: 109), there was 'nothing more natural' than worshipping in a forest, and Schama (1996) recounts the counseling of missionary Mellitus by Pope Gregory in 601 A.D. about the establishing of churches on the site of pagan groves in England:

> When this people see that their shrines are not destroyed, they will be able to banish error from their hearts and be more ready to come to places they are familiar with, but now recognizing and worshipping the true God. (Schama 1996: 216)

But over the centuries, this message was twisted and lost, such that there is an attitude of distrust and blame from those outside established religions, a view held by many reflected in the position established by Toynbee (1961), White (1967) and McHarg (1969). It would be difficult to say that if a person with an attitude of disdain toward religion could come and visit, possibly meditate or pray at the Green Cathedral, then he/she would change their overall view. A pagan may well be quite bemused by the development, as one interviewee noted 'it's not overly surprising to see it start to come back full circle [...] to the natural surroundings because that's basically where it all started from in the first place' (Pagan, male, 40–49, rural NSW).

There is an obvious simplicity at the Green Cathedral that allows for deep contemplation, without the distractions of ornate man-made artifacts; and by being exposed to the 'soft winds' and the 'fresh moist ground,' the opportunity for 'conforming the order of our lives' is ever present. One simply needs to be sensitive to nature, where the destiny of the Earth is synchronous with the destiny of humanity and an 'intimate, reciprocal emotional relationship' can be forged (Berry 1999: 175).

14.4 Implications

Admittedly the case studies are all in rural locations; it appears inevitable that in order to fully encapsulate a sustainable lifestyle especially with regard to supporting food production and native habitats, that a rural setting is best. There is much value to be found in the way these case studies highlight the concepts of integrating a worldview into a place of worship that also upholds principles of environmental sustainability. The implications, therefore, are considerable, where a working example can demonstrate new possibilities that can be developed further or replicated elsewhere. The question then becomes a fundamental issue of 'can this be franchised?' Just as a business model can emulate the success of a McDonald's restaurant franchise, so too can the examples of the case studies be replicated.

The catch, however, is the religious belief. There are many who would not find the lifestyle of the Twelve Tribes appealing, in some cases it would be tantamount to sacrilege. Indeed, changing one's belief is the hardest challenge because of its strong links to language, culture, identity, family, and society. Thus, to franchise these various systems of living sustainably throughout other parts of Australia and beyond is extremely difficult. Certainly, the Twelve Tribes model has only been in existence for less than 50 years and is still quite an isolated island of activity. The St Francis monastery is based on a several hundred year-old system that has been exported throughout the world but not everyone is willing to forego the conveniences and luxuries of modern living to live in a monastic environment. So, with the Green Cathedral, perhaps this is the best model to franchise; to simply provide a space that is as close to the natural elements as possible and make it open to all (another similar example – Bexhill Open Air Cathedral – exists near Lismore, NSW, maintained by a different Christian denomination). Despite its rural location, the possibility of 'exporting' it to the urban environment is not impossible; nature reserves, public parks, and botanical gardens are already in existence throughout urban and regional centers. The task then is to 're-enchant' the space, to set aside a space that can be considered sacred. Again, the problem is in the belief; as managers of these spaces will be hard pressed to establish exactly where that space will be and how to ensure and maintain its sacredness. Examples of sacralizing public space can be found in urban environments; but these are diffuse and somewhat surreptitious occurrences, where pagans meet in urban public reserves and in national parks to celebrate seasonal rituals and more progressive members of the UCA conduct sermons in botanical gardens (the latter is publicly advertised).

14.5 Conclusion

In conclusion, not one of the case studies could be highlighted as being the true way to holistic environmental sustainability, as that would be too simple and naive; however, there are elements within all of them that can be used. These consist of conserving native habitats, supporting natural farming techniques, recycling materials, preserving cultural property and providing a natural setting for contemplation.

Given that attitudes and behaviors can change to a favorable outcome for the environment we nonetheless see that the causal relationship between attitudes and behaviors is still not clear. Stern and Dietz (1994) have remarked that addressing values, beliefs and worldviews as being important because of the power of 'social amplification'; strong values will lead to behavioral change. This is highlighted in the framework of environmental concern by Stern et al. (1995). Simply put, religions are in the business of values and thus can spearhead that loop of change. Gardner (2002, 2006) is promoting their efforts through such catch-phrases as 'invoking the spirit' and 'inspiring progress,' and Berry (1999) acknowledges the vast challenge that is being met by redefining our values and knowledge system, labeling it 'the great work.'

The findings here are but a snapshot of that great work unfolding; understandably, there is much more that needs to be done. Goricanec (2008) has described the difficulties in integrating alternative pathways but also that there is a pressing need for the approach. At this stage, it appears that the awareness of environmental issues is quite high and the intention of engaging is very present.

Acknowledgments Firstly, the authors thank the School of Life and Environmental Sciences and the Environmental Sustainability Research Group at Deakin University for providing the funding to be able to undertake the study. We also thank the study participants for their valuable contributions. The study was undertaken in accordance with the requirements set by Deakin University's Human Research Ethics Committee (project no. EC54-2009).

References

Ajzen, I. (1991). The theory of planned behavior. *Organizational Behavior and Human Decision Processes, 50*, 179–211.
Ajzen, I., & Fishbein, M. (1977). Attitude-behavior relations: A theoretical analysis and review of empirical research. *Psychological Bulletin, 84*(5), 888–918.
Anglican Consultative Council. (2011). Mission – The five marks of mission. Retrieved May 1, 2011, from www.anglicancommunion.org/ministry/mission/fivemarks.cfm
Armitage, C. J., & Christian, J. (2003). From attitudes to behaviour: Basic and applied research on the theory of planned behaviour. *Current Psychology, 22*(3), 187–195.
Australian Bureau of Statistics. (2005). 1266.0 – Australian Standard Classification of religious groups, 2005. Retrieved February 2, 2010, from www.abs.gov.au/AUSSTATS/abs@.nsf/DetailsPage/1266.02005?OpenDocument
Berkes, F. (2008). *Sacred ecology: Traditional ecological knowledge and resource management.* New York: Routledge.

Berry, T. (1988). *The dream of the earth*. San Francisco: Sierra Club Books.
Berry, T. (1999). *The great work: Our way into the future*. New York: Bell Tower.
Black, A. (1997). Religion and environmentally protective behaviour in Australia. *Social Compass, 44*(3), 401–412.
Blumer, H. (1955). Attitudes and the social act. *Social Problems, 3*(2), 59–65.
Bouma, G. (2006). *Australian soul: Religion and spirituality in the twenty-first century*. Port Melbourne: Cambridge University Press.
Carey, H. M. (1996). *Believing in Australia: A cultural history of religions*. St Leonards: Allen & Unwin.
Cary, J., Webb, T., & Barr, N. (2002). *Understanding landholders' capacity to change to sustainable practices: Insights about practice adoption and social capacity for change*. Canberra: Bureau of Rural Sciences.
Chaiklin, H. (2011). Attitudes, behavior, and social practice. *Journal of Sociology & Social Welfare, 38*(1), 31–54.
Davis, R. A. (1985). Social structure, belief, attitude, intention, and behavior: A partial test of Liska's revisions. *Social Psychology Quarterly, 48*(1), 89–93.
de Silva, P. (2002). *Buddhism, ethics and society: The conflicts and dilemmas of our times*. Clayton: Monash Asia Institute.
DeWitt, C. B. (2002). Spiritual and religious perspectives of creation and scientific understanding of nature. In S. R. Kellert & T. J. Farnham (Eds.), *The good in nature and humanity: Connecting science, religion, and spirituality with the natural world* (pp. 29–48). Washington, DC: Island Press.
Doob, L. (1947). The behavior of attitudes. *Psychological Review, 54*, 135–156.
Douglas, S. (2009). Religious environmentalism in the West. II: Impediments to the praxis of Christian environmentalism in Australia. *Religion Compass, 3*(4), 738–751.
Earth Charter International Council and Secretariat. (2010). *Earth charter initiative handbook*. San José: Earth Charter International Secretariat.
Figà-Talamanca, I. (1972). Inconsistencies of attitudes and behavior in family-planning studies. *Journal of Marriage and Family, 34*(2), 336–344.
Frazer, J. G. (1993). *The Golden Bough: A study in magic and religion*. Ware: Wordsworth Editions.
Fuller, R. C. (1992). *Ecology of care: An interdisciplinary analysis of the self and moral obligation*. Louisville: Westminster/John Knox Press.
Gardner, G. (2002). *Invoking the spirit: Religion and spirituality in the quest for a sustainable world*. Washington, DC: Worldwatch Institute.
Gardner, G. (2006). *Inspiring progress: Religions' Contributions to sustainable development*. New York: Norton.
Goricanec, J. (2008). How do we (all of life) live sustainably on out fragile planet. In M. Clarke, P. Connors, J. Dillon, M. Kelly, & S. Kenny (Eds.), *Community development and ecology: Engaging ecological sustainability through community development: an international eco community conference*. Melbourne: Deakin University.
Griffin, D. R. (Ed.). (1988). *The reenchantment of science: Postmodern proposals* (Constructive postmodern thought). Albany: State University of New York Press.
Hageneder, F. (2005). *The meaning of trees: Botany, history, healing, lore*. San Francisco: Chronicle Books.
Haught, J. F. (1993). *The promise of nature: Ecology and cosmic purpose*. Mahwah: Paulist Press.
Herring, H. (2009). Sufficiency and the rebound effect. In H. Herring & S. Sorrell (Eds.), *Energy efficiency and sustainable consumption: The rebound effect* (pp. 224–239). Houndmills: Palgrave Macmillan.
Holland, R. W., Verplanken, B., & Van Knippenberg, A. (2002). On the nature of attitude–behavior relations: The strong guide, the weak follow. *European Journal of Social Psychology, 32*(6), 869–876.
Jupp, J. (2009a). *The encyclopedia of religion in Australia*. Port Melbourne: Cambridge University Press.
Jupp, J. (2009b). The social role of religion. In J. Jupp (Ed.), *The encyclopedia of religion in Australia* (pp. 28–40). Port Melbourne: Cambridge University Press.

Jupp, J. (2009c). Time, place and social status. In J. Jupp (Ed.), *The encyclopedia of religion in Australia* (pp. 41–52). Port Melbourne: Cambridge University Press.
Kellert, S. R. (1984). Assessing wildlife and environmental values in cost-benefit analysis. *Journal of Environmental Management, 15*(4), 355–363.
Keynes, J. M. (1963). *Essays in persuasion*. New York: Norton.
Khisty, C. J. (2006). Meditations on systems thinking, spiritual systems, and deep ecology. *Systemic Practice and Action Research, 19*, 295–307.
Kohn, R. (2006). True stories. *The spirit of things*, ABC Radio National.
LaFreniere, G. F. (1985). World views and environmental ethics. *Environmental Review: ER, 9*(4), 307–322.
LaPiere, R. T. (1934). Attitudes vs. actions. *Social Forces, 13*(2), 230–237.
Lawson, J. T., & Miller, K. K. (2011). Green revelations in a country of drought, flood and fire: A case study of Abrahamic faith communities and sustainability. *International Journal of Environmental Studies, 68*(6), 965–979.
Liska, A. E. (1984). A critical examination of the causal structure of the Fishbein/Ajzen attitude-behavior model. *Social Psychology Quarterly, 47*(1), 61–74.
Madden, T. J., Ellen, P. S., & Ajzen, I. (1992). A comparison of the theory of planned behavior and the theory of reasoned action. *Personality and Social Psychology Bulletin, 18*(1), 3–9.
Manfredo, M. J. (2008). *Who cares about wildlife?: Social science concepts for exploring human-wildlife relationships and other issues in conservation*. New York: Springer.
McFague, S. (1993). *The body of God: An ecological theology*. London: SCM Press.
McHarg, I. L. (1969). *Design with nature*. Garden City: Natural History Press.
Nash, R. F. (1990). *The rights of nature*. Leichhardt: Primavera Press.
Nisker, W., & Gates, B. (2000). The third turning of the wheel: A conversation with Joanna Macy. In S. Kaza & K. Kraft (Eds.), *Dharma rain: Sources of Buddhist environmentalism* (pp. 150–160). Boston: Shambhala.
Pinn, J. (2003). Restor(y)ing a sense of place, self and community. In J. Cameron (Ed.), *Changing places: Re-imagining Australia* (pp. 38–47). Double Bay: Longueville Books.
Porter, M. (1990). *Land of the spirit: The Australian religious experience*. Geneva/Melbourne: World Council of Churches Publications/The Joint Board of Christian Education.
Posey, D. A. (Ed.). (1999). *Cultural and spiritual values of biodiversity*. London/Nairobi: Intermediate Technology Publications/United Nations Environment Program.
Prabhupāda, A. C. B. S. (2001). *Bhagavad-gītā as it is*. Los Angeles: The Bhaktivedanta Book Trust.
Rasmussen, L. (1992). Ecocrisis and theology's quest: Today's theologies must include a cosmology and ethic worthy of the name. *Christianity and Crisis, 52*(4), 83–87.
Ruether, R. R. (1992). *Gaia and God: An ecofeminist theology of earth healing*. San Francisco: HarperSanFrancisco.
Schama, S. (1996). *Landscape and memory*. Hammersmith: Fontana Press.
Schneider, D. J. (2004). *The psychology of stereotyping*. New York: Guilford Press.
Schwartz, S. H. (1992). Universals in the content and structure of values: Theoretical advances and empirical tests in 20 countries. *Advances in Experimental Social Psychology, 25*, 1–65.
Sessions, G. (Ed.). (1995). *Deep ecology for the 21st century*. Boston: Shambhala.
Stern, P. C., & Dietz, T. (1994). The value basis of environmental concern. *Journal of Social Issues, 50*(3), 65–84.
Stern, P. C., Dietz, T., & Guagnano, G. A. (1995). The new ecological paradigm in social-psychological context. *Environment and Behavior, 27*(6), 723–743.
Tacey, D. J. (2000). *Re-enchantment: The new Australian spirituality*. Sydney: HarperCollins.
The Ecostery Foundation of North America. (2004). Philosophy & vision of TEFNA & the ecostery movement. Retrieved March 1, 2012, from www.ecostery.org/vision.htm
The Twelve Tribes – The Commonwealth of Israel. (No Date-a). Sixties movement. Retrieved September 13, 2014, from http://twelvetribes.com/category/topics/history/sixties-movement
The Twelve Tribes – The Commonwealth of Israel. (No Date-b). Timothy Leary's dead. Retrieved September 13, 2014, from http://twelvetribes.com/articles/timothy-leary

Thurstone, L. L. (1928). Attitudes can be measured. *American Journal of Sociology, 33*(4), 529–554.
Toynbee, A. J. (1961). *A study of history*. London: Oxford University Press.
Triandis, H. C. (1971). *Attitude and attitude change*. New York: John Wiley & Sons.
Tucker, M. E., & Grim, J. A. (2007). Daring to dream: Religion and the future of the Earth. *Reflections – The Journal of the Yale Divinity School, Spring*: 4–9.
White, L. (1967). The historical roots of our ecological crisis. *Science, 155*, 1203–1207.
Wicker, A. W. (1969). Attitudes versus actions: The relationship of verbal and overt behavioral responses to attitude objects. *Journal of Social Issues, 25*(4), 41–78.

Chapter 15
"Let My People Grow." The Jewish Farming Movement: A Bottom-Up Approach to Ecological and Social Sustainability

Rachel Berndtson and Martha Geores

15.1 Introduction

Sustainability is a complex concept, encompassing both ecological and social systems. Religious groups have long been concerned with the social sustainability of their member base, but have also increasingly addressed the ecological sustainability of the natural environment. The Jewish religion is embracing social and ecological sustainability through the Jewish farming movement. The movement addresses social sustainability by uniting Jews based on individual interests rather than hierarchical boundaries, through the provision of a grassroots, pluralistic forum for Jewish identity and community to accommodate a changing twenty-first century Jewish population. The movement addresses ecological sustainability by addressing internal and external boundaries to adopting a sustainable lifestyle, through the reinterpretation of environmental concepts and land-use practices from a Jewish lens and the physical application of these concepts in Jewish farms and gardens. This chapter presents a case study on the Jewish farming movement in Baltimore County, Maryland, using a critical theory approach to uncover how a grassroots, religious movement impacts the social and ecological systems surrounding it. The authors conclude that: (1) Jewish farming serves as an interest-based, non-hierarchical form of Jewish participation and identification, which impacts participants' Jewish identities, and: (2) Jewish farming creates a community field for its participants, which eases several barriers to pro-environmental living and impacts participants' sustainable lifestyles (Fig. 15.1).

R. Berndtson (✉) • M. Geores
Department of Geographical Sciences, University of Maryland,
College Park, MD 20742, USA
e-mail: rberndts@umd.edu; mgeores@umd.edu

Fig. 15.1 The Kayam Farm is a 2 ha (5 ac), non-profit Jewish educational farm located at the Pearlstone Conference and Retreat Center in Reisterstown, Maryland (Photo by Rachel Berndtson 2012)

15.2 The Jewish Farming Movement

Farming Jews are often associated with biblical figures or Israeli kibbutzim,[1] but since 2003 American Jews have been challenging those stereotypes by engaging in small-scale, sustainable farming across the country. As a movement directly influenced by the Jewish environmental movement, the Jewish farming movement (JFM) is a phenomenon that practices sustainable agriculture with messages rooted in Jewish values and ecological and social-wellbeing (Jewish Farm School 2011; Kayam Farm 2011). The movement operates through several 0.4–4 ha (1–10 ac) non-profit and educational organic farms and community gardens (Adamah 2011; Eden Village Camp 2011; Jewish Farm School 2011; Kayam Farm 2011; Urban Adamah 2011). Although funds do come through the sale of produce and programing revenue, many of the movement's Jewish farms and gardens rely on external donations and grants from the institutionalized American Jewish community (Adamah 2011; Jewish Farm School 2011; Kayam Farm 2011).

Contemporary Jewish farming is part of a recent revival in social justice action amongst American Jews. American Jewish community leaders believe social action will do good for others as well as for young American Jews to reinforce their own identity (Kaplan 2009). Much of this social action stems from the Jewish ideal of "*tikkun olam*," which means "repair the world" (Cohen and Eisen 2000; Heilman 2004; Kaplan 2009). Although variations exist between organizations, the goals and missions of each Jewish farm can be generalized under two larger purposes: using hands-on education to deliver/reinvigorate (1) Jewish values and (2) ecological

---

[1] A kibbutz is an agriculturally based Israeli collective community providing food, shelter, clothing, education and health care to its members, who receive small work stipends (Sosis and Ruffle 2003).

values. The following sections review the impacts of Western individualism on Jewish cultural sustainability and ecological sustainability, the recent bottom-up approaches to reviving each, an overview of religious environmental movements and a case study on the impacts of the Kayam Farm (a Jewish farm in Baltimore County, Maryland) on its participants' Jewish identities and sustainable lifestyles.

15.3 Transitions

15.3.1 From Jewish Boundaries to Jewish Individualism

Modern influences have affected the processes through which identity is established and communities are maintained, resulting in a decline of identities and networks based on socialization within consistent groups and a rise of those formed through individualism (Taylor 1991). The prominence of individualism in contemporary Western society has impacted the processes of Jewish identity formation and maintenance. Although a rich history of toward Jewish individualism dates back to European Enlightenment (Diner 1992; Sorin 1992; Bloomfield 1999; Cohen and Eisen 2000), this paper picks up with the trend in twentieth century America.

During the early to mid-twentieth century, urban American Jews were still members of the greater Jewish community "if only through osmosis" (Kaplan 2009: 21). However, the 1950s brought a significant American Jewish exodus from urban environments to the suburbs, diminishing the geographic and social closure amongst Jewish networks and thus the power of cultural norms[2] to dictate social control. During the suburban exodus, many American Jews either lost their sense of Jewish connection or saw a need to actively reevaluate their Jewish identity (Kaplan 2009). The suburban exodus led many American Jews to find other means to sustain a Jewish ethnic identity, such as historical relations of Jewish survival with the Holocaust, attachment to the state of Israel, and membership to Jewish organizations such as Jewish Community Centers (JCCs) and philanthropic groups (Herman 1989; Selengut 1999; Poll 1998; Cohen and Eisen 2000; Lipset 2003; Kirshenblatt-Gimblett 2005). However, today the relevance of those measures is declining, forcing American Jews to once again reinterpret ethnic identity (Etzioni-Halevy 1998; Sharot 1998; Cohen and Eisen 2000; Teutsch 2003; Kirshenblatt-Gimblett 2005; Beck 2009).

With the dissipation of old ethnic enclaves and increased residential mobility away from family homes, American Jews are no longer socialized in spaces that carry consistent unifying orders, thus impacting Jewish identification (Teutsch 2003; Groeneman and Smith 2009). Jewish identities and lifestyles are more fluid in terms of social and physical boundaries, giving Jewish individuals the opportunity to choose and re-choose identity based on individually made decisions (Cohen and Eisen 2000; Kelman 1999 in Horowitz 2002: 23; Heilman 2004; Kirshenblatt-Gimblett 2005; Windmueller 2007; Ellenson 2009; Grossman 2009; Heller 2009;

---

[2] Norms powerful enough to dictate society stem partially from the high degree of social capital maintained in such societies based on high network closure (Coleman 1988, 1993).

Herring 2009). Many contemporary American Jews want to be Jewish based on personal meaning rather than communal obligation or historical destiny (Azria 1998; Cohen and Eisen 2000; Horowitz 2000; Heilman 2004; Kirshenblatt-Gimblett 2005; Windmueller 2007). The decline of religious communal obligations and the desire to be Jewish on an individualistic basis has led to Jewish identification through the "sovereign self," a trend classified by Steven M. Cohen and Arnold Eisen explaining religious identification and practice that emerges from the significance of personal meaning, gives authority to the individual and transpires primarily through private and intimate spaces (2000: 184). This form of Jewishness has been described as an "elective ethnicity," based on its voluntary and non-compulsory characteristics (Azria 1993 in Azria 1998: 29), and leads to cultural identities forming from "consent" rather than "descent" (Sharot 1998: 87; Kirshenblatt-Gimblett 2005). The shift in Jewish identity construction from tribal to individual exemplifies the shift from an encompassing narrative of Jewish peoplehood and Jewish survival to a series of local narratives reflecting individual needs and wants in a Jewish ethnic identity (Cohen and Eisen 2000).

15.3.2 The Fragmentation of Networks and Decline of Social Capital

The aforementioned individualistic societal trends have also led to a transition in social structures and social capital, which ultimately influenced methods of local response to contemporary societal issues, including cooperative action toward ecological sustainability. James Coleman classifies the shift in social organizations as one from "primordial" to "purposively constructed" (Coleman 1993: 2). "Primordial" social organizations are constructed through family and clan ties, operate through closed networks and rely on normative power for social control (Coleman 1993). "Purposively constructed" social organizations are created between "positions" and/or offices rather than persons, and rely on rules, laws and formal incentives for social control (Coleman 1993).

Today's purposively constructed social organizations have changed the way many people access resources, which has in turn affected the strength of communal social capital. Individuals develop a series of loose relationships to secure a complete set of goods and services that they no longer receive from a single community (Wellman 1999; Tuan 2002). Each specialized tie connects the individual to a different community resulting in multi-layered communities providing separate resources for separate needs (Fischer 1982 in McMillan and Chavis 1986: 19). This series of weak,[3] specialized ties, however resourceful, does not provide as solid a

---

[3] Weak ties are connections between acquaintances that provide information and resources beyond one's immediate social circle, but are less readily available and are of less assistance than strong ties. Strong ties are connections between friends and family that are easily accessible and provide deep measures of assistance (Granovetter 1983).

support system as strongly tied, centralized networks (Tuan 2002). Tighter networks made of individuals' "whole selves," (rather than "fragments of selves") enable the creation and maintenance of social capital, which empowers local groups to successfully embark on future cooperative actions (Wellman 2001: 244; Coleman 1993; Putnam 1994), rather than relying on the top-down actors to develop policies and regulations.[4] Social capital is essential for cooperative action towards ecological sustainability (Putnam et al. 1993 in Dale 2005: 26; World Bank 2003 in Dale 2005: 26, 2005).

Attempts at stimulating pro-environmental behavior have largely stemmed from purposively constructed social organizations (such as national governments), using top-down, regulatory approaches. Beginning after the 1992 and 2002 United Nations Earth Summits, national governments have implemented financial disincentive policies[5] aimed at stimulating individual-level pro-environmental behavioral changes. However, such approaches have been criticized for relying too heavily on the individual actor, overlooking social, economic, psychological and cultural factors and choosing a strategy stemming from superficial rather than morally intrinsic reasons for pro-environmental behavioral changes (Barr 2003; Dobson 2003; Seyfang 2005, 2006a, b). Some argue rather than solely promoting "greener" consumption, approaches inspiring sustainable lifestyles should emphasize value changes, guided by intrinsic moral motivations such as those surrounding environmental justice, ecological citizenship and "new economics" (Seyfang 2006a, b; Matti 2008; Jagers et al. 2009). However, even if pro-environmental behavior stems from value-based (rather than solely fiscal) motivations, a number of barriers to adopting a sustainable lifestyle remain. Barriers preventing a sustainable lifestyle may be internal (such as level of environmental knowledge and awareness and perceived amount of control) or external (such as institutional, economic and cultural). Table 15.1 summarizes each barrier.

Community-based organizations, such as local religious groups, may have the ability to ease some of the barriers to adopting sustainable lifestyles (Defra 2005 in Middlemiss 2008: 78; Gardner and Stern 2002 in Middlemiss 2008: 78; Jackson 2005 in Middlemiss 2008: 78; Middlemiss and Parrish 2010).

---

[4] Although this chapter focuses on local collaboration for ecological sustainability, effective sustainability efforts require actors at many scales from both strongly tied networks (for trust-building, "place-specific social ecological information" capture, emotional support, effective communication, resource provision and successful deliberation) and weakly tied networks (for information diffusion, adaptation to new situations, prevention of groupthink, diversification of knowledge and actor bridging) (Prell et al. 2009; Barthel et al. 2010 in Ernstson et al. 2010: 32; Granovetter 1983 in Ernstson et al. 2010: 32; Oh, Chung, and Labianca 2004 in Ernstson et al. 2010: 32, Scheffer and Westley 2007 in Ernstson et al. 2010: 32; Wasserman and Faust. 1994 in Hinrichs 2000: 19; Burt. 1992, 2000, 2001 in Hinrichs 2000: 19).

[5] Many governments have adopted the Organization for Economic Cooperation and Development's (OECD) suggestion that market failure will lead to unsustainable action, and therefore prices and regulations should encourage pro-environmental consumer choices (OECD 2002 in Seyfang 2005: 293).

Table 15.1 A review of external and internal barriers to adopting a sustainable lifestyle

External barriers	Internal barriers
Institutional factors include the provision and upkeep of necessary infrastructures (such as bike paths, recycling centers, and public transportation systems) to adopt pro-environmental behaviors and the provision of appropriate policies to enforce behavior (Kollmuss and Agyeman 2002; Barr 2003; Redclift and Hinton 2008)	**Environmental knowledge and awareness** is the extent to which one knows of environmental issues and problems associated with unsustainable human activity (Kollmuss and Agyeman 2002; Barr 2003)
Economic factors include the ability of the individual to afford adopting a sustainable lifestyle (Kollmuss and Agyeman 2002)	**Perceived amount of control** is the extent to which one believes his/her behavioral change will impact the current unsustainable conditions (Newhouse 1991 in Kollmuss and Agyeman 2002: 255; Maiteny 2002; Eden 1993 in Barr 2003: 231; Hinchliffe 1996 in Barr 2003: 231). If individuals become too overwhelmed with their lack of capacity to solve environmental problems, they may respond with an "unconscious act of denial" (Maiteny 2002: 300). Individuals may feel so disheartened by their individual inability to sustain pro-environmental behavior and so overwhelmed by the "enormity of the task and the apparent futility of their behaviour" that they remove themselves completely (Maiteny 2002: 301)
Cultural factors include norms that may prevent pro-environmental behavior (Chan 1998, 2001 in Barr 2003: 230; Georg 1999; Tucker 1999 in Barr 2003: 230; Kollmuss and Agyeman 2002; Middlemiss 2008; Redclift and Hinton 2008)	

Source: Rachel Berndtson and Martha Geores

15.4 Religious Groups and Environmental Action

Aligning with the trends of Western individualism, America's religious climate has been increasingly categorized by a focus on selective, individual spirituality rather than top down observation of obligation from religious authorities (Petersen 2009). As such, many religious groups have incorporated constituents' interest in the environment with achieving spiritual satisfaction. Faith-based groups may be particularly successful at shaping the ecological beliefs and actions of adherents, because they provide "contextual connection" by framing environmental issues from a value-based theological perspective, thus giving those issues a greater meaning (Maiteny 2002: 305; Brockelman 1997; Rockefeller 1997; Jacobs 2003; Macnaghten 2003; Middlemiss 2008; Smith and Pulver 2009). Pro-environmental behavior is more likely to occur if the individual finds personal meaning and sense of well-being in the sustainable lifestyle (DeYoung 1996 in Barr 2003: 230; Maiteny 2002; Macnaghten 2003). Additionally, religious groups in particular present a "promising strategy" for forming sustainable agricultural networks, due to their pre-existing shared common values and spaces for action (Feenstra 1997: 34). By situating ecological sustainability within meaningful frameworks, religious groups can alter the normative group culture surrounding the environment (Petersen 2009; Middlemiss 2010). According to Ann Dale, without meaning or purpose from consistent shared values, contemporary individuals and communities will fail to respond to the "dynamic centrifugal forces of this era" (2005: 14).

According to David Ehrenfeld and Philip Bentley, many people fail to realize that Judaism is "one of the first great environmental religions" (2001: 125). Judaism is built on the notion that God is the sole creator of the universe, and the natural world, as part of this creation, relies on God for maintenance and sustainability (Tirosh-Samuelson 2001). Therefore those following a religious Jewish tradition respect and revere the natural world not for its own sake, but for its connection to God (Tirosh-Samuelson 2001). American Jewish environmentalism emerged at a large scale in the 1960s, in defense of the religion against "accusations that the Judeo-Christian tradition is responsible for Western society's destruction of nature" (Jacobs 2003: 450). Lynn White Jr.'s article, "The historical roots of our ecological crisis" (1967), prompted many defensive responses from across the Judeo-Christian spectrum, resulting in the eventual creation of religious-environmental groups (Gendler 1997; Tirosh-Samuelson 2001; Jacobs 2003). Ellen Bernstein's creation of the non-profit, *Shomrei Adamah* (Keepers of the Earth), was an early religious-environmental organization. From 1988 to 1996, Shomrei Adamah operated on a mission to "explore and illuminate the ecological roots of Jewish tradition and make them accessible to wide audiences" by organizing educational outdoor trips, holiday celebrations and other Jewish environmental events (Ellen Bernstein 2011; Jacobs 2003: 453). In 1992 Carl Sagan pursued eco-Judaism through political advocacy by partnering with Al Gore and several Jewish political leaders in Washington to found the Coalition on the Environment and Jewish Life (COEJL) (Jacobs 2003: 457). COEJL is a non-profit organization that seeks to "enact a distinctively Jewish programmatic and policy response to the environmental crisis" (COEJL 2011). Other Jewish institutions following the eco-trend include the Union of American Hebrew Congregations, the National Jewish Community Relations Advisory Board, the United Synagogue of Conservative Judaism and the Rabbinical Assembly (Jacobs 2003: 455).

A major product of the Jewish environmental movement was the realization of environmentalism's power and effectiveness as a means for Jewish education. The leading organization in the development of Jewish environmental education is the Teva Learning Center. Created in 1994, the Teva Learning Center uses a philosophy of immersive environmental education to help students "develop a more meaningful relationship with nature and their own Jewish practices" for "personal growth, community building, and a genuine commitment to Tikkun Olam, healing the world" (Teva 2011). Teva runs Jewish environmental education programs for groups across the Jewish denominational spectrum, hosted at several Jewish retreat and environmental centers across the country (Teva 2011).

The successful and growing Jewish environmental movement spawned a sub-movement around ecologically sustainable and socially-just food, later to be dubbed the "new Jewish food movement." Developed between 2004 and 2005, the new Jewish food movement focuses on issues of food access, food justice and food production domestically and abroad, from a Jewish moral lens (Feldstein 2011). The new Jewish food movement, like the Jewish environmental movement, also serves as a new approach for reengagement in Jewish life. According to Dana Kaplan "Perhaps the most interesting of these new approaches to traditional observances is eco-kashrut" (2009: 74). The Eco-Kosher Project was developed in 1990 by Rabbi Arthur Waskow, co-founder of the Alliance for Jewish Renewal (ALEPH), in order to "reevaluate the observance of kosher laws" (Kaplan 2009: 75; Jacobs 2003). According to Waskow, "What the Eco-Kosher Project implies is that we can strengthen our Jewish distinctiveness and serve the needs of the earth as well; that we can strive to heal ourselves by helping to heal the earth, and help to heal the earth by healing ourselves" (1995 in Kaplan 2009: 75). In this way, eco-kashrut makes a traditional element of Jewish life relevant to a new generation of Jews concerned with environmental issues (Kaplan 2009).

The leading organization in the new Jewish food movement is "Hazon," a national non-profit whose goal is to bring change through, "transformative experiences for individuals and communities, thought leadership in the fields of Jewish and environmental knowledge, and support of the Jewish environmental movement in North America and in Israel" (Hazon 2011). The organization seeks to involve Jews from across geographic scales, generations and religious denominations, in order to use "food as a platform to create innovative Jewish educational programs… touch people's lives directly…strengthen Jewish institutions…and create healthier, richer and more sustainable Jewish communities" (Hazon 2011). Hazon's Jewish food programs include establishing CSAs (Community Supported Agriculture), hosting and organizing national Jewish food conferences, maintaining the "Jewish Food Education Network (JFEN)," sponsoring the "Shmita Project" and providing the Jewish food blog, "The Jew & the Carrot" (Hazon 2011).

The Jewish farming movement came as a natural evolution from the Jewish food movement, with the first Jewish educational farm founded in 2003 at the Isabella Freedman Retreat Center in Falls Village, Connecticut (Adamah 2011). Jewish agricultural educators originally trained at Adamah have established other farms in different locations across the United States. However, "The goal of the Jewish farm-based schools is not to churn out farmers but to make gardening and farming normative practice within the wider Jewish community. The leaders of these programs say they look forward to the day when every Jewish community center, synagogue and day school will have its own garden" (Fishkoff 2009). The farms of today's JFM use Jewish traditions and values to exemplify ecological education and behavior, and use ecological connections to enliven Jewish community and identity.

Notions of environmental protection and appreciation are apparent throughout Jewish culture, with origins in religious texts. The Torah and the Talmud[6] describe both general notions of Jewish environmentalism (such as stewardship to God and *bal tashchit*) and specific laws on the social justice surrounding agricultural societies and land conservation (such as *peah, leket, shichecha, shvi'it, orlah* and *kilayim*), which are summarized in Table 15.2.

---

[6] The Talmud is a compilation of "Jewish wisdom, and the oral law, which is as ancient and significant as the written law (the Torah)….legend, and philosophy, a blend of unique logic and shrewd pragmatism, of history and science, anecdotes and humor." Although the compilation can be used as a source for Jewish law, it "cannot be cited as an authority for purposes of ruling" (Steinsaltz 2006: 4).

Table 15.2 Explanation of Jewish environmental and agricultural concepts and laws

Concept	Explanation	Citation in Torah/Talmud/Mishnah[a]
Stewardship to God	God establishes stewardship as He describes Himself as The Creator and assigns man as the humble steward	1. "And…God…put him into the garden of Eden to till it and to keep it" (Genesis 2: 15 in Ehrenfeld and Bentley 2001: 129)
		2. "The land shall not be sold forever: for the land is Mine; for you are strangers and sojourners with Me" (25: 23 in Ehrenfeld and Bentley 2001: 129)
Bal tashchit	A commandment from God requiring that humans "do not destroy" (Ehrenfeld and Bentley 2001: 131)	"When in your war against a city you have to besiege it a long time in order to capture it, you must not destroy its trees, wielding the axe against them. You may eat of them, but you must not cut them down. *Are trees of the field human to withdraw before you under siege?* Only trees which you know do not yield food may be destroyed; you may cut them down for constructing siegeworks against the city that is waging war on you, until It has been reduced" (Deuteronomy 20: 19–20, New Jewish Publication Society Translation in Schwartz 2001: 231, author's emphasis)
Peah	Leaving corners of one's field to be left for the poor and needy (Manela and Silverstein 2010)	"One should not give pe'ah less than one-sixtieth [of the field], and even though they said pe'ah has no measure, everything depends on the size of the field, the number of the poor, and the extent of the crop" (Mishna Pe'ah 1: 2 in Manela and Silverstein 2010: 7)
Leket	Allowing the dropped harvest to be left for the poor (Manela and Silverstein 2010)	"And when you reap the harvest of your land, do not wholly reap the corner of your field, and do no gather the gleaning of your harvest. And do not glean your vineyard, and do not gather the fallen fruit of your vineyard; rather leave them for the poor and for the stranger: I am the Lord you God." (Leviticus 19:9–10 in Manela and Silverstein 2010: 9)
Shichecha	Leaving harvested crops that have been forgotten in the fields for the poor (Manela and Silverstein 2010)	"When you reap your harvest in your field, and you forget a sheaf in the field, do not go back to fetch it; it shall be for the stranger, for the fatherless, and for the widow; that the Lord your God may bless you in all the work of your hands." (Deuteronomy 24: 19 in Manela and Silverstein 2010: 10)
Shvi'it	Letting the earth go fallow for an entire year, every seventh year (Manela and Silverstein 2010)	"But in the seventh year shall be a Sabbath of solemn rest for the land, a Sabbath unto the Lord; do not sow your field, nor prune your vineyard. That which grows of itself of your harvest do not reap, and the grapes of you undressed vine do not gather; it shall be a year of solemn rest for the land…" (Leviticus 25:4-7 in Manela and Silverstein 2010: 39)
Orlah	Prohibits harvesting of fruit trees for their first three years (Manela and Silverstein 2010)	"When you enter the land and plant any tree for food, you shall regard its fruit as forbidden. Three years it shall be forbidden for you, not to be eaten. And in the fourth year all the fruit shall be holy, for giving praise unto the Lord" (Leviticus 19: 23–24 in Manela and Silverstein 2010)

(continued)

Table 15.2 (continued)

Concept	Explanation	Citation in Torah/Talmud/Mishnah[a]
Kilayim	Forbidden mixtures of crops and species	1. "You shall keep my statues. Do not interbreed your cattle; do not sow your field with two kinds of seed; and do not wear a garment made of two kinds of material mixed together" (Leviticus 19: 19 in Manela and Silverstein 2010: 15)
		2. "A path of six handbreadths by six handbreadths may be sown with five varieties – four on the four sides of the path, and one in the middle. If it has a border one handbreadth high, it may be sown with thirteen – three on each border, and one in the middle. One may not plant turnip-heads in the border because it would fill up. Rabbi Yehudah says, six in the middle" (Kilayim Mishnah 3: 1 in Manela and Silverstein 2010: 15)

Source: Rachel Berndtson and Martha Geores
[a]The Mishnah is (together with the Gemarah) a book of the Talmud, containing rabbinic discussions dating from roughly a century prior to year 1 up until the 500s. Scholars of later generations reorganized the rabbinic discussions into new texts, with (what many believe to be) the most authoritative as the Schulchan Aruch, compiled in the 1500s (Blecher 2007)

Such laws were applied in historical Jewish farming communities and are reinterpreted today through the contemporary Jewish farming movement. Farming was a way of life for Jews living in ancient Israel, but was limited for the nineteenth and early-twentieth century Jews of Eastern Europe due to restrictions on landownership (Sorin 1992; Eisenberg 1995). In America, individual Jews farmed as early as the colonial era, and collective American Jewish farming came about (although ultimately dissipated) in the late nineteenth century (Eisenberg 1995; Lavender and Steinberg 1995). Until the recent emergence of the contemporary American Jewish farming movement, the "urban complexion of American Jewry disguises, even from itself, a people rooted in agrarianism" (Goldberg 1986: xxiii).

15.5 Bottom Up Approaches

15.5.1 Bottom Up Approaches to Jewish Identity

As noted above, the sovereign self has become the "principle authority" in identity formation for many American Jews, given the declining practice of religious norms and communal affiliations and the decay of externally imposed identities (Cohen and Eisen 2000: 2; Sharot 1998; Selengut 1999; Horowitz 2000). American Jews have created their own authentic selves from various aspects of Judaism rather than stepping into an "inescapable framework" of identity based on familial, communal or traditional prescriptions (Cohen and Eisen 2000: 2; Poll 1998; Horowitz 2000; Windmueller 2007; Ellenson 2009; Grossman 2009; Starr 2009). Sources of Jewish identity have shifted from externally bounded, top-down religious norms to internally driven, bottom-up, personally meaningful concepts (Horowitz 2000, 2002; Woocher 2009). As such, Jewish individualized identities fit less neatly into hierarchically established denominational labels such as "Conservative" or "Reform," making nondenominational and transdenominational movements an attractive option (Kirshenblatt-Gimblett 2005; Ellenson 2009; Heller 2009; Grossman 2009). American Jews, and particularly young American Jews, are more likely to form networks through loose social ties rather than formal ties (Heilman 2004; Cohen and Kelman 2005 in Kirshenblatt-Gimblett 2005: 6). These networks of porous boundaries, contextual relationships and connections without affiliation are not attained through Jewish Community Centers and traditional Jewish organizations (if their programming remains unchanged), but rather through "intermittent involvements structured around common interests" (Cohen and Kelman 2005 in Kirshenblatt-Gimblett 2005: 6; Grossman 2009). According to Zachary Heller, many American Jews seek to establish "independent, collaborative religious settings that meet their needs," which may fall outside existing congregations or traditional practices (2009: 5). A suggested approach in making Judaism relevant to many American Jews is for synagogues and other Jewish organizations to reinterpret Jewish values and apply them to contemporary concerns and questions (Herring 2009).

15.5.2 Bottom Up Approaches to Ecological Sustainability

Community-based organizations may enable pro-environmental behavior by easing internal and external barriers overlooked by top-down actors, including delivering informational resources relevant to local contexts, providing "meaningful experiences"[7] with the environment, offering the necessary facilities and infrastructure through which to act sustainably and creating community norms that "stimulate a culture in which sustainable consumption is acceptable" (Middlemiss 2010: 87; Georg 1999; Maiteny 2002; Barraket 2005). Internal and external barriers are addressed through the creation of a community field. A community field[8] is a "purposefully organized sub-network" that "provides the structural or institutional framework for the development and/or mobilization of social capital" (Sharp 2001 in Barraket 2005: 78, 80). The field serves as a structural base upon which to organize local action, mobilize social capital within and amongst local communities, and address issues of cultural, social, environmental and economic development (Barraket 2005). The difficulty of addressing the internal and external barriers to pro-environmental

---

[7] Meaningful environmental experiences are more likely to result in long-lasting pro-environmental behavior than behavior changes made solely in response to "externally imposed regulations and incentives" (Maiteny 2002: 304).

[8] The term "community field" originates from K.P. Wilkinson's notion of an "emergent structure for collective action that cuts across a range of specific social fields, or interest areas" (1991 in Barraket 2005: 78).

behavior is magnified by today's highly dispersed and fragmented networks. However, by establishing community fields, community-based organizations centralize the location at which members may access ecological information, guidance and the necessary infrastructure through which to act sustainably (Barraket 2005).

By uniting previously socially-separated group members over a common goal of ecological betterment, community-based organizations re-embed social interactions at the local level and enhance community social capital (Georg 1999), which is considered a "prerequisite" and "critical" for success in cooperative action towards ecological sustainability (Putnam et al. 1993 in Dale 2005: 26; World Bank 2003 in Dale 2005: 26, 2005: 14). Such local engagement may lead to the creation of "social-ecological memory," which is described as social ties that form over ecological issues due to collected lived experiences (Barthel et al. 2010 in Ernstson et al. 2010: 39). Sustainable farming initiatives in particular have been associated with reembedding[9] social relations at the community level (Feenstra 1997; Wells et al. 1999; Cone and Myhre 2000; Hinrichs 2000; Lacy 2000; Hendrickson and Heffernan 2002; Feagan 2007; Feagan and Henderson 2009).

15.6 Case Study: The Kayam Farm

> We're trying to change a food system and the face of what Judaism is or can be (Kayam Farm Interviewee 2, 2011)

The aforementioned themes of contemporary Jewish identity and ecological sustainability align with the goals and programming of many Jewish farms and community gardens in the United States. We use a case study of the Kayam Farm in Baltimore County, Maryland to present the impacts of Jewish farming on participants' Jewish identities and sustainable lifestyles. Data were collected from January 2011 to April 2012, using 9 in-depth open interviews, 28 structured phone interviews and 58 in-depth online surveys with individual (persons) and institutional (Jewish day schools, synagogues, Jewish organizations) Kayam Farm participants.

Kayam is a 2 ha (5 ac) educational Jewish farm located in Reisterstown, Maryland. Although it sells produce through a 50-share Community Supported Agriculture (CSA), at farmers markets and to wholesale buyers, the Kayam Farm is a nonprofit institution, financially supported by individual donors, family foundations and the Jewish Federation of Greater Baltimore. The farm offers Jewish engagement structured around reinterpretations of the environmental, ecological and agricultural concepts embedded in the Torah and Talmud (Kayam Farm Interviewees 1, 2, 5, 6, and 8, 2011). Ecologically grounded methods of Jewish engagement are enacted through the Kayam Farm's educational programming, farm structure and operations, and ritual activities.

---

[9] Embeddedness is a term originally used by Karl Polanyi in his description of economic institutions (1957 in Hinrichs 2000: 296).

Fig. 15.2 The Gan Alef Beit (Alef Beit Garden), located on a section of the 2 ha (5 ac) farm, contains boxes that each represent the letters of the Hebrew alphabet. The Gan Alef Beit is used for Kayam Farm Jewish environmental educational programming (Photo by Rachel Berndtson)

The farm's experiential educational programming is an alternative method of Jewish education to conventional Hebrew school. The Kayam Farm's place-based Jewish agricultural education provides students with a meaningful Jewish connection and an alternative to an "emotionally semi-involved participant who was sent to Hebrew school and pushed through a 'bar mitzvah factory' as part of childhood rites of initiations" (Kaplan 2009: 104) (Fig. 15.2).

The Kayam Farm's physical structure and operations are grounded in Jewish ecological concepts. The Kayam Farm reinterprets biblical land-use laws and applies them to a contemporary understanding and application of sustainable agriculture (Kayam Farm Interviewees 1, 2, 3, and 5, 2011). For example, the farm reinterprets and physically implements the *kilayim, peah* and *leket* laws, which dictate what kinds of crops can be planted where and how to harvest them (Fig. 15.3).

Jewish ritual engagement takes on a new meaning when framed through an ecological lens at the Kayam Farm. Examples include Sustainable *Simcha* workshops and holiday celebrations. Kayam's Sustainable *Simcha* workshops offer ways to "green" one's Bar or Bat Mitzvah by reducing carbon footprints during and after the event. The farm's monthly *Rosh Chodesh* ("head of the month") celebrations are structured around an ecological interpretation, as they are meant to "pass through time in a grounded ways, connected to land, seasons and Jewish culture" (Kayam Farm 2011) (Fig. 15.4).

Fig. 15.3 The Kayam Farm Patriarch's Vineyard, located on a section of the 2 ha (5 ac) farm, contains three grape varieties to represent the three patriarchs from the Torah: Abraham, Isaac and Jacob (whose artistic busts sit atop the three posts leading into the Vineyard). The Patriarch's Vineyard serves as a tangible example of several Talmudic agricultural laws including orlah (specifying the harvesting of fruit trees) and kilayim (specifying forbidden crop mixtures and appropriate crop spacing) (Photo by Rachel Berndtson)

The Kayam Farm offers new forms Jewish engagement by reinterpreting and reapplying traditional Jewish customs through ecologically grounded meanings (Fig. 15.5).

15.6.1 Kayam Farm Impacts on Participants' Jewish Identities

> It's a reawakening and an enlivening of the story of Jewish community, Jewish vibrancy, Jewish purpose. (Kayam Farm Interviewee 2, 2011)

Kayam Farm programming strives to instill a renewed sense of Jewish identity. As described by a Kayam Farm employee, "[the Kayam farm] is a Jewish community with Jewish values and Jewish way of life – a way of being Jewish in the world that is *relevant to us*" (Kayam Farm Interviewee 2, 2011, emphasis added). Many interviewees described farming as a way to reconnect the previously unengaged Jewish population to Jewish identities, lifestyles and community (Kayam Farm Interviewees 1–4, 8, and 9, 2011). A shared interest in the environment acts as a bonding mechanism from which to enter a Jewish realm, serving as an "access point" back into Judaism for the previously un or under engaged (Kayam Farm Interviewees 3 and 4, 2011).

Fig. 15.4 The Kayam Farm Hebrew Calendar Garden, located on a section of the 2 ha (5 ac) farm, contains 12 raised beds arranged in a circle to represent each Hebrew month. Kayam Farm participants gather at the Calendar Garden at each Rosh Chodesh (head of the month) to recite the Rosh Chodesh blessing, review the past month and welcome the new. The heirloom perennials planted in each month's bed represent the Jewish spiritual energy and Jewish holidays of that month (Photo by Rachel Berndtson)

Those interested in reinvigorating their Jewish identity may feel "[environmentalism] is a shared common value of the people in this community, and maybe I'll also be able to make those Jewish connections and feel strengthened in Jewish identity" (Kayam Farm Interviewee 3, 2011). Although many of today's "new" Jewish initiatives originally create "weak ties" over intermittent commitments, these may morph to strong ties if sustained for longer, more consistent periods of time (Kirshenblatt-Gimblett 2005: 5). Rabbi David Teutsch explains such shifts as interest-based, intermittent Jewish organizations changing from Jewish "communities of convenience," to "communities of care," to "communities of commitment" (2003). Once Judaism "feels relevant again" through agricultural and ecological experiences, Kayam Farm participants "may say 'well if my Jewish identity is strengthened by this environmental Jewish connection, what else may Judaism have to say about other issues?'" (Kayam Farm Interviewee 3, 2011). The Jewish farming movement in Baltimore "is not just about farming and connections between Jews and farming, but about exploring Jewish identity, and spiritual practice and pluralistic communities" (Kayam Farm Interviewee 3, 2011).

The researchers sought to gauge the impacts of Kayam Farm participation on individual Kayam Farm participants' Jewish identities. Specific indicators of Jewish identity (listed in Table 15.3) were extracted from open interviews and from literature

Fig. 15.5 Etrog trees grow in the Kayam Farm Greenhouse. Etrogs are citrus fruit used for ritual activities during the Jewish harvest holiday of Sukkot (Photo by Rachel Berndtson)

Table 15.3 Specific indicators of Jewish identity

Type of indicator	Specific indicators
Cognitive	Exploration of Jewish religion
	Exploration of Jewish culture
Affective	Feeling of connection with the Jewish peoplehood
	Feeling of connection with the Baltimore Jewish community
	Feeling of connection with the Baltimore Jewish environmental/agricultural community
	Sense of personal Jewish identity
Behavioral	Personal interactions/connections with other Jews
	Professional interactions/connections with other Jews
	Participation in Jewish events/activities/organizations

Source: Rachel Berndtson and Martha Geores

reviews and used in online surveys to measure cognitive, affective and behavioral (as all three are important in understanding contemporary Jewish identities (Horowitz 2000, 2002)) impacts of Kayam Farm participation on Kayam Farm participants' Jewish identities.

Survey takers were asked to rank the level of impact they felt occurred to each specific indicator of Jewish identity, based on their Kayam Farm participation. Survey responses were tallied and merged into three levels of impact: no impact to very little impact, little impact to moderate impact, and great impact to very great impact. Figure 15.1 represents Jewish identity impact levels separated by each specific indicator of Jewish identity ranked by survey takers (Fig. 15.6).

Participants' Jewish identities were impacted at different levels (no impact to very great impact) and in different impact area (cognitive, affective and behavioral) based on their participation with the Kayam Farm. The data show no distinct trends based on impact levels or impact areas. Impact levels and impact areas vary based on the specific indicators of Jewish identity with no clear pattern. More data collection and analysis are necessary to determine if Jewish identity impact levels and areas vary based on Kayam Farm participant characteristics or other factors.

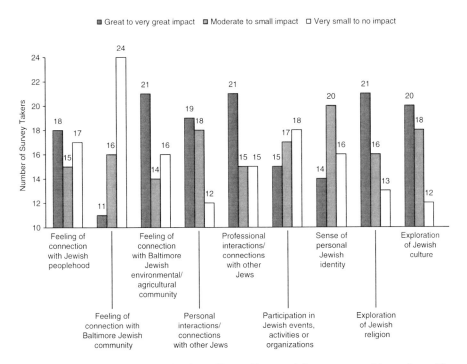

Fig. 15.6 Jewish identity impact levels on Kayam Farm participants separated by each specific indicator of Jewish identity (N=51) (Source: Rachel Berndtson and Martha Geores)

15.6.2 Kayam Farm Impacts on Participants' Sustainable Lifestyles

> I think the more people are educated about it, the more they see the benefits. (Kayam Farm Interviewee 6, 2011)

Although the Kayam Farm aims to build Jewish identity, encouraging sustainable lifestyles is a complementary focus. A principal goal of the Kayam Farm is, "empowering and mobilizing the Jewish community to be more sustainable and make healthier choices and eat local food" (Kayam Farm Interviewee 2, 2011). Agricultural lessons are framed through a Jewish lens to incorporate ecological values into participants' daily lives (Kayam Farm Interviewees 3, 5, and 8, 2011). Ecological impacts on Kayam Farm participants' may come in cognitive, affective or behavioral forms.

The Kayam Farm impacts cognitive and affective facets of its participants' sustainable lifestyles, by addressing the internal barriers to adopting a sustainable lifestyle. The amount of ecological knowledge one holds impacts (although is not completely responsible for (Kollmuss and Agyeman 2002; Maiteny 2002; Barr 2003)) his/her ability to act in a pro-environmental manner (Barr 2003; Middlemiss 2010). The farm circulates environmental information in many ways, including field trips, seminars, conferences and learning series (Kayam Farm Interviewees 2, 4, and 8, 2011). According to a Kayam Farm volunteer, participating on the farm can "influence people to learn more. Even what you don't get here you can choose to look more into it" (Kayam Farm Interviewee 7, 2011). The farm empowers participants to continue learning about agriculture, ecology, food justice and the environment after they leave the physical farming site. Participants' ecological awareness has been increased in a variety of topic areas, including runoff prevention, local growing season cycles, pesticide use, recycling, composting, healthy food choices, garden techniques, farming practices and agricultural education models (Kayam Farm Interviewees 3, 4, 7, and 8, 2011). In serving as a community field, the Kayam Farm provides a centralized resource for individuals to "tap into" to gain more ecological information (Kayam Farm Interviewee 4, 2011).

The Kayam Farm also affectively impacts participants. According to a Kayam Farm employee, "The goal is for [participating] institutions to *feel* comfortable and confident in embodying the values we stand for here" (Kayam Farm Interviewee 3, 2011, emphasis added). Although they are not entirely responsible for behavioral changes, affective changes to environmental attitudes impact individuals' pro-environmental behaviors (Kollmuss and Agyeman 2002; Maiteny 2002). As one interviewee notes, "I encountered a tremendous amount of fear and anxiety about … changing your behavior. Even if [non-sustainable behavior is] costing them more money or is bad for their health, people are much more likely to remain status quo and not change because it's easier" (Kayam Farm Interviewee 8, 2011). The Kayam Farm alters educational programming to make ecological knowhow more accessible and tailored to individual situations and is thus breaking the barrier between lack of knowledge, feelings of fear and motivations to act (Kayam Farm Interviewee 8, 2011).

Until recently, the Jewish farming movement in Baltimore was limited to a single rural location, the Kayam Farm, which is not only physically inaccessible to many, but also poses a psychological barrier to adopting a sustainable lifestyle. When asked, "Which of the following are barriers to your maintained/increased participation with the Kayam Farm (check all that apply)?", 62.5 % of individual participants noted "geographic distance to Jewish farming movement sites."

The physical distance to the Kayam Farm also posed a psychological barrier to adopting a sustainable lifestyle. As an interviewee explains, such a displaced location was

> ...almost propagating that idea of 'you can do [farming and pro-environmental activities]... out there.' 'You can do this if you have 100 acres'...'you can do this if you don't have to live within walking distance of a synagogue'...and 'you can do this 'if this,' 'if that'... 'those people over there do that'... 'if you go out there you can do that, but in our home it's status quo.' (Kayam Farm Interviewee 8, 2011, emphasis on original)

However, the recently created Kayam Farm Community Gardening Initiative is a program helping Jewish institutions across Baltimore to create and maintain their own Jewish community gardens, and is thus enabling participants to engage in ecological activities more frequently and in a familiar, spatially proximate, home location. As more Kayam Farm participants *feel* as though they are able to engage in pro-environmental tasks on a regular basis, behavioral changes may follow.

Lastly, Kayam Farm participants are impacted behaviorally, which is shaped by the cognitive and affective realms. The Kayam Farm addresses external barriers to sustainable lifestyles by providing the necessary infrastructure (the farm, gardens, seeds, tools, etc.) for action, as well as creating a cultural climate for action by linking Judaism and agriculture. The Kayam Farm Community Gardening Initiative has enabled pro-environmental behaviors to occur that may not have otherwise taken place due to the lack of infrastructural capabilities. Examples of behavioral changes that Jewish Baltimore has begun to internalize at the individual and institutional levels include: recycling, composting, gardening, joining a CSA, attending farmers markets, shopping within the local economy, eating healthier food, using renewable energy sources and improving land-use practices (Kayam Farm Interviewees 1–5 and 8, 2011).

The researchers sought to gauge the impacts of Kayam Farm participation on individual Kayam Farm participants' sustainable lifestyles. Specific indicators of sustainable lifestyles (listed in Table 15.4) were extracted from open interviews and from literature reviews and used in online surveys to measure cognitive, affective and behavioral impacts of Kayam Farm participation on Kayam Farm participants' sustainable lifestyles.

Survey takers were asked to rank the level of impact they felt occurred to each specific indicator of sustainable lifestyles, based on their Kayam Farm participation. Survey responses were tallied and merged into three levels of impact: no impact to very little impact, little impact to moderate impact, and great impact to very great impact. Figure 15.7 represents sustainable lifestyle impact levels separated by each specific indicator of sustainable lifestyles ranked by survey takers.

Table 15.4 Specific indicators of sustainable lifestyles

Type of indicator	Specific indicators
Cognitive	Exploration of other pro-environmental causes/knowledge/issues
	Level of pro-environmental knowledge increase
Affective	Level of anxiety/fear of environmental issues/situations decrease
	Level of pro-environmental attitude increase
Behavioral	Making "greener" personal/institutional food choices (ex: local, organic, non-packaged, low petroleum)
	Making "greener" personal/institutional energy choices (ex: public transportation, biking/walking, shutting off electricity, lower water use)
	Making "greener" personal/institutional purchase choices (ex: less plastic packaging, more recycling, more reusing, less non-recyclable purchases)
	Making "greener" personal/intuitional land-use choices (ex: composting, creating rain gardens, decreasing impervious surfaces, decreasing non-native species) (*behavioral*)
	Level of pro-environmental advocacy (ex: spreading the word, donating, fundraising, signing petitions, protesting)

Source: Rachel Berndtson and Martha Geores

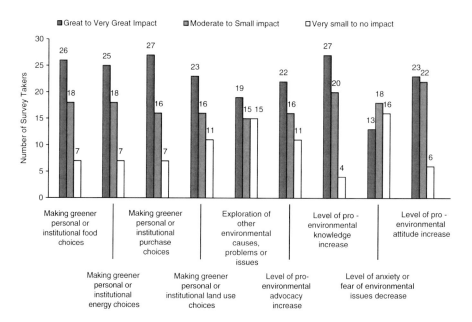

Fig. 15.7 Sustainable lifestyle impact levels on Kayam Farm participants separated by each specific indicator of sustainable lifestyles (N=51) (Source: Rachel Berndtson and Martha Geores)

Similarly to the Jewish identity impact results, survey responses show that Kayam Farm participants' sustainable lifestyles were impacted at different levels (no impact to very great impact) and in different impact areas (cognitive, affective and behavioral) based on their participation with the Kayam Farm. However, in contrast to the Jewish identity impacts, in most cases the highest proportion of survey takers felt they were greatly or very greatly impacted by their participation on the Kayam Farm. This trend holds true for 8 out of the 9 specific indicators of sustainable lifestyles.

15.7 Conclusion

The farms and gardens surrounding the Jewish farming movement use bottom-up approaches to address issues of Jewish cultural sustainability and ecological sustainability. As an interest-based, non-hierarchical form of Jewish participation and identification, Jewish farming is a new method for individual Jewish identification. Although these new methods initially form weak, intermittent ties, the relationships developed may advance into strong ties, shifting from Jewish "communities of convenience" to "communities of commitment." Weak ties morph to strong ties as interactions are deepened and sustained. In serving as community fields, Jewish farms and gardens centralize the locations through which individuals may access ecological resources and mobilize social capital. Although bottom-up approaches lay the framework for community mobilization on environmental issues, they alone are not a solution to environmental problems. The mobilization of local level social capital across vertical and horizontal networks will strengthen the coordination of action within and beyond a community. Interest-based Jewish involvement and community-based ecological organizations cannot be taken as a panacea to delivering immediate change in either realm, but, as exemplified through the Kayam Farm case study, offer promising approaches to beginning it.

References

Adamah. (2011). Isabella Freedman Jewish Retreat Center. Retrieved April 4, 2011, from http://isabellafreedman.org/adamah/intro

Azria, R. (1993). France and the United States, the promised land of the Jews? *Archives de Sciences Sociales des Religions, 84*, 201–222.

Azria, R. (1998). The diaspora-community-tradition paradigms of Jewish identity: A reappraisal. In E. Krausz & G. Tulea (Eds.), *Jewish survival: The identity problem at the close of the twentieth century* (pp. 21–32). New Brunswick: Transaction Publishers.

Barr, S. (2003). Strategies for sustainability: Citizens and responsible environmental behaviour. *Area, 35*, 227–240.

Barraket, J. (2005). Enabling structures for coordinated action: Community organizations, social capital, and rural community sustainability. In A. Dale & J. Onyx (Eds.), *A dynamic balance: Social capital and sustainable community development* (pp. 71–86). Vancouver: University of British Columbia Press.

Barthel, S., Folke, C., & Colding, J. (2010). Social–ecological memory in urban gardens: Retaining the capacity for management of ecosystem services. *Global Environmental Change, 20,* 255–265.

Beck, J. (2009). Challenges to Jewish community. In Z. I. Heller (Ed.), *Synagogues in a time of change: Fragmentation and diversity in Jewish religious movements* (pp. 189–193). Herndon: The Alban Institute.

Bernstein, E. (2011). Retrieved April 4, 2011, from http://ellenbernstein.org/about_ellen.htm

Blecher, A. (2007). *The new American Judaism: The way forward on challenging issues from intermarriage to Jewish identity.* New York: Palgrave MacMillan.

Bloomfield, D. M. (1999). Jewish survival: Operation educate. In C. Selengut (Ed.), *Jewish identity in the postmodern age* (pp. 259–271). St. Paul: Paragon House.

Brockelman, P. (1997). With new eyes: Seeing the environment as a spiritual issue. In J. E. Carroll, P. Brockelman, & M. Westfall (Eds.), *The greening of faith* (pp. 30–43). Hanover: University Press of New England.

Burt, R. S. (1992). *Structural holes: The social structure of competition.* Cambridge: Harvard University Press.

Burt, R. S. (2000). The network structure of social capital. In B. M. Straw & R. I. Sutton (Eds.), *Research in organizational behavior* (pp. 345–423). Greenwich: JAI Press.

Burt, R. S. (2001). Structural holes versus network closure as social capital. In N. Lin, K. Cook, & R. S. Burt (Eds.), *Social capital, theory and research* (pp. 31–56). New York: Aldine de Gruyter.

Chan, R. Y. K. (1998). Mass communication and pro-environmental behaviour: Waste recycling in Hong Kong. *Journal of Environmental Management, 52,* 317–325.

Chan, R. Y. K. (2001). Determinants of Chinese consumers' green purchasing behaviour. *Psychology and Marketing, 18,* 389–413.

Coalition on the Environment and Jewish Life. (2011). Retrieved April 4, 2011, from http://www.coejl.org/~coejlor/about/history.php

Cohen, S. M., & Eisen, A. M. (2000). *The Jew within: Self, family and community in America.* Bloomington: Indiana University Press.

Cohen, S. M., & Kelman, A. Y. (2005). Cultural events and Jewish identities: Young Jewish adults in New York. National Foundation for Jewish Culture.

Coleman, J. S. (1988). Social capital and the creation of human capital. *The American Journal of Sociology, 94,* 95–120.

Coleman, J. S. (1993). The rational reconstruction of society: 1992 presidential address. *American Sociological Review, 58*(1), 1–15.

Cone, C. A., & Myhre, A. (2000). Community-supported agriculture: A sustainable alternative to industrial agriculture? *Human Organization, 59,* 187–197.

Dale, A. (2005). Social capital and sustainable community development: Is there a relationship? In A. Dale & J. Onyx (Eds.), *A dynamic balance: Social capital and sustainable community development* (pp. 11–30). Vancouver: University of British Columbia Press.

Defra. (2005). Securing the future: Delivering UK sustainable development strategy. Sustainable-Development.gov.uk [online].

De Young, R. (1996). Some psychological aspects of reduced consumption behavior: The role of intrinsic motivation and competence motivation. *Environment and Behavior, 28,* 358–409.

Diner, H. R. (1992). *A time for gathering: The second migration 1820–1880.* Baltimore: The Johns Hopkins University Press.

Dobson, A. (2003). *Citizenship and the environment.* Oxford: Oxford University Press.

Eden, S. (1993). Individual environmental responsibility and its role in public environmentalism. *Environment and Planning A, 25,* 1743–1758.

Eden Village Camp. (2011). Eden Village Camp. Retrieved April 4, 2011, from http://edenvillage-camp.org/vision/

Ehrenfeld, D., & Bentley, P. J. (2001). Judaism and the practice of stewardship. In M. D. Yaffe (Ed.), *Judaism and environmental ethics: A reader* (pp. 125–135). Lanham: Lexington Books.

Eisenberg, E. (1995). *Jewish agricultural colonies in New Jersey, 1882–1920.* Syracuse: Syracuse University Press.

Ellenson, D. (2009). Thoughts on American Jewish denominationalism today: Culture and identity. In Z. I. Heller (Ed.), *Synagogues in a time of change: Fragmentation and diversity in Jewish religious movements* (pp. 73–79). Herndon: The Alban Institute.

Ernstson, H., Barthel, S., Andersson, E., & Borgström, S. T. (2010). Scale-crossing brokers and network governance of urban ecosystem services: The case of Stockholm. *Ecology and Society, 15*, 28.

Etzioni-Halevy, E. (1998). Collective Jewish identity in Israel: Towards an irrevocable split? In E. Kraysz & G. Tulea (Eds.), *Jewish survival: The identity problem at the close of the twentieth century*. New Brunswick: Transaction Publishers.

Feagan, R. (2007). The place of food: Mapping out the 'local' in local food systems. *Progress in Human Geography, 31*, 23–42.

Feagan, R., & Henderson, A. (2009). Devon Acres CSA: Local struggles in a global food system. *Agriculture and Human Values, 26*, 203–217.

Feenstra, G. W. (1997). Local food systems and sustainable communities. *American Journal of Alternative Agriculture, 12*, 28–36.

Feldstein, S. (2011). So, what's "the new Jewish food movement?" A movement for justice. American Jewish World Service. http://blogs.ajws.org/blog/2011/08/24/so-whats-the-new-jewish-food-movement/.

Fischer, C. S. (1982). *To dwell among friends*. Chicago: University of Chicago Press.

Fishkoff, S. (2009). Farming the land, Torah in hand. Retrieved October 2, 2011, from http://www.jewishjournal.com/articles/item/farming_the_land_torah_in_hand_20090204/

Gardner, G., & Stern, P. (2002). *Environmental problems and human behaviour*. Boston: Pearson.

Gendler, E. (1997). Join the chorus, recapture the rhythms. In J. E. Carroll, P. Brockelman, & M. Westfall (Eds.), *The greening of faith: God, the environment, and the good life* (pp. 67–78). Hanover: University Press of New England.

Georg, S. (1999). The social shaping of household consumption. *Ecological Economics, 28*(3), 455–466.

Goldberg, R. A. (1986). *Back to the soil: The Jewish farmers of Clarion, Utah and their world*. Salt Lake City: University of Utah Press.

Granovetter, M. (1983). The strength of weak ties: A network theory revisited. *Sociological Theory, 1*, 201–233.

Groeneman, S., & Smith, T. W. (2009). *Moving: The impact of geographic mobility on the Jewish community*. New York: The Jewish Federations of North America.

Grossman, L. (2009). Denominations in American Judaism: The dynamics of their relationships. In Z. I. Heller (Ed.), *Synagogues in a time of change: Fragmentation and diversity in Jewish religious movements* (pp. 25–34). Herndon: The Alban Institute.

Hazon. (2011). Retrieved April 4, 2011, from http://www.hazon.org/

Heilman, S. C. (2004). American Jews and community: A spectrum of possibilities. *Contemporary Jewry, 24*, 51–69.

Heller, Z. I. (2009). Introduction. In Z. I. Zeller (Ed.), *Synagogues in a time of change: Fragmentation and diversity in Jewish religious movements* (pp. 1–10). Herndon: The Alban Institute.

Hendrickson, M., & Heffernan, W. (2002). Opening spaces through relocalization: Locating potential resistance in the weaknesses of the global food system. *Sociologia Ruralis, 42*, 347–369.

Herman, S. N. (1989). *Jewish identity: A social psychological perspective*. New Brunswick: Transaction Publishers.

Herring, H. (2009). Synagogue renewal in an age of extreme choice: Anything, anyone, anytime, anywhere. In Z. I. Heller (Ed.), *Synagogues in a time of change: Fragmentation and diversity in Jewish religious movements* (pp. 111–138). Herndon: The Alban Institute.

Hinchliffe, S. (1996). Helping the earth begin at home: The social construction of socio-environmental responsibilities. *Global Environmental Change, 6*, 53–62.

Hinrichs, C. C. (2000). Embeddedness and local food systems: Notes on two types of direct agricultural markets. *Journal of Rural Studies, 16*, 295–303.

Horowitz, B. (2000). *Connections and journeys: Assessing critical opportunities for enhancing Jewish identity: A report to the Commission on Jewish Identity and Renewal*. UJA-Federation of New York, Continuity Commission.

Horowitz, B. (2002). Reframing the study of contemporary American Jewish identity. *Contemporary Jewry, 23*, 14–34.

Jackson, T. (2005). Motivating sustainable consumption: A review of evidence on consumer behaviour and behavioural change. Sustainable Development Research Network [online].

Jacobs, M. X. (2003). Jewish environmentalism: Past accomplishments and future challenges. In H. Tirosh-Samuielson (Ed.), *Judaism and ecology: Created world and revealed world* (pp. 449–480). Cambridge: Harvard University Press.

Jagers, S., Martinsson, J., & Matti, S. (2009). On how to make the theoretical concept of ecological citizenship empirically operational. In *Climate change politics and political theory workshop*. Uppsala University.

Jewish Farm School. (2011). Jewish Farm School. Retrieved April 4, 2011, from http://www.jewishfarmschool.org/

Kaplan, D. E. (2009). *Contemporary American Judaism: Transformation and renewal*. New York: Columbia University Press.

Kayam Farm. (2011). About. Retrieved April 4, 2011, from http://www.kayamfarm.org/?page_id=7

Kayam Farm Interviewee 1. (2011, January). Interview by Berndtson, R.

Kayam Farm Interviewee 2. (2011, January). Interview by Berndtson, R.

Kayam Farm Interviewee 3. (2011, June). Interview by Berndtson, R.

Kayam Farm Interviewee 4. (2011, June). Interview by Berndtson, R.

Kayam Farm Interviewee 5. (2011, June). Interview by Berndtson, R.

Kayam Farm Interviewee 6. (2011, June). Interview by Berndtson, R.

Kayam Farm Interviewee 7. (2011, June). Interview by Berndtson, R.

Kayam Farm Interviewee 8. (2011, July). Interview by Berndtson, R.

Kayam Farm Interviewee 9. (2011, July). Interview by Berndtson, R.

Kelman, H. C. (1999). The place of ethnic identity in the development of personal identity: A challenge for the Jewish family. In P. Medding (Ed.), *Studies in contemporary Jewry*. Oxford: Oxford University Press.

Kirshenblatt-Gimblett, B. (2005, *December 6–7*). *The "new Jews": Reflections on emerging cultural practices*. Paper delivered at Re-thinking Jewish communities and networks in an age of looser connections, Wurzweiler School of Social Work, Yeshiva University, and Institute for Advanced Studies, Hebrew University. New York City. Online at http://www.nyu.edu/classes/bkg/web/yeshiva.pdf

Kollmuss, A., & Agyeman, J. (2002). Mind the gap: Why do people act environmentally and what are the barriers to pro-environmental behavior? *Environmental Education Research, 8*, 239–260.

Lacy, W. (2000). Empowering communities through public work, science, and local food systems: Revisiting democracy and globalization. *Rural Sociology, 65*, 3–26.

Lavender, A. D., & Steinberg, C. B. (1995). *Jewish farmers of the Catskills: A century of survival*. Gainesville: University Press of Florida.

Lipset, S. M. (2003). Some thoughts on the past, present and future of American Jewry. In S. M. Lyman (Ed.), *Essential readings on Jewish identities, lifestyles and beliefs: Analyses of the personal and social diversity of Jews by modern scholars* (pp. 28–37). New York: Gordian Knot Books.

Macnaghten, P. (2003). Embodying the environment in everyday life practices. *The Sociological Review, 51*, 63–84.

Maiteny, P. (2002). Mind in the gap: Summary of research exploring 'inner' influences on pro-sustainability learning and behaviour. *Environmental Education Research, 8*, 299–306.

Manela, J., & Silverstein, Y. (2010). Chai ve'Kayam: A curriculum manual for Jewish agricultural education. In *Unit 2: Jewish garden education*. Baltimore: Kayam Farm at Pearlstone.

Matti, S. (2008). From sustainable consumers to ecological citizens: Elucidating attitudes towards individual environmental action in Sweden. Sharp Research Programme.

McMillan, D. W., & Chavis, D. M. (1986). Sense of community: A definition and theory. *Journal of Community Psychology, 14*, 6–23.

Middlemiss, L. K. (2008). Influencing individual sustainability: A review of the evidence on the role of community-based organisations. *International Journal of Environment and Sustainable Development, 7*, 78–93.

Middlemiss, L. K. (2010). Community action for individual sustainability: Linking sustainable consumption, citizenship and justice. In D. Pavlich (Ed.), *Managing environmental justice* (pp. 71–91). Amsterdam: Rodopi.

Middlemiss, L. K., & Parrish, B. D. (2010). Building capacity for low-carbon communities: The role of grassroots initiatives. *Energy Policy, 38*, 7559–7566.

Newhouse, N. (1991). Implications of attitude and behavior research for environmental conservation. *The Journal of Environmental Education, 22*, 26–32.

OECD. (2002). Policies to promote sustainable consumption: An overview. In *ENV/EPOC/WPNEP (2001) 18/FINAL*. Paris: OECD.

Oh, H., Chung, M. H., & Labianca, G. (2004). Group social capital and group effectiveness: The role of informal socializing ties. *Management Journal, 4*, 860–875.

Petersen, R. L. (2009). American dissonance: Christian communities in the United States and their cultural context. In Z. I. Heller (Ed.), *Synagogues in a time of change: Fragmentation and diversity in Jewish religious movements* (pp. 49–70). Herndon: The Alban Institute.

Polanyi, K. (1957). The economy as instituted process. In K. Polyani, C. M. Arensberg, & H. W. Pearson (Eds.), *Trade and markets in the early empires: Economies in history and theory* (pp. 243–270). Glencoe: Free Press.

Poll, S. (1998). Jewish identity in the twenty-first century. In E. Kranz & G. Tulea (Eds.), *Jewish survival: The identity problem at the close of the twentieth century* (pp. 145–161). New Brunswick: Transaction Publishers.

Putnam, R. (1994). Social capital and public affairs. *Bulletin of the American Academy of Arts and Sciences, 47*, 5–19.

Putnam, R., Leonardi, R. R., & Nanetti, R. (1993). *Making democracy work: Civic transitions in modern Italy*. Princeton: Princeton University Press.

Redclift, M., & Hinton, E. (2008). *Living sustainably: Approaches for the developed and developing world*. London: Progressive Governance Summit.

Rockefeller, S. C. (1997). The wisdom of reverence for life. In J. E. Carroll, P. Brockelman, & W. Westfall (Eds.), *The greening of faith: God, the environment, and the good life* (pp. 44–61). Hanover: University Press of New England.

Scheffer, M., & Westley, F. (2007). The evolutionary basis of rigidity: Locks in cells, minds, and society. *Ecology and Society, 12*(2), 36.

Schwartz, E. (2001). Bal tashchit: A Jewish environmental precept. In M. D. Yaffe (Ed.), *Judaism and environmental ethics: A reader* (pp. 230–249). Lanham: Lexington Books.

Selengut, C. (1999). Introduction: The dilemmas of Jewish identity. In C. Selengut (Ed.), *Jewish identity in the postmodern age* (pp. 1–9). St. Paul: Paragon House.

Seyfang, G. (2005). Shopping for sustainability: Can sustainable consumption promote ecological citizenship? *Environmental Politics, 14*, 290–306.

Seyfang, G. (2006a). Ecological citizenship and sustainable consumption: Examining local organic food networks. *Journal of Rural Studies, 22*, 383–395.

Seyfang, G. (2006b). Sustainable consumption, the new economic and community currencies: Developing new institutions for environmental governance. *Regional Studies, 40*, 781–791.

Sharot, S. (1998). Judaism and Jewish ethnicity: Changing interrelationships and differentiations in the Diaspora and Israel. In E. Krausz & G. Tulea (Eds.), *Jewish survival: The identity problem at the close of the twentieth century* (pp. 87–105). New Brunswick: Transaction Publishers.

Sharp, J. (2001). Locating the community field: A study of interorganizational network structure and capacity for community action. *Rural Sociology, 66*, 403–424.

Smith, A. M., & Pulver, S. (2009). Ethics-based environmentalism in practice: Religious-environmental organizations in the United States. *Worldviews, 13*, 145–179.

Sorin, G. (1992). *A time for building: The third migration 1880–1920*. Baltimore: The Johns Hopkins University Press.

Sosis, R. H., & Ruffle, B. J. (2003). Religious ritual and cooperation: Testing for a relationship on Israeli religious and secular kibbutzim. *Current Anthropology, 44*, 713–722.

Starr, D. B. (2009). History as prophecy: Narrating the American synagogue. In Z. I. Heller (Ed.), *Synagogues in a time of change: Fragmentation and diversity in Jewish religious movements* (pp. 13–23). Herndon: The Alban Institute.

Steinsaltz, A. (2006). *The essential Talmud*. New York: Basic Books.
Taylor, C. (1991). *The ethics of authenticity*. Cambridge: Harvard University Press.
Teutsch, D. A. (2003). Technological, organizational and social turbulence: Contemporary Jewish communal challenges. *Contemporary Jewry, 24*, 70–81.
Teva Learning Center. (2011). Retrieved April 4, 2011, from http://tevalearningcenter.org/
Tirosh-Samuelson, H. (2001). Nature in the sources of Judaism. *Daedalus, 130*, 99–124.
Tuan, Y. (2002). Community, society, and the individual. *Geographical Review, 92*, 307–318.
Tucker, P. (1999). Normative influences in household recycling. *Journal of Environmental Planning and Management, 42*, 63–82.
Urban Adamah. (2011). Urban Adamah. Retrieved April 4, 2011, from http://urbanadamah.org/
Waskow, A. (1995). *Down-to-earth Judaism: Food, money, sex, and the rest of life*. New York: Morrow.
Wasserman, S., & Faust, K. (1994). *Social network analysis: Methods and applications*. Cambridge: Cambridge University Press.
Wellman, B. (1999). From little boxes to loosely-bounded networks: The privatization and domestication of community. In J. L. Abu-Lughod (Ed.), *Sociology for the twenty-first century: Continuities and cutting edges* (pp. 94–114). Chicago: University of Chicago Press.
Wellman, B. (2001). Physical space and cyberplace: The rise of personalized networking. *International Journal of Urban and Regional Research, 25*, 227–252.
Wells, B., Gradwell, S., & Yoder, R. (1999). Growing food, growing community: Community supported agriculture in Iowa. *Community Development Journal, 34*, 38–46.
White, L., Jr. (1967). The historical roots of our ecological crisis. *Science, 155*, 1203–1207.
Windmueller, S. (2007). The second American Jewish revolution. *Journal of Jewish Communal Service, 82*, 252–260.
Woocher, J. (2009). Jewish education: Postdenominationalism and the continuing influence of denominations. In Z. I. Heller (Ed.), *Synagogues in a time of change: Fragmentation and diversity in Jewish religious movements* (pp. 139–160). Herndon: The Alban Institute.
World Bank. (2003). *Social capital for development*. Washington, DC: World Bank. http://www.worldbank.org/poverty/scapital/index.htm

Chapter 16
Religious and Moral Hybridity of Vegetarian Activism at Farm Animal Sanctuaries

Timothy Joseph Fargo

16.1 Introduction

Moral concern for the well-being of farm animals, especially those used for human consumption such as cows, pigs, turkeys, chickens, goats, and sheep, and an ideal of small-scale, non-industrial plant-based agriculture have influenced proponents of vegetarianism and animal welfare to form farm animal sanctuaries. These are places where farm animals that have been abandoned or considered abused and removed from agricultural settings are brought to live out their lives. Moreover they are places where activists educate the public about the treatment of animals in agriculture. Farm animal sanctuaries are transformative places, made by activists as moral tools to challenge views of farm animals and to convert humans to vegetarianism by having animals serve as "ambassadors" to humans unfamiliar with farm animals and their lives in agricultural settings. Farm animal sanctuary activists have been strongly influenced by Western and Eastern religious ideals as well as moral concepts from nontheistic philosophers. They draw upon those beliefs to form complex personal rationales for activism characterized by moral hybridity.

16.2 Farm Animal Sanctuaries in the United States

Farm animal sanctuaries are a subset of animal sanctuaries, which number in the hundreds in the United States. GreenPeople, the largest directory of eco-friendly and holistic businesses and organizations, lists 265 animal sanctuaries (GreenPeople 2006).

T.J. Fargo (✉)
Department of City Planning, City of Los Angeles, 200 N. Spring Street,
Los Angeles, CA 90012, USA
e-mail: timfargo@hotmail.com

An animal sanctuary is defined by The Association of Sanctuaries (TAOS), an accrediting agency, as "a place of refuge where injured, abused, or displaced animals are provided with appropriate lifetime care, or, when possible, rehabilitated and returned to the wild." Additionally, "farmed[1] and companion[2] animals can be adopted to carefully screened homes." Sanctuary animals commonly have been displaced from "natural" or "appropriate conditions," according to TAOS. Three primary categories of animal sanctuaries accredited by TAOS are wildlife, farm animal, and companion animal sanctuaries (TAOS 2006a). Some sanctuaries focus on one of these three groups of animals. Others take in a wide variety from all of the categories. This research concerns sanctuaries that concentrate exclusively or heavily on farm animals, especially farm animals used for food. There are 37 such sanctuaries for farm animals in the United States examined in this study (Lightfoot 2003; TAOS 2006b; ASA 2006; Animal Place 2006). The sanctuaries are located throughout the country with concentrations on the outskirts of major cities in the Pacific Northwest, California, the Northeast, and the Mid-Atlantic, with a few in the Midwest and Southwest (Fig. 16.1). The locations of these sites roughly correlate to areas of liberalism in the United States. Those who organize sanctuaries may start one where they live or may organize ones in other areas for specific reasons, such as to reach target human populations. Many of these locales are near strong vegetarian and animal welfare communities in nearby cities, from which the sanctuaries recruit members and volunteers, derive revenue, and can more easily reach out to new supporters.

The farm animals that are brought to farm animal sanctuaries have escaped from farms, fallen off transportation vehicles, are abandoned and left to die but are rescued, or come from farms that have been closed for legal violations or struck by natural disasters. Employees and volunteers locate and transport animals to the sites and then look after them. They feed them, allow them to get adequate exercise, clean up after them, and provide them with medical care. A map of the grounds of the 300-acre Farm Sanctuary shelter in Orland, California in Fig. 16.2 illustrates how one sanctuary is laid out, including a "Visitor Barn" with educational displays for visitors.

Sanctuary activists seek to convert non-vegetarians to vegetarianism at farm animal sanctuaries. At one sanctuary, a sign conveys the negative aspects of the lives of pigs as they are raised in factory farm conditions (Fig. 16.3). Many of those who visit farm animal sanctuaries do convert to vegetarianism (Fig. 16.4).

---

[1] Sometimes farm animals are referred to as "farmed" animals by subjects and sources in order to destabilize the naturalization of the term "farm animal" and make obvious the human role in raising animals as sources of food.

[2] Companion animal is a synonym for pet and is the preferred term for animal advocates who consider "pet" to be demeaning. Dogs, cats, rabbits, birds and any other animals considered to be "pets" are examples of companion animals.

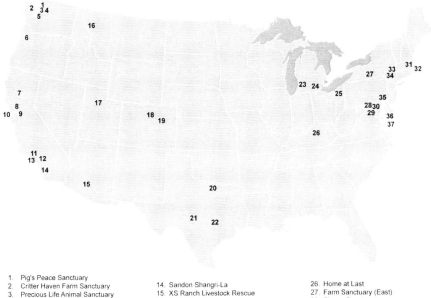

1. Pig's Peace Sanctuary	14. Sandon Shangri-La	26. Home at Last
2. Critter Haven Farm Sanctuary	15. XS Ranch Livestock Rescue	27. Farm Sanctuary (East)
3. Precious Life Animal Sanctuary	16. Montana Large Animal Sanctuary	28. Pigs, a Sanctuary
4. Pasados Safe Haven	17. Ching Farm Animal Rescue	29. Poplar Spring Animal Sanctuary
5. BAAHAUS Animal Rescue Group	18. Bleating Hearts Sanctuary	30. Star Gazing Farm
6. Lighthouse Farm Sanctuary	19. Peaceful Prairie Sanctuary	31. Veganpeace Animal Sanctuary
7. Farm Sanctuary (West)	20. Serenity Springs Sanctuary	32. Winslow Farm
8. Animal Place	21. Safe Harbour Animal Refuge	33. Catskill Animal Sanctuary
9. Harvest Home Animal Sanctuary	22. Dreamtime Sanctuary	34. Woodstock Farm Animal Sanctuary
10. Humane Farming Association	23. Grateful Acres Sanctuary	35. Chenoa Manor
11. Animal Acres	24. Sasha Farm	36. Eastern Shore Chicken Sanctuary
12. Hooves and Paws	25. Happy Trails Farm Animal Sanctuary	37. United Poultry Concerns
13. The Gentle Barn Foundation		

Fig. 16.1 Location of farm animal sanctuaries in the United States (Map by Timothy Joseph Fargo)

Fig. 16.2 Map of the 300-acre Farm Sanctuary shelter in Orland, California (Map by Kari Hallenburg, Farm Sanctuary, used with permission)

Fig. 16.3 Sign at Farm Sanctuary depicting adverse health conditions for pigs in factory farms (Photo by Timothy Joseph Fargo)

Fig. 16.4 Bumper sticker on a car at Animal Acres event (Photo by Timothy Joseph Fargo)

16.3 Moral Vegetarianism

According to the Vegetarian Resource Group, who compiled surveys from Gallop, Time/CNN, Zogby, and the National Restaurant Association, between 5 and 9 % of the adult population (between 10 and 17 million adults) in the United States are semi-vegetarian, true vegetarian, or vegan (VRG 2006). Semi-vegetarians rarely eat meat, self-identifying as "almost vegetarian." There are an estimated five million true vegetarians in the United States, those who never eat meat, poultry, or fish, and sometimes forego clothing made from animal products. True vegetarians include two million vegans, who never eat meat, poultry, fish, diary, eggs, or honey, and do not use animal products such as leather, wool, and silk. Vegetarianism is popular in many Western societies and is not a uniquely American phenomenon. In the United Kingdom, the only other country for which data is available, 5–7 % of the population are believed to be vegetarians (Hamilton 2000), indicating a similar percentage as the United States.

Nearly all subjects in this research working at farm animal sanctuaries (95 %) were vegetarians of various classifications or consumed animals or animal products solely due to health concerns. Thus I discuss sanctuary activists in the context of holding vegetarianism and veganism in particular as ideal lifestyles. Of the 37 participants, 68 % were fully vegan (n=25), 5 % were nearly completely vegan (n=2),[3] 14 % were non-vegan vegetarians (n=5), 3 % were semi-vegetarian (n=1),[4] and 11 % were non-vegetarians (n=4).[5]

In broader society, some consume predominantly vegetarian fare for economic reasons due to the greater expense of animal products. People with economic means who choose to become vegetarians, however, do so for many different reasons including health, religion, animal welfare, the environment, and social justice. For an individual, any one or a number of these factors may be of greater importance. Vegetarianism can be understood as a solution settling anxieties about modern food production including nutrition, food safety, the environment, and ethical concerns for animals (Beardsworth and Keil 1991; Hamilton 2000). Whether a person becomes a vegetarian for a single reason or multiple reasons, he or she often acquires additional reasons as time passes, becoming a vegetarian "for all reasons" with an increasing degree of moral certitude (Amato and Partridge 1989; Hamilton 2000). This accumulation of rationales forms a system of beliefs linking views of animals, views of humans in relation to nature such as environmentalism, and ideals of egalitarian social organization (Twigg 1979; Amato and Partridge 1989; Beardsworth and Keil 1991; Rozin et al. 1997; Jabs et al. 1998; Hamilton 2000; Lindeman and Sirelius 2001; Fessler et al. 2003).

Vegetarians concerned for the well-being of animals, such as farm animal sanctuary activists, often are true vegetarians and vegans, motivated equally by health, religion, ethics, and environmental values (Rozin et al. 1997; Fessler et al. 2003; VRG 2006). Animal rights organizations such as People for the Ethical Treatment of Animals (PETA), for example, often promote vegetarianism. The conventional use of the term "moral vegetarian" is one with an ethical concern for animals, but it may also describe vegetarians with other religious, environmental, or social values. In many religious traditions there are rationales for adopting vegetarianism that provide strong motivation for vegetarian devotees such as the concept of ahimsa, or nonharming, in Jainism, Buddhism, and Hinduism, and the mandate that humans be vegetarians in the Garden of Eden of the Abrahamic traditions (Genesis 1: 29).

16.4 Research Methodology

Dietary practices such as vegetarianism are psychologically complex mechanisms through which individuals express their personality, ideals, and identity (Lindeman and Sirelius 2001). The process of becoming vegetarian is rooted in the sense of the

---

[3] This includes one who only ate eggs from chickens personally raised by the subject.
[4] This subject only occasionally ate meat from animals personally raised by the subject.
[5] This includes two subjects who were not vegetarians solely for health reasons.

individual self and how the self fits in with the world (McDonald 2000), reflecting a perspective that is personal as well as geographic (Tuan 1974). This in-depth qualitative study of activists sheds light on how individuals develop their attitudes and materialize their beliefs through place-making. Few studies in this line of qualitative research on vegetarianism have been undertaken (Beardsworth and Keil 1991; Jabs et al. 1998; McDonald 2000; Boyle 2007). None has considered the particular role of place-making as a tool to further moral causes.

The 37 farm animal sanctuaries identified in preliminary research were contacted by telephone and email. I requested an official sanctuary contact to send a recruitment email to employees and volunteers who were informed to contact me if they wished to take part in the study. Eligibility was determined by enlisting all participants through the official contacts of the sanctuaries to ensure that the subjects had experience working or volunteering at farm animal sanctuaries. There was no initial exclusion of specific participants for any reason.

Questionnaires were emailed to the 45 employees and volunteers who said they were willing to participate. Subjects completed the questionnaires and returned them via email and the postal service. The questions asked pertained to becoming vegetarian, experiences with animals, religious and ethical beliefs, views of nature, views of society, as well as beliefs and experiences regarding the operation of farm animal sanctuaries and their relation to the public. Participants indicated whether they wished to answer follow-up questions over email. In the follow-up emails, I looked for clarification of responses and greater depth.

Of the 45 potential subjects who requested questionnaires, 36 returned them completed, for a response rate of 80 %. The response rate was improved by sending up to three reminders. Questionnaires were received from the following 13 sanctuaries: Animal Place (Vacaville, CA): 8; Bleating Hearts Animal Sanctuary (Boulder, CO): 1; Chenoa Manor (Avondale, PA): 5; Eastern Shore Chicken Sanctuary (Princess Anne, MD): 2; Farm Sanctuary (Watkins Glen, NY and Orland, CA): 7; Grateful Acres Animal Sanctuary (Otsego, MI): 1; Happy Trails Farm Animal Sanctuary (Ravenna, OH): 1; Pasado's Safe Haven (Sultan, WA): 1; Peaceful Prairie Sanctuary (Deer Trail, CO): 4; PIGS, a Sanctuary (Shepherdstown, WV): 3; Poplar Spring Animal Sanctuary (Poolesville, MD): 1; VeganPeace Animal Sanctuary (Sherborn, MA): 1; Woodstock Farm Animal Sanctuary (Woodstock, NY): 1. Respondents to the questionnaires were 81 % female (29) and 19 % male (7). Additionally, I visited three California farm animal sanctuaries: Animal Acres (Acton, CA), Farm Sanctuary (Orland, CA), and Animal Place (Vacaville, CA). At these sanctuaries, five interviews were conducted in person. All names were changed to protect anonymity.

16.5 Religious-Moral Hybridity at Farm Animal Sanctuaries

A variety of religious and philosophical perspectives regarding eating animals have influenced farm animal sanctuary activists, especially the call of the Abrahamic traditions (Judaism, Christianity, and Islam) for dominion over animals, the Eastern religious concept of ahimsa, or, nonharming, and Enlightenment-based notions of utilitarianism, the categorical imperative, and rights as extended to animals. For some there is a desire to recover or create a state akin to the Garden of Eden. For others there is an explicit rejection of the Garden of Eden ideal. Often many of these perspectives were held simultaneously, leading to the formation of complex personal moral justifications for vegetarianism and activism at farm animal sanctuaries.

The breakdown of religious affiliation of respondents (n=36) was as follows: 22 % Christian (n=8), 6 % Buddhist (n=2), 3 % Jain (n=1), 3 % Wiccan (n=1), 8 % undefined spiritual (n=3), and 58 % nonreligious (n=21) including 17 % (of the overall) atheist (n=6). I did not ask how views differed depending on Christian denominational adherence or upbringing. Generally respondents had been exposed to many religious perspectives although most had a Christian upbringing. Although there were denouncements of religious perspectives by some atheist respondents, Christian, Eastern, and Enlightenment thought converge at sanctuaries, which have become places marked by religious-philosophical moral hybridity. Their disparate beliefs were united, however, in sharing a critique of anthropocentrism. Despite individual religious and philosophical differences, sanctuary employees and volunteers held a common system of interrelated beliefs regarding vegetarianism and its relation to moral concerns such as health, world hunger, the environment, and economic efficiency.

The beliefs that activists at sanctuaries seek to change in others are fundamental ideas about the relationship of humanity to nature and the purpose of animals in relation to humans. These perspectives are connected to religious beliefs about the human relationship to the rest of the universe. The religious aspect of vegetarianism is usually downplayed since vegetarians come from many faiths, each providing some basis for a vegetarian lifestyle. Indeed, one of the most interesting aspects of vegetarianism in the United States is that there is not only a strong basis in Western religious and nonreligious thought (ancient, Christian, and modern) for vegetarianism, but that Eastern thought has had a significant impact as well (Hindu, Buddhist, and Jain concept of ahimsa), making for a cultural-religious hybridity, based on the common principles of nonviolence, reducing suffering, and reverence for life. Researchers have explored the religious and ethical underpinnings of vegetarianism, but each position is usually addressed in isolation (Dombrowski 1984; Gregerson 1994; Spencer 1995; Rosen 1997; Walters and Portmess 1999, 2001; Sapontzis 2004). Some research subjects did hold specific religious perspectives regarding animals, but others drew on diverse traditions to forge new personal belief systems with hybrid worldviews. Vegetarianism itself was cited to be a primary framework through which many participants understood their world.

One of the places in which the quasi-religious aspect of vegetarianism becomes clear is the farm animal sanctuary. The concept of a sanctuary itself has religious underpinnings, seen as bounded sacred place, as a refuge in a landscape of decay or violence. Farm animal sanctuaries challenge the conventional agricultural relationship between humans and animals, manifesting a moral stance on the relationship of humanity to nature, reclassifying farm animals, and creating a new place for their existence in relation to humanity at least superficially in the image of the Garden of Eden.

Even though vegetarianism is not explicitly religious for most, it shares many elements with religion and can be examined from a perspective drawing inspiration from classic geographic work on religion. David Sopher considered the importance of religion in shaping plant and animal species to human purposes (Sopher 1967). While religion may have been central to the domestication of plants and animals, sanctuary activists, in contrast, use religious and moral arguments to reject the shaping of animals for human purposes. Taboos regarding foods are strongly associated with religious beliefs. Examples from contemporary mainstream religions include prohibitions against pork and alcohol in Islam as well as not mixing certain foods in Judaism. Jains and Manichaeans require(d) strict vegetarianism (Twigg 1979).

Religious prohibitions such as taboos on food directly affect the land, according to Sopher (1967). These can be described as negative expressions of religion on the landscape, due to the absence of certain features that might have been present otherwise. One of the primary reasons given for vegan vegetarianism is that it is beneficial to the environment because it requires less land for the production of grain, most of which is fed to animals, and no land for animal husbandry.

Sopher described religion in so-called "simple ethnic systems" as a ritualization of ecology, "the medium whereby nature and natural processes are placated, cajoled, entreated, or manipulated in order to secure the best results for man" (Sopher 1967: 17). He argued that as society and economy increase in complexity, so do the symbolization and abstraction of ecological matter. Universal religions, according to Sopher, are systems of belief that are considered by their adherents to be proper for all humanity, have mechanisms to facilitate their transmittal, have broken through restrictions to a special place or group, and have been established as dominant religions on at least a regional scale. Vegetarians often consider their beliefs proper for all humans and actively proselytize, although they are not localized to any particular region due to the universalizing nature of the belief system. They often hold normative moral views of agriculture including stewardship of the environment and efficiency of production. Likewise, their behaviors often are intended to achieve the idealized ecological results they seek to establish.

The importance of mythical religious geographies such as the Garden of Eden, Valhalla, Elysian Fields, and Hades also were identified by Sopher (1967). Arguments have been made that vegetarians, like pacifists and naturists, seek to deny the so-called "fallen" state of humanity by acting in ways that are considered to be as pure as living in the Garden of Eden (Clark 2004). Instead of idealizing a mythical utopia, sanctuary activists, in contrast, attempt to realize a new relationship between humans and animals through present-day places such as farm animal sanctuaries, with the goal of transforming relations between humans and animals on a global scale.

16.6 The Western Religious Origins of Vegetarianism and the Role of Edenic Narratives

The historical lineage of vegetarianism in the United States reveals a strong religious heritage. Reverend William Cowherd (1763–1816), formerly a Swedenborgian,[6] founded the Bible Christian Church in England in 1807, requiring his congregation to be vegetarians. He believed it was "the duty of mankind to abstain from the use of animal food … as a religious requirement" (Iacobbo and Iacobbo 2004: 10). William Metcalf (1788–1862) and 40 members of the Bible Christian Church emigrated from England to the United States in 1817, establishing a foothold in Philadelphia and winning converts to vegetarianism such as Sylvester Graham (1794–1851) and William Alcott (1798–1859). Graham, Alcott, and Metcalf were champions of vegetarianism in nineteenth century America. They formed the American Vegetarian Society in 1850, the official beginning of organized vegetarianism in the United States, believing vegetarianism to be a Christian duty.

Nineteenth century Christian vegetarians were motivated by the wish to control the passions, especially sexuality, and to promote the purity of the body, linking vegetarianism with morality, cleanliness, rationality, and spirituality. They believed that ingesting meat made humans more violent through the ingestion of animality, symbolically linked with lust and brutality. Associating animal flesh with evil had been present in minor traditions of Christianity for over a thousand years. Early Christianity co-existed with and was perhaps influenced by the dualistic religion of Manichaeanism c. 300–400 C.E. Manichaeanism itself was perhaps influenced by earlier Christian Gnosticism. In Manichaeanism, flesh was considered evil, connected with the body, animality, and sexuality (Twigg 1979). Manichaeans practiced vegetarianism as a rejection of the body and darkness in favor of the spirit and light.

Twentieth and twenty-first century vegetarians have shifted in the past 200 years from rejecting animal passions associated with the consumption of meat to a rejection of animal suffering caused by the consumption of meat. An example of this new way of thinking, Christian theologian and philosopher Albert Schweitzer (1875–1965) developed an ethic of "reverence for life," encompassing more than human life, that has been highly influential to both Christian and non-Christian vegetarians. Schweitzer may have never become completely vegetarian, although he held vegetarianism in highest esteem. Reverence for life "considers good only the maintenance and furtherance of life," branding "as evil all that destroys and hurts life, no matter what the circumstances may be" (Schweitzer and Joy 1950: 189). Humans must decide on a case-by-case basis how far they can remain ethical and to what extent they must destroy or harm life out of necessity, recognizing a fundamental truth that "I am life that wills to live, in the midst of other life that wills to live" (Schweitzer and Joy 1950: 30). He desired a new enlightenment of humanity with greater

---

[6] Swedenborgians believe that the Second Coming is currently taking place in the present reality as a return of spirit and truth as opposed to a physical manifestation of the Lord (The Swedenborgian Church of North America, 2008).

awareness of our place in the universe (Schweitzer 1949). Often reverence for life is equated with the Eastern concept of ahimsa, yet when Schweitzer studied the concept he criticized Indian religious and philosophical traditions for conceiving of it as merely avoiding harm rather than taking positive action to help others (Schweitzer and Russell 1936).

In Christian thought, ideas of the Garden of Eden strongly have influenced the development of vegetarianism. The intellectual lineage of Bible Christian Church founder Cowherd included English Quaker and vegetarian Thomas Tryon (1634–1703) and German mystic Jakob Böhme (1575–1624) (Spencer 1995; Antrobus 2002). Tryon believed human beings had been herbivores in the Garden of Eden and that eating animal flesh made humans more prone to violence. Benjamin Franklin, an early American vegetarian, had been a Tryonist in his youth. The significance of Tryon was his belief that conditions in the Garden of Eden such as vegetarianism should be replicated in the present day. Böhme, by contrast, believed it was necessary for humanity to be expelled from the Garden of Eden in order to redeem itself and to evolve to a new and more perfect state of harmony apart from God. Among the Christian respondents in my research, many did seek to restore conditions akin to the Garden of Eden. Non-Christian research subjects, however, following Böhme more than Tryon, did not expect to restore the actual Garden of Eden, but to create a new human relationship with nature.

Carolyn Merchant argued that the story of modern Western civilization in the Americas can be thought of as a grand narrative of fall, or decline and loss of innocence, and recovery, with humans seeking to reinvent the whole earth in the image of the Garden of Eden through labor (Merchant 1996). She identified two versions of the narrative of origins, Genesis 1 (Chap. 1), recovery through domination, which is strongly interventionist, and Genesis 2 (Chaps. 2 and 3), recovery through stewardship, which involves softer human management of nature (Merchant 2003). I adapted the two versions of the Edenic narrative she identified to relate specifically to vegetarianism and sanctuary activism. The adapted Genesis 1 version emphasizes that animals should fear humans and that humans should eat animals and treat animals as harshly as they see fit. In the adapted Genesis 2 interpretation, humans should seek to re-create Eden in the present-day, be vegetarians, and have animals solely as companions.

In the Genesis 1 version, which is the dominant interpretation of the relationship between humanity and animals, humans were made in the image of God and instructed to "fill the earth and subdue it" (Genesis 1: 28 [RSV]). They were given "dominion over the fish of the sea and over the birds of the air and over every living thing that moves upon the earth" (Genesis 1: 28 [RSV]). In this version of Genesis, dominion is interpreted to mean domination, giving license to humans to use animals as means to any ends. It should be noted, however, that the passage often cited as supporting a strict vegetarian diet Genesis 1: 29 [RSV]: "I have given you every plant yielding seed which is upon the face of all the earth, and every tree with seed in its fruit; you shall have them for food." This mandate does not hold strength for long, however, as Genesis 9: 2–3 [RSV] redefines the relationship of humanity to animals following the flood, as God says to Noah: "The fear of you and the dread of you shall be upon every beast of the earth, and upon every bird of the air, upon everything that creeps on the ground and all the fish of the sea; into your hand they are delivered. Every moving thing that lives shall be food for you; and as I gave you the green plants, I give you everything."

In the Genesis 2 version, man was created from dust and God put man in the garden "to till it and keep it" (Genesis 2: 15 [RSV]). It is a portrayal of the Garden in which humans care for plants and animals as a sacred task. In this version of Genesis, dominion means kind stewardship. Animals are understood to be the original companions of man, as in Genesis 2: 18–19 [RSV]: "It is not good that the man should be alone; I will make him a helper fit for him. So out of the ground the Lord God formed every beast of the field and every bird of the air, and brought them to the man to see what he would call them." In accordance with Genesis 1: 29–30, not only are humans to be vegetarians, but all animals are to consume only green plants. Several later prophecies envision a return to a vegetarian world, such as the passage envisioning wolves and lions dwelling and feeding with lambs and calves when "the earth shall be full of the knowledge of the Lord" (Isaiah 11: 6–9 [RSV]).

Self-professed Christian research participants primarily interpreted the ideal state of humans to animals in accordance with the kind stewardship of the Genesis 2 interpretation. They believed animals to be roughly equivalent in value to humans. One claimed animals are as sacred in God's eyes as humans, saying "no God of Love could want us to torture [animals] for luxuries. … all sentient creatures have rights and are sacred" (Wyatt 2006), adding "[e]ven if God gives us permission to eat animals when we are in need, it would never give us an excuse to torture and kill them for luxury (as pretty much all meat-eating in the U.S. is)" (Wyatt 2006). This subject spoke about the role of Jesus in relation to dominion, explaining that "I do think of Jesus' words as sacred and he taught ideals of loving, protecting, and self-sacrifice. I easily extend this sentiment to the non-human animals" (Wyatt 2006), continuing "it would be very difficult for me to believe that a teacher was sent by God if she or he did not have the proper respect for all sentient creatures" (Wyatt 2006). She emphasized the belief that God placed great trust onto human beings by awarding them dominion over animals:

> I think that "dominion" is quite a charge. The God of Genesis has dominion over us, and yet chooses to treat us with great love, watching over us, grieving for us, and taking pains to save us from ourselves. If we are called to have dominion over the other animals, we have quite a responsibility to them. I don't think this God would ever approve of us raising animals in squalid conditions, then butchering them in gruesome ways, for mere taste. In fact, I think this God would disapprove of our killing animals all together, except for survival. (Wyatt 2006; original emphasis)

Other Christian respondents also interpreted dominion to mean that God wants human beings to care for animals and the environment in a nurturing manner: "My personal belief is that we do have dominion over the animal kingdom and through good stewardship we are to treat all creatures with care and compassion" (Zundel 2006). She extended the concept beyond the treatment of animals to include "good stewardship to the environment, to others, [to] try to do my best not to contribute to pollution, [and to] reuse and recycle things I no longer use" (Zundel 2006).

The Christianity activists espoused aimed to attain the kind stewardship of Eden in the here and now. They sought to live lives as they were believed to have been originally intended with animals more as companions than food:

> If one looks at the Biblical story as a time line we were intended to live in harmony with all living things, then came the first fall of man, then the second fall. At the end of the epoch we are intended to live once again in harmony with all creation. In my mind our best choice is to live our lives as close to the perfection that God intended rather than give in to our barbaric past and base desires to live in dominion [Genesis 1 interpretation] as it was given to Adam and Eve. That is why I try to create Eden here and now every moment of every day. That is the most responsible thing I can do for creation and myself. (Warner 2006)

Another observed "I believe God's original plan was for us not to eat animals. It was only because of man's sin that caused us to eat them" (Baxter 2006). These words were based on her belief that "God gave us animals for companion[ship], not for food, entertainment, etc." (Baxter 2006). She contended that "animals do go to heaven and all the 1,000's of animals I have rescued and saved will be in heaven and greet me there" (Baxter 2006). Her belief was substantiated by "biblical verses, especially in Revelations, [that] mention animals in heaven next to God's throne" (Baxter 2006). In this view, God considers animals to have a place alongside humans as companions even in the afterlife.

The belief in intended purposes for humanity and animals in Eden and heaven extends to present-day personal teleological objectives. The Christian in the last paragraph explained that tending to animals was a personal duty, saying "saving/rescuing animals is my calling, a gift given to me by God" (Baxter 2006). "God has created us and He gave us all a purpose. But none of us deserve to be neglected, that is no one's/animal['s] purpose in life" (Kellar 2007), said a subject who integrated views on her own life and those of animals. She continued "I am a disciple of God and I have a need to nurture, care for, and love those less fortunate. I choose to help His animals because they don't ask to have owners that don't feed them. … I do believe that humans were supposed to see/use the animals for a purpose. … [A] sanctuary is a chance to fulfill the purpose God created them for" (Kellar 2007). When asked more about the idea of purpose, she identified how each animal, like each human, has unique talents and gifts that may or may not be useful under certain circumstances. She argued that humans have a responsibility to see where those talents are to be put to best use and facilitate that animal in achieving its purpose:

> Each animal has its own unique purpose (just like us humans) [and] some may have to travel a long and hard road to get to why they were created. … Some animals, [for] example a young draft horse, your hopes would be they will grow big and strong to plow fields since that must be their purpose God intending for such a big strong animal. Then to find a) it didn't grow to be as strong to work all day in the big fields but it can pull logs [and] b) it has torn a tendon and was sent down the road to an old farmer as a companion for his other old horse. Their true purpose is unknown until it just happens (God's will, we don't ask why when we are called to do his will we just thank him for the opportunity). … Another reference is what good would a 30+ horse with 4 teeth be that had been starved for most of his life, he really has no special skills and he has an enlarged right front knee that doesn't completely bend because it was not treated when it was fractured? Well the answer is I was in need of that kind of horse and he is now serving his purpose as a pal to my old horse. True miracle. (Kellar 2007)

In this view, humans at sanctuaries act toward animals in the place of God, who fulfills intended purposes for humans by locating them in certain places to make the best use of their gifts and talents.

Equal treatment of humans and animals formed a cornerstone of Christian vegetarian beliefs in this research. Accordingly, one of the major themes in responses was the concept of the Golden Rule, or, "do unto others as you would have them do unto you," as extended to animals. One subject made clear that "God says to 'treat those as you want to be treated'! and I will do just that" (Kellar 2007). Self-identified as a follower of Jesus, another explained "[m]y religious upbringing as a child was about 'do unto others…' and 'what you do to the least of them you do to me.' Given that, I guess it's easy for me to feel more in partnership with other creatures rather than believe that my role is to dominate, use and abuse" (Watts 2007). Extending the same treatment to animals as humans also was reflected by comments of those who extended the fifth commandment to animals. One participant who had been raised Catholic, but was no longer practicing replied "I believe that 'thou shalt not kill' should apply to animals too" (McFadden 2007).

Farm animal sanctuaries are superficially similar to Gardens of Eden based on the Genesis 2 narrative in the views of Christian subjects, places where activists have attempted to restore peaceful relationships between humans and animals through the building of "paradise for farmed animals and the people who love them," the motto of one farm animal sanctuary that does not advocate a religious perspective but does appropriate Edenic language (Fig. 16.5). The sanctuaries also have become loci in the transformation of the earth in the image of the Garden, especially for Christian respondents for whom ideals of the Garden of Eden were particularly influential. The ultimate goal for a majority of subjects, however, 58 % of whom were nonreligious, was the complete end of human dominance of animals and thus the elimination of domesticated animals. Thus, sanctuaries are not ideal end states for most, but steps along that path.

Fig. 16.5 Welcome sign at Animal Acres (Photo by Timothy Joseph Fargo)

Christian respondents who believed that their advocacy for animals was consistent with their religion were counterbalanced by those who had been raised Christian and became dissatisfied when taught that these same concepts did not apply to animals. These subjects often spoke solely about being taught Genesis 1 conceptions of dominion, criticizing them and identifying them as primary reasons for disillusionment with religion in general. In one example, a person who had been raised a strict Jehovah's Witness said "I was taught it was ok to eat meat and that animals were provided to humans for eating. But then again I was also taught not to kill (I guess that applied to humans only) and to love and have compassion" (Millard 2006). She continued, "[it] doesn't really make sense to me now. Compassionate killing doesn't make sense …" (Millard 2006). This demonstrates how both practicing Christians and disillusioned former Christians readily viewed animals and humans as members of the same community, subject to the same moral laws regarding their treatment.

When presented with Genesis 1 interpretations, many questioned their religion. In one case, a respondent from a family of Methodist ministers asked "if God wanted us and expected us to eat animals, why did he give them personalities, emotions, etc.?" (Caruthers 2006). Another wondered how people who consider themselves religious could disregard other living creatures and how God could let humans destroy animals and the environment:

> I question religion constantly because of how I feel about animals. I question how a loving and just God can stand by and watch the human race destroy the animals and environment supposedly created by that same loving and just God. I question how people can make decisions based solely on their happiness, with a complete and total disregard for other living creatures, while professing to be so-called God fearing people. … I have to question how the religious have yet to incorporate the animals into their philosophies. I have spoken ad nauseum about this with my parents and others who are very conservative Christians. All mumble the same invalid excuses, i.e., it has been going on since the beginning of time, people have to eat, etc., etc. (Huston 2006)

A participant who now identified as an atheist argued that Christians are not sensitive to the concerns of animals, saying that "[t]he Ten Commandments were never taught as if they applied to anyone other than other humans. I have frequently heard Christians state that animals don't have 'souls,' so it is no big deal to kill them, eat them, and that they were 'put on earth to serve us'" (Rosas 2006). Another echoed this thought and added: "These [Genesis 1] ideals have only made me more convinced that religion is evil and must be eliminated if we are ever going to truly change how we treat other creatures" (Cromwell 2006). Genesis 1 interpretations of dominionism were seen, in this case, as an impediment to achieving the goals of the vegetarian movement.

The activist in the following passage argued that dominentrism in general is a faulty way of thinking because it is based on anthropocentrism and therefore Christianity is not able to address concerns of animals and the environment:

> Dominionism, the idea that it is humans' God-given right to rule other animals is absurd to me. I see no valid justification for it beyond what I consider ill-conceived theology. 'Thou shalt not kill' is probably meant only for humans to other humans in most people's eye.

> This, again, seems to have no valid justification. I think religion can be useful if it breeds compassion and openness, and any anthropocentric doctrine fails to do that with the world in general. (Buck 2006)

Both forms of dominion, domination and stewardship, are fundamentally anthropocentric concepts. Accordingly, most subjects, especially the nonreligious, rejected these ideas, desiring instead non-anthropocentric equality between humans and animals. The goal would be a new earth where either an egalitarian partnership between humans and animals would be achieved or humans would live in separate existence from wild animals, the only remaining animals, which would have minimal interaction with humans. In the following passage, another subject extended the criticism of Christianity to all religions, finding them archaic and seeking to use modern legal systems instead to determine what behaviors are inhumane and should be illegal:

> I have found it unfortunate that people utilize archaic texts to excuse inappropriate behavior. There is no rhyme or reason to treat animals the way we do: historical and religious texts support behavior that modern legal systems have deemed inhumane, barbaric and illegal. If there are areas we can improve upon, we should, and we should not rely so heavily on historical tenets from religious texts. I, of course, support any and all core truths that promote compassion, respect and dignity. ... I will add that some Eastern traditions (Buddhism, Jainism) have core truths and ideas that mesh more with my own beliefs than many Western traditions. But again, they have the same fundamental problems as Western traditions do, draconic [sic] wordage abused and misused by people. (Woodley 2007)

16.7 Ahimsa and the Eastern Religious Influence on Western Vegetarianism

An understanding of the influence of Eastern religions on Western vegetarianism is critical to understanding the religious-moral hybridity at farm animal sanctuaries. In Great Britain, the contact of Christian vegetarians with Indian religions, especially from the 1880s onward, stimulated and reinforced the vegetarian movement (Twigg 1979), primarily through the concept of ahimsa, signifying abstinence from killing or harming other living beings. It is an important principle in Hindu and Buddhist thought but holds the greatest significance for Jains for whom ahimsa pervades every aspect of the religion. Jains hold cloths over their mouths, engage in no agricultural production, and avoid travel, as these activities may harm insects and other beings. Scholars contend that in the case of the Jains, the ultimate goal of ahimsa is non-action and life-negation in order to avoid karma completely, ending the cycle of rebirth (Schweitzer and Russell 1936; Dundas 2002). The traditional notion of ahimsa thus is morally individualistic, aiming for the deliverance of the individual soul from rebirth and lacking concern for others, human or nonhuman, apart from this (Gupta 1992). Vegetarianism in India is historically embedded in social caste and is not universalistic, whereas in the West it is strongly associated with an egalitarian, anti-hierarchical ethic (Twigg 1979).

The initial differences between Eastern and Western vegetarianism were significant, yet through contact with the West, Mohandas Gandhi helped to transform the concept of ahimsa into one more suitable to the Western mind. His contact with British vegetarians was formative, developing a passion for vegetarianism only after reading *A Plea for Vegetarianism* by English activist Henry Salt. Influenced by Hindu scripture, the life of Jesus, as well as the works of Thoreau, Ruskin, and Tolstoy, Gandhi redefined the concept of ahimsa into an ideal of constructive nonviolent action for social change (Gupta 1992). Many now understand ahimsa to represent a positive virtue, promoting universal love and identification with nature and life (Jhaveri 1985). This new conception became a central theme of the American Vegan Society, founded in 1960. Ahimsa subsequently has been used as an underlying principle in vegetarian and vegan literature.

While most surveyed sanctuary activists had converted to vegetarianism in their lifetimes, one subject was a Jain whose family had been vegetarian for over 500 years. She explained "[w]e practice active non-violence in words, thoughts and deeds. All forms of life are sacred and part of the divine. … Ahimsa is the only way of life I have known. In my personal life, I strive to live with greater degrees of non-violence every day" (Sahu 2007). I asked further about her perceptions of the meaning of ahimsa given the externally oriented interpretations of Jhaveri and the individualistic interpretations of Dundas and Gupta. She explained "I feel that ahimsa as practiced by most Jains is more in keeping with Dundas and Gupta's perception. Some of us practice the active form as stated by Jhaveri" (Sahu 2007).

A few activists were influenced by other Eastern religions. Now an atheist, being raised by practicing Hindus influenced one subject at least in the habit of vegetarianism although "the religion of my parents did not influence my philosophy toward animals" (Patel 2006). One participant who reported being most influenced by Jesus and Christianity was interested in Eastern religious concepts such as karma and ahimsa, but only because they already matched similar ideas in Christianity, explaining "[m]y beliefs about animals came first, and so, though I do believe in karma and respect ahimsa, I am drawn to them because of my belief in a respect for life. Elimination of suffering is a great way to put it. There is no reason to cause suffering when survival and self-preservation are not on the line" (Wyatt 2006). This is a case of affirming certain beliefs by finding support from different religious faiths and moral positions. Another who had been raised a Christian believed strongly in karma, emphasizing the consequences of the human treatment of animals:

> I am a firm believer in karma, and someday, we may all face the animals who have come across our path, whether they were the neighbor's starving dog we refused to help or the dead cow on our plate. We will be held accountable and we will see retribution for what we have done directly via our own actions, or indirectly via our inaction. (Huston 2006)

Buddhist respondents argued that the concept of suffering was especially meaningful to them, often interpreted in a utilitarian sense of the least amount of suffering to the greatest number. For example, one said "I devote my life to the alleviation of suffering wherever it may exist" (Warner 2006). At least one Buddhist was a former Christian disillusioned with Christianity. Like other Christian participants, he said "[w]hen I was practicing Catholicism I applied 'the Golden Rule' to the animals, too. However, my gradual conversion to Buddhism was a result of disappointment in the Catholic Church" (Radcliffe 2006). The participant in the following response spoke of being fundamentally agnostic, but engaged in many of the practices of the Buddhist religion, especially the concept of mindfulness applied to the consumption of food:

> I practice some teachings from Buddhism such as meditation, compassion, tonglen, and mindfulness. A good example of how my spiritual practices have fit with my views of animals is the practice of mindfulness. Mindfulness has made me more aware of how my choices affect animals, such as with what I wear and what I eat. … Compassion and openness are important principles for Eastern religion, and can be extended to count for all living things. … My compassion for all sentient beings has excluded me from consuming their flesh. I don't want to contribute to an inherently cruel industry. (Buck 2006)

As seen in many of these examples, sometimes religions or philosophies are drawn upon as cultural support or justifications for personal moral beliefs rather than as the initial rationales for beliefs toward animals. This is especially evident in cases where disillusionment with Genesis 1 interpretations of Christianity led to drawing upon Eastern religious concepts perceived to be more supportive of their beliefs. For many of the activists, compassion towards others was the cornerstone of their beliefs regardless of religious or philosophical background, demonstrating the importance of temperament and personal experiences:

> I suppose there are philosophical perspectives that mesh nicely with my own beliefs. However, my initial beliefs, those core truths I have always believed in, have always come from my heart and mind. That they coincide with so many other philosophical and moral perspectives is a beautiful coincidence (it's nice to know more eloquent beings share the same ideas as me!). (Woodley 2007)

The following passage demonstrates how one used religion to support individual perspectives and gain credibility rather than as an overall guide for life:

> Although I do not practice any particular religion, I always enjoy being able to refer to various religious principles in support of vegetarianism, kind of a way to gain "credit" on the matter via God or other religious mandates of higher beings. I am influenced by these various teachings in that they further confirm my previously drawn beliefs and have provided me spiritual comfort when I finally got around to considering some of them and found that I am making my way in life pretty good just going on what I decide to be right and just. I had, in fact, actually avoided religious teachings for a long time because my early, little exposure to Christian religions led to me believe that it was all simply ridiculous justifications for doing bad. It is only in the last few years that I have taken a stronger interest in various religions, and I admit to having a strong preference to certain Eastern religions …. (Whitney 2006)

16.8 Ethics of the Enlightenment: Utilitarianism and Animal Rights

In addition to religious ideas, ethical arguments for animal rights, based on Enlightenment value systems, such as utilitarian ethics, have contributed an additional layer of moral significance to the vegetarian and animal rights movements (Singer 1975; Dombrowski 1984). A slight majority of research subjects (58 %) reported that they practiced no religion, with nearly a third of those stating that they were atheists. Most of the nonreligious cited philosophical perspectives in their thought regarding animals.

Utilitarianism-influenced animal rights philosopher Peter Singer credits Jeremy Bentham as being one of the first modern philosophers to denounce the human domination of animals. In 1780 Bentham made the argument that the capacity of animals to suffer and enjoy are prerequisites for having interests and the key elements making it morally necessary to take them into consideration (Singer 1975: 7). Bentham stated regarding animals, "[t]he question is not, can they reason? Nor can they talk? But, can they suffer?" This principle combined with modern scientific knowledge that animals feel pain has become central to the movement for animal rights since previously animals were believed to be machine-like, lacking ability to sense pain. Bentham compared the position of animals with that of slaves and looked to a day "when the rest of the animal creation may acquire those rights which never could have been withholden from them but by the hand of tyranny" (Singer 1975: 7). Influenced by such thought, social movements advocating humane treatment of animals began in the nineteenth century and grew in strength resulting in improvements in conditions for animals through the establishment of laws against abuse, especially from the Victorian Age (Ritvo 1987) through the present day.

Employees and volunteers at sanctuaries acknowledged utilitarianism, a leading Enlightenment ethics whose origins are credited to Bentham, but were popularized in the animal rights community by Singer, to be a primary influence. The greatest good for the greatest number, including animals, and the widespread diminishment of suffering, formed cornerstones of their beliefs: "My current attitude is merely born of a desire to not contribute to increasing suffering in this world, an attitude I think I have always held" (Patel 2006). This participant reported having been influenced by utilitarianism through the books of Peter Singer while another specifically mentioned his work as a supplemental influence, stating that "Singer's utilitarianism was important in developing my commitment to veganism, but I was already on my way" (Wyatt 2006). Credit was given to John Stuart Mill by another subject who stated that utilitarianism "seemed so simple, was clearest, and made the most sense. Even though there are complex situations where it fails, for common, everyday actions, it is mostly correct" (Jenson 2006). This respondent summarized the importance of utilitarian arguments to animal rights:

> Utilitarianism does … force you to think about 'everyone,' and not just you (and not just about people that are immediately around you). It also boils morality down to suffering and

happiness, and in drawing the scope of concern around all sentient beings. It's a philosophy that leads one right into the hands of animal rights and environmentalism. (Jenson 2006)

Recognizing the limitations of utilitarianism, however, if animals are not considered subjects of moral concern, he admitted "without the realization that animals are deserving of being included in the equation, utilitarianism doesn't do anything for them" (Jenson 2006).

Utilitarianism can be expanded, however, to include animals through a critique of anthropocentrism, which is credited as having prevented humans from extending ethical guidelines to include animals. Arguments made by those influenced by both religious and philosophical teachings shared this critique of anthropocentrism. In the following passage, a participant sought to go beyond utilitarianism as a sole basis for ethics:

> I am especially drawn to Peter Singer's work. He shows in a convincing manner that one cannot justify harm done to animals successfully through a philosophical argument. He also shows the grand fallacy committed by nearly all ethicists: anthropocentrism. Utilitarianism is appealing to me, the basic concept of 'doing the most good for the most amount of people,' but I am more drawn to Immanuel Kant's categorical imperative. One of its facets is that a person cannot be used as a mere means to another's end. If the principle were extended to animals it would forbid such acts as animal experimentation, because even if it does benefit many people (which utilitarianism would seek to justify) it is still violating a principle (because it uses an animal as [a means to] an end). (Buck 2006)

This subject identified Kant's categorical imperative as more important to arguments in support of animal rights than utilitarianism alone. It must be pointed out, however, that Kant said, as quoted by animal rights philosopher Tom Regan, "[s]o far as animals are concerned, we have no direct duties. Animals are not self-conscious, and are there merely as a means to an end. That end is man" (Regan 1983: 177). This demonstrates how an ethical concept could be taken from its original source and reinterpreted to be extended to animals. Kant did believe, however, that we ought to treat animals without cruelty because cruelty would harden us in our interactions with other humans: "[i]f a man shoots his dog because the animal is no longer capable of service, he does not fail in his duty to the dog, for the dog cannot judge, but his act is inhuman and damages in himself that humanity which it is his duty to show toward mankind" and thus "we have duties towards the animals because thus we cultivate the corresponding duties towards human beings" (Regan 1983: 178).

Animal rights theory maintains that some amount of legal rights previously reserved solely for humans be extended to animals. When asked what philosophical perspectives mattered most, one subject replied that it was "[t]he rights of animals as thinking, feeling beings with the same will to live and survive as any other creature" (Carson 2007). Another respondent expressed the core of what it seemed most subjects believed, that animals should be free of human domination, arguing "I believe that animals, as sentient beings, have the right to live free from the tyranny of humans, that is, to paraphrase the famous PETA quote, they are not ours to eat, to experiment upon, to use for our entertainment, etc." (Albright 2007), echoing the idea that animals should not be means to an end. For many participants it was about humans not using animals for human purposes, such as "I just believe that we don't have the right to use animals for any reason" (McFadden 2007).

The movement for animal rights is linked to other progressive causes based on the common goals of attempting to undo devaluation, domination, and violence of all kinds. The following quotation demonstrates how some of these perspectives came together for one activist:

> Animal rights theory primarily [has influenced my beliefs on the relation of humans to animals] and also the study of language/semiotics, history, and sociology in general plus feminism, sexuality studies, and gender, race and class studies influence my feelings about human/nonhuman relations. I see parallels all over the place and do believe that it is essentially the same thought processes (or interruption of thought processes via social conditioning) which allows us to harm other humans, other animals, damage the environment, etc., by devaluing them first based on social constructs and then taking advantage of or enacting violence upon them. The following quote from Theodor Adorno resonates strongly with me: "Auschwitz begins whenever someone looks at a slaughterhouse and thinks: They're only animals." (Tenney 2007)

Activists thus sought the progressive elimination of negative forms of dominance. Many were as concerned with issues of human subjugation as animal subjugation. One respondent subscribed to the idea of the progression of ethics as an expanding circle, a concept attributed to Singer, arguing "[w]e are here to promote the moral evolution of humans. Humans have started to understand why racism is intellectually untenable, and morally repugnant. We are starting to catch on to sexism, and culturalism, and the many prejudices that have been taught [to] us" (Wyatt 2006). There is a teleological quality of progress to these arguments in that ethics are seen to unfold to include wider and wider circles. The same subject placed the role of the farm animal sanctuary as central to this moral progress, saying "[t]he role of the sanctuary is to help humans take the next step, to see other species as deserving of moral consideration, just like those who fought for moral consideration for non-whites, women, etc." (Wyatt 2006). The goal evident from responses was a progressive enlightenment, characterized by "respect for all life forms, an ideology necessary for an evolved society" (Watts 2007). The philosophy of this new civilization would be a secular mixture of the reverence for life of Schweitzer and the progressive externally-oriented ahimsa of Gandhi. Often these ideas were equated and formed the cornerstone of their moral vision for the future:

> [U]nless humans and all of humanity develops a reverence for all life, you know, ahimsa, … it is going to be impossible to have any kind of symbiotic relationship with humans and nonhumans because humans are in a position where [we] can exploit, and are doing so at great lengths and at great harm. So if we could develop the human consciousness, you know, to an enlightenment, where [we] really did see animals as kin, where [we] saw and [we] did not use animals at all for any purpose, then yes it could be a beautiful world where humans and nonhumans co-existed in nature. I don't know that we'll ever see that. I mean, one would hope that the human consciousness will evolve to that … but I don't know if humans are … capable of that. … I can only tell you what I see now is that we're pretty far off from that and yet I have to hope that we are evolving that consciousness, that enlightenment, and that we'll get there before we blow up the world. (Root 2006)

16.9 Conclusion

Farm animal sanctuaries are morally hybrid places, composed of individuals of differing religious and philosophical backgrounds, united through a politics of affinity based on the common goal of reducing the suffering of farm animals. Edenic narratives influenced most sanctuary activists and their views of the relationship of humanity to animals, some in a positive way and some in a negative way. Christian subjects held that dominion meant kind stewardship and sought to achieve a new relationship with animals based on the model of the Garden of Eden. Many former Christians specifically credited their religious disillusionment to Genesis 1 interpretations of dominion. Most rejected the anthropocentrism of both the Genesis 1 and Genesis 2 narratives. Jain and Buddhist subjects believed in ending suffering as well as the progressive conception of ahimsa, including nonviolence towards others and the self. Atheist participants often believed in utilitarian reduction of suffering and conceptions of animal rights, such as that animals should not be used as means to human ends. Adherents of religions or philosophies often drew upon supporting concepts from other religions and philosophers forming hybrid worldviews. The key moral principles they identified regarding animal welfare included a rejection of anthropocentrism, reverence for life, progressive ahimsa, animal rights, and the utilitarian reduction of suffering.

References

Albright, C. (2007). Bleating Hearts Animal Sanctuary. Written questionnaire received by Fargo, T. J., Santa Monica.
Amato, P. R., & Partridge, S. A. (1989). *The new vegetarians: Promoting health and protecting life*. New York: Plenum Press.
Animal Place. (2006). Animal Place's Sanctuaries.org. Retrieved February 14, 2006, from www.sanctuaries.org/
Antrobus, D. (2002). Vegetarianism: Neither phoenix nor fad. *35th World Vegetarian Congress: Food for all our futures*. Heriot Watt University, Edinburgh.
ASA. (2006). "Accredited farmed animal sanctuaries." Retrieved February 14, 2006, from www.asaanimalsanctuaries.org/Sanctuaries/Categories/Farm.htm
Baxter, L. (2006). PIGS, a Sanctuary. Written questionnaire received by Fargo, T. J., Santa Monica.
Beardsworth, A. D., & Keil, E. T. (1991). Health-related beliefs and dietary practices among vegetarians and vegans: A qualitative study. *Health Education Journal, 50*, 38–42.
Boyle, J. E. (2007). *Becoming vegetarian: An analysis of the vegetarian career using an integrated model of deviance*. Ph.D. dissertation, Virginia Polytechnic Institute.
Buck, K. (2006). Peaceful Prairie Sanctuary. Written questionnaire received by Fargo, T. J., Santa Monica.
Carson, A. (2007). VeganPeace Animal Sanctuary. Written questionnaire received by Fargo, T. J., Santa Monica.
Caruthers, A. (2006). Farm Sanctuary. Written questionnaire received by Fargo, T. J., Santa Monica.
Clark, S. R. L. (2004). Vegetarianism and the ethics of virtue. In S. F. Sapontzis (Ed.), *Food for thought: The debate over eating meat* (pp. 138–151). Amherst: Prometheus Books.

Cromwell, L. (2006). Eastern Shore Chicken Sanctuary. Written questionnaire received by Fargo, T. J., Santa Monica.
Dombrowski, D. A. (1984). *The philosophy of vegetarianism*. Amherst: The University of Massachusetts Press.
Dundas, P. (2002). *The Jains*. London: Routledge.
Fessler, D. M. T., Arguello, A. P., Mekdara, J. M., & Macias, R. (2003). Disgust sensitivity and meat consumption: A test of an emotivist account of moral vegetarianism. *Appetite, 41*, 31–41.
GreenPeople. (2006). Animal sanctuaries and wildlife shelters and rehabilitation centers. Retrieved February 14, 2006, from www.greenpeople.org/sanctuary.htm
Gregerson, J. (1994). *Vegetarianism: A history*. Fremont: Jain Publishing Company.
Gupta, V. K. (1992). *Ahimsa in India's destiny: A study of the ethico-spiritual Ahimsa, its roots in ancient Indian history, and its role as a political weapon during the Gandhian era*. Delhi: Penman Publishers.
Hamilton, M. (2000). Eating ethically: 'Spiritual' and 'quasi-religious' aspects of vegetarianism. *Journal of Contemporary Religion, 15*(1), 65–83.
Huston, M. (2006). Peaceful Prairie Sanctuary. Written questionnaire received by Fargo, T. J., Santa Monica.
Iacobbo, K., & Iacobbo, M. (2004). *Vegetarian America: A history*. Westport: Praeger.
Jabs, J., Devine, C. M., & Sobal, J. (1998). Model of the process of adopting vegetarian diets: Health vegetarians and ethical vegetarians. *Journal of Nutrition Education, 30*(4), 196–202.
Jenson, R. (2006). Animal Place. Written questionnaire received by Fargo, T. J., Santa Monica.
Jhaveri, S. S. (1985). Ahimsa: Best solution for all problems. In S. K. Jain & K. C. Sogani (Eds.), *Perspectives in Jaina philosophy and culture* (pp. 49–50). New Delhi: Ahimsa International.
Kellar, A. (2007). Happy Trails Farm Animal Sanctuary. Written questionnaire received by Fargo, T. J., Santa Monica.
Lightfoot, V. (2003, July/August). The VegNews guide to farmed animal sanctuaries. *VegNews: North America's Vegetarian Newspaper*.
Lindeman, M., & Sirelius, M. (2001). Food choice ideologies: The modern manifestations of normative and humanist views of the world. *Appetite, 37*, 175–184.
McDonald, B. (2000). "Once you know something, you can't not know it": An empirical look at becoming vegan. *Society & Animals: Journal of Human-Animal Studies, 8*(1), 1–23.
McFadden, V. (2007). Poplar Spring Animal Sanctuary. Written questionnaire received by Fargo, T. J., Santa Monica.
Merchant, C. (1996). Reinventing Eden: Western culture as a recovery narrative. In W. Cronon (Ed.), *Uncommon ground: Rethinking the human place in nature* (pp. 132–159). New York: W. W. Norton & Company.
Merchant, C. (2003). *Reinventing Eden: The fate of nature in western culture*. New York: Routledge.
Millard, J. (2006). Farm Sanctuary. Written questionnaire received by Fargo, T. J., Santa Monica.
Patel, A. (2006). Chenoa Manor. Written questionnaire received by Fargo, T. J., Santa Monica.
Radcliffe, A. (2006). Chenoa Manor. Written questionnaire received by Fargo, T. J., Santa Monica.
Regan, T. (1983). *The case for animal rights*. Berkeley: University of California Press.
Ritvo, H. (1987). *The animal estate: The English and other creatures in the Victorian age*. Cambridge: Harvard University Press.
Root, S. (2006). Animal Acres. Interview by Fargo, T. J., Santa Monica.
Rosas, K. (2006). Chenoa Manor. Written questionnaire received by Fargo, T. J., Santa Monica.
Rosen, S. (1997). *Diet for transcendence: Vegetarianism and the world religions*. Badger: Torchlight Publishing, Inc.
Rozin, P., Markwith, M., & Stoess, C. (1997). Moralization and becoming a vegetarian: The transformation of preferences into values and the recruitment of disgust. *Psychological Science, 8*(2), 67–73.
Sahu, R. (2007). Animal Place. Written questionnaire received by Fargo, T. J., Santa Monica.
Sapontzis, S. F. (Ed.). (2004). *Food for thought: The debate over eating meat*. Amherst: Prometheus Books.

Schweitzer, A. (1949). *Out of my life and thought, an autobiography*. New York: H. Holt.
Schweitzer, A., & Joy, C. R. (1950). *The animal world of Albert Schweitzer; jungle insights into reverence for life*. Boston: Beacon.
Schweitzer, A., & Russell, L. M. R. (1936). *Indian thought and its development*. New York: H. Holt and Company.
Singer, P. (1975). Animal liberation: A new ethic for our treatment of animals. *New York Review*: Distributed by Random House, New York.
Sopher, D. E. (1967). *Geography of religions*. Englewood Cliffs: Prentice-Hall.
Spencer, C. (1995). *The Heretic's feast: A history of vegetarianism*. Hanover: University Press of New England.
TAOS. (2006a). TAOS – The Association of Sanctuaries – About Us. Retrieved February 14, 2006, from www.taosanctuaries.org/about/index.htm
TAOS. (2006b). TAOS – The Association of Sanctuaries – Farmed Animal Sanctuaries. Retrieved February 14, 2006, from www.taosanctuaries.org/sanctuaries/farmed.htm
Tenney, C. (2007). Peaceful Prairie Sanctuary. Written questionnaire received by Fargo, T. J., Santa Monica.
Tuan, Y. F. (1974). *Topophilia: A study of environmental perception, attitudes, and values*. Englewood Cliffs: Prentice-Hall.
Twigg, J. (1979). Food for thought: Purity and vegetarianism. *Religion*, 9(Spring), 13–35.
VRG. (2006). The market for vegetarian foods. Retrieved February 14, 2006, from www.vrg.org/nutshell/market.htm
Walters, K. S., & Portmess, L. (Eds.). (1999). *Ethical vegetarianism: From Pythagoras to Peter Singer*. Albany: State University of New York Press.
Walters, K. S., & Portmess, L. (Eds.). (2001). *Religious vegetarianism: From Hesiod to the Dalai Lama*. Albany: State University of New York Press.
Warner, M. (2006). Farm Sanctuary. Written questionnaire received by Fargo, T. J., Santa Monica.
Watts, D. (2007). Animal Place. Written questionnaire received by Fargo, T. J., Santa Monica.
Whitney, E. (2006). PIGS, a Sanctuary. Written questionnaire received by Fargo, T. J., Santa Monica.
Woodley, R. (2007). Animal Place. Written questionnaire received by Fargo, T. J., Santa Monica.
Wyatt, A. (2006). Farm Sanctuary. Written questionnaire received by Fargo, T. J., Santa Monica.
Zundel, D. (2006). Animal Place. Written questionnaire received by Fargo, T. J., Santa Monica.

Part III
Sacred Spaces and Places

Chapter 17
Religions and Ideologies

Paul Claval

17.1 Introduction

Most persons think that religions differ profoundly from ideologies. On reflection, however, the two terms offer many similarities (Dubuisson 1998; Duméry 1985). The French dictionary Robert defines *religion* as a "system of beliefs and practices, involving relations with a superior principle, and specific to a social group" and *ideology* as a "system of ideas, philosophy of the World and life." Definitions in English dictionaries show similarly the proximity of the two words: for instance, *religion* is "a set of beliefs, values, and practices based on the teachings of a spiritual leader" and *ideology*, "a set of doctrines or beliefs that form the basis of a political, economic, or other system" (*American Heritage Dictionary* 2011). Our purpose here is to compare religions and ideologies in order to understand the role they play in the organization of space and the life of individuals and social groups.

17.2 Similarities and Differences

17.2.1 Giving a Significance to Existence

We may summarize the notions of religion and ideology in the following way (Claval 2008): (i) Religions appear as sets of beliefs dealing with the significance of cosmos, nature, life, and offering salvation to women and men. (ii) Ideologies appear as sets of beliefs dealing with the significance of nature, life, society and promising fulfillment and happiness in this world to human beings. Both definitions

---

P. Claval (✉)
Department of Geography, University of Paris-Sorbonne, Paris, France
e-mail: p.claval@wanadoo.fr

give a sense to the existence of individuals and/or human groups. Those who believe in a religion or adhere to an ideology express their convictions through specific forms of behaviour. Religious women and men pray, meditate, attend services in temples, churches, synagogues or mosques, give money to their churches and participate in processions. They become pilgrims to visit holy places. Religions practice sacrifices of animals, but also, in the past, of men or women, or are reminded of them through their rituals, as is true for Christians.

Those who adhere to an ideology do not pray nor meditate; they do not visit temples, churches, synagogues or mosques, but attend meetings where orators remind them of the main tenets of their ideology and apply them to social problems – here and now, or in other places and other times. They participate in walks, which are in many ways similar to religious processions. They sometimes visit the places where the ideology was born, where its first exponents lived or where militants were killed when fighting for their cause, which is another form of pilgrimage. Ideologies consider that the sacrifice of animals and human beings for the Glory of God, or gods, is an abomination, but revolutions are one of their main rituals. In these times militants do not hesitate in sacrificing thousands or millions of people to eradicate the classes which exploit the proletariat, or the groups which have treacherous links with foreign countries.

Religions and ideologies are not identical, but their manifestations offer many analogies.

17.2.2 Salvation of the Soul vs. Individual or Collective Fulfillment and Happiness

The aims of religions and ideologies are akin, but not identical: personal salvation for most religions; a better life and happiness in this World – now, or in the future – for ideologies. Ideologies differ also from religions by their greater emphasis on collective destiny.

Religions are based on a dual conception of human beings. On one side the body is made of matter, and on the other, the soul appears as an immaterial principle, a breath, a spirit. The interpretation religions propose of human life is a consequence of this dualism. The death deprives the body of life, but does not destroy the soul. Religions reflect on the destiny of the soul after death – and sometimes, before birth. It offers ways to insure its salvation.

Ideologies do not accept this dualism. When they developed in seventeenth century Europe, epicurean atomism was one of the main references for the new scientists. It was the root of the sensualist conceptions of Thomas Hobbes and later, John Locke. The mind did not pre-exist to the body. It was not born as a full-fledged entity. It developed progressively thanks to the senses and their impact on the brain.

Ideologies ignore problems of salvation. They deal only with bodies and minds, and look for the individual or collective fulfillment of life in this World. They speak about happiness, not salvation.

17.2.3 A Different Role in History

Another difference is linked to the role of religions and ideologies in history. Religion was, and is, present in all societies; the prehistoric paintings preserved in caves were certainly linked to the worship of gods and the cult of the dead. Ideologies have a shorter history. They were unknown in the Antiquity and the Middle Ages. They appeared with Thomas More's *Utopia* in 1512, but became mainly influential during the seventeenth and eighteenth centuries with the rise of the idea of Progress and the success of philosophies of history. Their role grew in the nineteenth century. In many ways they appeared as dominant in the twentieth.

The decline of the philosophies of history reduced their audience in the late decades of the twentieth century. After the Fall of the Wall, in 1989, the socialist regimes of Eastern countries, which were grounded in the socialist ideology, disappeared in Europe; they survived only in Cuba, North Korea, Vietnam, and, under a much altered guise, in China.

Elsewhere, the influence of ideologies was reduced by the renewal of old religions and the development of new ones. This new success of religious beliefs – old or new ones – relies on a move towards a simpler interpretation of Revelation and purer forms of practice (Berger 1999). It can be achieved through a move back to the primitive expression of traditional religions, as in fundamentalisms, or a new emphasis on the Holy Spirit, as in Pentecostal and neo-Pentecostal churches and charismatic movements, or syncretic forms of Eastern religions and Western ideologies, as in many sects.

Are ideologies doomed to disappear? No, since they are taking new forms. They have ceased to rely on philosophies of history and the idea of progress. They are grounded in the exploration of the unconscious or in a new faith in the power of Nature.

On the whole, ideology appears as a Western venture. Some oriental "religions" present analogies with ideologies, since they do not invoke a unique God or a plurality of gods, which was the case of Confucianism, but favored the cult of ancestors, which was common to many traditional religions.

In the last two centuries, most of the ideologies which met success in the East came from the West. Among the two variants of the Philosophies of Progress, the liberal one seduced many indigenous elites during the nineteenth century, as shown by the intellectual milieu in which Gandhi grew up, for instance. Later on, the socialist variant was preferred, since it was at the same time a Western Ideology and a way to struggle against the Imperial Western hegemony. Some forms of national identities existed outside Europe in the past, but their role in the building of States was a limited one, except perhaps for some Eastern countries like China, Japan and Korea and to a lesser degree Vietnam, Thailand and Burma. The Japanese people were the first to graft the Western idea of a nation on an older indigenous one – already modernized and theorized, however, in the eighteenth century by Motoori Norinaga, whose ideas were influential at the time of the Meiji revolution.

17.3 Religions and the Topologies of Next Worlds They Are Built On

17.3.1 The Truth Comes from Another World

Religions and ideologies present important differences. The main one stems from the sources of authority they rely on. Religions refer to God, or gods or a superior principle; ideologies only deal with terrestrial realities.

Religions rely on a "belief in one or more gods," "in a divine power" or "in a supernatural power or powers that control human destiny." They show "belief in and reverence for a supernatural power or powers regarded as creator and governor of the universe." They express "belief in, worship of, or obedience to a supernatural power or powers considered to be divine or to have control of human destiny."

How this God or these gods were – and are – known? (i) It is (was) sometimes through the mythical tradition of an immemorial time when animals and human beings communicated freely with gods (Ricœur 1985). These traditions reveal(ed) also the existence of sacred places where hidden forces or spirits are (were) present. (ii) Religions may (might) rely on God's Word as revealed by one (or several) prophet(s); they recognize(d) the sacredness of the places where the Revelation occurred, where the prophet(s) was(ere) born and lived, and where God was – and is – present through the Eucharist or because the faithful are praying and singing in His/Her name. All religions rely in this way on the belief in the existence of Next Worlds, of Beyonds; human beings cannot visit them, but thanks to tradition or revelation, they know about their existence.

The relation of these Next Worlds, of these Beyonds, to our World is not always the same. They may be *transcendent* and reside in the Heaven – as in most Revealed religions. They may also be *immanent*: they are then made of forces or spirits who lay inside the terrestrial things or beings (Claval 1999).

17.3.2 Why Geographers Have to Study the Next Worlds of Religions

In the past, geographers only dealt with material space. Today, they are also interested in the representations people have of it. They analyze these other spaces. In this respect, two aspects have to be more specifically stressed.

(i) The analysis of the notion of "horizon" explains the way Next Worlds may be conceived and imagined, as shown by Jean-Marc Besse:

> … the horizon expresses … much more than the existence of faraway worlds. This term has an ontological as much as epistemological significance. It refers to the part of invisible which is located within all that is visible, to this incessant fold of the world which transforms definitely the real into an uncompleted space, an open space, which can not be totally categorized. The horizon is the name given to this power of overflowing of the being (débordement de l'être) which is present in the landscape. (Besse 2009: 53)

In this way, geographical experience goes beyond the real. It opens the way to next worlds.

(ii) In some places, the invisible "other spaces" get in contact with the material world and transform it. In the overwhelming profane environment where people live, there are points or areas which differ by their nature: heterotopias, in the sense defined by Michel Foucault (1984/1985).

These places present to human experience the appealing and frightening character of sacredness, since they communicate with the Next World. Mircea Eliade (1949/1975: 316) was the first to underscore this major fact: sacredness has to do with the terrestrial incidence of Next Worlds. It is often present where the axis of the world passes through the surface of the Earth, since it favors the relations between the Earth and the Beyonds.

In this respect, a major difference exists between revealed religions and polytheisms and animisms. For the first ones, terrestrial space is mainly profane; sacredness is an exception. For animisms, it is much more frequent, since many things or beings are inhabited by supernatural forces or beings. With the passage of polytheism to a unique God, the disenchantment of the world began, but it never was complete, since God had manifested himself/herself where the Revelation occurred. He is also present through the sacrifice of the Mass or because the faithful are gathered in order to pray and sing in God's name.

17.4 The Next Worlds of Ideologies

Ideologies reject the idea of a supernatural reality, which is central to religions,+ either traditional or revealed. They try to explain our world through terrestrial forces or factors only. Just like religions, their construction relies, however, on Next Worlds – but they are terrestrial ones.

17.4.1 Terrestrial Next Worlds

The construction of ideologies relies on the belief in the existence of terrestrial beyonds (Claval 1999); they serve as models to understand reality and conceive the ways of transforming it. They pertain to two types. First, they are made of real but unattainable places located in the past of the Golden Age, the future of an Utopia or the Earth without Evil of a faraway country. Second, they are located in the unconscious forces at work in the depths of nature, human beings or social systems.

The way these beyonds are known differ widely. Thomas More (1512) conceived his *Utopia* as a Country without Evil. The narrator had met in Antwerp Raphael Hythloday, who just came back from a journey to the newly discovered America, where he visited a former peninsula, transformed into an island by its inhabitants and who wished to sever all contacts with the rest of the World in order to build there a perfect society. As described by More, *Amoraute* was a wonderfully organized country and ignored the evils which plagued the Christian Kingdoms of Europe. A curious coincidence: at the time when Thomas More wrote *Utopia,* Portuguese and Spanish discoverers were meeting the Tupi-Guaranis of Southern America. From time to time, one of their tribes left its village and moved under the leadership of a shaman towards the Land without Evil he had just discovered in his dreams.

Utopia has always been located on the Earth, but very soon, it ceased to exist in the present. In the fifteenth and sixteenth centuries, the majority of humanists situated it in a Golden Age of the Past, in the texts and ruins of Rome and Greece. By the beginning of the seventh century, Utopia had become a feature of the future (Manuel and Manuel 1979).

A novel like *Utopia* was a good way to criticize the weaknesses of Western societies, but something else was required in order to give credence to the Next World it revealed.

17.4.2 Ideology and Myth

All the Next Worlds explored in Antiquity were not religious ones. In the myth of the cave, Plato compared the way human beings discover the world to that of prisoners living in a cave where the only thing they could observe was the shadows projected on the wall in front of them by a fire located behind them. This experience explained that their world was only a pale projection of the real one. The lesson for ordinary human beings was a simple one: their world was only a pale projection of that of Ideas. Hence the role played in the Western tradition by a non-religious beyond, that of metaphysics, which contributed much to the enrichment of the Christian tradition in the Antiquity and during the Middle Ages (Claval 2008).

In the Modern Times, the construction of non-religious Next Worlds did not result from the activity of philosophers, but from social thinkers (Claval 1980, 2001). Social sciences began do develop during the seventeenth century, when scholars started to study social realities according to the principles of the new physics and the new interest in epicurean atomism. Their approach was, however, different from that of physicists. They could not rely on experiences performed in the real world. Their only possibility was to imagine them. Hobbes was the pioneer in this field; for him, in the prehistory of humanity, there was no society. Each individual relied only on his/her forces to be secure. Nobody protected him/her. He/she was like an atom in a gas. According to Hobbes (1651), in this state of nature, man (generically speaking) was a wolf for man. People lived in permanent warfare; they ignored security. This condition was unbearable once people became conscious of this fact. In order to break with this situation, they decided to renounce the right they had to use physical violence against their enemies. They signed a Covenant, the Social Contract, which transferred it to the Leviathan (the State) and which insured security for all.

The story of the Social Contract was not a scientific demonstration. It offered a plausible version of an imagined event in the past. It was a myth, just like the myth of the Cave of Plato (Claval 1980, 1985). It was thus through a tale that the Next World of ideology gained a wide audience. The story appeared as grounded in Reason; it looked scientific, even if it was not built on a demonstration. Hobbes talked about a moment of the past, but his analysis invites us to replicate it. Hobbes' vision could be read as a recipe to build a better future for humanity. Utopia ceased to be a literary device. It became a projection that guided human action.

The Social Contract knew a fantastic success in the second half of the seventeenth century and in the eighteenth. As used by John Locke (1690), it was signed in a society which suffered all the most from insecurity that its members were already engaged in a process of enrichment through their labor: the social contract appeared as a condition for the betterment of everybody's material life. For Jean-Jacques Rousseau (1762), humanity had a long history, with positive as well as negative aspects: positive ones, since it had entered into a process of civilization; negatives ones also, since this process was responsible for the depravity of women and men. The signature of the Social Contract was the only way to give human history a right orientation and to insure happiness for all.

17.4.3 From Myth to "Scientific" Explanation

By the early nineteenth century, the standards of scientific research in the social field had improved. Myths could no more appear as convincing arguments, even if their lessons seemed rational. Auguste Comte (1822) was conscious of this fact, but he did not succeed in giving a firm statute to his Law of the Three States of Humanity. It was just another tale about the mythic prehistory and history of societies.

Karl Marx chose another way to present his own philosophy of progress and history. He relied on the only social science which already had a well-defined methodology, economics. The first book of *Capital* (Marx 1867) analyzed the genesis of profits through the dialectics of the market- and labor-value of human work. The exploitation of the proletariat appeared as the result of a hidden process. The suppression of the capitalist regime through a social and political revolution would stop human alienation. Societies would then move out of history to enter into an era of happiness.

Marx's *Capital* marked a major turn in the construction of ideologies. From then on, theory began to be "scientifically" built and anchored in what appeared to be the most convincing form of scientific demonstration. Marx presented a new version of the philosophies of history and progress, but at the same time, he introduced something new. He was not a builder of *utopias*: everyone knows his condemnation of the utopian socialisms of his time. His thesis was based on the analysis of contemporary realities; it relied on the discovery of a hidden mechanism in economic life, which remained unconscious for most of the population and liberal economists.

It means that for the first time, the Next World upon which an ideology relied was not built on a historical narrative, but on the "scientific" study of a hidden process (Claval 2008). In the second half of the nineteenth century, other forms of "unconscious" were successively explored by social and human sciences. Sigmund Freud stressed the significance of the unconscious in mental processes and the role of inner forces in the human sciences. Linguists explored in a similar way the unconscious of language; a speaker does not always express his/her personal views. When using ready-made expressions, he/she reflects his/her social position and the opinions of the groups he/she pertains to.

The discovery of unconscious mechanisms was exploited in two ways. First, scientifically, through the deciphering of new causal relations, and then, ideologically, through the use of these unconscious mechanisms to explain for what reasons human beings remained unfulfilled since they suffered from repression. Freud himself did not condemn systematically the role of the superego as a regulatory social mechanism, but for those who developed an ideological interpretation of his thought, all form of repression was bad; the truth of women and men was located in their depth, in their instincts. Hence there developed an ideology of liberation, which differed from the past ideologies of history, since what it described was located in the individuals, not in the society as a whole, and might happen now.

During the nineteenth twentieth century and the first half of the twentieth ideologies were either based on philosophies of history and progress or the liberation of the forces repressed by unconscious social mechanisms. During the last 50 years, the ideologies of history regressed, since the future of humanity appeared increasingly threatened by arms of massive destruction and the ruin of nature. Simultaneously, ideologies of the unconscious, which locate their Next World in the depth of things or beings, developed. Nature itself was increasingly considered as an entity whose inner mechanisms are threatened by human action. Hence the rise of green ideologies, for which the truth of the World does no more lay in humanity, but in the depth of nature.

17.4.4   *Instead of Prophets, Intellectuals*

In revealed religions, beliefs came from the testimony of prophets, who received the Divine Word and diffused it among the multitudes. To complete their work, religions rely on the existence of groups of theologians, *ulemas,* priests, ministers, imams, monks, preachers et al.

The situation of ideologies is not so different in this respect; they were born out of the activity of social thinkers, who imagined stories of Social Contract, or stressed the revolutionary consequences of the discovery of unconscious processes in the mind of people, in social life or in Nature. Their ideas were diffused by professors, journalists, novelists or film-makers through the publication of "scientific" articles and books, novels, the realization of films. Religions have their clerics; ideologies, their intellectuals.

17.5 The Institution of a Normative Order Thanks to the Next Worlds

17.5.1 Heavenly Next Worlds, Divine Law and Morals of Love and Fear

Even if the nature of the Next Worlds they believe in is not the same for religions and ideologies, their functions present similarities. The Next World of religions is not earthly. It is the World of God, or the gods and spirits, and possibly, as Paradise, the home of souls after death – if they had not been thrown down in the subterranean World of Hell after the Judgment.

The Heaven is the place of Integrity and Truth. God, or gods, provide women and men with rules to respect in order to achieve salvation. In this way, they impose a normative and moral dimension to human life (Claval 1988). People have to conform to the divine Law, worship the Lord, obey God's rules, honor their parents, respect one's word. They have to refrain from killing their foes, committing adultery, thieving etc. From one religion to another, the list of moral obligations is not the same, but there is always one. People have several incentives for complying with the divine Law, viz., the fear of the Divine wrath, the respect and love of the Lord and the perspective of salvation, which is generally linked in a way or another to the compliance of moral obligations.

17.5.2 Terrestrial Next Worlds, Rational Behavior and the Pursuit of Happiness or Fulfillment

The earthly next worlds of ideologies do not play exactly the same role; they are not inhabited by a supernatural power. For the ideologies of progress and history, they just offer a view of terrestrial environments, but environments which differ much from those in which human beings live here and now. They offer more harmonious landscapes, more peaceful relations between human beings and more prosperous economies and societies. Hence, for rational people, a simple conclusion: in order to achieve happiness, the best recipe is to take these earthly next worlds as models (Claval 1988).

The next worlds of the ideologies of unconscious are located in the inner strata of the self, society, economy or nature, which reveal the real forces at work and the obstacles they are confronted with. Here also, the lesson is evident. In order to insure their own fulfillment, women and men have to subvert all forms of repression and liberate themselves. In order to create a society of justice, they have to struggle against the capitalist system. In order to restore Nature in its entirety, they have to limit human impact on local, regional and global environments.

Next Worlds, those of religions and those of ideologies, had similar functions, viz., to give a normative dimension to human action, through the revelation of a more real and true World (the Heaven), or a more harmonious one, that what we shall experience here below (utopia). For religions, the stress is on fear and love; for ideologies, it is on reason and computation.

17.6 The Relations Between Religions and Ideologies: Complementarity and Competition

Relations and ideologies share many characteristics. Both are systems of beliefs, both are built on a dual interpretation of the World and the existence of an elsewhere – heavenly or terrestrial. Both give a sense to life – salvation in the Next World for religions, happiness and human fulfillment here below for ideologies. At the same time, they deeply differ. Religions generally rely on a dual conception of human nature, when ideologies stress its unity and its materiality. Religions often equate God and metaphysic Reason, but give room to mystery. Ideologies stress the role of scientific reason.

Because of these similarities and differences, religions and ideologies share a similar aim – to give a sense to human life – but along different lines. They are at the same time complementary and competitive.

17.6.1 The Complementarity of Religions and Ideologies

In a way, religions and ideologies appear as complementary. Religions are mainly concerned with the salvation of the soul and its life after death in a heavenly Paradise or a subterranean Hell. Ideologies tell what happens, or will happen, here below in the future of Utopia. They do not speak of salvation, but of happiness and for the ideologies of History and Progress, of the happiness that our children or great children will enjoy in the future.

As a result, people may believe at the same time in a religion and an ideology. Religion explains what each individual has to do in order to save her/his soul; the ideology shows how, thanks to their political involvement, the hard life of people today will give way to a better condition for their descendants.

This position explains why, since the seventeenth or eighteenth century, ideologies of Progress and History had coexisted peacefully with Christianity in the Western World. At the same time women and men labored hard for the salvation of their soul and contributed to the betterment of their lives and those of their children and grandchildren in this World below. It was for the same reason that Western ideologies appeared as seductive to a part of the colonized populations. They had the feeling that it was possible to remain faithful to the religion of their fathers, either Buddhism, Confucianism or Islam, and to believe in a better future for their families.

In such a context of complementarity, religions appeared more as systems of personal relief, and ideologies as a basis for collective action – for politics.

17.6.2 *The Competition Between Religions and Ideologies*

Between religions and ideologies, competition came along with complementarity. There were several reasons for that. The first one was clear: the perspective of a bright future in this World below reduced the seduction of the salvation of the soul in the Next World. The success of ideologies was conducive to a progressive restriction of the religious field to the personal endeavor for salvation and to the substitution of ideologies to religious believes. This position is the well-known thesis of the secularization[1] of modern societies – or expressed in another way, in the thesis of the disenchantment of the World (Weber 1904–1905; Cox 1965; Wilson 1966; Luckman 1967; Bellah 1970; Berger 1971; Gauchet 1985; Bruce 2002). It explains the decline of attendance to religious services, the Eucharist, the communion, and in the belief in a Next World.

This interpretation knew a great success in the second half of the twentieth century. It was recently revamped in France by Marcel Gauchet (2003).

During the last generation, the thesis of the secularization of modern societies lost, however, a part of its credibility (Berger 1999; Kepel 1991; Martin 1991). The ideologies of Progress and History suffered from the threats on the future of humankind and nature. Religions knew a renewal linked to their move back towards their origins, especially for fundamentalist groups, or to the creation of new forms of religious expression, on market segments (Lenoir 2003) through charismatic or evangelical movements as expressed by the multiplication of new denominations.

At the same time, new ideologies grew and replaced the philosophies or History and Progress. These are based on the idea of liberation of the self and the fulfillment of individuals or on the protection to be given to nature in order to save our planet.

Instead of speaking of the competition between religions and ideologies as conceived by the secularization thesis, people in the Western World are increasingly conceiving religions and ideologies as beliefs offered on a market where believers choose the set which appears the most convenient for them. For Rodney Stark and Laurence Iannacone (1994), as summarized by Knippenberg (2005: 44) "religion is subject to similar mechanisms as economy: people have religious needs, which result in a demand for certain activities, services and institutions. In this way, a religious market can direct the supply of their particular religious goods and services to their specific target groups or market segments. It is not surprising at all that this line of thought originated in the United States." Knippenberg mentions that the comparison of religion and the economy was introduced by Finke and Starke (1988).

---

[1] For Jose Casanova (1994), the word secularization has three meanings: (i) the separation between the religious and the secular fields; (ii) the decline of religious practices and beliefs; (iii) the negation of religion in the private sphere.

17.6.3 How Religions and Ideologies Interact

A deeper analysis stresses also the interaction between religions and ideologies. Religions have often borrowed some features to the ideologies with which they competed. Roman Catholicism developed in this way a growing interest in the social dimension of Christ's message since Leon XIII's Encyclical *Rerum Novarum* of the 15th of May 1891. It culminated in the theologies of liberation which flourished mainly in South America in the 1950s, 1960s and 1970s. This conversion of a part of the Roman Catholic Church to the rhetoric of social struggles and a lesser interest in the salvation of individual souls, had a double effect on Roman Catholics. It seduced a part of the working class, but left unsatisfied those who were endeavoring for the salvation of their souls.

The interaction between religions and ideologies is now taking new forms. The philosophies of History and Progress appear no more as a good platform of beliefs. New religious movements and sects (Wilson 1990) draw a part of their inspiration from the ideologies of self-fulfillment or of those of nature's protection (Corten 1997). In Brazil, for instance, the Universal Church of the Kingdom of God, founded in 1977 by Edir Macedo, owes a part of its success to the incorporation of new themes into old Christian ones. Its theology underscores the omnipresence of Evil forces, from which one frees oneself through practices of exorcism. It is at the same time a theology of prosperity. Poverty is diabolical, and God is portrayed as a rich parent who loves his children and wishes all of them prosperity Success is a proof of the divine presence in the life of people. The Theology of Prosperity restores the body to favor, as explained by Ricardo Mariano

> A perfect health, a material prosperity and the felicity are the rights of the Christians as announced in the Bible [...]. Through the mediating sacrifice of His Son, God has already done all that He could do for humanity, since his favors have become from then on affordable for men and women. They have to decree, require, vindicate, be faithful in their quest and generous in their offerings, as prescribed by God, in order to enjoy the benedictions they are entitled to. (Mariano 1998: 225)

And Mariano goes on:

> The theology of Prosperity is promoting a strong inversion into the axiological Pentecostal system. It emphasizes the move back to the faith in this life, instead of giving priority [...] to the abrupt apocalyptic end of this World and the bliss of the blessed elected to the celestial paradise. It contrasts with the traditional biblical terms of self-sacrifice [...] asceticism, negation of carnal pleasures and Worldly things. Instead of that, it values the faith in God as a mean to obtain physical wellbeing, wealth, happiness and power. (Mariano 1998: 226)

17.6.4 Instead of Secularization, a Dynamic Relation Between Systems of Beliefs

The above interpretation of religious evolution in terms of secularization does not fit the most recent phase of history. The joint approach of religions and ideologies we propose offers a more satisfactory interpretation of contemporary changes.

Paraphrasing Knippenberg (2005), we may say: people have spiritual needs, either religious or ideological, to give a sense to their life. This results in a demand for certain activities, services and institutions. In this way, a market of beliefs, either religious or ideological, can direct the supply of their particular religious or ideological goods and services to their specific target groups or market segments (Hervieu-Léger 2001).

17.7 Conclusion

All systems of beliefs, either religious or ideological, are based on similar, and different, topologies, viz., the existence, beyond or below observable realities, of a more real world, which serves as a model for the people living here and now, and introduces norms and objectives in individual and collective life.

This broader conception of systems of beliefs is conducive to: (a) a better interpretation of Western civilization and its religious dimension and correlates since the seventeenth century and (b) an understanding of the spread of Western ideologies all over the World as a form of globalization, and (c) an explanation of the contemporary syncretism between religions and ideologies. Since they do not rely on a set of beliefs fixed by a Revelation, but on reflection, meditation and self-discipline, Eastern Religion can more easily mix with modern ideologies than Christianity or Islam: hence the incorporation of Eastern components in many contemporary sets of beliefs, those of New Age for instance (Ferreux 2000).

References

American Heritage Dictionary of the English Language. (2011). Boston: Houghton-Mifflin.
Bellah, R. N. (1970). *Beyond belief: Essays on religion in a post-traditional world*. New York: Basic Books.
Berger, P. L. (1971). *La religion dans la conscience moderne: essai d'analyse culturelle*. Paris: Editions du Centurion.
Berger, P. L. (Ed.). (1999). *The desecularization of the world: Resurgent religion and world politics*. Grand Rapids: B. Eerdmans Publishing.
Besse, J.-M. (2009). *Le goût du monde. Exercices de paysage*. Arles/Versailles: Actes Sud/ENSP.
Bruce, S. (2002). *God is dead. Secularization in the West*. Oxford: Blackwell.
Casanova, J. (1994). *Public religion in the modern world*. Chicago: University of Chicago Press.
Claval, P. (1980). *Les mythes fondateurs des sciences sociales*. Paris: PUF.
Claval, P. (1985). Idéologie et sciences sociales: Quelques points de vue. *Cahiers de Géographie de Québec, 29*(47), 185–192.
Claval, P. (1988). Notre monde et les autres mondes: la pensée normative et l'espace. *Revue des Sciences Morales et Politiques, 143*(1), 7–18.
Claval, P. (1999). Les religions et leurs substituts. Réflexions géographiques sur la transcendance et sur l'immanence, Colloquio Internacional "Geografia de las religions," Santa Fe, 11 al 15 de Mayo de 1999 (pp. 21–56). Santa Fe: Universidad Catolica.
Claval, P. (2001). The geographical study of myths. *Norsk Geografisk Tidsskrift, 55*(3), 138–151.

Claval, P. (2008). *Religions et idéologies. Perspectives géographiques.* Paris: PUPS (Publications de l'Université de Paris-Sorbonne).
Comte, A. (1822). *Plan des travaux nécessaires pour reorganiser la société.* Paris, s. n. New edition, Paris: Aubier-Montaigne (1970).
Corten, A. (1997). *Le Pentecôtisme au Brésil. Emotion du pauvre et romantisme idéologique.* Paris: Karthala.
Cox, H. (1965). *The secular city. Secularization and urbanization in theological perspective.* New York: Macmillan.
Dubuisson, D. (1998). *L'Occident et la religion. Mythes, sciences et idéologies.* Bruxelles: Complexe.
Duméry, H. (1985). Religion et idéologie. *Encyclopædia Universalis, 15,* 3754–3758.
Eliade, M. (1949). *Traité d'histoire des religions.* Paris: Payot.
Ferreux, M.-J. (2000). *Le New Age. Ritualités et mythologies contemporaines.* Paris: L'Harmattan.
Finke, R., & Starke, R. (1988). Religious choice and competition. *American Sociological Review, 63,* 761–766.
Foucault, M. (1985–1986). Other spaces. The principle of heterotopia. *Lotus International,* 48–49.
Gauchet, M. (1985). *Le désenchantement du monde. Une histoire politique de la religion.* Paris: Gallimard.
Gauchet, M. (2003). *La condition historique.* Paris: Stock.
Hervieu-Léger, D. (2001). *La religion en miettes, ou la question des sectes.* Paris: Calmann-Lévy.
Hobbes, T. (1651). *Leviathan.* London.
Kepel, G. (1991). *La revanche de Dieu. Chrétiens, juifs et musulmans à la reconquête du monde.* Paris: Le Seuil.
Knippenberg, H. (Ed.). (2005). *The changing religious landscape of Europe.* Amsterdam: Het Spinhuis.
Lenoir, F. (2003). *Les métamorphoses de Dieu. La nouvelle spiritualité occidentale.* Paris: Plon.
Locke, J. (1690). Two treatises on government. P. Laslett (Ed.). Cambridge: Cambridge University Press, 1961.
Luckman, T. (1967). *The invisible religion. The problem of religion in modern society.* New York: Macmillan.
Manuel, F. E., & Manuel, F. P. (1979). *Utopian thought in the Western world.* Oxford: Blackwell.
Mariano, R. (1998). Eglise Universelle du Royaume de Dieu. De la banlieue de Rio de Janeiro à la conquête du monde. *Cahiers du Brésil Contemporain, 35–36,* 209–229.
Martin, D. (1991). The secularization prospect issue: Prospect and retrospect. *British Journal of Sociology, 42,* 465–474.
Marx, K. (1867). *Capital.* London: Penguin, 1967; trans: Fowkes, B., Vol. 1.
More, T. (1512), *Utopia.* New York: Norton, 1975; translated and edited Adams, R. N.
Ricœur, P. (1985). Mythe. L'interprétation philosophique. *Encyclopædia Universalis, 12,* 883–890.
Rousseau, J.-J. (1762). *Le Contrat social.* In Halbwachs (Ed.). Paris: Aubier, 1943.
Stark, R., & Iannacone, L. (1994). A supply-side reinterpretation of the 'secularization of Europe'. *Journal for the Scientific Study of Religions, 31,* 230–252.
Weber, M. (1904–1905). *The Protestant ethic and the spirit of capitalism.* London: Routledge, 2001; trans: Parsons, T.
Wilson, B. R. (1966). *Religion in a secular society: A sociological comment.* Harmondsworth: Penguin Books.
Wilson, B. R. (1990). *The social dimensions of sectarianism: Sects and new religious movements in contemporary society.* Oxford: Clarendon Press.

Chapter 18
Sacred Space and Globalization

Alyson L. Greiner

> *The present epoch will perhaps be above all the epoch of space.*
> (Michel Foucault, Of Other Spaces)

18.1 Introduction

Not long ago I found myself at the Detroit Metropolitan Airport with a few hours of layover time. On previous stopovers at this airport I had noticed signs to a religious reflections room. These signs had piqued my curiosity and I had regretted not being able to glimpse what this space was like. Although I had noticed signs for airport chapels at other airports over the years, I had not given much thought to them. But I knew that Detroit had a sizable Muslim population, and I wondered to what extent the prayer room would be an ecumenical, multi-faith space particularly with respect to airport space itself, which uneasily straddles the boundary between public and private space.

As I emerged from the underground tunnel into the main terminal building and confirmed the details for my connecting flight, I spotted the sign (in English and Chinese) indicating the direction of the religious reflections room. I followed the arrow on the sign, which led me in the direction of restrooms and toward a narrow corridor with a very institutional feel to it. The signage ended. I immediately felt out of place and retraced my steps to make sure I had taken the proper direction. I had. Still unsure of myself, I continued on. I heard people approaching me from behind, and moved to one side to let them pass. They were two male security agents and as they walked by one looked me in the eyes and asked me where I was going. "To the religious reflections room," I said. "Take the elevator to the third floor," he directed as he passed by me.

Beyond a bend in the hallway I found the elevator and took it to the third floor. When I stepped out of the elevator, another sign pointed me towards the religious reflections room. After a few more steps, I reached the room. A notice on the door admonished airport employees not to use this space for their work breaks. I opened

---

A.L. Greiner (✉)
Department of Geography, Oklahoma State University, Stillwater, OK 74078, USA
e-mail: alyson.greiner@okstate.edu

the door and stepped inside a square, dimly-lit room. I immediately spotted a compass rose on the floor, adjacent to the entrance. Each wall also carried a label indicating the cardinal direction, creating a slight incongruity with the directional labels on the compass rose. Most of chairs were bunched together against the west wall. A few other chairs, including one stacked with several prayer rugs, lined the south wall. I took a seat and contemplated this distinctively Muslim space.

This experience prompts many questions about mobility, globalization, space, place, and religion. How did this space come to be? What is the story behind the labeling of the walls and the placement of the compass rose on the carpet? Who uses this space, how often, and to what extent does it function as an ecumenical religious space as the name "religious reflections room" suggests? Marc Augé (1995) writes of airports as non-spaces—social domains which cater to and promote highly prescribed individual behavior. Think of the protocol of moving within airports, not to mention all the steps we follow (ritually!) to get through airport security. Is Augé's assessment appropriate or do we find that there is a sense of community held by those who use this space, or perhaps even that a community of airport workers use this space? Do people perceive it as a sacred space? What does the practice of religion in this space mean for our understanding of the relationships among private, sacred, public, and secular space?

Using these questions as a point of departure, this chapter provides a critical review of some facets of the intellectual history of sacred space. The literature on this topic spans the work of geographers, philosophers, religious historians, theologians, and many others. This review is necessarily partial and selective and it seeks to provide a concise distillation of the major trends, debates, and current research directions influencing the study and analysis of sacred space. It proceeds in a largely chronological fashion. In the process, the chapter reclaims a place for some of the key early works on sacred space and addresses the intersection of sacred space with the rise of new spiritualities, cyberspace, transnationalism and commodification.

18.2 Approaching Sacred Space

Among scholars there exists a tendency to use the term "sacred space" without defining it. Whether this is the result of assuming that the reader will intuitively understand the concept and not need a definition or because the term is fraught with a messiness and complexity that is easier to elide remains unclear. Often the practice of considering sacred place and sacred space as virtually synonymous prevails. For example, the entry on sacred space in the *Encyclopedia of Religion* begins with this observation: "A sacred place is first of all a defined place, a space distinguished from other spaces" (Brereton 1987: 7978). At other times, sacred space is treated rather generically, as a space for the conduct of any religious practice (O'Leary 1996). The propensity to broaden what is meant by sacred space leads Peter W. Williams to question the utility and value of the term, and consider jettisoning it in favor of "ritual space" (2002: 607).

18.2.1 Foundations: Van der Leeuw and Eliade

Although Mircea Eliade, for many years a historian of religion at the University of Chicago, is most closely associated with the term "sacred space," the Dutch religious scholar and philosopher of religion Gerardus van der Leeuw provided one of the early treatises that specifically addresses sacred space as a meaningful component in the experience of religion.[1] Van der Leeuw offers significant insights in his two-volume book, *Religion in Essence and Manifestation*, originally published in German in 1933 under the title *The Phenomenology of Religion*. Van der Leeuw (1963) focuses much of his study on an examination of the topics of object, subject, and their interconnections. By object he means the "*highly exceptional and extremely impressive 'Other'*" (van der Leeuw 1963: 23; italics in original). His discussion of sacred space comes in the section on object-subject interrelations.

Van der Leeuw defines sacred space as a zone where people perceive power to be situated. This is an auspicious or divine power and he recalls the Latin phrase "*liberatus et effatus*"—a reference to the process by which augurs in Roman antiquity would "set free" and, therefore, consecrate a space and delineate it with metaphorical or physical boundaries (1963: 393). Thus, sacred space in his view is a perceived and bounded space often marked by a threshold. Sacred space is also a refuge. But sacred space is not monolithic. Rather, it is a zone of differential power, where some positions are perceived as being more sacred than others. Sacred space may exist in auspicious as well as mundane spaces. Arguably, van der Leeuw's most significant contribution to the concept of sacred space comes in his recognition that sacred space can be a real, material space as much as it can be an imagined space, though he does not use these exact terms. As he shows in his discussion of pilgrimage, a sacred space can be a spiritual, emotional, or personal destination (1963: 402). This dimension of sacred space gained resonance for me in my own research on the Dunkard Brethren, a conservative anabaptist Christian community for whom this world is a profane space (Greiner 1991). To them, sacred space is the beyond, the other world associated with everlasting life in the Kingdom of Heaven.

The sacred/profane binary specifically, and spatial awareness more generally, provide important anchors to the argument in Mircea Eliade's book, *The Sacred and the Profane: The Nature of Religion* (1959). In that work, Eliade not only describes the sacred in contradistinction to the profane, but does so in an explicitly geographical way, making reference to the spatial and territorial aspects of sacred space. For Eliade, "*sacred* and *profane* are two modes of being in the world, two existential situations assumed by man in the course of his history" (1959: 14).[2] The development of these modes of being is bound up with the ways in which people

---

[1] I exclude Durkheim who identifies the sacred and profane and speaks of "sacred things," (1915: 52ff) but does not explicitly address sacred space.

[2] Eliade credits Otto for advancing the concept of the sacred in his 1917 book, *Das Heilige*, subsequently published in English as *The Idea of the Holy*. Even so, Eliade's sacred/profane binary has much in common with Durkheim's (1915) work.

and societies create worlds or generate social order from their experience, a process from which they derive ontological and existential security. Eliade identifies secularization as one of the forces responsible for transforming what was once a sacred world into a profane one (1959: 51). In his view and prior to the modern period, the identification of sacred space provided a means by which *homo religiosus*—Eliade's term—perceived the world. As Eliade argues, the construction of sacred space involves homologizing processes—behavior that repeats the work of the gods and sanctifies space. Sacred space develops where a "hierophany" or manifestation of the sacred occurs (1959: 20–21, 31). This manifestation consecrates the space—giving it form, order, and meaning—and distinguishes it from profane space. Thus, sacred space introduces and is characterized by spatial variation and heterogeneity whereas profane space remains homogeneous and undifferentiated (Eliade 1959: 20).

18.2.2 Arguments Against and for Eliade

In a strident and insightful critique of Eliade's work, Larry Shiner (1972) presents three basic complaints. First, he takes issue with Eliade's conceptualization of profane space as homogeneous, pointing out that "we do not natively experience space as a kind of container in which we find ourselves along with a collection of objects" (Shiner 1972: 427). In his view, Eliade would have been better served to speak of "lived space" for "we are not 'in' space as shoes are in a box … Through our bodies we are intimately intermingled with our surroundings" (Shiner 1972: 427). So-called profane space is not experienced as homogeneous space at all, he argues. A second complaint of Shiner's centers on Eliade's tendency to explain modern religious behavior as an unconscious "survival" or residue that speaks to the legacy of our much more religious past.[3] As Shiner explains,

> …much of the centering and differentiation of zones of varied spiritual significance which occurs in the modern period is neither a direct continuation nor a mere vestige of the past but a new expression of our fundamental human spatializing activity. (1972: 431; italics in original)

Shiner's third complaint draws attention to the inability of the sacred/profane dichotomy to capture the political dimensions of sacred space, and the blurring that sometimes occurs when a sacred center becomes transformed into a political center as well.

In addition to these criticisms, Eliade's work has suffered numerous other condemnations. Perhaps first and foremost among these involves the simplistic dualism between the sacred and profane; a dualism that belies the complexity and nuances not just of sacred space, but of sacredness more broadly. A second

---

[3]This appears to be a reference to anthropologist E. B. Tylor's, concept of "cultural survivals"—beliefs, practices, or artifacts that had their origins in an earlier time but persisted as anachronisms. Eliade was familiar with Tylor's work.

major criticism focuses on the essentialist and elitist aspects of Eliade's work. Although Eliade studied sacred space from a cross-cultural perspective, he did so in order to distill the essence of sacred space. He writes, "But for our purpose it is not the infinite variety of the religious experiences of space that concerns us, but, on the contrary, their elements of unity" (1959: 63). In addition, he showed some disdain for the "little religions," his term for new spiritual groups and movements (1959: 206).

There are still other criticisms of Eliade's work,[4] but it seems worthwhile to draw attention to several of his contributions, which often go unrecognized. Significantly, Eliade does associate the act of consecrating space with mechanisms for taking possession of it, though he does not pursue the political ramifications of this. As he explains (1959: 32), "a territory can be made ours only by creating it anew, that is, by consecrating it …. The Spanish and Portuguese conquistadores, discovering and conquering territories, took possession of them in the name of Jesus Christ." Similarly, he identifies the settlement process itself as a form of consecration: "… to organize a space is to repeat the paradigmatic work of the gods" (1959: 32). It is fair to say that with these observations Eliade recognizes that the making of sacred space involves possession, dispossession, and political motivations. His work also opens the door to studies of the processes of sacralization as well as desacralization.

From a geographic perspective, Eliade brings to his work an awareness of the importance of scale and he shows that sacred space is constructed at the level of the country, city, and basilica or temple. Moreover, he recognizes the importance of creating sacred space to the community and individual. Building a home or a hearth, for example, incorporates understandings of sacred space and orientation and serves a symbolic purpose by repeating the sacred acts of cosmicization (Eliade 1959: 32). Clearly, the idea that sacred space is a social construction is at the core of Eliade's work. What unsettles academics today, however, seems to be the teleological nature of his thought and language, as, for example, when he writes that people "… are not free to *choose* the sacred site … they only seek for it and find it by the help of mysterious signs" (Eliade 1959: 28). When reading Eliade it helps to keep in mind that his statements apply to an idealized, and not unproblematic, representation of premodern society where myth and religion saturated one's existence and where human agency was not denied, as Nelson suggests (2006), but set within teleological parameters.

Philosophically, what are we to make of Eliade's work? It has been characterized as both ontological (Lane 2001), and structural (della Dora 2011). These characterizations are certainly appropriate, but Eliade writes from the standpoint of a historian of religion and his aim is also phenomenological because he endeavors to understand the experience of sacred and profane space. His approach, however, is strongly guided and informed by essentialism. Eliade states, "Our primary concern is to present the specific dimensions of religious experience, to bring out the

---

[4] For example, Smith (1978) faults him for focusing too heavily on centers as opposed to the peripheries in his discussion of sacred space.

differences between it and profane experience of the world" (1959: 17). Indeed, the intellectual roots of phenomenology and essentialism are closely intertwined. As Mohanty reminds us, "phenomenology started out with the program of describing essences and essential structures of … phenomena" (1997: 1). Even Van der Leeuw sees part of the role of the phenomenologist as one who "assigns names: sacrifice, prayer saviour, myth, etc." (1963: 688). Although the study of sacred space began as an essentialist and phenomenological project, it has long since shed its essentialism for more grounded, reflexive, and alternative ways of knowing.

18.2.3 Early Geographical Works

Among North American geographers, Yi-Fu Tuan contributed to early discussions of sacred space in his essay, "Sacred Space: Explorations of an Idea." Tuan (1978) uses a series of categories, sometimes based on oppositional terms, to structure his essay and reflect on the qualities that people associate with sacred space: order and wholeness, light and darkness, structure and anti-structure, power and purity, and so on. Consequently, the style of his essay recalls the work of both Van der Leeuw and Eliade. One of Tuan's most noteworthy contributions comes toward the end of the essay when he, too, challenges Eliade's sacred/profane binary and suggests that we should be open to the possibility that sacred space might exist in typically secular spaces such as wilderness, suburbs, and the state.

Two other geographers, Richard Jackson and Roger Henrie, define sacred space as "that portion of the earth's surface which is recognized by individuals or groups as worthy of devotion, loyalty or esteem" (1983: 94). Although they do not appear to have been familiar with Shiner's work, they offer some similar observations. Like Shiner, they reject Eliade's characterization of how people experience profane space and point to the importance of what they call "mundane space," which is analogous to Shiner's "lived space." The main objective of Jackson and Henrie, however, consists of presenting a typology of sacred space. They identify three categories of sacred space: mystico-religious, homelands, and historical. Their approach is strongly functional—reflecting the behavioral turn within geography at the time—and somewhat reductive in the way that they use subcategories such as the level of sanctity, degree of permanence of the site or space, and its areal extent, for example, to "neatly" classify the different kinds of sacred space. With such a rigid typology, however, they fail to acknowledge that these kinds of sacred space are not mutually exclusive. That is, homelands can have mystico-religious and even historical characteristics. They also claim that the number and importance of sacred spaces vary by scale. They write, "an individual recognizes the greatest extent of sacred space at the local level in his present home, ancestral home and church, as well as important historic sites in the local community and historical or religious events important to him personally" (Jackson and Henrie 1983: 97). Their neglect of diasporic communities (somewhat ironic given the fact that they include Mormon examples) diminishes the relevance of this observation. It is not just the extent or number of sacred spaces that matters, but also the intensity of feeling, attachment, and as Van der Leeuw or Tuan would say, their power.

18.2.4 Re-thinking Sacred Space

Remarkably few geographers have engaged with the work of the eminent historian of religion, Jonathan Z. Smith. Of particular relevance are his books, *Map is Not Territory* (1978) and *To Take Place: Toward Theory in Ritual* (1987). The former consists of a collection of several of his previously published articles; the title comes from his lecture on the topic in 1974. That essay probably should be required reading for all scholars interested in religion. The essay is not directly about sacred space, but matters to a discussion of sacred space because it addresses the importance of cognitive maps, or as Peter Jackson (1989) has called them, "maps of meaning."

The premise of Smith's essay is that scholarship on religion can be defined by two kinds of cognitive maps: *locative maps* and *utopian maps*. The locative map "is a map of the world which guarantees meaning and value through structures of congruity and conformity" (Smith 1978: 292). With this metaphor, Smith recognizes that scholarship on religion—particularly so-called native religion—has been blinded by its rather single-minded focus on essences, commonalities, and Western conceptual frameworks that are imposed on other societies and traditions. The utopian map, in contrast, "perceives terror and confinement in interconnection, correspondence and repetition" (1978: 309). It represents resistance to such totalizing worldviews. To resolve this tension, Smith proposes a third map which he does not name, but describes as opening up the opportunity for more creative thought rather than shoe-horning religion into Western conceptual categories.

For Smith, these ideas of locative and utopian maps have spatial parallels and appear to have taken shape as he studied changes in the nature of Jewish sacred space, a topic subsequently taken up by Baruch M. Bokser, a rabbinical scholar. Bokser (1985) specifically explores how the creation of sacred space changes from a locative to non-locative activity. He proceeds by first defining sacred space as "the extended zone of the sacred" (1985: 279). Then, using a hermeneutical approach, he shows that through a process of redefining the sacred, post-biblical scholars effectively de-centered religious practices. This had the effect of allowing sacred space to be spatially extended to larger areas beyond the Temple Mount and made it possible for people to create temporary sacred spaces through their activities and ritual behavior. As Bokser explains, "the loss of a sacred center yields not only the negative experience of 'exile,' but positive efforts to locate the sacred in other places and to provide new centering structures" (1985: 288). In his view, "the rabbis preserved the idea of sacred space in a manner that enabled the group to function without a single center" (1985: 299). As suggested by Smith, and developed by Bokser, and in strong contrast to the work of Jackson and Henrie, sacred space is not necessarily a spatially fixed domain. Moreover, Bokser draws important attention to the purposeful reinvention of sacred space.

If Smith and Bokser highlight the lack of fixity of sacred space, then Chidester and Linenthal (1995) argue that sacred space is the product of contestation and differential inflections of power. In fact, they provide not only one of the most thorough reviews of sacred space, but also one of the most significant reconceptualizations of it. Through a combination of deconstruction and pastiche, they turn Eliade's modernist truths, if you will, about sacred space on their head. In contrast to Eliade, they take the view that "the most significant levels of reality in the formation of sacred space are not 'mythological' categories, such as heaven, earth, and hell, but hierarchical power relations of domination and subordination, inclusion and exclusion, appropriation and dispossession" (1995: 17). They think that Eliade's views so distort the way the world works and neglect the considerable effort individuals and groups devote to making sacred space that these misrepresentations have the effect of completely missing "the symbolic violence of domination or exclusion that is frequently involved in the making of sacred space" (1995: 18).

Although Chidester and Linenthal are quick to dismiss Eliade's work, they credit Van der Leeuw with providing the conceptual foundations of their theoretical framework. That is to say that they construct their understanding of sacred space as contested space on some observations that Van der Leeuw makes about sacred space. These observations include understanding sacred space as a kind of property; as a space defined by practices of inclusion and exclusion; as a space associated with deep emotional attachments and, when those attachments are severed, a sense of exile; and as a space created by intentional political decisions about the value of a certain site or area (Chidester and Linenthal 1995: 7–8). Thus, they see the study of sacred space as anchored to two main epistemologies that they refer to as a "poetics" and a "politics," the former associated with the experiential aspects of sacred space, and the latter associated with its contested quality (1995: 6).

Of the many valuable studies in this collection, Chidester's own chapter charts a new and perceptive direction in the study of the production of sacred space that deserves more attention. Through an examination of symbolic sacred space that builds on Smith's contrast between locative and utopian cognitive maps, he shows how America has been represented as a sacred space in South Africa. More specifically, he tracks the representation of sacred space as "spaces of liberation" and shows how "through local, African initiatives, these specific sacred sites were relocated and grounded in South African space" (1995: 273). With the press of ongoing globalization, the process of transposition—both symbolic and real—may likely have continued relevance in our understandings of sacred space, its production, and re-creation.

For about two decades now, Lily Kong has influenced the contours and direction of the geographical scholarship in the areas of religion, sacred space, and identity (1993, 2001a, b, 2002). She has been a key figure in the re-theorization of sacred space not just as contested space, but also as an arena for negotiation with and even resistance to state actions or policies affecting sacred spaces (Kong 1993). Her research has also refocused attention on places and spaces that she calls the "unofficially sacred" (Kong 2002: 1573).

Following Kong's lead, Julian Holloway (2003) extends the focus on unofficially sacred spaces through a study of New Agers. Like Chidester and Linenthal, Holloway marshals evidence that destabilizes some of Eliade's claims about sacred and profane space. For example, he shows that for New Agers the sharp dichotomy that separates sacred and profane space effectively dissipates as they experience mundane events. These events function as hierophanies that enchant or sacralize everyday or profane space. Of particular significance to Holloway's work is his consideration of "… the sensuousness and embodiment of sacred space" (2003: 1963). For New Agers, embodied practice and performance are necessary steps that make possible the attainment of spiritual knowledge. Thus, the body itself plays a role in the construction of what Holloway terms a "sacred space-time" (2003: 1967). In this way, Holloway highlights the salience of nonrepresentational theory to the study of sacred space, pointing out the importance of paying attention to what is heard, felt, or sensed in addition to that which is simply visual. With a nod to actor-network theory he illustrates how the profane objects around and in the midst of these New Agers become part of their spiritual space. As he explains, "enacting the sacred space-times involves a collective agency where heterogeneous elements are coinvestors in sanctification" (2003: 1972). His subsequent research on nineteenth century spiritualism and the séance advances the conceptualization of sacred space as a vitalized space (Holloway 2006).

For Belden C. Lane (2001), vitality infuses place and space. This vitality stems in part from the ways in which we encounter them: as numinous, storied, contested, magical and mundane. He, too, recognizes the material context as a "participant in the process of perceiving and experiencing sacred places" (2001: 44). As Lane makes clear, attention to these reciprocities between the body and its surroundings is needed to avoid one of the pitfalls of social constructionist approaches: the tendency to focus exclusively on groups and individuals in conflict and, in doing so, ignore the importance of place (2001: 44).[5]

Lane writes from the perspective of a Christian who sees (and seeks) evidence of God in nature, who cherishes the human experience of the mystery of place, and advocates a Christian "theology of place" (2001: 242–255). His scholarship reveals that spatial awareness has deep roots in Christian thought and this prompts him to call for a more deliberate respatialization of theology that directly recognizes the immanence of God rather than a more generalized divine transcendence, is explicitly built on "a deep respect for place," and incorporates "questions of poverty and eco-justice" (2001: 247). Lane brings to the study of sacred space a conscious integration of awareness of the social and natural environment. In doing so he presses for a sacred ecology, to borrow Fikret Berkes' (2008) term, within a Christian imaginary. It is a surprisingly short step from such imaginaries to the idea of globalization itself as a kind of transcendental force.

---

[5] Lynne Hume's important work on Wiccan sacred space also shows that not all sacred space is contested space and that this idea "… is untenable when speaking about a witch's sacred circle" (1998: 310). The sacred circle is the physical and imaginary space that Wiccans use to enable communication between the material and metaphysical worlds.

18.3 Globalization and Sacred Space

Those who frame globalization within a religious context tend to position the market as a kind of omnipotent force and situate its theology in neoliberalism, its salvific message in capitalism, and its institutional apparatus in the IMF and other global financial organizations (Cox 1999; Hopkins 2001; Csordas 2007). Secularists among us may gravitate to such views, but more meaningful for this study is an approach that incorporates an understanding of globalization as a set of integrated social, economic, and political processes which alter secular and sacred space. Globalization remains bound up with technological change and technology affects the pace of globalization. More specifically, globalizing processes make possible the opening up of new sacred spaces, including for example, new virtual sacred spaces. Globalizing processes also contribute to the stretching of sacred space through transnational flows. Thirdly, globalizing processes bring to the fore the importance and impact of the often intertwined processes of the commodification and hybridization of sacred space.

18.3.1 Sacred Space and Virtual Space

With respect to new virtual spaces, research on the relationship between computer-mediated communication and sacred spaces remains nascent. Surprisingly little research has been conducted on the popular virtual world Second Life, where avatars form the basis for interaction with other people and places, and participate in business activities and other events.[6] Second Life includes a significant religious dimension such that one can walk the Via Dolorosa, go on a virtual hajj, visit a Coptic Christian church or a Jewish synagogue, or spend time in a tower of meditation. Numerous questions about the representation and experience of sacred space in Second Life abound, not unlike many of those posed at the start of this chapter.

Another set of intriguing questions surrounds the maintenance, alteration, or even annihilation of socio-spatial boundaries in online environments. Do our understandings of inclusion, exclusion, and access to sacred sites or sacred spaces need modification? In an insightful study, Heinz Scheifinger (2009) shows that online activities can give certain groups who might otherwise be excluded from sacred space access to a computer-mediated version of that space. Scheifinger studies the online presence of the Jagannath Temple, a major pilgrimage destination and sacred center in eastern India, in relation to *darshan*. In Hinduism, the practice of darshan constitutes the reciprocal act by which a worshiper sees a deity and simultaneously is also seen and blessed by that deity. A reason for the growth of the Jagannath Temple as a pilgrimage site stems from the desire of Hindus to receive darshan there. However, the temple grants entry only to ethnic Hindus. In spite of this, the growth of services on the

---

[6] For an introduction to Second Life see Radde-Antweiler (2008).

Internet potentially provides a means by which all Hindus, regardless of ethnicity, can receive darshan. Of course, some people maintain that virtual access to the temple and image of the deity is not the same as visiting the temple in person. Although Scheifinger does not incorporate the views of non-ethnic Hindus who do receive darshan online or pursue the broader implications of access to or exclusion from sacred space within the context of religious and political identities, his work suggests that we still have much to learn about the construction of sacred space on the internet, the extent to which the provision of such spaces might be used to justify or rationalize real-world exclusionary practices, and even the issue of jumping scale—reframing the scalar context of sacred space for political or ideological purposes.

18.3.2 Sacred Space and Transnationalism

There now exists a sizable and growing literature on transnationalism and its impacts on livelihood, identity, and religion. The work by Vasques and Marquardt (2003) remains a very important contribution, but see also the special issue in *Environment and Planning A* (2006). Another special issue appearing in *Anthropological Theory* (2007) treats religion and globalization through the perspective of "transnational transcendence" (Csordas 2007: 266), but Elaine Peña's (2008) work expressly addresses sacred space. More specifically, she traces the production and meanings of a transnational sacred space among Guadalupanas/os in a Chicago suburb. The construction, use, and performance of activities at a "Second Tepeyac"—a replica of the site where the Virgin of Guadalupe is believed to have appeared to Juan Diego—produces a sacred space that creates a "politics of integration" which reinforces a sense of identity among immigrants not just from Mexico, but other parts of the Americas as well (Peña 2008: 724). Most tellingly, at certain times during the year part of the building that the community uses as a chapel becomes discreetly transformed into a space for volunteers to provide information about the U.S. naturalization and citizenship process. As Peña explains, "… this type of activity at the Second Tepeyac evinces its dual role as religious sanctuary and political safe haven" (2008: 740). Her research highlights at least two important facets of sacred space in a transnational context. First, a shared sacred space may provide a basis for community among diasporic peoples, and second, sacred spaces in transnational contexts may function in multivalent ways, serving much more than a religious or spiritual purpose.

A final example of the relevance of understandings of sacred space to transnationalism centers around development and the role of religious NGOs. Based on fieldwork in Aceh, Indonesia which was devastated by the 2004 tsunami, Andrew McGregor (2010) finds that transnational development networks tend to distance themselves from the religious needs of local communities during the relief and rebuilding process even though many of the NGOs are faith-based organizations. As a consequence, this distancing can deprive people of spiritual support and lead to the (inadvertent) secularization of the built environment. Concerns about the perception of proselytizing often come into play in ways that work to keep NGOs from rebuilding religious structures. Moreover, there are sometimes latent expectations that it is more appropriate for Christian NGOs to build mosques or meunasahs (village buildings for daily prayer) in Islamic communities and for Islamic NGOS to build churches in Christian communities. McGregor concludes that "access to sacred spaces and the right to faith in Aceh are not only negotiated within local and national communities and authorities, but within international development networks as well" (2010: 743). This admirable project brings to light both the relevance of sacred space and religion within the context of international development.

18.3.3 Sacred Space and Commodification

Like the spread of the Internet and the expansion of transnationalism, commodification is intimately associated with globalization. Indeed, it is now possible to have a prayer placed in the Wailing Wall (donations to support this free service accepted), and Hindus can pay to have a puja serviced performed at a temple on a specific date on their behalf. So, in 2011 when reports circulated in the media that the Catholic Church had approved an app for confession via the iPhone, they seemed startling, but within the realm of possibility. In reality, an iPhone app called "Confession" was released that helps a person prepare for confession but does not enable them to make confession.

The commodification of the sacred, including sacred space, remains a contentious issue. To some, the sale of holy water at pilgrimage sites constitutes a kind of profaning of the sacred. Others are disturbed at having to pay to enter religious sites, or find that the high numbers of tourists passing through shrines, memorials, and other sacred spaces lessens the sanctity of these spaces. There is often a fine line between commodification and exploitation. As we continue to theorize the process of commodification in relation to sacred space, a consideration of the development of themed spaces or event spaces may be helpful. Aaron K. Ketchall (2007) teases out some of the relations between the perception of the Ozarks as a sacred space and the emergence of the Christian-influenced Silver Dollar City theme park in Branson, Missouri. In the process, he highlights the disjunction between the discourse and image of the Ozarks as sacred space and the ongoing alteration of the landscape for the sake of progress. The idea of event space is implicit in Doron Bar's (2009) discussion of the development of pilgrimage routes, festivals, and other events at sites of holy graves in Israel. The identification of these new sites and the consumption of them through their development as significant destinations helped extend Israel's sacred space in the first two decades after statehood.

In an innovative study of prayer space as sacred space at the Millennium Dome in Greenwich, England, Sophie Gilliat-Ray (2005) investigates the making of a sacred space in a public institution and popular tourist site. Since this is a shared space that simultaneously belongs to everybody and no one, it becomes a kind of ambiguous space that facilitates personal and informal expressions of spirituality. Through a study of the use of the space, including notes and prayers written in the visitor notebooks, she shows that the space becomes sanctified through acts of contemplation and reflection; that is, "… not simply on account of the work of those who wrote the entries, but also by the efforts of those who read them and those who offered them for prayer" (2005: 368). Her research suggests that studies of sacred space in public facilities may provide insights about the growth of new, more personal spiritualities in contrast to conventional or mainline religious practices. More broadly, her work links the making of sacred space to forms of consumption such as leisure and tourism.

Commodification, cultural diversity and sacred space converge in Gordon Waitt's (2003) research on the construction of the Nan Tien Buddhist Temple in a New South Wales suburb. His study highlights not only the contested views about what does and does not belong in an area, but also raises questions about the use of sacred space as an expression and measure of cultural capital as well as a means for creating a tourist attraction. As Waitt explains, some of the opposition to the temple dissipated when local residents recognized the economic benefits that might be attained by virtue of having a Buddhist temple there (2003: 231). With this in mind, we might ask to what extent the marketing of sacred space creates a façade of cultural diversity or is used to present a certain image of a place or people.

18.4 Conclusion

This chapter has explored some of the recent research that brings together sacred space and globalization, highlighting the importance of new virtual spaces, transnationalism, and commodification. In the process, it has also attempted to reclaim a place for some key foundational works by Eliade, Smith, and Bokser, in part because some scholars—notably geographers—have been somewhat remiss in acknowledging their geographical perspectives and insights. In a recent editorial, Yorgason and della Dora ask, "how helpful, for instance, is Mircea Eliade's (1959) rigid structural opposition between sacred and profane space? Does it still make sense to think of religion through binary thought" (2009: 633). Although they do not directly answer these questions, I would suggest that Eliade's work is helpful as an important starting point.[7] Both the weaknesses and contributions of his conceptualization should be acknowledged, keeping in mind the advantages of hindsight. As Holloway (2003) shows, Eliade's concept of hierophanies has a certain salience for understanding sacred space as an enchanted, vitalized space, particularly within the context of New Age spiritual seekers. If we think about it, there is really not such a great distance between Eliade's discussion of homologizing behavior that repeats the work of the gods, and what Judith Butler refers to as "the stylized repetition of acts" through which people construct and constitute their identities (1988: 520).[8]

---

[7] Della Dora (2011) provides a very good example of ways of engaging scholarship, including Eliade's work, on sacred space, in a field course.
[8] Butler speaks specifically to the constitution of gendered identity, but the importance of repeated, performative acts as a basis of identity and meaning remains clear.

Eliade's rigid binary also remains helpful because both the practice of recognizing and the discourse of erecting such a sharp divide continues to be common. To give just one example, in 2010 Ross Workman the president of the Laie Temple of the Church of Jesus Christ of the Latter Day Saints in Hawaii, presented a devotional talk on sacred space that uses this very binary. As he made clear, "when Lucifer was cast out of heaven into the earth, the earth became profane space. It is the dwelling of Satan … The sacred sanctifies; the profane contaminates.... One cannot take the profane into the sacred and vice versa." Moreover, sacred space is identified specifically as something that is God-given (Workman 2010). If we as scholars fail to recognize the resonance that such a strict binary (and ontology) has for certain religious groups and individuals, then we diminish the value and relevance of our work. In addition, it gives the appearance that our scholarship places more weight on the imposition of our own etic categories, frameworks, and theories rather than building from emic ways of being and knowing.

A broader question still remains: what makes the concept of sacred space useful for our intellectual enterprise, or should we jettison it in favor of "ritual space" as suggested by Peter W. Williams (2002)? First, the concept of ritual space is a useful one. Indeed, Smith (1987) has shown how ritual contributes to the construction of sacred space. Even though the concept of ritual space may be meaningful for academics, one wonders what kind of traction the term would have within the wider public. Following from the emic/etic discussion above, it seems unlikely that most people would describe or interpret spaces that they perceive as immanently powerful, enchanted, or even dreadful as ritual space. Another reason for retaining the term sacred space has to do with the fact that the performance of some rituals is not necessarily associated with the sacred or sacred space. Countless ritual acts, such as convocations, commencements, and inductions into organizations, to give just three examples, may create and make use of ritual space but may not involve the sacred or create a sacred space. Simply stated, sacred space may be easier to comprehend and more versatile. It remains our challenge to tease out the nuances in the ways different groups and individuals understand the term.

To date, there has been surprisingly little work on the intersection of gendered identities and sacred space. Across the United States and in parts of Canada, Australia, and Mexico, the growth of "Cowboy Churches" presents some tantalizing research possibilities. These churches follow a conservative Christian theology anchored to an idealization of cowboy heritage and values. Church services may be held in rodeo arenas, barns, or other facilities, and baptisms often take place in stock tanks or troughs. A number of these churches emphasize a come-as-you-are openness and informality that welcomes workaday attire including jeans, cowboy boots and hats. Broadly speaking, cowboy values include privileging a strong work ethic, doing what is morally "right," favoring a staunch independence, and embracing a rugged individualism that is laced with a tinge of libertarianism and variously expressed as disdain for the government. Curiously, cowboy values are simultaneously gender neutral yet strongly masculinized, with much of the imagery showing men on horseback. Sometimes specifically "cowgirl" values are stressed, raising many questions about the nature of these gendered constructions within the discourse of Cowboy churches and, related to this, the ways in which mundane spaces become transformed into sacred spaces. Similar gender-based issues might be pursued with respect to the spread of Pentecostal and Renewalist churches in Latin America and Africa.

In addition, there are many possible ways that scholars could contribute to the ecologies of sacred space, and within them, constructions of nature. What churches, religious organizations, or social movements deploy narratives of sacred space when articulating their positions vis-à-vis the environment or resource use? A study of Santo Daime, a syncretic religion which originated in Brazil and has connections with narratives and programs associated with environmental sustainability, might shed light on some of these issues.

Taking a different perspective on sacred ecologies, we know, for example, that sacred groves are often associated with ecodiversity and constitute a kind of environmental conservation, but we know much less about the social consequences of converting such sacred spaces into parks or heritage areas, particularly with respect to indigenous peoples. Ironically, the creation of a heritage site can sometimes alter access to or management of a sacred space such that local or indigenous people inadvertently suffer from diminished access to it.

With respect to sacred space, civil religion also affords an avenue for additional research. How do monuments or memorials architecturally and aesthetically construct sacred space? At the Oklahoma City National Memorial sacred space and time intersect via the designation of two gates, one marked with the time 9:01, and the other with the time 9:03. These gates construct a sacred space centered on a reflecting pool and together they surround and simultaneously set apart the sacred time when the Murrah Building was bombed. Following Foucault (1986), this memorial constitutes a heterotopia—that is, a paradoxical place. The paradigmatic heterotopia for Foucault was the mirror because, as he explains, "I see myself where I am not…" (Foucault 1986: 24).

Returning to the Oklahoma City memorial, it functions as a heterotopia in the same way that a cemetery does: it gives adjacency to the end of life as well as eternal remembrance, and it simultaneously sets the site apart while making it accessible. The idea of sacred space as a Foucauldian heterotopia deserves additional attention. Indeed, an important subtext of Foucault's work centers on his view that these heterotopic paradoxes are "nurtured by the hidden presence of the sacred" (1986: 23). In sum, we have much to learn about sacred space as the "powerful, cognitive space" (Nelson 2006: 2) that it is, and the role that it plays in our geographical imaginations.

References

Augé, M. (1995). *Non-places: Introduction to an anthropology of supermodernity* (J. Howe, Trans.). London: Verso.

Bar, D. (2009). Mizrahim and the development of sacred space in the state of Israel, 1948–1968. *Journal of Modern Jewish Studies, 8*(3), 267–285.

Berkes, F. (2008). *Sacred ecology* (2nd ed.). New York: Routledge.
Bokser, B. M. (1985). Approaching the sacred. *Harvard Theological Review, 78*(3–4), 279–299.
Butler, J. (1988). Performative acts and gender constitution: An essay in phenomenology and feminist theory. *Theatre Journal, 40*(4), 519–531.
Brereton, J. P. (2005). Sacred space. In L. Jones (Ed.), *Encyclopedia of religion* (2nd ed., Vol. 12, pp. 7978–7986). Detroit: Macmillan Reference. Retrieved October 31, 2013, from http://go.galegroup.com/ps/i.do?id=GALE%7CCX3424502693&v=2.1&u=stil74078&it=r&p=GVRL&sw=w&asid=5d1757392b121bb0e21c2599668648a8
Chidester, D. (1995). "A big wind blew up during the night:" America as sacred space in South Africa. In D. Chidester & E. T. Linenthal (Eds.), *American sacred space* (pp. 262–312). Bloomington: Indiana University Press.
Chidester, D., & Linenthal, E. T. (Eds.). (1995). *American sacred space*. Bloomington: Indiana University Press.
Cox, H. (1999). The market as God. *Atlantic Monthly, 283*(3), 18–23.
Csordas, T. J. (2007). Introduction: Modalities of transnational transcendence. *Anthropological Theory, 7*(3), 259–272.
della Dora, V. (2011). Engaging sacred space: Experiments in the field. *Journal of Geography in Higher Education, 35*(2), 163–184.
Durkheim, E. (1965 [1915]). *The elementary forms of the religious life* (J. W. Swaim, Trans.). New York: Free Press.
Eliade, M. (1959). *The sacred and the profane: The nature of religion* (W. R. Trask, Trans.). New York: Harcourt Brace and World.
Foucault, M. (1986). Of other spaces. *Diacritics, 16*(1), 22–27 (J. Miskowiec, Trans.).
Gilliat-Ray, S. (2005). "Sacralising" sacred space in a public institution: A case study of the prayer space at the Millennium Dome. *Journal of Contemporary Religion, 20*(3), 357–372.
Greiner, A. (1991). *Geography, humanism, and "Plain People" in Missouri: The case of the Dunkard Brethren*. Thesis, University of Missouri-Columbia.
Holloway, J. (2003). Make-believe: Spiritual practice, embodiment, and sacred space. *Environment and Planning A, 35*, 1961–1974.
Holloway, J. (2006). Enchanted spaces: The séance, affect, and geographies of religion. *Annals of the Association of American Geography, 96*(1), 182–187.
Hopkins, D. N. (2001). The religion of globalization. In D. N. Hopkins, L. A. Lorentzen, E. Mendieta, & D. Batstone (Eds.), *Religions/globalizations: Theories and cases* (pp. 7–32). Durham: Duke University Press.
Hume, L. (1998). Creating sacred space: Outer expressions of inner worlds in modern Wicca. *Journal of Contemporary Religion, 13*(3), 309–319.
Jackson, P. (1989). *Maps of meaning: An introduction to cultural geography*. London: Unwin Hyman.
Jackson, R. H., & Henrie, R. (1983). Perception of sacred space. *Journal of Cultural Geography, 3*(2), 94–107.
Ketchall, A. K. (2007). *Holy hills of the Ozarks: Religion and tourism in Branson, Missouri*. Baltimore: Johns Hopkins University Press.
Kong, L. (1993). Negotiating conceptions of 'sacred space': A case study of religious buildings in Singapore. *Transactions of the Institute of British Geographers, New Series, 18*(3), 342–358.
Kong, L. (2001a). Mapping 'new' geographies of religion: Politics and poetics in modernity. *Progress in Human Geography, 25*(2), 211–233.
Kong, L. (2001b). Religion and technology: Refiguring place, space, identity and community. *Area, 33*(4), 404–413.
Kong, L. (2002). In search of permanent homes: Singapore's house churches and the politics of space. *Urban Studies, 39*(9), 1573–1586.
Lane, B. C. (2001). *Landscapes of the sacred: Geography and narrative in American spirituality* (Expanded ed.). Baltimore: Johns Hopkins University Press.
McGregor, A. (2010). Geographies of religion and development: Rebuilding sacred spaces in Aceh, Indonesia, after the tsunami. *Environment and Planning A, 42*, 729–746.

Mohanty, J. N. (1997). *Phenomenology: Between essentialism and transcendental philosophy.* Evanston: Northwestern University Press.

Nelson, L. P. (Ed.). (2006). *American sanctuary: Understanding sacred spaces.* Bloomington: Indiana University Press.

O'Leary, S. D. (1996). Cyberspace as sacred space: Communicating religion on computer networks. *Journal of the American Academy of Religion, 64*(4), 781–808.

Peña, E. (2008). Beyond Mexico: Guadalupan sacred space production and mobilization in a Chicago suburb. *American Quarterly, 60*(3), 721–747.

Radde-Antweiler, K. (2008). Virtual religion: An approach to a religious and ritual topography of Second Life. *Online–Heidelberg Journal of Religions on the Internet, 3*(1), 174–211. Retrieved January 6, 2012, from http://archiv.ub.uni-heidelberg.de/volltextserver/portal/relinternet/

Scheifinger, H. (2009). The Jagannath temple and online darshan. *Journal of Contemporary Religion, 24*(3), 277–290.

Shiner, L. E. (1972). Sacred space, profane space, human space. *Journal of the American Academy of Religion, 40*(4), 425–436.

Smith, J. Z. (1978). *Map is not territory: Studies in the history of religions.* Leiden: Brill.

Smith, J. Z. (1987). *To take place: Toward theory in ritual.* Chicago: University of Chicago Press.

Tuan, Y.-F. (1978). Sacred space: Explorations of an idea. In K. W. Butzer (Ed.), *Dimensions of human geography: Essays on some familiar and neglected themes* (pp. 84–99). Chicago: University of Chicago. Department of Geography.

Van der Leeuw, G. (1963). *Religion in essence and manifestation* (J. E. Turner, Trans.). New York: Harper and Row.

Vasquez, M. A., & Marquardt, M. F. (2003). *Globalizing the sacred: Religion across the Americas.* New Brunswick: Rutgers University Press.

Waitt, G. (2003). A place for Buddha in Wollongong, New South Wales? Territorial rules in the place-making of sacred spaces. *Australian Geographer, 34*(2), 223–238.

Williams, P. W. (2002). Sacred space in North America. *Journal of the American Academy of Religion, 70*(3), 593–609.

Workman, R. (2010). *Sacred space.* Devotional talk given at Brigham Young University–Hawaii. October 5, 2010. Retrieved May 28, 2012, from http://devotional.byuh.edu/script/sacred-space

Yorgason, E., & della Dora, V. (2009). Geography, religion, and emerging paradigms: Problematizing the dialogue. *Social & Cultural Geography, 10*(6), 629–637.

Chapter 19
Dark Green Religion: Advocating for the Sacredness of Nature in a Changing World

Joseph Witt

19.1 Introduction

In recent decades, global environmental issues have received increasing attention from politicians, scientists, and the broader public. Issues such as anthropogenic climate change, electronic waste and pollution, equitable access to clean water and other resources, and global food security pose some of the most pressing concerns for communities around the world. Increasingly, religious leaders and communities have entered these debates and reconsidered their theologies and practices, envisioning specifically religious responses to environmental problems. Advocates of the Christian "Creation Care" movement, for example, argue that environmental stewardship is a biblical mandate for observant Christians and they have led many efforts to address environmental concerns around the globe (see, for example, DeWitt 1998; Sleeth 2006; and McDuff 2010). In England, some Muslim groups such as The Islamic Foundation for Ecology and Environmental Sciences have developed educational initiatives to promote awareness about environmental problems and propose sustainable development projects from Islamic perspectives (Gilliat-Ray and Bryant 2011). Some religious studies scholars, as well, have led the push for the "greening" of religion by illuminating theological, historical, and cultural resources from the world's religions as especially appropriate for addressing environmental problems. The Forum on Religion and Ecology (FORE), for instance, has led a groundbreaking effort toward this end. Following some organizational conferences in the 1990s, the FORE published a ten volume "Religions of the World and Ecology" book series which brought together numerous scholars to explore the

J. Witt (✉)
Department of Philosophy and Religion, Mississippi State University, Mississippi State, MS 39762, USA
e-mail: jwitt@philrel.msstate.edu

environmentally positive and negative facets of different world religions. The FORE's work has continued into the twenty-first century with regular conferences and publications. Some of the group's founders, like Mary Evelyn Tucker and John Grim, have become internationally recognized authorities on the place of religious communities in developing environmentally sustainable policies. Further examples of faith-based and scholarly efforts at understanding and resolving environmental problems could be listed at length, but it is clear that these communities are increasingly concerned with solving environmental crises and adding their voices to policy debates. Given the important place of religion in public life, particularly in North America, it is very likely that religious communities will continue to weigh in on environmental problems into the future.[1]

Examining world religions and faith-based organizations, however, paints only part of the picture of contemporary religious responses to environmental problems; also theoretical changes within the field of religious studies point to new directions for consideration. Several scholars have begun to criticize what they consider to be the problematic colonial history of the so-called "world religions." Jonathan Z. Smith (1978), Talal Asad (1993), and Tomoko Masuzawa (2005), among others, have encouraged a critical reappraisal of previously comfortable concepts such as "religion" and "world religions," situating the terms in their broader political and historical contexts. As Smith argued, "'Religion' is not a native term; it is a term created by scholars for their intellectual purposes and therefore is theirs to define" (1998: 281). In a similar vein, others have worked to cross the borders of religions as distinct, bounded entities, focusing instead on the hybridity, negotiation, and contestation of religious values, concepts, and identities on the ground. For example, Robert Orsi advocated for a "lived religion" approach to religious studies, meaning that scholars could not ignore the complex interrelations between religious values and concepts and the practices of common people. He argued that "all religious ideas and impulses are of the moment, invented, taken, borrowed, and improvised at the intersections of life," and that religions cannot be extracted as abstract concepts from the "material circumstances in which specific instances of religious imagination and behavior arise and to which they respond" (Orsi 1997: 7–8). For Orsi, scholars should not reify typological boundaries, but instead attend to the complexly integrated relationships between values and specific practices. Thomas Tweed, as well, acknowledged the shifting, contested nature of religious ideas, defining religions as "confluences of organic-cultural flows that intensify joy and confront suffering by drawing upon human and suprahuman forces to make homes and cross boundaries" (2006: 54). Rather than firmly defined categories, for Tweed,

---

[1] There is an extensive body of literature on faith-based environmental efforts around the world. A good starting place for those interested in learning more is the FORE's website (http://fore.research.yale.edu/), which includes numerous faith statements on environmental problems and scholarly resources for further study. Roger Gottlieb's Oxford Handbook of Religion and Ecology (2010) and Ecospirit, by Laurel Kearns and Catherine Keller (2007), are also useful introductions to the field. See also Kalland (2005) and Taylor (2005) for more critical perspectives on the FORE and other similar academic projects.

religions are networks of ideas, practices, and emotions that intertwine with other cultural and environmental forces. Voices such as those of Tweed, Smith, Orsi, Asad, and Masuzawa have become more pronounced in recent decades, and although they should not be seen as composing a unified project (indeed, there are significant theoretical and methodological disagreements between them), taken together, they do present something of a different direction for the study of religions and environmental concern. In this context of religions and the environment, when scholars move away from official theological pronouncements and categories, focusing instead on religious negotiation and articulation among specific communities on the ground, what might they find?

One answer comes from the 2010 work of Bron Taylor on a set of religious phenomena he termed "dark green religion." This includes forms of religiosity that posit an inherent sacredness and value to the natural world and propose spiritual methods for addressing environmental problems. Rather than working from within a defined religious or theological tradition to discern appropriate responses to environmental problems, dark green religious practitioners build upon multiple spiritual and scientific ideas to express their feelings of connection to the natural world and advocate for dramatic lifestyle changes. Instead of a concrete tradition, dark green religion can be understood as a thread tracing through many contemporary religious and environmental negotiations at the individual, community, and international levels. For Taylor, these themes of sacredness and inherent value infuse much contemporary environmentalist discourse and they sometimes serve to bridge gaps between movements and groups. As complex anthropogenic environmental crises continue to proliferate, Taylor concluded, dark green religious themes will continue to inform public environmental discourse and decision-making.[2]

This chapter explores dark green religion as one pathway to understanding the changing world religion map in light of increasing environmental concerns. As a theory, dark green religion allows scholars to attend to the ongoing hybridity and dynamism of religious values, cultures, and practices. It thus calls scholars to reflect back upon what "religion" means in a continuously changing world. Beyond its theoretical implications, however, dark green religion also describes a diverse set of beliefs and practices shared by environmentalist and activist communities around the globe. Therefore, this chapter provides direct evidence of dark green religion in practice among activist groups in the Appalachian Mountains of North America. Adding dark green religion to the collection of religious responses to environmental problems helps us to better understand contemporary religious complexities and the interrelations between places, politics, and identities in the twenty-first century.

---

[2] I am not suggesting that the FORE and the "World Religions and Ecology" series should be seen as competing research projects to dark green religion, or that all of the hundreds of contributors to FORE projects share the same theoretical and methodological groundings. Adding the consideration of dark green religion only increases the number of approaches available to study religions and the environment.

19.2 Dark Green Religion in Theory

For centuries philosophers, scholars, authors, and theologians have contemplated the relationships between humans and their environments. In his influential study of environmental thought, for example, Clarence Glacken (1967) argued that many contemporary ideas concerning the relationships between humans and the natural world trace back at least 2,500 years to the philosophical works of the ancient Greeks. While researchers remain unclear on whether and how religious and philosophical values influence cultural practices, these ideas have likely influenced human behaviors toward nature, both in efforts to preserve natural environments and in justifications for continued exploitation.[3] Some modern commentators point directly at religious and philosophical systems, such as the Western philosophical disconnection between human spirits and physical nature, as influences on contemporary environmental problems (White, Jr. 1967; Merchant 1980). Others have looked to the natural world as a source of spiritual inspiration, leading them to work to preserve natural ecosystems and landscapes. Natural themes are interwoven through Western intellectual history, whether a continued dependence upon almanacs and natural portents among early North American colonists, or views of connections between nature, bodies, and the cosmos among New Age spiritual movements of the twentieth century (Albanese 1990; Hall 1990). In the North American context, these ideas have persisted from the writings of early colonists such as Anne Bradstreet and Thomas Morton to more recent environmental luminaries such as John Muir, Aldo Leopold, and Rachel Carson (Nash 2001; Gatta 2004). These themes of interconnection and sacredness have entered into international environmental policy advocacy as well. The Earth Charter, for example, was drafted in the 1990s as a voluntary international policy through which signatories agreed to respect life and ecological thriving in its many forms. The Earth Charter offered a "holistic understanding of what constitutes a sustainable way of living and sustainable development," and was grounded in the belief that "all beings are interdependent and all life forms have value regardless of their worth to people" (Rockefeller 2005: 517). Whether in the philosophy of ancient Greeks, modern art and literature, or contemporary international policy, ideas about the relationships between humans and the natural world and arguments over nature's value remain present.

The theory of dark green religion offers a way to understand how this language of sacredness, interconnection, and inherent value has been interwoven with religious and spiritual concerns through history and how it continues to adapt and change in modern circumstances. In his work, *Dark Green Religion* (2010: 13), Taylor defined his term as a value system that is generally:

---

[3] For example, environmental ethicist Anna Peterson problematized simplistic causal connections between value traditions and environmental practices. She argued, "values and practices, desires and structures, are always interacting with and transforming each other; to single one out as the dominant or sole factor in any social process reflects a deep failure of understanding" (2009: 131). See also Heberlein (2012).

19 Dark Green Religion: Advocating for the Sacredness of Nature in a Changing World

(1) based on a felt kinship with the rest of life, often derived from a Darwinian understanding that all forms of life have evolved from a common ancestor and are therefore related; (2) accompanied by feelings of humility and a corresponding critique of human moral superiority, often inspired or reinforced by a science-based cosmology that reveals how tiny human beings are in the universe; and (3) reinforced by metaphysics of interconnection and the idea of interdependence (mutual influence and reciprocal dependence) found in the sciences, especially in ecology and physics.

Most basically, dark green religion combines spiritual insights with scientific understandings of the natural world, leading to moral principles of responsibility, kinship, and humility. These moral principles are then often enacted among practitioners in various types of ritualization or direct actions aimed at stopping specific activities seen as environmentally exploitative (for example, blocking logging in the North American Pacific Northwest, or strip mining in Appalachia). While not exclusive to environmentalist communities, these dark green religious themes can be most often found among those who strive to change what they perceive to be environmentally-destructive practices and policies.

It is within this "environmentalist milieu" that much of the creativity, negotiation, and hybridity of dark green religion occurs (Taylor 2010: 13–14). Individuals within this countercultural milieu (akin to the "cultic milieu" described by Colin Campbell (2002)) share perspectives and ideas. Drawing upon multiple foundations, such as Asian religious traditions, indigenous worldviews and ritual practices, scientific descriptions of biological evolution, and theories from physics that seem to support ideas of the interconnectedness of all matter, some creatively construct distinct religious and ethical perspectives. Taylor employed French anthropologist Claude Lévy-Strauss's (1967) term "bricolage" to describe the constructive process of dark green religious development, in which diverse themes are combined to form novel approaches (Taylor 2010: 14). Dark green religion points to these multiple strands that remain in negotiation among different communities and the ongoing religious creativity that is directly related to changing environmental conditions and policies.

One philosophical strand that has been particularly influential in developing and supporting environmentalist worldviews is deep ecology. Most basically, deep ecology is the philosophical perspective developed by Norwegian philosopher Arne Naess positing intrinsic value to all living things (Taylor and Zimmerman 2005: 456). Naess distinguished deep ecology from shallow ecology—anthropocentric arguments to preserve and manage resources for the continued thriving and growth of human communities. For Naess, this philosophy was practiced through "ecosophy-T," a personal philosophy emphasizing self-realization, diversity, and symbiosis between living beings. While ecosophy-T was Naess's personal ecosophy (named after Tvergastein, one of his favorite mountains), he believed other individuals could adopt different personal ecosophies that would support the deep ecology movement (Bender 2003: 422–423). Some environmentalists in the 1970s and 1980s took Naess's philosophy and expanded it into a spiritual worldview, complete with different rituals such as the Council of All Beings, meant to break down anthropocentric assumptions among participants and foster deeper connections to the earth (Taylor 2010: 21–22). In 1985, George Sessions and Bill Devall provided a popular overview of the deep ecology perspective, including what they termed sources of the perspective from Asian and indigenous religious traditions (Devall and Sessions 1985). By the middle of the 1980s, deep ecology represented the bricolage Taylor found so characteristic of dark green religiosity, combining sometimes disparate religious and philosophical ideas into a unique, earth-revering system.

Given its multiple sources and the continued hybridity between communities, dark green religion is by its nature a diverse set of ideas and practices. Still, Taylor broke versions of dark green religion into four basic forms, acknowledging that the boundaries between them remain porous. The categories sat on a scale between supernaturalistic and naturalistic (considering whether or not there exists some sort of spiritual plane beyond the natural world as explained by science), and with foci on either pluralistic or holistic elements (considering whether individual creatures are considerable, or whether broader ecosystems, the earth itself, and cosmic processes deserve consideration) (Taylor 2010: 14–16). Taylor employed the term "animistic" to describe elements on the individualistic side of the spectrum. The discussion of "animism" originates with the early anthropological work of E.B. Tylor. In his *Primitive Culture*, Tylor defined animism as "the belief in controlling deities and subordinate spirits, in souls, and in a future state," typically found in indigenous cultures (1871: 386). Tylor argued that the origins of human religiosity may be found in the animistic beliefs and practices of so-called primitive cultures, and that animism was "the groundwork of the Philosophy of Religion, from that of savages up to that of civilized men" (1871: 385). Due to his connections to the European colonial enterprise and his dismissive attitude toward indigenous cultures, many modern indigenous religious practitioners, and the scholars who study them, have abandoned "animism" as an overly problematic term. Despite this troubling history, though, others have worked to positively reclaim the term in the modern context. Graham Harvey, a scholar of global indigenous cultures, found a popular reclamation of "new animism" among modern indigenous communities and "nature-venerating religionists" (2006: 3; see also Snodgrass and Tiedje 2008). Even with the term's negative history, Harvey argued that it still reflects a positive identification among some communities and so can remain an appropriate term for a specific set of beliefs, practices, and modern identities. In *Dark Green Religion*, Taylor sided with Harvey and defined animism among dark green religious communities as involving "a shared perception that beings or entities in nature have their own integrity, ways of being, personhood, and even intelligence" (2010: 15). In practice, dark green animism often encompasses forms of paganism and indigenous religious practice; or in other words, any spiritual tradition that emphasizes the plurality of spirits and intelligences in nature.

At the other end of Taylor's spectrum lies holism, or the emphasis upon a total spiritual unity to the cosmos. Taylor termed this "Gaian earth religion," which "understands the biosphere (universe or cosmos) to be alive or conscious, or at least by metaphor and analogy to resemble organisms with their many interdependent parts" (2010: 16). The use of "Gaia" for this collection of dark green elements refers to the ancient Greek earth goddess of the same name; but it also derives from the work of British scientist James Lovelock. In his 1979 book *Gaia: A New Look*

19 Dark Green Religion: Advocating for the Sacredness of Nature in a Changing World

at Life on Earth, Lovelock argued that the entire biosphere comprised one self-regulating system, and that the entire earth with all of its life could be thought of as a singular, living entity. Lovelock did not originally intend his theory to support spiritual worldviews, but many Neo-pagan and New Age religious practitioners took his concept as a useful referent for their feelings of interconnection between nature, animals, and humans (Pike 2004: 23; Taylor 2010: 35–36). In practice, Gaian earth religion refers to those forms of dark green religion that emphasis a fundamental unity among living beings, and for more spiritual versions, an all-encompassing spirit within the cosmos.

Along with the spectrum between pluralism and holism, Taylor argued that examples of dark green religion sit on a spectrum between spiritualism and naturalism, or claims of a spiritual, mystical presence in the universe versus reverence for the natural world as described by the sciences alone. While practitioners on the spiritual side of Taylor's spectrum may draw upon numerous religious traditions for inspiration, religious naturalists tend to emphasize the complex wonders described by physics, geology, biology, and other sciences as sufficient ground for reverence. Philosopher Donald Crosby (2002) has written extensively on religious naturalism, but perhaps one of the best known examples of this form of religiosity comes from the Epic of Evolution. Inspired by the work of Thomas Berry (Berry 1988; Swimme 1994), proponents of the Epic of Evolution argue that the story of the emergence of life and complexity in the universe, as described by the physical sciences, is a narrative that can function to unify humanity into a common respect for life. Without appealing to supernatural forces, yet frequently drawing upon religious metaphors and language of the sacred, religious naturalists find sufficient mystery in the universe to support their concerns for environmental preservation and continued research (Taylor 2010: 16).

Given all these diverse influences, it is clear that dark green religious theory entails a more expansive understanding of religion itself. Rather than focusing on distinct traditions and theologies, Taylor emphasized a "polyfocal approach" to the study of religion, examining "the widest possible variety of beliefs, behaviors, and functions that are typically associated with the term" (2010: 2). Inspired by the work of anthropologist Benson Saler and philosopher Ludwig Wittgenstein, this approach de-emphasized rigid definitions of religious boundaries and traits, and focused instead upon the explanatory power of considering "religion-resembling phenomena" among broader networks of values and practices (Taylor 2010: 3). Taking a wider view of religion allowed Taylor to include forms of religious naturalism within dark green religion, but it was also a point of contention among other religious studies scholars and scientists who were not so willing to challenge older, concrete definitions of the term. Richard Dawkins, the prominent evolutionary biologist and outspoken atheist, bristled at other attempts to describe scientific naturalism as a form of religious expression (Taylor 2010: 177). As Taylor noted, however, the study of dark green religion was meant to "rattle assumptions as to what counts as religion in order to awaken new perceptions and insights" (2010: 4). By expanding the definition and challenging previously comfortable categories, dark green religion allowed the inclusion of beliefs, perspectives, and practices that could have been previously overlooked by religious studies scholars interested in the interconnections between values and environmental practices. The theory of dark green religion is not only an approach to the study of religions, though; it describes the beliefs and practices of specific communities as well. How successful is dark green religion as a descriptive category?

19.3 Dark Green Religion in Practice

If dark green religion is a phenomena associated largely with environmentalist communities, as Taylor argued, it should be evident wherever those communities thrive. My own research has been among opponents of mountaintop removal coal mining (a form of surface mining resulting in dramatic changes to the local landscape). In the first years of the twentieth century, Appalachia became a focal point for national environmental policy with the increased use of mountaintop removal. Efforts to stop the practice escalated into the second decade of the twenty-first century, and the resistance movement brought together radical environmental activists (some with extensive experience among radical communities in the North American west coast) with local residents and community activist groups. The merging of these different communities initiated a local development of dark green religious hybridity, blending local Native American lore, cultural values of place and community, Christian creation care, and other spiritual traditions into a localized set of claims regarding the sacredness of the Appalachian Mountains. Of course, not all activists (radical or mainstream) expressed religious connections to the earth—some activists, for example, identified as atheist or agnostic. Others identified primarily as Christians, and while they respected the beliefs of others, they generally argued that sacredness was not inherent in the land itself, but in the transcendent God of the Bible. Still, in the Appalachian environmentalist milieu of the early twentieth century, dark green religion was an evident thread among religious responses to mountaintop removal.[4]

In interviews and oral histories, several activists expressed spiritual connections to the land and people of Appalachia. One middle-aged West Virginia activist who had lost his home to a flood associated with deforestation from a strip mine, who I will call "James,"[5] described his religious view as "a love for the land. If you want to call that spirituality—which I choose to do—then, there is a spirit-based, cultural part of that love for the land that I would put firmly in religion and faith" ("James" 2010). Other well-known activists expressed similar affective connections to the land. Larry Gibson (Fig. 19.1), who defended his home on Kayford Mountain, West Virginia, from an encroaching mountaintop removal mine until his death in 2012,

---

[4] For lengthier introductions to mountaintop removal coal mining and its resistance, see Montrie (2003), Burns (2007), and Scott (2010).

[5] Some of my interview subjects preferred to remain anonymous. I have given these individuals pseudonyms, which are indicated by quotation marks.

Fig. 19.1 West Virginia activist Larry Gibson describes the strip mine surrounding his ancestral home on Kayford Mountain, West Virginia (Photo by Joseph Witt, July 3, 2009)

explained that Appalachian people retained special connections to their land that other U.S. citizens often failed to understand. He said in an interview, "mountain people didn't live *on* the land, they lived *with* it" (Gibson 2009). Reflecting his sense of connection and dependence upon his land, Gibson continued, "my mother gave me birth, but the land gave me life" (Gibson 2009). While Gibson did not use words like "spiritual" or "religious," his feelings of connection to the mountains, combined with a critique of the coal industry, clearly formed an important motivation for his continued activism. These local dark green religious themes trace further back into history as well. Joe Begley, a deputy sheriff and anti-strip mining activist from Kentucky, offered a naturalistic statement on his own sense of connection to Appalachia. In an oral history, he said, "my religion is the stream, the timber, the wildlife, the animals and the people. I'm not worried about where I'm gonna go—I don't care where I'm gonna go in the end. It's here now, now's when I'm living, and I want to preserve that, I want to have that" (Begley 1987). For each of these activists, the dark green theme of kinship and connection to the natural world was clearly woven through their broader attitudes toward mountaintop removal coal mining.

Individual connections to places, however, make up only one facet of dark green religious hybridity in Appalachia. Taylor noted that dark green religious practitioners are frequently influenced by indigenous cultures, and this is certainly true among Appalachian environmental activists as well (Taylor 2010: 75). Appeals to perceived environmentally sustainable beliefs and practices among indigenous peoples go back at least to the writings of Jean-Jacques Rousseau, and the literary trope of the "ecologically noble savage" has persisted through the writings of popular authors such as James Fennimore Cooper and into modern iterations such as the infamous 1971 "crying Indian" commercial sponsored by the Keep America Beautiful campaign (see Deloria 1998; Krech 1999). For some non-native authors, indigenous peoples represent romantic exemplars of a pristine moral tradition and social structure, and spiritual movements of the nineteenth and twentieth centuries have appropriated indigenous religious practices in efforts to reclaim that perceived connectedness to the earth. Among several native commentators, this romanticism of native cultures and appropriation of indigenous practices is a continuation of exploitative attitudes and practices, dating back to the colonial era. For example, Cherokee activist and scholar Andy Smith argued that New Age religious practitioners who appropriate indigenous practices, regardless of their good intentions, "are in fact continuing the same genocidal practices of their forebears" (1991: 44). Adding to the critique of romantic portrayals of indigenous peoples as "first ecologists," some historians have shown that different indigenous cultures have dramatically impacted the natural environment at various points in time and that certain seemingly indigenous ideas, like "Mother Earth," have complex histories associated with colonization, exploitation, and romanticism (Cronon 1983; Gill 1987; Krech 2005). To critique romantic assumptions and explore the complex histories of indigenous ideas and practices, however, is not necessarily to reject the idea that indigenous cultures may offer useful knowledge about the natural world. Numerous anthropologists have pointed to the intimate connections between certain indigenous cultures and their surrounding environments, and others like Fikret Berkes have suggested that western environmental management practices could benefit from the shared knowledge of indigenous peoples (Berkes 1999). The relationship between indigenous cultures and nature is complex, and dark green religious practitioners wrestle with this complexity as well.

While some locals cite indigenous heritage (most frequently Cherokee, one of the most populous tribes of the southern Appalachian region), any overt appropriation of indigenous practices is generally downplayed within anti-mountaintop removal communities. In interviews, some activists expressed sincere respect for native cultures, but did not claim to be direct descendants of indigenous spiritual traditions, and I have not seen typically indigenous elements such as sweat lodges or sage smudges at the anti-mountaintop removal events that I have attended. For example, when asked about her thoughts on the impacts of mountaintop removal mining, one Catholic activist said, "we are so interdependent with the rest of creation, our Native American brothers and sisters got it right, we're all related. What we do to the environment, it's definitely gonna come back on us" ("Carolyn" 2009). While this activist approved of what she considered a specifically indigenous perception about interrelatedness in nature, she was not herself a native religious practitioner and did not attempt to recreate indigenous practices in her own spiritual life.

In an effort to include indigenous perspectives that they deem necessary voices for the movement, while simultaneously avoiding inappropriate misrepresentation and romanticism, opponents of mountaintop removal have included native leaders in protests and other direct actions. For example, on June 23, 2009, activists organized a rally and protest at Marsh Fork Elementary School, in Raleigh County,

West Virginia. At the time, Marsh Fork was a focal point in the anti-mountaintop removal movement because it sat immediately below a slurry impoundment (a giant reservoir of toxic mine sludge, retained only by an earthen dam). Activists also cited increased health problems for Marsh Fork students, especially increased rates of asthma, due to their proximity to an active strip mine. In 2010 the Annenberg Foundation added a generous donation to the years of effort by local citizens and construction of a new school was finally set into motion; but in 2009, these issues remained unsettled. The June 23 rally drew hundreds of mountaintop removal opponents as well as counter-protestors—miners, coal company employees, their families, and supporters. In the tense environment on the school grounds, where the rally was held, several critics of mountaintop removal attempted to speak to the assembled crowd over the shouts, air horns, and other distracting noise from counter-protestors. Among the speakers in opposition to the mine was Matt Sherman, a Blackfoot Indian activist from West Virginia (Fig. 19.2). Holding a hawk feather the entire time,

Fig. 19.2 Matt Sherman addresses the crowd gathered near Marsh Fork Elementary School. Notice the coal silo and loading ramp in the background. Sherman's shirt reads, "Sacred Ground" (Photo by Joseph Witt, June 23, 2009)

Sherman called upon those assembled to recognize their connections to the land and work together under a common cause of saving the land and people of Appalachia. He said, "whether we're with Mountain Justice [an anti-mountaintop removal activist group] or Massey Coal, the blood of our relatives is buried in these mountains." Within this message of unity, Sherman argued that native peoples (especially native women) held special power to restore order to the mountains. Pointing to Maria Gunnoe, the Goldmann Prize winning West Virginia activist, Sherman announced, "the souls of the Indian women will bring an end to mountaintop removal … The soul of the Indian is strong in the mountains of West Virginia."

Along with pointing to the indigenous presence in Appalachia, Sherman drew upon distinctly indigenous spiritual themes in his speech. As many of the anti-mountaintop removal activists spoke, one counter-protestor regularly interjected with the question, "how did you get here today?" The implication of his question was that, if the environmental activist had arrived by car (which of course they all had), then he or she was a hypocrite for denouncing a resource upon which he or she depended. Most speakers ignored the counter-protestor, but Sherman responded, to exuberant cheers from the crowd, "I soared in here today on the wings of eagles, sent by the Creator, to stop mountaintop removal!" As Sherman continued, a lone hawk began to circle in the sky above. As the crowd slowly recognized the bird, I overheard one older woman say, "it's an omen, nature approves." When Sherman noticed the hawk, he paused in his speech to acknowledge its presence. After a few seconds of silence, he said to the crowd, "those are our ancestors, come to join us … they are with us in these mountains today, and they are proud of what we are doing."[6] In his brief address to the crowd, Sherman clearly articulated dark green and indigenous themes of connection to the guiding power of natural spirits.

While Sherman drew upon his indigenous religious tradition to express his opposition to mountaintop removal, other activists turned to other sources for dark green inspiration. One example came with Kentucky artist Jeff Chapman-Crane's sculpture, "The Agony of Gaia." The life-sized sculpture was completed in 2004 and has since travelled to numerous activist meetings and protests around Appalachia. The piece focuses on a woman (Gaia) lying on her side with her hands covering her crying eyes. The figure is surrounded by a miniature landscape, representing the Appalachian Mountains. At her head, the landscape is green and forested; but moving toward her feet, her body turns gradually into a strip-mined mountain, surrounded by miniature bulldozers and dump trucks excavating the coal of her midriff. Finally, Gaia's feet have become terraced blocks, much like the remnants of a mountaintop removal mine. Quite graphically and literally, the sculpture portrays the destruction of the feminized earth from surface coal mining. The themes of destruction, connection, and redemption are expressed directly in a poem, "Cry of the Unreclaimed," written along the base of the sculpture. Composed by Chapman-Crane, the poem begins by reflecting upon the pristine wildness of Appalachia: "only the clarion cry of woodland bird, the song of mountain stream, should fill the silence

---

[6] Sherman's quotes and other details from the rally come from the author's field notes, Marsh Fork Elementary, Sundial, West Virginia, 23 June 2009.

19 Dark Green Religion: Advocating for the Sacredness of Nature in a Changing World

Fig. 19.3 Jeff Chapman-Crane's sculpture "The Agony of Gaia." Notice the verdant landscape at Gaia's head, compared to the strip mine at her feet. Small bulldozers and dump trucks are visible excavating her midriff (Photo by Joseph Witt, with the artist's permission, at Chapman-Crane's studio in Eolia, Kentucky, June 4, 2010)

here." The poem continues, describing the devastation of mountaintop removal: "But no. Lust would never have it so…Stripped to the bone and ripped apart she grieves, her heart is left to die alone." The poem concludes with a hopeful tone: "But know. Her seeds are left for us to sow" (Fig. 19.3).

I asked Jeff Chapman-Crane about his motivations in constructing "The Agony of Gaia" at the 2010 Mountain Justice Summer Camp, an educational and planning meeting for college environmental activists in the region. He explained that he hoped the striking image would evoke strong emotions in viewers, inciting them to stand against mountaintop removal. He said,

> for a long time in this country, in our culture, [there has] been the attitude that the earth is just there for us to exploit for resources with no responsibility towards it at all, no thought to the consequences, the environmental impacts. It's just simply something for us to exploit. I wanted to create a piece that showed that the earth is a living creature, that it feels what we're doing to it, and that there is profound impacts to doing what we're doing. And it's not really just coal mining here, you know, it's much larger than that. It's a world-wide problem. We simply treat the earth as this source of raw materials to fuel our energy consumption. Our greed, basically. So I wanted to do something that really showed the opposite view. (Chapman-Crane 2010)

Like other environmentalists and dark green religious practitioners, Chapman-Crane critiqued the anthropocentrism of American culture, calling viewers instead to consider the earth as a living being worthy of reverence and respect. He drew upon the image of Gaia, he explained, because he thought metaphors outside of the dominant Christian worldview of Appalachia were necessary to invoke the creative and critical thinking needed to break the lengthy historical connection between Appalachians and the coal industry. The conscious deployment of organic, holistic metaphors in his artistic production makes Chapman-Crane's work a clear example of Gaian earth religion, which "relies on metaphors of the sacred to express its sense of the precious quality of the whole" (Taylor 2010: 16).

Taken together, the spiritual connections to the landscape of Appalachia articulated by residents such as Larry Gibson, the indigenous religious perspectives offered by Native activists such as Matt Sherman, and the Gaian spirituality expressed by Jeff Chapman-Crane, reveal the diverse set of beliefs and practices that make up the local Appalachian dark green religious milieu. Although each of these individuals may not agree on spiritual values or the appropriate response to mountaintop removal, their work draws together dark green themes of kinship, interdependence, humility, and critiques of anthropocentrism. To be sure, these dark green religious themes are not the only (or perhaps even the most dominant) form of religious responses to mountaintop removal. For example, Christian groups such as Christians for the Mountains (an Appalachian community of Christian social and environmental activists), the Lindquist Environmental Appalachian Fellowship (based in Knoxville, Tennessee), and Restoring Eden (a national evangelical environmental group) have been very active in the opposition to the practice. As the evidence above reveals, however, dark green religious themes are present in the movement and influential for some activists in their work. Activists have drawn upon diverse spiritual and ethical themes in their responses to mountaintop removal, and focusing on dark green religion helps scholars to see this ongoing hybridity in action. As the topographic map of Appalachia changes due to surface mining, so too does the religious map.

19.4 Conclusion

Mountaintop removal is only one of many contemporary environmental concerns around the world; and if the predictions of climatologists are accurate, more climate-related crises will continue to emerge into the future. Toward the end of his work, Taylor predicted that examples of dark green religious production would continue to develop as global debates over environmental issues continued. He even believed dark green themes could form the basis for a "terrapolitan Earth religion," which could motivate collective global efforts to solve environmental problems (Taylor 2010: 180). The accuracy of this prediction, of course, remains to be seen; but in the meantime, dark green religion presents one useful approach to understanding how complex and diverse values and identities are negotiated, debated, and constructed in response to these issues. It is also one avenue through which religious studies scholars may continue to revise and debate the core terms defining their field. Among the Appalachian environmentalist milieu, indigenous cultures, personal feelings of kinship and connection to the natural world, and holistic visions of the sacredness of nature have combined to support efforts against mountaintop removal mining. Turning to a global perspective, scholars of religion must strive to understand similar interconnections between places, identities, and politics to help illuminate the shifting role of religions in public life. Their insights may contribute to the continued efforts toward sustainability and social justice as humanity faces the unforeseen challenges and opportunities of the twenty-first century.

References

Albanese, C. (1990). *Nature religion in America: From the Algonkian Indians to the new age*. Chicago: University of Chicago Press.
Asad, T. (1993). *Genealogies of religion: Discipline and reasons of power in Christianity and Islam*. Baltimore: Johns Hopkins University Press.
Begley, J. (1987). *Blackey, Kentucky oral history project, 2000OH02.1a*. Frankfort: Kentucky Historical Society. Kentucky Oral History Commission.
Bender, F. (2003). *The culture of extinction: Toward a philosophy of deep ecology*. Amherst: Humanity Books.
Berkes, F. (1999). *Sacred ecology: Traditional ecological knowledge and resource management*. Philadelphia: Taylor and Francis.
Berry, T. (1988). *The dream of the earth*. San Francisco: Sierra Club Books.
Burns, S. (2007). *Bringing down the mountains: The impact of mountain removal on southern West Virginia communities*. Morgantown: West Virginia University Press.
Campbell, C. (2002). The cult, the cultic milieu and secularization. In J. Kaplan & H. Lööw (Eds.), *The cultic milieu: Oppositional subcultures in an age of globalization* (pp. 12–25). Walnut Creek: AltaMira Press.
"Carolyn." (2009). Interviewed 8 July, Charleston.
Chapman-Crane, J. (2010). Interviewed 4 June, Eolia.
Cronon, W. (1983). *Changes in the land: Indians, colonists, and the ecology of New England*. New York: Hill and Wang.
Crosby, D. (2002). *A religion of nature*. Albany: SUNY Press.
Deloria, P. (1998). *Playing Indian*. New Haven: Yale University Press.
Devall, B., & Sessions, G. (1985). *Deep ecology: Living as if nature mattered*. Salt Lake City: Gibbs M. Smith.
DeWitt, C. (1998). *Caring for creation: Responsible stewardship of God's handiwork*. Grand Rapids: Baker Publishing Group.
Gatta, J. (2004). *Making nature sacred: Literature, religion, and environment in America from the Puritans to the present*. Oxford: Oxford University Press.
Gibson, L. (2009). Interviewed 5 July, Kayford Mountain.
Gill, S. (1987). *Mother earth*. Chicago: University of Chicago Press.
Gilliat-Ray, S., & Bryant, M. (2011). Are British Muslims "green"? An overview of environmental activism among Muslims in Britain. *Journal for the Study of Religion, Nature and Culture, 5*(3), 284–306.
Glacken, C. (1967). *Traces on the Rhodian shore: Nature and culture in western thought from ancient times to the end of the eighteenth century*. Berkeley: University of California Press.
Gottlieb, R. (2010). *Oxford handbook on religion and ecology*. Oxford: Oxford University Press.
Hall, D. (1990). *Worlds of wonder, days of judgment: Popular religious belief in early New England*. Cambridge: Harvard University Press.
Harvey, G. (2006). *Animism: Respecting the living world*. New York: Columbia University Press.
Heberlein, T. (2012). *Navigating environmental attitudes*. Oxford: Oxford University Press.
"James." (2010). Interviewed 26 September, Washington, DC.

Kalland, A. (2005). The religious environmentalist paradigm. In B. Taylor (Ed.), *The encyclopedia of religion and nature* (pp. 1367–1370). London: Continuum Press.

Kearns, L., & Keller, C. (Eds.). (2007). *Ecospirit: Religions and philosophies for the earth.* New York: Fordham University Press.

Krech, S., III. (1999). *The ecological Indian: Myth and history.* New York: W.W. Norton and Co.

Krech, S., III. (2005). American Indians as "first ecologists". In B. Taylor (Ed.), *The encyclopedia of religion and nature* (pp. 42–45). London: Continuum Press.

Lévy-Strauss, C. (1967). *The savage mind.* Chicago: University of Chicago Press.

Lovelock, J. (1979). *Gaia: A new look at life on earth.* Oxford: Oxford University Press.

Masuzawa, T. (2005). *The invention of world religions.* Chicago: University of Chicago Press.

McDuff, M. (2010). *Natural saints: How people of faith are working to save God's earth.* Oxford: Oxford University Press.

Merchant, C. (1980). *The death of nature: Women, ecology and the scientific revolution.* San Francisco: Harper.

Montrie, C. (2003). *To save the land and people: A history of opposition to surface coal mining in Appalachia.* Chapel Hill: University of North Carolina Press.

Nash, R. (2001). *Wilderness and the American mind* (4th ed.). New Haven: Yale University Press.

Orsi, R. (1997). Everyday miracles: The study of lived religion. In D. Hall (Ed.), *Lived religion in America: Toward a history of practice* (pp. 3–21). Princeton: Princeton University Press.

Peterson, A. (2009). *Everyday ethics and social change: The education of desire.* New York: Columbia University Press.

Pike, S. (2004). *New age and neopagan religions in America.* New York: Columbia University Press.

Rockefeller, S. (2005). Earth charter. In B. Taylor (Ed.), *The encyclopedia of religion and nature* (pp. 516–518). London: Continuum Press.

Scott, R. (2010). *Removing mountains: Extracting nature and identity in the Appalachian coalfields.* Minneapolis: University of Minnesota Press.

Sleeth, J. M. (2006). *Serve God, save the planet: A Christian call to action.* Grand Rapids: Zondervan.

Smith, J. Z. (1978). *Map is not territory.* Chicago: University of Chicago Press.

Smith, A. (1991, November/December). For all those who were Indian in a former life. *Ms. Magazine*, pp. 44–45.

Smith, J. Z. (1998). Religion, religions, religious. In M. Taylor (Ed.), *Critical terms for religious studies* (pp. 269–284). Chicago: University of Chicago Press.

Snodgrass, J., & Tiedje, K. (2008). Guest editor's introduction: Indigenous nature reverence and conservation—Seven ways of transcending an unnecessary dichotomy. *Journal for the Study of Religion, Nature and Culture, 2*(1), 6–29.

Swimme, B. (1994). *The universe story: From the primordial flaring forth to the Ecozoic Era—A celebration of the unfolding of the cosmos.* New York: HarperOne.

Taylor, B. (2005). Religious studies and environmental concern. In B. Taylor (Ed.), *The encyclopedia of religion and nature* (pp. 1373–1379). London: Continuum Press.

Taylor, B. (2010). *Dark green religion: Nature spirituality and the planetary future.* Berkeley: University of California Press.

Taylor, B., & Zimmerman, M. (2005). Deep ecology. In B. Taylor (Ed.), *The encyclopedia of religion and nature* (pp. 456–460). London: Continuum Press.

Tweed, T. (2006). *Crossing and dwelling: A theory of religion.* Cambridge: Harvard University Press.

Tylor, E. B. (1871). *Primitive culture: Researches into the development of mythology, philosophy, religion, art, and custom.* London: John Murray.

White, L., Jr. (1967). The historical roots of our ecologic crisis. *Science, 155,* 1203–1207.

Chapter 20
Reinventing Agency, Sacred Geography and Community Formation: The Case of Displaced Kashmiri Pandits in India

Devinder Singh

20.1 Introduction

The sudden displacement of an ethnic group can be visualized as a catastrophe which sets in motion the process of rebuilding, with what is left and what can be managed in the new circumstances. Such social rebuilding involves utilization of material and non-material resources an ethnic group has accumulated through its journey to forge a new community life in view of its present needs, future aspirations and conditions of the new environment. As a review of Foucault's work demonstrates, a culture cannot understand itself without first understanding its implicit connection and development within the constructs of religious belief and practice (Carrette 1999: 33). Considering religion as a universal ingredient of culture James Proctor (2006: 188) raises a stimulating question: Is religion some sort of cultural natural kind? Echoing the same conceptualization Brace et al. (2006: 29) argue that, in order to understand the construction and meaning of society and space, it is vital to acknowledge that religious practices in terms both of institutional organization and of personal experience are central not only to the spiritual life of society, but also to the constitution and reconstitution of that society.

Religious and spiritual matters form an important context through which the majority of the world population live their lives and forge a sense (indeed an ethic) of self, and make and perform their different geographies. Religious beliefs are central to the construction of identities and the practice of people's lives, from the habitual (the food that is eaten, the clothes people wear, the routines of daily prayer), to structuring the "vital" events of births, deaths and marriage (Holloway and Valins 2002: 6). Despite a great deal of recent attention paid to the practices and politics of identity formation, geographers have been slow to fully acknowledge the place of

---

D. Singh (✉)
Department of Geography, University of Jammu, Jammu, Jammu and Kashmir 18006, India
e-mail: devinder_jmu@yahoo.co.in

© Springer Science+Business Media Dordrecht 2015
S.D. Brunn (ed.), *The Changing World Religion Map*,
DOI 10.1007/978-94-017-9376-6_20

religion alongside such an axis of identity as race, class, nationality and gender in their analyses (Kong 2001: 226). Religion, Ceri Peach (2002: 255) emphasizes is not only a key to unravel ethnic identity, but is even perceived by many ethnic groups more than other ethnic markers in the conception of self. Geography faces a paradox in this regard as the construction of identity has become one of the central issues in contemporary geography (Keith and Pile 1993) and geographers still arguably know little about how communal identities in specific places are built around a sense of religious belonging (Brace et al. 2006: 29).

This study broadly touches the fertile terrain of the interface between the religion and the geographies of mobility and explores the processes through which an ethnic group rediscovers its religious world and creates religious spaces to (re)produce religious rituals in a quest to perpetuate distinct identity and continuity of spiritual life. It demonstrates the way (re)produced religious life implicates space, place and religious landscape in the new milieu. This focus constitutes one of strands recently identified by Lily Kong (2010: 761) at the interface of human mobility and religion. Through the case study of displaced Kashmiri Pandits[1] in Jammu city and the dynamics of cultural memory, landscape and performance, it illustrates the intimate interweaving of religion and identity. It further highlights how religion can become a bedrock, not only for reshaping and rebuilding ethnicity, but as a durable cushion during the crucial period in the life of a community. The study draws largely on fieldwork conducted from 1992 to 2012.

20.2 Kashmiri Pandits: Belief Systems and Sacred Geographies of Homeland

The Kashmiri Pandits are unique among all of their Brahman counterparts in the rest of South Asia in that they form one single group, the *Kasmira Brahmanas*, without any real subdivisions; they are equally unique in their language, customs, and traditions (Witzel 2008: 37). Their religious beliefs and traditions have evolved over centuries under a distinctive natural setting and historical circumstances. The *Nilmatapurana*, a seventh century local Sanskrit text, opens a broad picture of existing mythologies,[2] socio-religious life and sacred geography of the ancient Kashmir.

---

[1] Kashmiri Pandits form a distinct cultural group and associated with the Kashmir valley in the western Himalayas. Saraswat Brahman by caste, they occupy the highest echelon in the traditional Hindu social structure. The word Pandit in the traditional sense means a "learned" person or priest. Lawrence ([1895] 2005: 296) mentions their population at the end of nineteenth century as 52,576 persons. And, of the total, 28,695 were residing in Srinagar and small towns, while 23,881 were scattered far and wide in the valley.

[2] The *Nilamatapurana* mentions that the land of Kasmira was occupied by a vast lake called Satisar and inhabited by a demon Jalodbhava. The Vishnu (the supreme god in the Vaishnava tradition) along with other gods and goddess ordered Ananta to drain off the lake so as to eliminate Jalodbhava-invincible in waters. But Ananta made an outlet with a plough and, thus valley came into existence (Kumari [1968] 1988: 16).

Fig. 20.1 A destroyed temple built by King Avantivarman (AD. 855–833) about 30 km from Srinagar City (Photo by Devinder Singh)

The Kalhana's *Rajatarangini* (river of kings), compiled in the twelfth century, is another source of information on both Kashmir and its surrounding regions. Though confused in its early chronology (Witzel 2008: 46), it helps one to reconstruct the shaping of sacred geographies associated with Buddhism and Hinduism in the pre-Islamic period.[3] Some of the imposing imprints of ancient sacred landscape still dot the Kashmir valley in the form of relics (Fig. 20.1).

The destruction of the Kashmiri temples is universally attributed both by history and tradition to the bigoted Sikander (1489–1513), whose idol-breaking zeal procured him the title of *Bud-shikan*, or "iconoclast" (Lawrence [1895] 2005: 166). Abul Fazl at the end of sixteenth century, notes that the "whole country is regarded as holy ground by the Hindu sages" and enumerates 45 shrines dedicated to Mahadeva (one of many names of Lord Shiva), 64 to Vishnu, 2 to Brahma, and 22 to Durga (one of the forms of goddess). He adds, in 700 sacred places there are graven images of snakes which they worship and about which wonderful legends are told (Jarrett [1927] 2006: 356). We get a nuanced picture of the last quarter of the nineteenth through the scholarly works of George Buhler (1877), Aurel Stein (1894, 1900), and Walter Lawrence (1895). In the words of Lawrence, "there is hardly any river, spring or a hill-side in Kashmir that is not holy" and "structures of Hindus old temples defy time and weather" ([1895] 2005: 297, 161). In a recent work,

---

[3]The Kashmir was ruled by the Muslims for about five centuries. It was ruled by the Kashmiri rulers (1339–1586), the Mughals (1587–1753) and the Afghans (1753–1818).

Walter Slaje (2012: 12) illustrates "how remoteness of the valley and insuperable barriers in the past to visiting the original sacred places in India caused the development of nominal surrogates which in the course of their sanctification became suitable for serving the same sacred purpose as the remote original locality." Not surprisingly, Kashmiri Pandits theology shows a strong interweaving between local terrestrial elements and cosmic conceptions of Hinduism. And, this interweaving has not only imbued the landscape with cosmic meanings, but has profoundly shaped and crystallized their religious practices. The *Vitasta* (River Jhelum) in the Kashmir Valley, for example, is considered as a reincarnation of goddess Uma and their religious practices have strong connection with the river.

Historically speaking, Shaivism and Vaishnavism, the two dominant traditions of Hindu theology, have flourished in varying degrees in different periods and shaped the making and remaking of successive sacred spaces (Fig. 20.2). For example, Stein ([1900] 1989: 288) has recorded how political disturbed conditions of Upper Kisanganga Valley during the later Mughal and Afghan rule caused the neglect of one of the most central Sarada[4] *tirtha* (place of pilgrimage) and the development of a substitute for this ancient *tirtha* within the valley itself. Obviously, for infusing

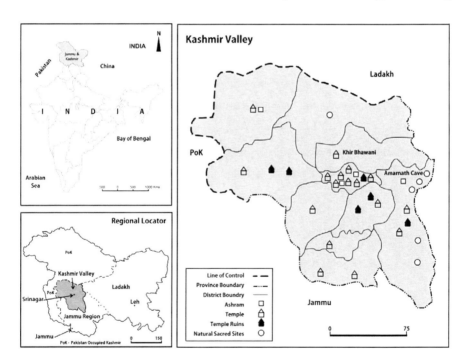

Fig. 20.2 Distribution of sacred places in the Kashmir Valley (Map by Devinder Singh)

---

[4]The ancient temple of Sharda is now located in Pakistan Occupied Kashmir. It is around 125 km (76 mi) north of Muzaffarabad. In Sanskrit, the word Sharda denotes both goddess Saraswati and goddess Durga.

Fig. 20.3 The original ambience of Khir Bhavani, the most venerated shrine of the community at Tulmul village in the Kashmir Valley. The goddess was worshipped in the form of spring initially. A temple was built in this spring by one of the Hindu Dogra rulers in the late nineteenth or early twentieth century (1885–1926) (Photo By Unknown Photographer)

spiritual life into new sacred space, it must have been necessary to produce and reinforce new discourses. Wangu (2008: 259) has shown in her research how from the late nineteenth to the early twentieth centuries, the social, religious and political situation in the valley of Kashmir was such that it helped to create a unique goddess popularly known as Khir Bhavani (Fig. 20.3). Lawrence ([1895] 2005: 296) also records that Khir Bhavani is their favorite goddess and perhaps the most sacred place in Kashmir is the spring of Khir Bhavani at the mouth of the Sind valley. This shrine has now acquired a Mecca-like status for the community. In the words of Chowdhary (2003) "the Khir Bhawani is part of life, existence and consciousness." Originally, constituting a serpent and spring worship at the site and reflecting a strong animistic character of their religion, a temple was built by one of the Hindu Dogra[5] ruler who ruled from 1885 to 1926 (Fig. 20.4). There is a belief that the water of the spring changes color symbolizing the future course of events.[6] It is a core Vaishnav shrine where unlike other Shivites shrines offering of meat is a taboo.

---

[5] The Kashmir valley came under the Hindu Dogra ruler of Jammu in 1846. They popularized vaishnavism by building Lord Rama temples and patronizing certain religious festivals which were not observed in Kashmir Valley (See Stein [1900] 1989; Wangu 2008; Madan 2008).

[6] Lawrence visited this shrine in the last quarter of the nineteenth century and writes, "When I saw the great spring of Khir Bhavani at Tula Mula, the water had a violet tinge, but when famine or cholera is imminent the water assumes a black hue" ([1895] 2005: 296).

Fig. 20.4 The present form of Khir Bhavani in the Kashmir Valley. This photograph was taken in 2011, when a large number of Kashmiri Pandits, pilgrims largely from Jammu and other parts of India, came to celebrate the *Jath Ashtami*, one of two yearly religious festivals associated with the shrine (Photo by Devinder Singh)

The Khir Bhawani is one of the *iesht devies (*presiding deities) of Kashmiri Pandits. Kashmiri Pandits theology links her to Sri Lanka. There are few legends about the circumstances which made her change her abode from Sri Lanka to the Kashmir valley. The most popular narrative holds that she was called Shama in Sri Lanka. When Rama (the main hero of the Hindu Epic *Ramayana* and considered incarnation of the Vishnu, and one of the three main gods in Hindu religious philosophy) defeated Ravana (the main villain in the Epic), *Shama* showed her desire to establish her new abode in Kyashav Bhumi (Kashmir Valley, the land of *Kyashav* seer). It was Hunumana, the dearer devotee of Rama, who lifted her in the form of water in a pitcher and transported her to the valley. Within the Kashmir valley, Hunumana made stoppage at Khanbarn and Manzgam before finally landing her at Tulmul, the village where Kheer Bhavani temple is located. It is further believed that she came along with 360 serpents. So initially, and until the ending of nineteenth century, water of the spring was the only object of veneration. Even after the establishment of a temple in the spring, the religious significance of water has not been affected. Even today when somebody returns after paying a visit to the shrine, Raina (2011) discloses that we ask him *darshan kasa tha*, which means what did the water look like?

Like Benares, the Kashmir has remained an ancient seat of learning and has contributed distinctively to Hindu philosophy, Abhinavgupta (950–1020 AD) being the foremost exopounder of Kashmir Shaivism. The spiritual philosophy of the Kashmiri Pandits accords supreme position to Lord Shiva which makes them essentially Shaivites. The Shivaratri or *Haerath* is one of the most sacred and ancient festivals of the community. Manvati (2011: 21) reveals that there are philosophical, mythological, religious, socio-cultural, and historical dimensions attached to this great and biggest festival of the community. This centrality of Shiva in their life resonates in their daily prayer in the form of a Sanskrit hymn composed by Abhinavagupta "*chinamayam ekam anantam anadim*" which translates as "Shaiva is the embodiment of knowledge, unique, without beginning and end." However, the goddess in the form of Sharika, Ragyani, Bhagwati, Uma, Baderkali and Sharda also forms an integral part of their spiritual beliefs and practices. Their devotion to a living or non-living saint is well described by Walter Lawrence:

> Every Kashmiri believes that "saint will aid if man will call," and they think that a dead saint is more efficacious than the living priest. The Kashmiri are called by foreigners *Pir Parast*, that is, saint worshippers, and the epithet is well deserved. ([1895] 2005: 286)

The rhythmic religious life of the community follows the rhythmic movement of moon; the *Ashtami,* the eighth day of the waxing moon considered the most auspicious day for religious performances. Like *Losar* of the Tibetan Buddhist, *Nevrah,* the new moon of *Chetr* (March or April) represents New Year for the community. It is one of the most sacred times for Kashmiri Pandits to begin a new affair and is celebrated at individual and community levels. The following lines summarize an insider's view of what religion mean to Kashmiri Pandits:

> To them the religious places of their motherland are more sacred than the holy shrines of the outside Hindu world. A Pandit adores his native Ganges, his own Prayag[7] and worships his own ancestral gods. A history of the heroes of the ancient Kashmir send a thrill of joy through his whole being and he is never weary of extolling the greatness of Shaivism, the school of philosophy which was born in his own land. (Bazaz 1941: 280)

20.3 Political Turmoil in Kashmir: Displacement and Resettlement

One of the tragic consequences of the turbulent political history of Kashmir has been the recent exodus of Kashmiri Pandits. In a short period from late 1989 to early 1990 an astounding number[8] (over 58,000 Kashmir Pandits) families fled their

---

[7] The term Paryag is used for the confluence of the Ganga, Yumana and legendry Saraswati at Allahabad. It is one of the significant places of pilgrimage of the Hindus.

[8] The figures given for the displaced Kashmiri Pandits are varying and confusing. The India Today, a widely read and circulated weekly magazine (15 July 1992) wrote that of the 125,000 Pandits who once inhabited the Valley, only 3,000 remain. On 22 August 2009, the opposition party BJP leader complained in the Jammu & Kashmir Assembly about the contradictory figures (33,499 and 56,000) of Kashmiri Pandits registered migrant families within a gap of 1 year (Early Times, 26 August 2009). On 15 May, 2012 the Indian Minister of State for Home M Ramachandran told Indian Parliament in a written reply that there are 58,697 Kashmiri migrant families registered with

homeland and resettled largely in Jammu city, the winter capital of the state of Jammu and Kashmir[9] and few other Indian cities,[10] particularly Delhi, Faridabad, Bangalore and Pune. This recent migration was the result of the atmosphere of insecurity and fear that erupted by the beginning of insurgency in the Kashmir valley in 1989. The Muslim insurgents were demanding Azadi or freedom. In some places the Kashmiri Pandits were implicitly and explicitly threated to either support or leave the valley. Reflecting on the issue, the present Union Minister and former Chief Minister of the State Farooq Abdulla admitted that a "mistake led to exodus of Kashmiri Pandits" (*Greater Kashmir*, 21 February 2011).

It is noteworthy that there are legends and historical evidence of their migration in the past. Witzel (2008: 70) mentions the emigration of large groups of Brahmans under some of the Kashmiri sultans, notably Sikandar (1489–1513), and especially during the Afghan occupation (1752–1819) when they spread all over India. Their greater preference for Jammu city, this time, may be linked to a community historical connection,[11] occupational structure, and government resettlement policy,[12] Initially, it was a gigantic task to provide accommodation to these uprooted people

---

respective relief authorities. These include 38,119 families in Jammu, 19,338 in Delhi and 1,240 in other places (Daily Excelsior 16 May 2012). The Relief and Rehabilitation Organization, Government of Jammu & Kashmir (2011: 1) shows the figure at 60,000 migrant families. Surprisingly, research literature too has provided different figures for the total numbers of migrants. For example scholars like Evans (2002: 23–26) mentions it was between 155,000 and 170,000 persons; however Madan (2008: 25) reports 200,000 and for Duschinski (2008: 46) the total number was 100,000; and for Slaje (2012: 26) 400,000.

[9] The multireligious and multicultural state of Jammu & Kashmir forms one of the 35 states of India. This state is popularly known to the outer world as Kashmir. It has three major natural regions, namely, the Kashmiri valley, Ladakh and the Jammu Region. Kashmir and Ladakh experience severely cold winters. The southern parts of Jammu which adjoin the Punjab Plains experience sub-tropical climate and undergo very hot summers. The Jammu & Kashmir is the only Muslim dominated (66.9 % in 2001 Census) state in India. In the same Census year, other major religious groups percentages were Hindu 29.6 %, Sikh 2.0 %, Buddhist 1.1 %, and Christian 0.20 %. It is noteworthy and relevant that the Kashmir region is overwhelming dominated by Muslims (over 97 %) and Jammu region has Hindu domination (65.2 %). And, within Jammu region, Jammu District has 86.0 % Hindus, 5.7 % Muslims, 7.40 % Sikh, 0.01 % Christians, and 0.12 % Janis.

[10] Henny Sender (1988; see also Pant 1987) has elaborately discussed their historical connection with different cities in the nineteenth and early twentieth centuries.

[11] The state of Jammu & Kashmir has a practice of having a winter (Jammu) and summer (Srinagar) capital since the second half of the nineteenth century. Kashmiri Pandits have traditionally been associated with Government service (see Bazaz 1941: 282–283) and also with migration. Besides, a good member of Kashmiri Pandits in other government jobs, had established links with Jammu city during their stay in the city. The Kashmiri Pandit Sabha, one of the oldest organisation of the community established its office and a temple in Jammu in 1914. Until the 1960s the Kashmiri Pandits residences were dispersed and confined to areas close to the old secretariat in the northeast of the old city. Subsequently, two Kashmiri Pandits localities (Mohinder Nagar and Bhagwati Nagar) emerged in the 1960s in the southwest of the old city.

[12] A majority of Government organized relief camps were established in and around Jammu city. A new department (Relief and Rehabilitation Organization [Migrant]) was set up to look after different relief and rehabilitation aspects of the migrants.

in Jammu city. Some were accommodated in tented camps; some occupied incomplete government buildings and a majority of them found rented accommodation in and around Jammu city. This temporary arrangement initiated the process of search for a permanent residence. A directional bias was seen in their intraurban mobility; the western parts of the old city being the most preferred destination. This spatial redistribution was embedded in the political, economic, and cultural circumstances. The segregated spaces, initially created by the state in the form of tented settlement, was followed by one room tenements and then by two room flats which were located in this part of the city, which created a cultural affinity for this area. Secondly, the availability of inexpensive uncultivated and agricultural land was another gravitating force. At present, almost all the old and new religious, social, cultural, and political institutions or organizations of the community are located in this western part of the city.

20.4 Collective Memories and New Sacred Geographies

Our continuity of life is tied to space and time. The rootedness of people in space gives shape to place. The bonds between the place and inhabitants become so intimate through time that any voluntary or involuntary spatial shift constitutes merely the movement of the bodies while the self continues to be in touch with the lived places. This remembrance of the lived life, condensed with ideology, spatial practices and performances lead to shared memories or cultural memories. Lowenthal (1975: 12) notes that the collective past is no less precious than the personal; indeed, one is an extension of other. Beginning with the work of French sociologist Maurice Halbwachs ([1951] 1992), many scholars including geographers have come to see memory as a social activity as both an expression and a binding force for group identity (Crang and Travlou 2001; cf Hoelscher 2003: 660). Awareness of the past, Lowenthal (1979: 103) suggests, is essential to the maintenance of purpose of life. Without it we would lack all sense of continuity, all appreciation of causality, all knowledge of our own identity. Each is essential to the being of others. This empirical research makes an attempt to unfold the processes through which the collective memories of a displaced community are crystallized through cultural performances and landscapes, the articulations considered unusually important by Steven Hoelscher (2003: 661).

Materially, the western section of Jammu does not provide an easy answer to how community has reconfigured the geography of the city. However, a closer look reveals the ways community has scripted a new chapter of its autobiography by creating new sacred spaces for enacting spiritual life in a meaningful way. The most striking feature of the new religious landscape constitutes the creation of the replicas of the shrines located in the Kashmir valley. These replicas include temples as well as *ashrams*, a Hindu shrine associated with some living and non-living saints (Fig. 20.5). Unlike some documented cases (Wood 1997; Dunn 2001) where construction of religious landscapes has faced substantial local opposition, the emergence of around

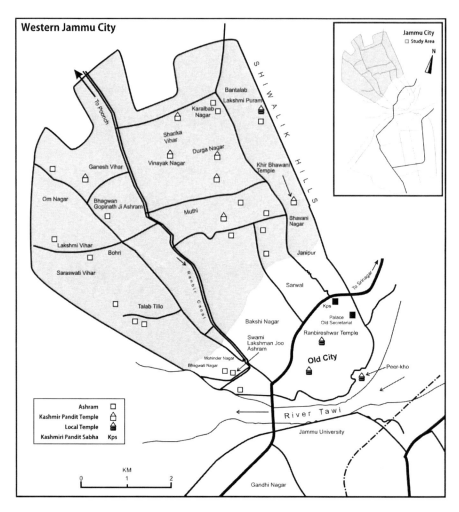

Fig. 20.5 Distribution of sacred places built by the migrant Kashmiri Pandits in Jammu City since the 1990s (Map by Devinder Singh)

25 Kashmiri shrines in Western Jammu has gone smooth. This reminds us of what Schein (1997: 660) suggests that interpreting the cultural landscape is a geographically specific exercise that requires interrogating the role of landscape in social and cultural reproduction, as well as understanding landscape within wider social and cultural contexts. Jammu, popularly known as the "City of Temples" largely because of the imposing temples built by the Hindu Dogra[13] rulers (1822–1947), has historically offered a conducive social environment to the community.

---

[13] The Dogras comprise a distinct ethnic group and are spread over parts of the states of Jammu and Kashmir, Himachal Pradesh, and Punjab.

"Centrality," a key element of space, has already appeared in the sacred space produced by the community. The *Bhagavaan* Gopi Nath (1898–1968) *ashram*, has undoubtedly assumed the central position (see Fig. 20.5). The spatial and social forces which have been at work to give it a supreme position bring up another interesting theme for investigation. Making its beginning from a rented accommodation in the south Jammu, it soon shifted to the present location in 1993. Since then it has grown both in area and activities. Its six times horizontal expansion has pushed away the residential area. Considered nerve centre (Wattal 2012) of the community, its premises include: *Param Dham* (heavenly abode); two huge halls one as *Havan Shalla* (place for scared fire) and other for collective hymn singing and preaching; the Heritage Hall, displaying personal belongings of the saint, his history and natural milieu of the Kashmir Valley; office (only shrine with computers and internet facilities); a community kitchen, store, dispensary, spacious lawn; and a shop for selling souvenirs, audio cassettes of kashmiri hymns and religious books. The two main yearly functions of the shrine attract more than 10,000–15,000 devotees. Not surprisingly, *ashram has* an active look every evening and the *Param Dham* with a capacity of over 100 people is full during evening *arti*, collective hymn singing. People often bring both home cooked or purchased items which are distributed after the evening rituals. Kashmir tea in the form of *parshad,* (sacred eatable gift) is served to the devotees in the spacious lawn.

Surprisingly, the replicas of the Khir Bhawani (Fig. 20.6), the most central and highly venerated shrine of the community in the Kashmir valley and the *ashram* of Swami Lakashman Joo (1907–1991), a great exponent of Kashmir Shaivism, have not been able to attract a large number of devotees on the daily basis. The replica of

Fig. 20.6 Replica of Khir Bhavani in Jammu City. The migrant Kashmiri Pandits are celebrating Jath Ashtami (Photo by Vishal Sharma)

Khir Bhavani only turns into a vibrant sacred place on two yearly festivals: one on *Jath Ashtami,* the eighth day of the waxing moon (May or June) and other on *Har Ashtam,* the eighth day of the waxing moon (June or July). *Ashram* congregations are confined to the birth and death anniversaries of Swami Lakashman Joo.

Other shrines dispersed in the westerns section of city may not be popular at the community level, but their ethnic anchoring or unifying role and creation of sense of unity cannot be underestimated. In fact these shrines have been the citadels for place making, which to Wood (1997: 58) is a continual process of shaping identity and expressing social relations. Among the migrant community in Jammu city, the domestication of local features is another striking feature highlighting performances of the agency to preserve its identity rooted in religion. The river Chenab, located around 25 km (15.5 mi) from Jammu city and the Ranbir canal (see Fig. 20.5) are recreated as the *Vitasta* (River Jhelum) to celebrate the *Veth Truwah,* the birth of the goddess on the 13th day in the month of *Bhadon* (August/ September). A new feature of Shivratri celebration is the visit to local Shiva temples particularly the Ranvireshwar temple, Apshambu, and Peer Kho (see Fig. 20.5). It is interesting that the celebrations in Kashmir were generally confined to home and neighborhood spaces.

An exploration of the "place names" in the city discloses the way the Kashmiri Pandits have struggled, negotiated, and compromised in reconfiguring the symbolic landscape across the western parts of the city. Lowenthal (1975: 9) has illustrated the significance of a portable symbol in aiding continuity. What has prominent in this new symbolic space is its sacred character. The community has deeply imbued it with names of gods (*Ganesh Vihar* and *Vinayak Nagar*), goddesses (*Saraswati Vihar, Lakshmi Vihar, Sharika Vihar, Durga Nagar, Bhavani Nagar* and *Lakshimipuram*) and saints (Gopinath Vihar and Karalbab Nagar) (see Fig. 20.5). It is interesting that this symbolic landscape opens another productive site to explore the relationship between periodic trends of naming and renaming and the overall atmosphere in which distinct toponym come into existence. It is often seen that people generally turn towards religion and spirituality in a time of crises and a toponymic terrain of that period should have a touch of religiosity in it. In *City of Some: The Hidden History of Jerusalem,* Meron (1996): 68; but also see Azaryahu and Golan (2001) describes naming as an act of taking possession. Evidences from the field also reveal how sacred names can act as a double edge knife for creating both identity and a sense of territoriality. For Massey (1995: 187) it is the resonance of the past in the present as a self-conscious building-in of "local character." Even at the individual level a sacred name may signify a double symbolic meaning. Piarey Hatash's (2012) description of his residence's location is insightful:

> You will find a house named Satisar [14] opposite Doordarshan (Television Station). Why have I kept this name Satisar? So that after my death people will know that he was a Kashmiri who had been driven out of Kashmir.

---

[14] The word Satisar is a combination of two words, that is, Sati means goddess and sar is lake. The *Nilmatapuran* mentions Satisar occupying the Kashmir Valley before it was drained off by divine intervention.

The year 2007 marked the beginning of a new phenomenon in the resettlement when for the first time a few temples were visited by migrant Kashmiri Pandits in the Valley largely due to the normalization of militancy and the proactive policies of the state. Though trend has grown over the years, with some breaks in between, when the political situation in the valley has not been conducive and when the Amarnath land controversy[15] erupted in the year 2008 and violence in 2010.[16] On May 2012, 65 buses were arranged by State of Jammu and Kashmir to take pilgrims to different shrines and sites of pilgrimages in their homeland.

20.5 Dialectic of Religious Landscape: Original Versus Replicas

Migration to a new environment engenders both continuities and discontinuities of past religious practices and inventions of the new traditions. Obviously, the additions and omissions in the religious life of a migrant community manifest connections between the group and the new social and physical environment. The blending of the old and new practices is mediated through the creation of new religious landscapes as well as in some cases the adoption of existing ones. The new practices manifest adaptations which an ethnic group makes in negotiating a new journey in the alien setting. For example, the sharp differences in the natural environment of Kashmir and Jammu have brought about changes in the religious practices.[17]

Three pertinent questions are important in this context: why are replicas built, how does community respond to these duplicate cultural objects, and what purpose do they serve? Lowenthal (1975: 9) argues that some who sunder ties with past landscapes find continuity in furnishing their landscape with real and symbolic replicas. To Hoelscher (2003: 661), since the original experiences of the past are irretrievable and forever unstable, we can only grasp them through their remains, that is, through objects, images, events and representations. Such creations for Dipanker Gupta (2000: 41) are *vicarious* spaces, the spaces recalled in the diasporic setting. They are not only integral to cultural identity, but also shape place attachment because of what they represent and they can serve as powerful sources of memory (Hayden 1997: 116). As already illustrated, cultural memories and the urge to preserve ethnicity have been the potent forces for the (re)production landscapes and practices.

---

[15] In the year 2008 a decision of the state government to handover some land to the Sri Amarnathji Shrine Board to create the infrastructure for the pilgrims' visiting the Amarnath Shrine in Kashmir Himalayas led to agitation in the Kashmir Valley. When the government reviewed the initial decision, there was a mass agitation in Jammu province and in some other parts of India which continued for 61 days and ended on 31st August.

[16] A teenager was killed in a cross-firing in Kashmir which led a prolonged violent agitation in the valley.

[17] Anne Buttimer (2006: 197) has shown her fascination with the role of geography, the biophysical world, its seasonal rhythms of climate, wildlife, and livelihood, shaping religious sensitivities and practices, liturgies, and ways of life for most communities of the world.

Empirical evidence with regard to the second question reveals a sense of duality prevailing particularly among the adult and aged.[18] They still carry the memories of lived experiences. A community member who lived in Jammu responded:

> I do not support this (building of replicas). Can you bring Panja Sahib[19] here? It is just to console the heart. We may put up Lord Shiva's photograph at home but cannot bring cave[20] here. A shrine stays where it exists. (Nath 2003)

The original shrine of Khir Bhawani is steeped in philosophical background which provides it both antiquity and natural aura. The replica not only lacks this, but does not match the architecture, size and natural surroundings of the original. The last factor has even affected practices associated with the original. The offering of lotus and rose has been replaced by the marigold and milk offering which has been substantially reduced.

The response of a senior member of the migrant community provides some glimpses of community's dilemma:

> They have made it on a mountain. They should have chosen a plain area for the shrine.[21] There (in case of original), as we entered the premises, we would get into heaven and mind would remain greatly tranquil. In the case here (the replica) mind only turns to the Bhagwati.[22] (Bhat 2011)

This quote brings home the point that the original can never be replicated in totality. Not only the construction, but also the additions and omissions in the replicas produce divergent discourses. The replica is like a flower whose fragrance always lies in the original. For example, in response to the replica of the goddess Shirika in Faridabad, near the Indian capital of New Delhi, Moti Lal Khemu (2008) said, "they have made a nice replica, but it is very hot there." Vimla Raina (2012), showed her displeasure for depiction of Led Ded, a fourteenth century poetess and saint as worshiping the snow *linga* (symbol of Lord Shiva), at the site of replica (Fig. 20.7). She said, "it does not fit well as she (Led Ded) was not a worshipper of idol and that her notion of Shiva was omnipresent" (Raina 2012).

---

[18] An intergenerational survey exploring the faith in original sacred places in the Kashmir Valley and those built in Jammu city after migrations was conducted in 2010. An overwhelming majority (87 %) of the migrants in the age above the 40 years identified their most venerated shrine in the Kashmir Valley. For 36 %, the Khir Bhawani was the most respected sacred place. Surprisingly, 61 % informants who had come of age after migration and were below the age 25 years at the time of survey, mentioned different religious places outside the Kashmir valley as number one in their spiritual life. Only 10 % mentioned Khir Bhawani as their highly venerated spiritual place.

[19] The Panja Sahib is a Sikh Shrine associated with their first Guru Nanak and located in Pakistan.

[20] Cave here refers to the Amarnath Cave, one of the most venerated natural sites of pilgrimage associated with Hindu god Shiva and located at 13,000 ft (3,962 m) above sea level in Kashmir Province. The route leading to it is highly treacherous and receives heavy snowfall. More than 630,000 pilgrims visited the shrine in 2011.

[21] The original shrine of the Khir Bhawani is located amidst a spacious wooded plain in the Kashmir Valley. The temple of Khir Bhawani in Jammu city is located in a congested locality and one has to negotiate over 100 steps to reach the main premises.

[22] The Bhagwati is one of the names of the Goddess. Kashmiri Pandits call the presiding deity of Khir Bhawani as mother Ragyani or Bhagwati.

Fig. 20.7 A tableau representing Lal Ded, a fourteenth century saint and poetess worshipping snow linga (symbol of the Lord Shiva) at Amarnath Cave in the Kashmir Himalayas. This representation is part of the spacious replica of the Sharika temple built at Faridabad near New Delhi. The original temple is located on the Hari Parbat hill at Srinagar city in the Kashmir Valley (Photo by Devinder Singh)

This discussion does not belittle the significance of replicas and gives lead to a third question related with the significance of these duplicates. The replicas according to Dipanker Gupta (2000: 41) authenticate cultural enactment and practices. The monthly and yearly festivals which were part of life in the Kashmir Valley are celebrated at the replicas built in Jammu city. Some of the duplicates of ashrams have gained a tremendous following in the migratory setting which may be attributed both to the spatial concentration of the community and strong urge to preserve the community culture. In sum, these scared places often become sites where community members get together to celebrate both religious and social functions and pass on their cultural traits to their next generation.

These duplicates open another productive but neglected aspect of geographies of religion and belief systems which Holloway (2006: 182; also see Kong 2001) calls sensuous, vitalistic, and affectual forces through which spaces of religious, spiritual, and sacred are performed. Arjan Dev Vishan (2012), a founder member of the organization (*Ardh Ratri Maharagyani Save Sanstha*) which resumed the tradition of a monthly congregation of the original Khir Bhavani at a private residence after migration in 1990 and which finally created this replica in 1993, narrated how at different stages of the evolution of the replica of Khir Bhavani, they felt the presence and guidance of divinity at the site. Having left Jammu in 1997 and now living in Noida, he was here along with his wife to participate in one of two the yearly festivals on *Har Ashtami,* the eighth day of the waxing moon.

20.6 Conclusion

This empirical case study confirms the notion of religious beliefs and practices as key elements or key tools of what Swindler (1986: 273) terms as the 'tool kit' of the culture. The experience of Kashmiri Pandits unfolds the processes through which a displaced community rebuilds its lost world with the inexhaustible fuel of collective memories, landscapes, and practices in the original or modified forms, in a quest to perpetuate distinct identity and continuity of spiritual life. In this process a new identity and a new sacred geography is taking roots. In such a situation one can think of extending what Doreen Massey (1995: 186) tells us about places identity to communities identity that "think them as temporal and not just spatial: as set in time as well as place." The sacralization of space by means of sacred topomyms, replicas, and domestication of local features for recreating homeland rituals and practices discussed here reactivated John K. Wright concept of "geopiety." However, it must be remembered that attachment to the past landscapes or spiritual imaginations of the past can also lead to disenchantment of the present (re)production particularly among the elders in the community which this study has highlighted. And finally, the study supports Anne Buttimer's (2006: 198) thought of wisdom that geography, in the broadest sense, as a lived reality and academic subject, affords a fertile ground for research on religion and environment.

Acknowledgement I am grateful to Professor Saraswati Raju, CSRD, Jawahar Lal Nehru University, New Delhi for her encouragement. I also wish to thank Dr Suresh Babu, Department of Sociology, University of Jammu, for his fruitful comments. I extend my regards and gratitude to innumerable members of the Kashmiri Pandit community for their cooperation and hospitality.

References

Azaryahu, M., & Golan, A. (2001). Re(Renaming) the landscape: The formation of the Hebrew map of Israel 1949–1960. *Journal of Historical Geography, 27*, 178–195.
Bazaz, P. N. (1941). *Inside Kashmir*. Srinagar: The Kashmir Publishing Co.
Bhat, J. L. (2011, April 11). *Priest and shopkeeper*. Interview by author. Sharika Vihar, Jammu.
Brace, C., Bailey, A. R., & Harvey, D. C. (2006). Religion, place and space: A framework for investigating historical geographies of religious identities and communities. *Progress in Human Geography, 30*(1), 28–43.
Buhler, G. (1877). Detailed report of a tour in search of Sanskrit MSS made in Kashmir, Rajputana, and Central India. *Journal of Bombay Branch of the Royal Asiatic Society of Great Britain and Ireland*, Bombay Branch, 12, extra number 34 A.
Buttimer, A. (2006). Afterword: Reflections on geography, religion, and belief systems. *Annals of the Association of American Geographers, 96*(1), 197–202.
Carrette, J. (1999). *Religion and culture by Michel Foucault*. Manchester: Manchester University Press.
Chowdhary, K. L. (2003, February 8). *Physician and author*. Interview by author. Bantalab, Jammu.
Crang, M., & Travlou, P. S. (2001). The city and topologies of memory. *Environment and Planning D: Society and Space, 19*, 161–177.

Daily Excelsior. (2012, May 16). 58,697 Kashmiri migrant families in India, p. 1.
Dunn, K. M. (2001). Representation of Islam in the politics of mosque development in Sydney. *Tijdschrift voor Economische en Sociale Geografie, 92*(3), 291–308.
Duschinski, H. (2008). Survival is now our politics: Kashmiri Hindu community identity and the politics of homeland. *International Journal of Hindu Studies, 12*(1), 41–64.
Early Times. (2009, August 26). Statistics again embarrass government in assembly, p. 1.
Evan, A. (2002). A departure from history: Kashmiri Pandits, 1990–2001. *Cotemporary South Asia, 11*(1), 19–37.
Greater Kashmir. (2011, February 21). Mistakes led to exodus of Kashmiri Pandits, p. 3.
Gupta, D. (2000). *Culture, space and nation-state.* New Delhi: Sage.
Halbwachs, M. ([1951] 1992). The social frameworks of memory. In *On collective memory* (pp. 37–189). Chicago: University of Chicago Press.
Hatash, P. (2012, May 7). *Retired professor of Hindi.* Interviewed by author. Jammu University.
Hayden, D. (1997). Urban landscape history: The scene of place and politic of space. In P. Growth & T. W. Bressi (Eds.), *Understanding ordinary landscapes* (pp. 111–133). New Haven: Yale University Press.
Hoelscher, S. (2003). Making place, making race: Performances of whiteness in the Jim Crow South. *Annals of the Association of American Geographers, 93*(3), 657–686.
Holloway, J. (2006). Enchanted spaces: The séance, affect, and geographies of religion. *Annals of the Association of American Geographers, 96*(1), 182–187.
Holloway, J., & Valins, O. (2002). Editorial: Placing religion and spirituality in geography. *Social and Cultural Geography, 3*(1), 5–9.
India Today. (1992, July 15). Kashmiri Pandits: Living on the edge, p. 75.
Jarrett, H. S. ([1927] 2006). *The Ain-i –Akbari by Abul Fazal Allami* (Trans. Vol. 11). New Delhi: D.K. Publishers.
Keith, M., & Pile, S. (1993). Introduction. Part 1: The politics of place. In M. Keith & S. Pile (Eds.), *Place and the politics of identity* (pp. 1–21). London: Routledge.
Khemu, M. L. (2008, April 24). *Writer and director.* Interview by author Apna Vihar, Jammu.
Kong, L. (2001). Mapping 'new' geographies of religion: Politics and poetics in modernity. *Progress in Human Geography, 25*(2), 211–233.
Kong, L. (2010). Global shifts, theoretical shifts: Changing geographies of religion. *Progress in Human Geography, 34*(6), 755–776.
Kumari, V. ([1966] 1988). *The Nilmatapurana: A cultural and literary study.* Srinagar: J & K Academy of Art, Culture & Languages.
Lawrence, W. ([1895]2005). *The valley of Kashmir.* Srinagar: Gulshan Books.
Lowenthal, D. (1975). Past time, present place: Landscape and memory. *The Geographical Review, 65*(1), 1–35.
Lowenthal, D. (1979). Age and artifact: Dilemmas of appreciation. In D. W. Meinig (Ed.), *The interpretation of ordinary landscape: Geographical essays* (pp. 103–128). New York: Oxford University Press.
Madan, T. N. (2008). Kashmir, Kashmiris, Kashmiriyat: An introductory essay. In A. Rao (Ed.), *The valley of Kashmir: The making and unmaking of composite culture* (pp. 1–34). New Delhi: Manohar.
Manvati, R. (2011, March 21–23). *Haerath- traditional feature.* Koshur Samachar.
Massey, D. (1995). Places and their past. *History Workshop Journal, 39,* 180–192.
Meron, B. (1996). *City of some: The hidden history of Jerusalem.* Berkeley: University of California Press.
Nath, B. (2003, March 23). *A Kashmiri Pundit migrant.* Interview by author. Borhi, Jammu.
Pant, K. (1987). *The Kashmiri Pandits: Story of a community in exile in the nineteenth and twentieth centuries.* New Delhi: Allied.
Peach, C. (2002). Social geography: New religion and ethnoburbs: Contrasts with cultural geography. *Progress in Human Geography, 26*(2), 252–260.
Proctor, J. (2006). Religion as trust authority: Theocracy and ecology in the United States. *Annals of the Association of American Geographers, 96*(1), 188–196.

Raina, T. N. (2011, April 24). *Community member*. Interview by author. Muthi, Jammu.

Raina, V. (2012, February 16). *Exponent of Shaivism and writer*. Interview by author, Gandhi Nagar, Jammu.

Relief and Rehabilitation Organization (M) Government of Jammu and Kashmir. (2011). Rebuilding life step by step. Jammu, p. 1.

Schein, R. H. (1997). The place of landscape: A conceptual framework for interpreting an American scene. *Annals of the Association of American Geographers, 87*(4), 660–680.

Sender, H. (1988). *The Kashmiri Pandits: A study of cultural choice in North India*. New Delhi: Oxford University Press.

Slaje, W. (2012). Kashmir minimundus: India's sacred geography en miniature. In R. Steiner (Ed.), *Highland philolog* (pp. 9–34). Wittenberg: Universitätsverlag Halle.

Stein, A. (1894). *Catalogue of the Sanskrit manuscripts in the Regunath Temples Library .... Kashmir State Council*. Bombay: Nirnaya Sagara Press.

Stein, A. ([1900] 1989). *Kalhana's rajatarangini: A chronicle of the kings of Kashmir* (Trans. with an introduction, commentary, and appendices, Vol. II). New Delhi: Motilal Banarsidass.

Swidler, A. (1986). Culture in action in action: Symbols and strategies. *American Sociological Review, 51*, 273–286.

Vishan, A. D. (2012, June 27). *Founder member of the Khir Bhavani organization in Jammu*. Interview by author, Bhavani Nagar, Jammu.

Wangu, M. B. (2008). Maji Khir Bhavani: The Kashmir Kuladevi. In A. Rao (Ed.), *The valley of Kasmir: The making and unmaking of composite culture* (pp. 259–300). New Delhi: Manohar.

Wattal, S. (2012, May 15). *Key caretaker of the Bhagawan Gopinath Ashram*. Interview by author. Gopinath Vihar. Jammu.

Witzel, M. (2008). The Kashmiri Pandits: Their early history. In A. Rao (Ed.), *The valley of Kashmir: The making and unmaking of composite culture* (pp. 37–95). New Delhi: Manohar.

Wood, J. (1997). Vietnamese American place making in Northern Virginia. *The Geographical Review, 87*(1), 58–72.

Chapter 21
Symbiosis in Diversity: The Specific Character of Slovakia's Religious Landscape

Juraj Majo

21.1 Introduction

Exploring the relationship between landscape, society, and religious systems is one of the fundamental research topics in the geography of religion. This paradigm is based on a model designed by M. Büttner, and interacts between three spheres – the spiritual, the social, and the level of visible indicators which is a part of the environment (Büttner 1980: 102) (Fig. 21.1). For human geography, or religious and cultural geography respectively, today's interest is not so much in the positivist tracking or *functional and descriptive efforts*, including the distribution of buildings and other sacred elements, but rather tries to find and explain the importance for the community or larger communities (such as places of pilgrimage) or the deconstruction of a specific symbology that remains often imperceptible by non-contextual observation (Kong 2005: 367). The religious significance can be of a certain sacred building for a community member which may be have a different meaning for non-members. An important aspect is, therefore, the process of constructing the meaning of a sacred building and its importance over time. Therefore, the implication of shifting interest in cultural geography from geography to culture (Peach 2002: 257) could to some extent also be applied to religious geography. The strong relationship between place and religions is also suggested by another author (Stump 2008: 221), claiming that there is a strong meaning for the concept of place, that is, places that symbolize the expressions and extensions of the faith of adherents with ascribed transcendent and religious significance. An important aspect of a sacred place according to C. Park (Park 2003: 210–211) is the stability of a sacred place (most often a church building or structure) over time which is influenced either by local factors

---

J. Majo (✉)
Department of Human Geography and Demography, Faculty of Sciences,
Comenius University in Bratislava, Bratislava, Slovak Republic
e-mail: majo@fns.uniba.sk

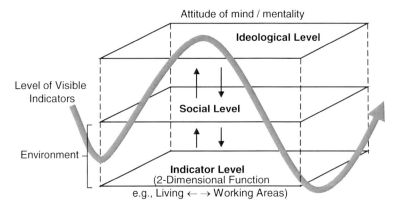

Fig. 21.1 Interaction between religion, society and landscape (Source: Manfred Büttner 1980, reprinted with permission of Taylor & Francis, Ltd.)

(local migration movements) or broader factors (major demographic and religious trends), which together result in either the growth or decline of the church. In Slovakia, we face both types of dynamics, from which the specific suggests hints of secularized rural regions, such as in Gemer in South Central Slovakia (Majo 2009). The region with an above average level of the share of non-church affiliated population is somewhat of a unique situation in Slovakia, which is generally considered as a country with high religiosity level especially in rural regions.

Researching religious landscapes is one of the most important topics in the so called "new geography" of religions (Kong 2005: 367), especially as seen in the conflict between the sacred and profane (cultural and political) which may be a cause for dualism. Constructing a sacred place, according to Kong (2005), is both a *politics of place* and a *poetic of place*. Within the sacred field is an equally interesting exploration; it is the relationship between the sacred objects of different confessions and religions and vice-versa. Even in the context of Büttner's model outlined above, it is important to look for guidance in exploring and developing the state of religious communities in different contexts. Many tendencies can be very appropriately applied to the territory and within the context of Slovakia.

Sacred objects with their local and wider significance in place are a demonstration of religious identity Its conceptual deployment reflects not only the religious structure of the population, but also encompasses the overall historical development of the relationship between denominations. These relationships in the past were often influenced by political interference. The local level often reflects the higher levels of this relationship.[1] So the politics of place seems to be a more important factor than

---

[1] There are several decades in history of Slovakia where the political situation stimulated hostility of clergy between various denominations. For example P. Kónya states (Kónya 2010: 63) that at the end of seventeenth century in time of culminating Counter-Reformation in Austrian Kingdom, "Lutheran ministers, doomed to live at the periphery were always under the threat, attacks and humiliation from the authorities, Catholic clergy and soldiers." Other historically salient events

the poetics of place in the geographical representation of inter-denominational relation in history in Slovakia. Though it might seem too rational and too profane in its essence, the "politics of space" created a specific atmosphere where the minority denomination had to survive, maintain its identity, and develop its own cultural codes and signs, that altogether make up the religious landscape of Slovakia unique.

Signs of religious identity are evident not only in the case of religious buildings, but also cemeteries, which are directly related to religious life. This fact is very evident in the variety of religious identities that exist in this part of Europe, which are inevitably reflected in the spatial differentiation of burial methods and the different types of tombstones which reflect the natural conditions of the region as well as the social status of those buried and others who played major roles in a community.

21.2 Believing and Belonging in Slovakia

Confessional identity is an important type of identity that in Central Europe, and especially Slovakia, is perceived almost as "hereditary," at least in rural areas (Beňušková 2004: 17). It has significantly shaped social relations, and the place of each denomination, for example, Lutheran and Catholic, is manifested and expressed uniquely. In terms of inter-denominational problems, it is also an interesting insight into the dynamics of the level and diversification of religious life and their manifestation in the community's social life, especially where the size of the dominant denomination in a village is changing, or if under the influence of external impulses the denominational body is undergoing a process of weakening or strengthening social boundaries with other faiths. The most common example of the influence of inter-confessional relations in the past was the intensity of endogamous marriages, and after 1950 it was the effects of secularization and the atheisation of society were mainly expressed in several communities and several denominations. There are two specific aspects of maintaining religious identity that take place not only in Slovakia but are evident throughout Central Europe, which are the diversity of religious structure and inter-confessional relations.

---

were conflicts and hostility against Jewish communities during WW II. In recent history, very peculiar were conflicts between Greek Catholics and Orthodox Church in 1990s in many communities in Eastern Slovakia. Greek Catholic Church was banned in 1950–1968 and its property was overtaken by Orthodox Church. After the ban came to an end in 1968, and especially after political changes in 1990, conflict based on discrepancies in ownership of land and estates became an inter-denominational issue. Beňušková (2004: 114) provides an example from Ladomírová (northeastern Slovakia) where conflict started as a Greek –Catholic parish was owner of a parcel on which there was a path to an Orthodox Church. Representatives of the Greek Catholi parish, as owner, demolished an historical Orthodox gate on this path. Conflict was broadcasted in national TV and especially influenced people in confessionally mixed marriages and caused arguments in families.

21.2.1 Diversity of Religious Structure

The first significant factor is the diversity of the religious structure in the population in Slovakia. The roots of diversity have a long historical record. In modern history, it is mainly the aftermath of the complicated processes of the Reformation, Counter-Reformation, Re-Catholization, and the Greek-Catholic unions, all which formed the basis of diversity until the issue of the Patent of Toleration in 1781.[2] Even after 1781 a typical demographical regulation of religiously exogamous marriages (Catholic marrying a Protestant) was an act prescribing the wedding ceremony by a Catholic priest, which resulted in the disadvantageous positions for Protestant spouse (Šoltés 2009: 72) as almost all children born to the spouse had to be baptised as Catholics. The results of the oppression of non-Catholic denominations before 1781 can be found in a number of buildings, and their appearance in particular in relations with Protestantism and Judaism which together formed also distinct architectural diversity of religious landscape.

Even though there were solid boundaries between religious groups until the middle of the twentieth century, they were not isolated from each other. Historically, only a few dozen municipalities in Slovakia could be described as mono-confessional, that is, with a share of one denomination over 90 %.[3] That level in the modern history of Slovakia, according to census data does not exceed 50 %. Most villages in Slovakia, therefore, have a mixed religious structure. That situation is subsequently reflected in the structuring of the local religious landscape, different forms and intensity of demonstrating their identity, and the strength to maintain ethnic boundaries between them. The variety of religious structures can be expressed in several ways, one of which is the index of ethnic fragmentation[4] (Figs. 21.2 and 21.3). Both reference years of 1880 and 2001 show an increase in diversification from north to south and from west to east. Here it is confirmed that religious diversity is indeed strongly associated with ethnic diversity. Liszka (2003: 347) states that in south-eastern Slovakia (Zemplin region), where Slovaks, Hungarians, Ruthenians and Romas live together, a village with higher status was considered to be one in which there were more denominations and, therefore, more church buildings and

---

[2] The Patent of Toleration issued by the emperor Joseph II in 1781 removed discrimination against non-Catholic denominations (Lutherans, Reformed Church, Orthodox Churches, and Jews) in Habsburg lands (including today's Slovakia). Non-Catholics were allowed to conduct their services freely, organize congregations in communities with at least 100 families and built their houses of prayer, however, with several architectural restrictions (no main entrance on street, no belfry and bells) and so on. From the point of view of civil rights Protestants and Orthodox became equal citizens to Catholics, although from the religious point of view, these denominations were just "tolerated" and the Catholicism still preserved its status of a state church (Bartl 2002: 320).

[3] The share of municipalities with almost exclusively one denomination in Slovakia reached the proportion of 52 % in 1880 and 38 % in 2001 (Data were calculated according to censuses 1880 and 2001 (A Magyar korona… 1882, and Štatistický úrad SR 2002)).

[4] Index of Ethnic Fractionalism was used by Yeoh (2003: 28). The index values varies from 0 to 1, where 0 represent completely homogeneous region and 1 represents completely heterogeneous region.

21 Symbiosis in Diversity: The Specific Character of Slovakia's Religious Landscape 419

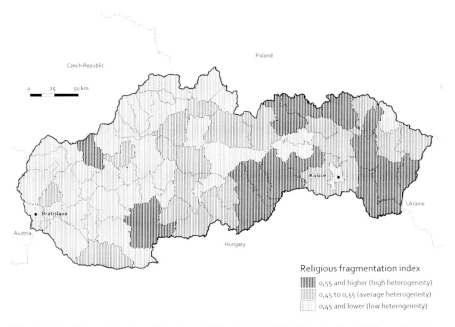

Fig. 21.2 Religious diversity of districts of Slovakia in 1880 (Map by Juraj Majo, calculations based on population census 1880)

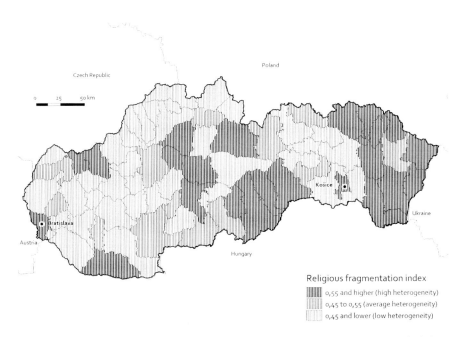

Fig. 21.3 Religious diversity of districts of Slovakia in 2001 (Map by Juraj Majo, calculations based on population census 2001, Štatistický úrad SR 2002)

established congregations. If the village was mono-denominational, it was considered less tolerant, and had therefore lower status. Population in more religiously diverse villages was more open to accept the others either as neighbors, friends or even family members.

21.2.2 Interdenominational Relations

Similarly, an appropriate sign for the establishment of a confessional identity in a village is the number of churches. Historically, from this aspect the most interesting case seems to be eastern Slovakia where we can find the highest number of villages with three church buildings established in one village (most often it is a combination of Roman Catholic, Greek Catholic, and Reformed or Lutheran church) (Fig. 21.4). In this area it is not unusual to attend the church of another denomination, where there is mutual participation of religious feasts and other events. Eastern Slovakia is notable for an interesting slight diffusion of customs from one church to another at the level of folk religion, although such may not be in accordance with the teachings of the official denomination. There are examples including the use of consecrated water by Reformed women in acts of folk healing, or perceiving the

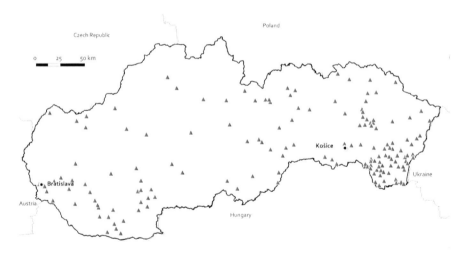

Fig. 21.4 Municipalities with three and more denominations with church buildings in Slovakia (Map by Juraj Majo, calculations based on Statistics of Urban and Rural Communities, Mestska a obecna statistika, 2008)

Holy Communion administered to a Protestant who is about to die as an extreme unction (Liszka 2003: 348). A similar model of attending the ritual of an opposite denomination in several villages in Slovakia that are mainly Lutheran and Roman Catholic. It is not unusual for Lutherans to attend Christmas Eve midnight Mass, even those without a Catholic spouse. In some cases of Slovaks living in northern Hungary (in Csömör) that is a recorded habit of decorating streets and houses by both Catholic and Lutherans during the Feast of Corpus Christi, although that holiday is not a part of the Lutheran liturgical[5] calendar (see Beňušková 2004: 95).

Much more difficult were the relations between Christian churches and Judaism. Even though Judaism had been established in Slovakia for a long time, it was usually perceived as a foreign element. Mutual prejudices occur, not only due to the religious origin that caused the Jewish community to isolate themselves. The majority had then seen such isolationism as animosity and hostility towards the Slovak national movements (Mlynárik 2005: 63) since nineteenth century when Slovaks as an ethnic group started to transform to a political nation. The concentration of the Jewish population in the past was determined by a settlement ban in Central Slovakis mining towns; this resulted in concentrations of Jews mostly in the western and eastern parts of the country (Botík 2007: 101). As a result of this deployment, the current expressions of Judaism place are mostly encountered in Western and Eastern Slovakia rather than Central Slovakia. In the course of gradual emancipation and equality in the nineteenth century however, most villages in Slovakia according to the 1880 census recorded the distribution of Jews and also the Roman Catholics (Jews were present in 93 % of municipalities, although the national proportion was 6 %) (Majo 2008: 110).

21.3 Church as a Symbol

Church buildings are the most typical structures representing the materialization of religious identity, as well as its manifestation of denomination established inside and outside a municipality. Most often a church building is the premise of a religious ceremonies, a place for daily or at least regular interaction between believers and the sacred place (Stump 2008: 330). Thus it is often the only place for mass gathering and maintaining identity in the community, which is usually wholly owned by denomination (at least this is the case in Slovakia)[4]. The church is one of the most important places where community members also undergo religious rites of passage (baptism, marriage and often burials) which are considered as a change in both ontological and social status (Eliade 1987: 185). Therefore, in many cases the importance of the church as the central sacral building in many ways crosses the

---

[5] Besides a church, a common liturgical place is frequently regarded as a house of mourning, which is respectively a place for rituals in a communal cemetery. In many towns and cities, it does not exclusively belong to a church, but rather is a building in secular (municipal) property, where funeral rites of other churches or civil funerals are also performed).

boundaries of the community, since the rites of passage are mainly also reflected in the sphere of non-religious individuals, families and communities (for example, changes in social status). Generally, the church building of an opposite denomination is also considered part of the municipality as it is one of its most important buildings by non-members too, yet it never cannot attain the level of own worship space of the main or central church building. Such perception of own and foreign is a historical and contemporary phenomenon and has always reflected the need for its own community in many of life's situations. These examples of activities between religious denominations are common inside one community and, of course, they do not occur in mono-denominational municipalities with just one church.

The relationship between village and church was extremely important in the past, and the church's location in the village was also a matter of prestige in relation to surrounding villages. The presence of a church assumed some degree of social consolidation within the village, and was also a sign of higher social status than churchless communities. A church in the village shifts the position of the village in imaginary village hierarchies, and represents a certain independence proportional to the size and style of a church building. An even higher status was attached to the village where there was a rectory or parsonage. The presence of a residence for the clergymen (historically just men) assumed greater social status resulting from the sufficient financial surplus with which the village could sustain a priest. Also from another point of view, the parsonage was a centre of intellectual life in the village, and the priest or pastor was an important representative personality in the village. Especially Lutheran parsonages and Roman Catholic rectories were considered notable centers of education and intellectual resources of Slovak national revival in the past (especially in nineteenth century).

If a village, or rather a town had several churches in the village of one denomination, a higher and central status was always awarded if it had a parish church. This is especially the case of the Roman Catholic Church. The importance grew when the parish included, besides a central community, also communities living in the surrounding subsidiary villages. There was, therefore, a sort of community subsidiaries that were subject to a parental center. Often this "mother – daughter" or subsidiary relationship, as the term is used, was also reflected in other areas in community life (such as prejudice, feeling of greater importance of inhabitants inside parish community, but also established positive forms of relationship including friendships, economical, family and marriage ties, and so on).

21.3.1 Location of a Church in a Community

The location of the church in the village expresses today, more intensively than in the past, a centripetal and the most important position in the village. The church, as the most important building in a medieval village, was a measure of the division of property and the setting of boundaries (Habovštiak 1985: 147). The position of churches followed so called "polar positions." The first was "symbolic" when the dominant churches were built in the center of an inhabited are, that is, on the most accessible and fastest approachable places in the village, although it was many times church construction was limited by the physical conditions of an area. In many communities the position of the church was reduced to a dominant, though not always the most accessible place (for example on a hill). The second approach was called an "emergency" position that had to accept a place in the circumstances and location of the village. Most often we encounter an effort by the ecclesiastical community a desire to locate sacred buildings in the middle of an inhabited area (Lukáčová et al. 1996: 163). This decision attempts to take into account the location of the church, including the importance of the church tower to profane life. Especially this is the case if the church had a tower, since towers in the past fulfilled an important role as a means of communication especially in the event of fires and floods, and the need for cooperation.

More possible sites for sacred buildings are found in urban areas of Slovaia, therefore, research on this topic is a bit more complicated than in villages. In cities there would be expected to be a general model for a church location on the main square in the city. This location could vividly emphasize its status not only to other sacred (also to other denominations), but also to secular buildings. Indeed, this model can be seen in many cities, where the central church lies in the main square. The most typical example is found in towns in Eastern Slovakia, where the churches are dominant and usually also a central part of the mainly spindle-square[6] (Košice, Prešov, Poprad, Spišská Nová Ves, Bardejov, etc.). Since most of these towns became religiously diverse especially after the Reformation, we are here dealing with the location of the Roman Catholic, usually parish church and the other dominant (often exclusively Lutheran) denominations in the square. After the Patent of Toleration this was an important attempt to show the cultural, economic, and social position of the formerly ostracized, yet important community of Protestants. Such dualities can be seen in the towns such as Spišská Nová Ves, Poprad, Levoča, Prešov, and in central Slovakia: Rimavská Sobota (uniquely with Reformed Church) Zvolen, and Partizánska Ľupča. Such a position was barely attainable by Jewish communities. There are very few examples of synagogues built on central squares of towns. An exception is Michalovce (Eastern Slovakia), where Jews were allowed to build it there, which reflected the important position of this community in the life of a town (Mešťan et al. 2002: 23).

An interesting example of the relationship between the church and city is in Bratislava, the capital city. The Main Square of the historical part of the town does not have a central church. The main church of the town is located on the west side of town, and in the past the western wall of today's St. Martin Cathedral was part of a town fortification. This model suggests that the church is older than the actual urban area (Mencl 1938: 188). The importance of the position of the church as a cult place in towns and cities also attracted other than religious functions. In the majority

---

[6] Spindle square are types of squares (especially in Slovakia) whose groundplan resembles a spindle. It is broadest in the center and is narrowing towards both sides. In the broadest parts of a square the most important churches were built in mentioned towns.

of towns and cities in Slovakia, we encounter intersections of these features in one place, which means that next to the church there was usually a pub, town hall, and market. In the Orava region (northern Central Slovakia), however, there are hints that the center of social life was transferred from the religious site to an economic place, creating two centers of community life (Langer 1990: 48). In others, especially the northern regions of Slovakia, we do not historically find such a strong form of polarization. In other parts of northern Eastern Slovakia we do not even find a trace of the economic centre.

21.3.2 Identity and Its Expression in Church Architecture

When comparing the position of sacred buildings, we are primarily focusing on the differences between Roman Catholic and Protestant churches. The Roman Catholic and Lutheran Church is the most usual combination in our settlements, and the differences between these structures shows some of the same characteristics, but there are also some regional peculiarities. Some examples will be added with information about Greek-Catholic, Orthodox and Jewish sacred buildings. A specific set of differences will be highlighted in cemeteries and gravestones.

The presence of several churches in the village clearly reflects the mixed religious structure of the population in the past. Not always, however, is such situation reflected to the present. Often under the influence of various external factors, such as the intense process of secularization and atheisation, relocation and resettlement by the population, and similarly by mechanical as well as structural population movement, the religious structure of a community can be altered. In Slovakia we have some examples as in a certain period of time the religious structure of one region changed collectively. Both examples relate to the Lutheran Church. The first example is the Spiš region (northern Eastern Slovakia). After WWII, the German population displacement changed the religious structure of the population. Many Lutheran churches remained empty and unused. Some were devastated with the help of the local socially excluded population (Lomnička), some were offered for use by the Greek-Catholic Church (Žakovce), and others completely changed function (Mlynica – museum of veterans).

Another example is the Gemer region (southern part of Central Slovakia). Under the influence of intensive processes of secularization in 1948, many Lutheran churches were unused or the intensity of their use dropped significantly. In many cases, this intensity of usage was accented by the depopulation of many villages or the increase of other population of other denominations (for example, Roma inhabitants). The intensity of depopulation will gradually also affect the northeastern region of Slovakia. Several mostly Greek Catholic churches no longer fully serve their purpose. Therefore, more often we can see the case that dwindling communities often highlight the physical condition of religious buildings in the village.

To distinguish the sacred buildings of a denomination there is a set of individual characters that may have some degree of variability and intensity of occurrence.

These signs are usually visible only after a thorough inspection of the exterior of the building. The establishment of denominations in a village or in several municipalities already indicates the presence of certain characteristics in the region, which is typical for the dominant faith. Such regional signs are usually mark the presence of roadside crosses, chapels and statues. Their frequent occurrence indicates the established and historical presence of Roman Catholics. Therefore we almost do not find such sculptures in the landscape of regions with a historically Protestant majority (Myjava region in Western Slovakia or Gemer in South Central Slovakia). Even if some Catholic community is preserved in such region, the landscape of such community is marked by the evident expression of its distinct identity. A stone crucifix usually denotes entry to the village (Ratkovská Lehota – small Roman Catholic enclave in the Gemer region).

Although small roadside chapels and crosses are not uncommon in our region, it is a relatively recent phenomenon that has been spreading since the Counter-Reformation. Together with festivals, pilgrimages, and other visual aspects, it worked as quite ostentatious religious, symbolical and spatial expression of non-Protestant identity in efforts to restore the Catholic forms of piety in re-Catholicized regions (Beňušková et al. 2005: 22). The evidence of such sculptures in the time of the Counter-Reformation period is its absence in regions that had been hitherto adhering mostly to Protestantism.

The location of sacred buildings in the country was not a random act at all. First of all they were placed in special-purpose sites. These sites were especially "crossed roads," where roads either crossed or diverged. Often they were in the area on the border line of villages, and in popular culture the boundary represented a magical significance of this place (meeting of witches, often a site for burying those who committed suicide) (Botík, Slavkovský et al. 1995: 277–278). As a sign of protection against evil forces, small religious buildings were built especially at these sites (Fig. 21.5). Often the motivation to build a sacred object was the expression of gratitude for protection during illnesses, natural disasters, or other life events, either of an individual or whole community (for example during epidemics). In urban areas, the most common plague pillars on town squares were built after the plague epidemics in the seventeenth and eighteenth century.

The set of features distinguishing churches is not as clear as in the case of small sacred buildings, which almost exclusively are ascribed to the Roman Catholic or Greek Catholic Church. Nevertheless, there are several important details and features upon which the denominational identity of the church building can be identified. The most important sign, which in Central Europe mostly works as a differentiating sign of identity, is the *symbol on the tower*. In the case of the Roman Catholic and Greek-Catholic Church, it is almost exclusively the symbol of a cross. It is common practice to use either a single Latin cross, or the Apostolic – Patriarchal Cross with two arms. The use of these two types of crosses does not reflect any ethnic or regional specificity. In Slovakia and Central Europe we can find the cross on the top of churches of Slovaks, Hungarians and Germans, and on the top of the churches of Lutherans as well. In the case of Greek Catholic and Orthodox churches, the temples are usually topped with an Eastern Cross that has three arms,

Fig. 21.5 Cross in the field in Oravská Lesná. The Orava region is a predominantly strong Roman-Catholic region in Northern Slovakia (Photo by J. Majo 2009)

where especially the third arm is slanted in the case of Orthodox Churches. An important sign of evident Catholic Church (either Roman or Greek Catholic) seems to be the presence of the "missionary cross" on one of the walls, or the statue of Calvary, patron, or piety, that is usually integrated into the facade too. In some cases, close to the church we can find other sculptures, mostly of the church patron, the Virgin Mary, or a symbol of Lourdes cave. Other characteristics, such as the Latin inscription above the portal, stained windows, more than one nave in the church or outlined presbytery, does not clearly indicate a clear affiliation to the Roman Catholic Church or Greek Catholic Church.

Lutheran churches have a large number of similar exterior features as the Catholic churches. Just like Catholic churches, Lutherans incline to the usage of a cross at the top of towers, either as a Latin or Apostolic cross. There is no common spatial pattern of distribution of that symbol. This factor is shared with Catholic churches too.

Compared with the Roman Catholic Church, however, Lutheran churches in some regions tend to be marked by other features that are typical signs of the Reformed Church – star, rooster, crescent, or a different geometric character with no explicitly Christian symbolical meaning. Such symbology is mainly used in the southern part of Central Slovakia close to the Hungarian border, which could indicate Calvinist influences. In the case of Reformed and Lutheran churches, we more often encounter churches without towers or separately standing bell tower or campaniles (typical for the Gemer region). Most often, it is the case of churches built shortly after the Patent of Toleration, and in cases where the community has failed to build a tower or belfry mostly due to financial reasons. In conclusion, the scope of symbols used at the top of Lutheran churches in Slovakia and Central Europe can be considered as the richest among all historically established denominations.

A typical character of a Reformed church and in some regions also of a Lutheran Church (Botík 1999: 34) is its resistance against the symbol of the cross from the time of Counter-Reformation, which is again a specific peculiarity of the Carpathian basin (Fig. 21.6) Reformed churches are therefore always crowned with a star, crescent, rooster, or other geometrical symbols, but never a symbol of cross. Significant differences between the Lutheran and Calvinist church are seen in the interiors only. Reformed churches never have an altar, altar cross, candlesticks, paintings or other elements that are found in Lutheran and Catholic churches. Also the orientation of a central place in the reformed church (pulpit and Lord's table) is usually on the side of the long wall of the church, and not in the shorter side of a rectangular shape.

Orthodox churches and Greek-Catholic churches share most signs, with the exception of the absence of sculptures in Orthodox Church art. Typical is the use of a two-dimensional painting of the crucified Jesus Christ on the cross instead of a sculpture (Fig. 21.7). Such a motif is used not only on their church buildings, but became a specific element of rural religious landscape too. Orthodox churches are usually newer, built only after 1950, due to "orthodoxification" which encompassed movements toward the "faith of our fathers" (Pešek and Barnovský 1997: 134) and the ban of the Greek Catholic Church that happened in many Eastern Bloc countries after 1950.

Special types of sacred architecture are synagogues. These were mostly built on territory shortly after the establishment of Jewish communities, one of the oldest Jewish communities was in Nitra (mentioned in the twelfth century) (Borský 2007: 143) and Stupava (from the sixteenth century) with a prayer house built in 1803 (Borský 2007: 134). Probably some of the oldest surviving synagogue buildings are located in Bardejov, Skalica, and Sobotiste (all from the second half of the nineteenth century), of which the two latter are in western part of Slovakia, close to the Czech Republic These three, just like almost all other preserved synagogues, no longer have a sacred use. They are often completely rebuilt, and from the original architecture there remains only fragments. Most of the synagogues were built in our country after the issue of the Patent of Toleration in 1781. Since then, several architecturally valuable buildings such as in towns Liptovský Mikuláš, Bratislava, Bardejov, Košice, Žilina and others were built. The specific of Orthodox synagogues apart from Reformed ones is especially in respecting architectural principles in the construction,

Fig. 21.6 The star and crescent at this Lutheran church in Brádno, Gemer region, South Central Slovakia, is a typical sign of many Reformed and some Lutheran churches in Slovakia (Photo by J. Majo 2009)

according to Halacha. Synagogue must, if possible, be higher than the surrounding buildings in the area, and the entrance to the nave of synagogues may not be located directly from the street. Orientation of *bima* must be to the east (towards Jerusalem) (Mešťan et al. 2002: 8). However, synagogue buildings have the reflection of the contemporary architectural style in which they were built, and especially Jewish Orthodox communities had been trying to make its exterior not resemble Christian churches, especially in not having towers or bells. Only the synagogues of Reformed Judaism were usually finished with a tower, often in the form of a bulbous roof. This type of architecture in the urbanism of Slovak and Central European towns represented a specific and valuable element that has unfortunately not been preserved for future generations in the whole extent.

Fig. 21.7 Orthodox church in Nižná Polianka, North Eastern Slovakia, built in 1992 (Photo by J. Majo 2009)

21.3.3 Church Buildings in Time and Space

The majority of churches in Slovakia were built by the end of the eighteenth century (Poláčik and Judák 2005: 68). Between eighteenth and twentieth century the majority of Roman Catholic churches were built (almost 80 %). Churches built in the eighteenth century, however, are dominant only in the former Bratislava – Trnava archdiocese (Poláčik and Judák 2005: 66). This is connected mainly to the culminating period of counter-reformation efforts in the country and the subsequent recovery of the Roman Catholic Church, as well as the emergence of new religious orders (the Jesuits) and the dissemination of their activities. The second most significant period of growth of Roman Catholic Churches was in the twentieth century. The rising number of churches is not just related to the growth of urban structures, but also reflects the rapid renewal of religious life after 1989 (742 new churches – almost one-quarter of the total number of church buildings). Regionally, most of the churches in the twentieth century were built in the Archdiocese of Košice in Eastern Slovakia. This region of Slovakia, according to indicators from several churches, experienced the most intense restoration of religious life in our country. In 2008 Slovakia had 2,547 Roman Catholic churches (Statistics of Urban and Rural Communities 2008).

Even in the Greek-Catholic Church, the majority of churches were built in the twentieth century (more than one-third) and in the nineteenth century (35 %)

(Poláčik and Judák 2005: 81) and almost all are located in Eastern regions of Slovakia (Prešov and Košice region exclusively). That the twentieth century is for this Church one of the most crucial and intensive periods of growth is a reflection of changes in society and a response to the ban of the church from 1950 to 1968 and the attempt to erase the image of the church from the religious face of Slovakia. Slovakia now has 484 Greek-Catholic churches.

The biggest variance in time issues related to the construction of churches has the Lutheran Church. Overwhelming the majority of churches was built in the eighteenth and nineteenth century. Especially after the formal renewal of church life after the issuing of the Patent of Toleration in 1781 until the end of the eighteenth century, in less than 20 years in Slovakia, 121 Lutheran churches were built, which represents more than one-fifth of all Lutheran churches in Slovakia (data according to Mlynár 1989). However, most churches were built in the nineteenth century, which is the period of the gradual full acceptance of non-Catholic denomination in Austria and Hungary. Currently, the Lutheran Church in Slovakia has 587 functional churches and prayer houses.

The other churches are, according to the Statistics of Urban and Rural communes in 2008, owned by Reformed churches (301), and Orthodox (140). Among the other churches with a relevant number of churches are Baptists (33), Church of Brethren (12), the Union of Jewish religious communities (15), and others. The total number of churches owned by other denominations was 90. The time period of building churches is followed by the periods of the emergence and establishment of denomination in society. For example, the Reformed Christian Church has similar time periods of intensive church building as the Lutheran Church, which means that most were built after the Patent of Toleration in 1781. Similarly, one of the oldest Orthodox churches was built for the Serbian community in Komárno in 1754 (Lukáčová et al. 1996: 124), however, the vast majority were mostly built in the twentieth century due to the late establishment and processes of "orthodoxification." Churches of minor Protestant denominations were also built mainly in the twentieth century, when that branch of Protestantism in our country began to be established. This data indicated the church buildings that are in the property of a denomination. In many cases, however, for financial reasons a community might only rent a place of worship, although it should be noted that even in the case of many smaller denominations, several were successful in building their own churches after the revival of religious life after 1989.

21.4 Cemeteries in Slovakia – Identity Embodied

Research of sacred places connected to death is a topic of the day in cultural geography. Geography of religions offers approaches in examining the spatial arrangement of funeral rites, and mostly focuses its interest in the cemetery as a major clash between the sacred and profane world. There is an interesting position of cemeteries to other elements in a country, considering not only practical but also the cosmological localization (Park 2003: 217). For example, the use of Feng Sui of China (Kong 1999: 3), architecture and the development of tombstones, and no less interesting is the issue of spatial segregation. This element is also an important topic in performing research on cemeteries in Slovakia, as the spatial distribution of tombstones and their appearance reflects the social status of the buried and often religious and ethnic identity. Not to mention are the special places for burying those who committed suicide and other socially less accepted individuals of the community. The importance of the cemetery itself lies in the similarities as well as differences from the other sacred buildings and places. On one hand is the intense site of convergence of the sacred and the profane and on the other hand the cemetery's significant sensual importance. It is, therefore, a symbiotic combination of both the functional and emotional in a single place (Francaviglia 1971: 501). What is also unique for the location of cemeteries is that across cultures have taken efforts to locate cemeteries in places distant from dwellings and in most traditional and contemporary cultures (Pitte 2004: 348), although the importance of that type of sacred place is substantial for the development of a community and its members.

In Slovakia for many centuries it has been the custom to bury the dead in a specially designated area. This was in concordance with animistic practices inherited from Slavic ancestors. Most often, a corpse was buried in a forest, under a tree in a field or grove (Bednárik 1972: 23). Only with the advent of Christianity did burial occur on consecrated ground, most often close to the church, but this rule was not strictly obeyed. In some regions, the practice of burial under trees lasted until the seventeenth century (for example, the Region of Upper Hron river in Central Slovakia) (Bednárik 1972: 23). Communities according to church dogmas distinguished between natural death and unnatural death. Special dignitaries were buried in crypts in churches, or at least their importance was pronounced in the architecture and position of the cemetery grave. A specific position in some cases can be seen with localities having a Roma population. Many Roma communities do not just encounter separation during life, but also when they die, since they use a specially designated area within the communal cemetery (Fig. 21.8).

From the sixteenth century onward it became more common to bury outside the village. Thus communities abandoned the practice of burying in and around churches. In 1876 in Hungary passed a general law prohibiting the establishment of new cemeteries near churches (Bednárik 1972: 15). Cultural evolution also caused growing importance of gravestone maintenance and decoration by the bereaved. In the past, graves were often unmarked, covered with highly stacked rocks or grass. It was not unusual to use the cemetery area for cattle grazing. Today's practice is to more or less formally visit the gravestone around All Saints' day. In some regions, families decorate the gravestone according to holidays, especially decoration with a Christmas Tree in some localities (Fig. 21.9). In this way families try to keep emotional and spiritual contact with the departed by sharing the atmosphere of important holidays.

Like other sacred objects, tombstones reflected the religious and social identity of an individual. Every denomination had to bury its members, thereby gravestones became, besides churches, the most common means of religious expression and

Fig. 21.8 Birdseye view of a cemetery in Muránska Dlhá Lúka in Gemer region in South Central Slovakia with spatial segregation of graves of Romas (*in circle*) (Source: Google Earth, 2012)

other associated identity. In the religious landscape, small religious buildings, for example, are specific only to non-Protestant denominations.

The most common feature of a Christian cemetery is the central cross. It occurs almost everywhere in Slovakia, except for some mono-denominational Protestant villages in the south of Central Slovakia. The cross has several significances and functions: if more denominations share the cemetery, it usually separates their areas with the graves of adherents (Fig. 21.10), since the confessional segregation is historically common pattern (Matlovič 2001: 222) in many usually rural communities and their cemeteries in Slovakia. The central cross is originally considered as a central place of worship in cemeteries especially for Roman Catholic rites. Centrality of the central cross is highlighted by the fact that graves located closer to the cross, or church, if it is part of the cemetery, were considered the most prestigious areas, and vice versa those at the fence were considered inferior and for socially excluded individuals, for example, suicide (Jágerová 2001: 26).

Tombstones and gravestones can be sorted in several ways, depending on the material used (the oldest are wooden and stone), followed by forms and symbols used to represent the buried. In particular, the shape and symbolism of drawings or

21 Symbiosis in Diversity: The Specific Character of Slovakia's Religious Landscape

Fig. 21.9 A grave in Oravská Lesná, Orava region, Northern Slovakia, is decorated with a Christmas tree, serving as a symbol of sharing the joy of festival even with the deceased members of family (Photo by J. Majo 2009)

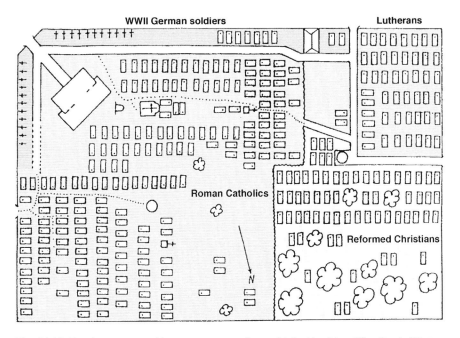

Fig. 21.10 Rural cemetery with separate space for multiple identities (Iža, South Western Slovakia), although the boundaries in cemeteries between denominations become nowadays more blurred due to exogamous marriages (Source: Adapted from Jozef Liszka 1998, used with permission)

Fig. 21.11 Roman Catholic cemetery in Detva, Central Slovakia, with carved crosses in first half of the twentieth century (Photo by Vavroušek 1929)

text demonstrates the religious identity of the buried, especially with distinct symbology codes in rural areas. Roman Catholics and Greek Catholics used the symbol of the cross in various sizes (with or without corpus, or with extra hagiographic decoration) while Lutherans, Calvinists, and Jews preferred boards and columns (Bednárik 1972: 46) (Fig. 21.11). Detailed sets of symbols were typical of each denomination. Catholics have inclined to use crosses, Lutherans and Calvinists chalice, a weeping willow symbol, stars and other geometric shapes. Besides these symbols, Lutherans also used the symbol of the cross, usually without a corpus. The Jewish symbolism contained mainly the symbol of the menorah (light – symbol of eternity), also a Jewish star pitcher and bowl for gravestones of Levites, and just like Protestants. Jews used the symbol of a weeping willow for a descendant of the priestly tribe of Cohens, the symbol of crossed fingers of two hands (Mešťan 2001: 164). (Fig. 21.12) Gravestones of all denomination are sometimes marked by a symbol of the profession of the deceased.

The modern phenomenon of burial practice is cremation. In Slovakia, that form of burial has been applied very slowly due to religious practice. Cremation in our country emerged after the formation of Czechoslovakia, where it was used especially among members of the Czech minority (Zelinová 2001: 207). The majority population

Fig. 21.12 Gravestone of a Lutheran teacher with a symbol of a weeping willow. Teplý Vrch, Gemer region, South Central Slovakia, from the first half of a twentieth century (Photo © J. Majo 2009)

refused cremation for a long time, the method of burying was opposed by Catholic Church. Since 1950 there was a gradual acceptance of cremation, which has become more popular inside the urban population contrary to the more rural conservative population. The urban cultural environment uses this method of burial mainly due to the result of more progressive secularization, just like those efforts to save space and free up inexpensive land for development within cities.

21.5 Conclusion

Can we find symbiosis in religious diversity in Slovakia? Yes, we definitely can. Such a level of diversity in a relatively small country required denominations to develop strategies to maintain mutual coexistence, no matter the relations between churches in higher levels. Common people with their spiritual leaders (usually priests and pastors) had to accept the essential rules of their own denomination, but in daily interaction with neighbouring denominations in the community they usually tried to avoid conflict, learned to accept otherness, and found a more or less reasonable and sustainable *modus vivendi*. Therefore, the system of solid and barely permeable religious boundaries served as a good "compass" in the social place. It helped the community to survive, to maintain their identity and traditions with regulated networks with members of an "out-group." This surprisingly, yet effective working system, was functional for centuries, creating unwritten rules and order in mutual relations. For that reason an outstanding mosaic of religiously diverse villages could survive throughout the major parts of Central Europe. That mosaic is an essential explanation of the diversity of religious identity imprints in Slovakia. The reflection of power and subordination embodied in various types of sacred buildings, tombstones, and small sculptures therefore works as a good history textbook, showing both us and subsequent generations the traces of interdenominational relationships and the consequences.

Finally, in the case of the religious landscape in Slovakia and partly across Central Europe, we can state that Slovakia is a place/country of much diversity in its religious landscape, a landscape that was created that from today's point of view developed into a harmony and symbiosis of the whole set of symbols and signs that depict the richness of the complicated history of religions in the country.

Acknowledgements The author gratefully acknowledges that the chapter was carried out under VEGA Scientific Research Agency Grant Nr. 1/0562/12 and APVV-00818-12.

References

A magyar korona országaiban az 1881. év elején végrehajtott népszámlálás föbb eredményei megyék és községek szerint részletezve 2. vol. [Main results from the 1881 Census in the countries of Hungarian crown sorted according to provinces and municipalities]. Budapest: Pesti Könyvnyomda – Részvény – Társaság, 1882.

Bartl, J., et al. (2002). *Slovak history. Chronology & lexicon*. Bratislava: Slovenské pedagogické nakladateľstvo.

Bednárik, R. (1972). *Cintoríny na Slovensku* [Cemeteries in Slovakia]. Bratislava: SAV.

Beňušková, Z. (2004). *Religiozita a medzikonfesionálne vzťahy v lokálnom spoločenstve* [Religiosity and inter-denominational relations in local community]. Bratislava: Merkur.

Beňušková, Z., et al. (2005). *Tradičná kultúra regiónov Slovenska* [Traditional culture of the regions of Slovakia]. Bratislava: VEDA SAV.

Borský, M. (2007). *Synagogue architecture in Slovakia. Towards creating a memorial landscape of lost community*. Bratislava: Jewish Heritage Foundation – Menorah.

Botík, J. (1999). *Tam zložili aj svoje kosti: kultúrnohistorické hodnoty náhrobných pomníkov zahraničných Slovákov* [They laid down to rest there. Cultural-historical values of the grave memorials of the Slovak emigrants]. Bratislava: Lúč.
Botík, J. (2007). *Etnická história Slovenska* [Ethnic history of Slovakia]. Bratislava: Lúč.
Botík, J., & Slavkovský, P. (Eds.). (1995). *Encyklopédia ľudovej kultúry Slovenska* [The encyclopedia of folk culture of Slovakia]. Bratislava: VEDA SAV.
Büttner, M. (1980). On the history and philosophy of the geography of religion in Germany. *Religion, 10*(1), 102.
Eliade, M. (1987). *The sacred and the profane: The nature of religion*. San Diego/New York/London: A Harvest Book, Harcourt, Inc.
Francaviglia, R. V. (1971). The cemetery as an evolving cultural landscape. *Annals of the Association of American Geographers, 61*(3), 501–509.
Habovštiak, A. (1985). *Stredoveká dedina na Slovensku* [Medieval village in Slovakia]. Bratislava: Obzor.
Jágerová, M. (2001). Slovenský pohreb (Slovak funeral). In J. Botík (Ed.), *Obyčajové tradície pri úmrtí a pochovávaní na Slovensku* [Customs connected with death and funerals in Slovakia] (pp. 13–32). Bratislava: Lúč.
Kong, L. (1999). Cemeteries and columbaria, memorials, and mausoleums: Narrative and interpretation in the study of deathscapes in geography. *Australian Geographical Studies, 37*(1), 1–10.
Kong, L. (2005). Religious landscapes. In J. S. Duncan, N. C. Johnson, & R. H. Schein (Eds.), *A companion to cultural geography* (pp. 365–381). Oxford: Blackwell.
Kónya, P., et al. (2010). *Konfesionalizácia na Slovensku v 16.-18.storočí* [Confessionalization in Slovakia between 16th and 18th century]. Prešov: Vydavateľstvo Prešovskej univerzity.
Langer, J. (1990). Spoločenské stavby v sídelnej štruktúre oravských dedín [Communal buildings in settlement's structure of the villages in Orava]. *Zborník SNM Etnografia, 31*, 39–51.
Liszka, J. (1998). Ungarische Friedhofe im nordlichen Teil der Kleinen Tiefebene (Sudwest-Slowakei). *Ostbairische Grenzmarken, 40*, 229.
Liszka, J. (2003). *Národopis Maďarov na Slovensku* [Ethnography of Hungarians in Slovakia]. Komárno/Dunajská Streda: Lilium Aurum.
Lukáčová, E., et al. (1996). *Sakrálna architektúra na Slovensku* [Sacral architecture in Slovakia]. Komárno: KT.
Majo, J. (2008). Niektoré priestorové vlastnosti sociálnej hranice [Some spatial aspects of social boundaries]. *Geographia Cassoviensis, 2*(1), 108–111.
Majo, J. (2009). Cobwebs in pews – Religion, identity, and space in Western Gemer region in Central Slovakia. Acta Facultatis studiorum humanitatis et naturae Universitatis Prešoviensis. *Folia Geographica, 50*(15), 55–68.
Matlovič, R. (2001). *Geografia relígií* [Geography of religions]. Prešov: Fakulta humanitných a prírodných vied Prešovskej university.
Mencl, V. (1938). *Středověká města na Slovensku* [Medieval towns in Slovakia]. Bratislava: Učená společnost Šafaříkova.
Mešťan, P. (2001). Pochovávanie v židovskej komunite [Burials in jewish community]. In J. Botik (Ed.), *Obyčajové tradície pri úmrtí a pochovávaní na Slovensku* [Customs connected with death and funerals in Slovakia] (pp. 163–172). Bratislava: Lúč.
Mešťan, P., Krivošová, J., & Kissová, D. (2002). *Architektúra synagóg na Slovensku* [Architecture of synagogues in Slovakia]. Bratislava: SNM – Múzeum židovskej kultúry.
Mestská a obecná štatistika. Vybrané údaje [Statistics of urban and rural communities, Selected data from an electronic database]. Trnava: Krajská správa Štatistického úradu Slovenskej republiky. 2008.
Mlynár, J. (Ed.). (1989). *Schematizmus Slovenskej evanjelickej cirkvi a.v. v ČSSR* [Directory of the Slovak Lutheran Church in Czechoslovak Socialist Republic]. Bratislava: Cirkevné nakladateľstvo.
Mlynárik, J. (2005). *Dějiny Židů na Slovensku* [History of Jews in Slovakia]. Praha: Academia.
Park, C. C. (2003). *Sacred worlds. An introduction to geography and religion*. London: Routledge.

Peach, C. (2002). Social geography: New religions and ethnoburbs – Contrasts with cultural geography. *Progress in Human Geography, 26*(2), 252–260.

Pešek, J., & Barnovský, M. (1997). *Štátna moc a cirkvi na Slovensku 1948–1953* [The State and the Churches in Slovakia 1948–1953]. Bratislava: Veda SAV.

Pitte, J.-R. (2004). A short cultural geography of death and dead. *GeoJournal, 60,* 345–351.

Poláčik, Š., & Judák, V. (Eds.). (2005). *Atlas Katolíckej cirkvi na Slovensku* [Atlas of the Catholic Church in Slovakia]. Bratislava: Lúč.

Šoltés, P. (2009). *Tri jazyky, štyri konfesie. Etnická a konfesionálna pluralita na Zemplíne, Spiši a Šariši* [Three languages, four denominations. Ethnic and confessional plurality in Zemplín, Spiš, and Šariš]. Bratislava: Historický ústav SAV.

Štatistický úrad SR. (2002). *Sčítanie obyvateľov, domov a bytov 2001. Bývajúce obyvateľstvo podľa pohlavia a náboženského vyznania* [The 2001 population and housing census. Population according to sex and religious affiliation]. Bratislava: Štatistický úrad SR.

Stump, R. (2008). *The geography of religion. Faith, place and space.* Plymouth: Rowman & Littlefield Publishers.

Vavroušek, B. (1929). *Kostel na dědině a v městečku* [Church in a village and in a small town]. Praha: Kvasnička a Hampl.

Yeoh, K. K. (2003). Phenotypical, linguistic or religious? On the concept and measurement of ethnic fragmentation. *Malaysian Journal of Economic Studies, 40*(1 & 2), 23–47.

Zelinová, H. (2001). Podiel Čechov na propagácii kremačného hnutia na Slovensku [The role of Czechs in spreading the cremation movement in Slovakia]. In J. Botik (Ed.), *Obyčajové tradície pri úmrtí a pochovávaní na Slovensku* [Customs connected with death and funerals in Slovakia] (pp. 207–216). Bratislava: Lúč.

Chapter 22
Religion Inscribed in the Landscape: Sacred Sites, Local Deities and Natural Resource Use in the Himalayas

Elizabeth Allison

22.1 Introduction

That religion shapes land use is a geographical truism: temples, churches, mosques, and other religious structures have a myriad of influences over the way people and materials interact with space. Religious sites and structures influence the flows of people, goods, and capital in their vicinity. They dictate acceptable practices, and shape the uses to which the nearby space may be assigned. They shape daily, weekly, and annual pilgrimage routes, and serve as magnets for tourism and associated commerce.

The converse of this assertion – that landscape shapes religious belief and expression – has been less thoroughly explored. Natural sites have been places of religious revelation and ecstatic experience for people from the Desert Fathers to contemporary Burning Man revelers. Throughout history, mystics and religious leaders, including Jesus, Moses, Siddhartha Gautama (the historical Buddha), John Muir, Thomas Merton, Henry David Thoreau, and Edward Abbey have retreated to the wilderness, to contemplate and seek enlightenment away from the distractions of human society and worldly cares. However, accounts of such spiritual experiences in the wilderness tend to depict nature as an empty vessel that contains the experience of the religious seeker, and through which spiritual enlightenment can come. Accounts tend not to emphasize the active role of natural sites in shaping the human spiritual imagination. The intention here is not to return to mid-century environmental determinism, but instead to take account of the role of nature in not only supporting and providing raw material for, but actively shaping, people's material livelihoods and perceptions. While natural systems provide a host of ecosystem

E. Allison (✉)
Department of Philosophy and Religion, California Institute of Integral Studies,
San Francisco, CA 94103, USA
e-mail: eallison@ciis.edu

services that support human material existence, providing everything from water purification to detoxification and decomposition of wastes, these processes often take a backstage to human-generated activity. Human culture is seen as acting on an inert non-human nature. However, human culture is equally shaped by non-human nature, which provides many of our literary and religious metaphors. Nature, environment, and place, as well as being both materially and socially constructed, are also sources of perceptions and ideas: the natural constructs the social (Watts and Peet 1996: 263).

Throughout the world, humans respond to the subtle energies in particular places that connect the spiritual and sacred to the phenomenal world. Chinese chi gong, Tai Ch'i, feng shui and acupuncture all represent means by which people recognize, interact with, and channel subtle energies, within the human body and in the surrounding world. The ancient Greek, Roman, German, and Nordic peoples recognized gods in their surroundings; contemporary African and Australian tribes identify the sacred in the landscape; the Japanese locate a kami spirit at the base of Shinto shrines; mountain peoples through the world revere singular peaks as homes of gods (Bernbaum 1990). All these traditions have expressed connection between people and the spirit world, through particular, powerful places on the landscape (Hayward 2000). Powerful places may be identified by their unusual landforms, patterns of active animal use, or particularly lush or vibrant vegetation. Human visitors may experience unusually vivid dreams or other types of visions linked to the place (Krippner et al. 2003). Pilgrims may travel to such areas out of a belief in their healing or restorative powers, leaving both destructive and restorative material traces on the landscape (Bernbaum 1997; Bernbaum and Purohit 1999; Huntsinger and Fernandez-Gimenez 2000). Religious monuments are often built on ancient power places (Gyatso 1987; Hayward 2000).

Rather than consigning nature to a background, static stage on which the historical and cultural activities of human affairs play out, this view brings the agency of nature into focus, suggesting that the materiality and activity of nature importantly shape the ways that people think about and work in it (Peet and Watts 1996; Plumwood 2006). In contrast to a strictly materialist viewpoint that seeks to improve human well-being through the accumulation of greater material comfort, a place-based spiritually-generated perspective places humans within networks to which humans and other living and non-living beings contribute and respond. Such a view sees forests, rivers, and soils not simply as natural resources to be used with maximal efficiency, but as beings with their own inherent value, purpose, meanings, and destinies, who resist being treated as objects, and retaliate against inappropriate actions. In a time when the Earth's capacities for generation and regeneration have been stretched to their maximum, and when human activity is leading to the reduction and extirpation of countless other species, a view that accords meaning, value, and purpose to non-human life, and even the abiotic elements of landscapes, offers a potentially curative alternative to the current human obsession with ever more rapid consumption of the Earth's bounty.

Claims about the "agency" of nature often run aground on the shoals of intention, where doubts about the intentionality of a rock or an earthworm rise up (Nash

2005). However, many non-Western peoples ascribe agency, and even intention, to the activity of nature, or to spirits that direct seemingly natural acts. For example, Himalayan villagers may attribute a landslide or hailstorm to deities who are angry because of the violation of some social rule. Adopting Bruno Latour's concept of actant, "whoever and whatever" that acts and has a "spokesperson" to represent it extends the realm of actors beyond that of the human, and its institutions and processes (Latour 1987: 84, 71). The concept of actants within a network of interactions helps reveal the ways in which non-human nature is always already acting on human processes, while removing the requirement that conscious, discernable intention guide action. As Latour shows (1987, 1993), we can trace the influence of actants on the other members of their network, while remaining agnostic about the actants' intentions. Thus, the important role – the agency – of a boulder or a pine tree in shaping a particular course of events can be identified without the necessity of attributing conscious human-style thought to these ecological elements.

The activity of nature in shaping the both the material and imaginal lives of humans can be examined through "environmental imaginaries." As defined by political ecologists Richard Peet and Michael Watts (1996), environmental imaginaries are the place-specific and regionally variable conceptions of nature that shape the ways that people perceive, discuss, work and play in nature, and reflect a community's values with respect to its environment. Though they may be unarticulated, these holistic conceptions underlie the decisions that people make with respect to use of their surrounding environments. Environmental imaginaries are often presented in narrative, religious, or spiritual terms. They encapsulate people's values with respect to their surroundings, defining actions that are morally appropriate and acceptable. This concept incorporates political-economic conditions together with social, cultural, and ideological threads that reflect and respond to the material conditions of particular surroundings, to envisage a regionally distinctive conception of nature, and of the appropriate human relation to non-human nature.

The moral and spiritual aspects of the local environmental imaginary are evident in the eastern Himalayas, where relations with non-human nature are mediated through construction of sacred sites, local beliefs about deities, and perception of sacred natural sites. Drawing on fieldwork conducted in Bhutan and Nepal on repeated visits since 2001, this essay describes the beliefs and practices through which Tibetan Buddhists engage with specific places on the landscape to create and nurture the habits of mind that recognize forests, soil, rivers, mountains, and wildlife as beings worthy of respect. Specific examples are drawn from interviews conducted in Thimphu, the capital city, as well as in districts across Bhutan, particularly Trashi Yangtse, in far eastern Bhutan. The prohibitions and requirements of autochthonous deities, believed to be the original inhabitants of the land, have shaped human settlement, land use patterns, and local custom, as well as the placement of religious structures. In the Tibetan Buddhist parts of Bhutan, religion shapes the landscape by dictating land uses and proscriptions in certain areas through (1) land use restrictions that may prohibit agriculture, trespass, or resource collection in places where deities are believed to live; (2) the construction of monasteries, temples (lhakhang), and monuments (chorten); and (3) prohibitions against construction that result from beliefs that deities reside in the land.

Fig. 22.1 Large mango tree, believed to be the home of a deity, in Trashi Yangtse, eastern Bhutan (Photo by Elizabeth Allison)

Simultaneously, the landscape shapes the local, indigenous religion by providing sites of holiness and reverence. Sacred natural sites in Bhutan are sites of on-going negotiation between local people and the landscape, in which features of the landscape, and the landscape itself, are understood as active beings whose needs and wishes must be respected. This agency of the landscape is personified by the deities and spirits believed to inhabit the landscape. In this tradition, unseen beings do not just inhabit features of the landscape, but actually shape its physical appearance through human interaction with the landscape (Huber 1999). The perception of a deity residing in a particular feature of the natural landscape shapes human response to this feature, thereby shaping the physical appearance of the feature itself, setting off a positive feedback loop that reinforces a particular set of human responses toward the feature. This dynamic can be seen in the local stance toward huge old trees believed to be inhabited by deities (Figs. 22.1 and 22.2). The belief that deities live within the trees shapes the human response of leaving the trees alone, allowing them to grow to great sizes, which reinforces the human belief that the trees are inhabited by deities, which reinforces the tendency to leave the trees undisturbed.

22.2 Spiritual Traces in the Landscape

The attitudes, practices, and beliefs of the eastern Himalayan peoples who follow Tibetan, or Vajrayana, Buddhism create a particular kind of space, in which the spiritual is grounded in and intimately connected with specific places in the phenomenal world. This practice reflects both canonical Buddhism, by venerating historical events associated with religious masters and particular landforms believed to offer

Fig. 22.2 Gnarled tree, believed to be the home of a deity, in Trashi Yangste, Bhutan (Photo by Elizabeth Allison)

soteriological salvation from this life, and indigenous spirituality, in honoring spirits – such as mountain deities and protector deities – believed to inhere in the landscape (Blondeau 1994; Blondeau and Steinkellner 1996; Diemberger 1998; Huber 1999; Karmay 1998; Pommaret 1996; Samuel 1993; Ura 2001b).

From its early days in Bhutan, Buddhism has been associated with particular features of the landscape. In the seventh century, a Tibetan king, Srongtsen Gampo, had monasteries constructed at Bumthang in central Bhutan (Jambay Lhakhang) and in Paro in western Bhutan (Kichu). These two were among the many temples

Fig. 22.3 Taktsang, the site of Guru Rimpoche's meditation, in western Bhutan (Photo by Elizabeth Allison)

that he sponsored throughout the Himalaya to pin down a demoness ravaging the region (Mills 2007). Despite the construction of these Buddhist monasteries, Bhutan was still known as an "unlit land," where enlightening Buddhism had not yet penetrated (Karmay 1998: 142).

Buddhism took hold after the great saint Guru Rimpoche arrived in western Bhutan, near Paro, the current location of the airport, in 747 CE. According to legend, Guru Rimpoche, also known as Padmasambhava, arrived in Taktsang, or Tiger's Nest, a perch high on a cliff wall, where he meditated so intensely as to have molded the rock around himself, according to local legend (Fig. 22.3). The sites of his meditations – including Kurjey ("body print") Lhakhang, in central Bhutan, and Gom Kora, in eastern Bhutan – are highly revered pilgrimage sites more than a millennium later (Fig. 22.4). Thus, from its beginning in Bhutan, Buddhism has interpenetrated the landscape.

Himalayan Buddhism had less success in eradicating the pre-existing deities than, for example, Christianity did in Europe (Schneider 1991). Instead, Himalayan Buddhism incorporated these deities into its pantheons and practices, creating a mechanism that mediates the relationships and strengthens the connections between people to their landscapes. As he spread the Buddha Dharma through the Bhutanese lands, Guru Rimpoche encountered unruly autochthonous spirits and deities, believed to be the original owners of the land, and subdued them, converting them to Buddhism and making them sworn protectors of the Dharma. Because the autochthonous spirits were and continue to be inextricably connected with particular places, their inclusion in the Vajrayana Buddhist pantheon of Bhutan tied religious and spiritual practice to the landscape.

Fig. 22.4 Political map of Bhutan, with sites discussed in this chapter (Map by Dick Gilbreath, University of Kentucky Gyula Pauer Center for Cartography and GIS; commissioned by the editor)

22.3 Historical Spiritual Sites: Contiguity with the Sacred Past

Built sacred sites, such as monasteries, temples (lhakhang), and reliquaries (chorten) highlight the sacred or spiritual aspects of the landscape, bringing historical events into current awareness and linking the practitioner into the practice lineage of a living tradition. The faithful receive blessed energy through proximity to the historical spiritual site. Travel to, and engagement with, Buddhist holy sites, including circumambulation and offerings of prayers and incense, cleanses practitioners of negative karma. Transfer of spiritual energy or blessings occurs through proximity to a sacred or spiritual object, site, or person. The degree of transfer depends on the ritual preparation and purity of the practitioner, as well as temporal and geographic proximity to the sacred site or person. After appropriate ritual preparations, and close engagement with the sacred site, the practitioner becomes imbued with the sacred energy of the source (Makley 2003). For this reason, locations associated with Guru Rimpoche's travels are popular pilgrimage sites. Guru Rimpoche's spiritual energy is believed to have permeated and altered the surrounding rock, such that his footprints and body prints remain as traces of past spiritual power. At Taktsang in western Bhutan (see Fig. 22.3), the temple encloses a rock shrine where Guru Rimpoche meditated and is thought to have left a body print on the rock. Near Gom Kora, a large temple in Trashigang, eastern Bhutan, oblong divots in the stone along the Drangme Chhu river are believed to be Guru Rimpoche's footprints immortalized in ancient mud. In the same valley, according to legend, Guru Rimpoche subdued a demon snake that had been menacing the villagers by cutting it into three parts and locking the head part behind a stone door. A giant hexagonal boulder near the National Highway is believed to be the middle section of the demon, petrified in rock. Nearby, a small temple encloses a shallow cave where Guru Rimpoche's meditation was so long and intense as to leave his body impression in the rock. Visits to these sites remind believers of the living power of Buddhism, and of the continuous spiritual presence of their revered founding saint, whom many regard as a second Buddha. These sites on the landscape where Guru Rimpoche mediated, subdued demons, and walked by the river immerse the Bhutanese in a vibrant and embodied spiritual tradition that remains present, both spatially and temporally. Lay people visit these sites to offer prayers, incense, butter lamps, and other talismans frequently, especially on the holy days of the Bhutanese calendar (the 8th, 10th, 15th, 25th, and 30th day of each month) when spiritual endeavors gain extra merit. That the places associated with their founding saint are embedded in the natural landscape gives the spiritual practice that much greater connection to place.

22.4 Chorten: Tamer of Unruly Energy

In addition to providing the location for blessings received through proximity, built sites also serve to ameliorate or disperse destructive energies. Ridgelines and passes are often locations of chorten, white-washed stone monuments that are filled with religious relics, often built to pin down destructive spirits or to control unruly energies. The constant wind and exposure of high ridgelines dissipates and disperses energy, while the solidity of the rock chorten, encasing religious artifacts, serves to ground and stabilize energy in these remote and treacherous places. A white-washed chorten is visible from a great distance, guiding the traveler toward and over an exposed ridge, assisting the traveler with both practical route finding aid, and spiritual sustenance through proximity to religious objects. Though apparently constructed by humans, chortens often have mystical origins recounted in traditional stories. For example, Chorten Kora (Fig. 22.5) in Trashi Yangtse, Bhutan, built to pin down a destructive spirit that was tormenting the people of the Kulong Chuu (River) valley, is built in the style of Boudanath, a large Buddhist stupa or chorten in Kathmandu, whose image was carved into a radish to be carried to Bhutan. Because the radish shriveled and shrunk in transit, Chorten Kora has different dimensions than Boudanath. When the chorten was nearly finished, a devout princess from Tawang, Arunachal Pradesh, just across the Indian border, is said to have entombed herself in the chorten to meditate on behalf of all sentient beings. A similar legend surrounds the construction of another large chorten, Chendebji Chorten near Trongsa, which is said to have been built to cover the remains of an evil spirit.

Smaller structures, in the shape of chorten, known as lu khang (Tib. klu khang) or ney khang (Tib. gnas khang) are built to house spirits, keeping them happy and placated within farms and fields (Fig. 22.6). By providing the spirits clean and respectable shelter, people are able to stay in their good graces and gain continual boons for their harvests and well-being. The lu, a spirit associated with water, brings prosperity and is believed to control a storehouse of underground treasure, like the

Fig. 22.5 Chorten Kora, in Trashi Yangtse, eastern Bhutan (Photo by Christopher J. Flynn, http://en.wikipedia.org/wiki/File:Chorten_Kora_080720.JPG)

Fig. 22.6 Lu khang near agricultural fields, Trashigang, eastern Bhutan (Photo by Elizabeth Allison)

Indian naga. If the lu is well-placated, this bounty will accrue to the family who maintains their habitat, through bountiful harvests and other good fortune. However, the lu is particularly vulnerable to physical and spiritual pollution, and will retaliate for this befoulment by causing illnesses, including boils and other skin infections. The lu's home must be kept clean with milk and water three times per month, on days designated by the tsipa (astrologer).

22.5 Identification of Sacred Natural Sites on the Landscape

In addition to the historical legends surrounding sacred sites on the landscape, sites may be identified by their physical characteristics or by the perception that deities or spirits inhabit the place. Bhutanese deity sites are known as phodrang (citadels) in Dzongkha, the national language. Throughout the world, sites felt to have special or unusual power – through observation of their landforms, patterns of animal use, or particularly lush or vibrant vegetation – are designated as sacred. New religions tend to build their monuments on these places, co-opting the sacredness and enfolding the worshippers of that place (Gyatso 1987; Hayward 2000).

The perception of deities of the landscape has often been read through a materialist lens as deification of material resources on which villagers depend. For example, mountain communities often deify the mountain from which their water flows. In recognizing that their fields, their cattle, their very life, are utterly dependent on the mountain, they adopt a worshipful attitude toward it (Blondeau and Steinkellner 1996). However, a less strictly materialist reading of this phenomenon suggests that the villagers are responding to inherent, energetic qualities of particular places on the landscape, such that the perception of a deity represents the crystallization of a relationship between people and the landscape. Buddhist cosmology allows an entrance into this hypothesis with its acknowledgment of six realms of sentient beings, which also makes room for the existence of unseen yet active beings. The idea that relatedness-of-beings can be manifest echoes the Buddhist idea of "interdependent co-arising," which states that all phenomena arise from combinations of factors, and no phenomena is entirely self-generated. In this notion, nothing has any individually existence, but is radically dependent on all other circumstances, factors, and beings that bring it into being. A human being is not a tightly delimited, atomistic individual, but exists in a network of relationships, exchanges, and dependencies. Through engagement with places on the landscape and their particular qualities, people come to express their relationships with these places through the personification of deities. This relationship unfolds like a conversation through which people come to know and understand their landscape through their interactions with it. Through conversation, multiple parties express their relatedness, creating meanings and personification that did not exist before the conversation began (Bird-David 1999). When this relationship is smooth and harmonious, and the needs of both sides are met, the deities are happy. When the relationship goes awry, personal and environmental calamities, understood to be the vengeance of the deities, reflect the disruptions in the larger system.

22.6 Physical Qualities of Sacred Natural Sites

Sacred natural sites are protected at a variety of scales, ranging in size from a few square meters of a water spring or single rock, to dozens or hundreds of acres making up a mountaintop, forest, or entire valley. Typically, the geographical features encompassing a sacred natural site are in 'natural' areas, relatively undisturbed by humans or livestock (although they may be surrounded by settlements or agricultural fields). Thus, a standing grove of tall trees in the middle of a settlement area, or a lake around which there is no human development might be identified as sacred natural sites. Religious or cultural restrictions prohibit the harvest, collection, or destruction of living plant and animal material from sacred natural sites.

Protector deities, who oversee entire valleys or regions, are thought to reside high in the mountains where they remain aloof from the spiritual and physical pollution of human activities below. Climbing the highest peaks is forbidden in Bhutan, for fear of offending these deities. Even the lower peaks are rarely broached, as many villagers observe a traditional practice known as la dam or ri dum that prohibits travel up the mountains and harvesting of timber or bamboo during the spring and summer growing seasons. Villagers will occasionally ascend lower peaks to conduct ceremonies for the local protector deities.

Particular qualities of the landscape may contribute to exceptionally dense, lush growth of forests. In interviews, villagers described the places associated with deities as being dark and damp, and having particularly tall trees (Fig. 22.7) (Allison 2004b). Sacred natural sites studied in eastern Bhutan were dominated by four species of trees Quercus griffithii (oak), Schima wallichii (needlewood), Alnus

Fig. 22.7 Deity citadel in Trashi Yangtse, eastern Bhutan (Photo by Elizabeth Allison)

nepaliensis (alder), and Juglans regia (English walnut), as measured by estimations of stem frequency (Allison 2004b). These species all have a great deal of utility in village life, suggesting that these deity citadels are important in sustaining human communities, perhaps by providing seed stock and local refugia for important species (Sharma and Rikari 1999).

Water springs, seeps, and mud were often found in deity citadels (Allison 2004b). A deity's protective influence may serve to monitor water usage necessary for human purposes. Because of the need to irrigate multiple rice paddies in close proximity to each other, the potential for conflict over water distribution and allocation is high. Religious means of water allocation have been shown to be more efficient than externally imposed structures in Indonesia (Lansing 1987). Like the Indonesian system, the control of water usage by means of the deity's restrictions may help allocate a scarce resource among multiple players.

22.7 Deities and Spirits of the Landscape

The Bhutanese recognize two broad categories of deities: those enlightened beings who dwell beyond the realm of karmic existence, and are envisioned during Buddhist meditation and prayer; and the "haughty and wrathful" deities, residing within the six realms of the mundane world, who are not enlightened (Ura 2001b). The latter group of "mundane" deities has a "more pronounced environmental significance in mediating between resources and people" (Ura 2001b). The mundane deities exist on the same continuum as humans, but at a different level. Though they have lifetimes of up to millions of years, the mundane deities are not immortal or eternal: like humans, they are caught in the Buddhist cycle of samsara and rebirth. They are sentient beings with particular names, personalities, and characteristics, who – though generally invisible – occupy physical space and perform actions with effects in the material world. Among the most common mundane deities in Bhutan are the tsan, gyalpo, neypo, yul lha, key lha, lu, tsomen, and dud. The lu may appear in the material world, or in dreams, in the form of a snake, often of white or green color. The tsan appears as a red-countenanced horseman, carrying weapons and other implements.

These mundane deities, broadly grouped into protector deities (tsan, gyalpo, neypo, yul lha, key lha) who are seen as the original owner of particular localities, and authochthonous spirits (lu, tsomen and dud), who may control small areas within the landscape, are believed to inhabit rocks, trees, lakes, forest groves and river sections, proscribing human use and resource extraction from these areas. Like a protective homeowner, deities resent intrusions onto their property, and punish those who trespass or harvest resources without permission with persistent personal or family illness or injury, crop failure, untimely or heavy rain, or other unusual misfortune. Belief in various local deities and spirits has been documented in most districts of the nation, with local variation in the types and qualities of deities and spirits (Allison 2004b; Ura n.d.).

Villagers believe that if they disturb a deity – by harvesting timber or clearing the area for crops – bad weather, crop failures and illness will follow. Depending on the particular requirements of the deity, the following actions may be proscribed, or proscribed at particular times: entering a forested area, collecting any material from the area, removing living timber from the area, making loud noises, excreting human waste, throwing rocks, cooking or burning meat, bringing meat to the area, visiting the area shortly after child birth or handling a dead body, and bathing (in a lake). In addition, deities require certain activities to maintain their good graces. The propitiations include, among others: creating pleasant smell through the burning of incense juniper and wheat flour; making offerings of milk, popcorn, and wheat flour; prayers and invocations; keeping the area clean; and building, maintaining, and cleaning small spirit houses called lu khang (see Fig. 22.6).

22.8 Deities and Modern Development

In recognizing the spatial and temporal primacy of autochthonous deities, Bhutanese villagers are sensitive to the potential for offending the deities through their actions. The concern for these unseen prior inhabitants affects the way some think about modern development in Bhutan. In the early years of the twenty-first century, some Bhutanese expressed concern about upsetting the deities with the telephone lines, electricity, Indian products, and new home construction that had infiltrated Trashi Yangtse, one of the districts in eastern Bhutan most remote from the national capital. In an interview, Phurba Tsering, the head gomchen (lay priest) of Trashi Yangtse, emphasized a need to maintain a balance and equilibrium with the deities, fulfilling the human side of the contract.

> Before, there was no development. No pollution, no noise pollution, no air pollution, none of these things. Before, there were not many cases of illness, no periods of uneven sunshine and rainfall. Now, we have air pollution, noise pollution, littering. The tsan is not pleased. More people are getting sick, we have more periods of hail: these are signals that such things should not happen. But the government promotes development, and the tsan has to bear with us. He has to tolerate the changes. The gomchens try to make offerings to counterbalance the increasing pollution from modern development. (Phurba Tsering, interview with the author)

Throughout the country, the belief in these spirits can slow infrastructure development, even when the planned construction would benefit local people, who are eager to have better access to roads, hospitals, and schools. Three key informants associated with construction and infrastructure development told of situations in which local belief in a spirit or deity abode had stalled construction progress. A road contractor, interviewed in 2007 about a specific situation in which local people had complained that a road would go through a deity abode, commented, "Only later, when the road building commences, people come out and complain that they road is going through a spirit's place." A British foreign development consultant, who had lived in Bhutan for several years, described a situation in which local deity beliefs interfered with hospital construction, because a spirit was believed to inhabit a large tree at the construction site. When many of the workers fell sick, a lama was called to conduct a ceremony to appease the spirit. In another area, local people first agreed to, and then objected to, the construction of a water treatment plant, because it would trespass on the home of a deity. The British development consultant felt that these objections were raised for political ends, simply to prevent the nearby construction, and did not reflect a sincere belief. In other cases, capitulation to development follows objection and negotiation. A Bhutanese town planner in the capital commented:

> I heard of a problem in Mongar. People didn't want the area to be developed because they thought there was a ghost in the area. I think we took it up anyway. There was some resistance from that side. They said 'we are not supposed to touch this area.' In some cases, we have to respect that viewpoint because it is the sentiment of the people. If there are alternatives, we try to check into them. (Bhutanese town planner, interview with the author)

Although the empirical reality of these deity citadels is difficult to determine with standard scientific methods, local people perceive their reality, and, to the extent that this perception then shapes development activities, the citadels have discursive and material reality. The above vignettes reveal conflicting environmental imaginaries among those who have been trained in Western-style empiricism and express doubt about the validity of the deity citadels, believing people were "making excuses" ("Only later … people come out and complain" and "I think we took it up anyway"), versus the beliefs of the local people ("they thought there was a ghost in the area.") Even among foreign-educated Bhutanese and international professionals, who have been trained in epistemologies that disallow such phenomena, the deity citadels shape and influence conservation and development initiatives because of their importance in the eyes of villagers.

Further, the negotiation of restricted and available space for the purposes of infrastructure development demonstrates the continual construction of space, as a process of negotiation between various human groups, the landscape, and the perceived deities. As the consultant suggested above, deity beliefs may be deployed – or seen to be deployed – for political purposes, such as preventing inconvenient development (the well-known NIMBY ["not in my back yard" syndrome]). At the same time, these beliefs carry sufficient power and valence as to be respected, even by those who subscribe to alternate or competing environmental imaginaries. Even the urban educated elite of Thimphu, the capital, maintains ties with their natal villages and with the practices described in this chapter. The "sacred" and "non-sacred" are not fixed terms or places, but fluidly, continually constructed, in negotiations between various human groups with the landscape and deities. In some places, spirits or deities are known to move – to flee a place that was unsuitable because of moral, spiritual, or material pollution, for a more appropriate place. For example, in Naykhar, of Zhemgang district in central Bhutan, a large lake, extending downward from the village temple, is believed to have disappeared because of offense to the local water deity, the companion of the area's protector deity. All that remains is a muddy spot next to the walking path (Fig. 22.8).

Fig. 22.8 Naykhar ancient lake, Zhemgang, central Bhutan. It is believed that this lake dried up after a deity became angry with the villagers (Photo by Elizabeth Allison)

22.9 Changes with Modernization and Globalization

Globalization is slowly and subtly shifting deity beliefs and associated practices on the landscape. Increased interaction between Bhutan and the world beyond the Himalayan and Tibetan regions is having the paradoxical consequence of increasing interest in sacred natural sites and spiritual landscapes, while intensifying the forces that jeopardize the cultural and ecological survival of these same landscapes. Globally, changes in social and economic conditions, including migration, academic education, and integration into the consumer economy have been shown to shake traditional religious and spiritual beliefs, leading to decreasing engagement with traditional practices (Anoliefo et al. 2003; Bhagwat and Rutte 2006; Sharma and Rikari 1999). Cultural assimilation with dominant surrounding cultures and increasing connectivity through infrastructure improvements and telecommunications both contribute to the erosion of beliefs and practice in Nepal (Sherpa 2005). New residents of an area may unwittingly trespass into a sacred natural site, and, experiencing no retribution, contribute to a decline in belief about the power of the place. Taboos may be relaxed, foreign visitors or outsiders may disregard or mock taboos, young people may lose the faith. The cultural norms and practices that maintain sacred groves and other sacred natural sites tend to be less widely practiced by the younger generations, educated in academic settings and prepared to participate in global capitalism (Allison 2004a; Fisher and Hillary 1997).

Economic development in Bhutan, one of the government's Four Pillars of Gross National Happiness, the guiding development paradigm, sometimes comes into conflict with the traditional cultural and religious beliefs that protect sacred natural sites. Economic development has brought improved material standards of living and has improved access to markets to the most remote areas, but has also introduced construction machinery and noise, non-biodegradable waste, and ideas that are at odds with traditional beliefs. In Bhutan, the older generation carries the knowledge of the locations and practices related to local and protector deities. The younger generation, educated in academic, rather than religious, schools, as in the past, is less familiar with the requirements of deities and religious practice (Tashi 2001). As young people move away from ancestral homes to pursue educational and economic opportunities in urban areas, they lose touch with the traditional beliefs and practices related to deities (Chhetri 2010). Young adults consult their parents or elders if they have questions about the location or rules of a deity citadel, but if the elders have passed away, or are otherwise unavailable, the practices can become attenuated. In addition, Bhutanese intellectuals have observed the "cultural cringe" that occurs when people who have grown up in quiet rural villages become seduced by the shiny lifestyles they see on cable television, assuming that everything Western is good and that people of developed countries are automatically more knowledgeable (Dorji 2001).

Interaction with international visitors and their relatively materially opulent lifestyles can exacerbate this problem, eliciting envy in local people, and cause them to doubt the value of their traditional ways, including adherence to prescribed rules for interaction with local deities and sacred sites. Most foreign visitors to Bhutan are wealthy by any standard, because tourism is limited not by a quota, but by a daily fee of $200–250 instituted by the government to ensure "high quality, low volume tourism." In the past few years, Bhutan has beefed up its image as an exclusive destination, adding two luxury lodges – with rooms costing up to US $1,248 per night – that offer services far beyond those of the standard clean-and-comfortable guesthouses (Cyr 2004: 50). The mandatory daily fee, which includes all food, lodging and ground transportation (excluding luxury accommodations), limits pleasure travelers to around 21,000 per year (Choden 2008), up from fewer than 8,000 annually in 2001 (Gyeltshen 2001), while still contributing US $30 million (in 2007) to the Bhutanese economy (Choden 2008).

Road construction is believed to disturb deities, potentially leading to their departure. During summer 2002, while a road was built toward a remote village in eastern Bhutan, some villagers expressed concern that the explosive charges used to blast the road up the rocky mountain would upset the local deities, and either bring harm or reduce the deities' power, causing them to flee, while others eagerly anticipated easier access to Trashi Yangtse town and its supply of material goods. One man described the lessening of negative effects from an area that had once been a highly restricted deity citadel. These days, he said, villagers could cut some trees nearby without serious effects. He thought this change was due to the "value of the area decreasing," and hypothesized that this decrease in value could be due to nearby blasting for road construction, changes brought by development, or extraction of resources.

Government efforts to improve rural sanitation have also had a paradoxical affect on deity beliefs. Government health workers and school teachers, often the only educated officials in the most remote areas, have instructed villagers to dig garbage pits in which to burn non-biodegradable garbage, which has become an issue only in the past 10–15 years. Some villagers are concerned that the foul smells released from burning old shoes and plastic sandals, items considered to be both materially and spiritually polluting, could offend deities, causing them to avenge the insult by striking the villagers with respiratory illness. Some suggested that the offensive smells could even cause the deities to vacate the area, with negative consequences for the well-being of the village.

While economic development practices can contradict traditional beliefs about sacred natural sites, park management practices sometimes embrace these beliefs as they are harmonious with park management goals. International interest in traditional ecological knowledge, sacred groves, and connections between cultural and biological diversity are increasingly highlighting traditional ecological beliefs and practices as valuable and worth preservation (Berkes 2008; Gadgil et al. 1993; Kellert and Farnham 2002; Posey 1999; Schaaf and Lee 2005). At Bomdeling Wildlife Sanctuary (BWS) in Trashi Yangtse, eastern Bhutan, park staff encourage the practice of la dam, a traditional practice that prohibits villagers from going to the high mountains during the early spring and summer to avoid disturbing the protector deities. Violation of the la dam rules can lead to a variety of environmental calamities, including hail, heavy or untimely rain, landslide, extreme wind or drought, all of which can damage the crops and lead to a poor harvest. The park staff at BWS sees the practice of la dam as a form of community-based natural resource management that protects trees and bamboo from harvest pressure during sensitive growing times. Forest guards remind villagers of the timing and rules of la dam each year as a way of preserving both ecological resources and traditional practices. Park staff has even encouraged villagers to revive the practice in locations where it appeared to be fading.

Noting the degree to which religious taboos protect some sacred natural sites, international organizations involved in biodiversity conservation have become interested in the potential of sacred natural sites to serve as a mechanism of indigenous community-based natural resource management (CBNRM). The World Wildlife Fund has been working for several years to establish a Sacred Himalayan Landscape, which encompasses parts of Bhutan, Nepal, and India, as a transboundary conservation initiative to leverage indigenous ethics, sacred natural sites, and local spiritual beliefs in service of the protection of endangered wildlife (WWF 2012; Gurung et al. 2006). Researchers in India have embarked on several ambitious projects documenting India's sacred natural sites to ensure the maintenance of this cultural form that contributes to the preservation of both biodiversity and indigenous livelihoods. This research has produced lists and compendia, but no comprehensive map or database of sacred natural sites in India has been produced to date. Though sacred natural sites have been recognized in diverse cultures throughout history (Chandran and Hughes 1997), Gadgil and Vartak (1976) brought scientific attention to the potential connections of sacred groves and biodiversity conservation with their early work on ethnobotany in India. Estimates of the number of sacred groves in India range from 13,720, based on published reports, to 150,000–200,000 sacred groves (Chandran and Hughes 1997; Gadgil and Vartak 1976; Gold and Gujar 1989; Malhotra et al. 2000, 2001; Polidor 2004; Wild and McLeod 2008). Surveyed sacred groves cover an area of at least 42,278 ha (163 sq miles) (Malhotra et al. 2001). While numerous studies have documented the sacred groves in particular Indian states or localities, there is no comprehensive inventory of sacred groves or sacred natural sites. In the early years of the twenty-first century, proposals were floated to establish a national inventory, or an electronic database, of sacred groves, modeled after the People's Biodiversity Registers program (Gadgil et al. 2000; Gaikwad et al. 2004; Ghate et al. 2004).

In 2010, the World Wildlife Fund organized a workshop on sacred natural sites and biodiversity conservation, held in Bhutan. The workshop's goal was to document the sacred natural sites of the eastern Himalaya, their role in the preservation of bio-cultural diversity, and the threats they face with climate change. The workshop also sought to engage the faith communities of the region in ecological conservation (WWF 2010).

More recently, Bhutan has taken up the notion of inventorying sacred sites, proposing that researchers at the Ugyen Wangchuck Institute for Conservation and Environment conduct a nationwide study to document sacred natural sites (Chhetri 2010). The inventory would help conserve the cultural heritage of Bhutan, while raising awareness of the role of sacred natural sites in biodiversity conservation and documenting the effects of climate change on such sites (Chhetri 2010). More complete understanding of the geographic distribution and extent, ecological characteristics, and socio-economic aspects of the sacred natural sites could contribute to greater appreciation and protection. In addition, the national inventory would suggest opportunities for legal protection of these sites, an idea that Dasho Karma Ura floated in a proposal in the early 2000s (Ura 2001a).

The national inventory and mapping project will need to proceed carefully, as some communities may not wish to reveal the precise locations of their sacred natural sites, for fear of unruly trespass by outsiders or causing offense to the deity. For example, during my research, one deity site in eastern Bhutan was considered to be so dangerous that villagers would not tell me where it was, nor let me visit it, for fear of offending the deity. The deities of the place – a tsan and a gyalpo – were thought to harm people by suddenly taking their souls away to the deity citadel, thereby causing death. According to one source, stones from outside the deity citadel were used in the construction of two houses near the citadel. Because of the power of the deities, one adult died in each house. In addition, mapping or inventorying the sites may make it easier to desecrate them for their natural resources. Overall, however, this effort will be an important step in documenting and analyzing the specific human-nature interactions that create and maintain sacred natural sites, and will further help shift ecological and geographical thinking toward a view that values the agency of non-human nature and the natural landscape.

References

Allison, E. (2004a). Religiously protected natural sites of Khumbu. In S. Subha & A.R. Sherpa (Eds.), *Sacred sites trails project brochure*. Kathmandu: The Mountain Institute (TMI).

Allison, E. (2004b). Spiritually-motivated natural resource management in eastern Bhutan. In K. Ura & S. Kinga (Eds.), *The spider and the piglet* (pp. 528–561). Thimphu: Centre for Bhutan Studies.

Anoliefo, G. O., Isikhuemhen, O. S., & Ochije, N. R. (2003). Environmental implications of the erosion of cultural taboo practices in Awka-south local government area of Anambra state, Nigeria: 1. Forests, trees, and water resource preservation. *Journal of Agricultural & Environmental Ethics, 16*(3), 281–296.

Berkes, F. (2008). *Sacred ecology*. New York: Routledge.

Bernbaum, E. (1990). *Sacred mountains of the world*. San Francisco: Sierra Club Books.

Bernbaum, E. (1997). Pilgrimage and conservation in the Himalayas: A model for environmental action based on cultural and spiritual values. The Mountain Institute.

Bernbaum, E., & Purohit, A. N. (1999). Badrinath: Pilgrimage and conservation in the Himalayas. In D. Posey (Ed.), *Cultural and spiritual values of biodiversity* (pp. 336–337). Nairobi: United Nations Environment Programme.

Bhagwat, S. A., & Rutte, C. (2006). Sacred groves: Potential for biodiversity management. *Frontiers in Ecology and the Environment, 4*(10), 519–524.

Bird-David, N. (1999). "Animism" revisited: On personhood, environment and relational epistemology. *Current Anthropology, 40*(SI), S67–S91.

Blondeau, A.-M. (1994). Tibetan mountain deities, their cults and representations. Papers presented at a panel of the 7th seminar of the international association for Tibetan studies. Graz 1995. In A.-M. Blondeau (Ed.), *International Association for Tibetan Studies Seminar* (Verlag Offentlichungen zur Sozialanthropologie, Bd. 3; Wien: Verlag der Osterreichischen Akademie der Wissenschaften).

Blondeau, A.-M., & Steinkellner, E. (1996). *Reflections of the mountain: Essays on the history and social meaning of the mountain cult in Tibet and the Himalaya*. Wien: Verlag Der Osterreichischen Akademie der Wissenschaften.

Capra, F. (1996). *The web of life: A new scientific understanding of living systems*. New York: Anchor Books.

Chandran, M. D. S., & Hughes, J. D. (1997). The sacred groves of south India: Ecology, traditional communities and religious change. *Social Compass, 44*(3), 413–427.

Chhetri, D. (2010, May 18). Towards a sacred natural site inventory. *Kuensel Online*

Choden, P. (2008, October 15). Tightening American belts may trim tourism-Bhutan. *Kuensel: Bhutan's National Newspaper*.

Cyr, C. (2004). Himalayan heaven. *Outside Magazine*, Santa Fe.

Diemberger, H. (1998). The horseman in red: On sacred mountains of la stod (southern Tibet). In A.-M. Blondeau (Ed.), *Tibetan mountain deities, their cults and representations. Papers presented at a panel of the 7th seminar of the international association for Tibetan studies* (pp. 43–55). Graz 1995. Wien: Verlag der Osterreichischen Akademie der Wissenschaften.

Dorji, T. (2001). Sustainability of tourism in Bhutan. *Journal of Bhutan Studies, 3*(1), 84–104.

Fallon, S., Connolly, P., & Enig, M. G. (2001). *Nourishing traditions: The cookbook that challenges politically correct nutrition and the diet dictocrats*. Washington, DC: NewTrends Publishing.

Fisher, J. F., & Hillary, E. (1997). *Sherpas: Reflections on change in Himalayan Nepal. With a foreword by Sir Edmund Hllary*. New Delhi: Oxford University Press.

Gadgil, M., & Vartak, V. D. (1976). Sacred groves of Western Ghats in India. *Economic Botany, 30*(2), 1521–1560.

Gadgil, M., Berkes, F., & Folke, C. (1993). Indigenous knowledge for biodiversity conservation. *Ambio, 22*(2/3), 151–156.

Gadgil, M., et al. (2000). New meanings for old knowledge: The people's biodiversity registers program. *Ecological Applications, 10*(5), 1307–1317.

Gaikwad, S. S., et al. (2004). Digitizing Indian sacred groves – An information model for web interfaced multimedia database. In V. Ghate, H. Sane, & S. S. Ranade (Eds.), *Focus on sacred groves and ethnobotany* (pp. 1232–1228). Mumbai: Prism Publication.

Ghate, V., Sane, H., & Ranade, S. S. (Eds.). (2004). *Focus on sacred groves and ethnobotany Mumbai*. Mumbai: Prism Publications.

Gold, A. G., & Gujar, B. R. (1989). Of gods, trees and boundaries, divine conservation in Rajasthan. *Asian Folklore Studies, 48*(2), 211–229.

Gurung, C. P., et al. (2006). The sacred Himalayan landscape: Conceptualizing, visioning and planning for conservation of biodiversity, cultures and livelihoods in the eastern Himalaya. In J. A. McNeely (Ed.), *Conservation biology in Asia*. Kathmandu: Society for Conservation Biology Asia Section and Resources Himalaya.

Gyatso, J. (1987). Down with the demoness: Reflections on a feminine ground in Tibet. In J. D. Willis (Ed.), *Feminine ground: Essays on women and Tibet* (pp. 33–51). Ithaca: Snow Lion.

Gyeltshen, T. (2001, March 10). The tourism trends. *Kuensel: Bhutan's National Newspaper*.

Hayward, J. (2000). Meeting the dralhas. In S. Kaza & K. Kraft (Eds.), *Dharma rain: Sources of Buddhist environmentalism* (pp. 271–277). Boston: Shambala Publications.

Hoffman, A. J. (2011). Sociology: The growing climate divide. *Nature and Climate Change, 1*(4), 195–196.

Huber, T. (Ed.). (1999). *Sacred spaces and powerful places in Tibetan culture: A collection of essays*. Dharamsala: Library of Tibetan Works and Archives.

Huntsinger, L., & Fernandez-Gimenez, M. (2000). Spiritual pilgrims at Mount Shasta, California. *Geographical Review, 90*(4), 536–558.

Karmay, S. G. (1998). *The arrow and the spindle: Studies in history, myths, rituals and beliefs in Tibet*. Kathmandu: Mandala Book Point.

Kellert, S. R., & Farnham, T. J. (2002). *The good in nature and humanity: Connecting science, religion, and spirituality with the natural world*. Washington, DC: Island Press.

Krippner, S., Devereux, P., & Fish, A. (2003). The use of the Strauch scale to study dream reports from sacred sites in England and Wales. *Dreaming, 13*(2), 95–105.

Lansing, J. S. (1987). Balinese water temples and the management of irrigation. *American Anthropologist, 89*(2), 326–341.

Latour, B. (1987). *Science in action: How to follow scientists and engineers through society*. Cambridge, MA: Harvard University Press.

Latour, B. (1993). *We have never been modern*. Cambridge, MA: Harvard University Press.

Makley, C. E. (2003). Gendered boundaries in motion: Space and identity on the Sino-Tibetan frontier. *American Ethnologist, 30*(4), 597–619.

Malhotra, K. C., Stanley, S., Herman, N. S., & Das, K. (2000). Biodiversity conservation and ethics: Sacred groves and pools. In N. F. Macer, & D. R. J. Macer (Eds.), *Bioethics in Asia* (pp. 338–345). Eubios Ethics Institute.

Malhotra, K. C., Gokhale, Y., & Chatterjee, S. (2001). *Cultural and ecological dimensions of sacred groves in India*. New Delhi: Indian National Science Academy.

Mills, M. A. (2007). Re-assessing the supine demoness: Royal Buddhist geomancy in the srong btsan sgam po mythology. *Journal of the International Association of Tibetan Studies*, (3), 47.

Nash, L. (2005). The agency of nature or the nature of agency? *Environmental History, 10*(1), 67–69.

Norgaard, K. M. (2006). People want to protect themselves a little bit: Emotions, denial, and social movement nonparticipation. *Sociological Inquiry, 76*(3), 372–396.

Peet, R., & Watts, M. (1996). *Liberation ecologies: Environment, development, social movement*. London/New York: Routledge.

Plumwood, V. (2006). The concept of a cultural landscape: Nature, culture and agency in the land. *Ethics and the Environment, 11*(2), 115–150.

Polidor, A. (2004). Sacred groves of India. Sacred Land Film Project www.sacredland.org/sacred-groves-of-india/%3E

Pommaret, F. (1996). On local and mountain deities in Butan. In A.-M. Blondeau & E. Steinkellner (Eds.), *Reflections of the mountain: Essays on the history and social meaning of the mountain cult in Tibet and the Himalaya* (pp. 39–56). Wien: Verlag Der Osterreichischen Akademie Der Wissenschaften.

Posey, D. A. (Ed.). (1999). *Cultural and spiritual values of biodiversity*. London: Intermediate Technology.

Samuel, G. (1993). *Civilized shamans: Buddhism in Tibetan societies*. Washington, DC: Smithsonian Institution Press.

Schaaf, T., & Lee, C. (2005). Conserving cultural and biological diversity: The role of sacred natural sites and cultural landscapes – international symposium. In T. Schaaf & C. Lee (Eds.), *Conserving cultural and biological diversity: The role of sacred natural sites and cultural landscapes* (p. 341). Tokyo: United Nations University and UNESCO.

Schneider, J. (1991). Spirits and the spirit of capitalism. In E. Wolf (Ed.), *Religious regimes and state-formation* (pp. 181–220). Albany: State University of New York.

Sharma, S., & Rikari, H. C. (1999). Conservation of natural resources through religion: A case study from central Himalaya. *Society and Natural Resources, 12*(6), 599–612.

Sherpa, L. N. (2005). Sacred hidden valleys and ecosystem conservation in the Himalayas. In T. Schaaf & C. Lee (Eds.), *Conserving cultural and biological diversity: The role of sacred natural sites and cultural landscapes* (pp. 68–72). Tokyo: International Symposium. UNESCO.

Tashi, K. P. (2001). Director, National Museum of Bhutan, Paro. Interivew with the author. Paro, Bhutan.

Ura, K. (2001a). *BhuNEAP conceptual proposal to protect the citadels of deities and spirits through legislative enactment*. Thimphu: Centre for Bhutan Studies.

Ura, K. (2001b, November 26). Deities and environment: A four part series. *Kuensel: Bhutan's National Newspaper*.

Ura, K. (n.d.). *Tables listing dzongkhag, deity, sex, abode of the deity, village, gewog*. Thimphu: Centre for Bhutan Studies.

Watts, M., & Peet, R. (1996). Conclusion: Towards a theory of liberation ecology. In R. Peet & M. Watts (Eds.), *Liberation ecologies: Environment, development, social movements* (pp. 260–269). London: Routledge.

Wild, R., & McLeod, C. (Eds.). (2008). *Sacred natural sites guidelines for protected area managers. Best practice protected area guidelines series*. Gland: IUCN.

WWF. (2010). W. W. F. sacred Himalayan sites bring together religious leaders, conservationists. www.worldwildlife.org/who/media/press/2010/WWFPresitem16324.html%3E.

WWF. (2012). Eastern Himalayas projects. www.worldwildlife.org/what/wherewework/easternhimalayas/projects.html%3E. Accessed 27 May 2012.

Chapter 23
Suppression of Tibetan Religious Heritage

P.P. Karan

23.1 Introduction

Religion and the web of Tibetan monasticism played a fundamental role in shaping the culture and society of the high plateau of Tibet which occupies approximately 1.5 million sq. miles (3.8 sq. km) with altitudes ranging from 4,000 to 20,000 ft (1,220–6,472 m) in central Asia. This "cultural Tibet" comprising areas where ethnic Tibetans comprise a significant part of the population which exceeds the boundaries of present political Tibet known as the Tibetan Autonomous Region of the Peoples' Republic of China. The barren nature to the north and west and the immense mountain ranges of the south (the Himalaya) served as isolating barriers and helped preserve the unique religious culture of the region. Tibetan religious institutions united many disparate groups, including nomadic pastoralists and urban craftspeople, slash-and-burn agriculturists and people engaged in international trade, serfs bound to the land, brigands and saintly mystics. Common ideology and common faith provided the basis for Tibetan culture.

23.2 Historical Background

For over 2,000 years Tibet has been an ethnically, culturally and linguistically distinct entity. In the eighth century Tibet emerged as the most powerful state in central Asia and conquered large parts of China. In the thirteenth century, however, Tibet came under the domination of the Mongol Empire and developed a priest-patron relationship with China's Emperors. This relationship is a uniquely Buddhist and

---

P.P. Karan (✉)
Department of Geography, University of Kentucky, Lexington, KY 40506, USA
e-mail: ppkaran@uky.edu

Central Asian religion-political institution formed as a personal bond consisting of the elements of protection by the Patron of his Priest and the Priests commitment to fulfill the Patrons' spiritual needs (Walt 1987: 123).

In 1720 Manchu troops entered Tibet, and again the priest-patron relationship became the basis for Tibet's association with the new invaders. Manchu influence over Tibet reached its peak after 1793 when Manchu officials in Tibet began to supervise, and in some cases conduct, Tibet's foreign affairs. Manchu influence steadily fell during the nineteenth century. Tibet was invaded, though not dominated, in 1842, 1856 and 1903 by the Dogras, the Gorkhas and the British respectively (Shakabpa 1988). In 1911 Tibet expelled all Chinese officials and was free of Chinese influence until 1950 when the People's Liberation Army overwhelmed an ill-equipped Tibetan force of only 8,000. The Tibetans, who had never experienced Chinese rule of their internal affairs, resisted the consolidation of Chinese power, which culminated in a national uprising in 1959. During the uprising, more than 87,000 Tibetans were killed by the Chinese army and the Dalai Lama narrowly escaped to India along with more than 100,000 followers.

Prior to 1950, Tibetans lived a manner unchanged since the Middle Ages. They had no electricity or mechanization, the economy was agricultural and nomadic. Valleys between 4,000 and 15,000 ft (4,575 m) were cultivated for barley, wheat, oats, peas, nuts, and fruit. One half of the population was nomadic, moving in fixed routes from pasture to pasture with yaks, yak-cow hybrids, goats, sheep and ponies as livestock. Despite the high altitude and dry climate, food was easily grown and herds could be fed where there was water and protection from the strong winds. The small population (3.5–4 million) could easily support itself in a primitive, but adequate standard of living.

Tibetans, spread across the isolated reaches of the plateau were commonly united by their faith and language, both imported from India in the seventh century A.D. The unique Tibetan form of Buddhism, molded over the succeeding centuries by indigenous beliefs and new influxes of mystical practices and monastic reforms from the border regions, had become the state religion by the eleventh century. Although separated from one another by great distances, the Tibetan people saw themselves unified as people of the faith, holding a common view of the supernatural as well as the earthly world.

The cultural landscape patterns which developed under the rule of successive Dalai Lama exhibited significant imprints of religion. Religious values, ideas and concepts permeated every aspect of Tibet's culture, economy and policy. The settlement morphology as well as the land use patterns provided visual manifestations of Tibetan's religious beliefs. Both political and socioeconomic affairs were viewed first and foremost in the light of their bearing on well-being of religion. Lamaist Buddhist ideology was decisive in imparting Tibet with the specific shape and direction it followed in economic and political affairs.

The imposition of Han Chinese Maoist ideology in 1950 marked the beginning of social, economic and political changes reflecting the Han ideology in all aspects of landscape and life of traditional Buddhist Tibet. Adherence to collectivism in the implementation of social and economic programs and an emphasis on integration of

Tibet into the Han Chinese communist system were the two key objectives of the communist ideology imposed on Tibet in 1950. These ideological values and attitudes have strongly influenced spatial processes and structures in Tibet during the last six decades. By 2010 Tibet had been transformed from a theocratic state into a land where communist ideology reveals itself in many geographic patterns. The communist/socialist doctrine has achieved remarkable spatial expression in political, economic and settlement patterns.

The obliteration of organized religion represents the profound impact wrought by communism in the cultural geography of Tibet. Only a handful of lamas and monks are reportedly left in major monasteries which at one time housed large numbers of religious communities. The role of monasteries as traditional centers of learning and culture has been eliminated and public education is under strict control of the party. The new secular structure of the Tibetan educational system indoctrinated with Marxist philosophy confirms the impact of communist ideology on the plateau.

23.3 Phases in China's Control of Religion

The socialist transformation of the Tibetan landscape and control of religion between 1951 and 2012 can be grouped under six phases. The first phase began in October 1950 with invasion and occupation of Tibet by the People's Republic. This initial phase culminated in 1954 with the extension of China's control over much of the plateau. The second phase, which began in 1955 and ended in 1959, marked a major extension of Chinese influence into the socioeconomic and religious aspects of Tibetan life. The third phase from 1960 to the eve of the Cultural Revolution in 1966 was marked by the abolition of the dual civil and religious governmental structure, obliteration of organized religion, and the intensification of socialist system. The fifth period marked the beginning of the massive influx of Han Chinese population in Tibet. The last and current phase of the Han penetration, which began in 1969, is characterized by an accelerated socialist transformation and intensified resistance by the Tibetans.

23.3.1 The First Phase 1950–1954: Laying the Framework for Extension of the Communist Ideology and Control of Religion

Following its successful invasion of Tibet, China pledged in the treaty of May 1951 not to alter the established status, functions and powers of the Dalai Lama. This pledge was observed on the surface with no formal attempt to alter the traditional institutions of Tibetan administration, but China began to lay the groundwork for concerted action to transform Tibetan society. To begin with China redefined the concept of "local government" as designated in the 1951 treaty. It began to consider

Fig. 23.1 Map of Tibet (Map by Dick Gilbreath, University of Kentucky Gyula Pauer Center for Cartography and GIS; commissioned by the editor)

Tibet not as a single political unit, but comprising three political units each having separate administrations. These three units consisted of: (1) the territory of central Tibet ruled by the Dalai Lama; (2) the area around Shigatse administered by the Panchen Lama; and (3) the strategic eastern-most area of Chamdo which formed the gateway to Tibet from China (Fig. 23.1). China interpreted its pledge to maintain traditional Tibetan institutions to apply only to the territory of central Tibet around Lhasa ruled by the Dalai Lama. To initiate reforms and changes in the rest of the Tibetan plateau outside central Tibet, China considered itself under no obligation to seek approval from the Dalai Lama's government at Lhasa. Although there is no historical evidence to support the partition of Tibet into three separate and equal entities, China cogently used the lack of the Dalai Lama's effective political control over the entire plateau for territorial division of the country in order to advance its own political control over Tibet.

The Chamdo region was first singled out for close integration with China because of its strategic location controlling major routes linking Tibet with China. As the most accessible city in Tibet to China, Chamdo began to emerge as a major political administrative center on the plateau as China began to move Tibetan offices to the city from outside the influence of the Dalai Lama's government in central Tibet.

23 Suppression of Tibetan Religious Heritage

As Han settlers began to arrive in Chamdo in the early 1950s, encroachments by the settlers on customary territory of Tibetan tribes in this area generated resentment and hostility towards Han. The Khampas, a border tribe, revolted against the Chinese. This defiance stood in the way of complete transformation of the Chamdo region. However, with the administration of public order in Chinese hands and full military dominance, China wielded complete authority in Chamdo by 1954.

In the Shigatse area China expanded its political control indirectly by securing the full loyalty of the Panchen Lama through elevating him on parity with the Dalai Lama and by exploiting the feud between the Panchen and Dalai and their respective entourage. Shigatse fell easily into the political control of China. Because of the Panchen Lama's high position and status as the head of vast land holdings in Shigatse having a large population, and spiritual eminence as head of the Tashilunpo monastery with vast wealth and affluence, the Dalai Lama's government generally refrained from direct interference in the internal administration of territories under the jurisdiction of the Panchen. However, the Lhasa government maintained provincial governors at Shigatse who were charged with the responsibility of coordinating the administration of the territory with the administrative organization responsible to the Panchen Lama. Historically, the Dalai Lama's political power and authority is regarded as paramount over all of Tibet, even though in reality the Dalai's effective control was circumscribed due to the country's feudal political pattern in which large landed estates were held in fiefdom by local lords who acted as administrative heads of various areas. The Panchen Lama, who stands below the Dalai Lama spiritually at the apex of monastic hierarchy, is not considered the secular equal of the Dalai. By treating the Panchen as sovereign within the Shigatse area, China effectively withdrew a part of Tibet from the Dalai Lama's secular control, and moved to establish its own control.

In contrast to the Chamdo and Shigatse areas where China encountered least resistance in establishing its domination, the traditionally strong Tibetan elements and institutions in the Lhasa region offered major obstacles to expansion of effective Chinese domination and socialist transformation of central Tibet. Under the circumstances China's policy was aimed at weakening the internal unity of the established Tibetan regime in Lhasa. It involved support of those indigenous elements such as the selected Tibetan hierarchy whose strength would destroy existing political balance, thereby facilitating the establishment of Chinese domination, and finally assist in bringing change in the existing socio-political organization of the Lhasa region. The policy called for downgrading the divine Godhood emblem of the Dalai Lama and undermining the awe and mystery surrounding the symbolic position of Dalai Lama as Tibet's God-King. A major effort was made to abolish the primacy of monks in Tibet's government. By attempting to eliminate monastic supervision over civil administration, China tried to "secularize" the traditional administrative system of Tibet. With extraterrestrial values deriving power from divine source of authority and possessing cohesiveness and unity invoked with religious sanction and sanctity of faith, monasteries in the Lhasa region occupied positions of great importance in the social and political fabric of the country. Through repeated efforts to curtail the public and semi-official privileges, prestige and influence of the major monasteries in the Lhasa region, China made an impressive attempt to reduce the importance of the major source of ideological values in Tibet.

Despite major attempts by China for influence and dominance in the Lhasa region, the power and prestige of the monastic Tibetan hierarchy remained supreme. However, China did succeed in bringing about some structural and functional changes in the Dalai Lama's government, particularly in secularizing the traditional administrating system of Tibet. By the end of 1954, China had consolidated its physical hold in all areas of Tibet except the Lhasa region.

23.3.2 The Second Phase 1955–1959: Development of New Political Structure to Advance Communist Ideology and Control of Religion

In the 1951 Sino-Tibetan treaty, China recognized the special and distinctive political status of Tibet. The special political position was downgraded when Tibet was made a standard autonomous region of China similar to other ethnically non-Han areas under the Constitution of the People's Republic of China. The concept of the Tibetan nation which the Dalai Lama's government had always advanced received a major setback when the new constitution was endorsed by Tibetan delegates attending the legislative session in Peking in 1954. The political relations between China and Tibet were not to be gained now under the para-international 1951 Sino-Tibetan treaty, but under Chinese domestic law. However, the regional autonomous status envisaged by the Constitution of the People's Republic was not endowed on Tibet in 1954 because the Chinese control was not fully secure in all parts of the country. The strong socio-political and religious groups in Lhasa represented major obstacles in execution of basic Chinese policies for communist transformation of Tibet.

A Preparatory Committee for Formation of a Tibetan Autonomous Region was established in 1955 to ready the country for autonomy as provided under the Constitution of the People's Republic. The Preparatory Committee comprised representatives of the Dalai Lama's government, the Panchen Lama's council, the People's Liberation Committee of the Chamdo area, the Chinese People's Government personnel in Tibet, and major monasteries and other organizations. With the Dalai Lama as the Chairman, the Panchen Lama and General Chang Kuo-hua as deputies and Ngabo Ngawang Jigme as the Secretary General, the Committee was to function as the local centralized administration of Tibet deriving its authority from and dependent on the State Council of the People's Republic in all respects.

The Committee established several subordinate agencies dominated by the Chinese personnel to facilitate administration. By creating agencies, run by Hans, on civil administration, finance, health, judiciary, agriculture, trade and industry, transportation and construction, China successfully established an institution in Lhasa to gnaw and seize the political power from Tibet's traditional leaders – a task in which they had not succeeded during the 1950–1954 period. The formal Chairmanship of the Committee by the Dalai Lama was of no value in restraining the Chinese from usurping political control of the country because most Tibetans appointed to the Committee were Chinese puppets. In any case the major decisions were made by the Committee of the Chinese Communist Party in Tibet.

The initiation of large scale secular education with a communist-oriented curriculum, the construction of hydroelectric stations, factories, experiment farms, and roads began to transform the geographical landscape of the country, its economy and people. However, the increasing Han control of Tibetan affairs began to generate mounting unrest among the people. Embittered by the Han management of their affairs through proliferating agencies of the Preparatory Committee and the rapid pace of development of local economic resources to strengthen China's position in Tibet, groups of Tibetans reacted with an armed uprising against the Chinese at various places on the plateau. The withdrawal of some Han personnel from Tibet in 1957 did not satisfy the Tibetans as the uprising continued in 1958 to which China responded by heightened repression. In early 1959, a stage was set for head-on collision between a determined China to gain political supremacy on the plateau and the Dalai Lama's government which sought desperately to preserve its traditional identity and religious institutions.

Defying Chinese regulations against carrying weapons, a crowd of about ten thousand Tibetans bearing arms marched on the Potala, the main palace of the Dalai Lama (Fig. 23.2), and the Norbu Lingka, the summer palace, to protect their sovereign God-King from the Chinese and to demonstrate against communist rule

Fig. 23.2 The Potala has been the residence of successive Dalai Lamas since 1649. It was also the seat of the Tibetan government. It has chapels, cells, schools for religious training as well as tombs for the Dalai Lamas. The Potala was shelled during the 1959 popular uprising against the Chinese. Government buildings and houses at the foot of the Potala were demolished by China to make room for the construction of wide boulevard seen in the foreground (Photo by P. P. Karan)

in Tibet. The Khampas from the border regions and people from the great monasteries near Lhasa led the insurgents into confrontation with the Chinese. The area of effective military confrontation between Tibet and China remained highly localized in Lhasa and its environs. Environmental difficulties and the lack of transport and communications prevented the uprising from spreading and assuming mass proportions. However, in a sense the revolt may be considered national since it involved Lhasa – the seat of government and the symbol of the spirit of the Tibetan people and the country's leadership in the struggle.

The revolt in Lhasa was effectively suppressed by the People's Liberation Army. Tibetan resistance collapsed although sporadic guerrilla warfare continued on the plateau. With the collapse of the insurgency in Lhasa and the decision of the Dalai Lama to exile in India rather than be a virtual prisoner of a foreign occupation, the way was paved for absolute Chinese control and communist transformation of Tibet (French 2003). In implementing Han policies in Tibet, China would have no longer to consider the reaction of the Dalai Lama and his considerable influence over his people.

23.3.3 The Third Phase 1960–1966: Monastic System Dismantled by China

The Han goal of molding the Tibetan society and policy into the "socialist" image of the People's Republic was greatly facilitated as a result of their triumph over the insurrection in Lhasa. China lost no time in establishing the required administrative system to execute programs of social, institutional and economic changes in Tibet to eliminate all remnants of the country's monastic system and autonomy and to substitute direct rule closely supervised by Chinese authorities (Fig. 23.3). These "reforms" deprived the monasteries of their land holdings, and entailed destruction of hundreds of monasteries and the imprisonment, execution and expulsion of tens of thousands of monks. They were undertaken, according to Chinese sources, to end the "exploitation of serfs" by the monasteries and disentangle religion from politics. Chinese sources credit the reforms with awakening the proletarian conscious of the monastic community. They contend that "the masses of the monks and nuns … felt so delighted and enthusiastic [they] demanded that they participate in labor" (Dreyer 1976: 169).

After the 1959 revolt a decree of the State Council of the People's Republic of China dissolved the local government of Tibet with the Dalai Lama as the head. At the same time the Preparatory Committee for the Tibetan Autonomous Region was made the principal instrument of Chinese rule with complete control of local administration. The Panchen Lama was named chair of the organization. The reconstituted Committee with new members replacing the "traitorous elements" undertook the task of reorganization of the religion and socio-political system of Tibet. The liquidation of the Dalai Lama's government and the purging of his supporters marked the elimination of the last surviving remnants of Tibetan governmental autonomy and

Fig. 23.3 The Jokhang is the most revered religious structure in Tibet. Its construction was initiated by King Songtsen Gampo in the seventh century. Much of the interior of Jokhang Temple was desecrated by Red Guards and many objects were removed and destroyed; several older residents of Lhasa reported that at one stage the monk's quarter was used as piggery. Since 1980 Jokhang has been restored and buildings in front of the temple have been demolished to create a wide plaza (Photo by P. P. Karan)

the establishment of a subordinate system of regional administration submissive to the People's Republic.

Major changes were made in the spatial organization of the administrative system to facilitate the effective control of the plateau by China. Tibet was divided into seven regions (Nagchu, Chamdo, and Lingtse. Gyangtse, Shigatse, Ari, and Lhasa) and further subdivided into 72 rural districts, four urban districts, and two suburban districts in the capital eliminating the former "feudalistic" division of territory for land tenure and civil administration into areas assigned to monasteries and nobility. Communist-indoctrinated Tibetans brought back from China were appointed to all levels of government in the new administration. These measures were designed to assure China's absolute control over the administration of Tibet.

A major land redistribution program was initiated, involving breaking up big estates, formerly owned by monasteries and nobility, and distribution of land to peasantry. Farmers' cooperatives were established to pave the way for eventual collectivization of Tibetan agriculture and the introduction of communes. Although major changes were accomplished in the transformation of agriculture, the vast pastoral areas of Tibet remained untouched by reforms. The Chinese concentrated their efforts on anchoring the pastoralists to permanent winter quarters to increase their control over the semi-nomadic population.

In the field of education, the role of monasteries, traditional centers of Tibetan learning, was eliminated with complete control of administration by the Chinese. An accelerated program of communist educational policies was adopted to replace the traditional educational system. Steady progress was made in developing secular education. Public education served as an indoctrinational weapon to convert the Tibetan people to communism.

With reforms in agriculture, advances in industrial development, road construction domination of commerce, and effective political and administrative control, the Chinese hold on Tibet was completed in the early 1960s. The abolition of the Panchen Lama's Council in 1961 marked the end of the last surviving political institution in Shigatse.

In 1965, after China's full political, administrative, and economic hold over Tibet was secured, the official designation of Tibet Autonomous Region was conferred on the country. Ngapo Ngawang Jigme, a Tibetan noble official who had become the leading collaborator with the Chinese since the 1950s, was named head of the new government of the Tibet Autonomous Region on September 9, 1965. His task was to guide Tibet's integration with China. The outbreak of the Cultural Revolution in 1966 pushed the new government headed by Ngapo Ngawang Jigme into obscurity before it could serve its ostensible purposes.

23.3.4 The Fourth Phase 1966–1968: Cultural Revolution in Tibet and Eradication of Religion

China's attempt to eradicate Buddhism in Tibet during the Cultural Revolution stands as one of the most macabre campaigns of the twentieth century. While Buddhism was an obvious target of the revolutionary effort, it was not explicitly identified. Rather, wholesale destruction of the monasteries was effected under the slogan "Smash the Four Olds" – old ideas, old culture, old customs, and old habits (Welch 1972). China's Cultural Revolution was set in motion by Chairman Mao Tse-tung to purge the Chinese communist party and government bureaucracy including the power base of Liu Shao-chi, a former chief of state, Red Guards (young militants) poured into Tibet to overthrow "capitalist power holders" among the communist party and government officials. For 2 years between August 1966 and August 1968, the Cultural Revolution in Tibet was marked by extensive house cleaning of local administration, dismissal of a large number of trained cadres, and elimination of "four olds" – old culture, old customs, old habits, and old thoughts. During this period thousands of young militant Red Guards smashed the monasteries, the party and the bureaucracy (Fig. 23.4). Not only was all religious activity strictly banned, but wearing Tibetan dress and hair styles were also forbidden. All religious items that were not hidden were destroyed – scriptures were burned, clay objects smashed and carved sacred stones were used for construction. Some monks were made to copulate in public and others forced to marry; thousands were executed or sent off to destinations, later found to be concentration camps, and never returned. Many committed suicide to escape their cruel fate.

Tibet replicated the confusion that characterized most of China. For a brief time upheaval resulting from activities of the Red Guards in Tibet and struggle of the revolutionary factions among themselves for supremacy loosened Chinese grip on the country and encouraged Tibetan resistance.

Fig. 23.4 Ganden monastery, the main seat of the Gelugpa Buddhist order, lies 25 mi (40 km) northeast of Lhasa. Ganden suffered most at the hands of the Red Guards during the Cultural Revolution. Destruction was caused by artillery fire and bombing in 1966. Large scale reconstruction of the monastic buildings is in progress (Photo by P. P. Karan)

General Chang Kuo-hua, Commander of the 1950 invasion forces into Tibet and the dominant military and political figure on the plateau, was accused by the radical Red Guards of dereliction of duty, "following the capitalist road" and "empire building" in Tibet. Apprehensive of the danger resulting from disorder and division among the Chinese in a country subject to outbreak of revolutionary violence by the native population, General Chang was determined to minimize the impact of the Cultural Revolution in Tibet. Despite his efforts to limit the activities of Red Guards in Tibet, the young militants transported from China by air launched their program of revolutionary activity on the plateau creating widespread confusion in Lhasa and outlying towns. On August 25, 1966, they sacked the Jo Khang, the main temple of Lhasa. Religious texts and paintings were set afire; images were destroyed and dumped into the river. All articles connected with the traditional ways of life were seized from private homes and destroyed. There were similar activities in Shigatse and other places in Tibet.

Chinese authorities insisted that it was Tibetan members of the Red Guard who pillaged and destroyed the monasteries. Reliable accounts indicate, however, that Chinese-staffed units were primarily responsible for the devastation (Dreyer 1976). Reports of precious religious objects being sold on the international market support Tibetan accounts of the systematic removal of monastic relics, rather than frenzied plundering by Tibetan members of the Red Guard units. In the late 1960s, numerous religious artifacts from Tibet's monasteries began appearing for sale in Chinese government-sponsored shops. During a visit to Tibet by a delegation representing the Government-in-exile, it learned that Chinese officials from the Mineral Department had removed all the precious stones from the statues, images and ritual objects housed in Lhasa area monasteries (Tethong 1982).

By the close of the Cultural Revolution, denunciation of the Dalai Lama reached an all-time high in Tibet. Chinese propaganda referred to the Tibetan's revered spiritual leader as the "chieftain of the Tibetan rebellious bandits, an executioner … with honey on his lips and murder in his heart" (Tungchou 1974: 11). It also contended that the "Dalai used 30 human heads and 80 portions of human blood and flesh each year as sacrificial offerings" when he held a religious service to curse the People's Liberation War.

In early 1967 Tseng Yung-pa was appointed Commander of the Tibet Military Region. In September 1968, he was also named chair of the Revolutionary Committee for the Tibet Autonomous Region. The Revolutionary Committees were formed during the Cultural Revolution to replace old governing bodies throughout China. The Revolutionary Committee became the effective governing apparatus in Tibet, as in most of China's administrative units, institutionalizing the outcome of the Great Proletarian Cultural Revolution. Although the Committee represented a political compromise, the army maintained a dominant role and decisive power.

23.3.5 Fifth Phase 1970–1990: Apparatus for Control and Implementation of Religious Policy

The physical destruction of the monasteries had mostly ended by 1970. The 20-year period (1970–1990) following the Cultural Revolution was characterized by intensification of repressive religious measures by the Chinese, a power struggle between contending factions among Han settlers in Tibet, and a growing military buildup on the plateau by China.

Religious policy in Tibet is managed and implemented by an array of departments in both the Communist Party structure and the government. China's highest party authorities, the Central Committee and Politburo, determine religious policy developed by the United Front Work Department. On the government side, the State Council is the highest authority and under it are the departments which actually implement religious policy. These departments are the Religious Affairs Bureau and the Tibetan Buddhist Association. They are closely supervised by the party. At the lowest level, religious policy is implemented by the "Democratic Management Committees" which have been set in all of Tibet's major monasteries. Religious policy is also implemented by cadres at the prefecture, county, and village level in close cooperation with the security forces.

All official documents regulating religious activities impose restrictions on permissible practices. The most authoritative and comprehensive statement issued by the central government on the permissible scope of religious freedom is the "Basic

Viewpoint and Policy on the Religious Question during Our Country's Socialist Period" reprinted in MacInnis (1989). The second important document is the "Rules for Democratic Management of Temples" enacted by the People's Congress of the Tibet Autonomous Region under the supervision of Hu Jintao, Communist Party chief of Tibet in the 1990s. These rules are the basis for the management of all temples in Tibet. The Tibetan Buddhist Association has urged all monks and nuns to study the regulations so as to take an active part in stabilizing the government and building a new socialist Tibet. There are many cadres in Tibet, both Tibetan and Chinese, who continue to thwart religion, and appear to enjoy support of their superiors in Lhasa and Beijing. Tibetans assert that religious policy is designed to let Buddhism wither away by circumscribing its essential components.

In 1990 the People's Congress also issued "Regulations on the Protection of Relics" which stipulates that all religious relics belong to the state. These regulations have authorized the expropriation of the religious wealth of Tibet. While there is certainly need to protect antique religious artifacts from leaving Tibet, these regulations represent a setback for Tibetan control over religious life.

Large numbers of primary and secondary schools were established during this period to ensure a continuation of the program of political indoctrination of young people. Educational institutions are effectively used in Tibet to meet the needs of socialist transformation. With Chinese culture taught in Tibetan schools and the People's Liberation Army commanders in key administrative roles, the Chinese began a program to root out the core of Tibetan culture which it perceives as a political threat. Each element of Tibetan self-identity – religion, customs, and culture – contains some seeds of anti-Han feeling, and until the Han succeed in exterminating the Tibetan culture, it will pose a political threat.

23.3.6 *The Contemporary Phase 1990–Present: Han Migration and Continuing Attempts to Consolidate Religious and Cultural Control*

China's outgoing President Hu Jintao was Communist Party Chief of the Tibet Autonomous Region during 1989, a time of political instability and growing demands from Tibet's people for cultural and religious freedom. Hu was responsible for a political crackdown in early 1989 that led to deaths of several Tibetan activists. Tibet became part of the central government strategy to develop the country's West in 2000. As a part of this program there was a major influx of Han Chinese into Tibet and growing restrictions on religious practices (Central Tibetan Administration 2000). In 2008 violence flared in Tibet after monks staged a protest on March 10, the anniversary of the failed 1959 uprising against China. Deadly burst of rioting in Lhasa, elicited a government response that sent hundreds of monks, nomads, students and shopkeepers to jail – and several of those accused of rioting, to their deaths. But unlike previous crackdown on dissent in Tibet, the current campaign since 2000 has deeply unnerved educated, middle-class and bilingual Tibetans.

Detentions, secret trials and torture accusations have prompted soul-searching and quiet resistance. Elderly Tibetan cadres have published memoirs on long-forgotten massacres by Communist troops. Middle-age functionaries have openly voiced qualms about their role in China's bureaucracy. Online, the young and the radicalized post provocative anti-Chinese comments. People are no longer hiding behind the tradition of self-censorship that comes from fear. Tragyal, a Tibetan writer known by his pen name of Shogdung, published *The Line between Sky and Earth* (2010), a book which is written indictment of Chinese rule in Tibet, called for a peaceful revolution against China's heavy-handed governing style. Tragyal was moved by the sight of so many monks marching in the streets and the stories about harsh punishments for the protestors. He was also moved by passive resisters like Runggye Adak, a nomad whose videotaped paean to the Dalai Lama earned him an 8-year prison term.

In 2010 acting jointly, China's Ministry of Public Security, the military and the United Front Work Department of the Central Committee of the Communist Party of China tightened up their control over the Buddhist monasteries in Tibetan areas of China (Fig. 23.5). It was decided that in order to ensure better supervision over the monasteries only competent Tibetan Buddhist monks and nuns who are "politically reliable" and widely respected should be selected to monastery management committees. The tightening of supervision over the Buddhist monasteries by the

Fig. 23.5 Tingri is a huddle of Tibetan homes in the backdrop of the Great Himalayan Range. It is a major military base along the southern border of Tibet (Photo by P. P. Karan)

Party as well as the Government indicates their continuing nervousness over the loyalty of the local monks to His Holiness the Dalai Lama and their reluctance to support the Panchen Lama nominated by the Party and the Government. The Ministry of Public Security and the United Front Work Department now hold regular re-education classes for the monks to stress the importance of patriotism and loyalty to the party. The Chinese are now facing a new threat in Tibet – a tweeting Dalai Lama; His Holiness started a direct dialogue with interested Chinese and Tibetan netizens in 2010. About 5,000 persons have been following his Tweets in which the Dalai Lama emphasizes the need for all ethnic groups in China to coexist amicably, and for Tibetans to protect their religion and culture.

23.4 The Future of Religion in Tibet

The future of religious and cultural freedom in Tibet is dim. Tibet today is a land which has been pushed into the communist world at a great cost to its people and culture (Borges 2012). There have been major achievements in economic development. But these have come at a high political and cultural cost. China has pushed a vigorous policy of assimilation and integration of Tibet into the Han culture where Tibetans might well preferred "benign neglect." The future of Tibetans under China in terms of the aspiration and prospects for cultural autonomy remains doubtful. Despite six decades of propaganda to portray China as a harmonious family comprising the Han, Manchu, Mongol, Hui, Uighurs and Tibetan ethnic groups, there is an uncomfortable gulf between China's Han majority and the minorities, particularly the Tibetans. Chinese consider "them as barbarians seeking to split the nation apart" (Jacobs 2012). The image of Tibetans as rebellious, uncultured, and unappreciative of government efforts to develop Tibet has been nurtured by the official propaganda. Gradually, the imposition of Chinese ideas and civilization are eroding one of the richest cultural heritages in the world, a heritage whose principal storehouse, the great monasteries of Tibet, have been largely destroyed. Only a few remain. Han Chinese are unmoved by the suffering and horrifying means of protest by self- immolations. Between March 2011 and November 2012 over 70 Tibetan monks including teenage monks have protested restrictions on religious freedom through self-immolation (Cumming-Bruce 2012). The Dalai Lama urged members of Japan's parliament to go to Tibet to examine the reasons for self-immolations (*Wall Street Journal*, November 14, 2012).

As China selected new leaders, Tibetans in Beijing were kicked out of their homes. The Tibetan wife of Wang Lixiong, a Chinese writer and novelist, was asked to leave Beijing by political police (*New York Times*, November 7 2012). The Communist Party views Tibetans as noxious. Wang's wife, also a writer, is not a terrorist. She has written about the fate of her fellow Tibetans, and for this reason the party has put her on a blacklist. She lost her job and was denied a passport. Wang writes "While Chinese people … are streaming into and out of Tibet by the thousands, Tibetans themselves have become outsiders in their own land" (Wang 2012).

The role that religion played in shaping both pre-1951 Tibet and the control of religion in post-1951 Tibet cannot be overemphasized. All of the new patterns reflect the Chinese communist ideology, just as the old patterns reflected the Buddhist religious ideology. Tibet is a good case study for the study of the impact of religious ideology on landscape.

Under present political situation the reintroduction of Buddhist learning and meditation practice in Tibet appears very difficult. Most of the great Tibetan lamas reside in exile, and the continuing political indoctrination of monks and nuns as well as a vigorous official campaign to marginalize the influence of the Dalai Lama among the Tibetans offer little hope for the revival of Tibetan religious cultural heritage. Social science research is difficult undertaking in Tibet today, but Tibet's unique cultural geography offers fascinating opportunities for further research.

Acknowledgement I appreciate assistance and good fellowship in the field during my travels in Tibet in the 1950s to G. R. Jani and C. K. Lee; and in the 2000s to Chou, Richard, Otis, Linda, Julie and Steve. I am also indebted to several Tibetans and Chinese residents of Tibet for facilitating field observations.

References

Borges, P. (2012). *Tibet: Culture on the edge*. New York: Rizzoli International Publications.
Central Tibetan Administration. (2000). *Tibet 2000*. Dharamsala: Gangchen Kyishong.
Cumming-Bruce, N. (2012, November 3). U.N. rights official criticizes China's limits on Tibetans. *New York Times*.
Dreyer, J. T. (1976). *China's forty millions*. Cambridge: Harvard University Press.
French, P. (2003). *Tibet, Tibet: A personal history of a lost land*. New York: Penguin.
Jacobs, A. (2012, November 19). Many Chinese intellectuals are silent amid a wave of Tibetan self-immolations. *New York Times*.
MacInnis, D. (1989). *Religion in China today*. New York: Orbis Books.
Shakabpa, T. (1988). *Tibet: A political history*. New York: Potala Publications.
Tethong, T. (1982). *Report on the second delegation to Tibet*. Dharamsala: Information Office of His Holiness the Dalai Lama.
Tungchou, S. (1974, 19 July). Emancipated serfs will never tolerate restoration. *Peking Review*, p. 11.
Van Walt, M. (1987). *The status of Tibet: History, rights, and prospects in international law*. Boulder: Westview Press.
Wang Lixiong. (2012, November 7). Unwelcome at the party. *New York Times*.
Welch, H. (1972). *Buddhism under Mao*. Cambridge: Harvard University Press.

Chapter 24
Archaeological Approaches to Sacred Landscapes and Rituals of Place Making

Edward Swenson

24.1 Introduction

The popular and often times sensationalized perception of archaeology as the study of ancient pyramids, tombs, and temples departs rather strikingly from the research prerogatives of much of the discipline during the past 50 years. Indeed, Hawke's famous ladder of inference placed religion on the highest and most inaccessible rung of archaeological interpretation, in contradistinction to economy and technology which could be more readily reconstructed from the analysis of patterned material remains (Hawkes 1954; see also Binford 1962). Hawkes certainly did not argue that religious institutions were inconsequential to historical process, he simply asserted that their complexity could never be adequately inferred from material traces (Hawkes 1954; see also Fogelin 2008: 129–130). However, beginning in the 1980s, archaeologists began to question the assumption that archaeological deposits are structured independently from the cultural schemas and meaningfully constituted worldviews of past communities. Technology or economy simply cannot be disembedded from historically contingent structures of practice, including ideological or religious constructions of reality (Pauketat 2001). Functional, cultural ecological, and evolutionary models were thus increasingly criticized for disregarding cultural factors and for reducing religion to ideology or to an epiphenomenal force of social integration and adaptation. This critique coincided with a new-found interest in the archaeology of sacred landscapes and ritual performance (Barrett 1993; Bradley 2005; Insoll 2004; Kyriakidis 2007; Renfrew 1985; Swenson 2008; Tilley 1994; Tilley and Bennett 2001; Bowser and Zedeño 2009). In fact, anthropological archaeologists are now increasingly concerned with interpreting meaningful places, an objective often dependent on investigations of the construction, experience, and

---

E. Swenson (✉)
Department of Anthropology, University of Toronto, Toronto, ON M5S 2S2, Canada
e-mail: edward.swenson@utoronto.ca

modification of sacred geographies. Therefore, perhaps ironically, constructivist and phenomenologically oriented archaeological research has moved closer to the popular imaginings of the discipline. However, this rapprochement is largely superficial, nor should it be viewed as a negative development. To be sure, ritual practice, defined heuristically and cross-culturally as highly formalized and symbolically charged acts, have dramatically shaped archaeological landscapes as diverse as households, agricultural fields, and monumental henges (Bradley 2005; Mills and Walker 2008; Plunket 2002).

This chapter will critically review archaeological approaches to religion and landscape arguing that archaeologists have much to offer geographers and others interested in the fundamental spatial mediation of religious experience. More specifically, the contribution will examine how ritual performances orchestrated within evocative built environments created charged thirdspaces, a process which engendered a critical consciousness of place and social identity (Lefebvre 1991; Soja 1996). The presentation of a particular case-study from the Moche culture of Peru (ca. AD 600–800) will serve to demonstrate the affective power of ceremonial constructions to create political subjectivities and generate plural social meanings and imaginings. The chapter will then conclude with an exploration of the thirdspace and heterotopic qualities of archaeological sites in contemporary northern Peru. This analysis will draw attention to the continued political efficacy of ancient ruins and to the potential ethical challenges facing archaeologists excavating sacred sites.

24.2 Archaeologies of Sacred Landscapes

The obsession with religion in anthropology and cognate disciplines, including more recently archaeology, is unsurprising given the widely shared view that ritual provides a window into the inner-workings of culture (Swenson 2010). This perspective holds for anthropologists of varying theoretical persuasions, including scholars who interpret ritual as an active force of structuration or those who view it in traditional Durkeheimian-Marxian terms as a reflection and rationalization of primary socioeconomic conditions (see Morris 1987; Swenson 2010). Archaeologists also increasingly privilege the analysis of the material signatures of past ritual events as a means to decipher power relations and ideological representations (mediations) of social organization and political economy (for a review see Swenson 2010). Although these approaches will continue to yield valuable insights, they tend to unwittingly downplay the conjunction of landscape and religion; sacred geographies are interpreted not so much to reconstruct past spirituality or the efficacy of ritually charged places, but rather to interpret implicitly "higher-order" and often essentialized (ahistorical) political forces. Of course, religion cannot be reified as a separate field of social practice, and an investigation of ritual performance is critical to understanding past political relations and ideological struggles (Fogelin 2008; Swenson 2008). Nevertheless, interpreting religious architecture simply as a signifier of sociopolitical conjunctures problematically relegates landscape to the status of epiphenomenon, a passive sign vehicle of social process that elides the historical contingency of sacred places.

Although archaeological approaches to built environments parallel broader divisions within the discipline, these different perspectives often continue to treat religious landscapes as representations, whether etic or emic, of primary social orders and economic structures. For our purposes, recent anthropological archaeological studies of sacred geographies can be conveniently divided into four or five camps: social constructivist, structuralist-symbolic, political-sociological (Marxian or otherwise), phenomenological, and practice theory approaches.[1] Of course, these are not mutually exclusive paradigms and many excellent studies eclectically incorporate theoretical insights from a number of such frameworks. Furthermore, recent research inspired by new developments in the study of landscape, Pericean semiotics, performance theory, social memory, and place-making have built on these perspectives and have moved beyond analyses of landscape as mere representation, demonstrating archaeology's exciting potential to contribute to theories on the intersection of place, religion, and the built environment (Bowser and Zedeño 2009; Harris and Sørensen 2010; Inomata and Coben 2006; Mills and Walker 2008; Preucel 2010; Swenson 2012; Van Dyke and Alcock 2003).

To be sure, examinations of past sacred geographies reveal that symbolism and representation constitute worthy subjects of study. Ruined temples, tombs, communal house structures, pilgrimage shrines, mountains peaks and so forth often materially symbolized culturally specific worldviews of death, the afterlife, ontological states, creation myths, and other theological and cosmological precepts. Of course, these meanings may have either affirmed or contradicted (misrepresented) particular social conditions (Miller et al. 1989). Inspired by the theories of Eliade, Geertz, and Wheatley, archaeologists researching pre-industrial urban settlements throughout the world have similarly interpreted ancient temples or even entire cities as cosmograms, simulacra of the hierophanic spaces of cosmogonic origins and creation (Ashmore 1989; Kolata 1993; Lilley 2004; Townsend 1982). Choay (1986) argues that built space was "deeply endowed with signification" and thus "hypersignificant" in pre-capitalist societies, given its homologous correspondence with cultural constructions of society and cosmos. She cites Levi-Strauss's famed study of the Bororo village to demonstrate the all encompassing semantic load of pre-modern spatial experience. In contrast, the market-driven parcelling of space in capitalist cities is described as "hyposignificant," and thus lacking in unequivocal signification.[2]

---

[1] Neo-evolutionary (selectionist) and cultural ecological studies of ceremonial architecture are also still undertaken in the discipline but are not considered here given their general disregard of religious meaning and experience (see Graves and Ladefoged 1995). However, archaeologists adopting historical ecological theories have successfully dispelled the myth that the environment constitutes an objective reality transcending the social context of its production and experience. At the same time, this perspective is equally critical of theories that reduce landscapes to a cultural construction and ignore the agency of anthropogenically transformed places (see Schann 2012).

[2] For similar analyses on the disconnect between the spatial signifier and the social signified in capitalist postmodern geographies see Baudrillard 1995 and Sorkin 1992.

However, can pre-capitalist cities and related landscapes be accepted simply as hypersignificant, a moniker that perhaps implies an essentialist and unifying sacrality? The problem with such a perspective is that monumental constructions do not conjure the same sentiments and meanings among different social actors, and rituals orchestrated in religiously charged arenas are often conceived (misconceived) in remarkably differing ways (Brück 2001). Indeed, there is danger in assuming *a priori* that Maya pyramids, Sumerian ziggurats, medieval cathedrals, Buddhist stupas, etc. were built first and foremost to materially inscribe celestial geographies. And it cannot be accepted as a given that their perceived microcosmic efficacy (as an *imago mundi* or *axis mundi*) dictated their construction, varied uses, or experience (see Carl 2000; Smith 1987). Indeed, traditional structuralists models have been criticized for universalizing unconscious classificatory principles, while social constructivist interpretations of religious architecture have been critiqued in turn for presupposing that higher symbolic (theological) orders inscribed in monuments impose "unequivocal" meanings and inculcate uniform affective responses. Rather than reducing the significance of a religious edifice to a singular indigenous theory of place (say from the vantage point of a high priest), archaeologists would be better served in following Lefebvre (1991) and examine struggles over the meaning of space and consider how this translated to the persistence or transformation of architectural forms through time (Swenson 2011).

In contrast to constructivist approaches, archaeologists prioritizing sociological or Marxian theoretical frameworks analyze religious buildings as signifiers of social, political, and economic processes. Thus, kivas in the American southwest are not interpreted in terms of their cosmological significance and mythological import but are examined as socially "integrative facilities" (Adler and Wilshusen 1990; Nelson 1995). Hence, the changing size and number of kivas are calculated to reconstruct population fluctuations and to explain how ritual performances served as mechanisms of cultural adaptation to shifting demographic and environmental conditions. In a similar, but less functionalist manner, Moore has investigated different traditions of mortuary architecture in the ancient Andes as reflecting the varied "grid" (level of rank distinction) and "group" (the degree of social cohesiveness as mandated ideologically) characterizing indigenous sociopolitical orders (Douglas 1972; Moore 2004) To be sure, a majority of archaeologists survey religious architecture and sacred geographies primarily as a means to reconstruct past ideologies, political agency, and power relations. For example, proponents of energetic and related approaches estimate the time and labor energy required to build religious monuments as a measure of past inequality and social complexity (Abrams 1989; Trigger 1990). A multitude of access patterns studies of sacred buildings, including proxemic and space syntax analyses of entry points, viewsheds, and the architectural prescription of movement and communication have also been mobilized to interpret instituted power asymmetries (Hillier 1996; Hillier and Hanson 1984; for an excellent application in the Andes see Moore 1996). That is to say, the interplay of exclusive and inclusive space within religious edifices is often deemed expressive of both the degree of social control and hierarchy characterizing a given community (and the degree to which religious knowledge was effectively monopolized) (Feldman 1987;

Fung Pineda 1988). Although Foucault argues for a fundamental break between pre-modern strategies of power based on ritualized spectacles of authority (including especially corporal punishment) and modern disciplinary technologies of surveillance, his theories on the panoptican have been employed to interpret ancient sacred landscapes as instruments of "normativization" and repressive social engineering (Graves and Van Keuren 2011; Heckenberger 2005: 313; Moore 1992: 97).

The last two decades have also witnessed important contributions by archaeologists adopting phenomenological methods and theories, an approach popular with British archaeologists studying the henges, barrows, and curses of Neolithic and Bronze Age England. Indeed, theorists inspired by Merleau-Ponty, Heidegger, and others have challenged the constructivist emphasis on the ideational, foregrounding instead issues of the sensual, experiential, affective, and immanently material (Barrett 1993; Ingold 2000; Tilley 1994; Thomas 2001; see also Moore 1996, 2005). However, critics of phenomenology warn that interpreting bodily experiences as prescribed by particular built environments runs the threat of imposing the subjective views of the archaeologist or that of the dominant faction within a past society. In other words, reconstructing the sensual experience of particular spaces most often privileges the ideological *intent* of those who designed and built the monuments. Thus a singular phenomenology (in which engagement with built forms elicits shared affective responses and forges homogenous subjects) is problematic and ignores potentially diverse and conflicting experiences within a given social and spatial milieu (Alcock 1993; Brück 2001; Johnson 2006a; Fleming 2006; Swenson 2008). The focus on "dwelling" (adopted from Heidegger) as a "sense of continuous being which unites the human subjects with their environments" (Thomas 1994: 28) discounts the existence of diverse modes of experience which transcend the monolithic architectural determination of place. In contrast, de Certeau (1984), Rapoport (1982) and others describe the creative and alternative modes by which agents "spatialize" encounters with the social and physical worlds, thereby subverting the constraints of hegemonic built environments. Criticism that phenomenology is overly individualized, subjectivized, and even romanticized is of serious concern for archaeologists interested in applying such frameworks. Therefore, reconstructing the potentials of bodily movement in ritualized spaces should not be reduced to totalizing phenomenologies (Swenson 2008).

To be sure, inspired by practice theory and phenomenological perspectives, recent archaeological analysis of architecture rightly acknowledges its power to shape cultural attitudes and habitual dispositions. However, despite the explicit recognition of "architecture's agency," studies of this kind commonly revert to traditional semiotic interpretations, examining how landscapes *reflect* religious values, social structure, and political authority (but see Lumsden 2004; Moore 1996, 2005; Pauketat and Alt 2005; Preucel 2010; Smith 2003). Although the word "constitute" is often strategically used in place of the more passive verbs "reflect" or "express," archaeological investigations of the built environment still tend to reaffirm the primacy of *a priori* worldviews and social institutions in moulding place and political subjectivity. In light of the above discussion, the reduction of the spatial to the communicative (representational) is understandable, and perhaps unavoidable as a starting point of analysis, given the nature of the archaeological record. Nevertheless, such an approach elides the subtle political and experiential dimensions of place, and the agency of architecture is reduced to its hegemonic and conservative potential to enforce political domination. That is to say, the built environment is most often interpreted as prefiguring the social by physically ensuring its maintenance and reproduction (Bourdieu 1977; Foucault 1979; Tilley 1994). This perspective is problematic, for the creative power of architecture is diminished to simply symbolizing and physically reproducing the status quo, and space remains paradoxically epiphenomenal (Lefebvre 1991; Preucel 2010; Swenson 2011, 2012). By extension, the politics of place are unwittingly homogenized and largely shorn of their cultural and historical particulars.

24.3 The Efficacy of Religious Space

As made clear in the above critique, theoretical frameworks that interpret religious landscapes simply as proxies of some other phenomenon elide the vitalism of place and the transformative and structuring potential of ritual performance as historically contingent and inherently spatialized acts. In fact, the recent "archaeological turn" in the social sciences demonstrates that praxis is inconceivable if divorced from its particular spatial and material context (see Dawdy 2010). These trends parallel developments in "non-representational" or "material-relational theories" in human geography that stress understandings of space and material culture not as passive backdrops or symbols, but as efficacious, vitalistic, and even agentive (for archaeological perspectives see Knappett and Malafouris 2008; Meskell 2005; Miller 2005). Proponents of material-relational theories locate the "making of meaning and signification in the 'manifold of actions and interactions' rather than in a supplementary dimension such as that of discourse, ideology, or symbolic order" (Anderson and Harrison 2010: 2; see also Deleuze and Guattari 1987; Latour 1993; Thrift 2008). Anderson and Harrison further explain: "humans are envisioned in constant relations of modification and reciprocity with their environs, action being understood not as a one way street running from the actor to the acted upon, from the active to the passive or mind to matter...." (Anderson and Harrison 2010: 5). These perspectives advance the program of practice theorists that interpret performance and social action as creative "presentation" and transformation (literal "culture-making") and not simply as representation (Pauketat and Alt 2005). Thus proponents of the "material turn" argue that experience and signification are not simply dictated by *a priori* symbolic codes, but are in a perpetual process of becoming as mediated by embodied interactions of social actors with objects, landscapes, and the natural environment.

Material-relational theories complement the recent "spatial" turn in the social sciences affirming the importance of theorizing the multifaceted dimensions of spatial experience and meaning as exemplified in Soja's notion of thirdspace or Foucault's concept of heterotopia. Soja deconstructs spatial experience into the three heuristics of first, second, and thirdspace, equivalent to Lefebvre's notion of space as *perceived*, *conceived*, and *lived*. Conceived space (secondspace), equated with Lefevbre's concept of the "representation of space" refers to the built environment as planned, idealized, and engineered, the purview of architects, city-planners, and politicians. In truth, archaeologists have been predominately concerned with Lefebvre's "representations of space," paying less attention to the two other dimensions of his spatial trilectics. Perceived space (firstspace) designates the built world as embodied and experienced in practice (the purview of phenomenological studies), while lived space refers to a heightened attunement to place as engendered through such perceptions (Casey 1997, 1998; Lefebvre 1991; Shields 1999; Soja 1996). Although its definition varies according to the theorist in question, lived or third space broadly refers to the emotional evocation or critical apprehension of place through *praxis*, the spaces where place-making is realized and subjectivity negotiated. In this formulation, place retains its definition as a historically contingent and meaningfully constituted locale (Ashmore and Knapp 1999; Basso 1996; Van Dyke 2003: 180). However, place as thirdspace implies a construal of the spatial as equally creative in the refashioning of subjectivity and social memory (Casey 1997, 1998; Merrifield 1993; Smith 2003; Swenson 2011). Lived space, then, is essentially "meta-space," a radically open place of potentially intense introspection/embodiment, where space is at once perceived and conceived, intensely material and imagined (see Soja 1996: 34). It is in such thirdspaces, comparable to Foucault's heterotopias, that cultural categories of place, time, and identity come potentially into heightened focus and scrutiny (Casey 1998; Foucault 1986; Hetherington 1997; Lefebvre 1991; Soja 1996). Lefebvre equated "lived space" with what he called "spaces of representation," which he contrasted with the conceived "representations of space" that constitutes the purview of most archaeological reconstructions. The former refers to how experiences in space actually creates and transforms representations of place—where new and possibly conflicting meanings, memories, and outlooks are generated within particular spatial encounters (see Lefebvre 1991; Shields 1999; Soja 1996).

In fact, the literature on Lefebvrian thirdspace strikingly resembles studies on the intensely political nature of ritual space, a correspondence that has remained largely under-theorized (but see Hetherington 1997; Kapferer 2004; Humphrey and Laidlaw 1994; Köpping 1997; Shields 1990; Swenson 2012). Similar to Soja's thirdspace, symbolically charged ceremonial loci are commonly interpreted as conflating the imaginative and real, the performative and introspective, the past and the present, the emotional and contemplative, the intimately personal and the intensely social. Of course, as a potentially liminal nexus of metamorphosis and heightened consciousness, ritually constructed places constitute the ultimate of thirdspaces, where subjects are bodily and spiritually reconfigured (Turner 1967). Foucault (1986) categorized pre-modern sacred spaces as *crisis heterotopias*, built environments that simultaneously compress, reflect, distort, or invert the many other places comprising a particular society. In fact, Soja has argued that the concept of heterotopia, as spaces of extreme otherness disruptive of the familiar and normative, shares much in common with Lefebvrian theories of thirdspace (Soja 1996: 154–163 see also Johnson 2006a, b; Hetherington 1997). Samuels notes (2010: 68): "Temples make appealing heterotopia … delineating sacred from profane space to act as foci around which the social and spiritual worlds is ordered, a place in which things were juxtaposed in such a way as to prompt reflection on the nature of other places." Therefore, ritualized acts often imbue place with thirdspace qualities, for they demarcate performative arenas of intense, alternative experience, the meaning of which is potentially fluid and variable. Although ceremonial architecture is generally interpreted as representing "controlled environments" in the extreme (Smith 1987), it is in precisely such places where identity, personhood, and one's place in the world fall under intensified scrutiny.

Of course, certain modes of ritual may be as habitual and taken-for-granted as house cleaning, refuse disposal, or food preparation. In other words, routinized religious observances are as ingrained in social "taskscapes" as other quotidian activities (Ingold 2000). Landscapes of the everyday have been interpreted as powerful instruments of social reproduction precisely by naturalizing cultural practices (ritual or otherwise) within the realm of the spatially pre-programmed and unreflexive (Bourdieu 1977). In contrast, exceptional, heterotopic spaces are argued to hold the potential to reveal the arbitrariness of socio-spatial realities and thus provide a fulcrum for their alteration. Indeed, it is within the sensual and inherently spatialized field of spectacular ceremonial events that social orders are especially amenable to reification, misrepresentation, or even transformation (Bell 1997; Inomata and Coben 2006; Swenson 2011). Over-engineered and aesthetically charged architectural constructions are sites of emotive remembering and imagination. As Hodder notes (2010: 17), rituals contexts including burials constitute "a marking event, and it can be called religious not because it is separate from everyday life, but because it focuses attention, arouses, refers to broader imaginings and deals with the relationship between self and community." Admittedly, theorizing past ceremonial constructions as "thirdscapes" as opposed to "taskscapes" poses the danger of repackaging and imposing problematic sacred/profane dichotomies onto past social worlds. It cannot be simply assumed that high ritual theater staged in magnificent monumental space stimulated heightened consciousness, strong emotions, and a critical awareness of place and memory ("ideological spaces") in any singular manner, while landscapes of the everyday simply reproduced the temporally habitual and mundane ("doxic space") (Johnson 2007: 131–133). In truth, comparative spatial categories are of interpretive value only if they can make sense of the vastly different agencies of past and present religious landscapes. Deleuze and Guattari's striated and smooth space, or Bachelard's notion of felicitous space recognize the historical specifics of spatial production and the irreducibility of place to clean categorical antinomies. They refer to qualities of space as opposed to rigid and ahistorical essences (Bachelard 1994; Deleuze and Guattari 1987). With this caveat in mind, evocative spaces were activated—and were activating in turn—strictly within historically contingent conjunctures of landscapes and practice.[3] These spaces may

---

[3] Such a perspective resonates with Jones's notion of "ritual-architectural events… occasions in which specific communities and individuals apprehend specific buildings in specific and invariably diversified ways" (Jones 2000: xiii).

have been scenes of rare, imagistic religious celebrations or loci of frequent but politically contested ritual performances (see Whitehouse 2004). The following case-study from pre-Inka Peru demonstrates how such heuristics, sensitive to the vicissitudes of place, can permit an archaeological approximation of the creative agency of distinct sacred geographies.

24.4 Sacred Landscapes of Becoming: Huaca Colorada, a Moche Site in Northern Peru

The Moche ceremonial center of Huaca Colorada is located in the dry desert of the Jequetepeque valley in northern Peru (AD 600–800). Moche designates a political and religious ideology propagated throughout the desert North Coast of Peru during the Andean Early Intermediate and Middle Horizon Periods (AD 100–800) (Bawden 1996; Quilter and Castillo 2010; Shimada 1994; Uceda and Mujica 1994, 2003). Archaeologists have argued that Moche society was defined by unprecedented social stratification and formed one of the earliest state polities in the Americas (Bawden 1996; Billman 2002; Shimada 1994). However, recent investigations have questioned the existence of territorial Moche state(s), and it seems increasingly apparent that Moche political organization varied considerably from region to region (Castillo and Donnan 1994; Quilter 2002; Quilter and Castillo 2010).

As has been well documented, Moche political theology was defined by cycles of warfare, prisoner capture, and human sacrifice that likely conformed to poorly understood cosmogonic myths and ideologies of legitimate religious authority (Alva and Donnan 1993; Bourget 2006; Donnan 1978; Swenson 2003). Human sacrifice and the aestheticization of violence defined the heart of Moche political theology, and the iconographic corpus suggests that ritually controlled death and destruction reciprocally enabled creation and was thus deemed generative of life, cosmos, time, authority, and ultimately place itself (Bourget 2006; Swenson 2003, 2012). Thus, death was not understood as an end but as the ultimate nexus of transformation and becoming. In fact, the religious and political landscapes of the Moche are fruitfully understood in terms of this specific sacrificial ontology, and the prevalence of architectural dedication and termination rites in Moche centers suggests that Moche conceptions of place resembled tenets of "material-relational theories" in contemporary human geography (Anderson and Harrison 2010; Thrift 2008). In other words, the Moche perceived space and architecture as alive, imbued with personhood, and infused with vital agency.

Of course, analyses of the platform mounds and plazas comprising Moche ceremonial architecture have played a critical role in interpreting the religious and political ideologies of this archaeological culture. The multi-tiered Huaca de la Luna in the eponymous Moche Valley has been the subject of intensive archaeological investigations in recent years, and it evidently served as the ceremonial nerve center of the great conurbation of Huacas de Moche (Uceda 2001; Uceda and Mujica 1994,

2003). This complex is characterized by a massive monumental plaza and series of elevated interior chambers and platforms. Painted adobe reliefs of predatory animals, warriors, captured prisoners, sacrifice, fanged divinities, and cosmic landscapes grace the temple walls, suggesting that this structure served as the central stage for the sacrificial rituals which underwrote the theocratic ideologies of Moche polities. Satellite ceremonial centers, modeled after Huaca de la Luna, were founded in the lower portions of supposedly conquered valleys, such as Cao Viejo in Chicama, attesting to the possible territorial expanse and political influence of the premier Moche site (Bawden 1996; but see Quilter 2002). These pyramidal platform mounds have been interpreted as simulating the form and sacred power of mountain peaks, the traditional nexus of supernatural forces, deified ancestors, and generative forces common to many Andean cosmologies (Bawden 1996; Uceda 2001).

Due to space limitations, a brief description of the smaller ceremonial site of Huaca Colorada will focus on archaeologically identifiable ritual practices staged within the ceremonial nucleus of the site in order to demonstrate how the complex in question can be fruitfully interpreted as a peculiar thirdspace, a place of creative signification and subject formation. Huaca Colorada is situated in the south bank of the Jequetepeque Valley, approximately 100 km (61 mi) north of the aforementioned Huacas de Moche. The settlement occupies an area of approximately 24 ha (59 acres) and is dominated by an elongated platform structure built on a modified hill of sand. This central construction measures 390 m × 140 m and rises 20 m at its central highest point (1,280 × 459 × 66 ft) (Fig. 24.1). This ample elevation is differentiated into three principal sectors which delimit Huaca Colorada's principal ceremonial and manufacturing/residential areas. It was occupied during the Late Moche Period (AD 600–850), an era marked by environmental perturbation and the diversification of Moche religious culture (Bawden 1996; Shimada 1994). Late Moche Jequetepeque witnessed demographic expansion in rural areas and the popularization of Moche ritual practices as evidenced by the construction of numerous stone platforms in fortified hillside settlements distributed throughout the countryside (Castillo and Donnan 1994; Dillehay 2001; Swenson 2006). The emergence of a polycentric settlement system is further indicative of political decentralization and internecine conflict. The so-called priestess cult was established during the Late Moche Period at the site of San José de Moro in the north bank of the Valley (Castillo 2001, 2010), likely the region's most prominent center headed by elite females buried in lavish tombs. The female figures appear to have assumed leadership of the Moche sacrificial complex in Jequetepeque as suggested by the rich iconographic corpus.

Huaca Colorada is the largest Late Moche settlement located in the southern bank of the valley, and the elaborate adobe constructions dominating the central and highest prominence of the pyramid points to the presence of high status individuals (for a detailed description of this complex see Swenson et al. 2011, 2012 and Swenson 2012; Swenson and Warner 2012). Indeed, it is possible that Huaca Colorada served as the ceremonial and political headquarters of a powerful polity that dominated the southern Jequetepeque valley. However, the unique layout, location, and artifactual associations of the site suggest that it was constructed to stage

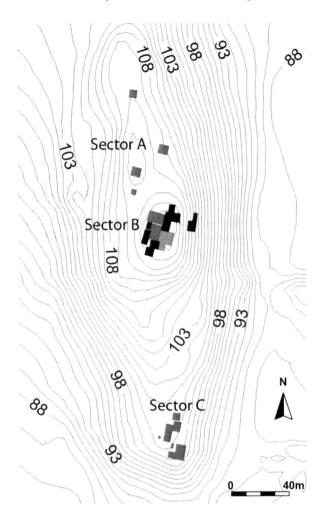

Fig. 24.1 Topographic map of the ceremonial site of Huaca Colorada (Map by Ed Swenson)

specific kinds of ceremonies unique to the region. In fact, beyond foundation sacrifices and the discovery of a few household burials in the lower production zones, no elite tombs have been unearthed in the monumental sector, and it is clear that the settlement did not function as a mausoleum as did San José de Moro and other prominent Moche centers. The recovery of copper ornaments, tools, and production detritus in makeshift domestic constructions indicates that metallurgy constituted an important pursuit of the settlement's inhabitants (Swenson and Warner 2012). The discovery of offerings of prills and finished copper objects in conjunction with foundation sacrifices in the ceremonial zone of the site might even suggest that Huaca Colorada elites were skilled artisans who based their authority on their guarded ritual knowledge of the metallurgical arts. However, most evidence of production derives from the non-elite sectors, and it is possible that peripatetic artisans visited the huaca during specific seasons under the patronage of the site's leaders. Copper items may have been produced and exchanged during fairs and grand religious celebrations. The high quantity of decorated feasting vessels unearthed in the domestic and ceremonial sectors also strongly suggest that artisans, pilgrims, and full time residents of the site were compensated with corn beer, a sacred beverage in the ancient Andes (Swenson 2012). It is also possible that Huaca Colorada served as the locus for the (religious) training and supervision of part-time artisans who engaged in agriculture during other seasons of the year. This scenario could also account for the cyclical abandonment and re-occupation of the extensive domestic areas of the site (see Swenson and Warner 2012).

The principal religious constructions of the monumental core of Huaca Colorada consist of six daises or altars, all of which were ritually interred under floors or construction fill. One of the more prominent daises formed the focal point of a sunken and spacious chamber. It supported two beautifully stuccoed columns and may have functioned in part as a stage for the presentation and consumption of comestibles including corn beer (Fig. 24.2). As argued by Wiersema, this architectural construction is also commonly associated with a number of key ritual practices in Moche iconography, including the arraignment of captives, the sacrifice of warriors, and the presentation of the goblet to the Warrior Priest (Wiersema 2010: 175–182). The different platforms were not all in use simultaneously, but are associated with distinct construction and occupation phases. In fact, the ceremonial precinct of the huaca appears to have been in a constant state of renovation, and it is evident that there was a religious expectation to ritually terminate and rededicate altars, rooms, and platforms, perhaps as dictated by a religious calendar or festival round. The discovery of foundation sacrifices of young women or children associated with both the closure and re-dedication of three different altar platforms corroborates this hypothesis. Moreover, copper objects were offered in conjunction with the human sacrifices, evidence with points to an ontological continuum of people, places, and things in Moche worldview (Fig. 24.3) (Sillar 2009; Swenson and Warner 2012). It also deserves mention that one of the major reductions of the central ceremonial chamber was also commemorated with the sacrifice of a woman thrown haphazardly into the construction fill (Swenson 2012; Swenson and Warner 2012).

Evidently, the ritual specialists and other community members of Huaca Colorada perceived sacred space as living, kinetic, and dynamic. Therefore, the underlying sacrificial ontology defining Moche worldview was not just mirrored in the built environment but was made palpably real in the architectural renovations of Huaca Colorada's ceremonial structures (Swenson 2012). That is to say, sacrificial conceptions of time and being were instantiated through repeated architectural interventions that inculcated Moche religious values through embodied ritual performance. In the end, the creation of place, as consummated through ritual performance and repeated architectural renovation, was tantamount to the creation and re-creation of subjectivities. Indeed, the sacrifice of space rendered physically real the abstract, dialectical understandings of time, life, and even personhood peculiar to the Moche. It could even be argued that for many of pilgrims or part-time inhabitants of Huaca Colorada, a critical consciousness of such conceptions was only fully realized through embodied acts of place-making. Huaca Colorada's monumental core was a

Fig. 24.2 An adobe platform with stucco columns dominates the principal ceremonial chamber at Huaca Colorada. The Moche fineline depiction of a roofed dais closely resembles the excavated platform (Photos and drawing by Ed Swenson, after Donnan 1978)

literally *enlivened* thirdspace; the complex appears not only to have induced a heightened awareness of Moche religious values specific to Middle Horizon Jequetepeque but was complicit in the actual creation and negotiation of these ideals (see Dawdy 2010: 773). Just as Huaca Colorada's built environment was in a state of constant flux and becoming, individuals moving through its precincts or involved in ritual rebuilding were no doubt transformed in body and mind by the kinetic power of the space in question.

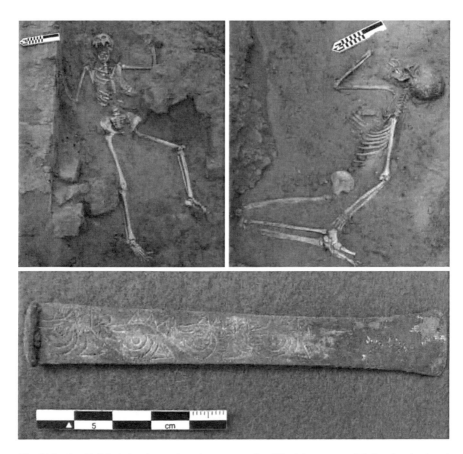

Fig. 24.3 Sacrificial victim thrown into the construction fill of the ceremonial chamber (*top*) and an engraved copper spatula (*bottom*) offered in association with a human foundation sacrifice deposited to commemorate the construction of an altar-platform (Photos by Ed Swenson)

In fact, Huaca Colorada can be fruitfully interpreted as exhibiting heterotopic qualities, an *other* place that at once juxtaposed and possibly disrupted the many sites and temporal rhythms of Moche life, a kaleidoscopic condensation of the lived places of the Moche-Jequetepeque world. Foucault (1986) argued that fairgrounds were exemplars of heterotopic spaces, places where diverse peoples converged to partake in convivial festivals removed from quotidian practices and temporalities. Huaca Colorada was similarly a locus of congregation and spectacle, where different communities engaged in copper production, exchange, feasting, ritualized architectural renovation, and emotionally charged human sacrifice. Of course, Huaca Colorada obviously departs from Foucault's understanding of (non-crisis) heterotopias as "de-centered anti-sites" of prisons, brothels, or asylums, the *other* places of marginalization and deviance, which at once reaffirm but inevitably distort and disrupt the utopian norms of modernist societies. Nevertheless, it is significant that diverse heterotopic places are commonly sites of (potential) violence and heightened performance, arenas of intense biopolitical control and struggle. Although the sacrificial pyramid, prison, or fairground obviously instantiate the liminal and heterotopic in markedly different ways, they nonetheless reveal how such evocative spaces could serve as instruments of subject formation as realized in the literal deconstruction and reconstitution of subjects. One might protest that foregrounding the heterotopic or thirdspace synergies of specific built environments overstates their agency. However, the affective force of the initiation hut, reliquary, or sacrificial stone is difficult to downplay; buildings saturated with violence and other-than-human powers induce intense if varied emotions (fear, deference, awe, joy), while emanating a gravitas and spatial/conceptual depth that powerfully reorients behavior, attitude, and subjectivity (Caton and Zacka 2010: 209).

Of course, it could be argued then that the creative agency of the huaca simply reinforced the authority of the site's elites. Although the apparent synchronization of architectural renovation with ritual bloodshed, copper production, and feasting events most likely fostered alternate understandings and heightened emotions, these sentiments appear to have legitimated the presiding ideological order. In other words a "smooth space" of excess and exception was framed within the decidedly hierarchical and striated space of the temple itself, as exemplified by the prevalence of sacrificial violence (Deleuze and Guattari 1987). Nevertheless, Lefebvrian conceptions of thirdspace stress that new meanings and often unorthodox perceptions of the socio-spatial order can be engendered in charged places defined as heterotopic or otherwise. Such exceptional or evocative spaces have been analyzed in terms of their redemptive and liberating potential or as smooth spaces of creative openness and transgression (Deleuze and Guattari 1987; Giles 2006). To be sure, it is in precisely thirdspace-like locales that participants are best positioned to exploit the creative and polysemic potential of ritual practice, especially of a violent nature. Das (2007: 86) has argued that in "occupying the edges of human experience violence often provides a space for the recognition and experimental transformation of culture." Of course, ritual violence did not erupt from the edges but was dramatically at the "center" of religious life at Huaca Colorada. However, it is not difficult to imagine the potentially unpredictable agency of such a thirdspace charged with sacrificial violence.

The proliferation of informal graffiti on the walls of the ceremonial precinct at Huaca Colorada points to the appropriation and even subversion of the intended meaning and functions of the structure during specific phases of its occupation (Swenson 2012). The graffiti suggest possible contradictions between ideologically mandated "representations of space" and lived spatial practices and perceptions. In fact, the informal etchings demonstrate that the monumental sector was at times remarkably open and accessible. Most of the instances of graffiti appear to have been executed during the later use-phases of the main ceremonial chamber. They may even have been drawn as part of the decommissioning rite prior to the ritual closure of this precinct dominated by the pillared platform. The depictions consist of crudely rendered warriors, litter bearers, birds, lizards, landscapes, maritime scenes, and

Fig. 24.4 Graffiti depicting litter bearers, warriors, animals, and architectural and landscape scenes etched on the walls of the monumental sector of Huaca Colorada (Source: Ed Swenson)

running figures (Fig. 24.4). Many of the representations resonate with Moche religious themes centered on warfare and death, and the ritual context of their execution seems probable. Therefore, the informal sketchings depart from contemporary connotations of graffiti as irreverent, subversive, or transgressive.

Nevertheless, the elites of Huaca Colorada clearly did not commission the artwork as integral to the overall architectural design and aesthetic plan of the monument, and the representations are haphazardly applied on the walls and blocked entrances of the main complex. Instead, individuals appear to have informally commemorated ritual activities staged within the ample ceremonial chamber. The figures express movement and transience, while their hurried applications are equally expressive of the momentary passage of the artists themselves, possibly analogous to religious tourists wishing to forever memorialize their fleeting encounters with the edifice. The etchings are remarkable, for they demonstrate the perduring thirdspace effects of the huaca which continued to evoke strong but possibly conflicting memories, emotions, and political stances. The actual engraving of roofs, structures, and other architectural elements identified on the northern and eastern walls of the main precinct confirms beyond a doubt that the huaca stimulated a critical consciousness of place—a quintessential thirdspace of becoming and introspection enabled by the intimate experience of the ceremonial chamber (Swenson 2012) (see Fig. 24.4). At least in certain phases, the huaca was open and accessible, a decidedly smooth and powerful space complicit in the reformation of political subjectivities.

As a nexus of intense biopolitics, the graffiti are suggestive of the creativity of the monument to generate new and conflicting meanings which perhaps contradicted the intended religious rationale of the periodic spectacles of ritual homicide documented at the site. Huaca Colorada was a thirdspace of multiple becomings that defies reduction to a single religious or political ideology.

24.5 Archaeological Sites as Thirdscapes in Contemporary Peru

The graffiti etched on the walls of Huaca Colorada radiate a spectral or phantasmagoric quality, an uncanny "absent-presence" characteristic of archaeological ruins (Edensor 2008: 324; Holloway 2006; Maddern and Adey 2008). Of course, wall foundations, pot sherd scatters, graffiti and other such remains constitute the *present* physical traces of absent people, and archaeological research itself has been described as a quintessentially third-space experience given the practitioner's sensitivity to the layered meanings, whether real or imagined, inscribed in abandoned places (Blake 2002). In fact, the foundation sacrifices, coinciding with the ritual closure of altars, ramps, and chambers sealed under floors and tons of sand, indicate that the Moche of Huaca Colorada were keenly aware of the power of the invisible, but immanently present agents incorporated into architectural constructions. The combined foundation sacrifice of finished copper objects and female victims perhaps served to transfer the life-force and youthful vigour of the offerings to the sacred space in question. It could even be argued that Huaca Colorada formed part of a "sacramental landscape" involving the mystical consubstantiation of humans, copper, and architecture. Memory of buried structures, copper ornaments, and people imbued extant altars and precincts with a vitality and haunting authenticity, creating a thirdspace of intense imaginings and emotion. Indeed the palimpsest of dedication and termination rites saddled the monument with a spectral weight that was no doubt effective in *conjuring* powerful sentiments of the sacred specific to Moche religious culture (see Mock 1998).

It is at once intriguing and unsurprising that archaeological ruins such as Huaca Colorada continue to exert a spectral thirdspace force in contemporary Peruvian society. In fact, many scholars recognize the heterotopic aura of ruins, places of alterity variably associated with decidedly *other* times and peoples (see Dawdy 2010). In the Peruvian context, archaeological ruins are flashpoints of political struggle and fulcra of guilt, fear, hope, desire and contested social identity (Smith 2005). Often located at the margins of agricultural fields or in deserted and ghostly pampas, archaeological loci represent rends in the fabric of space-time, where space and time are uncannily enfolded or disjointed (Maddern and Adey 2008: 292). Indeed, archaeological sites conform to Benjamin's notion of the "dialectical image," evocative places that invite contemplation on everything from the evanescence of life to the brutality and legacy of Spanish colonialism (Benjamin and Tiedemann 1999; Buck-Morss 1989; Dawdy 2010; Stoler 2008: 194). Pre-Columbian ruins are frequently spaces of a compounded *double othering*; foreign investigators or *crillo* Limeños commonly haunt archaeological sites, and researchers are held in suspicion by local communities as either greedy prospectors hunting for hidden treasure or as alienating scholars who express more interest in the dead than in living communities (Smith 2005).

The heterotopic quality of prehistoric settlements is exemplified by the remarkable diversity in the perceptions and experiences of ruined places. Therefore, the evocative and signifying power of pre-Columbian monuments defy reduction to any singular discourse. Certainly, socioeconomic background and cultural dispositions account in part for the kaleidoscopic meanings of archaeological sites. Traditional North Coast *curanderos* venerate ruined pyramids and burials grounds, establishing a divinatory relationship with the presiding spirit or *dueño* (steward) of a particular ruin in order to discover power-items which are incorporated into their curing mesas (Gündüz 2001). Curanderos also frequently stage healing and divination rites within the confines of archaeological sites which they perceive as portals to the spirit world. In contrast, many inhabitants of the Jequetepeuqe region (both farmers and urban dwellers) differently conceive of the sacredness of the numerous ruins in the valley. Looting is widespread, and syndicates of *huaqueros* have sacked thousands of cemeteries and pyramids scattered throughout the valley in search of fine pottery and gold objects that can be sold on the illicit antiquities market. Archaeological cultures are commonly labeled "*los gentiles*" ("gentiles"), and their non-Christian belief systems accentuate a perceived social distance legitimating the actual desecration of sites.

However, ruins in the Jequetepeque Valley are also the subject of myth and legend, and they are commonly spaces of enchantment, intense memories, and haunting specters. For instance, Huaca Colorada has fortunately been little disturbed by looters. Local laborers working at the site and villagers from the neighboring pueblo of San Lorenzo de Jatanca explain that Huaca Colorada's relatively pristine condition is due to the fact that it is haunted by malevolent spirits. Workers recount stories of the huaca literally eating looters and trespassers. According to local *campesinos*, the site and surrounding landscape emit chilling screams at night, and visiting the site can result in illness and troubling dreams. Unusual occurrences are also interpreted by the field laborers as bad omens. For instance, a mouse scurrying across the site was feared as a possible incarnation of shape-shifting shaman who was in the process of hexing the excavators (according to the labourers, mice cannot survive in the barren desert where the site is located). The workers are especially unsettled by the discovery of burials and the foundation sacrifices. As a result, a ceremony (*pagapu*) is conducted by a local shaman at the end of the field season to dispel the illnesses and traumatizing nightmares induced by working at the site (Fig. 24.5). To be sure, narratives of fear, death, and violence commonly underscore the mythologies of many of the archaeological ruins in the Jequetepeque valley and we often struggle to reconcile these beliefs with our community outreach program.

Haunted landscapes "quiver with affective energy," but spectral geographies of this kind, as with heterotopic places in general, are polysemic in their emotional and political effects (Holloway 2006: 182). Of course, religious monuments are commonly

Fig. 24.5 A local curandero presides over a purification ceremony (pagapu) in the excavated chamber of Huaca Colorada to placate spirits harming laborers participating in the excavations (Photo by Ed Swenson)

embroiled in politics and social violence. One needs only to consider the heated conflicts over heritage sites in Jerusalem or the destruction of Buddhas of Bamiyan in Afghanistan and the demolition of the mosque of Babri Masjid in Ayodhya and the deadly riots that ensued (see Abu-El-Haj 2001; Meskell 2003; Wallace 2004: 156–157). Indeed, a persistent difficulty of community archaeology campaigns that encourage active participation of locals in the excavation and interpretation of archaeological sites is the inevitable problem of reconciling diverse and contradictory understandings of the past (Dawdy 2009). Such programs have been criticized for homogenizing indigenous perspectives and for essentializing the identities of both archaeological cultures and their presumed descendants (Smith 2004). Despite the shortcomings of community archaeology, it would be wrong to simply defend the epistemic privilege of professional archaeologist as beyond reproach, and researchers have an ethical responsibility to accommodate and comprehend the varied interrelationships of local peoples with their surrounding archaeological landscapes (Bergquist 2001). Nevertheless, archaeological research is ideally equipped to transcend some of the conflicts and ethical dilemmas surrounding religiously significant places, precisely by identifying and historically contextualizing the thirdspace efficacy of particular monuments—in both *past* and *present* societies.

24.6 Conclusion

Huaca Colorada is productively investigated in terms of how it expressed tenets of Moche cosmology and politico-religious ideology (either idiosyncratically or expectedly). However, as an examination of the foundation sacrifices and graffiti intended to show, an aura of indeterminacy and fluidity underwrote processes of signification at the site. At the same time, the sensual encounter with the huaca played a critical role in the creation or re-formation of political subjectivities. Wallace (2004: 153–155) describes how certain tourists visiting iconic holy sites in Jerusalem, including the Holy Sepulchre, have suffered unexpected psychological breakdowns. Denoted the "Jerusalem Effect," individuals who were moderates in their religious convictions are suddenly compelled to don bed sheets and zealously preach in public. The sensory experience and overwhelming affective force of storied religious sites are thought to be among the principal causes of this particular disorder. Although admittedly an extreme example, it nonetheless demonstrates the often unpredictable agency of religiously charged historical locales and confirms that higher order ideological and symbolic structures are not the sole determinates of a sacred place's social significance and political instrumentality.

Of course, the local laborers who have been participating in the excavation of Huaca Colorada acquire news understandings of the site. The involuntary memories or emotions the monument may provoke after research has ceased, including meditations on friendships with North American archaeologists, will undoubtedly depart from those of local *huaqueros* and *campesinos*. To be sure, excavating the huaca constitutes an act of place-making in and of itself. However, the "network of material resistances" (Shanks and Tilley 1990) presented by the adobe constructions and attendant offerings plays an equally important part in such processes of place-making.

In light of the third-space sensitivities of archaeologists, it could be argued that an often misrecognized and under-theorized "sanctity" underlies much of archaeological practice. Such a claim would seem contradicted by the atheistic leanings of a good number of academic archaeologists or by the discipline's traditional concern with long-term social transformations prioritizing economic and ecological change, a focus which has treated religion as either inaccessible or inconsequential to understanding historical processes of the *longue durée* (see Wylie 2002). However, the growing realization that economy, technology, and other such analytics of modernist scholarship are often embedded in cosmological and related cultural mediations of the world demands that archaeologists address the religious forces complicit in the creation of place. Analyses of landscapes to interpret past domestic economies, irrigation systems, and human-environment relations, to provide just a few examples of what would commonly be perceived as not-ritual pursuits, cannot disregard the religious meanings that no doubt mediated such practices and the places they were constantly re-creating. Although a great deal has been written on the difficulties of empathizing with past subjects or in understanding the experiences of extinct communities (the classic challenges of the double-hermeneutic—see Barrett 2001;

Hodder and Hutson 2003), a recognition that past landscapes were at once variably meaningful and agentive renders a historicist epistemology theoretically inevitable and ethically imperative. In the end, by privileging the material and spatial conditioning of social practice, the insights of non-representational theory discussed in this chapter provide a strong basis for both "hyper-historical" reconstructions of past social realities and cross-cultural comparative studies of ancient religious landscapes.

To conclude, since much of what archaeologists investigate (dwelling foundations, relic fields, hearths and campsites, etc.) was potentially imbued with historically specific ritual significance (consecrating a house, harvesting corn as divine gift, recounting origins myths over a campfire), it is perhaps ironic that the imbrications of religion and archaeology have remained largely under-theorized (but see Bradley 2005; Fowles 2013; Insoll 2004). At the same time, the fact that archaeological ruins (whether ceremonial "thirdscapes" or everyday "taskscapes") are often considered sacred by local communities further heightens and complicates the ambivalent and poorly examined interface of the religious and archaeological. As Dawdy (2010: 768) notes in praising Benjamin's archaeological sensibilities, abandoned sites and artefacts "acquire more intensely affective, complex meanings as they age and become archaeological." Indeed, "ancient" relics and ceremonial spaces, imbued with timeless authority and deep social memory, foster feelings of reverence and encourage contemplation of bygone people, places, and even omnipresent though invisible (mystical, primordial, etc.) other-than-human powers. It is precisely this "absent-presence" dialectic (Edensor 2008: 324) underwriting the archaeological imagination and defining the discipline's distinct methodological challenges that can account for the spectral aura commonly underwriting archaeological research. By materially indexing past and alternate ways of living, being, and believing, archaeological ruins are quintessentially heterotopic and heterochronic spaces that have long excited the religious imagination.

References

Abrams, E. M. (1989). Architecture and energy: An evolutionary perspective. *Archaeological Method and Theory, 1*, 47–89.

Abu-El-Haj, N. (2001). *Facts on the ground: Archaeological practice and territorial self-fashioning in Israeli society*. Chicago: University of Chicago Press.

Adler, M., & Wilshusen, R. (1990). Large-scale integrative facilities in tribal societies: Cross-cultural and southwestern U.S. examples. *World Archaeology, 22*(2), 135–146.

Alcock, S. E. (1993). *Graecia capta: The landscapes of Roman Greece*. Cambridge: Cambridge University Press.

Alva, W., & Donnan, C. B. (1993). *Royal tombs of Sipán*. Los Angeles: Fowler Museum of Cultural History, University of California.

Anderson, B., & Harrison, P. (2010). The promise of non-representational theory. In B. Anderson & P. Harrison (Eds.), *Taking-place: Non-representational theories and geographies* (pp. 1–36). Burlington: Ashgate.

Ashmore, W. (1989). Construction and cosmology: Politics and ideology in lowland Maya settlement patterns. In W. F. Hanks & D. E. Rice (Eds.), *Word and image in Maya culture: Explorations in language, writing and representation* (pp. 272–286). Salt Lake City: University of Utah Press.

Ashmore, W., & Knapp, B. (1999). *Archaeology of landscapes: Contemporary perspectives*. Malden: Blackwell.

Bachelard, G. (1994). *The poetics of space*. Boston: Beacon.

Barrett, J. C. (1993). *Fragments from antiquity: An archaeology of social life in Britain, 2900–1200 BC*. Malden: Blackwell.

Barrett, J. C. (2001). Agency, the duality of structure, and the problem of the archaeological record. In I. Hodder (Ed.), *Archaeological theory today* (pp. 141–164). New York: Polity Press.

Basso, K. (1996). *Wisdom sites in places*. Albuquerque: University of New Mexico Press.

Baudrillard, J. (1995). *Simulacra and simulation: The body, in theory: Histories of cultural materialism* (S. F. Glaser, Trans.). Ann Arbor: University of Michigan Press.

Bawden, G. (1996). *The Moche*. Cambridge: Blackwell.

Bell, C. (1997). *Ritual: Perspectives and dimensions*. Oxford: Oxford University Press.

Benjamin, W., & Tiedemann, R. (1999). *The Arcades project*. Cambridge, MA: Belknap.

Bergquist, A. (2001). Ethics and the archaeology of world religion. In T. Insoll (Ed.), *Archaeology and word religion* (pp. 182–192). New York: Routledge.

Billman, B. R. (2002). Irrigation and the origins of the southern Moche state on the north coast of Peru. *Latin American Antiquity, 13*(4), 371–400.

Binford, L. R. (1962). Archaeology as anthropology. *American Antiquity, 28*(2), 217–225.

Blake, E. (2002). Spatiality past and present: An interview with Edward Soja. *Journal of Social Archaeology, 2*(2), 139–158.

Bourdieu, P. (1977). *Outline of a theory of practice*. Cambridge: Cambridge University Press.

Bourget, S. (2006). *Sex, death and sacrifice in Moche religion and visual culture*. Austin: University of Texas Press.

Bowser, B. J., & Zedeño, M. N. (2009). *The archaeology of meaningful places*. Salt Lake City: University of Utah Press.

Bradley, R. (2005). *Ritual and domestic life in prehistoric Europe*. New York: Routledge.

Brück, J. (2001). Monuments, power, and personhood in the British Neolithic. *The Journal of the Royal Anthropological Institute, 7*(4), 649–667.

Buck-Morss, S. (1989). *The dialectics of seeing: Walter Benjamin and the Arcades project*. Cambridge, MA: MIT Press.

Carl, P. (2000). Were cities built as images? *Cambridge Archaeological Journal, 10*(2), 327–365.

Casey, E. S. (1997). *The fate of place: A philosophical history*. Berkeley: University of California Press.

Casey, E. S. (1998). The production of space or the heterogeneity of place: A commentary on Edward Dimendberg and Neil Smith. In A. Light & J. M. Smith (Eds.), *Philosophy and geography II: The production of public space* (pp. 71–80). Lanham: Rowman and Littlefield Publishers.

Castillo, L. J. (2001). The last of the Mochicas. In J. Pillsbury (Ed.), *Moche art and archaeology in ancient Peru* (pp. 307–332). New Haven: National of Gallery of Art & Yale University Press.

Castillo, L. J. (2010). Moche politics in the Jequetepeque valley: A case for political opportunism. In J. Castillo & J. Quilter (Eds.), *New perspectives in Moche political organization* (pp. 1–26). Washington, DC: Dumbarton Oaks.

Castillo, L. J., & Donnan, C. B. (1994). Los Mochica del norte y los Mochica del sur. In K. Makowski & C. B. Donnan (Eds.), *Vicús* (pp. 143–181). Lima: Colección Arte y Tesoros del Perú, Banco de Credito.

Caton, S. C., & Zacka, B. (2010). Abu Ghraib, the security apparatus, and the performativity of power. *American Ethnologist, 37*, 203–211.

Choay, F. (1986). Urbanism in question. In M. Gottdiener & A. P. Lagopoulos (Eds.), *The City and the sign: An introduction to urban semiotics* (pp. 241–258). New York: Columbia University Press.

Das, V. (2007). *Life and words: Violence and the descent into the ordinary*. Los Angeles: University of California Press.

Dawdy, S. L. (2009). Millennial archaeology. Locating the discipline in the age of insecurity. *Archaeological Dialogues, 16*(2), 131–142.

Dawdy, S. L. (2010). Clockpunk anthropology and the ruins of modernity. *Cultural Anthropology, 51*(6), 761–793.

de Certeau, M. (1984). *The practice of everyday life*. Berkeley: University of California Press.

Deleuze, G., & Guattari, F. (1987). *A thousand plateaus*. Minneapolis: University of Minnesota Press.

Dillehay, T. D. (2001). Town and country in late Moche times: A view from two northern valleys. In J. Pillsbury (Ed.), *Moche art and archaeology in ancient Peru* (pp. 259–284). New Haven: National of Gallery of Art & Yale University Press.

Donnan, C. B. (1978). *Moche art of Peru*. Los Angeles: Museum of Culture History, University of California.

Douglas, M. (1972). *Natural symbols: Explorations in cosmology*. New York: Vintage Books.

Edensor, T. (2008). Mundane hauntings: Commuting through the phantasmagoric working-class spaces of Manchester, England. *Cultural Geographies, 15*(3), 313–333.

Feldman, R. A. (1987). Architectural evidence for the development of nonegalitarian social systems in coastal Peru. In J. Hass, T. Pozorski, & S. Pozorski (Eds.), *The origins and development of the Andean state* (pp. 10–15). Cambridge: Cambridge University Press.

Fleming, A. (2006). Post-processual landscape archaeology: A critique. *Cambridge Archaeological Journal, 16*(3), 267–280.

Fogelin, L. (2008). Delegitimizing religion: The archaeology of religion as … archaeology. In D. S. Whitley & K. Hays-Gilpin (Eds.), *Belief in the past: Theoretical approaches to the archaeology of religion* (pp. 129–142). Walnut Creek: Left Coast Press.

Foucault, M. (1979). *Discipline and punish: The birth of the prison*. New York: Vintage Books.

Foucault, M. (1986). Of other spaces. *Diacritics, 16*, 22–27.

Fowles, S. (2013). *An archaeology of doings: Secularism and the study of Pueblo religion*. Santa Fe: SAR Press.

Fung Pineda, R. (1988). The late preceramic and initial period. In R. Keatinge (Ed.), *Peruvian prehistory* (pp. 67–96). Cambridge: Cambridge University Press.

Giles, J. (2006). *The spaces of violence*. Birmingham: University of Alabama Press.

Graves, M. W., & Ladefoged, T. N. (1995). The evolutionary significance of ceremonial architecture in Polynesia. In P. A. Teltser (Ed.), *Evolutionary archaeology: Methodological issues* (pp. 149–174). Tucson: University of Arizona Press.

Graves, M. W., & Van Keuren, S. (2011). Ancestral Pueblo villages and the panoptic gaze of the commune. *Cambridge Archaeological Journal, 21*(2), 263–282.

Gündüz, R. (2001). *El mundo ceremonial de los Huaqueros*. Lima: Universidad Ricardo Palma, Editorial Universitaria.

Harris, O. J. T., & Sørensen, T. F. (2010). Rethinking emotion and material culture. *Archaeological Dialogues, 17*(2), 145–163.

Hawkes, C. (1954). Archaeological theory and method: Some suggestions from the Old World. *American Anthropologist, 56*(2), 155–168.

Heckenberger, M. J. (2005). *The ecology of power: Culture, place and personhood in southern Amazonia AD 1,000–2000*. New York: Routledge.

Hetherington, K. (1997). *The badlands of modernity: Heterotopia and social ordering*. London: Routledge.

Hillier, B. (1996). *Space is the machine: A configurational history of architecture*. Cambridge: Cambridge University Press.

Hillier, B., & Hanson, J. (1984). *The social logic of space*. Cambridge: Cambridge University Press.

Hodder, I. (2010). Probing religion at Catalhoyuk: An interdisciplinary experiment. In I. Hooder (Ed.), *Religion in the emergence of civilization; Catalhoyuk as a case study* (pp. 1–31). Cambridge: Cambridge University Press.

Hodder, I., & Hutson, S. (2003). *Reading the past: Current approaches to interpretation in archaeology*. Cambridge: Cambridge University Press.

Holloway, J. (2006). Enchanted spaces: The séance, affect, and geographies of religion. *Annals of the Association of American Geographers, 96*(1), 182–187.

Humphrey, C., & Laidlaw, J. (1994). *The archetypical actions of ritual: A theory of ritual illustrated by the Jain Rite of worship*. Oxford: Oxford University Press.

Ingold, T. (2000). *The perception of the environment: Essays in livelihood, dwelling and skill*. New York: Routledge.

Inomata, T., & Coben, L. S. (2006). Overture: An invitation to the archaeological theater. In T. Inomata & L. S. Coben (Eds.), *Archaeology of performance: Theaters of power, community, and politics* (pp. 11–46). New York: Altamira.

Insoll, T. (2004). *Archaeology, ritual, religion*. London: Routledge.

Johnson, M. (2006a). On the nature of theoretical archaeology and archaeological theory. *Archaeological Dialogues, 13*(2), 117–182.

Johnson, P. (2006b). Unravelling Foucault's "different spaces". *History of the Human Sciences, 19*(4), 75–90.

Johnson, M. (2007). *Ideas of landscape*. Malden: Blackwell.

Jones, L. (2000). *The hermeneutics of sacred architecture: Experience, interpretation, comparison* (Vol. 2). Cambridge, MA: Harvard University Press.

Kapferer, B. (2004). Ritual dynamics and virtual practice: Beyond representation and meaning. In D. Handelman & G. Lindquist (Eds.), *Ritual in its own right* (pp. 35–54). New York: Berghahn.

Knappett, C., & Malafouris, L. (2008). *Material agency: Towards a non-anthropocentric approach*. New York: Springer.

Kolata, A. (1993). *The Tiwanaku: Portrait of an Andean civilization*. Cambridge: Blackwell.

Köpping, K. P. (1997). The ludic as creative disorder: Framing, de-Framing, and boundary crossing. In K. P. Köpping (Ed.), *The games of God and man: Essays in play and performance* (pp. 1–39). Hamburg: Lit Verlag.

Kyriakidis, E. (2007). *The archaeology of ritual*. Los Angeles: Costen Institute of Archaeology, University of California, Los Angeles.

Latour, B. (1993). *We have never been modern*. Cambridge, MA: Harvard University Press.

Lefebvre, H. (1991). *The production of space*. Malden: Blackwell.

Lilley, K. D. (2004). Cities of God? Medieval urban form and their Christian symbolism. *Transactions of the Institute of British Geographers, 29*(3), 296–313.

Lumsden, S. (2004). The production of space at Nineveh. *British Institute for the Study of Iraq, 55*, 187–197.

Maddern, J., & Adey, P. (2008). Editorial: Spectro-geographies. *Cultural Geographies, 15*, 921–925.

Merrifield, A. (1993). Space and place: A Lefebvrian reconciliation. *Transactions of the Institute of British Geographers, 18*(4), 516–531.

Meskell, L. (2003). Memory's materiality: Ancestral presence, commemorative practice, and disjunctive locales. In R. M. Van Dyke & S. E. Alcock (Eds.), *Archaeologies of memory* (pp. 34–55). Malden: Blackwell.

Meskell, L. (2005). *Archaeologies of materiality*. Malden: Blackwell.

Miller, D. (2005). *Materialities*. Durham: Duke University Press.

Miller, D., Rowlands, M., & Tilley, C. (1989). *Domination and resistance*. New York: Routledge.

Mills, B. J., & Walker, W. H. (2008). *Memory work: Archaeologies of material practice*. Santa Fe: SAR Press.

Mock, S. (1998). *The sowing and the dawning. Termination, dedication, and transformation in the archaeological and ethnographic record of Mesoamerica*. Albuquerque: University of New Mexico Press.

Moore, J. D. (1992). Pattern and meaning in Prehistoric Peruvian architecture: The architecture of social control in the Chimu state. *Latin American Antiquity, 3*(2), 95–113.

Moore, J. D. (1996). *Architecture and power in the Ancient Andes: The archaeology of public buildings*. Cambridge: Cambridge University Press.

Moore, J. D. (2004). The social basis of sacred space in the prehispanic Andes: Ritual landscapes of the dead in Chimú and Inka societies. *Journal of Archaeological Method and Theory, 11*(1), 83–124.

Moore, J. D. (2005). *Cultural landscapes of the ancient Andes: Archaeologies of place*. Gainesville: University of Florida Press.

Morris, B. (1987). *Anthropological studies of religion: An introductory text*. Cambridge: Cambridge University Press.

Nelson, B. A. (1995). Complexity, hierarchy, and scale: A controlled comparison between Chaco Canyon, New Mexico, and La Quemada, Zacatecas. *American Antiquity, 60*(4), 597–618.

Pauketat, T. R. (2001). Practice and history in archaeology: An emerging paradigm. *Anthropological Theory, 1*(1), 73–98.

Pauketat, T., & Alt, S. M. (2005). Agency in a postmold? Physicality and the archaeology of culture making. *Journal of Archaeological Method and Theory, 12*(3), 213–236.

Plunket, P. (2002). *Domestic ritual in ancient Mesoamerica* (Monograph 46). Los Angeles: The Costen Institute of Archaeology, University of California Press.

Preucel, R. (2010). *Archaeological semiotics*. Malden: Blackwell.

Quilter, J. (2002). Moche politics, religion, and warfare. *Journal of World Prehistory, 16*, 145–195.

Quilter, J., & Castillo, L. J. (2010). *New perspectives on Moche political organization*. Washington, DC: Dumbarton Oaks.

Rapoport, A. (1982). *The meaning of the built environment: A nonverbal communication approach*. Tucson: The University of Arizona Press.

Renfrew, C. (1985). *The archaeology of cult: The sanctuary at Phylakopi* (Supplementary Vol. 18). Athens: The British School of Archaeology at Athens.

Samuels, J. (2010). Of other spaces; Archaeology, landscape, and heterotopia in fascist Sicily. *Archaeologies: Journal of the World Archaeology Congress, 6*(1), 62–80.

Schann, D. P. (2012). *Sacred geographies of ancient Amazonia*. Walnut Creek: Left Coast Press.

Shanks, M., & Tilley, C. (1990). Archaeology into the 1990s. *Norwegian Archaeological Review, 22*(1), 1–12.

Shields, R. (1990). *Places on the margin: Alternate geographies of modernity*. London: Routledge.

Shields, R. (1999). *Lefebvre, love and struggle: Spatial dialectics*. London: Routledge.

Shimada, I. (1994). *Pampa grande and the Mochica culture*. Austin: University of Texas Press.

Sillar, B. (2009). The social agency of things? Animism and materiality in the Andes. *Cambridge Archaeological Journal, 19*(3), 367–377.

Smith, J. Z. (1987). *To take place: Toward theory in ritual*. Chicago: University of Chicago Press.

Smith, A. T. (2003). *The political landscape: Constellations of authority in early complex polities*. Berkeley: University of California Press.

Smith, A. T. (2004). The end of the essential archaeological subject. *Archaeological Dialogues, 11*(1), 1–20.

Smith, K. L. (2005). Looting and the politics of archaeological knowledge in northern Peru. *Ethnos, 70*, 149–170.

Soja, E. (1996). *Thirdspace: Journeys to Los Angeles and other real-and-imagined places*. Malden: Blackwell.

Sorkin, M. (1992). *Variations on a theme park: The new American city and the end of public space*. New York: Hill and Wang.

Stoler, A. L. (2008). Imperial debris: Reflections on ruins and ruination. *Cultural Anthropology, 23*, 191–219.

Swenson, E. R. (2003). Cities of violence: Sacrifice, power, and urbanization in the Andes. *Journal of Social Archaeology, 3*(2), 256–296.

Swenson, E. R. (2006). Competitive feasting, religious pluralism, and decentralized power in the Late Moche Period. In W. H. Isbell & H. Silverman (Eds.), *Andean archaeology III: North and south* (pp. 112–142). New York: Springer.

Swenson, E. R. (2008). The disjunctive approach to the archaeological analysis of ritual politics. In L. Fogelin (Ed.), *Religion in the material world* (pp. 237–261). Carbondale: Center for Archaeological Investigations, Southern Illinois University Press.

Swenson, E. R. (2010). Revelation and resolution: Anthropologies of religion, cognition, and power. *Reviews in Anthropology, 39*, 173–200.

Swenson, E. R. (2011). Stagecraft and the politics of spectacle in ancient Peru. *Cambridge Archaeological Journal, 21*(2), 285–315.

Swenson, E. R. (2012). Moche ceremonial architecture as third space: The politics of place-making in ancient Peru. *Journal of Social Archaeology, 12*(1), 3–28.

Swenson, E. R., & Warner, J. (2012). Crucibles of power: Forging copper and forging subjects at the Moche ceremonial center of Huaca Colorada, Peru. *Journal of Anthropological, 31*(3), 314–333.

Swenson, E. R, Chiguala, J., & Warner, J. (2011). *Proyecto de Investigación de Arqueología, Jatanca-Huaca Colorada, Valle de Jequetepeque 2010*. Lima: Technical report submitted to the Instituto Nacional de Cultura.

Swenson, E. R, Chiguala, J., & Warner, J. (2012). *Proyecto de Investigación de Arqueología, Jatanca-Huaca Colorada, Valle de Jequetepeque 2011*. Lima: Technical report submitted to the Instituto Nacional de Cultura.

Thomas, J. (1994). The politics of vision and the archaeologies of landscape. In B. Bender (Ed.), *Landscape: Politics and perspectives* (pp. 19–48). Oxford: Berg.

Thomas, J. (2001). Archaeologies of place and landscape. In I. Hodder (Ed.), *Archaeological theory today* (pp. 165–186). New York: Polity Press.

Thrift, N. (2008). *Non-representational theory: Space, politics, affect*. London: Routledge.

Tilley, C. (1994). *A phenomenology of landscape: Places, paths and monuments*. New York: Berg.

Tilley, C., & Bennett, W. (2001). An archaeology of supernatural places: The case of West Penwith. *Journal of Royal Anthropological Institute, 7*(2), 335–362.

Townsend, R. F. (1982). Pyramid and Sacred Mountain. In A. F. Aveni & G. Urton (Eds.), *Ethnoastronomy and archaeoastronomy in the American tropics* (Annals of the New York Academy of Science 385, pp. 37–62). New York: New York Academy of Science.

Trigger, B. (1990). Monumental architecture: A thermodynamic explanation of symbolic behavior. *World Archaeology, 22*(2), 119–132.

Turner, V. (1967). *The forest of symbols: Aspects of Ndembu ritual*. Ithaca: Cornell University Press.

Uceda, S. (2001). Investigations at Huaca de la Luna, Moche Valley: An example of Moche religious architecture. In J. Pillsbury (Ed.), *Moche art and archaeology in ancient Peru* (pp. 47–68). New Haven: Yale University Press.

Uceda, S., & Mujica, E. (1994). *Moche: Propuestas y perspectiva*. Lima: Travaux de l'Institut Français d'Etudes Andines.

Uceda, S., & Mujica, E. (2003). *Moche hacia el final del milenio* (Vols. 1 & 2). Lima & Trujillo: Fondo Editorial, Pontificia Universidad Católica del Perú and Universidad Nacional de Trujillo.

Van Dyke, R. M. (2003). Memory and the construction of Chacoan society. In R. M. Van Dyke & S. E. Alcock (Eds.), *Archaeologies of memory* (pp. 180–200). New York: Blackwell.

Van Dyke, R. M., & Alcock, S. E. (2003). Archaeologies of memory: An introduction. In R. Van Dyke & S. E. Alcock (Eds.), *Archaeologies of memory* (pp. 1–14). New York: Blackwell.

Wallace, J. (2004). *Digging the dirt: Archaeology and the Romantic imagination*. London: Duckworth.

Whitehouse, H. (2004). *Modes of religiosity: A cognitive theory of religious transmission*. Walnut Creek: Alta Mira Press.

Wiersema, J. (2010). *Architectural vessels of the Moche of Peru (C.E. 200–850): Architecture of the afterlife*. PhD dissertation. College Park: University of Maryland, Department of Anthropology.

Wylie, A. (2002). *Thinking from things: Essays in the philosophy of archaeology*. Berkeley: University of California Press.

Chapter 25
Sacred Caves of the World: Illuminating the Darkness

Leslie E. Sponsel

25.1 Introduction

In the contemporary Western world caves are principally the site for research by scientists, most of all for archaeologists, biologists, and geologists. The scientists may or may not be adventurers called cavers or spelunkers. Speleology, the scientific exploration and study of caves, has only been developing since the mid-eighteenth century. Eventually specialized organizations were established, such as the National Speleological Society in the United States in 1941, currently with a membership around 12,000, and the International Union of Speleology in 1965 now with 60 member countries. Caves are also visited by tourists, especially show caves that are maintained as attractions for public curiosity and entertainment by government or private agencies. Thus, caves are sites for exploring the unknown as well as for recreational and aesthetic experience. They may even be viewed as a kind of wilderness (White and Culver 2005).

Throughout the world many people consider caves, or at least particular ones, to be sacred places, and in some instances inhabited by spiritual beings or forces. Yet caves may even be recognized as a spiritual or mystical place apart from any particular religion as revealed, for example, by the reactions of the first explorers visiting the Cueva de los Cristales (Cave of Crystals). It is located south of Chihuahua in northern Mexico and was only discovered as recently as the year 2000. This awesome cave contains 170 selenite crystals up to 37.4 ft (11.4 m) long that have been growing under extraordinary chemical conditions for more than 600,000 years. One of the explorers, Juan Manuel Garcia-Ruiz, "… called it the Sistine Chapel of crystals noting that in both cathedrals and crystals there's a sense of permanence and tranquility that transcends the buzz of surface life. In both there is the suggestion

---

L.E. Sponsel (✉)
Department of Anthropology, University of Hawaii, Honolulu, HI 96822, USA
e-mail: sponsel@hawaii.edu

of worlds beyond us" (Shea and Peter 2008: 77). The reflection by Garcia-Ruiz is reminiscent of the prominent historian of religion Mircea Eliade's (1959: 11) statement: "Man becomes aware of the sacred because it manifests itself, shows itself, as something wholly different from the profane." (Also see Hurd 2008).

Sacred caves are especially extraordinary places as spaces at the interface between the natural and supernatural. Accordingly, sacred caves are also extraordinary places for science and scholarship in geology and biology on the one hand, and on the other religion, culture, and history. Thereby sacred caves facilitate a holistic perspective and integrative synthesis ranging among the natural sciences, social sciences, and humanities, instead of the usual isolation of more or less arbitrary academic compartmentalization (Sponsel and Natadecha-Sponsel 2004).

Here, pursuing a holistic and interdisciplinary conceptual framework, it is helpful to first consider caves in general before surveying sacred caves in particular, the latter from prehistory to the present with examples from most of the main lived religions of the world. Then cave research and conservation are discussed briefly, and, finally, the most useful resources for further pursuing aspects of this subject are identified.

25.2 Caves

Caves are unusually attractive and fascinating in many ways as subterranean environments that are alien and mysterious. For instance, often caves contain unusual or beautiful shapes of mineral deposits called speleothems including stalactites and stalagmites. (The former grow from the ceiling, the latter from the floor). Also, troglobites, animals that live exclusively in the permanent darkness of the interior of the cave usually have lost skin pigment and eyesight. Some are referred to as extremophiles because of the extreme conditions to which they must adapt to survive, such as nutrient scarcity. Caves are island-like refuges with a relatively isolated and limited area, low biodiversity, and high endemicity. They may be considered as natural experiments in biological evolution and adaptation (for example, Krajick and Littschwager 2007).

In general, a cave is any natural cavity in the ground large enough for a human to enter and mostly without any natural light, except at its entrance and in a subsequent twilight zone before the total darkness of the deeper interior (White and Culver 2005). However, some caves are not entirely natural, humans having carved cavities into the rock in the side of a hill or mountain or made other modifications. There are many different kinds of caves as classified by their geological origin and the type of rock surrounding them, including mechanical processes (for example, talus caves); differential erosion and scour (for example, sea caves, aeolian caves, and rock shelters); volcanic (for example, lava tubes, vents, and blisters); glacial (for example, ice caves); and solution. A large variety of solution caves is recognized, depending on the type of rock as well as the source and chemistry of the water that generates the dissolving. Most solution caves are formed in limestone (calcium carbonate) or dolomite (calcium magnesium carbonate), although some form in deposits of gypsum (calcium sulfate) or salt (sodium chloride) (Palmer 2007; White and Culver 2005: 83–85). Karst is a widespread but distinctive landscape formed by the solution of the bedrock, usually limestone, and characterized by a particular suite of landforms such as towers and pinnacles of rock, blind valleys, and dolines (depressions). The bedrock may be exposed over large areas while drainage is mostly underground through caves and other cavities (Ford and Williams 2007). Solution caves develop from the chemical dissolution of the bedrock by circulating ground water. Interestingly, because solution caves are fed by water seeping down from the surface, significant changes in the climate and vegetation on the surface may leave their record in the growth bands of the speleothems (White and Culver 2005: 82). In environmental and other respects, such as the archaeological record, caves may serve in effect as time capsules (for example, Burney 2010; Burney and Kikuchi 2006; Katz 2003; Zhang et al. 2008). Also, karst landscapes with high levels of endemic species have been recognized as "arks" of biodiversity (Clements et al. 2006).

For many thousands of years, and for over a million in some areas such as southern Africa, humans and their ancestors have used caves for shelter and habitation (Bonsall and Tolan-Smith 1997). Natural resources are exploited in caves such as guano for fertilizer or gunpowder, bats and honey for food, and swifts for bird nest soup, a delicacy in Chinese cuisine. Caves have been used for storage, some by herders for livestock. They may also serve in essence as a kind of art gallery.

Different parts of the same cave might have different uses, and/or the uses might change over time (for example, Brucker 2005; Ustinova 2009). In the area of central Kentucky, Mammoth Cave was used by the ancestors of Native Americans for around 2,400 years starting in 2250 BC. They went several miles into the cave using torches made of cane and dry weeds. They visited it seasonally for shelter and also mined selenite, mirabilite, and gypsum minerals. It has been suggested that the indigenes may have used mirabilite as a laxative and possible food seasoning, gypsum as an ingredient for paint, and selenite crystals for rituals. (Today selenite is popular among adherents of New Age religions for various purposes). Much later Euroamericans mined saltpeter, a component of gun powder, from the cave. During the War of 1812 seventy slaves were used in this mining. This cave is famous for a variety of stalactites, stalagmites, columns, and flowstones. Tourism started as the population and roads expanded westward into Kentucky from 1795 to 1840, and the number of tourists increased markedly with the construction of the Louisville & Nashville Railroad in 1858. Mammoth Cave National Park was instituted in 1941. The associated Cave Research Foundation was established in 1957 (see Brucker 2005; Sears 1989; Watson 1987).

Show caves are specifically relegated to tourism (Cigna 2005; Show Caves 2012). At least as early as the seventeenth century a fee was charged for entrance into the Vilenica Cave in Slovenia. It is estimated that each year worldwide there are around 150 million visitors to caves and they generate a $2.3 billion business with some 100 million salaried employees. In the United States, for example, the three most famous show caves, Carlsbad Caverns in New Mexico, Mammoth Cave in Kentucky, and Wind Cave in South Dakota, are visited annually by an estimated 2.5 million people (Gigna and Burri 2000).

25.3 Sacred Caves

Sacred places are diverse sites and landscapes with extraordinary qualities that may evoke spiritual or mystical experiences (Bellows et al. 2008; Brockman 2011; Holm and Bowker 1994; Sponsel 2007a; Swan 1990). Attributes of the spiritual may involve an experience variously described as awesome, cosmic, deep, ecstatic, epiphanic, extraordinary, high, ineffable, infinite, mysterious, numinous, oneness, peak, powerful, primal, serene, sublime, tranquil, transcendent, transformative, unforgettable, unity, unreal, ultimate, and/or wonder (Grassie 2010; Oliver 2009; Saint-Laurent 2000; Sponsel 2012; Swenson 2009).

Sacred places are complex phenomena that can be viewed usefully as varying along several continua ranging from natural (or biophysical) to anthropogenic (or socio-cultural); prehistoric to historic, recent, or newly created; secret or private to public; single culture (or religion) to multi-cultural (or multi-religious); intrinsic to extrinsic in value; uncontested to contested; and protected to endangered. Particular sacred places variously tend toward one pole or another of some combination of these continua. It is also noteworthy that in many cases the same place may be experienced as sacred by persons from very different historical, cultural, linguistic, religious, and/or national backgrounds suggesting to some observers that the sacredness is something inherent in the place itself (Sponsel 2007a, b; Swan 1990).

Whether or not it is considered sacred, a cave is a truly extraordinary environment situated in mountains where earth and sky meet; surface and underground interface; darkness and silence prevail; and the environment is variously alien, awesome, mysterious, and even potentially dangerous. For many people entering a cave involves crossing a threshold from the profane into the sacred. As Eliade (1959: 37) observes: "a sacred place constitutes a break in the homogeneity of space … this break is symbolized by an opening by which passage from one cosmic region to another is made possible… from earth to the underworld." This crossing into the subterranean world may be characterized as liminal, a transitional state from the place of ordinary above ground activities into a place of extraordinary underground activities (Turner 1969). The cave becomes a gateway into the spiritual dimension of human experience, or even into the supernatural world. Accordingly, various religious uses of caves encompass, but are not limited to, pilgrimage sites; natural sanctuaries for retreat, contemplation, prayer, worship, and/or meditation; resonance chambers for singing or chanting; and/or reliquaries and tombs. It should be noted that the area around as well as inside a cave may be considered sacred (Moyes 2012; Sponsel and Natadecha-Sponsel 2004).

25.3.1 Prehistoric

Shanidar Cave in the Zagros Mountains of northern Iraq is likely among the earliest sacred caves. There intentional burials of Neanderthals were discovered and dated at between 60,000 and 80,000 years ago. Some archaeologists have interpreted them as a funerary ritual. Routine soil samples from around the Shanidar IV grave yielded the pollen remains of several species of colorful flowers including Cornflower, Bachelor's Button, St. Barnaby's Thistle, Ragwort or Groundsel, Grape Hyacinth, Joint Pine or Woody Horsetail, and Hollyhock Yarrow. Many of these species are used to this day for their medicinal qualities. However, archaeologists disagree as to whether the flowers were intentionally placed over the grave or were introduced by natural agents such as wind or rodents (Solecki 1975).

Archaeologists have discovered nearly 200 caves in Europe with prehistoric art associated with *Homo sapiens sapiens* from the Upper Paleolithic around 35,000–12,000 years ago. The paintings consist mostly of images of large animals and geometric symbols, but some are a composite of human and animal forms (therianthropes). For more than 20,000 years caves in Europe were used for various purposes, but most solely as ritual sites. Various religious uses may have included sympathetic magic for enhancing hunting success or prey population fertility; vision quests and shamanic trances or spirit possession with ritualized altered states of consciousness; and ceremonies for rites of passage at critical junctures of the human life cycle. The art is usually deep in the interior of the cave, instead of near the entrance or in the twilight zone. Commonalities in the art have been interpreted to reflect shared human neurology and psychology. (See Clottes and Lewis-Williams 1998; Galanti 1998; Herzog 2011; Lewis-Williams 2002; Lewis-Williams and Clottes 2008; Whitley 1998, 2009).

Some of the earliest archaeological evidence of art and religion is found in caves like Chauvet Cave in France dated around 32,000 years ago (Herzog 2011). Apparently a creative explosion occurred in this region during the Upper Paleolithic (Heyd and Clegg 2005). The most famous caves are Altamira in northern Spain and Lascaux and Chauvet in southwestern France. Discovered in 1940, Lascaux has been called the Sistine Chapel of the Paleolithic with some 600 paintings and 1,500 engravings deep in its interior (Davidson and Gitlitz 2002). One archaeologist, Geri-Ann Galanti (1998: 1), remarked after visiting Lascaux that it was "easily the most profoundly moving experience of my life." (See Arias 2009; Aujoulat 2005; Bahn 2007; Beltran 1999; Chauvet et al. 1996; Clottes 1998, 2003, 2010; Curtis 2007; Desdemaines-Hugon 2010).

One of the most famous prehistoric caves in the world is Niah in Borneo which humans have used regularly for as long as some 45,000 years. It contains 200 burials from the Neolithic period 5,000 to 2,500 years ago. The earlier burials are extended skeletons in log coffins, while the later ones are in ceramic, bamboo, or basketry containers. They are all located in the twilight zone of the cave (Barker 2005; Barker et al. 2005).

25.3.2 Indigenous

To this day caves may still be used for funerary and other purposes by relatively traditional indigenous societies. In the remote highlands of the Sepik River basin of Papua New Guinea, the Meakambut people, isolated semi-nomadic foragers, use some of more than a hundred natural caves as shelters, refuges from enemies, funerary sites with ancestral skulls, in rituals like male initiation rites, and for art such as negative hand prints. The hand is placed against the wall of the cave while red or black pigment is blown or painted over it like a stencil. Each of the caves is named and ownership is inherited from father to son. Often there are legends associated with a cave. One of the caves is where the Meakambut believe that their ancestors originated (Jenkins and Toensing 2012).

Most of the research on indigenous uses of caves has been conducted in the Maya region of Mexico and Central America since 1985. Most of this research has been done by archaeologists, even though considerable continuity is apparent in traditions of cave use from prehistory into the present. Even Maya pyramids have been viewed as sacred mountains with a cave underneath, usually human-made (Heyden 1975). Indeed, by now caves are recognized as a regular component of the sacred landscape of the Maya. Because they are so integral to the culture, society, religion, and history of the Maya, caves are also an important determinant of settlement pattern. However, most caves are too damp to serve for refuge, shelter, or habitation.

In the cosmology and mythology of the Maya, caves are where the sun, moon, gods, humans, animals, plants, water, and weather originate. As portals to the supernatural, caves are places where humans interact with the powerful forces that animate the universe. They are sacred places for burials, mummies, ossuaries, and cremations. There animals and humans may be sacrificed, such as to the rain deities. Caves are also art galleries connected to religion; sources of "virgin" water for ceremonies such as rites of passage; places for healing rituals; sanctuaries for the performance of magic for weather management and the farming cycle; sources of water for drinking and other uses; and depositories of discarded ceremonial objects. In addition, they may be sites for the practice of witchcraft. With exposure to Christianity, some speleothems are interpreted as Christian symbols.

Maya cave ceremonies are rarely mentioned in the literature let alone described in detail. Accounts are rare because the presence of an unsanctioned foreigner might cause misfortune, death, or even an earthquake or other catastrophe. Consequently the caves used for ritual are located in remote areas and kept secret from outsiders. Yet in 1968, Jaroslaw Theodore Petryshyn became the first eye witness of a cave ritual, one performed by the Lacandon of Chiapas in southern Mexico. He observed several human skulls on a natural shelf in a cave together with implements including a censer bowl representing a deity, and square boards with handles and balls of copal. (Copal is an aromatic tree resin used for incense). A shaman prayed and then chanted as he fashioned the copal and burned it while ritually moving a wrapped palm leaf over a fire. Thereby the shaman appealed to the cave god to protect his cornfield from damage by heavy rains (Petryshyn and Colas 2005). (Further material on Maya caves can be found in Awe 2006; Bassie-Sweet 1996; Brady 1997; Brady and Prufer 2005; Chladek 2011; Moyes 2012; Stone and Brady 2005).

Australian Aborigines still create and ritually use art on rock surfaces like cave walls as an integral component of their sacred geography. They believe that caves are created by spirits traveling through rock. Furthermore, they believe that some caves are inhabited by spirits that live within the rock and that the spirits materialize as art along the walls. The world inside the rock is thought to be similar to the one outside with various plants, animals, and landscapes. Some art depicts therianthropes. Numerous sacred myths and legends are associated with their rock art, some of which is more than 10,000 years old. For example, for thousands of years Aborigines have visited rock art in the many caves at the base of the red sandstone monolith Uluru (Ayers Rock) in the central desert of Australia. This is the most sacred site of all for Aborigines, a place where over 60 of their sacred routes of dream tracks or song lines converge. They feel the walls to contact their ancestral spirits and obtain their blessing. Uluru also attracts several hundred tourists annually, other Australians as well as foreigners, many of them New Agers (Davidson and Gitlitz 2002; Layton 1986; Morwood and Hobbs 1992; Tacon 2005).

In the Hawaiian islands there are more than a thousand sea caves and lava tube caves. Many were used by ancient Hawaiians as sites for defense, refuge, storage, burials, concealing religious artifacts, or rituals (Bollt 2005; Kennedy and Brady 1997). For example, Makauwahi Cave at Maha'ulepu on the island of Kaua'i is associated with oral traditions extending back to the fourteenth century. A kahuna (shaman) is known to have used the cave as a place of divination. People would visit him inside the cave. The kahuna kindled a fire, extinguished it, and then from the patterns created by the smoke and ashes read answers to their questions. At Makauwahi in recent decades the cave is a popular place for picnics, sightseeing, and contemplation by locals and tourists (Burney 2010; Burney and Kikuchi 2006).

25.3.3 Abrahamic

Sacred caves are associated as well with the Abrahamic religions, although much less so than with the previously mentioned Maya and, as discussed later, Buddhism and Hinduism. Many caves are cited in verses of the Bible as shelters, tombs, refuges, and sites for worship (Gospel Hall 2012). After being expelled from the Garden of Eden, Adam and Eve are supposed to have taken refuge in the Cave of Treasures. The altar beneath the Church of the Nativity in Bethlehem on the West Bank is one of the holiest places of Christianity. It is built around a cave where Jesus is believed to have been born. Quarantal, the Holy Grotto of the Temptation of Christ, is located in Jericho. After Jesus died he is said to have been carried to a cave for burial. The Chapel of the Holy Trinity on the summit of Mount Sinai in Egypt encloses a cave where Moses is supposed to have waited to receive the Ten Commandments. The Cave of the Apocalypse on a hilltop in the Dodecanese Islands of Greece is where St. John allegedly wrote the Book of Revelations. David is believed to have taken refuge and composed many of the Psalms in Adullam Cave (Bellows et al. 2008: 101, 106, 226; Holm and Bowker 1994; Steward 2005).

Caves have been used throughout Christian history. Saint Francis of Assisi frequented caves for prayer and contemplation (Francke 2005; Spoto 2002). Also especially interesting are the vast monastic complexes of rock-cut churches with wall frescoes from the tenth through the twelfth centuries at Cappadocia in Turkey (Goreme Open Air Museum 2012). In more recent history one of the most famous of all Christian pilgrimage sites is the Grotto of Our Lady of Lourdes in southwest France. (Our Lady of Lourdes is a title of the Blessed Virgin Mary). In Catholicism more than a dozen separate apparitions of Mary are said to have transpired at Lourdes since 1858. The spring water is believed to have healing qualities by Protestants as well as Catholics. Claims of miraculous cures have been investigated since 1861 by a special committee and many have been certified as not having any natural or scientific explanation. By now Lourdes encompasses a huge complex of sacred sites and receives 2–4 million visitors yearly. This world famous Grotto of Our Lady of Lourdes has been replicated by Catholics elsewhere such as in Skaro, Alberta, Canada (Davidson and Gitlitz 2002). Incidentally, at Lourdes and in many places in Europe including Rome there are catacombs beneath religious and other buildings entombing saints and other personages that form in essence sacred caves.

The famous Wieliczka Salt Mine is located in the town of Wieliczka, part of the Krakow metropolitan area in Poland. The mine has been in operation since the thirteenth century. A Salt Cathedral was carved out of the deposits by the miners in part of the mine. It contains statues, chandeliers, and other remarkable art work carved from salt. Other cathedrals carved out of salt deposits by miners are located in Spain and Colombia.

A second Abrahamic religion, Islam, has only a few famous sacred caves. The Cave of Hira near Mecca in Saudi Arabia is where Muhammad is said to have first received the word of God. The Temple Mount, or Temple Rock, is a famous cave where it is believed that Muhammad ascended to heaven. In Ethiopia, the Omar Caves are supposed to be where Allah revealed himself to Sheikh Sof Omar in the twelfth century. The cave was used as a mosque by the sheikh and his followers. It contains natural columns, buttresses, domes, vaults, and pillars eroded from the rock. To this day it remains a meeting site for local Muslims.

In the third Abrahamic religion, Judaism, the Cave of Machpelah (Cave of the Patriarchs) identified in Genesis 23 of the Bible is located on Mt. Carmel in Hebron in the West Bank. It is supposed to contain the tomb of the Jewish patriarchs and matriarchs, among them Abraham, Sarah, Isaac, Jacob, and Rebecca. It is also thought to be the passage to the Garden of Eden. Visited by more than 300,000 people every year, Machpelah is the second holiest site for Jews after the Temple Mount in Jerusalem. Unfortunately, like many sacred places, it is a contested site, in this case between Jews and Muslims (Bellows et al. 2008).

From 1947 to 1956 the 11 caves of Khirbet Qumran on the West Bank yielded the famous Dead Sea Scrolls comprising 972 mostly parchment texts from the Hebrew Bible in Aramaic, Greek, and Hebrew. As the oldest surviving copies of such documents from 150 BC to 70 AD, the scrolls are of tremendous historical and religious importance (Freund 2004; Ullmann-Margalit 2006).

25.3.4 Asian

In Buddhism the earliest record of the use of caves dates from the Emperor Ashoka (268–239 BC), the first monarch in India to pursue that religion. For more than a thousand years the use of caves for meditation and rituals continued in the subcontinent. There the most famous sacred caves are the Adjanta and Ellora complexes together comprising some one thousand caves carved out of cliffs from 200 BC to 700 AD. These were major pilgrimage sites for about a thousand years.

The Ajanta Caves in Maharashtra encompass a group of 29 Buddhist temples and monasteries from the second century BC to 480 AD. Both the Theravada and Mahayana traditions are represented. These rock-cut caves are remarkable feats of engineering with some halls without any supports reaching about the size of a basketball court. Cave number 26 is a prayer hall housing a giant carved statue of the reclining Buddha, representing his death with mourners below. The Ajanta caves contain the oldest known Indian wall paintings of historic times, many depicting legends associated with the Buddha. They were forgotten for 1,300 years until their discovery in 1819. Many European and American artists became obsessed with making copies of the paintings and sculptures. In 1983, UNESCO designated Adjanta as a World Heritage Site. (See Behl 1998; Caswell 2000; Dhammika 1998; Jamkhedkar 2009; Klein and Klein 2012; Malaandra 1993; Mitra 2004; O'Neill and Behl 2008; Spink 2009; Srinivasan 2007; Tsian 2010; Tulku 1994; Winchester 2006).

Elsewhere in South Asia there are many sacred caves as well. According to legend, Saptaparna Cave near Raja Guha in what is now called Nepal was the site of the first council of 500 followers after the Buddha's death in 543 BC. The council was held to reach consensus on a standard version of the Dharma, the Buddha's teachings. In Sri Lanka there are more than 2,000 sites of ancient monastic caves, although some with only one or two caves. Many have been used intermittently for more than 2,000 years and some continuously to the present (Dhammika 1998).

Sacred caves associated with various aspects of Buddhism can be found throughout many parts of Southeast Asia. In Thailand, for example, over a hundred natural caves have been identified as sacred among more than 4,000 in the 18 % of the country covered by karst terrain. Archaeological evidence reveals that some caves and rock shelters have been used for thousands of years for habitation, burials, painting, and rituals. In the case of Buddhism in Thailand, for well over a thousand years caves as natural sanctuaries have been used for meditation, acquiring merit, spiritual experiences, resonance chambers for chanting, and reliquaries and tombs. Most religious objects are located in the entrance and twilight zones for natural lighting, but some are hidden deep in the dark interior and in alcoves. Typically a large statue of the Buddha, and/or a stupa or chedi containing human remains, are positioned on a raised platform facing the main approach. This is the most sacred area of a cave. Other figures may include holy personages, animals, and mythical creatures. Some speleothems may be interpreted as religious symbols. Water dripping from stalactites and from sacred pools as well as mud may be used for healing and magical purposes. Monks use caves to overcome personal physical and mental fears as well as to develop their powers over ego, one condition for ultimately achieving enlightenment. For centuries caves, especially those recognized as sacred, have served as pilgrimage and tourist sites. Devotees commonly make offerings of incense, candles, flowers, prayers, or chants, and/or meditate silently. As elsewhere, various parts of the same cave may have been used for different purposes at the same time and/or through time. Some caves have temple complexes constructed adjacent to them (Munier 1998; Sidisunthorn et al. 2006; Sponsel and Natadecha-Sponsel 2004).

As Buddhism declined in India the centers for carved caves moved to northern China along the famous Silk Road until Islam gained power in some areas. The Silk Road was a historic corridor for the flow of ideas as well as trade goods between the West and East. Today the most popular sacred cave site is the Dunhuang or Mogao Caves along the border between Mongolia and Tibet in Gansu province of China just southeast of the city of Dunhuang. It is on the margins of the high desert of Gobi. A mile long sandstone cliff is honeycombed with some 800 caves carved into the rock, 492 of them ornately decorated with painted murals and 2,300 sculptures of Buddhist art. These caves were developed from the fourth through the fourteenth centuries. They are time capsules illustrating various aspects of the history of the evolution of Buddhism and Chinese art, among other phenomena. The caves were used for shelter, meditation, and occasionally also for burial. Some housed temples and lecture halls. Many of the caves contain sumptuous paintings covering almost everything, some with murals illustrating the Jatakas, tales from the Buddha's past lives. Cave 96, including a nine-story pagoda, houses a magnificent 116 ft (35.3 m) tall statue of the Buddha. Discovered in 1900, the sealed Cave 17, the Library Cave, conserved tens of thousands of silk and paper scrolls of paintings, banners, religious, and secular books in Chinese, Sanskrit, Tibetan, Mongolian, and other languages. One of the ancient documents was the famous Diamond Sutra. Miraculously somehow the caves escaped Chairman Mao's Cultural Revolution of 1966 which destroyed so many other Buddhist structures and artifacts throughout most of China (Cotter 2008; Larmer and Law 2010). (Also see Caswell 2000; Jinshi 2010; Palmer 1996; Whitfield 2000; Whitfield et al. 2000).

In Tibet much of the high plateau in the vicinity of Lhasa is limestone. In this terrain natural and anthropogenic caves and rock shelters have been used in various ways for at least 2,000 years, and especially by Buddhist monks and other holy personages, an integral component of the sacred geography of Tibet. For instance, at Piyang more than a thousand caves have been carved into the mesa, some for habitation while others for hermitage, meditation, and/or ritual purposes. Spiritual beings and forces are believed to dwell in landscape features, especially in mountains, in both the Buddhist and pre-Buddhist or Bon religions of Tibet. Caves acquire power through their association with famous religious personages and secular rulers as well as spiritual beings and forces. For instance, the Guru Rinpoche cave complex in the Yarlung Tsangpo river area incorporates deep natural caves with altars, statues, and wall paintings constructed over the past eight centuries. In the eighth century A.D., a tantric master from India, Guru Rinpoche, also called Padmasambhava, was asked by the Tibetan king Trisong Detsen to destroy the demons of the plateau. In their quest for enlightenment subsequent generations of holy persons frequented such caves in order to absorb some of the spiritual power. Caves are also pilgrimage sites for lay Buddhists. Pilgrims may take stones, soil, or water from a cave because of the spiritual energy they impart and place the specimen in their home shrine. In Lhasa the Potala Palace, the most sacred site of Tibetan Buddhism and the official residence of the succession of Dala Lamas for centuries, was constructed on the site of the meditation cave of Songsten Gampo by Tibet's first Buddhist king. (See Aldenderfer 2005; Buffetrille 1998; Davidson and Gitlitz 2002; Gutschow et al. 2003; National Geographic Society 2009).

A rare account of the recent religious use of a cave was published by Vicki McKenzie (1998). In 1976, Diane Perry, a young English woman from London, by then a convert to Buddhism with the Tibetan name Tenzin Palmo, secluded herself in a remote cavern only about 6 ft (2.7 m) square at some 13,200 ft (4,023 m) high in the Himalayan mountains. There for the next 12 years she dedicated herself to intense meditation with the primary aim to achieve enlightenment as a woman. She survived the bitter cold of winter, an avalanche, wild animals, near starvation, and almost complete solitude. Palmo grew her own vegetables seasonally. About once a year in the summer basic food staples and other supplies such as firewood and kerosene for a small stove were delivered from a village in the valley below. Her diet consisted of tsampa, rice, lentil flour, dried vegetables, cooking oil, salt, milk powder, tea, sugar, and apples. (Tsampa is a staple food of Tibetans made from roasted barley or wheat flower commonly mixed with butter tea).

Palmo meditated and slept in a meditation box only 2.5 ft (0.7 m) square never really laying down. Her regular daily schedule encompassed 3 h of meditation each in the morning, afternoon, and evening. Rising at 3:00 am. and retiring at 10:00 pm., she only ate breakfast and lunch. Palmo remarks that actually she was never lonely or bored. Her seclusion and meditation in the cave resulted in intense experiences. At times she felt that her body melted away and flew; then she felt great awareness and clarity, visions, and bliss, detachment, inner peace and freedom; and her warmth, mental sharpness, humor, and equanimity were enhanced.

Palmo reports that: "The advantage of going to a cave is that it gives you the time and space to be able to concentrate totally" (McKenzie 1998: 198). "In a cave you face your own nature in the raw, you have to find a way of working with it, dealing with it…." (McKenzie 1998: 4). She observes that: "One goes into a retreat to understand who one really is and what the situation truly is. When one begins to understand oneself then one can truly understand others because we are all interrelated. It is very difficult to understand others while one is still caught up in the turmoil of one's emotional involvement- because we're always interpreting others from the standpoint of our own needs." (McKenzie 1998: 144). Finally, she mentions that: "It's a poverty of our time that so many people can't see beyond the material. In this age of darkness with its greed, violence and ignorance it's important there are some areas of light in the gloom, something to balance all the heaviness and darkness. To my mind the contemplatives and solitary meditators are like lighthouses beaming out love and compassion on to the world. Because their beams are focused they are very powerful. They become like generators- and they are extremely necessary." (McKenzie 1998: 196; see Palmo 2012 and also Oostergaard 2010 and Rangdrol 2001).

Incidentally, Buddhism and other religions of Asia have spread into many other parts of the world. An especially interesting example that includes Buddhist meditation retreats in caves is EcoDharma (2012). It is inspired by Buddhism and by Arne Naess's deep ecology. It is located in the Pyrenees mountains near Catalunya, Spain.

In another Asian religion, Hinduism, adherents have long revered caves as places of habitation, retreat, and meditation. The famous Ellora site is in Maharashtra state in India. It comprises a complex of 34 monasteries and temples carved into a volcanic rock cliff, including 12 Buddhist temples and monasteries (630–700 AD) and five Jain temples (800–1000 AD) as well as 17 Hindu shrines (550–780 AD). Cave numbers 6 and 10 have images from both the Buddhist and Hindu religions. This and the overlap in time of the caves from these three different religions suggest centuries of inter-religious tolerance. In the Hindu caves pilgrims remove their shoes, ring the bells, and carry a variety of offerings to present to Shiva and other deities such as milk, holy water, ghee (clarified butter), flowers, fruits, coconut, sandalwood, and money (Davidson and Gitlitz 2002).

Another well-known site, the Elephanta Caves, is on Gharapuri Island near Mumbai in India. The seven caves were carved out of stone along a hillside from the fifth to the eighth century AD. They include five Hindu shrines and two Buddhist temples. These caves were covered with plaster and then painted in bright colors, although today only traces remain. Interestingly, even ancient temples were constructed with an undecorated inner most sacred sanctuary resembling a dark cave.

Amarnath is one of the most remarkable sacred caves. It is at a height of 12,760 ft (3,889 m) on the slopes of Mount Amarnath in the Himalayas of Jammu and Kashmir. One of the most sacred sites for Hindus, the cave contains several linga (holy phallic symbols) of stalagmites composed of ice that continue to grow from a natural spring. The largest is supposed to represent the god Shiva. Amarnath is visited every July and August by tens of thousands of pilgrims. They sing religious songs, make offerings of food or small clay lamps, and appeal to Shiva for blessings (Davidson and Gitlitz 2002; Steward 2005: 407; for more on Hindu caves see Housden 1996 and Religious Portal 2012).

As a last example from Asian religions, sacred caves are also found in association with Shinto in Japan. In this religion caves are believed to have inherent power as the junction between the physical and spiritual worlds, thus they are the focus of religious and ascetic devotion. They may be considered as the residence of deities. A specific case is the Cave of the Sun Goddess Amaterasu at Amano Yasukawara Shinto shrine near Takachicho in Japan (Holm and Bowker 1994: 190–193).

25.4 Research

Sufficient examples have been sited to indicate the extent and variety of sacred caves throughout the world. However, the coverage of this subject in the literature is very uneven and fragmentary in scope and content. The majority of the literature is on Maya and Buddhist caves. Most of the research on caves has been conducted by specialists in archaeology, art history, biology (especially ecology, zoology, and chiroptology), geography, and geology (mainly geomorphology and hydrology).

Ethnographic research within cultural anthropology through participant observation with interviews to document the beliefs and activities of people in sacred caves and their surrounding environment is sorely needed. This deficit of ethnography is revealed in the literature and in the proceedings of national and international conventions of organizations specializing in speleology (for example, Northrup et al. 1998). All aspects of the culture or subculture associated with caves from the material to the symbolic need to be described. Particular sacred caves also need to be considered as part of a wider geographical, historical, cultural, and religious landscape. Cultural, social, and religious aspects of sacred caves include the insider (emic) as well as the scientific outsider (etic) perspectives which can be complementary (see Feinberg 2003 and Moyes 2012 for cases, and for especially appropriate methods in research see Grimes 1995 and Wind 1997).

25.5 Conservation

More research and action are important for the purposes of the conservation of caves throughout the world, especially show caves which have the largest number of regular visitors (Cigna 1993, 2005; Cigna and Burri 2000; Elliott 2005; Show Caves 2012). Fortunately, among the current 936 UNESCO World Heritage Sites, 29 are caves such as the Longmen Grottoes, Mogao Caves, and Yungang Grottoes in China; the Ajanta Caves, Elephanta Caves, and Ellora Caves in India; and Altamira Cave in northern Spain.

In show caves conservation is a special issue because the large numbers of visitors may seriously alter the environment. A walking person emits about as much energy as a 200-W light bulb at a temperature of around 37 °C (55 °F); thus, the total heat energy released by hundreds to thousands of daily visitors can have a detrimental impact on the environment of a cave. Another source of disturbance is the lighting system. Measures to minimize human impact include high-efficiency lamps and limiting the number of visitors and time spent as well as exiting through another route rather than retracing steps from the entrance. Hair, flakes of skin, dust from shoes, and lint from clothing can also accumulate to cause deterioration of pristine speleothems reducing them to a black mess. Visitors also exhale carbon dioxide which can increase acidity reacting with the spelothems. The famous Lascaux cave in southwestern France had to be closed in 1963 because the increased carbon dioxide from visitors was causing the deterioration of the prehistoric paintings. Instead a museum with replicas was constructed for the public (Cigna 2005: 497–498).

The Dunhuang Academy, formerly established as the Dunhuang Research Institute in 1944, is in charge of stabilizing the structures in the Magao caves, conserving the invaluable paintings and sculptures, monitoring and restricting access by visitors, and training specialists in technical aspects of cave conservation. This institution now has a full-time staff of some 500 people and from the Chinese government $38 million dollars forming 70 % of its annual budget. An international advisory board has been established while international art organizations like the Getty Conservation Institute and the Mellon International Dunhuang Archive have also helped with technical assistance. A visitor center is being developed to first orient the hundreds of thousands of tourists regarding the history of the caves, provide a virtual tour, and show simulated restorations before they go to selected caves. Today the site is only a 3-h airplane ride from Beijing and most of the visitors are Chinese. This is probably the largest sacred cave research and conservation program in the world.

Previously some of the treasures of painted murals, sculptures, and manuscript scrolls from Dunhuang were pilfered by unscrupulous foreigners to end up in museum and private collections. The caves are also endangered by nature, especially the windblown desert sands and in some water seepage. However, increasingly the primary conservation challenge is to develop sustainable tourism, the number of visitors reaching half a million by 2006. For example, humidity fluctuations caused by mass tourism have discolored some of the paintings, while mold has grown in some areas as well. Today the number of tourists, the particular caves that they can visit, and their time in the caves are all restricted (Agnew 1997, 2004; Cotter 2008; Friends of Dunhuang 2012; Larmer and Law 2010).

A second major initiative in conservation for the Dunhuang caves is the multinational and multi-institutional high quality digital reconstructions of the murals, sculptures, and texts by the Mellon International Dunhuang Archive including a team from Northwestern University. By now the collection has 75,483 images comprising 95 % of the art works recorded for posterity (International Dunhuang Project 2012; Mellon International Dunhuang Archive 2012).

Most sacred caves are associated with karst terrain and often it is exploited in harmful or even destructive ways. In Southeast Asia, for instance, quarrying limestone is a primary threat. Improved land use planning is required to prevent karst resources from being exhausted in developing regions. The landscape needs to have a comprehensive assessment of its biological, historical, cultural, religious, and economic values. Improved legislation and enforcement for conservation is sorely needed. In addition, increased research, activities, and advocacy are necessary to promote more public awareness of the importance of karst terrain and its associated caves. To adequately protect a cave the adjacent landscape, and in some cases much more, needs to be protected as well (Clements et al. 2006; Elliott 2005).

Finally, it should be noted that caves in areas of violent conflict also require special attention. A case in point is the wanton and shocking destruction of the two giant carved statues of the standing Buddha by the Taliban in Afghanistan at Bamiyan in March 2001. It is not clear how many of the 750 Buddhist caves carved in the sandstone, including five seated Buddha statues, and some 50 caves with painted murals, have been damaged or destroyed in the warfare in this area (Higuchi and Barnes 1995; Shambhala Sun 2002).

25.6 Conclusion

As truly extraordinary places in many respects, caves attract religious and spiritual as well as scientific, adventurer, and tourist attention. The anomalous characteristics of caves contribute to the sacredness of some of them, although historical, cultural, and/or religious factors may be involved as well and likely more influential. While most religions have some sacred caves, by far the greatest number is associated with the Maya and Buddhism, judging from the available literature. Numerous caves have been recognized as World Heritage Sites because they are exceptional treasures of history, culture, and/or religion. Caves may be threatened by tourism and economic development, and some are contested sites, a few endangered, damaged, or even destroyed by warfare. Surely both sacred and secular caves merit far more research and protection throughout the world. In various ways and degrees the archaeology, biology, geography, geology, and history of caves have been pursued, but the ethnography of caves is embryonic at best, a new frontier for future research that might be called ethnospeleology.

25.7 Resources

For further information on caves see in particular Culver and White 2005; Gunn 2004; Moyes 2012; Northup et al. 1998; Steward 2005; Waltman 2008; Weinberg 1986. Two previous encyclopedia entries on sacred caves are by Heyden 2005 and MacRichie 1924, the former an especially useful survey. Articles may be of interest in specialized periodicals such as the *International Journal of Speleology* and the *Journal of Cave and Karst Studies*. Among special issues on caves in other journals are *Asian Perspectives* (Spring 2005, 44(1), 1–245), *Expedition* (2005, 47(3), 8–42), *Geoarchaeology* (1997, 12(6), 501–750), and *World Archaeology* (2009, 41(2), 191–344). In addition, the *National Geographic* magazine has published superb popular articles on caves and the National Geographic Society has produced some extraordinary documentary films about them. There are also websites and newsletters of international, national, state, and regional organizations of cavers that can be located through Google.com. The Cave Research Foundation is an American private, non-profit group dedicated to cave exploration, research, and conservation. The work of the Hoffman Environmental Research Institute (2012) at Western Kentucky University in Bowling Green encompasses a B.S. degree in the Department of Geology and Geography with a concentration in Karst Geoscience. Websites for specific show caves can be located through searching Google.com, and, in some cases, videos are available on YouTube. Beyond the websites already mentioned above, of special relevance are these selected websites, all accessed on May 31, 2012:

- Cave Research Foundation www.cave-research.org/
- International Union of Speleology www.uis-speleo.org/
- National Cave and Karst Research Institute www.nckri.org/index.htm
- National Geographic Traveler, 2009, Ten Sacred Caves http://traveler.national-geographic.com/books-excerpts/ten-sacred-caves-text.
- National Speleological Society (U.S.A.) www.caves.org/
- Sacred Caves and Cave Temples www.sacred-destinations.com/categories/sacred-caves
- Show Caveswww.showcaves.com

References

Agnew, N. (Ed.). (1997). Conservation of ancient sites on the Silk Road. In *Proceedings of an international conference on the conservation of grotto sites*. Los Angeles: The Getty Conservation Institute. www.getty.edu/conservation/publications_resources/pdf_publications/. Accessed 31 May 2012.

Agnew, N. (Ed.). (2004). Conservation of ancient sites on the Silk Road. In *Proceedings of the second international conference on the conservation of grotto sites, Mogao Grottoes, Dunhuang, People's Republic of China*. Los Angeles: The Getty Conservation Institute. www.getty.edu/conservation/publications_resources/pdf_publications/. Accessed 31 May 2012.

Aldenderfer, M. (2005, Winter). Caves as sacred on the Tibetan plateau. *Expedition, 47*(3), 8–13.

Arias, P. (2009). Rites in the dark? An evaluation of the current evidence for ritual areas at Magdalenian cave sites. *World Archaeology, 41*(2), 262–294.

Aujoulat, N. (2005). *Lascaux: Movement, space and time*. New York: Harry N. Abrams.

Awe, J. J. (2006). *Mayan cities and sacred caves: A guide to the Mayan sites of Belize*. Belize: Cubola.

Bahn, P. G. (2007). *Cave art: A guide to the decorated caves of Europe*. London: Frances Lincoln.

Barker, G. (2005). The Neolithic cemeteries of Niah cave, Sarawak. *Expedition, 47*(3), 15–19.

Barker, G., Reynolds, T., & Gilbertson, D. (Eds.). (2005, Spring). The human use of caves in peninsular and island Southeast Asia. *Asian Perspectives, 44*(1), 1–231.

Bassie-Sweet, K. (1996). *At the edge of the world: Caves and late classic Maya world view*. Norman: University of Oklahoma Press.

Behl, B. K. (1998). *The Ajanta caves: Artistic wonder of ancient Buddhist India*. New York: Harry N. Abrams.

Bellows, K., et al. (2008). *Sacred places of a lifetime: 500 of the world's most peaceful and powerful destinations*. Washington, DC: National Geographic Society.

Beltran, A. (Ed.). (1999). *The cave of Altamira*. New York: Harry Abrams.

Bollt, R. (2005). Tricks, traps, and tunnel: A study of refuge caves on Hawaii island. *Hawaiian Archaeology, 10*, 96–114.

Bonsall, C., & Tolan-Smith, C. (Eds.). (1997). *The human use of caves* (British Archaeological Reports international series 667). Oxford: Archaeopress.

Brady, J. E. (1997, September). Settlement configuration and cosmology: The role of caves at Dos Pilas. *American Anthropologist, 99*(3), 602–618.

Brady, J. E., & Prufer, K. M. (Eds.). (2005). *In the Maw of the earth monster: Mesoamerican ritual cave use*. Austin: University of Texas Press.

Brockman, N. C. (2011). *Encyclopedia of sacred places* (2nd ed.). Santa Barbara: ABC-CLIO, Inc.

Brucker, R. W. (2005). Mammoth caves system. In D. C. Culver & W. B. White (Eds.), *Encyclopedia of caves* (pp. 351–355). San Diego: Elsevier Academic Press.

Buffetrille, K. (1998). Reflections on pilgrimages to sacred mountains, lakes and caves. In A. McKay (Ed.), *Pilgrimage in Tibet* (pp. 18–34). Richmond: Curzon.

Burney, D. A. (2010). *Back to the future in the caves of Kaua'i: A scientist's adventures in the dark*. New Haven: Yale University Press.

Burney, D. A., & Pila Kikuchi, W. K. (2006). A millennium of human activity at Makauwahi Cave, Maha'ulepu, Kaua'i. *Human Ecology, 34*(2), 219–247.

Caswell, J. O. (2000). Cave temples and monasteries in India and China. In W. M. Johnston (Ed.), *Encyclopedia of monasticism* (pp. 255–263). Chicago: Fitzroy Dearborn Publishers.

Chauvet, J.-M., Deschamps, E. B., & Hillaire, C. (1996). *Chauvet cave: The discovery of the world's oldest paintings*. London: Thames and Hudson.

Chladek, S. (2011). *Exploring Maya ritual caves: Dark secrets from the Maya underworld*. Walnut Creek: AltaMira Press.

Cigna, A. A. (1993). Environmental management of tourist caves. *Environmental Geology, 21*, 173–180.

Cigna, A. A. (2005). Show caves. In D. C. Culver & W. B. White (Eds.), *Encyclopedia of caves* (pp. 495–500). San Diego: Elsevier Academic Press.

Cigna, A. A., & Burri, E. (2000). Development, management, and economy of show caves. *International Journal of Speleology, 29B*(1–4), 1–27.

Clements, R., Sodhi, N. S., Schilthuizen, M., & Ng, P. K. L. (2006, September). Limestone karsts of Southeast Asia: Imperiled arks of biodiversity. *BioScience, 56*(9), 733–742.

Clottes, J. (1998, March). The mind in the cave—The cave in the mind: Altered consciousness in the Upper Paleolithic. *Anthropology of Consciousness, 9*(1), 13–21.

Clottes, J. (2003). *Chauvet cave*. Salt Lake City: University of Utah Press.

Clottes, J. (2010). *Cave art*. New York: Phaidon Press.

Clottes, J., & Lewis-Williams, D. (1998). *The shamans of prehistory: Trance and magic in the painted caves*. New York: Harry N. Abrams.

Cotter, H. (2008, July 6). Buddha's caves. *New York Times* AR1. www.nytimes.com/2008/07/06/arts/design/06cott.html?pagewanted=1&_r=3

Culver, D. C., & White, W. B. (Eds.). (2005). *Encyclopedia of caves*. San Diego: Elsevier Academic Press.

Curtis, G. (2007). *The cave painters: Probing the mysteries of the world's first artists*. New York: Anchor.

Davidson, L. K., & Gitlitz, D. M. (Eds.). (2002). *Pilgrimage from the Ganges to Graceland: An encyclopedia*. Santa Barbara: ABC-CLIO, Inc.

Desdemaines-Hugon, C. (2010). *Stepping-stones: A journey through the ice age caves of the Dordogne*. New Haven: Yale University Press.

Dhammika, S. (1998, May). Sri Lanka's ancient cave monasteries. *The Middle Way, 73*(1), 31–36.

Eco-Dharma. (2012). Eco-Dharma. Catalunya, Spain. www.ecodharma.com

Eliade, M. (1959). *The sacred and the profane: The nature of religion*. New York: Harcourt Brace & Company.

Elliott, W. R. (2005). Protecting caves and cave life. In D. C. Culver & W. B. White (Eds.), *Encyclopedia of caves* (pp. 458–467). San Diego: Elsevier Academic Press.

Feinberg, B. (2003). *The devil's book of culture: History, mushrooms, and caves in southern Mexico*. Austin: University of Texas Press.

Ford, D. C., & Williams, P. (2007). *Karst hydrogeology and geomorphology*. New York: Wiley.

Francke, L. B. (2005). *On the road with Francis of Assisi: A timeless journey through Umbria and Tuscany and beyond*. New York: Random House.

Freund, R. A. (2004). *Secrets of the cave letters: Rediscovering a Dead Sea mystery*. Amherst: Humanity Books.

Friends of Dunhuang. (2012). www.friendsofdunhuang.org. Accessed 31 May 2012.

Galanti, G.-A. (1998, March). State of consciousness and rock cave/art. *Anthropology of Consciousness, 9*(1), 1–2.

Goreme Open Air Museum. (2012). www.goreme.com/goreme-open-air-museum.php. Accessed 31 May 2012.

Gospel Hall. (2012). www.gospelhall.org/bible-reference/bible-verses-about/bible-verses-about-caves.html. Accessed 31 May 2012.

Grassie, W. (2010). *The new science of religion: Exploring spirituality from the outside in and the bottom up*. New York: Palgrave Macmillan.

Grimes, R. L. (1995). *Beginnings in ritual studies*. Columbia: University of South Carolina Press.

Gunn, J. (Ed.). (2004). *Encyclopedia of caves and karst science*. New York: Fitroy Dearborn.

Gutschow, N., Michaels, A., Ramble, C., & Steinkeller, E. (Eds.). (2003). *Sacred landscapes of the Himalaya*. Vienna: Austrian Academy of Sciences Press.

Herzog, W. (2011). *Cave of forgotten dames: Humanity's lost masterpieces*. New York: MX Creative Differences Productions, Inc. (90 minutes).

Heyd, T., & Clegg, J. (Eds.). (2005). *Aesthetics and rock art*. New York: Ashgate Pub Ltd.

Heyden, D. (1975). An interpretation of the cave underneath the pyramid of the sun in Teotihuacan, Mexico. *American Anthropologist, 40*, 131–147.

Heyden, D. (2005). Caves. In L. Jones (Editor-in-Chief), *Encyclopedia of religion* (Vol. 3, pp. 1468–1473). New York: Thomson Gale.

Higuchi, T., & Barnes, G. (1995). Bamiyan Buddhist cave temples in Afghanistan. *World Archaeology, 27*(2), 282–302.

Hoffman Environmental Research Institute. (2012). Department of Geology and Geography, Bowling Green: Western Kentucky University. www.wku.edu/hoffman/

Holm, J., & Bowker, J. (Eds.). (1994). *Sacred place*. London: Pinter.

Housden, R. (1996). *Travels through sacred India*. San Francisco: Thorsons.

Hurd, B. (2008). *Entering the stone: On caves and feeling through the dark*. Athens: University of Georgia Press.

International Dunhuang Project. (2012). London. http://idp.bl.uk/. Accessed 31 May 2012.

Jamkhedkar, A. P. (2009). *Ajanta*. New Delhi: Oxford University Press.

Jenkins, M., & Toensing, A. (2012, February). Cave people of Papua New Guinea. *National Geographic, 221*(2), 126–141.

Jinshi, F. (2010). *The caves of Dunhuang*. London: Dunhuang Academy of London Editions.

Katz, N. (2003). The modernization of Sinhalese Buddhism as reflected in the Dambulla Cave Temples. In S. Heine & C. S. Prebish (Eds.), *Buddhism in the modern world: Adaptations of an ancient tradition* (pp. 29–44). New York: Oxford University Press.

Kennedy, J., & Brady, J. (1997). Into the Netherworld of island earth: A Reevaluation of refuge caves in ancient Hawaiian society. *Geoarchaeology, 12*(6), 641–655.

Klein, D. D., & Klien, A. (2012). Cave temples at Ellora, India. www.artstor.org/what-is-artstor/w--html/col-ellora.shtml

Krajick, K., & Littschwager, D. (2007, September). Discoveries in the ark. *National Geographic, 212*(3), 134–147.

Larmer, B., & Law, T. (2010, June). Caves of faith. *National Geographic, 217*(6), 124–145.

Layton, R. (1986). *Uluru: An aboriginal history of Ayres rock*. Canberra: Australian Institute of Aboriginal Studies.

Lewis-Williams, D. J. (2002). *The mind in the cave*. New York: Thames & Hudson Inc.

Lewis-Williams, D. J., & Clottes, J. (2008). The mind in the cave—The cave in the mind: Altered consciousness in the Upper Paleolithic. *Anthropology of Consciousness, 9*(1), 13–21.

MacRichie, D. (1924). Caves. In J. Hastings (Ed.), *Encyclopedia of religion and ethics* (Vol. III, pp. 266–270). New York: Charles Scribner's Sons.

Malaandra, G. (1993). *Unfolding a Mandala: The Buddhist cave temples at Ellora*. Albany: State University of New York Press.

McKenzie, V. (1998). *Cave in the snow: Western woman's quest for enlightenment*. New York: Bloomsbury Publishing.

Mellon International Dunhuang Archive. (2012). www.artstor.org/what-is-artstor/w-tml/col-ellon-dunhuang.shtml

Mitra, D. (2004). *Ajanta*. New Delhi: Archaeological Survey of India.

Morwood, M. J., & Hobbs, D. R. (Eds.). (1992). *Rock art and ethnography* (Occasional AURA Publication No. 5). Melbourne: Australia Archaeological Publications.

Moyes, H. (Ed.). (2012). *Sacred darkness: A global perspective on the ritual use of caves*. Boulder: University of Colorado Press.

Munier, C. (1998). *Sacred rocks and Buddhist caves in Thailand*. Bangkok: White Lotus Press.

National Geographic Society. (2009). *Secrets of Shangri-La: Quest for sacred caves*. Washington, DC: National Geographic Society (57 minutes). www.pbs.org/programs/secrets-shangri-la/. Accessed 31 May 2012.

Northrup, D. E., Mobley, E. D., Ingham, K. L., III, & Mixon, W. (1998). *A guide to the speleological literature of the English language 1794–1996*. Dayton: Cave Books.

Oliver, P. (2009). *Mysticism: A guide for the perplexed*. New York: Continuum International Publishing Group.

O'Neill, T., & Bhl, B. K. (2008, January). Faces of the divine. *National Geographic, 213*(1), 122–139.

Oostergaard, J. (2010). A topographic event: A Buddhist Lama's perception of a pilgrimage cave. *Social Analysis, 4*(3), 64–75.
Palmer, M. (1996). Buddhist caves and grottoes. In his *Travels through sacred China* (pp. 207–214). San Francisco: HarperCollins Publishers/Thorsons.
Palmer, A. N. (2007). *Cave geology*. St. Louis: Cave Books.
Palmo, T. (2012). Tenzin Palmo. www.tenzinpalmo.com
Petryshyn, J. T., & Colas, P. R. (2005). A Lacandon religious ritual in the cave of the God Tsibana at the holy lake of Mensabok in the rainforest of Chiapas. In J. E. Brady & K. M. Pufer (Eds.), *In the maw of the earth monster: Mesoamerican ritual cave use* (pp. 328–341). Austin: University of Texas Press.
Rangdrol, S. T. (2001). Retreat at hermit's cave. In his *The life of Shabkar: The autobiography of a Tibetan yogin* (M. Ricard, Trans., pp. 49–64). Ithaca: Snow Lion Publications.
Religious Portal. (2012). Sacred caves of India. www.religiousportal.com/SacredCaves.html
Saint-Laurent, G. (2000). *Spirituality and world religions: A comparative introduction*. Mountain View: Mayfield Publishing Company.
Sears, J. F. (1989). Mammoth cave: Theater of the cosmic. In his *Sacred places: American tourist attractions in the nineteenth century* (pp. 1–48). Amherst: University of Massachusetts Press.
Shambhala Sun. (2002, March). Buddhist treasures of Afghanistan. *Shambhala Sun, 10*(4), 36–43.
Shea, N., & Peter, C. (2008, November). Crystal palace. *National Geographic, 214*(5), 64–77.
Show Caves. (2012). Show Caves. www.showcaves.com
Sidisunthorn, P., Gardner, S., & Smart, D. (2006). *Caves of northern Thailand*. Bangkok: River Books Co., Ltd.
Solecki, R. S. (1975). Shanidar IV, A Neanderthal flower burial in northern Iraq. *Science, 190*(4217), 880–881.
Spink, W. M. (2009). *Ajanta history and development*. Boston: Brill.
Sponsel, L. E. (2007a). Sacred places and biodiversity conservation. In C. J. Cleveland, et al. (Eds.), *Encyclopedia of earth*. www.eoearth.org. Accessed 31 May 2012.
Sponsel, L. E. (2007b). Religion, nature and environmentalism. In J. Cutler Cleveland, et al. (Eds.), *Encyclopedia of earth*. www.eoearth.org. Accessed 31 May 2012.
Sponsel, L. E. (2012). *Spiritual ecology: A quiet evolution*. Santa Barbara: ABC-CLIO/Praeger. www.spiritualecology.info. Accessed 31 May 2012.
Sponsel, L. E., & Natadecha-Sponsel, P. (2004). Illuminating darkness: The monk-cave-bat-ecosystem complex in Thailand. In R. S. Gottlieb (Ed.), *This sacred earth: Religion, nature, environment* (pp. 134–144). New York: Routledge.
Spoto, D. (2002). *Reluctant saint: The life of Francis Assisi*. New York: Penguin.
Srinivasan, P. R. (2007). *Ellora*. New Delhi: Archaeological Survey of India.
Steward, P. J. (2005). Myth and legend, caves. In D. C. Culver & W. White (Eds.), *Encyclopedia of caves* (pp. 406–408). San Diego: Elsevier Academic Press.
Stone, A., & Brady, J. E. (2005). Maya caves. In D. C. Culver & W. White (Eds.), *Encyclopedia of caves* (pp. 366–369). San Diego: Elsevier Academic Press.
Swan, J. A. (1990). *Sacred places: How the living earth seeks our friendship*. Santa Fe: Bear & Company Publishing.
Swenson, D. S. (2009). *Society, spirituality, and the sacred: A social scientific introduction*. Toronto: University of Toronto Press.
Tacon, P. S. C. (2005). The world of ancient ancestors: Australian aboriginal caves and other realms within rock. *Expedition, 47*(3), 37–42.
Tsian, K. R. (2010). *Echoes of the past: The Buddhist cave temples of Xiantangshan*. Chicago: University of Chicago Smart Museum of Art.
Tulku, T. (1994). Cave temples of western India. In his *Holy places of the Buddha* (pp. 270–296). Berkeley: Dharma Publishing.
Turner, V. (1969). *The ritual process: Structure and anti-structure*. Chicago: Aldine.
Ullmann-Margalit, E. (2006). *Out of the cave: A philosophical inquiry into the Dead Sea Scrolls research*. Cambridge: Harvard University.

Ustinova, Y. (2009). *Caves and the ancient Greek mind: Descending in the search for ultimate truth*. New York: Oxford University Press.

Waltman, T. (2008). *Great caves of the world*. London: Firefly Books/Natural History Museum.

Watson, P. J. (1987). *Archeology of the Mammoth Cave areas*. St. Louis: Cave Books.

Weinberg, F. M. (1986). *The cave: The evolution of a metaphoric field from Homer to Ariosto*. New York: P. Lang.

White, W. B., & Culver, D. C. (2005). Cave, definition of. In D. C. Culver & W. B. White (Eds.), *Encyclopedia of caves* (pp. 81–85). San Diego: Elsevier Academic Press.

Whitfield, R. (2000). *Cave temples of Dunhuang: Art and history of the Silk Road*. Los Angeles: J. Paul Getty Trust.

Whitfield, R., Whitfield, S., & Agnew, N. (2000). *Cave temples of Mogao: Art and history on the Silk Road*. Los Angeles: The Getty Conservation Institute.

Whitley, D. S. (1998, March). Cognitive neuroscience, Shamanism and the rock art of native California. *Anthropology of Consciousness, 9*(1), 22–37.

Whitley, D. S. (2009). *Cave paintings and the human spirit: The origins of creativity and belief*. New York: Prometheus Books.

Winchester, S. (2006, November 6). In the holy caves of India. *New York Times*. http://travel.nytimes.com/2006/11/05/travel/05caves.html?pagewanted=all. Accessed 31 May 2012.

Wind, J. (1997). *Places of worship: Exploring their history*. Walnut Creek: AltaMira Press.

Zhang, P., Cheng, H., Edwards, R. L., Chen, F., Wang, Y., & Yang, X., et al. (2008, November 7). A test of climate, sun and culture relationships from an 1810-year Chinese cave record. *Science, 322*(5903), 940.

Chapter 26
Space, Time and Heritage on a Japanese Sacred Site: The Religious Geography of Kōyasan

Ian Astley

> *Place matters because among other things it is a repository of the past. (Yi-fu Tuan 2003: 878)*

26.1 Introduction

This contribution focuses on one area of the Kii Peninsula 紀伊半島 in west-central Japan which provides examples of a number of themes of religio-geographical significance in Japanese religions: the mountain precincts of Kōyasan 高野山, granted in 816 by Emperor Saga to the Buddhist monk Kūkai 空海 (774–835) and one of the most important religious sites in Japan (Londo 2002: 19). Primarily a Shingon 真言 Buddhist site,[1] Kōyasan developed from the late eleventh century as a place of goal-oriented pilgrimage[2] and of burial *ad sanctos*.[3] Faith in Kūkai is ubiquitous, both geographically and socio-religiously.[4] Beliefs which relate to his post-mortem existence are fundamental to his cultic ubiquity and to the layout of Kōyasan. Kūkai is believed not to have died but to have entered a state of contemplation, *nyūjō* 入定,

---

[1] Established in Japan in the early ninth century, Shingon is one of the Japanese forms of Buddhism often referred to as Tantric or Vajrayāna Buddhism, since it originates in the tantra class of Indian religious literature in a process often referred to as the "Hinduization" of Buddhism. "Esoteric Buddhism" is increasingly being used in the scholarly literature. The most up-to-date scholarship is to be found in Orzech et al. (2010).

[2] One may distinguish two types of pilgrimage: procession to a particular site (mairi 参り, mōde 詣で, and sankei 参詣); and circumambulatory pilgrimage (meguri 巡り, junrei 巡礼, and sanpai 参拝).

[3] The phrase was used by Gregory Schopen, following Brown, in his detailed analysis of Kōyasan (Schopen 1987; Brown 1981). Emperors, shōgun, prominent samurai, daimyō, poets, artists, and religious figures from a thousand years and more are at rest in Kōyasan's cemetery, Oku no In 奥の院.

[4] He is known throughout Japan as Kōbō Daishi 弘法大師, hence the term Daishi shinkō 大師信仰 "Daishi worship" to describe his cult; the phenomenon is dealt with extensively in Hinonishi (1988).

I. Astley (✉)
Asian Studies, University of Edinburgh, Edinburgh EH8 9LH, UK
e-mail: ian.astley@ed.ac.uk

a component of the path of meditation in Buddhist dogma and practice. For this reason, the site has become a focus of a kind of millennial belief, on the basis of awaiting the coming of the bodhisattva Maitreya when all beings attain enlightenment to the teaching of the Buddha's Greater Vehicle, the Mahāyāna, and of belief in the Pure Land, from which all beings attain enlightenment. Kōyasan has been seen variously as a Pure Land of Miroku [Maitreya] (弥勒), of Esoteric Splendor (密厳), and of the buddha Amida [Amitābha] (阿弥陀).[5] By extension it is also a *gokuraku* 極楽, a place of *post-mortem* bliss, a paradise. These presentations of Kōyasan thus lead on from orthodox Buddhist scholastic, cosmological and contemplative concepts, and mesh with Buddhist and popular millennial beliefs (Astley 2004).[6] Through the centuries, pilgrims have trodden paths through foothills and mountains punctuated by temples and shrines, ending at the mausoleum of Kōbō Daishi. By the 1930s, the modernization which had commenced in the wake of the Restoration of imperial power in 1868, brought travellers by rail to that Pure Land; although some few still do walk along the ancient footpaths up to the main gate at the western entrance of the precincts, the Daimon 大門, most take the luxury express. Many do so as part of a well organized package tour. We shall see that although the technology has changed, two important themes have not: the propensity to ascribe to the landscape a set of ideas and values, a propensity that is then acted out in practice; and the established religious authority's need to present those ideas in an attractive manner.

Many of the observations presented here that relate to Kōyasan, stem from time spent in the mountain community. My first visits were in spring 1982 and spring 1983, followed by a period of residence from August 1983 to December 1984, which included the 50-yearly *go-onki* 御遠忌 celebrations of Kūkai's death. These spanned the 49 days from the 21st March and various special events occurred throughout the year. I returned to Kōyasan on a number of occasions between 1989 and 2006 and my most recent information is from winter 2009 and spring 2012. Although there are many studies of the history of Kōyasan in Japanese as well as work in English,[7] a full anthropological study of Buddhism on the mountain, with extensive collection and analysis of statistics and the conducting of formal interviews, is lacking but falls outside the scope of this piece.

---

[5] These interpretations are found in publicity of all kinds, for example an exhibition of antique maps of Kōyasan held at its official museum, the Reihōkan, in 2005–2006 (www.reihokan.or.jp/tenrankai/list_tokubetsu/2005_12.html; retrieved 2012-10-15).

[6] Indeed, the latest Japanese nomination for UNESCO World Heritage Status, Hiraizumi 平泉 in north-east Japan (2011), was couched in precisely such terms: "Hiraizumi: Temples, Gardens and Archaeological Sites Representing the Buddhist Pure Land". The Japanese title is more specific, describing the site as both a "Buddha-land" (*bukkokudo* 仏国土) and a "Pure Land" (*jōdo*), http://whc.unesco.org/en/list/1277 and www.bunka.go.jp/bunkazai/shoukai/sekai_isan.html (both retrieved 2012-10-15; in English only a general description of the Agency's work on the preservation and utilization of cultural properties is available).

[7] These include Londo (2002), Hinonishi (2002), and a recent PhD thesis on pilgrimage to Kōyasan in the eleventh to twelfth centuries (Lindsay 2012).

Over-arching conclusions about the relationship between Japan's geography and the religious life of its people would involve over-ambitious extrapolations, but concentrating on Kōyasan, its geographic composition, historical resonances, and the activities of visitors can provide working perspectives, with indicators both to the issues that have wider significance for our understanding of Japanese religions (especially in the light of urbanization and demographic change) and for our appreciation of the way in which human individuals and societies relate metaphysically to their environment. An important lever for our understanding of these issues is Kōyasan's involvement in the election of sites in the Kii Peninsula to the register of World Heritage sites.

26.2 The Town of Kōyasan

Kōyasan is the central part of Kōyachō, a municipal authority that covers 137 sq km (53 sq mi) of mainly forested orogenic terrain in Wakayama Prefecture in west-central Japan, to the south of the Nara plain (Fig. 26.1). Historically and in terms of the geographic relationship to the ancient capitals, it represents the antithesis of court life, and has nostalgic associations of solitude and the beauty of the wild (Astley 2003). These still have resonance today and are a prominent feature in the marketing of this heritage site, whether for visits to the mountain or for museum exhibitions and publishing ventures. However, these efforts have been taking place against the background of population drain over the last decades, with younger people especially leaving to seek employment elsewhere. The student population, a significant part of the town's economic base when I first attended in the early 1980s, has also dwindled. The university, which is one of the seminaries that train the next generation of priests, has an undergraduate population of around 30 per annum as of April 2012 (the postgraduate figures are relatively healthy), which stands in stark contrast to the building projects of the mid-1980s. Population figures from the municipal authority show a population of 9,324 in 1960 (namely, the point when Japan was poised for its economic recovery after the Pacific War). By the end of the fiscal year 2010–2011 it had fallen to 3,797, 40.72 % of its 1960 level. A first significant slump occurred in 1965 (17 %). The 10 years from 1995 were also sobering: by 2000 the population had decreased by 16.1 % and the next 5 years saw a further 13.5 % reduction.[8] Thus the current population stands at 59.49 % of the 1995 level, the year of the Aum subway attack.[9]

Thus, in the 30 years or so that I have been making my observations, the town's population has halved. It is against this background that much of what relates to the

---

[8] These figures are published on the town's official website, www.town.koya.wakayama.jp/jinko.html (retrieved 2012-10-14).

[9] A causal connection would have to be corroborated by cross-reference to a number of other socio-religious and demographic factors but would not be inconsistent with other expressions of religious skepticism among the Japanese following that event.

Fig. 26.1 Kōyasan ezu (an illustrated map of Kōyasan, 1681). In this monochrome woodblock print, east is at the *top*, north to the *left*. Kūkai's mausoleum is at the *top left*. The pilgrim would arrive from the south at the Daimon (not shown) and proceed past the *garan* and into the cemetery. The perspective emphasizes the pre-eminence of the mausoleum (Source: Ian Astley's collection)

complex's current profile, with the emphasis on its heritage and a marketing strategy that targets cultural tourism, is to be understood. For example, internet search engines will typically return results for "Koyasan" that provide information on and booking procedures for tourist accommodation on the mountain, marketed in a highly professional manner for both domestic and non-Japanese consumers, with cultural themes and natural beauty to the fore.

The current capacity of the 52 *shukubō* 宿坊 on the mountain is just over 6,000, that is, almost twice the resident population of the whole authority. A *shukubō* is traditionally a monk's place of residence but the term now encompasses accommodation for any visitor to a temple; thus to all intents and purposes these are hotels, although the original idea was to give visitors an authentic sense of the monastic environment. A typical stay costs in the range of JPY20,000 to JPY30,000 per night for two people (approx. US$250 to US$380 as of Oct. 2012), comparable to very comfortable accommodation in major city hotels. Much is made of the food: *shōjin ryōri* 精進料理, traditionally the diet of an ascetic and containing no meat, poultry, fish, or dairy products. However, it is marketed as a plentiful gourmet offering and is intended as a high point of the stay, along with an impressive array of traditional, Japanese-style hot baths.[10] In short, although visitors to Kōyasan have long been a part of its economy, the tourist traffic is very much a lifeline for the economy of the town and the current tourist profile is directly related to the mountain's place in the UNESCO listing.

26.2.1 Kōyasan: The UNESCO World Heritage Program

Kōyasan is an important component of the complex of ancient sites on the Kii Peninsula which was inscribed under the UNESCO World Heritage program in 2004: "Sacred Sites and Pilgrimage Routes in the Kii Mountain Range".[11] The Inscription includes the Yoshino–Ōmine–Kumano complex (itself designated a national park as early as 1932) as well as the Kumano Sanzan (Moerman 2005, 2010), drawing on the historical role of the Kii Peninsula as a metaphysical adjunct to the centers of power in the Nara plains.[12] As such, it was international recognition of the remarkable history, institutions, ideas and practices which have built up over the centuries on the peninsula. Whilst no figures are given for the individual sites, UNESCO's information quotes 15 million visitors annually for the whole of the peninsula and for both religious and non-religious visits (although there is no evidence that this figure has been updated since the Inscription). This is a substantial number but the distinction is noteworthy: it implies firstly a realization that the

---

[10] A case in point is the discovery of hot mineral water beneath Kōyasan, the rights to which were bought by the largest *shukubō* on the mountain, Fukuchiin 福智院. Since 2005 this has defined the temple, which now markets itself very successfully as "Kōyasan Onsen Fukuchiin".

[11] Kii Sanchi no Reijō to Sankeimichi 紀伊山地の霊場と参詣道; ref. point 33.836944 [N 33 50 13] 135.776389 [E 135 46 35] (Hongū, Kumano); ref. no. 1142; http://whc.unesco.org/en/list/1142 (retrieved 2012-10-15); see UNESCO document 28COM 14B.28, 28 June–7 July 2004.

[12] Essential reading would include the special issue of the *Japanese Journal of Religious Studies* dedicated to pilgrimage, containing both specific articles on the Yoshino–Ōmine–Kumano route and key studies on pilgrimage in Japan (Reader and Swanson 1997).

religious attitudes of the Japanese have changed such that a significant sector of the population would find a religious stance off-putting; and secondly that non-Japanese visitors need to be attracted first and foremost by an emphasis on the cultural aspects of a visit to the mountain.

The Inscription brings to the fore a Japanese perception of the intimate relationship between the physical geography of Japan and religious attitudes and activities.[13] It also emphasizes non-religious travel to these ancient centers, indicative of the increased prevalence of secular attitudes as well as the current attention to environmental concerns of global resonance.[14] I do not wish to imply that Japanese sites of religious significance were never visited for non-religious purposes[15]: the point is that in Japan it has now become markedly more politic to make the distinction.

The primary focus on the sites' religious nature combines discrete elements of Japan's major religious traditions, Shinto and Buddhism. Shinto 神道, as one might expect, is far more complex than the nature-worship presented in the Inscription: for example, the Chinese systems of statecraft adopted from the seventh–ninth centuries CE are a fundamental part of Shinto, as are autochthonous ritual cycles and human responses to divine presence in the natural environment. Still, the more simplistic line taken by the Inscription sits more easily with the mainstream projection of an exotic Japan promoted by official bodies such as the Agency for Cultural Affairs (which collated and submitted the definitive documents) and the Japan Tourist Board (JTB).[16] The additional emphasis on secular aspects in the Inscription is indicative of changing Japanese attitudes to religion, which must be understood in terms of the marked and incontrovertible decline in attitudes to matters religious that has characterized the post-war period (Reader 2012a, b).[17]

It should be noted in parenthesis that the Bunkachō has a number of proposed World Heritage nominations. The Kii Heritage website, for example, states that the Shikoku regional authorities are preparing an application for inscription, under the title "Shikoku Hachijūhakkasho Reijō to Henromichi 四国八十八箇所霊場と遍路

---

[13] A comprehensive thematic and bibliographic survey of geography and environment in Japanese religions has been provided by Ambros (2006).

[14] These two aspects of religious travel—the pilgrim's and the tourist's, the devout and the ludic— have long been noted in other contexts (Morinis 1981, 1992).

[15] The Tokugawa period (1603–1868) for example saw a dramatic increase in pilgrimages, the ludic—even antinomian—element of which increased as the Tokugawa regime lost its grip on power and hence its control over people's movements. While the system never really achieved uniformity or full control over the movement of the populace, the intent was clear and its restrictions had to be negotiated (Vaporis 1994).

[16] The best introduction to Shinto is by Breen and Teeuwen (2010); more detailed studies are to be found in their earlier volume, Breen and Teeuwen (2000); for a discussion of the role of space and the nature of kami, see Nelson (2000), ch. 3. The Agency for Cultural Affairs is the official English name of the government office, Bunkachō 文化庁.

[17] The decline is not merely a consequence of the Aum attacks on the Tokyo subway system in 1995, as it is often presented, though that incident was a significant factor; cf. the slump in the population of Kōyachō, noted above.

道" [The Eighty-Eight Sacred Sites of Shikoku and the Pilgrimage Route].[18] This is confirmed by documents published by the Agency for Cultural Affairs, in which the Shikoku pilgrimage is listed among 23 tentative candidates as of 2008. Even a cursory reading of these documents will indicate the high level of official activity that has been assigned to the nomination of as many Japanese cultural assets as possible. This applies on the level of local government, too. The rewards, however, are significant, as evidenced by the changing demographics of visitors to Kōyasan.[19]

26.3 Kōyasan: Pilgrims' Maps and Publicity Materials

Below we will look at the image of Kōyasan presented to prospective visitors in maps and pamphlets: a pamphlet for the mountain-path that leads from the foothills at Kudoyama all the way up to Oku no In and famous for its Kamakura-period (1192–1333) way-markers, and one pamphlet for Oku no In itself; but we shall begin with some examples from the Tokugawa period.

26.3.1 Pilgrim's Maps in the Tokugawa Period

"Old maps" (*koezu* 古絵図), printed to help pilgrims find their way to and around the centers of worship, are a hive of information.[20] They were being circulated widely from the early Tokugawa period, by which point pilgrimage had become a phenomenon that spanned the gamut of social classes (Reader 2005: 113, 129–131) and was well on the way to becoming very much a secularized leisure activity (Vaporis 1994: ch. 6). Travel for individuals was also catered for, as may be seen in the existence of maps for those travelling alone.[21] Generally produced through the woodblock printing processes which were refined greatly as the era progressed, they show the various sites on the mountain and are often supplemented with pertinent information. One late-seventeenth

---

[18] The Shikoku pilgrimage is one of Japan's most famous and has long-standing connections to Kōyasan, although the way in which the sect has drawn Shikoku pilgrims to worship on Kōyasan, is a modern development (Reader 2005: 14, 108–111).

[19] See www.sekaiisan-wakayama.jp/index.html; www.bunka.go.jp/bunkashingikai/sekaibunkaisan/singi_kekka/besshi_8.html (both retrieved 2012-10-15). Despite the statement on the Kii World Heritage web-site, inscription-related activity has been put on hold since 2008, due perhaps to the fact that China, Korea and Japan, in mutual competition, were considering submitting a disproportionate number of sites for UNESCO inscription.

[20] Hinonishi (1983–8) is the prime resource for maps of Kōyasan. Guelberg (2009) has a useful online resource hosted at Waseda University (Tokyo), taken from Hinonishi's compendium, which gives a representative selection of maps from the early eighteenth century (www.f.waseda.jp/guelberg/halle2009/japomain.htm; retrieved 2012-10-14).

[21] See the map from 1734 at www.f.waseda.jp/guelberg/halle2009/e_iri_annai/annai.htm (retrieved 2012-10-14). Such maps generally contain quite a number of items in the more widely understood phonetic kana syllabary rather than the traditional Chinese characters.

century example in portrait orientation (Fig. 26.1) shows Kōyasan from the north, i.e. from the Kinki plains, and evinces a compressed perspective, with the River Kii and adjunct sites such as the Jison'in 慈尊院 at the bottom. The perspective is foreshortened to emphasize the pre-eminence of the Kōyasan complex, especially Oku no In and the mausoleum. Thus the prime cultic site dedicated to Kōbō Daishi assumes pride of place, with other places on Kōyasan subordinate to it and the related sites south of the river in clear hierarchical relationship to the precinct. Graphically the map is simple compared even to examples of a hundred or so years later, but it presents the viewer, the prospective pilgrim, with a clear portrayal of the metaphysical superiority of the mountain complex, whilst at the same time presenting the possibility of access. East is shown at the top of the illustration to give the pilgrim the impression of progress through the Daimon, past the *garan*[22] and into the cemetery.

By the nineteenth century a number of maps seem to have been in circulation, often printed in landscape orientation. For example, the *Kōyasan saiken ezu* gives a bird's eye view of the complex, with the main thoroughfare stretching from west to east.[23] The final approach to the mausoleum turns northwards, possibly an indication of the direction of the imperial capitals, possibly in imitation of the Buddha's *parinirvāṇa*, which in the Buddhist world is traditionally assigned to that direction. Another map which shares this graphic structure is that by Asai Kōei 浅井公英 from 1858 (Fig. 26.2), two copies of which are held in the Central Library at Edinburgh, as well as an abridged version by the same designer. Coloured prints published in Osaka, they detail the places to be visited and have vignettes of pilgrims going about their holy-cum-ludic business.

That Kōyasan was seen as part of a traveller's network, can be seen in a table with distances in *ri* 里 ("league"; *c*.3.965 km, 2.3 mi) to related pilgrimage centres, printed in the bottom left hand corner of the abridged version, roughly in ascending order of distance. Whilst major towns—where many of the pilgrims might originate or pass through—are listed, the principal concern is to list other major sites of pilgrimage in the Kinki and Kii areas: first, the Shinto shrine at Amano 天野神社[24] and the Buddhist temple Jison'in, already noted; then further afield there is Kokawadera 粉河寺 (by the River Kii between Hashimoto and Wakayama (Iwahana 2000)), Makiosan Sefukuji 槙尾山 施福寺 (in the Izumi 和泉 range between Hashimoto and Kawachi Nagano), Negoroji 根来寺 (founded by the medieval Shingon prelate Kakuban 覚鑁), Ki-Miidera 紀三井寺 (and its resonance with the

---

[22] This and other traditional maps show the Middle Gate (Chūmon 中門), the traditional entrance to the *garan* until it was destroyed in 1843. It is now under re-construction (as of April 2012) in anticipation of the thousand-year anniversary of the founding of Kōyasan (www.koyasan.or.jp/feature/news/110603.html; retrieved 2012-10-07). This may be seen as another example of considerable funds being used for monumental purposes (understandable in itself and part of the authorities' obligations under the World Heritage status), while the pastoral demographic is in decline.

[23] Tachibana (1813); the traditional Japanese way of reading a manuscript or illustration is from right to left, i.e. east to west in this picture. However, in this case pilgrims would have approached the complex from the west end of the town, arriving at the Grand Gate (Daimon 大門) and proceeding through the various sites of worship to the mausoleum.

[24] The spatial and religious relations between Amano Jinja and Kōyasan have been discussed by Wada Akio 和多昭夫 (1965; reprinted in Takagi and Wada 1982).

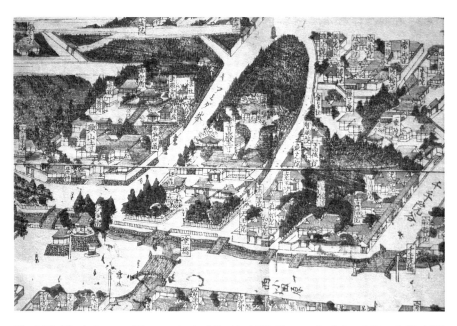

Fig. 26.2 Pilgrim's map of Kōyasan by Asai Kōei in 1858. The temples Seiganji (founded in 1593 by Hideyoshi to commemorate his mother) and Gōzanji were joined in 1869 to form the Kongōbuji, now the national headquarters of the Kōyasan branch of Shingon Buddhism. The *garan* with the two-storied pagoda, Shinto shrine and the Middle Gate can be seen to the *left* (Photo by Kevin Maclean, used with permission)

famous temple of Miidera (Onjōji 園城寺) by Lake Biwa, near Kyoto, which incidentally is affiliated with Kumano), and Hasedera 長谷寺 by Sakai, a leading commercial center during the Tokugawa period. Of importance for Kōyasan in its context of other traditional sites of pilgrimage, are the imperial retreat at Yoshino,[25] the main (Shinto) shrine at Kumano, the mountain center at Ōmine, and the Inner Shrine at Ise where Amaterasu, divine progenitor of the sovereign family ("The Sun Goddess"), is enshrined.[26]

These places fall into three categories: relevant urban centers (that is, practical information); centers which have a close connection to Kōyasan (Jison'in, Amano); and centers with a more general connection in the context of over-arching relations

---

[25] Yoshino is a long-standing topos in Japanese literature. One of Japan's best-loved writers, the peripatetic monk-poet Saigyō 西行 (1118–90), used Yoshino as one of his long-term bases— Kōyasan was another, due to his indebtedness to Kūkai. These places and the landscapes that constituted them, have long been woven into society's perception of itself, something which is brought out in a number of ways in the UNESCO inscription.

[26] This map seems to have enjoyed wide circulation: this and similar maps were re-carved and expanded and also seem to have attracted the attention of Westerners living or travelling in Japan at that time of transition and modernization. Apart from Tachibana's map mentioned above, a copy of a similarly composed work by Itō Ryūzan (surname first, in accordance with Japanese custom) is held in the British Museum (1894,0510,0.10AN590443); a further version is held in the C. V. Starr East Asian Library, University of California at Berkeley, illustrated in Proser (2010: pl. 42).

of a regional or even national significance (Yoshino, Ise, etc.). They are thus evidence of the persistence of this matrix of holy places in the middle of the nineteenth century and the popularity of travel between them as constituting a coherent group of religious sites. In the light of the UNESCO inscription, their persistence into the present day and the ratification of this matrix at government level shows us that although the permutations have changed, the place of a cultic center in a national—and now international—matrix is an important part of such sites' significance.

26.4 The Kōyasan Stone Marker Path

One of the appeals of Kōyasan is the opportunity to stroll past tokens of Japan's history. Indeed, one of the pamphlets advertising the Kōyasan Stone Marker Path (named explicitly in the UNESCO Inscription), issued for the foreign tourist by the Wakayama Prefecture World Heritage Kōya Region Association, proclaims (in English) "Hiking through History—Visit the Kōyasan Choishimichi". Issued by the tourist office of Wakayama Prefecture, it is available for free at a number of places in the area and is clearly part of a well thought out marketing strategy (Fig. 26.3). The context in which the Stone Marker Path is placed, describes the sites in terms of "the gods from the distant past (*tōi mukashi no kami-gami* 遠い昔の神々)", a phrase that evokes a

Fig. 26.3 Pamphlet of the Kōyasan Stone Marker Path for the modern hiker or pilgrim, presenting the path as part of the World Heritage site. The blurb is printed above the pilgrim's stamp for the Jison'in, typical of those the pilgrim is invited to collect at the designated centers along the way (Source: Wakayama Tourist Board public promotional pamphlet)

particular, monolithic and nostalgic view of Japan's history, a perspective that is commonly found in the Japanese media, too. The reverse contains a simple map that places the Kōyasan Stone Marker Path in the context of the pilgrim routes on the peninsula. Other pages give short explanations of the history, religious ideas, and practices associated with the region, and travel information.

The pamphlet even invites the visitor to try Shingon Buddhist meditation at one of the temples on Kōyasan (Fig. 26.4). The relevant page not only equates one's own body with the universe (as the body of Dainichi, the central Buddha in the Shingon understanding of the universe-as-sacred), in line with one of the core elements of Shingon dogma and ritual, it also presents the mysterious syllables from the Sanskrit script that have presented the exotic face of Esoteric Buddhism since it was introduced to Japan in the ninth century. The advertisement ends with an exhortation for world peace through simple Buddhist practices, thus completing the global integration of this remote center with the rest of the world in one common, spiritual quest.[27]

The front of the pamphlet consists of a clear map of the path between the foothills and Kōyasan, with all the stone markers indicated as well as important sites along the way: (1) Jison'in 慈尊院, (2) Niukanshōbu-jinja 丹生官省符神社, (3) Niutsuhime-jinja 丹生都比売神社, (4) Kōda-Jizō-dō 神田地蔵堂, (5) Yatate-jaya [tea-house] 矢立茶屋, (6) Dai-garan 大伽藍 (the central precinct on Kōyasan), and (7) Oku no In 奥の院. One of its intended uses is therefore to collect imprints from the named sites as a souvenir of the hike, much like stamping one's travel diary with the insignia provided at train stations in Japan as a record of one's journey, and conceivably resonant of the practice of having temple seals placed in a pilgrim's log.[28]

The Jison'in, said to be founded by Kūkai himself in 816, is an important site, being the first on the way up the mountain; it now boasts a substantial stone sculpture placed prominently in the main precinct in celebration of the Inscription. During a visit in spring 2012, I was informed by a temple incumbent that visitors still consist mainly of day-trippers from the Kinki region at weekends, mostly Sundays. In contrast to Kōyasan itself there are hardly any foreign visitors.[29] Whilst

---

[27] Interestingly, too, the word "spiritual" occurs in the heading, not as a normal Japanese word but as a transcription of the English. This sits strangely alongside the word o-susume, "our recommendation [for you, the customer]", which is often used to introduce special offers in stores and restaurants. This is typical of the clearly commercial techniques utilized in much of the marketing material for Japanese religious centers and activities.

[28] This custom originates in the Tokugawa period, when pilgrims were required to furnish evidence that they had in fact completed the journey for which the authorities had given permission; see Reader (2005: 22–23) for a succinct explanation and further sources; Vaporis (1994: chs. 4 to 6) has detailed information on the context in which pilgrims travelled in the Tokugawa period.

[29] This is conceivably exacerbated by the fact that foreign visitors are usually in Japan on tight schedules and can ill-afford the full day it would take to make the detour of this and the related sites on the Stone Marker Path. The priest had no precise figures for visitors but was adamant that the Inscription had made a difference.

Fig. 26.4 An invitation to Shingon Buddhist meditation, an esoteric Buddhist visualization technique, presented as part of the experience of making the ascent to Kōyasan. Other experiences, such as sūtra-copying, are also offered through the tourist office and the Teaching Center (Source: Kōyasan Tourist Board public promotional pamphlet)

this is probably no different to previous decades, it is clear that the Inscription has not actually done much to increase traffic to the subsidiary waypoints or to extend the clientele in line with the internationalizing intentions of the Inscription. Indeed, there is good reason to believe that focus moved away from these sites when the Nankai railway line was extended to Kōyasan prior to the 1934 *go-onki*. The latest move to bolster the pilgrim traffic to Kōyasan, fuelled by the recent inscription as a UNESCO World Heritage site, has only added to the rupture from sites on the traditional approach to Kōya such as Jison'in and Amano Jinja. The completion of the connection between Hashimoto and Kōyasan is in evidence on a map from 1934, *Kōya seizan chōkan zu* 高野聖山鳥瞰図, where it is highlighted in red.[30] Whilst Jison'in and Amano Jinja are still prominent on that map, since increased visitor numbers to Kōyasan were a consequence of the improved communications, it follows that over time it would alter the pattern of patronage of these subsidiary sites.

Car and motorcycle ownership might have been a factor in offsetting the effects of the direct rail connection to Kōyasan, but the building of a toll road from Kudoyama in the 1970s and a further toll road to the south, the Ryūjin Skyline, simply gave owners of private vehicles a further option for easy access to the mountain. The right of the construction companies to collect tolls has long since passed but the re-orientation of the mountain's profile has not seen visitors use the direct roads to the town in anything like the numbers which were common previously. In the early to mid-1980s, weekend traffic in particular swelled the town. Now, such casual visitor traffic has declined significantly and many long-standing restaurant and cafe businesses on the central thoroughfare have closed, patronage, it would seem, having been transferred to the confines of the *shukubō*.

To return to Jison'in, the temple showed evidence of regular activity. The main pagoda was under repair—indicative of a decent level of income for the temple, since the traditional methods necessary for maintaining such buildings are very specialized and hence relatively expensive. Also, the temple is a repository of items in the government's inventory of National Treasures and Important Cultural Treasures and is hence part of the institutional focus of modern Shingon Buddhism, as opposed to the pastoral crises dealt with by Reader. Further indications of continuing patronage were the presence of fresh or nearly-new offerings. There was a coach park for two coaches in the narrow street by the main entrance, indicating regular patronage from tour companies; and there were well-maintained items indicative of historical continuity: for example, permanent offerings (such as stone steles) from the Go-onki of 1984, not only here but also at the Niukanshōbu Jinja up the steep hill behind the temple.

Jison'in also functioned historically as a proxy for Kōyasan, especially for women, who were not allowed into the precincts until well into the Meiji period. Following the legendary example of Kōbō Daishi's mother, who is also said to have passed away at Jison'in rather than encroach upon the monastic precincts

---

[30] This "Bird's Eye View", a reproduction of which is given by Tarui (http://kouyasiki.sakura.ne.jp/oldmap/detail/map-17.html; retrieved 2012-10-15), was probably made for the Go-onki celebrations of that year.

to visit her son, women would make their offerings there.[31] The temple continues to feature women-related themes, a part of the precinct being dedicated to safe childbirth and bringing up young children, etc. The offerings there were both plentiful and newly added.

26.4.1 *Oku no In: A Conscious Encapsulation of History*

Over the centuries the cemetery at Oku no In[32] has spread out from the mausoleum of Kūkai, becoming in the process a microcosm of the religious, political and cultural life of Japan. Laid out as an extension to the eastern end of the main east–west thoroughfare which defines Kōyasan, it stretches for almost two kilometers into the cryptomeria-clad slopes. It has maintained its basic shape certainly since its revival in the eleventh century but expanded greatly with the popularization of pilgrimage in the Tokugawa period.

There are many old, historical graves to commemorate famous households, especially those of regional *daimyō*, and individuals (Fig. 26.5). Proximity to Kūkai's mausoleum represents a powerful point about space and spatial relations correlative to social and political standing. To the right hand side of the mausoleum, looking from Kūkai's place of rest, is the imperial tomb, Rekidai Tennōryō 歴代天皇陵.[33] Farther away from the mausoleum are generously sized precincts for Oda Nobunaga and for Toyotomi Hideyoshi. The latter's *Kōyasan monjo* 高野山文書 of 1585 gives an insight into the secular and military power that Kōyasan (along with many other estates) had acquired by the end of the Warring Provinces (Sengoku 戦国) period. In it Hideyoshi admonishes and threatens the monastic authorities should they continue to abuse their secular power (Lu 1997: 195–196). In that respect, the prominent presence of Hideyoshi's memorial at Oku no In must be seen as a very political statement. Interestingly, it is not uncommon to see cranes for peace hanging from the trees by that memorial. Also, the house of Tokugawa has a memorial temple (Daitokuin 大徳院) and a mausoleum (Tokugawa Reidai 徳川霊台) in the centre of the town and their crest can be found on other temples on Kōyasan. Thus, the "Three Unifiers" of Japan, who one after the other eventually brought an end to the destructive, one-and-a-half-centuries long warfare, have a prominent presence on the mountain. All this, and the consistent marking of these sites on the various

---

[31] There were also routes around the mountain itself that enabled women to visit the mausoleum of Kōbō Daishi; Jison'in was important for women who did not have the opportunity even to make the journey up the mountain.

[32] *Oku* means "distant" or "furthest reach" and in such contexts as Kōyasan indicates a distant site which is referred to by a cult in the pertinent centre; an *oku no in* (Buddhist) or *oku no miya* 奥の宮 (Shinto) would thus refer to a subsidiary precinct at a significant distance from the main temple or shrine. See Grapard (1982: 199–200), referring to Kageyama Haruki.

[33] On the 1681 map this is shown simply as the Tenshi Misasagi 天子陵, the newer name reflecting the changes in the imperial cult that occurred during the Meiji Restoration (1868).

26 Space, Time and Heritage on a Japanese Sacred Site: The Religious Geography... 537

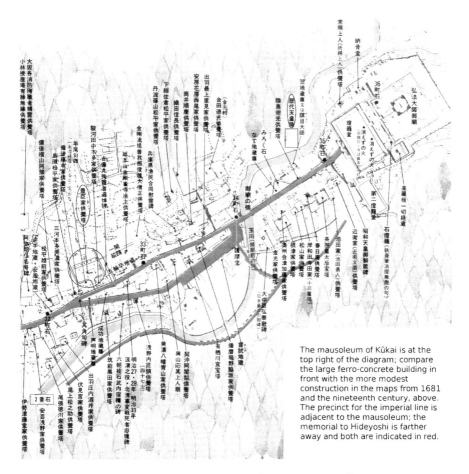

The mausoleum of Kūkai is at the top right of the diagram; compare the large ferro-concrete building in front with the more modest construction in the maps from 1681 and the nineteenth century, above. The precinct for the imperial line is adjacent to the mausoleum; the memorial to Hideyoshi is farther away and both are indicated in red.

Fig. 26.5 Diagram of Kūkai's mausoleum. Detail of Kōyasan tourist office pamphlet, showing Kūkai's mausoleum and its surroundings. Designed by Hinonishi Shinjō, a renowned scholar of the founder's life and his popular worship, it shows the location of most of the noteworthy grave markers in the main cemetery, including that of Hideyoshi (Source: Kōyasan Tourist Board public promotional pamphlet)

pilgrim's maps, indicates that Oku no In and the town as a whole have developed as a series of referents, markers, to society and the body politic.

In the course of the past four decades or so the cemetery has seen significant changes. Although it remained within its traditional confines until 1969, it had to be expanded that year, into the Daireien 大霊園. This lies to the south of the oldest part of the cemetery, on the other side of Route 480, which leads out of Kōyasan (Nakamaki 1995: 148). By the late 1980s there was very little space left there and a large area through a short valley to the north of Naka no Hashi was cleared for a further subsidiary cemetery. This last expansion was certainly fuelled by the throngs of people who visited Kōyasan in 1984 as the Shingon sect celebrated the 1,150th anniversary of Kūkai's passing. The sense of optimism, not to mention the income which the success of the *go-onki* generated, had led to a period of prosperity, typified by the expansion and modernization of the Kōyasan University campus in celebration of its own hundredth anniversary (its founding date was set to commemorate Kūkai's passing). When I visited the mountain in 1989, a striking rejuvenation of the town was in evidence: stores on the main street had been renovated and a modest ski and leisure resort had sprung up on the outskirts of the town, adjacent to the new cemetery extension.

However, the boom did not last and it became clear that the numbers of people and organizations choosing to be laid to rest on Kōyasan had begun to stagnate: in 1993, the Naka no Hashi extension, as one might expect, was hardly used. In the spring of 1999 I discovered that this extension, despite a comprehensive and tastefully executed advertising campaign on the railway, was still remarkably underpopulated. Even on my most recent visit (spring 2012), the numbers of graves in the new cemetery had increased only minimally. This may be simply an indication that the cemetery already housed more or less as many as could have been expected and that the temple administrators had misjudged the demand. However, it is more likely that rising costs of such sites coupled with the general fall in prosperity on the bursting of the Bubble, not to mention the changing values among Japan's youth, have caused a very significant fall in the symbolic value of this ancient site.

The expressions of faith in Kōbō Daishi at Oku no In include grave plots for many companies and other corporate organizations, making not only for an extension of corporate life beyond the mundane but also a tremendous variety in the types of grave to be seen there. These range from a professional photographers' association through medium-sized electronics companies to heavy industry involved in the making of rockets. The graves are designed to represent the companies' or associations' characteristic concerns or products in a very direct, this-worldly manner (Pollack 1993). Of particular relevance because of the long-term associations with Kōyasan, is the railway company one of whose founders in commemorated at Kōyasan, the Nankai Dentetsu 南海電鉄.

The Nankai Kōya Line 南海高野線 is part of a modern railway that serves commuters and tourists in the Osaka, Nara and Wakayama regions; it also runs ferry connections to the islands of Shikoku and Awaji. The company's roots go back to 1884 and one of its founders, Matsumoto Jūtarō 松本重太郎 (1844–1913), is the one whose commemorative stele was erected near Naka no Hashi in 1953. An annual memorial service is conducted every June (Nakamaki 1995: 150). The railway to Kōyasan was conceived in 1898, though the final part of the line, the funicular connection to Kōyasan itself, was not completed until 1930.[34] Most visitors to Kōyasan now take the luxury express, the Kōyagō 高野号, which since its introduction in time for the 1,150th anniversary celebrations (*go-onki*) in 1984 has gradually replaced the standard express as the prime means of through transport to Kōyasan. Since the UNESCO inscription some of these trains now include special

---

[34] The company publishes a wealth of information on its website, e.g. on its history, see www.nankai.co.jp/traffic/museum/muse/ji0001.html (retrieved 2012-10-14; in Japanese).

panorama carriages, which have the seating arranged such that passengers can enjoy the spectacular views of the valleys as the train climbs to Kōyasan. Taking the standard train (which does not incur a hefty surcharge for a seat reservation) now typically requires a change at Hashimoto on the River Kii, just before the intricate climb to Kōyasan itself.

The ramifications of the Nantetsu's involvement with traffic to Kōyasan extend beyond the railway. The company also serves Kansai International Airport (KIX), which opened on 4th Sept 1994. Whilst the airport was intended as relief for the already over-crowded airport at Osaka and was seen as a vital factor in the region's, especially Osaka's economic future,[35] the Nankai Railway also cast the airport as an additional point of access for traffic to Kōyasan, as evidenced by a significant level of advertising (occurring even 10 years prior to that, during the 1984 *go-onki*) and by its being a further factor in the tourist traffic in the wake of the UNESCO inscription. Over the years, the company has also provided sponsorship for cultural work, such as the recent exhibition on Kūkai and Shingon Buddhism at the Tokyo National Museum,[36] and the work of the photographer Toshinori Tarui, who produces beautiful work that complements the kind of image which attracts visitors interested in the cultural and aesthetic aspects of the mountain.[37]

26.5 Conclusion: Landscape and Heritage

Place can matter for many reasons. Historically place was very important in the earliest records of the Japanese polity. Moerman points out that although the official record of Japan's origins is mythological, crucial events, such as the death of Izanami after giving birth to the god of fire, were given a specific locus. Indeed, "Kumano served as both the birthplace and the sepulchre of Japan's royal traditions" (Moerman 2005: 44). Over time—and this is fundamental to the UNESCO Inscription—this assertion of the importance of place was reinforced practically, often by ritual action such as courtly or imperial pilgrimage and procession (Fujiwara nobles and retired emperors (*insei* 院政)), or by institutional presence (Kōyasan as a monastic center and as a powerful temple estate), or by personal practices (pilgrimages in the Tokugawa or modern-day recreational trips). So, underlying the UNESCO Inscription is a complex of ideas about space,

---

[35] It was, for example, being showcased in the New York Times' business section in 1994 (www.nytimes.com/1994/08/05/business/worldbusiness/05iht-airport.html (retrieved 2012-10-14)).

[36] 20 July–25 September 2011; the catalogue, in which Nantetsu is named as the only commercial supporter, is a prime example of the high-quality work that goes into the production of books and events that celebrate Japan's cultural heritage (Tokyo National Museum et al. 2011); www.tnm.jp/modules/ (retrieved 2012-10-15).

[37] Surname given second, as per his web-site, http://kouyasiki.sakura.ne.jp/profile/index.html (retrieved 2012-10-14). This site also contains high-quality scans of pilgrim's maps from 1732 to 1934, the last one shows the new railway to the top of Kōyasan quite prominently.

Fig. 26.6 The cemetery extension at Naka no Hashi. The last two decades have seen relatively little occupation of the latest extension to the ancient cemetery at Oku no In (Photo by Ian Astley)

more specifically human understanding and valorization of space; and the manner in which that understanding and valorization are expressed as practice. As evidenced by the Oku no In, the Japanese religious ethos lays some store by having one's life commemorated at a site of major religious significance, commemoration which can be individual or corporate (in the broadest sense—the early modern domains were also a kind of corporate entity). The commitment of time and financial resources indicates this quite clearly: the fact that people are now—in increasing measure—choosing not to do so, is rather significant. As Reader indicates, we have already entered a period where an important change has taken place, with serious consequences for religious behavior and the demographics of mortuary belief (Reader 2012b). The cemetery extension at Naka no Hashi is melancholy testimony to a significant change in the valorization of life and death (Fig. 26.6).

While Kōyasan is benefitting from the tourist interest generated by the Inscription, there are telling signs that the fabric of life on the mountain has changed over the last three decades and that although the aspects of commerce and marketing were readily obvious before, its traditional religious activities have decreased in intensity compared to the touristic elements that the Inscription has helped to promote. By its very nature, Kōyasan will always be a fascinating encapsulation of Japanese life over the past millennium, both with reference to the specific life of the mountain community and in the broader context of the practice of Japanese religions. It has attracted markers of events and personages from many walks of life, distillates of myriad stories, from rulers to poets, to major religious founders, to champions of

Fig. 26.7 Steles donated by Matsushita, founder of Panasonic. Erected in the 1930s and located close to the central two-tiered pagoda in the *garan*, this is one of a number of steles offered by leading industrialists in the period of Japan's imperial expansion; Matsushita's are the two in the *center*, the one on the *left* was donated by Kaijima Ta'ichi (Kyūshū zaibatsu) (Photo by Ian Astley, 2012)

Japanese industry (Fig. 26.7).[38] Many ordinary people are buried there, too, either with their own memorials or through identification with a larger community, such as the company they worked for. However, Kōyasan is not only a historic site but also

---

[38] There are two stone steles (c.150 cm high) donated by the founding president of the company widely known as Panasonic (Matsushita 松下) in the central *garan*: one stele was a personal gift and one was erected on behalf of his company. Panasonic has a prominent company memorial in Oku no In but these steles, on the eastern approach to the Grand Pagoda, stand side by side with a number of others donated by prominent figures in the industrial build-up to the Pacific War.

the headquarters of a religious community with dependant temples throughout Japan, each of which pays a substantial annual subscription for the privilege of affiliation. As Reader's fieldwork has shown, especially over the last 5 years or so, many parish temples are increasingly exposed to financial difficulty because of dwindling demand for their pastoral services. There is a real danger that the attention to institutional and monumental continuity will result in the demise of many local parish communities. This situation was exacerbated by the disclosure that the Kongōbuji headquarters lost 6.8 billion yen (US$70 million) (which included funds collected from dependent temples), due to errors in disbursement during fiscal 2012 (Asahi Shinbun 2013-02-26).

In a rather different context, though still in reference to issues of landscape and heritage, Tilley has stated, "Modernity is erased in favour of nostalgic reference to a lost past in an analogous way to the manner in which the official promotion of world heritage sites requires architecturally restoring the past in the present to project possibilities for a desirable future" (Tilley 2006: 14). The involvement of official bodies in the nomination for the Kii Peninsula, as for Hiraizumi and the applications that are now on hold, was very complex. At the highest administrative level was a state bureau (the Agency for Cultural Affairs (Bunkachō)), which was gathering information through delegation from a number of regional and local government agencies and religious and cultural bodies right down to the local level. The Bunkachō itself was operating at the behest of its government, which in turn was responding to the demands of an international body, UNESCO. The milieu in which this all occurred was also in effect contested by other sovereign states. In such a context a matrix of values is brought into play, either more or less consciously.

But however correct (or indeed irrelevant) all these statements may be from a scholarly point of view, the restoration (or construction) of a particular view of the past always runs the risk of abstraction from present concerns. The structural, one might even say the syntactical distance between the extremities of this official, highly bureaucratized activity mean that the values can become irrevocably abstracted from the environment of the practised religion, such that a rupture can occur that alienates the latter from the very monumentality that gains recognition. Of course, the human element is always emphasized in these enterprises and it would be foolish to deny that element. Yet, in the very act of inscribing there is the ominous risk of an irrevocable obliteration of the very life on-the-ground that was the original human impetus behind artefacts and activity.

Acknowledgements The author would like to thank Ian Reader (Lancaster University), Joachim Gentz (Edinburgh University), and Rosina Buckland (National Museums of Scotland) for their constructive comments; and Hill Williamson and Kevin Maclean (Central Library, Edinburgh) for access to Koyasan-related materials and the photography for Fig. 26.2, respectively.

References

Ambros, B. (2006). Geography, environment, pilgrimage. In P. L. Swanson & C. Chilson (Eds.), *Nanzan guide to Japanese religions* (pp. 289–308). Honolulu: University of Hawai'i Press.

Asahi Shinbun. (2013-02-26). Retrieved March 11, 2013, from http://www.asahi.com/national/update/0227/NGY201302260021.html

Astley, I. (2003). Kūkai's Kōyasan petition to the Court: Reclusive sentiments in a public undertaking. In Kōyasan University (Ed.), *Matrices and weavings: Expressions of Shingon Buddhism in Japanese culture and society* (pp. 217–230). Kōyasan: Kōyasan University.

Astley, I. (2004). Salvation and millennial expectations in the Japanese cult of Kōbō Daishi Kūkai. In K. Triplett, C. Kleine, & M. Schrimpf (Eds.), *Unterwegs: Neue Pfade in der Religionswissenschaft* [New paths in the study of religions] (pp. 287–296). Munich: Biblion.

Breen, J., & Teeuwen, M. (Eds.). (2000). *Shinto in history: Ways of the kami*. Richmond: Curzon Press.

Breen, J., & Teeuwen, M. (2010). *A new history of Shinto*. Oxford: Wiley-Blackwell.

Brown, P. (1981). *The cult of the saints: Its rise and function in Latin Christianity*. London: SCM Press.

Grapard, A. G. (1982). Flying mountains and walkers of emptiness: Toward a definition of sacred space in Japanese religions. *History of Religions, 21*(3), 195–221.

Hinonishi Shinjō (Ed.). (1988). *Kōbō Daishi shinkō* 弘法大師信仰. Tokyo: Yūzankaku.

Hinonishi Shinjō. (2002). The Hōgō (treasure name) of Kōbō Daishi and the development of beliefs associated with it. *Japanese Religions, 27*(1), 5–18.

Iwahana, M. (2000). Kokadera sankei mandara ni miru seiiki kūkan no hyōgen 粉河寺参詣曼荼羅にみる聖域空間の表現. In Ashikaga Kenryō Sensei Tsuitō Ronbunshū Hensan Iinkai (Ed.), *Chizu to rekishi kūkan: Ashikaga Kenryō Sensei tsuitō ronbunshū* 地図と歴史空間: 足利健亮先生追悼論文集 (pp. 428–435). Tokyo: Taimeidō.

Lindsay, E. (2012). *Pilgrimage to the sacred traces of Koyasan: Place and devotion in Late Heian Japan*. Ph.D. thesis, Department of Religion, Princeton University.

Londo, W. (2002). The 11th century revival of Mt. Kōya: Its genesis as a popular religious site. *Japanese Religions, 27*(1), 19–40.

Lu, D. J. (1997). *Japan: A documentary history*. Armonk: M. E. Sharpe.

Moerman, D. M. (2005). *Localizing paradise: Kumano pilgrimage and the religious landscape of premodern Japan* (Harvard East Asian monographs 235). Cambridge, MA: Harvard University Press.

Moerman, D. M. (2010). Outward and inward journeys: An introduction to Buddhist pilgrimage. In A. G. Proser (Ed.), *Pilgrimage and Buddhist art* (pp. 5–9). New Haven: Yale University Press and the Asia Society.

Morinis, E. A. (1981). Pilgrimage: The human quest. *Numen, 28*(2), 281–285.

Morinis, E. A. (Ed.). (1992). *Sacred journeys: The anthropology of pilgrimage*. London/West Port: Greenwood Press.

Nakamaki, H. (1995). Memorial monuments and memorial services of Japanese companies: Focusing on Mount Kōya. In Bremen & Martinez (1995), 146–158.

Nelson, J. K. (2000). *Enduring identities: The guise of Shinto in contemporary Japan*. Honolulu: University of Hawai'i Press.

Orzech, C. D., Sorensen, H. H., & Payne, R. K. (Eds.). (2010). *Esoteric Buddhism and the Tantras in East Asia* (Handbook of Oriental studies, Vol. 24 [Section four: China]). Leiden: E. J. Brill.

Pollack, A. (1993, September 8). Koyasan Journal: For Japan Inc., Company Rosters That Never Die. *New York Times*, p. 4.

Reader, I. (2005). *Making pilgrimages: Meaning and practice in Shikoku*. Honolulu: University of Hawaii Press.

Reader, I. (2012a). Buddhism in crisis? Institutional decline in Modern Japan. *Buddhist Studies Review, 28*(2), 233–263.

Reader, I. (2012b). Secularisation, R.I.P.? Nonsense! The 'rush hour away from the gods' and the decline of religion in contemporary Japan. *Journal of Religion in Japan, 1*, 7–36.

Reader, I., & Swanson, P. L. (Eds.). (1997, Fall). Pilgrimage in Japan, special issue. *Japanese Journal of Religious Studies, 24*(3/4).

Schopen, G. (1987). Burial ad sanctos and the physical presence of the Buddha in early Indian Buddhism: A study in the archeology of religions. *Religion, 17*(3), 193–225.

Tachibana Hoshun (1813). Kōyasan saiken ezu 高野山細見絵図 [Re-engraved detailed map of Koyasan], mounted by Iemon at Kōyasan from an original drawing by Tachibana Hoshun [Yasuharu] 橘保春. [Bunka 10.7:] July 1813.

Tilley, C. (2006). Identity, place, landscape and heritage. *Journal of Material Culture, 11*(1/2), 7–32.

Tuan, Y.-F. (2003). Perceptual and cultural geography: A commentary. *Annals of the Association of American Geographers, 93*(4), 878–881.

Vaporis, C. N. (1994). *Breaking barriers: Travel and the state in early modern Japan*. Cambridge, MA: Harvard University, Council on East Asian Studies.

Wada Akio. (1965). Kōyasan to Nyūsha ni tsuite [Kōyasan and the Nyū Shrine], *Mikkyō Bunka, 73*, 1–19. Reprinted in S. Wada & S. Takagi (Eds.) *Kūkai*. Tokyo: Yoshikawa Kōbunkan, 1982.

Chapter 27
Greening the Goddess: Sacred Landscape, History and Legislation on the Cāmuṇḍī Hills of Mysore

Caleb Simmons

27.1 Introduction

Within Hinduism's vast collection of mythology, the landscape of India plays a crucial role in the epic stories of divine struggle. The entire subcontinent becomes the playground and battlefield in which the *devas* and the *asuras* struggle for supremacy. As a result the landscape itself becomes an integral part of the myth and is established as a locus of the sacred. Often, however, the landscape lacks the same reverence that devotees pay to temples or shrines, despite its importance to the region and in the collective narrative of myth. Sacred centers that have become the focus of popular pilgrimage are often covered with trash discarded by the devotees. The sacrality of environment gives way to the sheer number of devotees because there is insufficient infrastructure to support the massive amounts of people. Often religious rhetoric is invoked to press devotees for the need to protect the sacred, which has met with mixed results. In Mysore City in the state of Karnataka secular discourse has been produced that has effectively implemented legislation to protect the group of sacred hills and popular pilgrimage destination called the Cāmuṇḍī Hills. However, upon closer examination the secular discourse constructs the significance of the hills because they are sites of Indian "heritage." To be considered a "heritage site," the landscape of the hills is interwoven into the history of the city through its connection to the epic battle between the Goddess and the buffalo-king. The connection between the Goddess narrative and heritage is only possible through a conception of history that places the mythological events of the landscape within the chronology of human history that resulted from a negotiation of Indian and colonial understanding of historiography. Therefore, though the rhetoric employed by those seeking to protect the hills appears completely secular, its foundations are built upon mythological narratives.

C. Simmons (✉)
Religious Studies Program, University of Arizona,
Tucson, AZ 85721, USA
e-mail: calebsimmons@email.arizona.edu

I was originally drawn to the region of Mysore because of Cāmuṇḍī Hills, a popular South Indian pilgrimage site. The Cāmuṇḍeśvarī Temple on the Cāmuṇḍī Hills had over 700,000 people tour the grounds in 2005 making it one of the largest pilgrimage sites in all of India, and the steady stream of devotees and tourists has risen every year (*City Development Plan* 2006). Therefore, I expected to find the same ecological situation that is so prevalent in many large sacred sites in India, viz., the temple grounds littered with refuse and the path to the temple lined by both human and animal excreta that accumulates due to inadequate infrastructure that could properly accommodate the vast number of pilgrims. At first, I believed that I was on the right track. I found reports that the lakes surrounding the Cāmuṇḍī Hills were extremely polluted and caused a number of cases of malaria and respiratory diseases that bordered on the epidemic scale (Learmonth 1961). These lakes or tanks had been formed from the naturally purifying watershed of the hills by the former Woḍeyar kings of Mysore who devoted them to various deities. The kings realized the purifying and practical role of the hills, which had served as a site of power and divine favor at least since the time of 'Bōḷu' (Baldy) Cāmarāja IV (r. 1572–1576 CE); so they chose the waters that naturally ran down from the hill as the primary water source for the city. Studies conducted on sacred sources of purifying water by Kelly Alley and David Haberman that explore the contrast of pure and clean, impure and polluted and the role of the religious discourse in ecological movements that seek to clean them up immediately came to mind and seemed to be quite relevant (Alley 2002; Haberman 2006). In these studies, rivers that are considered both sacred and the embodiment of goddesses, such as the Yamunā and the Gaṅgā, had become polluted because the goddesses were so inherently pure that devotees assumed that they remained pure no matter the amount of toxic waste that accumulated in their streams. Therefore, despite being polluted their waters remained pure. However, as I looked to more recent environmental studies, I realized that same rhetoric was never applied to the lakes and, as far as I could find, there was never any indication that the waters were inherently pure because they came from the abode of the goddess. Though the lakes are still considered polluted and non-potable because of high levels of phytoplankton caused by physico-chemical and biological waste, the lakes are now relatively clean compared to most lakes of the country because of movements to clean-up the city's water resources (Mahadev and Hosamani 2004, 2005; Mruthunjaya and Hosamani 2004). In light of the more well-known studies of environmental crises of North India and the charged religious rhetoric that is used, I found the dearth of such arguments extremely curious. This led me to look deeper into the history of Mysore City's modern ecological history, especially as it relates to the preservation of the Cāmuṇḍī Hills.

27.2 Mysore City, the Goddess, and Mapping "Historical" Myth

Before the intricacies of the environmental situation can be discussed, a history of the city and the sacred site must be explored. Mysore City is located in the Mysore District of the South Indian state of Karnataka 135 km/ 84 mi from the state capital of Bangalore. While Bangalore is the political and technological center of the state, Mysore is renowned for its educational and cultural heritage. It served as the capital of the Woḍeyar dynasty from the founding of the city in 1572 CE until 1610 when Rāja Woḍeyar seized Śrīraṅgapaṭṭaṇa from the Vijayanagara viceroy. The city was reestablished as the capital after the fall of Tipu Sultan in 1799 CE when the Woḍeyars were reinstated as the kings of Mysore. The city is home to several elaborate palaces that were built during the colonial period of princely rule, which now serve as major tourist attractions. In 2005 the Delhi Assembly passed a bill, which focused on the development of tourism trade over increasing urbanization, that sought to protect buildings over 100 years old that were not covered by India's Ancient Monuments and Ancient Site and Remains Act of 1959. This bill protected so many of the structures built during the Woḍeyars' rule, which are renowned for their historical and architectural significance, that the Government of Karnataka declared the Mysore City a "Heritage City" (Rao 2006).

The Cāmuṇḍī Hills are a network of hills that come together to form one large hill located just outside of Mysore City. The Hills are largely unpopulated except for small enclaves, which are predominately inhabited by small groups of religious professionals and agriculturalists. The Hills are covered with vegetation and for most of its history the apex, where ancient temples to Śiva and Cāmuṇḍī are located, was extremely hard to access until Dēvarāja Woḍeyar (r. 1659–1673 CE) commissioned the installation of 1,000 steps leading up the hill. Though there is a road that leads to the top that has regular bus service, many of the devotees choose to walk up the steps on their way to the temple. There are three primary spaces for business on the Hills. The largest is at the top, which is home to dozens of shops, a government guesthouse, and a post office. The second largest is at the base of the Hills, which primarily serves the needs of pilgrims, who are in need of offerings for the Goddess. At the 600th step, there is a large sculpture of Nandi, Śiva's bull attendant (Fig. 27.1). The image of Nandi draws many tourists for whom small souvenir shops and snack shops are set up, which also provides a nice break for the pilgrims on their journey up the hill.

Because of the regional importance of the city and the Woḍeyars, Mysore's history has been interwoven with epic and Purāṇic narratives, making it an important city for pan-Indian mythology. Mysore is the Anglicization of *Mysūru* (or one of its variations: *Maisūru, Mahisūru*, etc.), which is Kannada for "city of Mahiṣa (buffalo)." Over the centuries the city of buffalo has become connected with the widely known Mahiṣāsura (buffalo-demon), the mythic *asura* from Purāṇic lore that is most well-known from the narrative of his death at the hands of the Goddess in the *Devīmāhātmya* of the *Mārkaṇḍeya Purāṇa* and the *Devī Bhāgavata Purāṇa*. According to the pop narrative of the story, the physical location of the buffalo-demon's kingdom was at the same locale as the modern city of Mysore. The Purāṇic narrative elaborates that from his capital the *asura* grew in power due to an ill-advised boon from Brahmā (or Śiva) and was able to acquire the throne of heaven and dispelled the gods (*deva*) to earth. The gods were filled with rage but were impotent to defend themselves. Consumed with rage, they let their energies (*śaktī*) emerge from their foreheads, which united to form the Goddess (Devī). The Goddess, formed from combined powers of the displaced deities, immediately began destroying

Fig. 27.1 Nandi (Photo by Caleb Simmons)

the armies of Mahiṣa with great ease. After defeating many of his generals, an epic battle ensued between the Goddess and Mahiṣa. Despite Mahiṣa's ability to shape-shift and his haughty antagonism, the Goddess was victorious and returned the gods to their abodes. At the close of the episode, she offered a boon to return whenever she is remembered by devotees. The epic battle is one of the most popular in the Indian imaginary and can be seen at almost every turn represented in the iconography of "slayer of the buffalo" (*Mahiṣāsuramardinī*) in which the Goddess drives her trident into the flesh of the buffalo-demon (Fig. 27.2). The physical site of the battle is connected with Cāmuṇḍī Hills, on the top of which a cleft rock formation is said to have been formed from the sheer force of the fatal blow delivered by the Goddess. Throughout India, this battle is commemorated during the 10 day festival of Dasara, which had special significance for the establishment of kingly authority in South India since the Vijayanagara dynasty, who the Woḍeyars attempted to emulate. During Dasara, the Cāmuṇḍeśvarī temple and Mysore City are flooded with pilgrims celebrating the Goddess victory over evil. Because of the connection to the widely known narrative, Mysore and the Cāmuṇḍī Hills are sacred landscapes that are integral to the narrative for the people of the region and of great importance to many devotees of the Goddess.

The connection that is drawn between the physical space of Cāmuṇḍī Hills and the sacred narrative of the slaying of the buffalo-demon is part of a larger complex that has great implications on our understanding of Hinduism and its relation to the construction of history and heritage that, in turn orients, popular conception of space, geography, and ecology in India. Human conception of history has amazing

Fig. 27.2 The goddess slaying the buffalo demon (Photo by Caleb Simmons)

implications for our perception of the world around us. For many of us that are educated in the Euro-American tradition, our understanding of time is relatively linear in which we focus on chronological continuity of 'factual' events that were observed by human sources that we trust to be "neutral" or "objective," who focus on "mapable" territory and the deeds of men (and sometimes women where convenient). Armed with this 'scientific' understanding of history, many colonial historians of India, were struck by the lack of chronological histories in Indian society, especially in light of the vast amount of writing that they produced on other topics. Therefore, many of these scholars reported back to their home institutions and published findings that concluded that India "had no history." Instead, Indian culture was labeled "mystical," "spiritual" or "other worldly." However, this was not necessarily the case. There was history – just a different kind of history. The Indian view of history was through a wider lens that recognized the importance of kings and territory (and chronicled their actions quite extensively in epigraphic form), but they also understood the history of the cosmos as much more grand system of macro and micro cycles of existence that continually degrades until it is recreated, simultaneously creating the longing for a golden age and the hope for a brighter future. Due to the colonial pressure to conform to the Euro-American understanding of history, interesting conceptions of time and space emerged. Building upon the presentation of Indian history in James Mills's *The History of British India* published in 1806, which created the illusion of a golden-age of India that was united under Hinduism, many tales of the Purāṇas became viewed in a different manner (See Inden 2000 for a thorough discussion of Mills's reconstructionist history). The tales of the deeds of deities and sages were re-conceptualized as historical descriptions of factual events that had taken place on this earth, much like Indian kings had described the great kings of the epic literature for centuries.

That is not to say that physical space was not considered sacred prior to colonial intrusion. There exists an entire genre devoted to describing the sacrality of cities and holy sites called *sthala purāṇas* (site stories); however, there is always an ambiguous separation between the actual space and a divine space that mirrors the physical world in the abode of the gods (for example, the Vārāṇāsī narrative found in the *Kāśī Khanda* of the *Rāmāyaṇa* in which Śiva is able to remove the cosmic city from the earthly equivalent). These two worlds, the divine and the earthly, were considered coterminous yet remained distinct at these sacred sites. There was no vocabulary for understanding this relationship within the colonial historical imagination; they were either divine or natural. Therefore, in the negotiated linear either/or history that was emerging, the sacred sites lost their both/and identity. It seems that more often than not this reconfiguration of history and religious landscape has had a negative impact on people's concern for preserving the environment. The prevalent mentality that resulted has been, "If the site is divine, it has the power to purify itself."

27.3 Heritage as the Secular Sacred

However, in Mysore City, a very different attitude arose in relation to Nehruvian secularism in which the preservation of sacred space would be the preservation of the city's historical heritage. From the earliest interest in the preservation of Mysore, the entire Cāmuṇḍī Hills have been earmarked because of their historical importance, a designation that is usually only reserved for important structures or monuments. The first attempt to preserve the sacred landscape was made by Devaraj Urs. Devaraj Urs was born in Mysore City in 1915. There, he was trained as an educator. He rose into politics from this career path, becoming a member of a regional wing of the Congress party that was commonly called the "Syndicate." His influence quick grew, and by 1953 he was the Chief Minister of Education of Karnataka. After becoming a major force within the political sphere, he became interested in the environment and the preservation of natural and cultural heritage of his home. Thus, in 1966 in conjunction with his wife, Subramanya Raje Urs, a novelist, he devised the Nehru Loka Project. This Project was to secure land for farmers and laborers and to be used for "parks and open spaces" in Mysore City and its surrounding areas. Under Nehru Loka, 12,355 acres (5,000 ha) of land were to be used only for uncultivated recreational and aesthetic purposes. A major portion of the project was also to preserve the ancient Cāmuṇḍeśvarī Temple and the natural landscape of the Cāmuṇḍī Hills, pointing out the Hills provide a link to the ancient cultural heritage of the city.

Never in any of the documents, however, is there any religious rhetoric or sentiments expressed. Even the name Nehru Loka, named after Jawaharlal Nehru, India's first Prime Minister, who is famous for his promotion of secular India, suggests the secular nature of the Project. Terms such as heritage and culture are the only non-biological reasons for the protection of the region or the hills. Devaraj Urs' loyalty to the Nehruvian ideals and the Congress party surely influenced his rhetoric, but the mythology of the Goddess provided the necessary link to constitute the space as a heritage site. The project never became official nor was it ever passed into law because of national politics and the introduction of nationally recognized land reform legislation, but it was implemented in Mysore as if it were.

In 1971, Urs successfully won the election for Chief Minister in the State of Karnataka for Indira Gandhi's Congress Party, where he immediately surrounded himself with scientists and academics and began a push for national land reform. In 1976, a year before Indira Gandhi's President's Rule was enacted, Urs convinced Andhra Pradesh, Haryana, Gujarat, Himachal Pradesh, Karnataka, Maharashtra, Orissa, Punjab, Tripura, Uttar Pradesh, and West Bengal (later joined by Assam, Bihar, Madhya Pradesh, Manipur, Meghalaya, and Rajasthan) to adopt the Urban Land (Ceiling and Regulation) Act. The goal of the Act was to distribute land equally so that farmers were able to retain the lands that they had been farming against the constant push of urbanization. The Urban Land Act also implemented many of the same reforms that were in the Nehru Loka Project, limiting the encroachment of urban development into cultural sites in whose list the Cāmuṇḍī Hills were included.

In 1999, the Urban Land Act of 1976 was repealed by Karnataka among other states. However, the city of Mysore was not affected because of the implementation of the Nehru Loka Project of 1966 that protected these spaces. In the following years, there was an onslaught of criticism over the land reforms imposed by the Nehru Loka Project. The "antiquated" Nehru Loka reforms were said to be a "hurdle to the modernisation process" because they prohibited the sell of part or all of the land to industrial farmers or developers thereby limiting the production and revenue for the state (Ramoo 2001). Under such criticism, the land that had been protected by the Nehru Loka Project was de-notified in 2002 by the Government of Karnataka. By 2005, the city was facing a rampant wave of unrestricted urbanization. At this same time, a land grabbing scandal was unfolding over the entire country. It was discovered that in the early 1990s the same government officials who had repealed the Urban Land Reform Act of 1976 had been taking land from farmers that had been protected by the Act and selling it to corporations for development projects. In order to combat overdevelopment and to investigate the injustices, Mysore formed the Mysore Agenda Task Force (MATF). MATF was to assess the situation and to create a City Development Plan (CDP) in the context of Jawaharlal Nehru National Urban Renewal Mission (JNNURM) that had replaced the Urban Land Act in 1999. The Development plan had no real power, but could suggest measures by which the city could regulate its development.

> The CDP seeks to set in place the directions and principles, rather than aim at being a definitive and conclusive document. Primary emphasis is on principles, directions and reform, rather than on specific projects that the City needs to develop. Given the complex and consensual nature of the exercise, it is clear that while such a consultative process gives room for all the views to be articulated, it is certainly not possible to adopt every view point. The final vision will therefore reflect a preponderance of opinion, rather than be a unanimous view. A two-phased approach has been adopted to chart the direction of the City's

development. A consultative, normative approach to envision the future, complemented by a bottom-up approach of specific interventions in the City. (City Development Plan 2006)

MATF was chaired by one of Mysore's leading industrialists, R. Guru of Rangsons along with several other well-known engineers and architects. Therefore, it was no surprise that the plan that they offered further reduced the amount of land to be reserved for parks and open spaces in Mysore City and its surrounding areas. According to the CDP, the land designated for parks and open spaces was to be further lowered to 8.9 %. The percentage under the previously instituted Nehru Loka Project had been 18.05 %. The CDP also suggested a railway system that traveled the 13 km (8 mi) from the heart of Mysore City to the base of Cāmuṇḍī Hills in order to tap into the "revenue potential" of the site.

The CDP still recognized Cāmuṇḍī Hills as a cultural heritage site and allotted a 680 acre (275 ha) buffer zone around the Hills. But many still feared that the introduction of the railway system would irreparably harm the site, both physically and aesthetically. The emissions of the gas powered rail cars would cause greater pollution in the region. It would also lead to greater noise and the loss of the natural landscape. There was also a fear that the advent of the rail system would encourage developers to pursue the land. In fact, by 10 June 2005 Mysore's Urban Development Authority had proposed six to eight new development layouts between Mysore and the Cāmuṇḍī Hills (Revival 2005). This concern led to the formation of the Mysore Amateur Naturalists (MAN), a grassroots youth organization aimed to prevent unbridled urbanization and unregulated development. MAN began the "Save the Cāmuṇḍī Beṭṭa (Hills)" campaign, which enumerated an 11-point objective aimed at protecting the ecosystem of the Cāmuṇḍī Hills. The plan entailed introducing solid waste management, promoting research on eco-tourism, and educating the villagers of the Devikere village outside the Hills. The group also pushed for a 500 m (1,640 ft) buffer zone for any development including the railway. They were joined by many other NGO's and environmentalists in their push to preserve the Hills. Even MATF eventually joined their side of the debate arguing for the importance of the "green belt" region that encompasses Cāmuṇḍī Hills for lung safety and tree cover (Kumar 2005). Throughout the debates and protests, the conversation either revolved around tourist income, environmental concerns, or its heritage, always evoking the historical significance of the natural landscape.

Simultaneously, there emerged a strong ecological movement to preserve the heritage of the Cāmuṇḍī Hills by regulating the use of non-biodegradable products at the top of the hill. The goal of the movement was to prohibit the use of plastic containers and plastic bags by the vendors, who sell devotional products, souvenirs, or foodstuffs to pilgrims and sight-seers. Before any legislation was passed, K.B. Ganapathy, the Editor-in-Chief of the *Star of Mysore* wrote a vehement article calling for personal responsibility of pilgrims and tourists to respect the heritage site regardless of the presence or absence of a law. Speaking about other sites he had visited and seen people abusing the sites by littering, he said,

> I think that this kind of irresponsible citizen should be mercilessly dealt with and heavily fined, if not jailed, for their irreverence to nature and for vandalizing the environment. (Ganapathy 2005)

Fig. 27.3 Plastic free zone (Photo by Caleb Simmons)

Ganapathy uses the overtly secular rhetoric of the citizen and nature for a site that is also a renowned home of a goddess and compared the space to other heritage sites in south India that have no living religious significance. By October 2005 legislation was enacted that declared the Cāmuṇḍī Hills a plastic-free zone by law (Fig. 27.3).

While the prohibition of the use of plastic by retailers on the hill was a victory for the preservation of the Hills, development continued to be unbridled and unregulated. In April of 2007, Mukhyamanthri Chandru from the Joint Legislature Committee that had been established to investigate the land grabbing scam from the early 1990s submitted a report on the encroachment of the Cāmuṇḍī Hills' buffer zone. This document cited an earlier report from 2005 that was given to the MATF by the head of the Heritage branch of the Mysore Agenda Task Force (MAHTF). However, that was never made public or even addressed prior to the MAHTF report. The original report exposed a high level of encroachment by developers into the buffer region of Cāmuṇḍī Hills. The finding stated in the MAHTF report showed that nearly 500 acres (202 ha) of the 680 acres (275 ha) designated by the Mysore CDP to remain open and free of all development had been developed. This has led to a renewal of the controversy and many citizens and environmentalist calling for a revival of Devaraj Urs's Nehru Loka.

However, there is still no need to preserve the sacrality of the site or the incorporation of Goddess themes in the fight. It would seem to make sense for the movement to preserve the Cāmuṇḍī Hills and to employ the narrative of the Goddess slaying the buffalo demon to fight against governmentally sanctioned usurpation of the land. The imagery in the narrative seems to align with the fight for preservation of the site. The activist, like the gods, could implore the Goddess to come to their aid because the powerful ruler(s) have overstepped their bounds and take what was not theirs to take. This would actually not be that novel. The narrative of the Goddess and buffalo is commonly evoked in popular media (film, television, political speeches, etc.) when a righteous battle is implied. Yet it is absent at the site, which is believed by many to be the actual spot where the Goddess defeated the demon and promised that she would return to do the same when she was remembered by devotees.

The consensus of those writing about the issues is that the problem is situated in the structures of the state government and should be addressed through secular legislative means. However, it seems like the state government is the cause of many of these ecological issues for the Cāmuṇḍī Hills. The government was so obsessed with development and industrialization that it has begun to disregard the importance of the space. Politicians changed legislation to allow for the development of the heritage land that had been protected. Similar issues have arisen all over India. Anil Agarwal, who has years of experience as an environmental activist and author in South Asia, arrived at this conclusion while discussing the ability of India to address its environmental crisis. He argues that India's democratic setup is not capable of supporting its needs on the village level:

> First, India needs a major overhaul of its structures of state. A sincere effort must be made to create systems that engender respect for rules and regulations at the community level. Participatory, local democratic systems must be created to replace the current exclusive dependence on highly centralized electoral democracy. Gandhi had said that a free India should be a country of 560,000 village republics. In other words, each village should be a republic. Unfortunately, that has not happened. (Agarwal 2000: 177)

The democratic rule by the majority by its very nature excludes the minority. Individual villages become the minority when faced with problems that will benefit the majority of the state. In the race for development, those who stand in the way of modernization, more often than not, get bulldozed by the system. This sad democratic reality leads Agarwal to conclude,

> It [a solution to India's environmental crisis] will require striving to achieve a balance between the accumulation of material wealth and the protection of the bounteous beauties of nature…. A government can never effectively legislate these balances. Only people's faith and belief and public education [can]. (Agarwal 2000: 178)

The case of Mysore City and the Cāmuṇḍī Hills seems to disprove Agarwal's conclusions by continually resisting faith-based arguments. While faith, belief, and education are integral in the production of a personal environmental ethic, secular responses to government have been the primary means through which its citizens have sought for ecological legislation, and it has been effective, far more so than many environmental movements throughout India that have employed religious rhetoric. In fact, in July 2007 in a study conducted by an independent research company, Mysore was named the least polluted city in India and encroachment into the hillside has been curtailed (Mysore 2007). However, before we can firmly draw this conclusion or Agarwal's thesis can be dismissed, I must return to the idea of history and heritage discussed above.

The demarcation of the Cāmuṇḍī Hills as a heritage site defies convention for such legislation. In the list of heritage sites created by the CDP, the Cāmuṇḍī Hills are the only site that is not of historical significance. In fact, in the section that lays out the criteria for selecting the sites, which only refers to buildings or other large human-made structures, there are no criteria that the Hills meets. Furthermore, the hill system itself is relatively devoid of any link to Indian, Karnatic, or Mysorean human history. The only significant elements are those contained along the thousand step path, which makes up a very small percentage of the protected area. But, if one considers the epic events of the Purāṇic narrative of the Goddess slaying the buffalo-demon to be a historical event, the entire landscape of the battle becomes a heritage site like those commemorating significant battles in human history. Indeed this seems to be the case in popular understanding of the history of Mysore City. Almost any popular account of the city's history that can be found in the city will say that it was formerly the kingdom of Mahiṣa until he was conquered by Cāmuṇḍī. Therefore, it seems that foundation of the secular argument for the preservation of the Cāmuṇḍī Hills rests on an implicit recognition that the entire landscape is connected to the structures at its apex through actual events that took place on the Hills in a time far passed. However, this is actualized differently than most other movements, which have elevated the seats of religious significance into the transcendent world. The advocates of ecological responsibility on the Hills have collapsed the sacred into the natural world. The placement of the mythological narratives within human history provides a new vocabulary of history and heritage, which can be translated into secular legislation.

27.4 Conclusion

The city of Mysore and the Cāmuṇḍī Hills present an interesting case study in the field of Hinduism and Ecology or perhaps a better term, Indian Heritage and Ecology. The city and its citizens have maintained a secular discourse in its approach to environmental issues dealing with extremely religious sites by reconsidering the mythological narrative as human history. Not only can this case show the multiple strategies that can be employed to preserve sacred space, but it also shows how our ways of conceptualizing history shape our understanding of the world around us and gives us the vocabulary through which we are able to make sense of larger issues relating to transcendence and the natural world.

The activists in Mysore City, who have fought to maintain the Cāmuṇḍī Hills, have proven that there are multiple ways to articulate the need for preservation of sites that are important because of their role in mythological narratives in India. Their emphasis on heritage, instead of 'religion' avoided many of the pitfalls that have befallen other such movements in India that have been mired in opposing views of the relationship between humanity and things considered sacred within the Hindu community. Similarly, the focus on the heritage of the site as part of the history of the city side-steps the divisive politics associated with religious communalism that segregates causes as either Hindu or Muslim. By employing secular terminology, the movement to save the Cāmuṇḍī Hills removed the issue from one that is mired with the potential peril associated with religious traditions and into an issue of the dirty politics and the cultural wealth of the city. Through its innovative discourse, the movement has provided an effective model that other environmental organizations in India can emulate even when navigating the hazardous terrain of important pilgrimage sites.

References

Agarwal, A. (2000). Can Hindu beliefs and values help India meet its ecological crisis? In C. Chapple & M. Tucker (Eds.), *Hinduism and ecology* (pp. 165–179). New Delhi: Oxford University Press.
Alley, K. D. (2002). *On the banks of the Gaṅgā: When wastewater meets a sacred river*. Ann Arbor: University of Michigan Press.
City Development Plan for Mysore. (2006). Retrieved November 22, 2011, from www.jnnurm-mysore.in/pdf/CDP.pdf
Ganapathy, K. B. (2005, May 18). Plastic-free-zone and all that Bakwas! *Star of Mysore*.
Haberman, D. L. (2006). *River of love in an age of pollution: The Yamuna river of northern India*. Berkeley: University of California Press.
Inden, R. B. (2000). *Imaging India*. Bloomington: Indian University Press.
Kumar, R. K. (2005, August 8). Buffer zone imperative around Chamundī Hills. *The Hindu*.
Learmonth, A. T. A. (1961). Medical geography in Indian and Pakistan. *The Geographical Journal, 127*(1), 10–20.
Mahadev, J., & Hosamani, S. P. (2004). Community structure of cyanobacteria in two polluted lakes of Mysore City. *Nature, Environment, Pollution, and Technology, 3*(4), 523–526.
Mahadev, J., & Hosamani, S. P. (2005). Algae for biomonitoring of organic pollution in two lakes of Mysore City. *Nature, Environment, Pollution, and Technology, 4*(1), 97.
Mruthunjaya, T. B., & Hosamani, S. P. (2004). Application of cluster analysis to evaluate pollution in the Lingabudhi Lake in Mysore, Karnataka. *Nature, Environment, Pollution, and Technology, 3*(4), 463–466.
Mysore Least Polluted City. (2007, July 2). *The Tribune*.
Ramoo, S. K. (2001, March 9). Land reforms in state a failure? *The Hindu*.
Rao, C. H. G. (2006, March 26). Place for heritage and culture. *The Hindu Business Line*.
Revival of the Green Belt favoured. (2005, June 10). *The Hindu*.

Chapter 28
Pollution and the Renegotiation of River Goddess Worship and Water Use Practices Among the Hindu Devotees of India's Ganges/Ganga River

Sya Buryn Kedzior

28.1 Introduction

When Raj, an architect in the industrial city of Kanpur, makes his way down to the banks of the sacred river *Ganga* for morning *puja* he performs two successive actions. First, he dips his palm into the cool river and, reciting a prayer, washes his brow with the waters that cleanse his body and spirit in a reaffirmation of life and recognition of the sacred bond between the river and those who inhabit her shores. Then, Raj endeavors to wipe the water from his hand and face, using a cloth napkin and bottled water to wash the residue from his body before proceeding to work in his downtown office. Raj, like many people living in Kanpur and elsewhere along the Ganges River, are faced with the question of whether to alter practices of river water worship in the face of staggering pollution that represents a real threat to human and ecological health.

This chapter examines changing patterns of river water worship in the central Ganges River Basin (GRB), with emphasis on the city of Kanpur, Uttar Pradesh. Drawing on survey and interview data collected in 2008 and 2009, the author argues that water use activities are being renegotiated and that, as people change water use habits in response to high levels of pollution, many ideas about the sacred nature of the river and the appropriate relationship between Hindu devotees and the mother goddess *Ganga* are changing as well. These changing values of use and worship are significant because they illustrate not only the unfixed and changeable nature of religious practice and belief, but also because they draw attention to those spaces where the material and spiritual worlds meet and to the interrelationships between ecological health and human value systems.

S.B. Kedzior (✉)
Department of Geography and Environmental Planning, Towson State University, Towson, MD 21252, USA
e-mail: skedzior@towson.edu

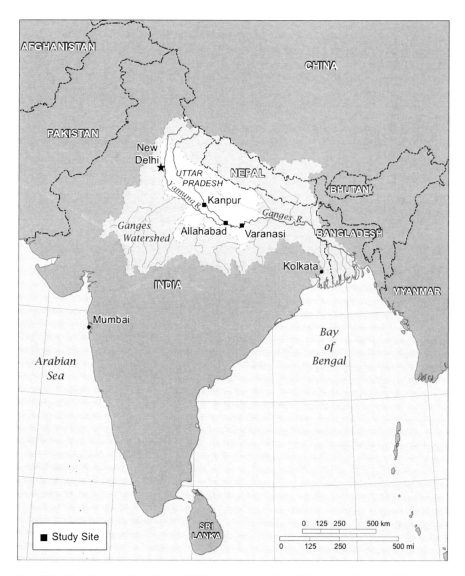

Fig. 28.1 The Ganges River Basin (GRB) and major cities (Map by Jeff Levy, University of Kentucky Gyula Pauer Center for Cartography and GIS; commissioned by the editor)

While this chapter attempts to tease out processes of religious change occurring along the river, discussion is limited to interview and survey findings from the city of Kanpur. Selected as a study site for its consistently low water quality and renown as a pollution-producing industrial center, Kanpur is also the largest city along the main stem of the Ganges before Kolkata[1] (Fig. 28.1). The city is located about

---

[1] Kanpur's population is approximately 6.3 million people (Census of India 2011).

400 km (244 miles) from the capital at New Delhi in the fertile Lower Doab between the Ganges and the Yamuna in the plains of Uttar Pradesh (UP), the largest and most populous state in India. UP lies at the center of the "Hindu belt" or "cow belt," and is a dominant center of Hindi language use and national politics.[2] But, the state also has some of the lowest literacy and life-expectancy rates in the country, a skewed sex ratio, and history of communal violence between the majority Hindu and minority Muslim populations. UP is also home to some of the most polluted stretches of the river, and to a majority of Ganga Action Plan (GAP)[3] implementation sites. While this imparts UP-State boards and officers with unusual influence over the governance of pollution and water quality in the basin, decision-making in terms of water quality governance and pollution abatement approaches is often carefully balanced between the central Ministry of Environment and Forests (MoEF), the Central Pollution Control Board (CPCB), the regional Ganga Project Directorate (GPD), the Uttar Pradesh Pollution Control Board (UP-PCB), and local municipal water authorities (Jal Nigam) in each of the cities incorporated into the plan.

Kanpur has been a growing industrial center since Independence. The city grew from a small village on the banks of the Ganges following the British establishment of a trading factory, and later a military station, at the site (Spate and Ahmad 1950). Subsequent demand for military-grade khaki, boots, and saddlery led to the development of a local leather processing industry. However, the "real rise" of the town came from the cotton mills built in the late nineteenth century to fill demand during the American Civil War (Spate and Ahmad 1950). The strategic location of the city in the agriculturally productive *doab* and on rail lines between Delhi and Kolkata (Calcutta), allowed for the rapid growth of the city's textile mills, sugar refineries, and tanneries, which soon began to dominate the region economically.[4]

Once dubbed by the British as the "Manchester of India" (Cumming 1994: 30), much of the textile sector has now left Kanpur, as factories became obsolete and new construction in the south became cheaper than updating Kanpur's aging infrastructure (Pandey 2008). Today, the city is best known for tanning, and is the largest exporter of finished and semi-finished leather products in the country, an industry which brings in about 2.75 billion Rs. (or around $61 million US dollars) annually (Mishra 2009). The industrial centers of the city are located along the southern bank of the river near civil lines (textile cluster), inland behind the central business district and military cantonment (munitions and metalworks cluster), and downstream from the city center in Jajmau (tannery cluster). Jajmau is known as a major leather production zone, housing around 305 tanneries that employ shared Common Effluent Treatment Plants (CETPs) for the treatment of toxic effluent that contains organic matter, chromium, sulfide, ammonium and salts used in stripping and curing hides as well as a number of trace

---

[2] Of India's 14 Prime Ministers, 8 were elected from Uttar Pradesh, including the powerful Nehru family. Leaders of the Indian National Congress (INC) Party also frequently come from UP.

[3] GPD is an interstate pollution abatement program implemented since 1985 in dozens of cities along the Ganges.

[4] On the eve of Independence, 42 % of factory workers and 62 % of textile workers employed in the state worked in Kanpur (Spate and Ahmad 1950).

metals (Beg and Ali 2008). In manufacturing the nearly 100,000 hides exported from Kanpur each year, these tanneries also produce around 1,500 tons of chromium sulfate discharge and 146,000 tons of solid waste (Schjolden 2000). This amount adds to 339 MLD of domestic wastewater from the city, agricultural runoff, and much of the uncollected and unburned trash that flows down to the river through the *nalas* or drains of the city (Fig. 28.2). The stretch of the river downstream from Kanpur frequently measures as the most polluted along the river (NCRD 2009), before downstream tributaries reduce the overall pollution load by adding to the water level.

In the following pages, we discuss how Kanpurites are responding to local changes in water quality by adjusting practices associated with traditional river water use and worship. Drawing on interview and survey data, this chapter emphasizes that people's perceptions of the river, and of relationships with the river on which water use and worship practices are based, are being renegotiated in the face of decreasing water quality. The nature of these renegotiations are unfixed and highly variable, with some water users changing the site at which they bathe in the river, others adjusting the physical acts associated with river worship, and still others reducing the frequency of bathing and drinking activities or even stopping them altogether. We begin this discussion by turning first to a brief introduction to the Ganges River, to the river goddess and her attendant practices of worship, and to a review of pollution causes and problems around Kanpur. Results of our study are then presented and discussed, with an emphasis on how water users make sense of the changing water quality in the basin and how they subsequently renegotiate practices associated with devotional river worship.

Fig. 28.2 A nala or drain empties runoff and debris into the Ganges River at Kanpur (Photo by Sya Buryn Kedzior)

28.2 The Goddess Ganga

The River Ganga is a goddess in the Hindu religious pantheon. Her shores are the genesis site of a number of the world's major religions (including Hinduism, Buddhism, and Jainism) and the stage for the most important legends and epics that constitute North Indian history and Hindu religious narratives (Bonnemaison 2005). Stories of the river and goddess Ganga are intertwined with those of the greatest Hindu gods and kings. References to the Ganga are found in nearly all major Hindu texts, starting with the Aryan Vedas of the sixteenth to nineteenth centuries BCE.[5] Ganga and her descendants play significant roles in the epic *Mahabharata* and an account of her birth is contained in the *Valmiki Ramayana*, wherein the Ganga and the Yamuna are described as sister rivers and daughters of the Lord Himalaya. Ganga, the elder of the daughters, is a "heavenly river undefiled," who was given to the immortals so that her waters might, "cleanse and save …. Purging all sinners to the sea" with "Her waves that bless and purify" (Ramayana I, translated by Ralph T.H. Griffith from the original Sanskrit 1870–1874, as quoted in Pant 1987: 3–4).

There are many creation myths associated with the formation of Ganga, her materialization from goddess to river, and her descent to earth. In most of these, the Ganga originates in the heavens and various gods and mythical figures are credited with the work done to bring her to earth. In these, she is frequently associated with *Vishnu*, especially his toes and feet, granting her the alternate name of Vishnupadi, meaning "originating from Vishnu's feet" (Gupta 1993: 107). In the *Bhagavata Purana*, Ganga is released to earth when Vishnu's large toe scratches a crack in the comic egg that encapsulates this world, releasing the river to wash the dirt from his foot and thereby washing away the sins of humanity and purifying the entire world (Alley 2002: 60). In another tale, Ganga is one of the three wives of Vishnu. During an argument with another of his wives, Saraswati, the two women curse each other to become rivers and descend to the earth (Gupta 1993: 108). In the *Puranas*, "The river issues from the foot of Vishnu above, washes the lunar orb and falls here from the sky, encircles the city [of Brahma, at the top of the Himalayas] and then divides into four mighty rivers flowing [in the four cardinal directions]" (quoted in Chapman 1995: 15). Thus, the origins of the Ganga and her sister tributary the Yamuna are divine, and not earthly, as with the other major rivers of the earth (Pandey 1984).

In addition to Vishnu, Ganga is also connected to other major Hindu gods, especially Shiva and Krishna, who are often credited with reigning in the river goddess' mighty power. In the *Bhagavad Gita*, the river is described as "The one that descends onto the earth from the heavens" (Kishore 2008: xiii) and is associated with Krishna, who reveals, "In rivers, I am the Ganga" (Pant 1987: 1). In the *Brahmavaivarta Purana*, Krishna's consort Radha drinks the river goddess in a fit of jealousy. Ganga takes shelter from her at Krishna's feet, removing all water from the world. After the gods pray to Krishna for her release, he ruptures his toenail and frees her back upon the earth (Gupta 1993).

---

[5] In the Rig Veda (seventeenth to twelfth century BCE), both the Ganga and the Yamuna are given priority over other rivers and described as divine (Pandey 1984).

Yet, the best-known legend associated with the descent of Ganga is that of the prolific King Sagara found in the *Ramayana*. The 60,001 sons of the King, an ancestor of Ram, are burned to death in battle with the sage Kapila, leaving their ashes scattered across the great plains of India. The descendents of Sagara appeal to *Ganga* to come to earth and release the sons from their earthly bondage. The descendants of Sagara eventually convince Brahma to release Ganga (or, in some versions of the story, convinced Ganga herself to descend). But, Shiva must contain Ganga's descent, as the power of her flow threatens to break the world apart. He stands at the top of the Himalayas, capturing Ganga in his hair and eventually setting her free across the plains, where she cleanses and releases the souls of Sagara's many sons (Pandey 1984; Alley 2002). It is this association of the river with the ability to purify bodies, ashes, and souls that becomes especially significant for practices of river worship and water use.

28.3 River Worship and Water Use

Although water, especially river water, is considered sacred in Hindu religious tradition, the Ganges is the holiest of all rivers in India, sent to earth to rid humankind of evil and impurity. In the *Mahabharata*, water from the Ganges, or *ganga jal*, is described: "As Amrita [the nectar of immortality] to the gods … even so is Ganga water to human beings" (XIII, translated and quoted in Darian 1978: 125). As with other rivers, the Ganges is feminized and "perceived to be nurturing (and sometimes judgmental) mothers, feeding, nourishing, quenching, and when angered flooding the earth" (Narayanan 2001: 193–194). Throughout the major Hindu texts, Ganga is repeatedly referenced as the most sacred river, revered for her purity, sanctity, and emancipatory power (Pandey 1984). To wade on her banks, swim in her streams, or even to take a drink of containerized Ganges water (*ganga jal*), is to be cleansed of material dirt (*gandagi*) and ritual pollution or contamination (*apavitra*). These actions allow devotees to attain *suddhata*, or religious purity. Placing the ashes or uncremated remains of loved ones into the river assists their release from the cycle of rebirth and ensures "attainment of heaven" of the dead (Pandey 1984: 21).

There are many rituals and methods of religious practice into which the use of *ganga jal* is incorporated. *Ganga jal* can be drunk or splashed as an element of *puja*, or worship. A vial or jug of *ganga jal* can be kept in the home as a blessing or for future use if a family member falls ill. But, above all else, bathing in the Ganga is the most favorable act for a Hindu devotee (Fig. 28.3). In an excerpt from the Mahabharata, a sage explains the sanctity of the Ganga and her waters:

> That end which a creature is capable of attaining by penances, by practicing celibacy, by sacrifice or by practicing renunciation, one is sure to attain by only living by the side of the Ganga and bathing in its sacred waters. Those creatures whose bodies have been sprinkled with the sacred waters of [Ganga] or whose bones have been laid in the channel of that sacred stream, have not to fall away from the heaven at any time. Those men, who use the waters of the Ganga in all their acts, surely ascend to heaven after departing this world …. Hundreds of sacrifices cannot produce that merit which men of restrained souls are capable of acquiring by bathing in the sacred waters of the Ganga… As cotton, when it comes into

Fig. 28.3 A worshipper bathes in the Ganga at Sirsaya Ghat, Kanpur (Photo by Sya Buryn Kedzior)

contact with fire, is burnt off without a remnant, even so the sins of the person that has bathed in Ganga become consumed without a trace. (quoted in Kishore 2008: 26–27)

While bathing in the Ganga at any time is surely beneficial, certain dates and hours are more auspicious than others. At sunrise and sunset each day, an *Aarti* is performed for the Ganga at numerous sites on the river. In most of these, an Aarti lamp is circulated during the recitation of prayers and then waved, while facing the river, in a circular pattern, to indicate that the Ganga is the goddess around whom all life and activity circulate (Fig. 28.4). Aartis for the Ganga also serve to "wake up" the goddess in the morning and to put her to "sleep" at sunset. After the performance of the Aarti, individual worshippers light candles suspended in a boat made of leaves (often lotus leaves) and filled with flowers, incense and other symbolic offerings. The boats are then placed in the river with a prayer and make their way downstream, presumably to the river delta and the ocean.

Pilgrimage to the Ganga, and especially circumambulation along her shores, is an important practice for many Hindus. The greatest number of *tirthas*, or pilgrimage sites, in India are located in the GRB, with significant clustering of these at the upper reaches of the river and in eastern Uttar Pradesh, around the *prayaga* at Allahabad (Bhardwaj 1983). Mass pilgrimage fairs are held regularly along the Ganges, with *Kumbh Melas*[6] drawing record-breaking crowds. Within these celebrations there are specific days and times during which bathing is particularly

---

[6] Kumbh refers to the "pitcher" or "urn" that contains Amrit, or the nectar of the gods and of life, while mela refers to a festival or fair.

Fig. 28.4 A tourist circulates the Aarti lamp during a ceremony at Rishikesh, Uttaranchal (Photo by Sya Buryn Kedzior)

rewarding. At these times, millions of pilgrims participate in ritual bathing simultaneously[7] (Fig. 28.5). These celebrations represent the largest gathering of people in the world and draw crowds so large that they can be viewed from space (BBC News 2001). The use of the river for mass bathing draws particular attention from religious officials, government agencies, and civil society organizations, who express concerns about bathers' exposure to pollutants and their production of non-point source pollution on a massive scale.

28.4 River Pollution

As much as the Ganges may be the spiritual heart of Hinduism, the river also serves as the demographic, political, and agricultural center of the nation. Because of this centrality, the river has long suffered from "pollution" of one sort or another.[8] Today,

---

[7] The most recent Purna Kumbh Mela, held in Haridwar between January and April 2010, was attended by tens of millions of people, with about 10 million of those bathing on April 14th (Yardley and Kumar 2010). Chapman (1995) reports that of the estimated 30 million pilgrims at the 1989 Kumbh Mela in Allahabad, 15 million bathed at daybreak on 6th of February. At the last Kumbh Mela to be held in Allahabad, in 2007, approximately 50 million people were present (Kishore 2008).

[8] Precolonial descriptions of the Ganga "are unanimous in their description of the Ganges as wholesome, clear, sweet, tasty, and digestive" (Markandya and Murty 2000: 222). But, in 1859, the

Fig. 28.5 Tent city at Sangam during the 2009 Mela, Allahabad (Photo by Sya Buryn Kedzior)

the waters of the Ganges River rank among the most polluted in the world (Ministry of Environment and Forests 2005). Yet, it is difficult to get a full picture of current water quality and sources of pollution largely due to the seasonal fluctuation of the amount of water in the river and noncomprehensive monitoring procedures. In the most general sense, water quality is highly variable along the course of the river, tending to decline gradually as the Ganges descends and reaches the plains, and becoming markedly worse in the stretch of the river from Kanpur to Varanasi. After Varanasi, water quality improves due to inputs from tributaries in the lower courses of the river.

According to a report published by the United Nations Environment Programme (UNEP) and World Health Organization (WHO), the main sources of pollution in the river are domestic and industrial waste (liquid and solid), agricultural runoff bearing pesticides and fertilizers, partially- or un-cremated human and animal

---

Mela festivities at Allahabad were nearly cancelled by local authorities concerned about the possibility that a cholera outbreak had been caused by the collection at the Sangam of "all the dejections and filth of this immense population" (Deputy Inspector General's Office, as quoted in Maclean 2008: 78). Concerns about the relationship between poor water quality and the outbreak of disease inspired the creation of Sanitation Police and later an office of North-Western Provinces Sanitary Commissions who were, on many occasions, charged with dispersing crowds, controlling mass bathing in the river, and even breaking up festivities during annual Mela celebrations (Deputy Inspector General's Office, as quoted in Maclean 2008). In 1896, Samuel Clemens, writing as Mark Twain, described the Ganges River at Varanasi as "nasty" and joined scores of other foreign travelers in complaining about the partially decomposed bodies found floating in the river (McNeill 2001).

remains, direct defecation into the river, and mass bathing and other practices associated with ritual worship of the river (Sharma 1997). Problems with water quality are compounded by dams and irrigation projects, which decrease the amount of water flowing into the main stem of the river and contribute to the rapid accumulation of silt on the riverbed. Because of these projects, the river has become innavigable in its upper stretches and pollution loads are concentrated during the dry winter season, when farmers compensate for lost rainfall by increasing water extraction. Unfortunately, it is at this time that human exposure is also high, as pilgrims flock to the banks of the Ganges for the winter/spring festival season.

Investments in increasing the river's irrigation potential have come as a response to state prioritizations of food security in the face of a rapidly growing population.[9] Before Independence, reliance on the monsoon contributed to cyclical famine. Development of canal infrastructure provided additional water resources for GRB farmers, especially those within reach of the Upper and Lower Ganges Canals. These canals act as the backbone of a multi-state irrigation network that waters about 1.8 mil acres (7.29 mill hectares) of farmland in Uttar Pradesh and Uttarakhand (Jain et al. 2007). Further investment in the extension of irrigation networks and expansion of lands under cultivation helped to improve regional food production in the postcolonial context. Adoption of Green Revolution technologies and strategies during the late 1960s and 1970s led to the development and distribution of high-yield rice and wheat varieties, subsidies and loans for farmers using chemical fertilizers and pesticides, and the continued expansion of irrigation infrastructure. Over the next decades, while grain production in north India increased markedly, so did the presence of chemical inputs, like synthetic fertilizers and pesticides, in agricultural run-off, and therefore in both ground and surface water,[10] including the canalized runoff that is returned to the Ganges.

While India's population remains, as a whole, relatively rural,[11] the central GRB is home to about a third of the nation's cities and another third of its urban residents (Kishore 2008). This population places significant pressure on the river not only as a

---

[9] At Independence, India's population numbered only 350 million people, with about a third of those living in the GRB (Chapman 1995). Over the past six decades, the population of India has grown to 1.21 billion people (Census of India 2011), with the population of the Indian GRB alone calculated between 400 million (Rashid and Kabir 1998) and 500 million (Sharma et al. 2008). Much of this population is clustered in the central basin states of Uttar Pradesh and Bihar, which claim some of the largest (199 and 103 million people, respectively) and densest (828 and 1,102 people per sq km (320 and 425 per sq mile, respectively, compared with 382 per sq km (147 per sq mile) nationally) populations in the country (Census of India 2011).

[10] GRB states alone consume nearly 10 million tons of chemical fertilizer each year (45 % of all fertilizers utilized nationally), of which 10–15 % are estimated to end up in surface water systems. The Ganges and her tributaries are believed to contain up 70 mg/l nitrogen and .05–1.1 mg/l phosphorus rates; much higher than 10 and 0.1 mg/l phosphorus rates that are considered unsafe in drinking water (TERI 2011). Pesticides, which have a greater toxicity than fertilizers are also used profligately throughout the GRB. About 21,000 tons of pesticides are applied to cropland in the GRB each year (47.6 % of those used nationally). However, it is difficult to get accurate data on the types and chemical composition of many pesticides, as they are often protected as intellectual property of their manufacturers.

[11] Only 29 % of India's population lived in urban centers in 2001 (Census of India 2011).

source of water for consumption, but also as a site of waste disposal. In most towns and cities along the Ganga there is a lack of infrastructure for water and sewage treatment, and where it does exist, facilities frequently fall far short of comprehensive treatment. From the 179 Class I basin cities for which information is available, about 11,100 million liters per day (MLD) of sewage is produced, of which an average of 24 % is subjected to treatment before being released into the river (Maria 2003). While reliable data on solid waste disposal are available for only five of the largest cities in the Indian GRB, none of these use processing facilities or sanitary landfills. Solid waste creation is estimated at 1,100 tons per day (TPD) in Kanpur alone (Maria 2003). But, few data exist on how much of this solid waste flows to the river or enters landfills, the size or capacity of current landfills, and their area of coverage. Non-point sources of pollution are equally difficult to measure, with estimates of human ash disposal running into several millions of tons at Varanasi alone—the product of 30 million bodies being burned each year in the holy city's official crematoria (McNeill 2001). Sources like detergents and bleaches from clothes washing at the *dhobi* ghats, solid and liquid waste from idol immersion and bathing, or organic pollution from direct urination and defecation into the river go unmeasured.

Industrial development has also clustered in major cities on the river, where factories take advantage of ready access to raw materials from both the plains and Himalayan foothills, unskilled urban labor, and proximity to the river for use in production and for waste disposal. Popular industries include metalworks, pharmaceuticals manufacturing, textile mills, distilleries, chemical plants (fertilizer and pesticide), tanneries, munitions factories, and paper and sugar mills. These industries are located near on the river because they are prolific water consumers, employing large tubs and washbasins to repeatedly treat or wash their products.[12] Many factories in the GRB specialize in production of low-quality, semi-finished goods, often produced in sub-standard labor and environmental conditions. Specialization in low-end goods means that factory owners frequently participate in a "race to the bottom" that involves reducing investments in manufacturing and treatment technologies, relying on underpaid or child labor and operating in sub-standard environmental conditions. Environmental protection legislation furthers local industrial agglomeration by encouraging factories to operate in clusters, whereby they share effluent treatment facilities, in addition to common labor pools and suppliers. According to one source, clustering allows factory owners to exert disproportionate political influence in the municipality and the state, and has promoted corruption of UP-PCB officials. Factory owners can use their economic clout to control the outcome of mandatory effluent testing and coerce employees not to vote for particular officials or political parties that have threatened to fine or shut down their factories. Regulations related to the treatment of industrial wastewater or effluent are reportedly easy for factory owners to ignore, because monitoring

---

[12]The textile industry serves as a good example of a high water consumption industry in the GRB. Almost every step of fabric production requires water, including scouring, bleaching and dyeing the product. As each stage of production, fabric is re-washed with fresh water in order to remove chemicals applied during previous stages, and is then often disposed of without treatment. Production of one kilogram (1 kg) of cotton fabric consumes between 272 and 784 kg of water, depending on processes and equipment used (AquaFit4Use 2010), and produces between 150 and 75 l of wastewater for every kg of fabric produced (Jacob and Azariah 1998/2008).

authorities have little desire to close those factories fueling regional and national economic growth, or that are themselves owned by the state.[13]

Under the current Best Use Designation scheme, the Ganges is a Class B river, indicating that water quality in the river *should* be safe for organized (mass) bathing. As Table 28.1 illustrates, the Ganges does not meet this standard at most testing sites and is most frequently classified as a Class D river, indicating that its waters are safe enough only for the propagation of wildlife and fisheries, and not human contact or consumption (MoEF 2009). Water quality monitoring data from 2009 (TERI 2011) indicate that average Biochemical Oxygen Demand (BOD) concentrations in the river were 5–7 mg/l (Class B ranking requires 3 or less), with higher levels recorded in the stretch of the river between Kanpur and Varanasi. The highest BOD values of 65.8 mg/l were reported during the pre-monsoon summer in Kanpur. Coliform levels were also significantly high at this time, with the monitoring station downstream from Agra reporting rates as high as 5.2×10^6 coliform counts per 100 ml of sampled water, which is far beyond both the CPCB standard of 500/100 ml for Class B rivers and the WHO recommendations of no more than 2/100 ml in recreational or bathing waters (NRCD 2009).

Despite the health risks posed by contaminates that regularly exceed national and international recommendations for safe use, the river continues to be a site where people gather water for watering livestock, washing clothes and dishes, and participating in the Hindu bathing rites believed to cleanse the soul of sin and impurity. These diverse water users are subjected to a number of health risks through their exposure to toxic chemicals and pollutants in the water, including diseases like cancer, endocrine disorders, cataracts, and kidney and liver disease (Maria 2003). Other effects include rashes and yellowed patches of skin, eyesight problems and heavy metal poisoning (Sengupta 2006; World Health Organization 2004). While these may sound more hazardous than the diarrhea and bacterial infections caused by E. coli and other pathogens introduced and spread through unsanitary water conditions, they affect fewer people and, therefore, are viewed as less of a potential threat to human health (Hosterman et al. 2009). Diarrhea alone causes an average of one death every minute in the Gangeatic region (Sampat 1996). While some direct water use is associated with daily material needs, devotees seek the waters of the Ganges because tradition identifies it as an infallibly pure river whose waters cure physical and spiritual ills. Many water users argue that these purificatory powers of the river actively prevent, rather than cause, the many illnesses associated with water pollution.

28.5 Pollution and the Renegotiation of River Water Worship

As described above, bathing and drinking of the Ganges River water is a common practice, especially among Hindus as an aspect of devotional worship. But direct observation and experience with river water pollution is changing how many devotees interact with the water, conduct practices of worship, and ultimately conceive

---

[13] Some of the most flagrant polluters in the GRB are government-owned industries, including the Ordnance Factory in Kanpur and the Diesel Locomotive Works in Varanasi (Krishna 2004).

Table 28.1 Classification of the Ganges River at various locations

Location	Desired class	Observed class (Critical parameter)					
		1997	1998	1999	2000	2001	2008
Ganga at Rishikesh	A	D (CF)	B (CF)	C (CF)	NA	C (CF)	B
Haridwar	B	C (CF)	C (CF)	C (CF)	NA	C (CF)	B
Garhmuktesar (UP)	B	B (BOD)	–	D (BOD)	NA	D (BOD, CF)	NA
Kannauj U/s UP	B	D (BOD, CF)	D (BOD)	D (CF)	D (BOD, CF)	D (BOD, CF)	C (CF)
Kannauj D/S UP	B	D (BOD, CF)	D (BOD)	D (CF)	D (BOD, CF)	D (BOD, CF)	C (BOD, CF)
Kanpur U/S UP	B	D (BOD, CF)	NA	D (CF)	D (CF)	D (CF)	D (CF)
Kanpur D/S UP	B	D (BOD, CF)	D (BOD)	D (CF)	D (BOD, CF)	D (BOD, CF)	D (BOD, CF)
Raibareilly UP	B	D (CF)	D (CF)	C (CF)	NA	NA	NA
Allahabad U/S UP	B	D (BOD, CF)	E (CF)	D (CF)	NA	NA	C (BOD, CF)
Allahabad D/S UP	B	D (BOD, CF)	E (CF)	D (CF)	NA	NA	D (BOD, CF)
Varanasi U/S UP	B	D (BOD, CF)	D (BOD)	D (CF)	–	D (CF)	D (BOD, CF)
Varanasi U/S UP	B	E (BOD, CF)	E (BOD, DO)	D (BOD)	NA	NA	D (BOD, CF)
Gazipur UP	B	D (BOD, CF)	D (BOD)	D (BOD)	D (CF)	D (CF)	NA
Buxar	B	D (BOD)	D (CF)	D (CF)	D (CF)	D (CF)	C (CF)
Patna U/S	B	D (CF)	D (CF)	D (CF)	D (CF)	D (CF)	NA
Patna D/S	B	D (CF)	D (CF)	D (CF)	D (CF)	D (CF)	NA
Rajmahal	B	D (CF)	D (CF)	D (CF)	D (CF)	D (CF)	D (CF)
Palta (WB)	B	D (BOD)	B	NA	D (BOD, CF)	D (BOD, CF)	D (CF)
Uluberia (WB)	B	D	B	NA	D	D (BOD, CF)	D (BOD, CF)

Data source: TERI (2011)

of the river as a purificatory goddess. Elsewhere, scholars have argued that belief in the goddess Ganga has prevented water users from acknowledging pollution in the river (see Alley 2002). However, surveys among water users at three urban sites in the central basin show that an overwhelming 78.5 % of respondents confirmed that there is indeed pollution in the Ganges. Rates of affirmative response were highest in Kanpur, where just over 90 % of respondents indicated that there is pollution in the river. Indeed, many water users expressed surprise at the idea that anyone living along the Ganges would not recognize that the river is polluted: "Everyone here knows the River is polluted. Have you met anyone who says it is not?" (male tannery consultant, personal communication); "Everyone should have knowledge of [problems of pollution in the Ganges] nowadays. This [the Ganges] is mother. This is God." (businessman, 2009, personal communication).

Among these Kanpuri respondents, personal observations associated with direct water use (as well as newspapers and family and friends) were reported as important sources of knowledge related to pollution. Compared to respondents in other cities, those from Kanpur reported with greater frequency that river water quality has declined in the city and that this decline is observable to the casual user. The river was variously described as "polluted," "dirty," and "not clean." Others reported visual evidence of deteriorating water quality, including dirtier looking, "blackened" water, a "whiteness" or a "pale yellow" color in the water, or increased garbage floating in the water. Some respondents' observations were more specific: "[Now there is] too much impurities by industries." (male teacher, 2009, personal communication); "[The water is] more dirty, because of tanneries." (male electrician, 2009, personal communication)

Compared with respondents at other sites, those from Kanpur were particularly adept at identifying sources of river water pollution. A majority 61.1 % of Kanpuri respondents indicated that pollution comes from the city itself, from the factories that line the banks and the sewage drains that empty directly into the river. In an interview with one worshipper, he argued that it is easy for people from the city to identify pollution, because it is more abundant and easier to see in Kanpur than in other cities. Numerous respondents pointed to the drains, visible especially during the dry low-water season, that empty colored effluent into the river in view of major water collection and bathing sites. Others pointed to the physical evidence of pollution: the debris that washes up on the shore, the black and green slime that collects on the *ghats* or stairs at bathing sites, and the rashes that now appear on the soft skin of bathers' stomachs and backs following a dip in the river at certain sites. These respondents drew attention to the idea that there is little room for plausible deniability when it comes to pollution in the Ganges: it can be seen, smelled, touched, and the effects of its existence experienced as illness in and on the body.

These observations and experiences with pollution have changed how worshippers in Kanpur conceive of their relationship with the river and have significant consequences for how water users interact with and make use of river water. Across the three study sites in the central GRB, 88.4 % of respondents indicated having bathed in the river at some point, with 84.3 % reporting that they had at one time drunk water, *ganga jal*, directly from the river. Among sites, Kanpur had the lowest rates of participation in both of these activities, with only 71.4 % of Kanpurites indicating that they have participated in either bathing or drinking activities at some time (compared with 94.7 % of respondents in Varanasi). Yet, many of these respondents report that they are bathing in and drinking from the river less frequently than they had before, and less frequently than water users at other study sites. Only 6.7 % of respondents at Kanpur reported that they bathe in the river on a daily or weekly basis. Nearly 40 % reported monthly bathing in the river, compared with more than 90 % of Varanasi respondents who reported daily bathing activities. This indicates that water users in Kanpur bathe in the river far less frequently than their counterparts in Varanasi. While this pattern may reflect lower rates of religious water use in Kanpur (compared with Varanasi which is a holy city), it also may be indicative of a more widely held apprehension that water quality in Kanpur is not suitable for bathing.

Only 38.1 % of study respondents at Kanpur agreed with the statement that, "It is safe to drink water directly from the river." Many people expressed concern that the safety of river water drinking was relative and varied according to the site where water was collected or the immunity of the individual involved. In terms of water treatment, 33 % of Kanpurites agreed with the statement that, "People should boil or treat *ganga jal* before drinking it." But, many water users argued that, while boiling or treating the water would be preferable, most people lack the means or resources to do so. Finally, 42.9 % of Kanpurites expressed some agreement with the statement that, "Drinking river water may make some people sick." It is, therefore, interesting to consider that in Kanpur, water users are nearly divided on whether they believe that drinking holy river water can lead to illness. One survey respondent pointed to inconsistencies between the contention that the purity of *ganga jal* prevents its drinkers from becoming ill and the real observations and lived experiences of those who have become ill from drinking river water: "According to religion, drinking [ganga jal] could not make one sick. But, according to experience, one can see otherwise" (serviceman, 2008, personal communication).

Follow-up interviews with survey respondents indicated that many Kanpurites are responding to decreasing river water quality by modifying bathing and drinking activities. Some worshippers no longer engage in bathing and drinking activities: "People shouldn't drink [*ganga jal*] at all" (male worshipper, 2009, personal communication); "No one in my family drinks from the river. I tell my children to drink only bottled water.... Bathing? No. We are educated people and do not need to bathe in the river in this manner" (male professor, 2009, personal communication).

But, rather than ceasing bathing and drinking activities altogether, most respondents indicated that they are changing where and how they engage with the river. One respondent said that yes, he bathes, but wanted to qualify that activity by stating that he only bathed on the far side of the river. A young woman explained that she bathes in the river, but is careful not to immerse her face in the water. A male technician reported that Kanpurites with the resources to do so will travel nearly 20 km (6 miles) to Bithoor, a site upstream from the factories and city of Kanpur, where bathing and drinking are thought to be relatively safer. Others, like the architect Raj with whom the chapter opened, are careful to minimize contact with the river water by rinsing their foreheads with a handful of water collected from the riverside before washing their face and hands clean with bottled water. One young male respondent enthusiastically exclaimed, "I try to bathe in the river nearly every day. But I run home immediately after, to shower and wash the filth from my body, so I do not fall ill" (student, 2009, personal communication).

These changes in bathing and drinking activities have significant implications both for worshippers' relationships with the river itself and for how water users portray the river as an inviolable goddess. Across the GRB, respondents reported that, against mounting visual and experiential evidence of pollution in the river, many worshippers are rethinking their ideas about the river as goddess and her divine powers to purify. One male military officer provided a clear case of a water user who had abandoned religious interpretations of the river's power to purify: "[The Ganga was] a pure river, but no longer. Its spiritual power is no more" (personal communication). Yet others are not as hasty to abandon a belief system that identifies the river as a "cleanser par excellence" (activist, 2009, personal communication). In order to account for apparent contradictions between this belief and their observations and experiences of pollution in the river, some water users have developed novel views, describing the power of the river as having a "limit" or "ceiling" that has been surpassed, or as being diminished by the placement of dams on the river and the subsequent reduction of water levels and flow rates. Others argue that the sacredness of the Ganga, her purificatory powers, were meant only for the Hindu religious practices performed on her shores, and that it is through the introduction of alternative forms of water use (for irrigation, sewage disposal and industry) that the river has become polluted. In other words, the river *does* continue to purify waste from ritual bathing and idol disposal, for example, but not from factory effluent, as she is a river sent to cleanse the bodies and souls of people, and not of industries. In this view, industries and other non-religious uses need to be removed from the river before the river can once again cleanse herself.

28.6 Conclusion

In the face of diminishing water quality and direct observations of both pollution and its effects, water users in the Ganges River Basin are changing the practices associated with river water worship and use. While some people are no longer participating in the bathing and drinking activities central to Hindu devotional practice, others are modifying these activities in order to reduce exposure to water pollution and limit the risks posed at sites perceived as having higher levels of pollution. These changes illustrate the unfixed nature of religious practice, as well as interlinkages between environmental and religious change.

But, it is important to be mindful that the modifications to water use and religious practice documented here are not occurring in a vacuum. In Kanpur and elsewhere along the river, civil society organizations actively lobby against not only pollution

Fig. 28.6 Protestors with the Save Ganga Movement hoisting a picket sign reading, "Save Ganga, Save the Country." (Photo by Rama Rauta 2012, http://vinacc.blogspot.com/2012/01/save-ganga-save-himalayas-meeting-12th.html; translation by Sya Buryn Kedzior, used with permission)

levels and unsuccessful state-run abatement programs, but also against the practices of water users and Ganga worshippers that contribute to poor water quality (Fig. 28.6). The Kanpur Eco-Friends (KEF), an holistic environmental organization, runs educational and media campaigns that aim to convince local people to stop worshipping the Ganga, to end practices of river bathing, and to substitute riverside burial for long-held practices of cremation and river immersion. While few survey respondents reported knowledge of the KEF or their anti-bathing campaign, state agencies and some local officials are also active in their efforts to reshape river water use practices.

Nonetheless, the relationship between water pollution and changed religious practice seems undeniable, especially as we turn to other sites on the river that are beginning to exhibit trends similar to those reported in Kanpur. In both Varanasi and Allahabad, river worshippers reported observations of decreased water quality and tied these to changes in religious practice: "Now I don't want to drink the water because it is so dirty" (male sweetmaker, 2009, personal communication); "We don't want to drink, but we drink out of faith … Religious [tradition] says we should [drink ganga jal], but it isn't safe" (retired male priest, 2009, personal communication); "It is so polluted now, we don't want to drink [the water]" (female homemaker, 2009, personal communication).

Frequencies of participation in both bathing and drinking activities were also much lower among younger generations, especially respondents in their teens and twenties, who reported little recollection of a once clean and pollution-free river.

But, while devotional water use practices change, belief in the Goddess Ganga may not, ultimately, fade. Kanpuri respondents who indicated that they no longer bathe in or drink from the river continued to flock to the river banks in order to participate in leisure activities or, more notably, for *dekne* (to look at or see the river) and *darshan* (to behold a deity). People of many religious creeds spoke of the river as an important site for finding peace, tranquility, and opportunities for meditation and self-reflection. For others, the idea that the holy river now carries too much pollution for direct worship activities motivates the desire for wider social change and the need for state and public intervention:

> Out of humanity, we should take care of Ganga. People should clean at the river every day just like they clean their homes every day. If you do not clean your own home, it will be dirty. The river is like this, your home. (farmer, personal communication)

Organization of devotees sharing these sentiments may ultimately lead to the formation of a sustained civic action campaign against pollution in the river and to the savior of the River Ganga by those who wish to continue her worship as an unadulturable goddess.

References

Alley, K. D. (2002). *On the banks of the Ganga: When wastewater meets a sacred river*. Ann Arbor: University of Michigan Press.
AquaFit4Use. (2010, April 13). Water quality demands in paper, chemical, food and textile companies. Retrieved June 22, 2011, from www.aquafit4use.eu/userdata/file/Public results/AquaFit4Use - Water quality demands in paper-chemical-food-textile industry.pdf
BBC News. (2001). Kumbh Mela pictured from space. Retrieved June 29, 2011, from http://news.bbc.co.uk/2/hi/science/nature/1137833.stm
Beg, K. R., & Ali, S. (2008). Chemical contaminants and toxicity of Ganga River sediment from up and down stream area at Kanpur. *Journal of Environmental Sciences, 4*(4), 362–366.
Bhardwaj, S. M. (1983). *Hindu places of pilgrimage in India: A study in cultural geography*. Berkeley: University of California Press.
Bonnemaison, J. (2005). *Culture and space: Conceiving a new cultural geography*. New York: Palgrave Macmillan.
Census of India. (2011). Census data summary. Retrieved June 23, 2011, from www.censusindia.gov.in/2011-common/CensusDataSummary.html
Chapman, G. P. (1995). Introduction: The Ganges-Brahmaputra basin. In G. P. Chapman & M. Thompson (Eds.), *Water and the quest for sustainable development in the Ganges Valley* (pp. 3–24). London: Mansell Publishing.
Cumming, D. (1994). *The Ganges delta and its people*. London: Wayland.
Darian, S. G. (1978). *The Ganges in myth and history*. Honolulu: University of Hawaii.
Gupta, J. (1993). Land, dowry, labour: Women in the changing economy of Midnapur. *Social Scientist, 21*, 74–90.
Hosterman, H. R., McCornick, P. G., Kistin, E. J., Pant, A., Sharma, B., & Bharati, L. (2009). *Water, climate change, and adaptation: Focus on the Ganges River basin [Working Paper]*. Durham: Duke University, Nicholas Institute for Environmental Policy Solutions.
Jacob, C. T., & Azariah, J. (1998/2008). Environmental and ethical Costs of t-shirts Tiruppur, South India. In N. Fujiki & D. R. Mauer (Eds.), *Bioethics in Asia: Proceedings of the UNESCO Asian Bioethics Conference* (pp. 191–195). Christchurch: Eubios Ethics Institute.

Jain, S. K., Agarwal, P. K., & Singh, V. P. (2007). *Hydrology and water resources of India.* Dordrecht: Springer.

Kishore, K. (2008). *The holy Ganga.* Kolkata: Rupa & Co.

Krishna, G. (2004). Uttar Pradesh: Ganga in crisis. In Panos Institute South Asia (Ed.), *Disputes over the Ganga: A look at potential water related conflicts in South Asia* (pp. 124–143). Kathmandu: Panos Institute South Asia.

Maclean, K. (2008). *Pilgrimage and power: The Kumbh Mela in Allahabad, 1765–1954.* New York: Oxford University Press.

Maria, A. (2003). *Costs of water pollution in India.* Paris: CERNA, Ecole Nationale Supérieure des Mines de Paris. Retrieved December 28, 2006, from www.cerna.ensmp.fr/cerna_globalisation/Documents/maria-delhi.pdf

Markandya, A., & Murty, M. N. (2000). *Cleaning up the Ganges: A cost-benefit analysis of the Ganga action plan.* New Delhi: Oxford University Press.

McNeill, J. R. (2001). *Something new under the sun: An environmental history of the twentieth-century world.* New York: W.W. Norton and Co.

Mishra, M. (2009). Leather tanning industry in Kanpur hit by ongoing global recession. ANI News. Retrieved March 22, 2009, from http://hamaraphotos.com/news/national/leather-tanning-industry-in-kanpur-hit-by-ongoing-global-recession.html

MoEF (Ministry of Environment and Forests). (2005). *State of environment report. Prepared by development alternatives.* New Delhi: Government of India.

Narayanan, V. (2001). Water, wood, and wisdom: Ecological perspectives from the Hindu traditions. *Daedalus, 130*(40), 179–206.

NRCD (National River Conservation Directorate). (2009). *Status paper on River Ganga: State of environment and water quality.* New Delhi: Ministry of Environment and Forests, Government of India.

Pandey, S. (1984). *Ganga and Yamuna in Indian art and literature.* Chandigarh: Prakashan.

Pandey, V. (2008). Kanpur textile mills going to seed. Business Standard, December 27. Retrieved January 19, 2009, from www.business-standard.com/india/storypage.php?autono=334461

Pant, P. (1987). *Ganga: Origin and descent of the river eternal.* New Delhi: Frank Bros. & Co.

Rashid, H., & Kabir, B. (1998). Case study: Bangladesh water resources and population pressures in the Ganges River basin. In A. de Sherbinin, & Dompka, V. (Eds.), *Water and population dynamics: Case studies and policy implications.* American Association for the Advancement of Science (AAAS). Retrieved May 12, 2010, from www.aaas.org/international/ehn/waterpop/bang.htm

Rauta, R. (2012). Save Ganga Movement. VINA News Blog. Retrieved April 21, 2012, from http://vinacc.blogspot.com/2012/01/save-ganga-save-himalayas-meeting-12th.html

Sampat, P. (1996). The river Ganges' long decline. *World Watch, 9*(14), 24–32.

Schjolden, A. (2000). *Leather tanning in India: Environmental regulations and firms' compliance* (FIL Working Papers, No. 21). Retrieved October 25, 2008, from www.cicero.uio.no/media/1677.pdf

Sengupta, R. (2006). The tragedy of the Ganges. *TerraGreen, 121.* Retrieved December 2, 2006, from www.teriin.org/terragreen/index.php?option=com_content&task=view&id=367&Itemid=3

Sharma, Y. (1997). Case study I: The Ganga, India. In R. Helmer, & Hespanhol, I. (Eds.), *Water pollution control: A guide to the use of water quality management principles.* New York: United Nations Environment Programme (UNEP). Retrieved March 26, 2008, from www.who.int/water_sanitation_health/resources/watpolcontrol/en/index.html

Sharma, B., Amarasinghe, U.A., & Sikka, A. (2008). Indo-Gangetic river basins: Summary situation analysis. Challenge program on water and food. Retrieved July 14, 2009, from http://cpwfbfp.pbworks.com/f/IGB_situation_analysis.PDF

Spate, O. H. K., & Ahmad, E. (1950). Five cities of the Gangetic plain: A cross section of Indian cultural history. *Geographical Review, 40*(2), 260–278.

TERI (The Energy and Resources Institute). (2011). *Environmental and social management framework (ESMF) Volume I: Environmental and social analysis*. New Delhi: National Ganga River Basin Authority. Retrieved June 24, 2011, from www-wds.worldbank.org/external/default/WDSContentServer/WDSP/IB/2011/03/24/000333038_20110324010819/Rendered/PDF/E26650v10REV011B0SAR1ESMF1P119085v2.pdf

World Health Organization (WHO). (2004). The environment: Where's the risk, and where are children safe? WHO Press Release [Online], June 22. Retrieved October 9, 2008, from www.who.int/mediacentre/news/releases/2004/pr43/en/

Yardley, J., & Kumar, H. (2010, April 14). Taking a sacred plunge: one wave of humanity at a time. *New York Times*. Retrieved January 14, 2011, from www.nytimes.com/2010/04/15/world/asia/15india.html?_r=1

Chapter 29
Privileged Places of Marian Piety in South America

David Pereyra

29.1 Introduction

Devotion to the Blessed Virgin Mary has been a centerpiece of Roman Catholic belief and piety for centuries. In the Americas it is a phenomenon that started in the early 1500s during the *Spanish Conquista* and continues into the twenty-first century.

Over the past 500 years many apparitions of the Virgin Mary (*Virgen María* as she is called by Latin Americans) have been reported in the Americas. These apparitions are often accompanied by various forms of evidence—tangible or symbolic—which serve as reminders of her visits. A well-known Latin American example is the image of Our Lady of Guadalupe which is reported to have been miraculously imprinted on the cloak of Saint Juan Diego. Although skeptics might say that these "reminders" could easily come from other sources, the stories that have been told throughout time about these objects or forms of evidence are valuable in understanding the nature and the history of belief within these communities.

To be officially recognized by the Roman Catholic Church, a Marian apparition can be granted approval either through the local Bishop from the direction of the Congregation for the Doctrine of the Faith or it may receive a direct approval from the Holy See. The apparitions are often given names based on the town where they appear and it is not uncommon for a Marian apparition to recur several times over an extended period of time (Marzal 2002: 243–264).

With some Marian apparitions a vision is reported to appear as a lady. In these cases the viewers often describe experiences that include visual and verbal interaction with the woman present at the site of the apparition. With other

D. Pereyra (✉)
Toronto School of Theology, University of Toronto, Toronto, Canada
e-mail: david.pereyra@utoronto.ca

apparitions, an image is reported, but there is no verbal interaction or conversation. Physical contact is hardly ever an element of a Marian apparition and in only the rarest of cases is a physical artifact reported. The expert in this matter is Father Rene Laurentin, a French theologian. He has followed with extreme interest almost every reported apparition, messenger, and other expressions over the past 40 years (Laurentin 1997).

In South America, some major Marian basilicas and traditions are based on legends that do not involve any specific apparitions, but do involve sacred objects that are assumed to have been associated with an apparition.

The last century and a half has seen a dramatic increase in Marian devotion worldwide. This resurgence of the *culto a la virgen* can be attributed to two primary factors. Firstly, Mary's already exalted status in the church was substantially enhanced by Catholicism's official acceptance of the Marian dogmas known as the Immaculate Conception (1854) and the Assumption (1950). Secondly, Mary's growth in popularity, especially among the laity, is not so much doctrinal as experiential as there has been an increased frequency in her alleged appearances since the nineteenth century. With an increase in various shrines dedicated to particular apparitions attracting millions of pilgrims each year, it is easy to see that this phenomenon is having a substantial impact on the intersection of religion, academics, politics and tourism, for the almost one-billion-member Roman Catholic Church.

This chapter will explore seven of the most famous locations in South America where apparitions of the Virgin Mary have taken place. Each of these locations has been officially recognized by the Roman Catholic Church. The geography and history of these sacred locations, the development of local pilgrimages, the architecture of sanctuaries and the subsequent rise of tourism will be considered. Ongoing conflicts between religious leaders and the local population will be noted. As well, theological aspects of popular piety, and the resulting religious practices that arise and occur outside of the official Catholic Church, will be highlighted. A map with the location of the different sanctuaries will be provided, together with images of the different churches and the religious pilgrimages and celebrations (Fig. 29.1).

The seven places we will examine are:

1. Bolivia: Nuestra Señora de la Candelaria de Copacabana, Lake Titicaca.
2. Ecuador: Nuestra Señora de la Presentación, Quinche.
3. Paraguay: Nuestra Señora de los Milagros, Caacupé.
4. Argentina: Nuestra Señora de Luján, Buenos Aires.
5. Venezuela: Nuestra Señora de Coromoto, Guanare.
6. Brazil: Nossa Senhora da Conceição Aparecida, São Paulo.
7. Colombia: Nuestra Señora de las Lajas, Ipiales, Andes Colombianos.

Fig. 29.1 South American religious sanctuaries discussed in the text (Map by Dick Gilbreath, University of Kentucky Gyula Pauer Center for Cartography and GIS; commissioned by the editor)

29.2 The Sanctuaries: Marian National Shrines

29.2.1 Bolivia: Nuestra Señora de la Candelaria de Copacabana, Lake Titicaca

In the Americas, *Nuestra Señora de la Candelaria* stands as one of the oldest advocations of the Virgin Mary, venerated in the Bolivian town of Copacabana (Cruz 1993: 51–56). Copacabana is located on a peninsula at the southeastern shore of Lake Titicaca, close to Isla del Sol and Isla de la Luna, islands sacred to the Aymaras and Incas.

The Dominicans evangelized the area from 1539 to 1574. At that time the inhabitants of Copacabana were divided into two groups: Anansayas (Inca newcomers) and Urinsayas (the traditional residents of the region). Despite conversion to Christianity, these two groups continued an attachment to their original religion. When poor harvests lead them to consider attracting favor from heaven through a new confraternity, the Anansayas resolved to venerate the Virgin Mary; the Urinsayas selected San Sebastian (Aramayo 1983: 10).

Francisco Tito Yupanqui, an amateur sculptor and a member of the Anansayas, decided to create an image of the Virgin, believing it would influence the population. Using clay, and assisted by his brother Felipe, Tito created the image of the Virgin. The sculpture was placed at the side of the altar by the pastor. When the pastor left, the priest Don Antonio Montoro assumed his position. Unhappy with the look of the coarse and disproportionate sculpture, he ordered that it be removed from the altar and placed in a corner of the sacristy. Tito was humbled and at the advice of relatives went to Potosí where there were outstanding teachers of sacred image sculpting. Studying in the workshop of Maestro Diego Ortiz, Tito gained expertise in sculpture and woodcarving. With his new skills he resolved to create an improved image of the Candelaria and searched through the churches of Potosí for an image of the Virgin to serve as a model. He found his inspiration at the Convent of Santo Domingo to the Virgen del Rosario. He studied it closely and held a Mass in honor of the Holy Trinity as a divine blessing for his work.

Eventually the Urinsayas accepted the establishment of the Virgin Mary confraternity, but they did not accept Francisco's new carving. However, when it reached Father Don Antonio Montoro, he decided to bring the image to the people. On February 2, 1583, the image of Mary was brought back (Alvarez del Real 1990: 82).

The Shrine From its beginning, the image gained a reputation for being an object of miracles. It was quite a radical carving in that the body of the image is carved in maguey wood and is gold laminated, and the clothes are those of an Inca princess. The Augustinians built their first chapel between the 1614–1618. Later, the Viceroy of Lima, Conde de Lemos, morally and financially supported the construction of a basilica to honor the Virgin. Construction of the Basilica of Our Lady of Copacabana began in 1668, was inaugurated in 1678 and completed by 1805. The venerated image was crowned during the pontificate of Pius XI. Subsequently, the faithful donated embellishments to the image, including valuable jewels, and the temple was filled with gifts and treasures.

When Bolivia gained independence in 1825, it was largely attributed to the faith of the population to the Virgin of Copacabana. However, its sanctity was defied for political purposes in 1826 when Marshal Antonio José de Sucre, the President of the Republic of Bolivia, expropriated all the jewels and colonial treasures at the Shrine of the Virgin, and used them to create the first coins from Bolivia (Fig. 29.2).

Festivity The Virgin of Copacabana is the patron saint of Bolivia. She is venerated in Bolivia during her feast day of February 2, the day of the Purification of Mary. She is also venerated on August 5 with her own liturgy and popular celebration.

Pilgrimage There are three occasions of mass pilgrimage to the shrine of Our Lady. During Easter, pilgrims from all over Bolivia and other countries come to the Sanctuary, most of them walking from La Paz. On August 5, a pilgrimage is marked by devotees who come mainly from Peru to rest at the foot of the Virgin and ask for her blessing and give thanks. Thirdly, on November 16, pilgrims come mainly from Cochabamba to visit the Virgin and bring flowers to the altar of the Lady.

Fig. 29.2 Basilica of Nuestra Señora de la Candelaria. Copacabana, Bolivia (Photo by Jimmy Gilles, http://commons.wikimedia.org/wiki/File:Church_of_copacabana.jpg)

29.2.2 Ecuador: Nuestra Señora de la Presentación, Quinche

Another of the earliest apparitions of the Virgin Mary reported in the Americas occurred in 1580 at Quinché, Ecuador, when a number of Oyacachi people reported seeing a woman in a cave who promised to assist them with a threat posed to their children by a bear (Cruz 1993: 66–68). This event occurred during a time when many Oyacachi people were being converted to Catholicism. Also at this time, a sculptor named Diego de Robles had been commissioned to carve an image of the Virgin with the baby Jesus. When it was completed the client refused to pay the agreed upon price, so in 1585, de Robles traded the statue to the Oyacachi people in return for some wood. When the statue arrived, the people were surprised to discover that it was an image of the beautiful lady they had seen in the cave. The face of the child Jesus had features resembling those of the mestizo children of those mountains. Mestizo—a synthesis of Inca and Spanish—was also the color of the lady. She quickly became very popular in Ecuador, especially among the Indians who affectionately referred to their protector in heaven as "La Pequeñita" (the little one).

The Shrine The statue remained under the care of the Indians for 15 years, (Alvarez del Real 1990: 180) but in 1604, the local bishop ordered it to be transferred to the town of Quinche, where Christian life had a firmer ground. It was placed in the parish church which became her new sanctuary and from which she took her name

(Salazar Median 2000: 24). Soon however, a larger church was needed and in 1630 the sacred image was moved to its new sanctuary. Over time the building underwent several modifications and after the earthquake of 1869 the temple was rebuilt. The last building of the temple dates back to 1905 and contains a dedication of 1928 (Fig. 29.3) (Sono 1883).

Festivity The image was crowned in 1943 and her feast is celebrated each year on November 21. The present shrine was declared a National Sanctuary in 1985. There are many songs in honor of the Virgin of Quinche, in many dialects of the region as well as in Spanish. These songs have been sung for three and even 400 years.

The Pilgrimage In the month of November the Virgin of Quinche draws pilgrims from the many and varied ethnic groups of Ecuador and in effect serves as an image of an ethnically united nation. Despite being located in an isolated village in the Ecuadorian Andes, 60 km from Quito, thousands of devotees undertake a walk to the Sanctuary to thank her or to ask for a special favor. The pilgrimage is among the oldest in South America, having taken place for over 400 years. The virgin of Quinche is attributed to countless miracles and favors that are portrayed in paintings and plaques from the devotees which hang on a wall of the church.

Fig. 29.3 Basilica of Nuestra Señora de la Presentacion del Quinche. Quinche, Ecuador (Photo by Mark Figuras, http://es.wikipedia.org/wiki/Archivo:Santuario_del_Quinche.JPG)

29.2.3 Paraguay: Nuestra Señora de los Milagros, Caacupé

Tradition says that in the town of Caacupé, founded in approximately 1600, there lived a Guarani sculptor who converted to Christianity from the Franciscan mission Tobatí. On one occasion, returning from the jungles of Ytú Valley, he was surrounded by the fierce Mbayáes, a tribe which refused to accept the Christian faith. In the forest, he sought refuge behind a massive tree trunk. He hid there asking for protection from his Mother in Heaven as the good friars had taught him to. There, he promised the Virgin that if he survived he would carve a beautiful image with the wood of the protective trunk. His persecutors continued on without discovering his presence and he escaped.

As soon as he could, he returned to the tree in the woods and used it to sculpt two images of the Holy Virgin. The larger one he gave to the church of Tobatí (Alvarez del Real 1990: 286). This statuette was miraculously saved from a great flood. Over time numerous miracles were ascribed to it and the people called it the Virgin of Miracles. Years later an unknown Indian moved with his family to this site and built a modest chapel. Like a magnet it attracted superstitious villagers and by 1765, the area was known as the Valley of Caacupé.

The Shrine The name Caacupé comes from the Guarani word "ka'a kupé," from "ka'aguy" which means "forest," and "ka'a" which means yerba mate. Caacupe—located 54 km (34 mi) from Paraguay's capital city of Asunción—has long been considered Paraguay's religious capital. It has the largest sanctuary in the country. Construction of the present church began in 1945 and it has been the sanctuary of the Virgin of the Miracles of Caacupé since 1980 (Fig. 29.4).

Festivity Paraguayans pay the pledge to their Holy Mother on December 8 each year. The popular remembrance begins 9 days in advance, during which they pray to the Rosary. On the hour, church bells, accompanied by loud explosions of bombs, rockets and fireworks proclaim the beginning of the Feast. The vigil is crowned with the traditional serenade to the Virgin, and dances such as the Galopera take place accompanied by Paraguayan harp. All the devotees sing together, greeting and paying homage to the Virgen Azul de Paraguay (the Blue Virgin of Paraguay) (Domínguez 1981: 77–80).

Pilgrimage For the great feast of "Maria de Caacupé" thousands of pilgrims congregate at the Virgin's sanctuary. They come on foot, on bicycles, or however they can, to show love and gratitude to their Mother.

29.2.4 Argentina: Nuestra Señora de Luján, Buenos Aires

The history of this shrine began in approximately 1630, (Cruz 1993: 1–5) when a peasant from Cordova, wishing to revive his neighbors' belief, commissioned two statues from Brazil—one of the Immaculate Conception, the other of Mary and her Son. Once the two statues were finished, the drivers placed them on a wagon and set

Fig. 29.4 Basilica of Nuestra Señora de los Milagros. Caacupé, Paraguay (Photo from http://es.wikipedia.org/wiki/Archivo:Caacupe.JPG)

off on their northbound journey. They stopped to spend the night on the shores of the Luján River, located approximately 60 km (37 mi) from Buenos Aires, Argentina. The next morning they were stunned to discover that the mules could not move the wagon; even after they removed some of the weight it was carrying. Only by strenuously removing the box that contained the statue of the Immaculate Conception, were they able to once again move the wagon. The drivers believed that this was a sign from Our Lady, that this was the location in which she wished to be venerated (Alvarez del Real 1990: 44–45).

People devoted to Our Lady soon began to relocate there, regarding it as a special site for children to receive the sacrament of Baptism and for people to perform penances. The site soon became a small village and took the name of The Village of Our Lady of Luján. In 1755 it was granted the status of a town. Devotion to Our Lady of Luján increased each year due to the miracles that occurred there and on October 23, 1730; the parish of Luján was established (Presas 2002: 1119–120).

The statue, made almost entirely of baked clay, has a dark, oval-shaped face and stands almost a foot high. Her feet rest over a half-moon, which rises above a cluster of clouds, surrounded by four open-winged cherubim. She wears a white tunic and a sky-blue mantle, and with her hands folded in front of her, is seen to be symbolic of the Argentinean flag. The devotion was adopted by the movement for independence in 1816.

The Shrine In 1677, her image was kept in a church built in her honor, until a larger one replaced it in 1763. On May 4th, 1890, the construction of the current basilica began by the impulse of Fr. Jorge María Salvaire, reaching its completion in 1904 when the image of Our Lady of Luján was solemnly transferred to this locale. Today, Nuestra Señora de Luján is one of the world's most famous shrines to Mary and has been honored by papal coronation (Hadad and Venturiello 2007: 34–39). The church was built in neo-gothic style from nineteenth century and has two 330 ft (100 m) foot high towers (Fig. 29.5). Since 1930 the cathedral has been raised to the rank of a basilica. Several religious orders, the Carmelites, the Dominicans, the Jesuits, the Franciscans and others, have established churches and monasteries around the shrine. Pope John Paul II visited the basilica on Friday June 11, 1982, in a very difficult time for Argentina (Falklands War). In Lujan, he gathered the largest crowd ever of pilgrims in the history of the sanctuary.

Festivity The virgin of Luján is the patron saint of Argentina, and Argentineans celebrate her feast on May 8.

The Pilgrimage The first general pilgrimage to the Shrine at Luján was organized on December 3, 1871. Since then, millions of people have come to visit each year. Every October large crowds of youth walk the 64 km (40 mi) from Buenos Aires

Fig. 29.5 Basilica of Nuestra Señora de Luján. Ciudad de Luján, Argentina (Photo by Luis Argerich, http://commons.wikimedia.org/wiki/File:Luján_-_Bas%C3%ADlica_de_Nuestra_Señora_de_Luján_-_200807e.jpg)

to Luján in an overnight pilgrimage of great Marian devotion (Hadad and Venturiello 2007: 39–42).

29.2.5 Venezuela: Nuestra Señora de Coromoto, Guanare

The story of our Lady of Coromoto is the corollary of a long process of conquest and evangelization of the Southwestern floodplains, near the Andes foothills, land of the Cospes people. When the city of Guanare was founded in 1591, Cospes people fled to the north jungle. According to oral tradition, 60 years hence a Lady appeared to Coromoto, the chief of the Cospes, in a river canyon and requested that he and his tribe be baptized (Maria 1996; xxxi–xxxiii).

This extraordinary event proved far more powerful than the work of the missionaries, and several Cospes converted and were baptized. But the chief refused because he feared that he would not be recognized as the legitimate chief under a new religion, so he fled, taking refuge in the forest. One year later, the lady appeared to him again. This time he tried to grab her and she vanished, but a small painting of her appeared in the bark of a tree. According to the legend, Coromoto was bitten by a poisonous snake, returned to Guanare wounded and near death, and begged to receive baptism. He was baptized under the name of Angel Custodio (Guardian Angel), became a faithful apostle, and evangelized the remainder of the rebel Cospes.

The Shrine The National Shrine of Our Lady of Coromoto is located 25 km (15 mi) from the city of Guanare. It is built on the site of her second apparition. Architect Erasmus Calvani began the project in 1975 and construction started in 1980. In February 1996, in the presence of more than two million devotees, the shrine was consecrated by Pope John Paul II. On August 12, 2006, Pope Benedict XVI elevated it to the dignity of Minor Basilica (Fig. 29.6).

Festivity Our Lady of Coromoto was declared the patron of Venezuela by Pope Pius XII in 1944. Venezuelans celebrate her feast three times a year on February 2, September 8 and 11. In preparation for the day of the celebration, the local Church celebrates Coromotana week, from August 29 to September 11, praying the Rosary and celebrating a mass every day in her honor. There is no formal pilgrimage to this sanctuary.

29.2.6 Brazil: Nossa Senhora da Conceição Aparecida, São Paulo

Tradition says that on October 12, 1717, when Brazil was still a colony of Portugal, three fishermen were out on the Paraíba River, between Rio de Janeiro and São Paulo (Brustolini 1998: 226). After no success catching fish, one of

Fig. 29.6 Basilica of Nuestra Señora de Coromoto. Guanare. Venezuela (Photo http://commons.wikimedia.org/wiki/File:Templo_Votivo_de_la_Virgen_de_Coromoto,_estado_Portuguesa_-_Venezuela_001.jpg)

them threw out his net and drew up the body of a small clay statue of Our Lady. Its head was missing. Later that day, in a different location on the river, he dropped his net and this time pulled up the head of the statue. After that, the fish filled the nets of the three fishermen.

As soon as the two parts of the statue were joined together it was venerated by the families and neighbors of the fishermen. It was considered to be an image of the Immaculate Conception that had appeared from the river waters. It soon became known as "Our Lady of the Conception Who Appeared from the Waters," which was shortened to Our Lady Aparecida. While it is not known why the statue was at the bottom of the river, it was eventually determined who the artist was Frei Agostino de Jesus, a Carioca monk from São Paulo. The image is of a black woman, stands less than three feet tall, was made around 1650 and had been underwater for many years.

Miracles of all kinds took place in her presence and devotion to her deepened and spread among the poor people and Afro-Brazilians, heightened in part because of her black Madonna status, and also because it was reported that one of the first miracles attributed to her image was performed on an enslaved young man.

Fig. 29.7 Basilica of Nossa Senhora da Conceição Aparecida. Sao Paulo, Brazil (Photo by Valter Campanato, http://en.wikipedia.org/wiki/File:Santuario_nacional.jpg#filelinks)

The Shrine For 15 years the image remained in the residence of one of the fishermen and many gathered there to pray. Later, they built a chapel. In 1888 a large colonial Basilica replaced the smaller chapel to host the pilgrims whose number had already reached 150,000 a year (Brustolini 1998: 90–92). In the mid-twentieth century, the construction of a much larger building to shelter the image became necessary. In 1955 work on a new Basilica of the National Shrine of Our Lady of Aparecida began. Architect Benedito Calixto designed a building in the form of a Greek cross, with an area of approximately 18,000 sq meters (194,000 sq ft). It can hold up to 45,000 people. The Basilica is the fourth most popular Marian shrine in the world. It receives eight million pilgrims each year, from the whole spectrum of Brazilian social and cultural diversity (Fig. 29.7).

Festivity Pope Pius XII proclaimed Our Lady Aparecida patron saint of Brazil in 1930 (Brustolini 1998: 186). Her feast day is celebrated on October 12 which has been observed as a public holiday since Pope John Paul II consecrated the Basilica in 1980 (Brustolini 1998: 223). The influence of the cult of Our Lady Aparecida on Brazilian Catholic society is incalculable. During a visit to the new Basilica in 2007, Pope Benedict XVI granted the shrine a Golden Rose.

Pilgrimage Every September 7, Brazil's Independence holiday, thousands of laborer pilgrims congregate at the new basilica. In recent years the official Church has promoted a National Family pilgrimage on April 28 and 29 (Pereira Böing 2007). Another is the "Caminho da Fé," (The Walk of Faith) inaugurated in 2003 inspired in the Santiago of Compostela Walk (Spain). Leaving from Aguas da Prata, "Caminho da Fé" runs 308 mi (496 km), of which 180 mi (292 km) across the Manitqueira mountain range, via small roads, trails, forest and pavement. At each stop the pilgrims contribute to the economy of the small cities and support the culture of the inhabitants.

29.2.7   Colombia: Nuestra Señora de las Lajas, Ipiales

Nuestra Señora de las Lajas' story began in 1754 when a native woman from the village of Potosi, María Mueses de Quiñones on a journey to Ipiales with her deaf-mute daughter Rosa, stopped to rest at a place called Las Lajas (The Rock Slabs) (Ocampo López 2006: 7–10). After a while, Rosa emerged from a cave shouting: "Mama, there is a woman in here with a boy in her arms!" Maria was shocked since her daughter could not speak. She ran into the cave, but could not see the figures her daughter claimed to have seen. On the way home, Maria shared what had happened to her employers, but they did not believe her. Later, Rosa returned to the cave where her mother found her in ecstasy contemplating the image of the Virgin. She took the girl to Ipiales and visited the priest's house. After hearing the story and seeing the girl talking, Dominican priest Gabriel Villafuerte began to ring the bells across the neighborhood. The entire town, armed with improvised torches, undertook the risky journey to the cave. This first pilgrimage arrived at the site early on the morning of September 16, 1754.

At the sight, they all saw a beautiful figure of the Virgin and Child, both crowned and both Mestizos, on one of the rock slabs. In the right hand of Our Lady hangs a rosary given to Santo Domingo de Guzman, the founder of the Order of Preachers or Dominicans and on the other side, a Child with head bowed offers the sackcloth girdle for St. Francis of Assisi.

The Shrine   Located on a bridge, which spans a spectacular gorge of the Guáitara River, the church is of Gothic revival architecture and was built from January 1, 1916 to August 20, 1949, with donations from local churchgoers. Interestingly, the church has been constructed in such a way that the rock (and image) is its high altar. In 1951 the official Church authorized the advocation of Nuestra Señora de Las Lajas Virgin, and in 1954 it was declared to be a Minor Basilica (Fig. 29.8).

Festivity   The festivities of the Virgen de Las Lajas take place from September 1–15, known as the eve of the holidays. People celebrate the apparition with traditional songs and dances. The entire region is involved during her feast, including the border cities of Ecuador.

Pilgrimage   Pilgrimages to the sanctuary are all year round, but the number of devotees increases during three seasons: September for the feast of the Virgin, on Holy Thursday, and during Christmas time and early January for Epiphany.

Fig. 29.8 Basilica of Nuestra Señora de las Lajas. Guáitara, Colombia (Photo by Martin St-Amant, http://en.wikipedia.org/wiki/Las_Lajas_Sanctuary)

29.3 Conclusion

Stories are our way of recording our experience of the world, of ourselves and of others, but how do the stories we tell help us understand ourselves and others? Can such stories, whether true or false, provide an entire society an identity? These seven extraordinary places, which all begin with their own unique story, are expressions of a meeting of different cultures. We see through their individual depictions of reality, various layers of meaning, such that our reading of these stories becomes endless. Every story alludes to, or suggests another narrative, so none are allowed to stand as the ultimate truth. As Alberto Manguel (2007: 77) says,

> Like other communal tasks, storytelling has the function of lending expression and context to private experiences, so that, under recognition by the whole society, individual perceptions (of space and time, for instance) can acquire a common, shared meaning on which to build learning.

For all native Latin Americans, the figure of Mary seems to be the only face of compassion that they found in their conquerors. The mother of the god of their conquerors, has the power to free them. This is the paradox. The language of these stories, which acknowledges the impossibility of accurately naming laws and dogmas, groups people under a common humanity while granting them self-revelatory identities. These stories are anchored in a place, and the basilicas which were subsequently built upon these sites have become icons of popular culture as well as places of beauty full of symbols understood by most devotees.

Socially, the popular expression of faith unites various social classes and cultural expressions. This congregation of the people is so powerful that even the politics of each country are deeply imbued with these sanctuaries. Due to uncertain origins, the Church was initially reluctant to acknowledge the shrines with even a seat of the local dioceses, until they reached a point of unquestionable appropriation and declaration of ownership, such as in Argentina, Brazil, and Paraguay. Across South America, the shrines we have examined have played an important political and historical role.

The sanctuary is always presented as a welcoming place to which pilgrims bring their concerns about survival and their questions regarding human dignity. In the 1970s, different social movements inside the Catholic Church brought politics to the sanctuaries and they became a space of social inquiry. The entire expression of their meaning is complex as several factors play a part in varying degrees of significance at different shrines—psychological, ecclesiastical manipulation, papal involvement, nationalist and political elements, the presence of something much older than Christianity, namely the worship of the goddess, and finally, the possible connection to New Age syncretism and neo-paganism. In Latin America today, a positive attitude towards cultural events predominates, but within the Christian faith there is a distrust of native religious resources and mixtures thereof. At the level of the official Church, little importance is given to popular forms of *living the sacred* or of reinterpreting Christianity. This is evident in the teaching of various episcopal conferences that neither interact with, nor are challenged by, autochthonous ways of believing. Another major problem is the tendency to qualify popular behaviors or any spiritual elaboration different from that prevailing in the West as animist and magical.

Marian apparitions answer a demand for integration between native people and their conquerors. The sanctuaries built in honor of the apparitions are privileged meeting places of the nation and the Church, populated and popularized by the people. The faith expressed by the people shows an attempt to overcome the social crisis that surrounds them, including economic problems, unemployment and illness, and a desire to create congruity in their ritual language that combines autochthonous religious features and Latin American reconstructions of the Christian message. The pilgrimages and participation in the Marian feasts provide people with evidence that there are forces of life in the cosmos that are in contact with their ancestors and are reflected in icons of popular piety such as Mary and God.

In fact, at present, Latin American Marian devotion as expressed through worship and images intersects and amalgamates various cultural forms, which helps to strengthen the mestizo identities. The cult of images shows, along with the intense piety and mysticism, the Latin American paradigm of reciprocity. Its meaning is not that of a sacred thing or a divinity (because people do not invoke the virgin as god). Rather, these signs and icons show how people feel that God specifically helps. More than sacred objects, they are icons. That is, they constitute a symbolic and inculturated presence of God. These images are then inculturations of the Christian faith.

References

Alvarez del Real, M. E. (1990). *Santuarios de la Virgen María apariciones y advocaciones*. Panamá: Editorial América.

Aramayo, A. (1983). *Breve historia de la Virgen de Copacabana*. La Paz: Editorial Don Bosco.

Brustolini, J. J. (1998). *História de Nossa Senhora Aparecida: Sua imagem e seu santuário*. Aparecida: Editora Santuário.

Cruz, J. C. (1993). *Miraculous images of our lady*. Rockford: Tan Books and Publishers.

Domínguez, R. (1981). *La religiosidad popular paraguaya: Aproximación a los valores del pueblo*. Asunción: Ed. Loyola.

Hadad, M. G., & Venturiello, M. P. (2007). La Virgen de Luján como símbolo de identidad popular: significaciones de una virgen peregrina. In R. Dri (Ed.), *Símbolos y fetiches religiosos II* (pp. 27–44). Buenos Aires: Biblos.

Laurentin, R. (1997). *Pilgrims, sanctuaries, icons, apparitions*. Milford: Faith Publishing.

Manguel, A. (2007). *The city of words*. Toronto: Anansi.

Maria, N. (1996). *Historia de Nuestra Señora de Coromoto*. Caracas: Ed. de la Presidencia de la República.

Marzal, M. M. (2002). *Tierra encantada: Tratado de antropología religiosa de América Latina*. Madrid: Editorial Trotta.

Ocampo López, J. (2006). *Mitors, Leyendas y Relatos Colombianos*. Bogotá: Plaza & Janés.

Pereira Böing, M. (2007). *Nossa Senhora Aparecida – A padroeira do Brasil*. São Paulo: Edições Loyola.

Presas, J. A. (2002). *Anales de Nuestar Señora de Luján: Trabajo histórico-documental, 1630–2002*. Buenos Aires: Editorial Dunken.

Salazar Medina, R. (2000). *El santuario de la Virgen de el Quinché: Peregrinación en un espacio sagrado milenario*. Quito: Ediciones ABYA-YALA.

Sono, C. (1883). *Historia de la imagen y del Santuario del Quinche*. Quito: Imprenta del Clero.

Chapter 30
Sacred Space Reborn: Protestant Monasteries in Twentieth Century Europe

Linda Pittman

30.1 Introduction

Monastic communities emerged like mushrooms after the rain throughout Protestant Europe during the twentieth century. After almost 400 years of a landscape empty of inhabited motherhouses and cloisters, there are now monasteries in the Lutheran and Calvinist heartlands. Since the Reformers heartily and uniformly disapproved of monks and nuns, their appearance begs the question, "Why?"

The monasteries are located in Sweden, Finland, Denmark, Norway, Germany, Austria, Switzerland, and France (Fig. 30.1). While no more than 1,200 individuals inhabit the communities, the number of persons influenced by the foundations through visits or the activities of members is much greater. The works of Mother Basilea Schlink of the Sisters of Mary in Darmstadt, Germany, for example, have been translated into over 30 languages and published in the millions of copies. Taizé, in France, has achieved worldwide fame as the first ecumenical monastery. Tens of thousands of visitors gather regularly to camp in its Burgundy fields.

A Protestant monastery is a community of Christians belonging to any denomination born of the Reformation and in which at least some members of the group live a life of poverty, celibacy, and obedience, the three traditional monastic vows. An expectation that members will remain with the community for life is also present. Members of the communities are called monks, referring to both men and women, or monastics, and brothers and sisters. Since the Reformation, monasteries have been banned from Protestant landscapes. Luther was adamant in his rejection of what he called "the fleas in God's coat" (Luther 1912: 125).

Sacred places disappeared with the monasteries. While "[w]eakly developed religious centers do appear in Protestantism," Protestants are most noted for their

---

L. Pittman (✉)
Department of Geography, Richard Bland College of the College of William and Mary, Petersburg, VA 23805, USA
e-mail: lpittman@rbc.edu

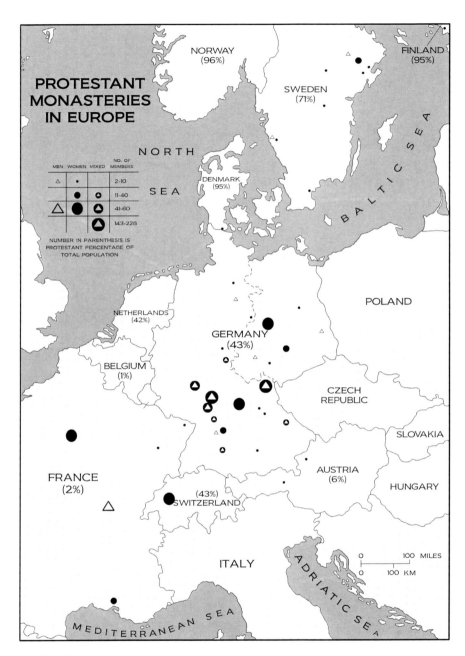

Fig. 30.1 Protestant monasteries in Europe (Map by Linda Pittman)

lack of sacred space. (Sopher 1967) The Protestant regions of Europe have traditionally displayed maps with far fewer dots noting sacred locations than their Catholic neighbors. Yet now the Protestant monastic community Taizé is counted as one of the top pilgrimage destinations on earth.[1]

The goals of this chapter are (1) to explain why monasteries disappeared and then reemerged in Europe's Protestant lands and (2) to explain why sacred space disappeared and reemerged with the monasteries.

30.2 Protestant Rejection of Monasticism

The notorious habits of sixteenth century monks and nuns are well known, and the Reformers were neither original nor alone in their antipathy for the monasticism of their time. The problem was that Protestants came to believe that the life many monks were living didn't succeed in delivering what it advertised – salvation.

As a former Augustinian canon, Luther was personally disappointed by monasticism. He failed to find peace with God through the Roman sacramental system. Luther's belief that Christians are saved by grace alone denied that human works play any part in accomplishing salvation. He held that vows of poverty, celibacy, and obedience did not get a monk into heaven. Since salvation was the primary reason many monks and nuns had entered monasteries, the most natural response, if one agreed with Luther, was to get out.

Luther put his opinions on the subject of monasticism in writing. In his 1520 *Address to the German Nobility,* mendicants were to be relieved of hearing confessions and preaching. The number of orders was to be reduced and there should be no irrevocable vows. The clergy should be permitted to marry. Luther suggested possible reforms, but, since most of the proposals were critical of the very idea of monasticism, his suggestions were generally not adopted. For Luther, the only real solution was to liberate the religious from their vows and empty the monasteries.

In comparison with Luther, Calvin wrote surprisingly little about monks. Not before 1559, when Calvin published the seventh and last edition of his *Institutes*, are there references specifically to the monastic life. When Calvin does deal with monasticism, he too addresses the problem of vows. Monastic vows require actions which cannot please God for they are incompatible with actions required elsewhere in the Bible. The vow of celibacy in particular is problematic because it shows a disdain for marriage, a divine institution. Marriage binds men and women mutually

---

[1] A 2012 Huffington Post article on Taizé records the number of visitors to the monastery as 100,000 annually. This is the number seen over and over again in almost all articles on Taizé as a pilgrimage destination. How many pilgrims actually visit remains undocumented. Where the visitors come from is also undocumented. Taizé is adamantly opposed to keeping records. This has been a thorn in the side of all those who have tried to write about the community. "It is no easy task to write a documentary history of a community that, at the end of each calendar year, burns almost all its documents" (Brico 1978: 9).

with an obligation that comes directly from God. No human authority can free men and women from their reciprocal obligation.

Most other theological leaders of the Reformation period do not offer anti-monasticism arguments differing significantly from those of Luther and Calvin. Zwingli, a monk from German Switzerland and the first theologian of the Reformed tradition, disagreed with other Reformers on a number of subjects, but monasticism was not one of them. As soon as he converted to the Reformed view of the world, Zwingli left his monastery. By the end of the sixteenth century only a few emasculated vestiges of what was formerly one of northern Europe's most powerful institutions remained.

30.2.1 Reformation Rejection of Monasticism Is Limited

Despite the critical nature of the Reformer's statements, they do not all seem to have been completely convinced that the monastery was a fundamentally bad place. In the 1536 Wittenberg Articles, Luther wrote:

> If certain men of outstanding character, capable of living a life under a rule, feel a desire to pass their lives in the cloister, we do not wish to forbid them, so long as their doctrine and worship remain pure, and, notably, so long as they consider the practices of monastic life as things indifferent. We are persuaded that numerous authentic Christians of irreproachable spirituality have passed their lives in convents. It is even certainly to be wished that such convents will exist, occupied by wise and fervent religious, in which the study of Christian doctrine can be pursued, for the greater good of the Church. These might be then a place where, by the practice of pious exercises of religious life, young people would receive not only an intellectual training, but a spiritual one as well. (Biot 1963: 63)

And in the Reformed Church, the *Confessio helvetica posterior*, published in 1568 in Zurich, contains this passage:

> Those to whom God has given the gift of celibacy, in such a way that they can continue wholly pure in heart and mind without grievously burning, must serve the Lord in this vocation as long as they feel endowed and protected with this heavenly gift, and so long as they do not therefore raise themselves up above other people, but assiduously serve the Lord in all simplicity and humility. Moreover, such individuals are better disposed to busy themselves with things divine than those who suffer the distractions of their families. (Biot 1963: 59)

This statement admits the possibility of a celibacy consecrated to the service of the Lord.

Twentieth century Protestant apologists for the renewal of the cenobitic life depend on these escape clauses in the Reformers' writings to justify the founding of monasteries. They also rely on the Reformers' generally positive attitudes towards early monasticism. Both Luther and Calvin hearken back to a pristine period when monks were true monks. Neither Reformer wanted to abuse the Apostolic Fathers. Nevertheless, the three centuries following the Reform witnessed continued dissolution of monasteries and the defamation of monasticism. They also witnessed the disappearance of sacred space from Protestant landscapes.

30.3 Sacred Spaces Disappear

"Religious space is powerful space" (Kilde 2008). But in some religious traditions it is more powerful than in others. Protestants are NOT known for the power of their religious spaces. In Jeanne Halgren Kilde's discussion of Christian architecture and worship, she argues that the power in religious space comes from three sources:

> (1) divine or supernatural power, or that attributed to God; (2) social power, or that pertaining to a variety of social, particularly clerical hierarchies; and (3) personal power, or the various feelings of supernatural empowerment that individuals derive from the experience of the divine. (Kilde 2008: 4)

In Christianity before Protestantism (and in every Catholic and Orthodox church today), every sanctuary was home to all three types of power. God's divine power was present in the substance of the body and blood of Christ. The social power of the priesthood with its ability to call the divine into existence through the celebration of the Eucharist was present. And individual empowerment was possible because each believer had access to the divine through partaking of the Eucharist, and through access to the saints, who were already in the presence of the divine.

As monasteries disappeared from Protestant territories, so did sacred places, for many of the same reasons. After the Reformation, Protestant church interiors were whitewashed, the medieval paintings of saints covered over. This was due in part to disagreements with the theology represented in the paintings, but it was also due to a changed understanding of the nature of the church building. Before the Reformation, churches were holy. They were home to the images of the saints, they were where the priest performed the great mystery of the Eucharist, and they housed the Blessed Sacrament. Christ was physically present and believers were reminded of the sacred presence by a constantly burning candle, and the glories of the art and architecture.

After the Reformation, the understanding of the sacrament changed. It was no longer the sacrament if it was not in the process of being consumed by believers. In Lutheran churches the action moved from transubstantiation to consubstantiation. The bread and the wine did not become the body and blood of Christ, they contained the body and blood of Christ. Luther likened the presence of God in the sacrament to the fire contained in the horseshoe while it was being forged in the blacksmith's flame. Once the Eucharistic celebration was over, the horseshoe cooled and returned to its state as horseshoe, not horseshoe and fire simultaneously. So too, the bread and the wine that had become the body and blood of Christ during the Eucharistic celebration returned to their original state of being simply bread and wine. The church building housed God only while the congregation was present. It didn't need to be big and showy. It needed to shelter the people of God while they worshipped. But the locus of holiness had shifted: it was the people who were holy, not the place.

For the followers of the Reformed tradition the changes were even greater. Zwingli saw the Eucharistic celebration as wholly spiritual. When Christians shared the bread and the wine they shared spiritual communion with Christ, not a ritual of sacrifice. The nomenclature changed and the ritual was called Communion. The altar came to be understood as a table for the Lord's Supper, a place to gather for

Communion with Christ. Calvin argued something other than Zwingli's purely spiritual communion. For him, the bread and the wine represented the body and blood of Jesus and he thought that an "exclusively spiritual communion neglected the human element of Christ" (Kilde 2008). But Calvin did not see the elements as supernatural the way Catholics and Lutherans did.

In addition to a changed understanding of the sacrament, the Reformation led to a changed understanding of the clergy. Luther preached the "priesthood of all believers." Prior to the Reformation, priests were viewed as having enormous power through their ability to facilitate the miracle of the Eucharist. Priests functioned as mediators between the human and the divine, between the earthly and the heavenly. Where the priest was, there was power. With the concept of consubstantiation, the mystery of the Eucharist was far less mysterious. A less mysterious priesthood combined with a universal priesthood added up to something significantly less sacrosanct than the pre-Reformation version. While Protestants may all have become priests, being one was no longer as much to brag about.

The nature of worship changed during the Reformation. For Protestants, divine power centered on God's Word. The Bible was God's effort to communicate with humankind. Protestants no longer had the option of communicating with God through the saints. The saints had been banished, eradicated, painted over, and quite literally defaced. The faithful knew God through Jesus alone, and Jesus was known through the Bible alone. The focus of Sunday morning changed from the Mass to the sermon.

For the same reasons that churches were sacred places in Catholic times, monasteries were doubly so. Not only was the monastery church with its lighted candle signifying Christ's sacred presence always open, it was often occupied with praying monks, many of whom were priests. Monasticism also offered the fast track to the divine. By following the rule, monks achieved salvation. After the Reformation, Protestants lost their sacred places because sacredness did not exist independently of congregational worship. Given the Protestant understanding of the omnipresence of God, who was thought to be present in all places and all times, meant that God was no longer **more** present in the Church than anywhere else. Bit by bit, Protestants changed their view of the church building. By the nineteenth century it had lost much of its power.

This is not to say that there were no sacred spaces in Protestant landscapes between 1600 and 1950. Cemeteries have served that function in most societies since time immemorial, and that was just as true in Protestant Europe as anywhere else. It would be impossible for many Finns to visit Helsinki's Lutheran Hietaniemi Cemetery, established in 1829, without experiencing the sacredness of the place in a way similar to that in which many Americans experience Arlington Cemetery. On a smaller scale, cemeteries in Scandinavia frequently celebrated All Saint's Eve (October 31), by putting lighted candles on the graves of loved ones. The tiny flames flickering in the darkness at the beginning of another long Scandinavian winter, symbolizing each soul's release from the grave, made churchyards sacred places. All Saints' Day was also Reformation Day in Lutheran Scandinavia, the day the Reformers were remembered and celebrated, so the holy day was a profoundly Protestant one.

Sacred space could also still be found in the churches, not all of which were whitewashed and cleansed of pre-Reformation elements. The great cathedrals continued to be used by Protestants, and, due to the nature of their space, it was common to experience a sense of awe and wonder when worshipping in them. Additionally, the church buildings where believers experienced the sacred mystery of the Eucharist were inevitably imbued with an extra bit of the *mysterium tremendum* associated with the divine presence.

But Protestant territories were far less likely to house pilgrims, shrines, hermitages, and even large and/or inspiring churches than Catholic and Orthodox territories. And Protestants themselves remained (in some cases, remain) uncomfortable with the elements that imbued pre-Reformation churches with holiness; images of any sort, stained glass, liturgical vestments, candles, incense, rood screens, crucifixes. Movements such as Puritanism that sought to complete the job the Reformation had begun pepper Protestant history. Iconoclastic controversies continue to break out, particularly in denominations inspired by Calvinism. And the root cause of the shift away from sacred space was Reformation theology. So for sacred space to once again become fully acceptable, theology would have to change. One theologian who helped initiate the transformation was Søren Kierkegaard.

30.3.1 *Kierkegaard*

Kierkegard was conscious of the fact that sacred places had been lost by the middle 1800s, and he missed them. He believed the Church lost its credibility the day it joined forces with the world. Only in complete opposition to the world could the Church be Christian. He called for radical Christianity, a serious and ascetical Christianity. He himself needed a monastery, but he did not have one because Protestants had outlawed them. There was no "place" for Kierkegaard to go to satisfy his spiritual needs.

From Kierkegaard's point of view, the ideal Protestant monastery was a place for those who did not fit into active society. "There is no doubt," Kierkegaard wrote,

> That the present, that Protestantism always, needs the monastery again, or that it should exist. 'The Monastery is an essential dialectical fact in Christianity, and we need to have it there like a lighthouse, in order to gauge where we are. (Kierkegaard 1951: 222)

One of the first strides in the direction of re-monasticizing Europe came through the Deaconess Movement of the nineteenth century.

30.4 The Deaconesses

In Europe between the revolutions of 1789 and 1848 women began to assume new roles in Europe. They started to work outside the home caring for the sick and needy. The French Revolution paved the way for this change in women's status in two ways: first, by fostering women's desire for greater freedom; and second, by making participation in the public workplace possible. The Napoleonic era slowed women's emancipation by supporting a return to more traditional social roles, but Napoleon did allow the Sisters of Charity, who were banned during the Revolution, to reorganize, focusing public attention on women who cared for the needy. The Sisters provided an example of the kinds of work women could do if allowed. Between 1815 and 1830 the order received nearly 200 novices a year (Christensen 1971). The Sisters of Charity were greatly admired by many Protestants, and gave rise to the wish that Protestants too would produce such sisters.

In 1822 a Lutheran pastor named Theodor Fliedner was assigned to the destitute parish of Kaiserswerth, Germany. The poverty of his small parish inspired in Fliedner the desire to help the needy. Eventually, he turned a house he and his wife owned into a hospital, and an unmarried parishioner named Gertrude Reichardt committed herself to helping the Fliedners care for the sick. Reichardt became the first modern deaconess and was soon followed by dozens of others (Weiser 1971: 23). The sisters, as they came to be called, committed themselves to running the various public services provided by the church; a shelter for the poor, an infirmary, and schools. The Kaiserswerth deaconess community saw itself as a revival of the ancient office of deaconess. The idea proved popular, and Kaiserswerth spawned deaconess homes and hospitals from Pittsburgh to Jerusalem. When Pastor Fliedner died in 1864 there were 1600 deaconesses and 32 motherhouses descending from the Kaiserswerth foundation.

As organized in Germany, the deaconesses took the form of a family-community acceptable to the social conventions of the day. The "Father" of the house was a pastor and the "Mother" was most often the pastor's wife. She was the directing deaconess. The "motherhouse," a term and concept borrowed from the Sisters of Charity, was originally designed as a place of training and protection for the Kaiserswerth deaconesses. When Pastor Fliedner founded Kaiserswerth, it was assumed that a fledgling deaconess would enter a motherhouse to receive her training, but once she had been properly trained, both vocationally and spiritually, she would move into the local church to undertake diaconal work on the parish level. In the parish, she would perform the tasks performed by the deaconesses in the Early Church. But the Kaiserswerth deaconesses stayed in the motherhouse stage. They were never incorporated into the local church in Germany. This made the German deaconesses quite different from the New Testament prototype. Deaconesses living in a motherhouse for life bore a far greater resemblance to the Catholic Sisters of Charity than to the Early Church's Phoebe. Except that deaconess motherhouses were never places for meditation. While chapel was an important part of the sisters' lives, it was their work as servants to Christ that was the centerpiece of their worship.

Deaconesses lived a communal life in the motherhouse. The poverty, celibacy, and obedience that became standard practice among the Protestant sisters was adopted not because of a desire to follow a monastic ideal (although such an ideal was present in many of the first deaconess communities), but because no other arrangement was practical. Without celibacy and obedience it was not possible to live the ordered existence demanded by the organization. Without the self-imposed poverty created by giving away all of ones earnings (the deaconesses gave the community any wages they earned), the motherhouse would not have had the finances to exist.

The rest of Europe was quick to follow Kaiserswerth's example. In Switzerland, in 1841, a Pastor Germond founded "The Lutheran counterpart of the Catholic Sisters of Charity" (Biot 1963). The first deaconess house was founded in Sweden in 1851, in Denmark in 1863, in Finland in 1867, and in Norway in 1868 (Lenhammar 1977). All used the motherhouse organizational style. Some of the foundations were Reformed in confession, some Lutheran, and some United. In time, the largest motherhouse in Europe, Bethel in Bielefeld, Germany, housed over 2,000 deaconesses.

30.5 Support of Monasticism by Twentieth Century Theologians

At the turn of the twentieth century, the monastic ideal began to come into vogue in Protestant circles on the continent. The Lutheran theologians Adolf von Harnack, Ernst Troeltsch, and Reinhold Niebuhr all recognized the need of some Christians for a withdrawal from the larger Christian community "so long as it is not divorced from the ongoing life of the Church" (Bloesch 1964). This need helped give rise to both monasticism and the retreat movement. During the early twentieth century the idea that Christians would separate themselves from their homes and workplaces to engage in a time of reflection and prayer became popular and the retreat movement developed. It helped spawn the monasteries that followed.

One of the first twentieth century theologians to espouse monasticism's cause was the German church historian and liberal, Adolf von Harnack. In Harnack's opinion, Luther had not wanted to eradicate monasticism. Rather, Luther was simply critical of the institution. His criticism destroyed the monastic way of life in Lutheranism, but, since that was not Luther's intent, there was no reason not to return to it. In fact, several good reasons existed to resuscitate it. According to Harnack, monasticism had a truth to it. He went so far as to say that an evangelically committed monasticism was not only an option, but a necessity. Cleansed of superfluous Roman Catholic trappings, monasticism could breathe new life into the Church (Harnack 1900). Harnack saw the deaconesses as a beginning, an introductory form of Protestant monasticism. But there had to be a contemplative aspect to complete the picture. Places dedicated to prayer and piety, holy places, needed to exist somewhere in the Protestant landscape.

But theologians seeing a need for monasteries did not mean that the monks would simply appear. Retreats, a period from a few days to a few weeks in which a believer removes him or herself from normal daily activities to concentrate on prayer and spiritual growth, were a step in that direction. Retreats eventually proved quite popular with the laity. Housing and feeding the retreatants became the means by which many founding members of communities discovered each other, and the primary means of livelihood for most Protestant communities even today.

30.5.1 High Church Movement

The high church movement also supported Continental retreats and monasticism. It was primarily Lutheran. In Germany it started with the "95 Theses" of theologian Heinrich Hansen, who was highly critical of what the Lutheran Church in Germany had become. He wanted to return to the orthodoxy of the seventeenth century. In Sweden, high church ideas arrived from Britain after several Swedish pastors spent time in England and were impressed by the liturgical and religious life there. The movement stressed the rediscovery and reaffirmation of things German or Swedish in church history, the German or Swedish saints in particular, and the high, holy places of the past, like Regensburg or Maria Laach in Germany and Saint Birgitta's Vadstena monastery in Sweden. To foster this interest, church societies that focused on a particular saint were founded. In Sweden in 1914 the St. Sigfrid Order was founded for clergy. In 1920 the St. Sigfrid Order gave birth to the St. Birgitta Order for pastors and lay men and women (Kilström 1970).

These orders were similar to Roman Catholic third orders, and often included a reference to "third order" or "tertiary" in their names. The Lutheran Franciscan Tertiaries, for example, started in Germany in 1927. Members took vows, followed a rule, and read daily from a breviary. Group spiritual retreats were held annually. Third orders were accused by the church's mainstream of cultivating monasticism for primarily "antiquarian" reasons. There is some truth to this. Many third order members wanted to bring back pre-Reformation rituals simply because they were old. The fact that monasticism had continued for most of the life of the Church was seen as sufficient justification for continuing its existence, no matter what Luther's and Calvin's opinions on the subject had been.

In France, where the Protestant population was small, the "third order" concept took on a slightly different cast. Pastor Wilfred Monod founded the Communauté des Veilleurs (The Watchers Fraternity) in 1923. Monod was known as a revivalist whose message focused on the Social Gospel. While the Watchers had high church elements such as a daily liturgy and vows, the emphasis was more on social justice and ecumenism than on high church ritualism. Monod was an ecumenical guru having masterminded the merger of several French Protestant denominations. The movement he supported was called Practical Christianity. Protestant denominations agreed to put aside some of their theological differences to concentrate on social issues. In the face of increasing secularization, this was deemed a wiser use of energies than continuing to fight old and increasingly irrelevant battles. The high church in France was not as antiquarian as that in Sweden and England because French saints and monasteries were still in good evidence in the dominant Catholic Church. They didn't need Protestant societies to recall and celebrate them. The Watchers had a profound effect on all of the French speaking communities. The liturgy they developed incorporated elements from several denominations. It was based primarily on the Beatitudes which were spoken at noon. This same liturgical practice, saying the Beatitudes at noon, was initially used by all of the French speaking communities – Taizé, Grandchamp, Pomeyrol and Reuilly.

These brotherhoods, sisterhoods, orders, fraternities, third orders, sodalita, or whatever one might choose to call them injected monastic vocabulary and custom into the Protestant body. The love affair with liturgy that is a fundamental part of the creation of monasteries must be credited to the high church movement. But the high church movement, despite its profound effects on twentieth century Protestant church life, never moved the theological mainstream. For many Protestants, the movement was too Catholic (Halkenhäuser 1982: 12). Despite protestations to the contrary, it continued to be seen by the majority as too "Romish." It was also, by and large, a product of the upper class and clergy. In the Europe of the 1920s and 1930s, this did not endear it to an increasingly politicized middle class. The brotherhoods, sisterhoods, orders and third orders remained exceptional. It was not until someone as solidly Protestant as Karl Barth emerged and affirmed the worth of monasticism that the spell cast by the Reformation was broken.

30.5.2 Karl Barth

Barth began by recognizing the correctness of the Reformers' rejections of monasticism at the end of the Middle Ages given its debased state. This approval of the Reformers' attitudes, however, did not mean that movements to reintroduce religious community life into the Church were unacceptable as long as they protected themselves against the abuses and distortions of the past.

In Volume III of *Church Dogmatics*, Barth raised a number of problems concerning monastic life and celibacy. For him, the fundamental question was whether or not the Reformers were in accord with the New Testament.

> With growing assurance – but also from a single viewpoint – Luther attributed to work in the fields, in the shop, or in the family household the value of a "service of God," which had been attributed in his day to the activities of the cloister. Luther's point was doubtless of very great importance. But did not Protestantism then and in the centuries to follow insist upon this point too much? Did it not over-rate marriage at the expense of celibacy? How far can the Word of God be brought to justify this preference?...Paul himself, in the perspective of the Lord's return, to correspond with the service required by the Lord in his spiritual army, understood celibacy as a call from God, and even gave all Christians the unequivocal advice that they should do as he did, for the Kingdom of Heaven. (Biot 1963: 145)

Karl Barth was the modern theologian willing to say that the Reformers were simply wrong. He wrote positively of the monastic life and urged Protestants to try it.

> What Monasticism, in its particular way, planned and undertook was a concrete method of following the Lord. Now following the Lord is something which in the Gospel we find required not only in general terms, but also described – in part at any rate, as a model – in extremely precise detail. Here, then, is what was planned and undertaken: individual and collective sanctification, a way of Christian life designed for a purpose, a concrete and organized fraternal life, and all this under the impulse and in service of a concrete and total love. (Biot 1963: 147)

Several monastic communities begun after World War II, especially those stemming from the Calvinist tradition, were inspired by Barth's arguments. Sister Antoinette Butte founded the community of Pomeyrol in Provence because of them. Roger Schutz (Brother Roger), founder of Taizé, also followed Barth's reasoning, and said that the Reformers were wrong to condemn the principles of monasticism.

What Barth did not do was address the fact that the principles of monasticism at the time of the Reformation were based on works righteousness, the doctrine that good deeds and clean living (works) could save the soul (righteousness). This was the doctrine Luther's theology was designed to refute, as Luther promoted the idea that salvation came through grace alone. Works righteousness was so forcefully and capably opposed that by 1922 when Barth published the completely rewritten second edition of his *Epistle to the Romans*, it seemed impossible for a Protestant to think that entering a monastery would save his or her soul. No Protestant monastery used salvation as a reason to enter the monastic life. The Reformation understanding of the Gospel was so deeply imbedded in Protestant theology that "sola gratia" was taken for granted. When Barth and the communities argued that the Reformers were wrong about monasticism because they could only focus on the abuses in the monastery, they failed to note the larger context of the Reformation argument. Luther would undoubtedly still argue that monasticism's evangelical worth is nothing unless the monk is fully aware that the monastery is not a strategic stop on the path to salvation. But 500 years after Luther's birth, it was possible for one of his German followers, Walther von Loewenich, to write these words:

> No intelligent person today should dispute the fact that the Reformation polemic against monasticism was a one-sided affair. This powerful manifestation of a heroic faith cannot be dismissed as a religion of works… The average modern Protestant is often culpably blind to the religious power latent in monastic piety and its ideal of sanctity. Of course, like everything else in the world, it has its dangers. But that should not lead us to overlook its positive side. Monasticism is the fount from which Catholic piety has been constantly renewed. (Loewenich 1959: 349)

The implication was that it would also be the fount from which Protestant piety would be renewed. The problem is that within Catholicism, monasticism was able to renew the church because of a steady stream of novices to fill the monasteries. And most novices were inspired to enter monasteries because they believed it would save their souls. Without salvation as an incentive, there simply are not many individuals who see monasticism as a particularly attractive way of life, despite the "religious power latent in monastic piety."

But even after the changes in theology that brought acceptance for monks, there were still aspects of the monastic life that just didn't feel right to Protestants. One of these was the idea of vows. Even Brother Roger, who never emphasized his Protestant roots, preferring to highlight the ecumenical nature of his community, did not see fit to use the word vows when he reintroduced them in Taizé on Easter night, 1949. The Reformers spoke forcefully and at length on the subject of vows. When Protestant communities sought to incorporate monastic life into Protestantism, they had to do so on terms acceptable to that tradition. Otherwise, what was the point of remaining Protestant? They could just as well have entered Catholic monasteries if they had not found something in Protestantism that they valued. The argument against vows was apparently too central to the tradition to be ignored. Thus, instead of vows, they made "commitments," "engagements," and "promises." In Brother Roger's words, "We prefer the expression *engagement* since it does not have the implication of merit that the word "vow" has" (Biot 1963: 125). While the new monks thus managed to create a monasticism that was truly Protestant, accruing no merit to those who chose the lifestyle, they also managed to become monks. This remains a balancing act, but in doing so they created a new type of Protestant clergy.

Some monks might argue that they are not clergy because they are not in official church leadership positions, but they are clearly religious professionals, which is another aspect of the definition of "clergy." It was this role as religious professional that proved so exciting to Protestants who flocked to the monasteries as they emerged because they wanted to see monks and nuns. With the advent of this new group of religious professionals the sacredness of monastic space was amplified. Power in religious space comes from three sources: (1) divine or supernatural power; (2) social power, or that pertaining to clerical hierarchies; and (3) personal power. Once they were established and accepted as monks and nuns, this new clergy satisfied the need for the social power that added to the sanctity of a place.

Acceptance as monks and nuns did not come overnight. The 40 Protestant monastic communities that developed after World War II all felt the need to explain themselves. They all knew that talking about monasticism put them on slippery ground. Some used high church arguments, or ferreted out positive remarks on monasticism in Luther and Calvin. Others used Barth's ideas. It was hard to ignore the harsh words against monasteries found in the Reformation texts. But it seems that the greatest justification for their existence was their existence. The fact that enough Protestants managed to get together to found the monasteries, that no one saw fit to challenge their legitimacy, that many devout Christians went to them for retreats, counsel, and confirmation classes year after year, could only mean that they had become acceptable because they were, in fact, accepted.

One of the measures by which the liveliness of Christianity is assessed is the rate of church attendance. European Protestant rates of church attendance remain low, the lowest of all Christian countries. In this, Europeans differ radically from North Americans, where about 40 % of the population (as opposed to Protestant Europe's average of 10 %) attend church regularly.[2] This aspect of culture is one of the few areas in which developed countries differ to such a high degree. In examining Protestant monasticism in Europe we are looking at something that is uniquely European, and by studying an example of radical faith in secularized societies, it may indicate the need to reassess the statistics by which secularization is understood.

---

[2] While church attendance statistics are notorious for their inflation, what is most significant for purposes of this paper is the fact that the attendance levels in Protestant Europe are so profoundly different from those in the United States. The Wikipedia entry on church attendance statistics lists the following numbers from the Gallup Poll of 2004: United States 43 %, Austria 18 %, France 12 %, Finland 5 %, Sweden 5 %, Denmark 3 %, Norway 3 %. No statistic is given for Germany or Switzerland.

While Protestant monasticism is European, interest in it is international and ecumenical. There are no Protestant monasteries in the Netherlands and only a few in the United States; but there are Dutch and American members in the European communities. Visitors to the monasteries come from all branches of Christianity and from all corners of the earth.

Several communities begun in the early 1940s and 1950s have failed. If a community does not grow quickly in its first decade, it is likely to remain small, usually ten or fewer persons. Half of the communities are small. This has been a particular problem in Scandinavia, where at least five communities have failed without ever achieving more than three members. Size remains a problem in Scandinavia where only one community, the Daughters of Mary whose primary foundation is in Vallby, Sweden, has more than eight members.

30.6 Founding Experiences

While the history of becoming a Protestant monastic community followed a distinct path for each group, there were common themes. The first is that the community rarely started off intending to become a Protestant monastic community. The second is that the intention to become a Protestant monastic community was a multi-step process, developing over many years and changing one step at a time.

Most Protestant monastic communities began with the "conversion" of or "call" to a charismatic leader. The life-changing religious experience is a familiar Christian phenomenon. Paul had one on the road to Damascus. Protestants, or at least certain denominations or groups among the churches of the Reform, have consistently sought and simulated the religious experience. A primary goal of tent meetings and television evangelists is the production of conversion experiences. In Europe, instead of tent meetings it was war, Girl Scouting, Pietism, he retreat movement, and the Oxford Group that brought about life changing experiences and stimulated community formation. All were catalytic in the lives of the first generation of community members. In the 1960s it would be the counterculture and the charismatic movement that served as catalysts.

The formative experience par excellence for monasteries was World War II. World War II was a cauldron for spiritual renewal. As small prayer and church groups formed to pray for sustenance through air raids and rationing, they experienced the fires that bond for life and began to look for forms for the common life to which they felt themselves called. In interviews with founding members of four of the largest communities, the Sisters of Mary, the Brotherhood of Christ, Casteller Ring, and Taizé, World War II was given as a cause for community formation. Even after 1945 the War continued to be a reason to join a community. Many young women in continental Europe were well aware that there could be no marriage in their futures. Far too many of their male age mates had died fighting. The demographic imbalance created by war made it clear that marriage would not be in the picture for many women, especially in Germany. This made it easier for some of the women to make life commitments to communities.

What the convert wanted was a group of like-minded souls. The main purpose for coming together was to do a better job living the Christian life. New Christians could keep each other from stumbling, and help each other if they did stumble. Strength came through numbers. But so did sacredness. Once a community had formed, the neighbors began talking about it. "What could cause a bunch of young people to spend so much time together praying if it wasn't the presence of God?," they asked themselves. When the neighbors talked, the curious showed up for a look and some of them wanted to join. More members meant more curious visitors. There were more conversion experiences and the perception of even greater sacredness.

Once like-minded individuals had committed themselves to living a Christian life together, choices had to be made. How were they to support themselves? How, where, and when should they worship? How should conflicts be resolved? The answers to these questions often depended on models already present in the church, or on elements already present in whatever group the first converts were part of when conversion occurred.

After a commitment to live a community life style was formulated, the first problem to solve was how to support the community. The most common means communities found to support themselves was the running of retreat centers. Spiritual retreats, which became popular with the high church movement, needed quiet, secluded places where retreatants could meet for meditation and prayer. The communities could support themselves by maintaining the retreat centers, providing clean rooms and simple meals for the guests while simultaneously housing and cooking for the community. Occasionally, the retreat center preceded community formation as in the case of Grandchamp and Taizé, and became a cause for formation. Scouting, the Oxford Group, and Pietism also served this function.

30.6.1 Scouting

Scouting is the largest youth movement in the world. It spread rapidly during the decade following Lord Baden-Powell's publication of *Scouting for Boys* in 1907, with Girl Scouts springing into existence in 1910. Scouting arrived in Germany shortly after its inception. Since Scouting was frequently sponsored by churches, it gained a strongly confessional aspect as it spread through Europe. Girl Scouting in particular was important to Protestant monasticism. Its goal of teaching twentieth century girls independence helped develop the leadership skills needed to start Casteller Ring. The first members of the Casteller Ring Community were a Girl Scout troop. Their present monastery is the site of their former summer camp. Antoinette Butte, the founder of Pomeyrol in Provence, also founded the Protestant Girl Scouts of France. Through Scouting, many members in the first communities discovered the existence of the groups they eventually joined.

Scouting imbued its members with an Acadian view of nature that would profoundly impact the formation of Protestant sacred spaces. The idea that living close to nature could save modern folks from the polluting and demoralizing effects of urbanization and modernization is one of the precepts on which Scouting is founded. It would be difficult to visit Casteller Ring in Bavaria or Pomeyrol in Provence and not remark upon the glorious beauty of their respective settings. A visit to either community means a trek to some of the more remote and scenic corners of Germany and France. The emphasis on the power of place to heal and strengthen the spirit became a cornerstone of the retreat movement that in turn became a cornerstone of Protestant monasticism. Both Casteller Ring and Pomeyrol are renowned for their role as retreat centers. They are visited by thousands of guests a year (Halkenhäuser 1982: 13).

30.6.2 Oxford Group Movement

The Oxford Group was begun by a Pennsylvania Lutheran pastor, Frank Buchman, in 1921. Influenced by the YMCA and English Pietism on a trip to Europe, Buchman underwent conversion at Oxford and distilled the experience into catchy formulas which he carried around the world. The European upper classes, with their well-developed sense of class difference, were especially receptive to Buchman's message because one of Buchman's key policies was to begin with "those at the top." Two Protestant monastic communities are now housed in the castles of Franconian nobility because the owners of those castles participated in the Oxford Group (Fig. 30.2). The founders of nine communities (among them three of the largest) were directly involved in Buchman's movement, some experiencing their own conversions at the house parties the Group sponsored.

Traditional Group practices include silent time (a period of time, usually in the morning, when Group members wait for guidance and keep a daily notebook of messages received), house parties (evangelistic pep-talks presented in the private homes of Buchman's followers to which the uninitiated were invited), a commitment to beginning with "those at the top," and the four absolutes: absolute honesty, absolute purity, absolute unselfishness, and absolute love. The practice of writing daily in a journal while in chapel (three communities still do this) comes from the Oxford Group connection.

Unlike the connection with Scouting which is frequently recalled and which the communities with scouting roots are glad to talk about, most communities are no longer aware that a number of their spiritual practices stem directly from contact with the Buchmanites. There are three reasons the Oxford Group connection has been forgotten. First, the Oxford Group is no longer part of the ecclesiastical environment. After Buchman's death in 1961, his influence faded rapidly. Second, the Oxford Group was not always accepted by mainline denominations. Even during its heyday before World War II, the Oxford Group received criticism from both within and without the church, especially because of the enormous wealth it amassed, including some exceedingly showy and expensive real estate. Third, after World War II the Oxford Group became increasingly political, eventually morphing into an organization called Moral Rearmament, from which many mainstream European churches wanted to distance themselves.

Fig. 30.2 Schwanberg Castle courtyard (Casteller Ring) (Photo by Dale Pittman)

30.6.3 *Pietism*

In general, the pietistic aspects of a community's history are the hardest to trace. Pietism started in Germany with Philip Spener and Hermann Francke in the eighteenth century. It was primarily a criticism of orthodox formalism. It promoted personal spiritual experience, often in an emotionally charged context. By the twentieth century, aspects of it had been institutionalized, but it continued to criticize the institutional church and was therefore regarded warily by members of the church establishment. Connections between communities and Pietism were often hard to uncover because they were the oldest influences and were thereby covered with the greatest number of other events and influences. But there was also reticence to speak of pietistic experiences because they were pietistic. Conversion experiences are not normal fare in modern European Protestantism. They are not PC.

But pietism could also be seen to have great truth and power in creating and maintaining community (Latourette 1975: 456). The Moravians, a Pietist group residing on Count Zinzendorf's estate along Germany's eastern border, was one of the few well-known "sacred places" in Protestant Europe. When they arrived on Count Zinzendorf's land in 1722, the Moravian Brethren established an intentional spiritual community as well as several successful industries and trading enterprises. Herrnhut, the name of the foundation, was home to a unique prayer and liturgical life. Community members committed themselves to 24/7 prayer and kept up the regimen for 100 years. Herrnhut became a center for the modern missionary movement. Moravian communities following the Herrnhut model were eventually established at 30 locations worldwide. The Moravians inevitably became the model for an intentional life in community for any of the groups with pietist roots.

30.7 Historical Summary

As Protestant communities aged, they changed. In the beginning, members of today's Protestant monasteries saw themselves or were seen as pariahs and sectarians, radicals and revolutionaries. If they managed to stay alive and healthy for a decade or two, they looked much less extreme. In part this was due to familiarity, the church and surrounding community accustomed themselves to the brothers and sisters (or when this was not the case and they were run out of town, they found someplace more welcoming and that place got used to them), but in part it was due to the fact that the community adapted itself to the needs and mores of that time and place. If they did not, they didn't survive. The community life settled down to a routine, displaying far fewer of the attributes of enthusiastic religion than it had when it began. This settling down, in addition to the fact that the church became accustomed to having communities nearby, helped make intentional communities more acceptable to the institutional church.

By 1949, theological developments and historical/religious events (deaconesses, high church movement, Scouting, Pietistic revivals, Oxford Group, ecumenism and World War II) led to the formation of groups of like-minded Christians willing to commit to life together in community, but they were not yet being called monasteries. The next step, the move to self-conscious monasticism, resulted when three parties – the communities, the leadership of the institutional church, and the laity – all discovered a common ground in monasticism. For monasteries to have become as popular as they became, all three parties were needed. Taizé could never have become the pilgrimage center it became had Protestant leaders and laity all agreed that monasticism was anathema. The movement towards monasticism reflected the fundamental longing for reconciliation common to many European Christians in the aftermath of WWII, and monasticism was seen as part of the reconciliation process.

During the first half of the twentieth century, sharp divisions could be seen within Protestant Europe. These differences cut across national, denominational, and confessional boundaries. There were three ways of being Christian. The first emphasized historic confessions, creeds and forms of worship (Traditionalists). The second way of being a Christian exalted the Bible as the inspired Word of God and stressed personal religious experience (Pietists). The third way of being a Christian tried to accommodate science and twentieth century philosophies and ideas (Modernizers). Groups that developed into communities were often inspired by one of the strains of Christianity and found themselves pitted against Church leaders and laity of another strain. After WWI ecumenism flowered, and after WWII it became a groundswell. There was a deeply felt desire to resolve the tension and to unite.

Protestantism internationalized this feeling through the birth of the World Council of Churches in 1948. At the local level ecumenism had the effect of inspiring the communities, the Church, and the laity to look for ways to resolve their differences. For the first time, the will was present to keep the peace rather than splintering, as had been Protestantism's practice when disagreements emerged in the past. The move toward monasticism helped Protestants keep the peace in three ways. First, elements of monasticism appealed to all three types of Christians so the peace was kept because everyone had come to agree that monasteries were OK. The Traditionalists identified with the fact that monasteries were part of the church's past, while Modernists were excited about the idea that monasteries were new to Protestantism. Pietists liked the idea that that the monastery was a place that encouraged personal holiness. Second, monasticism encouraged peace because it meant Protestants could have diversity within the denomination without splintering. Traditionally, monastic communities represented a means by which some believers could separate from the Church without becoming separatists. They could live and worship apart from the local church but still be part of the whole. They were the loyal opposition. Their existence told fellow Protestants that a commitment to Christ could mean 24/7 prayer, but their separation implied that the responsibility for 24/7 prayer belonged to those who felt called to it. Finally, the move toward monasticism helped keep the peace because it meant admitting that Protestantism had been wrong in the past. Admitting fallibility and accepting change were steps required to resolve all sorts of conflicts the denominations would be called upon to resolve at the meetings of the World Council of Churches. If Protestants had been wrong about monasticism, the possibility that specific denominations could have been wrong about other issues helped give way to an openness that characterized church dialogue in the aftermath of WWII.

As individual communities felt drawn towards monasticism, each began to adopt practices associated with monks. An example of one such practice was the wearing of habits (Fig. 30.3). Because the communities, especially after Taizé's 1949 public profession of poverty, celibacy and obedience, were much observed by the press, the change of habit was remarked upon. This meant more visitors came to check things out. That in turn, meant that novices were attracted to the community. With the success of each such innovation, that innovation diffused to other nascent monastic communities. These other communities took up the practice, modifying it to meet their own particular needs. Then they too became more curious about

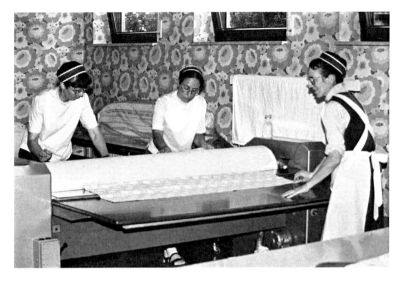

Fig. 30.3 Brotherhood of Christ sisters at work in 1989 (Photo by Dale Pittman)

monasticism. The more they read, the more they realized how much they shared with the monastic tradition. More monastic practices were then adopted, then popularized and diffused.

In order to better understand how this worked, we need to pick apart the idea of what makes a place distinctly monastic. What was it that a community had to do to be seen as a monastery? In general, a monastery is a place inhabited by people who make self-conscious commitments to a life of poverty, celibacy and obedience. Most of the groups were already doing that. They may not always have called them vows, but that was the life they were living. Adoption of a monastic vocabulary became part of the "monasticizing" process. Other elements of the monastic life include silence, enclosure, daily recitation of the liturgy, the use of the Rule, icons and crucifixes, meditation, and a rural location with a property large enough for a retreat house (Fig. 30.4). By 1950 only a smattering of Protestant communities had adopted some of these practices, but by 1990 these were all standard. And by 1990, the communities were called and were calling themselves, monasteries.

There is not space here to do the unique history of each monastic community justice, but it is also misleading to speak solely in generalities about the path monasticism took. It will be more helpful to look in depth at one monastery, at least. For this case study, I have chosen Taizé, in France.

30.8 Case Study: Taizé

Taizé is the most celebrated and publicized of all Protestant monasteries. It is the goal for over 100,000 pilgrims a year (See Note 1.) But while Taizé is the tip of the Protestant monastic iceberg, it is atypical of the movement as a whole. Taizé is

Fig. 30.4 Crucifix created for Casteller Ring's Chapel (Photo by Dale Pittman)

French while most Protestant monastic foundations are German, it is male while the majority population in Protestant monasteries is female, and it does not identify itself as Protestant. Taizé calls itself "ecumenical." It has no Protestant denominational affiliation. Most Protestant monastic communities have an official relationship with some branch of the local church. In the case of Taizé, the local church with which it has had the greatest affiliation is Catholic. But since no monastery is "typical," I have chosen Taizé because Taizé's impact on the monastic movement has been greater than that of any other community.

Taizé was founded by Roger Schutz, aka Frére (Brother) Roger. Born in 1915, as a youth Roger rejected Christianity. But as a young adult he experienced a conversion and decided to become a pastor like his father. He enrolled in the School of Theology of the Evangelical Free Church of Vaud, in Lausanne, Switzerland, where he soon gained popularity and was elected president of the Student Christian Association.

To allow the Student Christian Association to hold retreats regularly, in 1940 Roger Schutz purchased a run-down château near the demarcation line between occupied and free France. Because of World War II he was soon spending more time in France helping refugees fleeing south to escape the Nazis than he spent dealing with retreatants. By 1942 it became clear that the Taizé underground was known to the Nazis, so Roger returned to the relative safety of Geneva for the duration of the War.

During his time in France, Schutz had become acquainted with Abbé Couturier, the Catholic ecumenical pioneer from Lyon. The Abbé had visited Taizé and celebrated mass in the village church, a Catholic space the local bishop later allowed the ecumenical group to use as their place of worship. This meeting was crucial in Roger Schutz's monastic development. Schutz's growing interest in and commitment to community started to have a distinctly monastic cast after the encounter. Roger produced a small pamphlet detailing his ideas on community life. He distributed it to his friends and to those who might be in contact with others of like interests. One such locale was Grandchamp, a Swiss retreat house run by several sisters living in community. Grandchamp would later become Taizé's sister community when it adopted the Taizé Rule in 1953. Through Grandchamp, Max Thurian, Taizé's future theologian, found his way to Frére Roger. From 1942 on, they shared a Geneva apartment with two others interested in a life in community. After the War, all four returned to Taizé. By Easter 1949 they numbered seven. On that day they made lifelong commitments to poverty, celibacy, and obedience.

Word spread like wildfire – Protestants had taken vows! Frére Roger may have resisted calling them that, but that is how they were perceived. The community grew by leaps and bounds. Within a decade there were 39 brothers. They had come from all over the world to join the community. They farmed, painted, made pots, wrote books, and received retreatants for their livelihood. They did well enough at these tasks to support fraternities in other parts of the world. And unlike the other communities, they quite rapidly became fully committed to monasticism. By 1952 Brother Roger had written a Rule and the community had adopted it. In 1959 Max Thurian published a book supporting the celibate life for Protestants, so Protestants had theologically grounded celibacy. And, eventually, this ecumenical community had sacred space as well.

One of the ways in which sacred space is created is when a place is seen as being called into existence by supernatural forces. Stories of "coincidental" occurrences creating sacred acts often inaugurate a place's sacredness. Taizé is such a place and the following reminiscences are typical of the kinds of stories that accompany the formation of many Protestant monasteries.

> Roger signed the undertaking to purchase the property and the lawyer promptly sent a telegram to Lyon to the owners of the house. What the purchaser did not know then, but it is a story that has since been well substantiated, was that the very day on which she received the news that the property had been sold was the ninth day of a novena Madame de Brie had been offering that their house might find a buyer....Still, that woman had offered a novena which finished on 8 September, and on her return from mass that very day she received the telegram. She and her husband had nothing left to live on. They had one child, a boy who bore the Christian name of Roger. (Spink 1986: 36)

This story asserts that the hand of God was active in the sale of the property. Roger's arrival in Taizé and his procurement of the château were the answer to another Christian's prayers. Ergo, God was guiding Brother Roger.

Once established, Taizé attracted attention and crowds. The fact that Taizé self-consciously and publicly embraced monasticism gave other communities, communities that began living a monastic lifestyle before Taizé was even a twinkle in Brother Roger's eye, the courage to do the same. Why was Taizé the community that crossed the Rubicon? (1) It did not belong to an institutional church so there was no one to tell it that it could not to do so. (2) It was comprised exclusively of young men and (3) it grew very quickly. All these contributed to giving Taizé the freedom and the inclination to make what was still, to some degree, a radical break with the past. While the stage had been set for the event, and the needs of the times were calling Protestant monasteries into existence, the river needed to be crossed. Taizé was the community that did it.

During the 1950s the renovated château provided the brothers with all the shelter they and their guests required, and the small Romanesque church, once restored, provided a perfect setting for the community's liturgical life. But within a decade of the first vows it was more than obvious that a new church was desperately needed. There was no place for the ever-growing numbers of monks and visitors to worship together. But Brother Roger did not want to build a church. He had not imagined that the contemplative life would include the crowds filling the Burgundy hill, and was ambivalent about committing time and energy to structures that would only increase the number of visitors. One of the brothers was studying to be an architect and he wanted to design a church for the community. Frére Roger said he could go ahead and design it, but it would never be built, because, "We are not builders" (Spink 1986: 77). He was wrong.

At about the same time a German youth organization called Signs of Atonement (Sühnezeichen), whose members committed themselves to labor for at least a year on projects throughout Europe in atonement for German actions during World War II, approached Frére Roger with the offer to build a church they could see was desperately needed. In spite of himself, Roger Schutz found himself agreeing. The Church of Reconciliation was designed by the architect brother and built by teams of German students.

The Church of Reconciliation is cavernous and dark, a welcome change from the summer sun and a comforting warmth of a hundred candles on winter evenings. It has been said to resemble an airplane hangar, and there is truth to that (Whitgift 2007). It has room for 400 persons seated and for 500 standing. It contains no pews although benches line the walls. Upon entering the church, many Protestants are surprised by the décor. Not only is it much darker than a typical Protestant church, but it contains Catholic and Orthodox images. An icon of the Trinity hangs on the right wall at the front of the church, an icon of Mary with the Christ child on the left, and a Franciscan crucifix is situated more or less in the middle. All this is in the space that was originally designed as a Protestant sanctuary. In the crypt of the building are two small chapels: one is Roman Catholic, the other is Orthodox.

Eventually, even space for 500 persons proved insufficient. Eight years after the building was completed, when the community was expecting a large crowd for their Easter services, the back wall was knocked out and replaced with large folding doors. These are now opened whenever large crowds are expected. A marquee is extended to provide shelter for the congregation. Later, two onion domes were added to the roof of the Church (Fig. 30.5). By the 1990s Taizé was increasingly coming to be seen as Catholic. More and more visitors were coming from Catholic countries and they often didn't see a difference between Taizé and the monastic communities in their own churches. More of the brothers who entered the community were Catholic, and in 1988, Max Thurian, Taizé's theologian and subprior, converted to Roman Catholicism and was ordained as a Catholic priest. The addition of the onion domes served to stress the ecumenical part of the community's name.

When pilgrims arrive in Taizé, they see signs everywhere with the word "SILENCE" on them. But what the visitor initially hears is more likely to be a cacophony of tongues. Every day of the year at least half a dozen languages are

Fig. 30.5 Taizé's Church of Reconciliation (Photo by Maren Henke, http://commons.wikimedia.org/wiki/File:Taize05ostern20.jpg)

spoken here, and during the summer's crowds, that number quadruples. It is immediately apparent that Taizé is international. This is the only Protestant monastery where the visitor experiences this.

Where all the people come from has never been accurately determined because Taizé is adamantly committed to keeping no records. People arrive in buses and stuffed-to-the-brim cars and vans. They hitchhike and they walk. They take bicycles and trains. Most of them are young. Some bring their own tents, some sleep in community tents, some stay in the guesthouse. They eat simple meals in the open air. Most of the meals are prepared by volunteers who are also young. Often the volunteers were retreatants themselves a month or two earlier when they decided it would be a good thing to stick around and see what should happen next. The costs are low, but if low is still too high, the retreatant can talk to one of the Sisters of St. Andrew (Catholic) responsible for most of the hospitality and she'll see what can be done.

So what is the appeal? What makes Taizé so popular? What makes it such a spiritual center, such a sacred place? As a monastery, Taizé's message is the need simply to be. Not to do anything or say anything particular, simply to be and to keep on searching how to be. Taizé proclaims Kierkegaard's dialectical fact. A monastery's existence is to help the Church evaluate its existence and to keep it on the straight and narrow. Taizé's particular role in that process is seen as its commitment to being there, to listening, and to liturgical prayer. While that may sound too mystical/metaphysical for popular appeal, it has appealed profoundly to the 18–30 year old demographic for 40 years now.

Its liturgy is Taizé's the most widespread effect. Since Taizé is the center of a far flung pilgrim network, visitors have carried its simple, memorable liturgy back to home churches around the planet. The Taizé liturgy can be found almost anywhere there are Christians. At Amazon.com, when the name Taizé was entered in the search bar, 19 of the first 20 items listed with the Taizé name on them were collections or recordings of music. Only one item was a book about the place. The first answer visitors give to the question "What makes this place special?" is "The liturgy." It is simple, yet beautiful, easy to learn and easy to sing. Verses have been translated into numerous languages so one moment the congregation is singing in German, the next it is singing in Spanish. And the liturgy is being sung, together, by individuals from all over the planet. It is a powerful experience.

Another source of Taizé's popularity is its generic Christianity. It insists on the label ecumenical. But that term has traditionally had no doctrinal or liturgical content of its own. It means the cooperation between churches. At Taizé "ecumenical" has become a church in its own right. It is a mosaic church made of bits and pieces of a dozen traditions. This "mosaic" principle was used by the ecumenical leader Wilfred Monod when he created of the Watchers Liturgy on which Taizé's liturgy was originally based. Over the years the principle has expanded. Taizé's worship service was taken from a wide variety of Christian communities and includes a little bit of everything. Brother Roger believed in limiting the doctrinal elements of Christian worship. Because the Taizé liturgy was always multilingual, he preferred keeping the spoken word to a minimum, favoring song and liturgical prayer. The time for talk at Taizé is at the end of the evening service, when the brothers situate themselves around the perimeter of the room. Pilgrims who seek the brothers' counsel can wait for turns to talk with one of them, each of whom speaks several languages, and these conversations can sometimes go on long into the night.

Finally, Taizé's popularity itself generates mass appeal. Many visitors arrive just to see what the excitement is about. Frére Roger's stabbing death in 2005 in the Church of Reconciliation at the hands of a Romanian woman has only added to the curiosity.

30.9 Conclusion

Protestant monasteries are alive and well in Protestant Europe, as is Protestant sacred space. Monasteries are not numerous, but they are well known and frequently visited, and they are fully accepted within the bosom of the institutional church. The Church of Sweden actually lists its communities on the Church's website. Monasteries were helpful in reintroducing sacred space into the Protestant landscape in two ways. The first is through their existence. Each monastic community felt "called" into existence, making its presence God's doing and thereby sacralizing the space it occupied. Second, because the monasteries' existence caused such a

Fig. 30.6 Sisters of Mary Chapel in Darmstadt (Photo by Dale Pittman)

stir, it generated a discussion about and exposure to such high church practices as the Liturgy of the hours, crucifixes and icons, silent meditation, clerical vestments, habits and Rules. Eventually, many of these elements trickled down to the grass roots having been reinterpreted as ecumenical (Fig. 30.6).

Ecumenical, meaning the promotion of worldwide Christian unity, is not something most Protestant Christians were inclined to oppose. While it would be hard to argue that monks or sacred space are mainstream realities in European Protestant churches today, neither do they shock or astound most churchgoers. It seems that Protestantism has truly changed. Kierkegaard's wish has been realized.

The theological and organizational realities that conspired to allow the resacralization of churches and retreat centers, monasteries and pilgrimage destinations, rely on all three forms of religious power; (1) divine or supernatural power, (2) social power, (3) and personal power. When Protestant monasteries tell the stories of how they came into existence, the stories almost always include supernatural and miraculous elements. Once in existence, the inclusion of a new "clerical" element to Protestantism adds hierarchy to the traditionally flat Protestant social organization. Monks and nuns are a mystery made all the more mysterious by the fact that within Protestantism they have nothing to gain by living the ascetic lives to which they are vowed. It is not a fast track to salvation. Most Protestant visitors are amazed by the life choices monks and nuns make. Asceticism is rare in the Western world and therefore awe-inspiring. The third type of religious power, personal power, is the sense of empowerment that comes to individuals who encounter the divine. Sometimes this sense comes from proximity to the monks and nuns. It can also

Fig. 30.7 Interior of Casteller Ring's Church (Photo by Dale Pittman)

come from the monastic landscape, architecture, and art. Because of ecumenism, icons and crucifixes, stained glass and candles, incense and liturgy abound in Protestant monasteries. All of these elements serve to facilitate an experience of the sacred (Figs. 30.7 and 30.8).

The numbers of monasteries and monks has stabilized over the last two decades. New monasteries have started, others have disappeared. There has not been a monastic sweep in the Protestant heartland, with thousands of Calvinists and Lutherans flocking to live as Benedictines. But neither have the communities disappeared. There are young faces among the ranks in several monasteries and new work to be done. When the Iron Curtain fell and East and West Germany rejoined,

Fig. 30.8 Casteller Ring's Church (Photo by Dale Pittman)

Luther's Wittenberg Cathedral once again became the focus for pilgrimage. Hundreds of thousands of visitors traveled to Wittenberg during the last two decades. And for those who chose to, it was possible to stay in a Protestant monastery. What would Luther say?

References

http://www.huffingtonpost.com/2012/05/31/taize_movement_america_throwback_appeal_n_1544221.html.
Biot, F. (1963). *The rise of Protestant monasticism*. Baltimore: Helicon.
Bloesch, D. G. (1964). *Centers of Christian renewal*. Philadelphia: United Church Press.
Brico, R. (1978). *Taizé: Brother Roger and his community*. New York: Collins.
Christensen, A. (1971). The Sisters of Reuilly. In D. G. Bloesch (Ed.), *Servants of Christ* (pp. 50–70). Minneapolis: Bethany Fellowship.
Halkenhäuser, J. (1982). *Engagement und Zeugnis; Aufsatze Zum Christsein in Kommunitäten* (Schwanberg series, Vol. 8). Schwanberg: Casteller Ring Community.
Kierkegaard, S. (1951). *The journals of Søren Kiekegaard* (A. Dru, Trans.). London: Oxford University Press.
Kilde, J. H. (2008). *Sacred power, sacred space*. New York: Oxford University Press.
Kilström, B. I. (1970). *Societas Sanctae Birgittae's historiska bakgrund. In his Societas Sanctae Birgittae, 1920–1970* (pp. 3–17). Strängnäs: Societas Sanctae Birgittae.
Latourette, K. S. (1975). *A history of Christianity* (Vol. 2). New York: Harper and Row.
Lenhammar, H. (1977). *Genom tusen år*. Uppsala: Academic.
Luther, M. (1912). *Weimarer Ausgabe Tischreden*. Weimar: Hermann Bohlaus Nachfolger.
Sopher, D. E. (1967). *Geography of religions*. Englewood Cliffs: Prentice-Hall.

Spink, K. (1986). *A universal heart: The life and vision of Brother Roger of Taizé*. London: SPCK.
von Harnack, A. (1900). *Wesen des Christentums*. Leipzig: Hinrichs.
von Loewenich, W. (1959). *Modern Catholicism*. New York: St. Martin's Press.
Weiser, F. S. (1971). The origin of the modern diaconate for women. In D. G. Bloesch (Ed.), *Servants of Christ* (pp. 323–360). Minneapolis: Bethany Fellowship.
Whitgift, J. (2007). The Priory Church of Reconciliation, Taizé, France. Retrieved February 10, 2013, from http://shipoffools.com/mystery/2007/1472.html

Chapter 31
Cemeteries as a Template of Religion, Non-religion and Culture

Daniel W. Gade

31.1 Introduction

Each world's religion has rules, guidelines or preferences about how and where the living treats the deceased. Except in Hinduism and Zoroastrianism, cemeteries have been the primary places where dead people are honored and buried. Theological beliefs drive cemetery protocols only to a degree; equally important are cultural—usually expressed as traditional—values and economic imperatives. As one aspect of a geography of death and the dead, cemeteries present a particular kind of humanized landscape and cultural markers (Pitte 2004). Since the 1950s, geographers have explored the spatial settings and material manifestations of burial grounds (Pattison 1955; Kniffen 1967; Sopher 1967; Francaviglia 1972; Perret 1975; Zelinsky 1976; Jordan 1982; Meyer 1993; Harvey 2006).

With ideas and examples drawn from personal experience, I have chosen to elaborate how religious dimensions affect the appearance and organization of spaces for the dead, but also, as a counterpoint, how secularization intrudes into a domain that once was heavily sectarian. The dissonances of burial practices within one's own culture are revealed when comparisons are made with cemeteries of other cultures. My personal journey, in which religion lost its position as a final vocabulary and became instead a source of nostalgia for a transcendental vision, has influenced the topical and interpretive aspects of this essay. In Nietzschean terms, these remarks are not simply my interpretation or my perspective, but rather, as Grimm (1977) phrased it, that I am an interpretation and a perspective.

D.W. Gade (✉)
Department of Geography, University of Vermont, Burlington, VT 05405, USA
e-mail: daniel.gade@uvm.edu

31.2 North American Cemetery Ownership and Practice

The trauma of the end of a life guarantees that cemeteries have always called forth aspects of religion. The material and spiritual considerations involve matters of ownership, landscape, and burial practice. Who controls a cemetery normally determines who is included in it, but one is cautioned against making too many assumptions. Each country has its own burial ground protocols. In Sweden, for example, the state church owns and operates cemeteries; nevertheless, Swedish graveyards are considered to be public spaces open to all burials, including to members of other religions or of no religion. This practice offers a prime example of how the Protestant Reformation took responsibility for eliminating the church as a rival to the state. In North America, cemeteries may be owned by religious organizations, municipalities, private corporations that may be either non-profit or for profit or private family plots on private land. In France, local governments run all burial grounds and by law keep religion under a tight leash. In South America, local governments own most cemeteries, but for all intents and purposes they are burial grounds for Catholics who constitute by far most of the population. In Morocco, no distinctions between Islamic religion and Moroccan culture are even conceivable.

31.2.1 Cemeteries and Religion in North America

In the United States and Canada, civil law in all states and provinces has allowed any responsible entity to establish burial grounds. In most North American cemeteries burial concessions give the family holding a plot the right to keep the remains there in perpetuity. Some religious groups in North America actively created cemeteries for the interment of their own members and justified that by theological reasons or as a way to symbolize communal fellowship. Cemeteries under the auspices of a religious organization both complemented as well as competed with those that belonged to municipalities or to private entities. However, in the early history of North America, Protestants were often involved in creating public or private cemeteries rather than for their own religious organizations. Two Protestant denominations departed from that generalization. Anglicans/Episcopalians consecrated their burial grounds and in early decades often created graveyards adjacent to their churches. The more conservative Lutheran groups did not consecrate ground, yet they often maintained their own cemeteries, conveying the message that apartness in death reflected the ostensible theological purity by which they defined themselves (Fig. 31.1). More recently, Protestant parishes have laid out on their property memorial gardens to receive without an urn the ashes of dead members. Whatever kind of cemetery, individuals, either the deceased or their family, make decisions about a monument that defined them or their values in life. Mormons, for example, have monuments that often identify them as such (Fig. 31.2).

31 Cemeteries as a Template of Religion, Non-religion and Culture

Fig. 31.1 The inscription on this gravestone in Holy Ghost Cemetery, Bergholz, New York, reflects a literal approach to the sacred. Implied in the proclamation of Christ's resurrection is the expectation of a blissful eternal life for the couple. Beyond the deceased, this monument conveys the Lutheran emphasis on Biblical theology and music (Photo by Daniel W. Gade)

Two other faiths that were active in establishing their own cemeteries in North America have strong burial traditions. Jews, with have ancient rituals connected to death, are buried in cemeteries owned by holy societies that may or may not be connected to a particular synagogue. Catholics have had the most extensive systems of cemeteries held under religious auspices. Their graveyards date from the time when Catholics assiduously pursued their differences in the temporal realm and they

Fig. 31.2 This tombstone of a nonagenarian couple in St. George, Utah is indicative of the strong family emphasis in the Church of the Latter Day Saints. The importance of religion in their long wedded life is suggested by iconographic reference to their marriage in the local Mormon temple in 1923 when they were teenagers. The longevity of both man and wife suggests the attention given in Mormonism to healthful living (Photo by Daniel W. Gade)

provide sanctified ground for the dead as well as places of prayer and remembrance for the living. Protestants did not have these requirements, nor were they comfortable with the Catholic iconography adorning the common spaces and individual gravesites of Catholic cemeteries. Since nothing in U.S. or Canadian civil law prevented Catholic entities from doing so, it was a common practice for parishes and dioceses to lay out their own burial grounds. Lay Catholics also started some cemeteries, but the hierarchy most everywhere promoted church-owned burial grounds, for those under official church auspices could be tightly controlled. Catholic burial grounds sometimes divided towns when they would have benefited from community solidarity. For example, among the Czech immigrants who went as pioneers to Nebraska, some were free thinkers and others were Catholics, but the Catholic clergy refused to share cemetery space with the non-Catholic Czechs (Kiest 1993). Parishes had considerable autonomy in setting cemetery rules, including the kind of stone marker, its thickness and its lettering. On theological matters, cemetery managers had to defer to the Magisterium, which, for example, prohibited the burial of suicides. The *aggiornamento* of the Church in the 1960s changed that rule, as well as approving burials and services of non-Catholics in Catholic cemeteries.

In some places in North America, Catholics are buried in reserved sections of public cemeteries. For example, Roselawn Cemetery, which was founded in 1891 in Pueblo, Colorado as a non-profit privately-owned cemetery, established separate sections for Catholics, Freemasons, Elks, and veterans of the Grand Army of the Republic. The subsequent establishment of parish cemeteries in the area did not exclude Roselawn from receiving new Catholic burials; even a bishop was laid to rest there.

The diversity of North American cemetery practices explains the hundreds of thousands of burial grounds large and small that are spread over the continent. A modest sized town such as Keene, New Hampshire, with a population of 22,000, has ten cemeteries, nine of which are municipal and one that is Catholic. The perpetuity rule also plays a major role. Whereas French cities of Keene's size have only one burial ground, the cemetery inventory of a North American city must maintain areas that are fully occupied. Though Europe has a settlement record many times older than North America, it does not observe the rule of perpetuity and thus its cemeteries occupy one-tenth the area.

31.2.2 Montréal and Its Iconic Cemetery Division

Within North America, Montréal has a distinctive cemetery pattern that was created by the overlap of religion and culture. The combined linguistic and religious divisions between native French-speakers and English speakers created the so-called "two solitudes" that long characterized that city. Through most of its history, the religious separation into Catholic or Protestant was paramount in defining cemeteries and every other institution. Urban development obliterated the succession of small graveyards that each group had, which created a demand for a new burial space in the rapidly growing city. Cemetery developers in mid-nineteenth century Montréal favored the idea of a rural cemetery based on the grand model of Mount Auburn Cemetery near Boston, which was founded in 1831. Developers purchased land then outside the city on the northeast flank of Mount Royal and proposed to establish there one municipal cemetery for the whole city. That plan was not to be, for the Catholic authorities in Montréal insisted on having their own burial grounds. Thus two cemeteries were established: Mount Royal Cemetery, founded in 1852, for English-speaking Protestants and in 1854, the Cimetière Notre-Dame-des-Neiges for Catholics. Decades later, two small Jewish cemeteries, one Sephardic the other Ashkenazic, also found space on that side of the mountain.

The Mount Royal Cemetery Company, a private corporation that assumed a public function, laid out its 750 m^2 (8,073 sq ft) hilly expanse with a large variety of trees and shrubs in an English romantic tradition. Owners of individual plots often personalized the gravesites by installing railings and planting favorite flowers. When founded, this cemetery had as its clear mandate serving the one-third of Montréal's population that was Protestant. The Anglicans and Presbyterians who dominated the board of trustees until the mid-twentieth century had as their mission the burial of Protestants, almost all of whom were anglophones. Since this religious and linguistic group constituted Montréal's entrepreneurial elite, Mount Royal Cemetery acquired a reputation as being the rich man's place to be buried (Young and James 2003). The elite are always the manipulators of symbols in cemeteries and "the ultimate designers of the sociospatial arrangement of the dead" (Ericksen 1980). But such was the religious commitment of the cemetery company's directors, that poor people, if they were Protestant and of British ancestry, could count on burial there free of charge.

Religious exclusivity at Mount Royal started to break down early in the twentieth century. Mount Royal interred a fair number of Catholics, including suicides, retrogrades and Freemasons who were refused burial in the cemetery next door. In 1901 the cemetery built the first crematorium in Québec, which was approved by the provincial legislature with the proviso that only Protestants could be incinerated there. Eventually, some Catholics sought incineration in that facility, though, by doing so, their own Catholic cemetery denied burial of their ashes. The right to be buried in Mount Royal extended to atheists and agnostics as well as to an array of non-Catholic immigrants, including Chinese who, disconcertingly, sacrificed pigs near the graves of the Confucianist deceased. After World War II, Mount Royal accepted non-Catholic immigrants not of British stock, many of them Greek, Ukrainian or Russian Orthodox. Another barrier fell by the wayside when Reform Jews asked for and received a separate section for their dead.

Meanwhile, the Protestant workforce, that had defined the cemetery company before 1950 as had its Protestant board of trustees, disappeared. In the 1970s when Québec nationalist sentiment rose to a fever pitch, Mount Royal Cemetery became a symbol of the elitism of a privileged minority. Under the banner of francophone rights, nationalist outbursts in 1975 and 1976 overflowed into the cemetery, resulting in tombstone vandalism. On the defensive and with Montréal's native English-speaking population in sharp decline, the Mount Royal Cemetery Company introduced bilingual signage and provided an array of services in French.

Cimetière Notre-Dame-des-Neiges is twice as large as the adjoining Mount Royal Cemetery. Since its inception, Notre-Dame parish, operated by the *sulpiciens,* a wealthy order with roots in France, has owned and managed the cemetery. Since most parishes had no burial ground of their own, Cimetière Notre-Dame-des-Neiges (hereafter CNDN) served the sepulchral needs of all Catholics in the city, unless for some reason they were refused ecclesiastical burial. Unlike Mount Royal Cemetery, plots were leased for varying amounts of time and then renewed for an additional period. But with no pressure on plot space, disinterments were rare. CNDN became the largest cemetery in Canada with more than a million burials by the end of the twentieth century. However, as many as 80 % of those burials are not identifiable in the cemetery landscape, either because of plot turnover or the absence of a marker. On the grounds are a chapel, greenhouses, refrigerated corpse chamber, crematorium, and 10 communal mausoleums with crypts and four of them with columbaria. The invention of the backhoe that can dig open frozen ground made the winter corpse storage facility obsolete. Paved roads have facilitated the movement of visitors, who rarely arrive on foot to this vast sepulchral landscape, laid out in garden park fashion.

Until the mid-twentieth century, French Canadians, (now called Québécois), constituted by far the most numerous burials (Fig. 31.3). The linguistic and religious split in the city blocked dialogue between the CNDN and Mount Royal about mutual concerns. Not until after World War I did the Catholics and Protestants of Montréal jointly erect one monument to honor the servicemen who died in combat in Europe. Carved out of each cemetery was a military section at the mid-point of which a monument was erected. On one side, the inscription in English mentions

Fig. 31.3 Francophones dominate in Cimetière Notre-Dame-des-Neiges just as they have demographically dominated the city, if not the island of Montréal. This elaborate pierre tombale reflects a pre-Vatican II kind of Catholic piety now rare in more recent burials. As in the U.S., religious belief in Québec has now become more psychologized and internalized. The words at the base of the monument, "Souvenez-vous," form an injunction to remember, not just those buried there, but the whole historic saga of the French in North America (Photo by Daniel W. Gade)

the British monarch; the other side in French does not. Later the two cemeteries established a similar conjoined space to honor Montréal firefighters.

Since religion trumped culture, English-speaking Catholics, most of whom were of Irish descent, had the right to inhumation at CNDN. In the 1960s immigrants originally from Southern and Eastern Europe found burial there. The older parts of CNDN are ethnically mixed, which conveyed the message that being Catholic was more important than one's ethnicity. That began to change significantly in the 1970s when ethnic immigrant groups became more assertive in calling for their own enclaves of gravesites where a cluster of markers inscribed in their language and national symbols formed a definable community. Thus Polish-Canadian tombstones, many with Polish language inscriptions, display the Black Madonna of Czestochowa. Nearby, another section has tombstone inscriptions in Portuguese or French (the language group to which they assimilated) and images of Nossa Senhora de Fatima. Though abundantly represented in ground burials, Italo-Canadians later concentrated their presence in two mausoleums.

31.2.3 Non-religious Cemeteries in North America

Cemetery tradition in the United States and Canada stands between secular and religious assumptions. The guarantee of freedom of religion in both the United States and Canada permits religious cemeteries for those who want them, but they exist along with an abundance of religiously neutral cemeteries. An example of graveyard secularism, but one with ethnic bearings, occurs in Barre, Vermont, where more than 40 quarries yielding gray-colored igneous rock brought that small city its distinction as granite capital of the United States. First settled in 1788, Barre experienced an economic boom in the late nineteenth century prompted by a high demand for Civil War monuments, federal, sate and municipal governmental buildings and an increasingly affluent American population. that preferred granite tombstones over those of other kinds of stones. To meet the labor demand of those boom years, European immigrants arrived between 1880 and 1910 from Scotland (Aberdeenshire); Sweden (environs of Göteborg), Spain (around Santander); and Italy (Lombardy and Tuscany) to work in the quarries and granite sheds. As a group, the Italians stood out for their skills, passions, and demographic weight in the community. Stonecutters, artisans and sculptors, they were also literate, politically aware, and sensitive to workers' rights. Anti-monarchical and anti-clerical in their views, Italian immigrants to Barre were politically in one of two camps: anarchist or socialist. Strikes they called in the 1920s against the granite companies prompted quarry owners to bring in French Canadians from Québec who then became the next new ethnic influx to the Barre area.

The Italian immigrant community in Barre dominated a private company, which in 1895 laid out Hope Cemetery as an alternative to the newly established Catholic burial ground owned by St Monica Parish. Though its charter accepted Catholic or Protestant burials, religion received no special consideration and no religious symbols were permitted in its common space. From its beginning, Hope Cemetery became known much more than as a place to dispose of the dead. It became a showcase for those who quarried and wanted a place that highlighted Barre granite, the only kind of stone permitted for those monuments. Those who carved monuments saw themselves as individuals and wanted to honor and immortalize their artistic talent. This attitude differed from that of most gravestone carvers of Anglo-Saxon origin who remained anonymous (Hijha 1983). Many meticulously-carved tombstones in Hope Cemetery reflect the personal situations and secular world view of

31 Cemeteries as a Template of Religion, Non-religion and Culture 631

Fig. 31.4 This imposing monument in Hope Cemetery, Barre, Vermont, displays the well developed funerary art tradition that Italians brought to America. The specifics involve a historic local figure, Elia Corti, a skilled stone artisan, whose life had a tragic end in 1903 at age 34. His funeral, which took place without the presence of a priest, was a major local political event. This sculpture represents Corti as a thinker dressed in a suit and tie, though elsewhere on the stone are carvings of the mallet, hammer and calipers that he used in his art (Photo by Daniel W. Gade)

their carver. In one, Louis Brusa, deathly sick from silicosis, carved a dramatic self portrait in the arms of a woman. Another gravestone shows Albert Ceppi carving himself with the inscription "*sculptore supremo*." Among other markers is a married couple in a stone bed gazing in each other's eyes; a civil air patrol pilot in his biplane; a soccer ball; and a race driver's car. The most famous of the cemetery tombstones at Hope is that of Elia Corte, a politically active granite worker murdered in 1905 for his anarchist views (Fig. 31.4).

The secular spirit of Hope Cemetery presaged a major regional change in that direction throughout the State of Vermont. This polity now has the lowest per capita church membership and lowest per capita church attendance of any state. At the same time Vermont has become the most politically liberal of the 50 states (Worcester 2011). It was the first state to have civil unions and one of a handful that has same-sex marriage, a law that came from legislative overturning of a veto. A single payer system for health care is to be implemented. In that context, by the turn of the twentieth century, cremation had come to dominate disposal of its dead. The 1960s, which brought a substantial inflow of people from large eastern cities, was the defining decade in this social shift. Unlike the heavily in-migrant states of Arizona and Texas which have vaunted individualistic values, the strong communitarian thinking in Vermont found its models for public policy more in Western Europe than Western America. Vermont raises a big rhetorical question about the correlation between the religious and the political. Does rejecting religion make people liberal or, alternatively, do holding liberal values lead people to become apostates? In my judgment the second possibility takes precedence over the first.

31.2.4 Broad Trends in North American Cemeteries

Collectively, cemeteries manifest the attitudes of a religious affiliation, a subculture, but also a period in time. The Puritan view of death in the early seventeenth century contrasts with what emerged 150 years later. Judging from the constant epitaphic references to God, Jesus, Christ, and the Omnipotent, the afterlife was the major Puritan concern. Cemeteries were then in the center of villages and tombstone inscriptions, accompanied by the death's head, communicated to the living the importance of preparing for one's eternity (Stannard 1977). As material security improved, the early preoccupation about death as a spiritual passage evolved by the eighteenth century into a view of death as being part of a natural process. No longer did strong expressions of belief mark the words on tombstones; instead of the divine truths of revealed religion, the focus became sentimentalized on family loss. Émile Durkheim (1915: 62) observed that the social practices of a religion have been more central to many people than any dogma or belief. Among those who see death in religious terms, a wide range of interpretation is possible. As Michel Vovelle (1983: 713) has cautioned, no isometry ever existed between those who believe in God and those who believe in an afterlife.

The trend today toward secularization in North America either emphasizes religion as a social force or leaves religion behind. The same process occurred at least three decades earlier in Western Europe where the belief systems of both Catholics and Protestants have largely collapsed even as the churches themselves, though bereft of firm believers, manage to survive. In Germany the state pays salaries of the clergy and in France the state has responsibility for repairing religious monuments. In both the United States and Canada the downward spiral of religious involvement since the 1970s has affected Protestants and Catholics. Mainline American Protestants all reported membership losses in the order of from 0.5 to 2.0 % a year (Chaves 2011). Catholic sources do not annually publicly report defections from the fold, but it is generally believed that four American-born Catholics have left the Church for everyone who has joined it (Carlin 2003). Of those who still consider themselves Catholic, less than a fourth go to mass and many are not married by a priest in the Church. Hundreds of Catholic parishes have closed and more than 3,000 of them not closed now have no resident priest (Jones 2003).

Several groups, most notably Pentecostals, Mormons, and Jehovah's Witnesses, have grown in adherents, but their increases collectively do not offset the larger downward trend in religious affiliation and participation. The rise of people with no formal religious identity rose from 3 % in 1957 to 17 % in 2008. (Chaves 2011). In the 20-year period from 1990 to 2010, the number of Americans who claim no religion has nearly doubled. A fall in the belief in God of 1 % a decade since 1955 fits into this downward trend in religious adherence and practice (Chaves 2011). Uncontested belief in certain tenets of Christianity can no longer be assumed. Blaise Pascal's (1995) seventeenth-century utterance that "without faith in immortality, we are all doomed to utter despair" would not go uncontested even among assiduous churchgoers. The idea of heaven is, for many, counterintuitive, if only because, from a human perspective, the scenario of eternity is one of unremitting boredom. The kind of religious reflexes, both Protestant and Catholic, described in W. Lloyd Warner's (1959) classic book, *The Living and the Dead*, has disappeared. Perhaps as a consequence of its decline, religion has increasingly moved to the private sphere. No longer do churches have the Sunday-morning monopoly they had into the 1950s. Religion has lost its hold on social mores, so that, for example, "'living in sin' is so common that the phrase has become passé" (Bruce 2011).

The rise of secular attitudes in general and as they specifically relate to burial will lead to the decline of the cemetery as an institution. Although many religious people are now cremated, this practice reflects more than anything the growth of a secular attitude not just by individuals, but also by religious institutions. Since 1970 cremation in North America has increased at a rate of about 0.5 % a year. Shifts in religious thinking have fueled this trend, because cremation was once firmly opposed as it seemed to deny the possibility of corporeal restoration. In the Apostles Creed the allusion at the very end to "the life everlasting" comes right after the words "resurrection of the body." In the popular imagination, cadaveric presence and belief in its post-mortem resurrection went hand in hand for it is far easier to imagine an intact corpse coming back to life through divine intervention than the corporeal reconstitution of incinerated bits.

Nevertheless, already by the twentieth century, many Protestants had finessed that implied objection to cremation. The Catholic hierarchy continued to rail against it, which, in retrospect, was based largely on the practice's association with liberals, Freemasons, and communists. The realities of an acute shortage of cemetery space in many Catholic countries and the growing mobility of individuals made cremation an irresistible solution to the practicalities of disposing of the dead in modern society. Finally, in 1963, the reformist agenda spurred by the Vatican II Council gave Catholics the option of cremation in the Holy Office document *De cadaverum crematione*. A step further occurred in the 1990s when the Holy See authorized North American bishops to allow funeral masses in the presence of cremated remains. The lag that occurred in the actual implementation of cremation suggests that a decade was needed for people to get used to the idea. For example, the Cimetière Notre-Dame-des-Neiges in Montréal did not install a crematorium on its property until

1975 and even then the ecclesiastical authorities argued that its purpose was to dispose of body parts from Montréal hospitals. In 1966 a scandal had erupted when body parts were found in hospital garbage (Le Devoir 1975).

In 1990 in the United States, cremation accounted for 17 % of the disposal of the dead; in 2000 that figure rose to 25 % (Bryant et al. 2003). In 2010 it reached 40 %. In Canada the rate of cremation has been from 10 to 15 % higher than in the United States. The sharp increase in North American cremation has gone hand in hand with a decline in religious commitment. That parallel is even more evident in Great Britain, where more than 70 % of deceased persons are now cremated and formal religious involvement hovers around 11 %. Uncertainty about an afterlife is probably fueling this practice. Rather than deferring to the theological assertions of traditional Christianity, individuals increasingly make their own decisions about such matters.

Indeed, cremation represents an avenue for individual expression, breaking with the stereotypical thinking of the past. An individualistic ethos releases people to seek memorialization in ways other than by a cemetery gravestone. Pulverized into dust-like material in a kitchen blender, the bone fragments can be scattered to the wind, buried in the backyard, incorporated into fireworks display, floated in balloons, painted into portraits or compressed into synthetic diamonds. For a fee, remains can be scattered from helicopters. A relatively new reason for taking a cruise is to spread the ashes of a loved one at sea. As Hindus have always believed, scattering of ashes can have a profound religious meaning as an essential rite of purification. In 1985, the U.S. Department of Transportation approved a plan to send into orbit a flying mausoleum containing the remains of as many 15,000 people, each of whom could be encased in tube 50 mm [1.96 in.] long, which would stay in orbit for 63 million years.

31.2.5 The Future of the North American Cemetery

Cremation may eventually largely remove the need for cemeteries of traditional appearance. To be sure, ashes are still buried in cemeteries, where they require only tiny space compared to coffins. On a plot 10 m^2 (107 sq ft), 200 urns can be interred, compared to only four to six coffins. Furthermore, in or near large cities, cemeteries have walls and niches to receive large numbers of urns in a relatively smaller land surface. Meanwhile, burial perpetuity, which grants families the right to a cemetery plot from which the remains would never be removed, creates problems. That arrangement also usually includes perpetual care absolving families of the responsibility of caring for gravesites. That care, however, is often minimalist. Survivors rarely even visit their family plots in North America and it is no wonder that many look half abandoned, especially in comparison with those in other countries. In Italy, for example, graves are kept neat, clean and adorned with real (not artificial) flowers as a matter of family honor (Goody and Poppi 1994). A cultural more than religious difference distinguishes the family's role in death in Catholic Italy vis-à-vis Protestant America (Fig. 31.5).

Fig. 31.5 This simple but informative headstone in Limone sul Garda, Italy describes a septuagenarian as a "fervent and sincere Christian" who died in the "kiss of the Lord." However, more space on the epitaph was given to his earthly achievements: "an intelligent fisherman who knew the Lake (Garda) well" and whose efforts in organizing the fishing business was "welcomed and appreciated by everyone." (Photo by Daniel W. Gade)

In the years ahead, the North American trend toward secularization may also question whether the remains of one's loved ones must forever be preserved *in situ*. At its core, this idea arose from a spiritual belief and a material opportunity. In the former, the body is in an eternal sleep from which it must not be disturbed until resurrection day. By embalming the corpse, public exposition of the cosmetized body, and use of ground vaults a fiction was created that the person was not really dead. The American cultural unwillingness to accept the finitude of life has also been expressed among the living who undergo expensive surgical operations to prolong life well-beyond normal life expectancy. In the latter, this attitude has gone hand in hand with abundant and cheap land that made perpetuity contracts possible.

However, cost and mobility make this common arrangement ultimately untenable. The maintenance of expansive spaces with large plots and an abundance of vegetation needing frequent horticultural attention will become economically unsustainable. Secondly, North Americans tend to be very mobile and as people move away or die out, few will maintain a direct interest in a particular gravesite. Nothing is more ironic in this regard than the promise of eternal remembrance on a dilapidated and forgotten tombstone. North American cemetery developers did not take into account the role of time in the mourning process, which, in cemeteries, translates into an inbuilt "forgetting rate" of human burial. By correlating the year of death and the presence or absence of fresh floral offerings on graves, Hannon (1990) calculated that within 50 years a grave is normally thrown into total oblivion. In Europe, with its much longer history and higher population densities, the length of the human grieving period is more realistically assessed. There, after a few decades in the ground, remains are removed, burned, and the gravesite is recycled for another cadaver. It is a sustainable system compared to that of North American cemeteries, which are burdened by expansive spaces and the morass of forever contracts. Wider acceptance of the practice of cremation will free survivors from paying for plots in which, after a time, they lose interest. Meanwhile, the crisis of the traditional North American cemetery will come to ahead. Anticipating the need for changes, Québec altered in 1994, its civil code, removing perpetuity of cemetery burials for both maintenance and concessions. In its place, the province instituted a system of burial leases which are good for 25 years and renewable up to 100 years.

The dynamics of urban development impose additional pressures on cemeteries by demanding that land for other uses. Burial spaces in cities are not as untouchable as they once were when assumptions about "sacred ground" were universally respected and deferred to. The public relations departments of the two large Montréal cemeteries previously discussed vaunt their contributions to the ecological health of the urban environment and their role in the city's historical patrimony. Though dog walking and sunbathing are still banned, they permit jogging, bird watching and bicycling. Since cemetery companies do not pay property taxes, the land they occupy is likely to become a point of contention. At some point, cemeteries are likely to yield to pressures for either recreation or construction (Harvey 2006). Over the years, many cemeteries in the path of airport expansions or highway reconfigurations have been relocated. Prime candidates for land use conversion are old municipal graveyards no longer used for burials and located in densely populated areas whose living members have no dead relatives there. It is not inconceivable too those hard-pressed municipalities in their quest for tax revenues will at some point sell unused graveyards for urban development. By putting a premium on the living and their needs, such changes manifest the growing primacy of secular values. Some of this shift in attitude, which is still in its early stages, will come from the widespread acceptance of cremation as a practice. Incinerated human remains, like dust, are volatile, evanescent and untraceable. The psychological association that survivors have for cremation compared to an embalmed body are more abstract, remote, and secular. Photographic records of all grave markers may replace existing cemeteries as archives.

31.3 Cemeteries in the Andean Countries of Bolivia and Peru

For almost 500 years European Christianity has strongly influenced the Andean countries of South America, but cemeteries have been part of the cultural landscape for less than half of that time. Through most of the colonial period, most cadavers were placed in the basement (*bóveda*) of churches. Applying lime to the corpse hastened decay of the fleshy parts so that the bones could be compactly arranged in that space. Exceptions to burial inside the church occurred when accidental death precluded access to the last rites. For example in one small Bolivian village, I found 60 years of death records that included many who had died from a lightning strike. Those corpses had not received Extreme Unction and could be buried only in the atrium of the church (Gade 1983). Whether inside or outside the church building, the Catholic association with inhumation manifested the imposition of a new religious belief on the Indian population who had a different burial tradition. Well into the seventeenth century, indigenous people dug up the corpses of their relatives and mummified them in large ceramic containers or in caves. This custom conformed to an Inca mode of disposal of the dead, which considered earth burial to be an unacceptable smothering of the body.

In the late eighteenth century, an Enlightenment idea arrived in the Iberian colonies of the New World from France that pathological vapors emanating from church burials caused disease. That new assertion was a step beyond the common complaint about the mephitic odor from decaying flesh that seeped into the nave of the church from below. The faithful considered an attack on their sense of smell as a small price to pay for church burial that facilitated the breakout of the soul from purgatory and on to heavenly bliss. When official orders put baroque piety aside and prohibited burial in and around churches, the authorities in Andean cities and towns created cemeteries on the outskirts of settlements.

In 1789 the Peruvian city of Tarma constructed one of the earliest municipal graveyards and the Bolivian city of Potosi created one of the last in 1910. The municipalities, not the ecclesiastical establishment, took charge of these cemeteries, a transfer of power from the spiritual to the temporal. When I surveyed the

Fig. 31.6 The general cemetery of Potosí, Bolivia as it looked in 1980 lies in the cold puna 4,100 m (14,000 ft) above sea level. In Bolivia, many people of rural origins conceptualize gravesites as symbolic structures for the deceased. Those of a higher socio-economic class place their dead in a niche in the pavilions as shown on the right. The elite often have their own family mausoleums (Photo by Daniel W. Gade)

organization of the Potosi cemetery in 1980, it was the only cemetery in a city of ca. 70,000. Thirty-eight pavilions held 17,000 niches (Fig. 31.6). The municipality granted the niche free, but to seal it required paying a fee of 500 pesos (then about $US 25.00) for a 5-year lease. If the arrangement was not renewed for another 5 years, workers removed the coffin from the niche and either cremated the remains or threw them into a common pit. Restricted leases and short mourning periods guaranteed the rapid turnover of niches. In the 1980s, the general cemetery of Potosí also had 150 pavilions owned by associations, mainly trade unions and religious orders, whose members had rights to burial there. Another section of the cemetery was reserved for children up to 12 years of age. Burial space for this age group formed a cultural response to a Catholic theological notion—one uninformed by modern neuroscience—about the presumed impeccance of dead children who are likened to little angels ("angelitos").

About 120 km (75 miles) to the east of Potosí, the *cementerio general* of another Bolivian city, Sucre, followed the basic organization described above, except that in the latter city a well established elite constructed elaborate family mausoleums with legal rights in perpetuity (Gade 2010). As in Potosí, people holding certain occupations in life, among them taxi drivers, truck drivers, teachers and lawyers, were buried in pavilions belonging to their professional association. The corpses of indigents were thrown into a common pit with no marker of any kind. Dead children also were placed in a separate space. Family ties are close in the Andes, but, strangely, that has not led to keeping families together in death except among the wealthy. Municipal cemeteries in growing Andean cities have little space for all who want it, resulting in the short term concessions. New cemeteries ease pressure on municipal graveyards. A recent development in some Andean cities is the establishment of North American-style garden park cemeteries. Clandestine cemeteries, a response to the pressure on cemetery space elsewhere, also now occur. A cultural reticence toward cremation exists in the Andes; if and when that practice gains acceptance the pressure on cemetery space will be relieved.

A rural village in Bolivia and Peru typically has one cemetery on its outskirts. A surrounding adobe wall keeps out pigs and dogs that may disinter corpses. Individual graves typically are marked by a miniature adobe construction and a makeshift cross fashioned out of wood or tin plate. Inscriptions are rare and even the name of the deceased may not be present. The indigenous resistance to ground burial may have disappeared, but without burial perpetuity and with little identification of the interred, it is difficult to retrieve a village history by visiting its cemetery. Burial occasions in the Andes can be raucous, a custom of indigenous origin which includes eating, drinking, and playing games of chance (Fig. 31.7).

Fig. 31.7 Devoid of other-worldly sentiment, this glass-enclosed tombstone display in the cemetery in Urubamba, Peru is concerned with the material sustenance for the deceased in his afterlife. He may also have been someone especially fond of beer. In the Andes people who visit cemeteries sometimes bring food and drink for themselves and for relatives who have died. The idea behind the playing cards is to keep the deceased occupied in case he/she gets bored in the heavenly home (Photo by Daniel W. Gade)

31.4 French Cemeteries and Institutionalized Laicism

As an assertively secular country, France has a local cemetery protocol entirely different from that of the United States or Canada. France, the only Western democracy that places restrictions on religion, has for more than a century required every buried corpse to be interred in a space controlled by the commune or the municipality. This system reflects the strong anticlerical view of the French Revolution. After 1789, cemetery control was transferred from the Catholic Church to secular authorities and, through the decades, burial protocol was refined. A 1881 law forbad confessional divisions within cemeteries and a 1887 law guaranteed the right to religious symbols or words on grave markers. These refinements, together with a 1905 law prohibiting discrimination on the basis of faith, resulted in an unusual heterogeneity inside French cemeteries. No religious sections are permitted: a Catholic may be buried next to a Jew, Muslim, Buddhist or atheist (Fig. 31.8). Normally a burial concession comes with a 30-year lease, after which cemetery workmen disinter and incinerate the bones. In that way new burials can be accommodated. Pressure on burial space in France is further eased by the rise of cremation. Since the 1990s, *incinération* accounts for about one of two burials.

Fig. 31.8 This tombstone in the municipal cemetery in Carnoux, France communicates more than family sentiment. Its cryptic political message "to the memory of all our dead left in Algeria" refers to French people who had settled in North Africa, but who left Algeria in panic in 1962 without repatriating their deceased loved ones buried there. Mediterranean cultures have placed a strong emphasis on the materiality of bodily remains, the legacy of an ancient Catholic sensibility that death is a change of status more than it is a permanent absence from the world of the living (Photo by Daniel W. Gade)

Muslims have contested French burial democracy, which conflicts with several rules of Islamic inhumation. In France burials must be in a box placed in a deep hole. Civil authorities have sometimes accommodated the *qibla* requirement of orienting Muslim tombs southeastward toward Mecca, but they are under no legal obligation to do so. Mayors, who in France hold ultimate authority on such matters, have routinely rejected requests to negotiate longer burial leases and separate burial sections (*carré musulman*). The usual alternative for pious Muslims has been to ship the cadaver back to the country of origin. However, the second generation of French Muslims, mostly aware of the advantages of secularism in their lives, has rejected many Quranic proscriptions (Roy 2005). There is, however, an exception to secular cemeteries in France: the region of Alsace, where, since 1973, several Muslim sections were laid out. When they filled up, municipal authorities in 2010 constructed an exclusively Muslim cemetery. A legacy of its Germanic history, Alsace's legal framework ("*droit local d'Alsace-Moselle*") permits not only burial grounds according to religion or lack of it, but also the use of public funds for their construction.

Though named for the Jesuit confessor of Louis XIV, the Cimetière Père Lachaise in the very worldly city of Paris encapsulates the secular attitude that drives French public life. Even before the French Revolution, Paris was notably secular in its attitude, a condition that had its origin in a place that historically was never effectively converted to Christianity. In the early nineteenth century the municipality opened Père Lachaise, together with the two other grand cemeteries of Montmartre and Montparnasse to replace the unhygienic burial grounds that had proliferated around churches. In seeking to promote their new graveyard in a poor, semi-rural working class part of the city, the early managers of Père Lachaise transferred to it the bones of Molière (1622–1673) 144 years after he had died. The halo effect of such a cultural luminary attracted a long succession of notable painters, sculptors, actors, politicians and philanthropists who have sought burial there as well. With time, the stone monuments honoring those people left less and less space available for trees and shrubs. No longer a garden park, the Père Lachaise has become France's largest open-air sculpture museum. To meet its role as a tourist destination, Père Lachaise became the only cemetery in Paris to have its own curator.

Since its founding, more than two million people, most of whom had no monuments, have been buried there. Thanks to frequently renewed leases and public advocacy, the famous and the moneyed have perpetuity. Père Lachaise is one of the few graveyards where individuals with no relation to the deceased have brought flowers or cleaned graves on a regular basis. Its artsy cachet created a waiting list of those waiting to be buried there far in excess of the several hundred concessions that yearly come on the market. *Laïcité* in France is not to be confused with egalitarianism. However much *égalité* has been a pillar of French nationalist rhetoric, the elite long since learned how to manipulate the sociopolitical system in their favor. No place better exhibits the search for immortality, not through religion, but through the promotion of individuality which asserts itself in death as in life.

31.5 Islamic Cemeteries in Morocco

In contrast to the laic assumptions of French cemeteries, particularly apparent in the case of Père Lachaise, Moroccans have a burial tradition in which religion cannot be disengaged from culture. Although local governments now have the responsibility for working cemeteries in Morocco, the latter reflect Islam, which is not only the official state religion of the kingdom, but also tied to the notion of a divine kingship running the country. Moroccans with a secular worldview certainly exist, but, as in most Islamic societies, apostate ideas are not tolerated. If publicly expressed, these ideas constitute legal grounds for punishment. Consequently, non-Muslim religious minorities in Morocco organized their own burial spaces. The old Jewish quarter (*mellah*) of Moroccan cities possessed its own cemetery, which is now a fossilized relic. Most Jews left Morocco in 1967 right after the Six-Day War and even the cemetery caretakers are now likely to be Muslim. Likewise the 120 Christian cemeteries in the country have received few burials since independence in 1956. Because French colonizers started these burial grounds, responsibility for their maintenance has fallen on the French Embassy in Rabat.

Muslim cemeteries in Moroccan cities occupy large amounts of choice land wedged between the walled medina and the European-style new section (*ville nouvelle*). They replaced those graveyards within the walls of the medieval core when they filled up and could no longer accept new burials. The principle of perpetuity that regulates Moroccan cemeteries guarantees their eventual inability to accommodate any more burials. In Muslim belief, a dead person laid to rest in a grave will stay there until the day of resurrection when Allah will judge the person's deeds as deserving of either heaven or hell. Unlike the much less articulated notion of resurrection in Western Christianity, this impending eschatological event is believed to take place in the cemetery itself. Burial *in perpetuum* explains why Morocco has needed ever more cemetery space. Cremation remains a taboo in Islam.

Unlike large North American or European cemeteries configured to accept vehicles, those in Morocco are for pedestrians. A Christian funeral cortege in the United States proceeds in slow-moving vehicles, whereas in Islamic Morocco the procession moves with celerity and on foot, part of a burial process completed within 24 h of death. Quranic injunction requires the corpse to be put into the ground in a white shroud, with a board on top, but without a coffin. That sepulchral position is explained as enabling the resurrected person to sit up in his or her grave on the Day of Judgment. The requirement of a shallow grave is based on an old belief that the resuscitated body needs to be able to hear the call of the muezzin. On his or her first night in the grave, the dead person is said to be visited by black-skinned blue-eyed angels who, if they find fault with the life of the deceased, will flog the dead person. Blanket generalizations about mortuary practices in a country like Morocco where tribal traditions have been strong are risky, for Westermarck (1968) found considerable variation in ideas about death and burial.

The *qibla* requirement imposes a rough geometry in Islamic cemeteries. With the head to the north and the feet to the south, the face of the dead is turned toward Mecca. Tombs typically have an upright stone at the head and foot of the grave.

Many tombs are outlined by a concrete frame or are delineated by a ring of stones. Only some graves have the name of the deceased. On Fridays women and children, who are often banned from the inhumation, visit cemeteries in numbers. Beggars seeking alms or food roam the cemetery on that day too. A stark, unkempt appearance comes from the lack of landscaping and the neglect of many gravesites.

Cemeteries in Morocco reflect socioeconomic strata. In Rabat, Marrakech and other cities, the wealthy have selected certain cemeteries for their burials and have constructed family mausoleums there. Other cemeteries dominated by the poor started in an ad hoc fashion around the burial site of a holy man (*marabout*). The acute shortage of cemetery space in cities reflects their rapid growth from rural migration. Only since the mid-1970s have cemeteries in Morocco become integrated into urban master plans. Cemeteries no longer able to receive new burials have been seen as relieving the demand for recreational or social space in densely populated cities.

For example, in the medina of Salé, the 18-ha (444 acres) cemetery of Sidi Ben Acher, ran out of room for burials decades ago. It has become a de facto public space for social interaction, and the municipality of Salé has sought to turn the space into a public garden by removing the stones but not the graves (Philifert 2004). Such change requires intervention of the *ulemas* who must resolve conflicts between Islamic custom and societal demands. A look into the past, however, indicates that the shortage of cemetery space in Salé is not recent. Brown (1976: 43) discovered that a one cemetery outside the medina had layers of graves superimposed one upon the other, but that the stone markers were used to construct a wall of the medina.

31.6 Summary and Conclusion: Larger Meanings About Cemeteries as Markers of Religion and Culture

The cultural-historical tradition in American geography typically has draws its intellectual cues from the visible and the tangible in the material landscape. Instead of emphasizing theory, this subfield coaxes larger ideas from the factual and the particular. From my cross-cultural cemetery rambles I offer six first-order propositional reflections about graveyards as expressions of religion and non-religion. These remarks both summarize my ideas and provide a conclusion.

31.6.1 Individual Devotion Is a Continuum in Every Religious Manifestation

Markers erected in cemeteries reflect both individual and group differences either in religious beliefs or in fidelity to an institution. A feeling for the sacred contrasts with a search for emotional intimacy with the dearly departed, an indication of the variability in commitment which neither membership figures nor the presence of churches conveys. While it is certainly true that survivors, rather than the dead, decide the appearance and inscriptions of a cemetery monument, the personality of the deceased usually guides that decision.

31.6.2 Over the Past Half Century in North America Cemeteries Manifest a Convergence of Previously Divergent Religious Categories

Cemeteries provide evidence of changes in the practices of Protestants and Catholics in the United States and Canada. Once a defining social distinction in most places, the division has now faded into relative insignificance. Greater tolerance about where a person can be buried prevails almost everywhere. Gone are the times when the Catholic laity, intimidated by hierarchical power, kept apart. The separation of cemeteries into Catholic and non-Catholic had much to do with priests and bishops who ordered the establishment of their own church cemeteries. No theological necessity mandated that separation; rather it was one more way to distinguish themselves from Protestants. In a pluralistic America, the clergy had a major role in setting the personal identity of the faithful. But freedom of personal decision has not eliminated Catholic cemeteries. An inertia that allows the dead to rest in tax-free peace has allowed for graveyard survival at a time when other components of the Catholic institutional structure have fallen into a tenuous state. Catholic orphanages have disappeared, the majority of Catholic hospitals have been absorbed by others, and half the Catholic schools in existence in 1960 have closed.

31.6.3 Shifts in the Religious Commitment of a Society Are Not Necessarily Reflected in Cemeteries

Québec provides a dramatic example of the celerity with which a collective disaffection can occur in spiritual institutions. A controlling clericalism once had such a lock on most aspects of that society that public schools, hospitals, orphanages or cemeteries did not exist. Montréal was the only large North American city without a municipal library or a municipal cemetery. But in the half century since 1960, mass disaffection with the Catholic Church as an institution has emptied the parishes, shuttered convents, closed seminaries, laicized the universities and transformed the organization of elementary schools. By the turn of the twentieth century, the Church in Québec had dropped out of public life. For bishops to speak out on moral issues of the day risks vilification from secularists opposed to ecclesiastical influence. When people are asked to believe too much, they end up believing nothing at all.

On the surface, cemeteries seem not to have been much affected, but only because as landscapes of the dead, they are frozen in time and do not readily reveal the changes that have roiled the Church. Since 1970, hundreds of thousands of people in Québec have rejected Catholic burials. In the 1970s, linguistic nationalism replaced religion as the source of Québécois identity. Nationalism was sometimes blamed for crowding religion out of people's lives, but the larger trend toward secularization, evident in much of the Western world, has had a more fundamental effect in explaining Québec's rapid shift from devotion to impiety.

31.6.4 Cemeteries Convey the Meaning of Laicism

Secularity of space in French cemeteries is a reminder of how fundamental the idea of separation of religion and state can be in a society. It is manifested in the absence of a Catholic political party and in the lack of standing that religion has in the affairs of state. In the United States, where religion has long been viewed as a good thing, well organized and richly funded entities are able to push their agendas at various levels of government even though they represent small percentages of the population. In France, governments react only to the demands of individuals, not groups. That crucial political difference between France and the United States is explained by both the political power of American Christian evangelicals on the federal electoral process and the influence of the Israel lobby on American foreign policy. When French *laïcité* is analyzed in its contours, it is clear that the United States does not have the particular political and historical circumstances for the French system to ever be adopted.

31.6.5 Burial Practices Differ Partly on How Societies Separate or Conjoin Culture and Religion

Bolivia and Morocco both fuse religion and culture, whereas North America and Western Europe have societies that decouple the two elements. In the United States that separation predates the rise of secularism, whereas in Western Europe it did not. In both, certain cultural religious traditions, not only religions, influenced the rise of secular ideas.

31.6.6 No Universal Movement Toward Secularization Is Occurring That Might Account for Any Broad Trend in the World About Cemeteries or the Dead

Although secularization stands as a clear trend in Europe and North America, the same is not true for many other parts of the world. This divergence places doubt on the validity of the secularization paradigm, which posits the loosening of religious ties as a result of the triumph of science and technology (Bruce 2011). Moreover, the Zeitgeist can change to reverse the secularizing trends in Western countries that would once again legitimize religion among the masses. A series of catastrophic events may intervene to change human consciousness, but even more plausible is that technological society could become so oppressive that religion will emerge as a bulwark against its materialistic assumptions. The religious impulse has sustained humans for hundreds of millennia, not just to explain the universe or assuage insecurity, but also to escape the dreadful immanence of daily life that drains the human spirit.

References

Brown, K. L. (1976). *People of Salé: Tradition and change in a Moroccan city*. Manchester: Manchester University Press.
Bruce, S. (2011). *Secularization: In defense of an unfashionable theory*. New York: Oxford University Press.
Bryant, C. D., Edgley, C., Leming, M. R., Peck, D. L., Sandstrom, K. L. (2003). Death in the future: Prospects and prognosis. In C. D. Bryant (Ed.), *Handbook of death and dying* (2 Vols., pp. 1029–1040). Thousand Oaks: Sage.
Carlin, D. (2003). *The decline and fall of the Catholic Church in America*. Manchester: Sophia Institute Press.
Chaves, M. (2011). *American religion: Contemporary trends*. Princeton: Princeton University Press.
Durkheim, E. (1915). The elementary forms of religious life (J. W. Swain, Trans.). New York: Free Press.
Ericksen, E. G. (1980). *The territorial experience*. Austin: University of Texas Press.
Francaviglia, R. V. (1972). The cemetery as an evolving cultural landscape. *Annals of the Association of American Geographers, 61*, 501–509.
Gade, D. W. (1983). Lightning in the folk life and religion of the Central Andes. *Anthropos: International Journal of Ethnology and Linguistics, 78*, 770–788.
Gade, D. W. (2010). Sucre, Bolivia and the quiddity of place. *Journal of Latin American Geography, 9*, 99–118.
Goody, J., & Poppi, C. (1994). Flowers and bones: Approaches to the dead in Anglo-American and Italian cemeteries. *Comparative Study of Society and History, 36*, 146–175.
Grimm, R. H. (1977). *Nietzsche's theory of knowledge*. New York: Walter de Gruyter.
Hannon, B. (1990). The forgetting rate: Evidence from a country cemetery. *Landscape Journal, 9*(1), 16–21.
Harvey, T. (2006). Sacred spaces, common places: The cemetery in the contemporary American city. *Geographical Review, 96*(2), 295–312.
Hijha, J. (1983). American gravestones and attitudes toward death: A brief history. *Proceedings of the American Philosophical Society, 127*(5), 339–363.
Jones, K. C. (2003). *Index of leading Catholic indicators: The Church since Vatican II*. St. Louis: Oriens Publishing Co.
Jordan, T. G. (1982). *Texas graveyards: A cultural legacy*. Austin: University of Texas Press.
Kiest, K. S. (1993). Czech cemeteries in Nebraska from 1868: Cultural imprints in the Prairie. In R. E. Meyer (Ed.), *Ethnicity and the American cemetery* (pp. 77–103). Bowling Green: Bowling Green State University Popular Press.
Kniffen, F. (1967). Necrogeography in the United States. *Geographical Review, 57*, 126–127.
Le Devoir. (1975, 19 May). Montréal newspaper.

Meyer, R. E. (Ed.). (1993). *Ethnicity and the American cemetery*. Bowling Green: Bowling Green State University Popular Press.

Pascal, B. (1995). *Pensées/Pascal* (A. J. Krailsheimer, Trans.). New York: Penguin Books USA.

Pattison, W. (1955). The cemeteries of Chicago: A phase of land utilization. *Annals of the Association of American Geographers, 45*, 245–257.

Perret, M. E. (1975). Tombstones and epitaphs: Journeying through Wisconsin's cemeteries. *The Wisconsin Academy Review, 21*(2), 2–8.

Philifert, P. (2004). Rites et espaces funéraires à l'épreuve de la ville au Maroc: Traditions, adaptations, contestations. *Annales de la Recherche Urbaine, 96*, 35–42.

Pitte, J.-R. (2004). A short cultural geography of death and the dead. *GeoJournal, 60*, 345–351.

Roy, O. (2005). *Laïcité face à l'Islam*. Paris: Editions Stock.

Sopher, D. E. (1967). *Geography of religions*. Englewood Cliffs: Prentice-Hall.

Stannard, D. E. (1977). *The Puritan way of death: A study in religion, culture and social change*. New York: Oxford University Press.

Vovelle, M. (1983). *La mort et l'occident de 1300 à nos jours*. Paris: Gallimard.

Warner, W. L. (1959). *The living and the dead: A study of the symbolic life of Americans*. New Haven: Yale University Press.

Westermarck, E. (1968). *Ritual and belief in Morocco* [1926]. 2 Vols. New Hyde Park: University Books.

Worcester, T. (2011). http://www.huffingtonpost.com/thomas-worcester/vermont-green-mountain-conversion_b_1118945.html. Accessed 3 Dec 2011.

Young, B., & James, G. (2003). *Respectable burial: Montréal's Mount Royal Cemetery*. Montréal/Kingston: McGill-Queens University Press.

Zelinsky, W. (1976). Unearthly delights in cemetery names and the map of the changing American afterworld. In D. Lowenthal & M. Bowden (Eds.), *Geographies of the mind* (pp. 171–196). New York: Oxford University Press.

Chapter 32
Visualizing the Dead: Contemporary Cemetery Landscapes

Donald J. Zeigler

32.1 Introduction

In both the popular and academic worlds, visualization is in an ascendant mode, and cemeteries have not escaped the trend. After an era when a certain sameness settled upon cemetery landscapes, a new generation of cemetery arts has appeared to deliver highly personalized narratives of lives well lived. In the not-so-distant past, we knew what to expect when we looked down at a memorial marker. Now, we find ourselves surprised at the variety of textual and pictorial elements as they vary from one grave to another. Grave marking has come to reflect the looser norms of life in the wider world. Over the last quarter century, how we memorialize the dead has become more personal, more visual, and more secular. The themes of individualism and materialism have come to rival (1) religiosity, as it triumphed in churchyards, and (2) rigid conformity, as it triumphed in grassy memorial parks. The cemetery is more varied and more visual that ever, a reflection of changing cultural attitudes and technological capabilities.

Analysis of cemetery landscapes begins with zoning, the division of space into areal units that enable us to generalize about the nature and use of land that is used to both dispose of and memorialize the dead. Into four zones may we divide that space:

1. *The Burial Zone*: It is the space that envelops the mortal remains, whether corpse or ashes. The burial zone may take the form of a subterranean pit or above-ground vault. Although it is the cemetery's *raison d'être*, it is the least important part of the cemetery once interment has taken place. This part of the landscape is unseen, unknowable, and irrelevant to the lifeworld. Its edges are solid system boundaries; they permit no intercourse. What we know about the departed is what we find in the proximate visible landscape.

D.J. Zeigler (✉)
Department of Geography, Old Dominion University, Virginia Beach, VA 23453, USA
e-mail: dzeigler@odu.edu

2. *The Grave Marker*: Usually made of granite, marble, or bronze, the grave marker records information about the person whose remains are out of sight. It fixes the person in space and time, containing, at a minimum, the name of the deceased, plus years of birth and death. Once complete, it does not change; it is a snapshot that gives us a glimpse of the person and the times. It conveys a message of permanence. Various aspects of headstones (usually grounded in deep history) have been studied by Abby Collier (2003), Brown (1994), Hamscher (2002), McDannell (1987) and others. A unique study of pet cemetery grave markers has been done by Brandes (2009).
3. *The Contiguous Zone*: The surface space surrounding the grave and the grave marker may remain empty of anything but grass or gravel. More commonly, though, it is a zone of memorializing ephemera. Flowers (stones on many Jewish graves) are the most recognized components of this space. Unlike the grave marker itself, the contiguous zone is subject to change: from season to season and from generation to generation. Though localized in time and place, the most comprehensive study of the contiguous zone as an active space for ongoing memorialization is about cemeteries in the Appalachian Mountains west of Asheville, North Carolina (Jabbour and Jabbour 2010).
4. *The Land Use Zone*: Whether a family plot, a churchyard, or a larger cemetery, the land use zone is the acreage set aside for the burial or entombment of the dead. Its landscape architecture may evolve organically, or it may be comprehensively planned and managed, sometimes "in perpetuity." Over time, the names given to these land use zones have changed remarkably, always in the direction of creating mental images that deny death and attract sales. Before 1914, 97 % of all U.S. cemeteries carried the name "cemetery." Since then, names have been chosen to reflect "park-like edens." Memorial Park and Memory Gardens have led in popularity, with specific names attached, for example, Hillcrest, Roselawn, Resthaven, and Oaklawn (Zelinsky 1976: 178–182). The new generation of names, in fact, laid the foundation for the new generation of grave marking. These names have also been studied by Tarpley (2006). In addition, studies of cemetery locations and their cultural landscapes have been done by Darden (1972), Darnall (1983), Francaviglia (1971), Howett (1977), Jordan (1982), Kniffen (1967), Pattison (1955), Wright (2003), Zelinsky (1994), and, in the context of memorializing tragic and violent events, Foote (2003).

The grave marker and contiguous zone are the focus of this essay because they offer copious clues to the cultures that created them. What survivors put on the headstone and bring to the grave does memorialize the dead (Jackson 1967; McDannell 1987; Moore et al. 1991). But, really, the cemetery's petroglyphs and valuta reveal more about life than death. The stories they tell are conditioned by time and place. Their narratives comprise the final *résumés*, if not the *curriculum vitae*, of individual lives. Whereas the grave marker goes on forever (at least until lichens, frost wedging, gravity, acid rain, or vandalism bring it to an end), the contiguous zone is a space that responds continuously to the mourning and memorializing rituals of those left behind. Taken together, these two zones have led to the increasing personalization of cemetery space, especially over the last 25 years.

Whereas lavish memorials have always been an option for the wealthy, the middle class (urban and rural) increasingly demands a more personalized way of remembering the dead. The photographs in this essay reveal a post-modern reconstruction of how society views life and death. They were taken in the United States and Canada, and will be presented as gallery exhibitions, as if the reader were visiting a museum of the present. We will begin with two examples.

32.2 Posthumous Résumés

The grave of Robert Earl Crowe may be found in the Baptist Church Cemetery in Virgilina, Virginia (Fig. 32.1). What information about the departed's life does the black-granite grave marker offer? First, not only is a full legal name provided, but so is a humanizing nickname. We know him as a friend when we know him as "Crowe." Second, we are informed of the very day of his birth and his death. Third, by virtue of what might be a yearbook photograph, we know what he looked like. Fourth, we know he loved wrestling. The grave marker makes it his defining characteristic, and, in the etching, we see him as triumphant. Fifth, we read his final words (though probably put in his mouth by the closest of loved ones): "Remember me with a smile, for God has taken me home." These are the permanent records of his life; little do we know about his death. Those who knew him bring a personal narrative to the grave; those who didn't become symbolic archeologists challenged to construct a life story with only a few bits of evidence. To the permanent record, must be added the few items in the contiguous zone, but one can only imagine what it must have looked like in the few days after burial.

In Hagerstown, Maryland, surrounded by karst topography, the grave of Michael D. Brierly is not dominated by a tombstone (Fig. 32.2). Rather, a bronze plaque provides essential information, plus a textual and pictorial bonus. The deceased is honored as "father, son, brother, and friend," and we know he must have loved

Fig. 32.1 Grace Baptist Church Cemetery, Virgilina, Virginia, 2012 (Photos by Donald J. Zeigler)

Fig. 32.2 Rest Haven Cemetery, Hagerstown, Maryland, 2012 (Photos by Donald J. Zeigler)

skateboarding. It is in the contiguous zone, though, where we really get to know something of the person. Some of those who knew him best are having fun with this grave site; they are honoring the deceased with humor. They are adding personality to the being described on the bronze marker. Vinyl bats, a Cave Ave street sign, and other "five below" purchases upstage the plants and flowers that are normally a grave's star decorations. A small package of visuals gives a departed stranger a personality and reminds us that the cemetery is a repository of social memory at the circle-of-family-and-friends scale.

Contrast both images in this gallery with the mid-twentieth century's grave-marking practices. It was an era when, except for a few floral and scriptural embellishments, most grave markers epitomized the efficient use of space by noting only names, dates, and possibly relationship roles such as husband or mother. It was an era when the contiguous zone contained only flowers or stones. It was an era of the "organization man" when life templates were standard issue and social pressure was directed at conformity and fitting in, not standing out. Today, cemetery landscapes have moved beyond the industrial era's assembly-line mentalities. More and more frequently, the individuals interred and the survivors who tend their graves are intent on creating new norms for celebrating individual lives in the land use zones that are set aside for "disposing of" the dead (a phrase that sounds increasingly dated). "New life" has come to the post-modern cemetery, which continues to remind us that grave-marking is place-making and a discourse on what society finds acceptable at the intersection of religion and secularism.

32.3 Cemeteries as Museums of Biography

The space we reserve for the dead is inhabited by the daughters of memory: the Muses. Their influence took hold solidly in the 1990s as laser technology developed and as the economy pumped more disposable income into people's pockets.

In service to their mother, Mnemosyne, Greek goddess of memory, the Muses brought a wider and deeper variety of arts to the graveyard: photography, black-and-white and color etching, bas relief, sculpture, even stained glass. The cemetery muses are in the business of keeping the dead alive in survivors' memories. They do the job with textual and pictorial elements. The post-modern cemetery is essentially a biographical (and sometimes autobiographical) space where stories are spun to keep people "alive" for future generations. Cemeteries are essentially museums in the truest sense of the word: places (*-um*) where the Muses (*muse-*) reside as they tend to their individual specialties: epic, lyric, love, pastoral and sacred poetry; history and tragedy.

Memorializing stories presented on grave markers are sometimes told in text: names, dates, quotations, affiliations, poems, epistles, epitaphs, and special words. Other times, the elements are graphic. In fact, an increasing variety of design elements have found their way onto waiting *tableaux*, all designed to tell stories of lives well-lived. The themes around which these tombstone "biographs" coalesce are: (1) relationships, (2) faith and principles, (3) vocations and avocations, (4) secular associations, (5) tools and possessions, and (6) place and time (Fig. 32.3). These categories are presented in the matrix below, but to the matrix should be added one very large category called "miscellaneous."

Relationships, as memorialized on cemetery tombstones, are typically limited to the human domain; but, as pets have taken on an anthropomorphic presence in society, their images are also becoming a part of the graveyard scene. The rocking chairs in West Point, Virginia, and the two friends who are "forever free" in Mercersburg, Pennsylvania, both tell the story of loving relationships, even if one is about a woman and her dog. The rocking chairs are without bodily presence, but the departed spirits are symbolized by two free-flying butterflies. On the ground is a heart (symbol of devotion) with this inscription: JBA lvs RGA. The woman and her dog are shown twice on her headstone: on an earthly seashore and looking down from above. The ultimate message is not unique to these two grave markers. It may be the prevailing message of remembrance: the loving relationships of earth carry over into the hereafter, however conceptualized.

As subjects for pictorial interpretation, faith and principles are almost as important as relationships. In North Myrtle Beach, South Carolina, Maimonides speaks for a deceased rabbi, and his voice transcends eight centuries. "I believe" says he – in his native Hebrew and written in the script of the Scriptures. In Honolulu, Hawaii, a different alphabet and identifiably-Iranian names point toward a minority faith in the world today. The symbol at the top of the tombstone is Baha'i. Visits to cemeteries offer teachable moments about the value of faith and principles. Grave sites have always been places of asynchronous instruction where the dead offer lessons to the living and where the wisdom of one generation is offered to the next and the next and the next.

Spiritual though they be, people do have earthly lives. They have vocations and avocations with which they identify and which identify them. In Beckley, West Virginia, a coal miner of 50 years tells future generations how he fed his family and what he looked like when he went to work. Baseball, though not a vocation (for

Fig. 32.3 Graves representative of relationships, faith and principles, vocations/avocations, secular associations, tools and possessions, and place and time can be found across the United States. These graves are located in St. Theresa's Cemetery, West Point, Virginia; Church of the Brethren Cemetery, Mercersburg, Pennsylvania; Southern Palms Memorial Gardens, North Myrtle Beach, South Carolina; Oahu Cemetery, Honolulu, Hawaii; Sunset Memorial Park, Beckley, West Virginia; Mount Calvary Cemetery, Richmond, Virginia (Photos by Donald J. Zeigler)

32 Visualizing the Dead: Contemporary Cemetery Landscapes

Fig. 32.3 (continued)

most), can provide continuity and meaning to life, as it apparently did in Richmond, Virginia. Occupations, sports, hobbies: all are found with ever increasing frequency on the grave markers (and in the contiguous zones) of North American cemeteries. What a culture puts in our lives, we take to the "great beyond," almost as if we have returned to the ancient Egyptian practice of filling the tomb with treasures needed in the next life.

The most common secular association to be noted on or beside grave markers is service in the armed forces. In the case of a grave in Osakis, Minnesota, it was the U.S. Army. But, in the case of a grave in Sacramento, California, it was two state universities, where the departed had earned an undergraduate and a graduate degree, appropriately noted in a forever-open book. Also secular, and often represented in the cemetery are personal possessions. However, one imagines the tools on a headstone in Steinbach, Manitoba, not as possessions, but as extensions of the person. The same might be said of the full-bodied car in Barre, Vermont. Barre's Hope Cemetery displays more art than many art galleries and is a tourist attraction in its own right; right next door is the Rock of Ages granite quarry. It fought the trend toward lawn-type headstones, and led to an expansion wave of cemeteries as sculpture gardens.

Grave markings also reflect dimensions of place and time. Beside the church in Saint Jean Baptiste, Manitoba, farmstead and endless prairie mark the place as home. With its four layers (the land, the improvements, clouds in the sky, and heaven above), the image appears almost like a Geographic Information System. The Saint Jean Baptiste churchyard will forever preserve a "screen capture" of the Sabourin farm. Another landscape is captured on a headstone in Chloride, Arizona: a harsh landscape for those who had to find a livelihood there, but a beautiful landscape for those who could cruise through in an air-conditioned RV. Could these be "snowbirds," post-modern nomads, in their adopted home? Does the very portrayal of an RV suggest a particular era in the history of the arid West, a unique intersection of time and space?

32.4 Religious or Secular Narratives?

The cemetery is a place where human spirits live on after death, but the memorialization of those spirits may take on either religious or secular raiment, or both. Sacred and profane often compete for space on the same grave, making it possible to arrange cemetery memorials along a religiosity continuum.

Save for those of the rich and famous, traditional grave markers said little. But culture abhors a vacuum, so the culture of every era and every realm has found different ways of telling stories in graveyard environments. Some stories are told using religious symbols alone. In Serbin, Texas, husband and wife have their names etched into the Holy Book, which lays open next to a Christian cross (Fig. 32.4). Judging from the position of the bookmark, Scripture made them a better mother and father. Beneath the sacred symbols are the words "Asleep in Jesus." Even the

Fig. 32.4 St. Paul Lutheran Church Cemetery, Serbin, Texas, 2009; Catholic Cemetery, Granby, Québec, 2011 (Photos by Donald J. Zeigler)

wedding rings seem to signify spiritual togetherness without end. In Granby, Québec, another common element of grave marking is on display: a Bible verse, John 13: 34–35. In translation, it reads "Love one another as I have loved you," an essential Christian teaching. Both markers reveal Christianity's universalizing character, an attempt to reach the living and spread "The Word" even after death. They also lay the foundation for some asynchronous role modeling.

Then there are the life stories which carry but scant trace of religious conviction. In Lumberton, North Carolina, a man is portrayed playing golf while his wife is memorialized more abstractly, though flowering dogwood adds a hint of religiosity (Fig. 32.5). Added to the pictorial elements are the words "Family Record," direct

Fig. 32.5 Meadowbrook Cemetery, Lumberton, North Carolina, 2011; Pine Forest Cemetery, Homerville, Georgia, 2012 (Photos by Donald J. Zeigler)

and to the point, as if to say to generations yet unborn, "I know why you have come, so here's the information you wanted." In Homerville, Georgia, there is more recreation, in this case a fisherman who has reeled in a large-mouth bass as big as the boat. In both instances, a secular utopia is where you know the departed reside. Heaven seems to be imagined as earth. Both graves provide evidence of tombstone topophilia, the love of places translated into cemetery art. But in the post-industrial cemetery, a dichotomous classification system for grave markers does not work very well. Rather, both religious and secular elements often share the same *tableau noir* and the space around it. Exemplifying the hybrid model is the "Truckin for Jesus" headstone in Mercersburg, Pennsylvania (Fig. 32.6).

Fig. 32.6 Church of the Brethren Cemetery, Mercersburg, Pennsylvania, 2010 (Photos by Donald J. Zeigler)

32.5 Cemetery Art Imitates Life

The first modern technology to transform grave-marking was probably the black-and-white photograph. Black-and-white gave way to color, the technology of etching advanced, and computer-driven lasers transformed the ability to use polished granite as a canvas for portraits and other imagery. Photographic images usually show the departed at their best, never sick, and always looking content if not exactly happy. In Woodford, Virginia, a small headstone packs in an abundance of clues to the character of the deceased, including a color snapshot (Fig. 32.7).

Steinbach, Manitoba was settled by German Mennonites from Russia. Macy, Nebraska is located on the Omaha Reservation. In both cemeteries, etchings have replicated photographs, and other images have replicated local geography. Although tragedy is implied by the dates of Mr. Reimer's life, death-at-a-young-age is not the narrative chosen. Rather, than emphasize the meningitis that struck suddenly, he is shown fishing with his grandfather in a place where he is "safe in the arms of Jesus." And, on the Mitchell grave these words speak to survivors and unborn generations: "Look not mournfully into the past, it comes not back again. Improve the present – it is yours. Go forth to meet the shadowy future." Cemeteries are rich in ethnic history and contribute to what Harvey (2006) calls "ethnic memory." These three grave markers, in fact, represent three types of traditional ethnic cemeteries: African-American, rural ethnic, and Indian tribal. All three reveal a post-modern preference for personalizing the lives of departed loved ones using technologies that

Fig. 32.7 Lawn Cemetery, Woodford, Virginia, 2012; Heritage Cemetery, Steinbach, Manitoba, 2012; Omaha Tribal Cemetery, Macy, Nebraska, 2007 (Photos by Donald J. Zeigler)

make the dead come alive. Post-modern memorialization knows no ethnic bounds. New tombstones everywhere have been touched by national trends.

Cemetery art imitates individual lives, but it also imitates the culture of commercialization. Grave markers repeatedly tell us how much people loved the businesses they built and the jobs they held. It is said that "you can't take it with you," but some people try. They apparently want to take their work with them, a suggestion that we live in an era where our livelihoods often become our lives. In Chester, Virginia, the deceased owner of a Holiday Inn has brought the Great Sign with him to the grave (Fig. 32.8). It was the source of his identity in life, and it will be the source of his

Fig. 32.8 Sunset Memorial Park, Chester, Virginia, 2007; Hampton, Virginia, I-64 corridor, 1986 (not Mr. Whelan's Holiday Inn franchise) (Photos by Donald J. Zeigler)

identity long after death. Although the Holiday Inn Great Sign has disappeared from roadside America, it will never disappear from *la terre sacrée*. The cemetery becomes a museum of visual history.

Cemetery art does imitate life. In the case of this grave marker in Steinbach, Manitoba, the life lost was young (Fig. 32.9). From the clues etched on black granite, we may conclude that there were two loves in Mr. Wiebe's life: a secular love of motorcycles and a spiritual love for Jesus. Both sides of the memorial are used to tell his story. Two photographs and his signature add the personal touch. And, from the grave he speaks to the visiting stranger with words that are common to cemeteries across North America: "As you are now / So once was I / As I am now / So you will be." As the demand for personalized grave markers has risen, memorial dealers have offered customers an increasing abundance of copy art. This marker, though, is original and unique. It must be one in which the granite cutter takes great pride, one which he hopes will stimulate even greater demand for preserving personalities for posterity. Monument dealers have become an important part of the economy that services cemeteries. Just over the border from Manitoba, "EXPRESS YOURSELF FOR ETERNITY" has become the sales slogan for the Stennis Granite Company of Grand Forks, North Dakota. It's a slogan that should appeal to the pre-dead: become a stakeholder in your own memorial – buy before you die! Take charge of the narrative!

Municipal cemeteries, church cemeteries, private cemeteries, and ethnic cemeteries have all witnessed trends toward increasing secularization and personalization, with

Fig. 32.9 Heritage Cemetery, Steinbach, Manitoba, 2012 (Photos by Donald J. Zeigler)

Fig. 32.10 Dean Family Cemetery, Shenandoah National Park, Virginia, 2008 (Photos by Donald J. Zeigler)

words and pictographs chosen to customize each and every grave. These changes have also come to the family plots of Appalachia and other parts of the country. The Dean family cemetery has been completely enveloped by Shenandoah National Park, but it remains in family hands. Zelinsky (1976: 173) would classify it as "a sort of vestibule" into the Dean family home and their conception of the land of the departed. Two select tombstones from that cemetery illustrate the themes that have been spreading across the continent: personalization, earthly utopianism, and commercialism, in addition to traditional family values as they are represented by the photograph and the wedding rings (Fig. 32.10).

32.6 Ex Situ Humor

Hilarity is an unwelcome visitor to graveyards. Humor seems out of place. Few people think of cemeteries as comedy clubs. Even their names are "dignified and solemn" (Zelinsky 1976: 178). However, cemeteries which do offer a little levity can be very popular as tourist attractions and are often marketed by local communities. That is certainly the case of Boot Hill Cemetery in Tombstone, Arizona, where Lester Moore was interred, probably in the 1880s, with this epitaph: "Here Lies / Lester Moore / Four Slugs / From A 44 / No Les / No Moore" (Fig. 32.11). In Key West, Florida, B. P. Roberts was buried in an above-ground vault with a plaque attached. It reads: "I told you I was sick." And, in a rural cemetery in Boyce, Virginia, this inscription appears on the verso of a headstone: "She Stood Up For Her Principles / Even If She Stood Alone." When the grim reaper comes calling, however, few survivors are in the mood for what is otherwise one of the living spirit's best mood-managers, jocularity. Though humor may be one of the primary ways in which we exhibit our individuality, those who visit cemeteries bring with them two primary emotions, sorrow and solace, that often repress its palliative power (Bachelor 2004). Nevertheless, the trend toward pre-planned personal narrative may mean that macabre humor may have a brighter future in the expansion zones of American necropolises.

Fig. 32.11 Boot Hill Cemetery, Tombstone, Arizona, 1988; Green Hill Cemetery, Boyce, Virginia, 2012; Key West Cemetery, Key West, Florida, 1991 (Photos by Donald J. Zeigler)

Fig. 32.12 Beaufort National Cemetery, Beaufort, South Carolina, 2008 (Photos by Donald J. Zeigler)

32.7 The Case of National Cemeteries in the United States

Not even the U.S. National Cemeteries have escaped the trend toward personalization. Those who were taught conformity to military regimen during their time in uniform, have the opportunity to stand out (just a bit) as individuals in the country's military necropolises. Proud of their rank and service record are those interred; that is evident from their headstones. Notice of family relationships, religion, and some personal characteristics (for example, terms of endearment, titles, and nicknames) are also permitted.

The headstones in this gallery are located in the Beaufort National Cemetery in South Carolina (Fig. 32.12). The graphic element on each headstone (and the only one allowed except for the Medal of Honor) is called an emblem of belief, of which there are dozens approved by the U.S. Department of Veterans Affairs (2012). From left to right, this sample represents the Native American Church of North America, the Unitarian Church, Islam, and the United Methodist Church. Dozens of stylized Christian crosses are available to veterans, but there are also emblems for many other religions (for example, Buddhists, Sikhs, Hindus), even atheists and Wicca. Each memorial also carries a single line of personal characterization: "a gentle soul," "unforgettable," "loving memory," "super sparrow." The full context of each is known only to family and friends.

32.8 Toward a Post-cemetery Age

Cremation will soon be the method of choice for disposing of dead bodies in the United States. Sometimes the ashes will be inurned in cemeteries, sometimes they won't (Fig. 32.13).

Accompanying the trend toward cremation is the tendency to place memorial markers in the land of the living, rather than confining them to the space we have set

Fig. 32.13 Brunswick, Georgia, 2012 (Photo by Donald J. Zeigler)

aside for the dead. White crosses that mark tragic deaths along highways are examples of cemetery-like memorials diffusing into the world of the living. They may not take the place of grave markers, but they engender many of the same feelings and rituals. True examples of ex situ grave marking, however, are starting to appear. On the side of a building in Asheville, North Carolina, a bronze plaque, looking very much like a cemetery marker, commemorates an untimely death, in fact, a murder (Fig. 32.14). On city streets it will be seen; on cemetery lawn, it will not.

Another trend that imitates cemetery landscapes is the memorial plaza paid for with the sale of personalized bricks. Though small, such bricks tend to be in places of honor where people congregate: good places for memorializing rituals. "Looking for my brick" has become part of popular culture nationwide. In the case of Lauderdale-by-the-Sea, Florida, one person's brick will be her "permanent" marker; she plans to be cremated (see Fig. 32.14). Frequently, in fact, brick inscriptions will begin with the words "In Memory of." Strategies like these bring the specter of death back into our everyday lives, where it used to be when family cemeteries were on family farms and every church had a burial ground attached.

32.9 The [Dead] End

The cemetery of the twenty-first century has been transformed by the currents of change. Graves are more visually expressive; their narratives are more personal; their palettes more colorful; their symbolism more secular; their totems more materialistic; and their aura more artistic than ever. They are grist for the mill of visual geography, ripe fruit for the geohumanities; and potentially digestible by the enzymes we know as geospatial technologies. For local communities, there is now

Fig. 32.14 *Top*: Lexington Avenue, Asheville, North Carolina, 2009; *Bottom*: Melvin J. Anglin Square, Lauderdale-by-the-Sea, Florida, 2005 (Photos by Donald J. Zeigler)

more entertainment (read: tourism potential) in "cities of the dead" than in many uptown art galleries. Want to be inspired? Looking for some advice? Need a role model? Enjoy creative story-telling? Ready for a new narrative of your own? The cemetery may be the place for you. The dead end isn't so dead any more.

References

Abby Collier, C. D. (2003). Tradition, modernity, and postmodernity in symbolism of death. *Sociological Quarterly, 44*(Autumn), 727–749.

Bachelor, P. (2004). *Sorrow and solace: The social world of the cemetery*. Amityville: Baywood Publishing.

Brandes, S. (2009). The meaning of American pet cemetery gravestones. *Ethnology, 48*(Spring), 99–118.

Brown, J. G. (1994). *Soul in the stone: Cemetery art from America's heartland*. Lawrence: University Press of Kansas.

Darden, J. T. (1972). Factors in the location of Pittsburgh's cemeteries. *Virginia Geographer, 7*, 3–8.

Darnall, M. J. (1983). The American cemetery as picturesque landscape: Bellefontaine Cemetery, St. Louis. *Winterthur Portfolio, 18*(Winter), 249–269.

Foote, K. (2003). *Shadowed ground*. Austin: University of Texas Press.

Francaviglia, R. V. (1971). The cemetery as an evolving cultural landscape. *Annals of the Association of American Geographers, 61*, 500–509.

Hamscher, A. N. (2002). Scant excuse for the headstone: The memorial park cemetery in Kansas. *Kansas History, 25*(Summer), 124–143.

Harvey, T. (2006). Sacred spaces, common places: The cemetery in the contemporary American city. *Geographical Review, 96*(April), 295–312.

Howett, C. (1977). Living landscapes for the dead. *Landscape, 21*, 9–17.

Jabbour, A., & Jabbour, K. S. (2010). *Decoration day in the mountains: Traditions of cemetery decoration in the Southern Appalachians*. Chapel Hill: University of North Carolina Press.

Jackson, J. B. (1967). The vanishing epitaph: From monument to place. *Landscape, 17*, 22–26.

Jordan, T. G. (1982). *Texas graveyards: A cultural legacy*. Austin: University of Texas Press.

Kniffen, F. (1967). Necrogeography in the United States. *Geographical Review, 57*, 426–427.

McDannell, C. (1987). The religious symbolism of Laurel Hill Cemetery. *The Pennsylvania Magazine of History and Biography, 111*(July), 275–303.

Moore, J., Blaker, C., & Smith, G. (1991). Cherished are the dead: Changing social dimensions in a Kansas cemetery. *The Plains Anthropologist, 36*(February), 67–78.

Pattison, W. D. (1955). The cemeteries of Chicago: A phase of land utilization. *Annals of the Association of American Geographers, 45*, 245–257.

Tarpley, F. (2006). Naming America's graveyards, cemeteries, memorial parks, and gardens of memories. *Names, 54*(June), 91–101.

U.S. Department of Veterans Affairs. (2012). Available emblems of belief for placement on government headstones and markers. www.cem.va.gov/hm/hmemb.asp. Accessed 1 Aug 2012.

Wright, E. (2003). Reading the cemetery, "Lieu de Mémoire par Excellance". *Rhetoric Society Quarterly, 33*(2), 27–44.

Zelinsky, W. (1976). Unearthly delights: Cemetery names and the map of the changing American afterworld. In D. Lowenthal & M. J. Bowden (Eds.), *Geographies of the mind* (pp. 171–195). New York: Oxford University Press.

Zelinsky, W. (1994). Gathering places for America's dead: How many, where, and why? *Professional Geographer, 46*(1), 29–38.

Chapter 33
Sacred, Separate Places: African American Cemeteries in the Jim Crow South

Carroll West

33.1 Introduction

Segregated African American cemeteries were a fact of life in the post-Civil War American South (Aiken 1998; Litwack 1998). When historian Eric Foner reviewed the Reconstruction era across the South, he concluded that newly freed African-Americans rushed first to create three separate institutions—churches, schools, and cemeteries—within the rapidly segregating South (Foner 2002). The first two public spaces, churches and schools, are the subject of an extensive body of scholarship (Montgomery 1993; Butchart 1980). Cemeteries are comparatively ignored, even as segregated cemeteries for African Americans not only still exist by the hundreds across the southern landscape, but are still integral parts of today's African American neighborhoods. Their location helps to define the boundaries of historic African American neighborhoods while the gravemarkers and family plots enhance the cultural identity of these communities.

Two generations after the end of the Civil War, educator and historian W. E. B. DuBois observed,

> It is usually possible to draw in nearly every Southern community a physical color-line on the map, on the one side of which whites dwell and on the other Negroes. The winding and intricacy of the geographical color line varies, of course, in different communities. I know some towns where a straight line drawn through the middle of the main street separates nine-tenths of the whites from nine-tenths of the blacks. In other towns the older settlement of whites has been encircled by a broad band of blacks; in still other cases little settlements or nuclei of blacks have sprung up amid surrounding whites. Usually in cities each street has its distinctive color, and only now and then do the colors meet in close proximity. (DuBois 2007: 79)

C. West (✉)
Center for Historic Preservation, Middle Tennessee State University,
Murfreesboro, TN 37312, USA
e-mail: carroll.west@mtsu.edu

Historians have since documented the lines of segregation found in most southern places. Leon Litwack explains: "To find the black neighborhood in almost any town or city, one needed no map or signs. The streets in black districts were seldom if ever paved, and in rainstorms they were certain to turn into quagmires or mud. The housing was the least desirable, sometimes places discarded by whites" (Litwack 1998: 336). In extensive fieldwork in Tennessee African American places, supplemented by field research in Mississippi and Alabama, the author found that invariably historic African American cemeteries helped to define the edges between the white and black sections of a town or to separate the African American world from that of the dominant culture within a county. Typically located in close proximity to other community institutions as churches and schools, the large open space conveyed by a cemetery spoke to the history, culture, and separation of African Americans in a Jim Crow South (West 2008).

As they developed fully in the late nineteenth century, African-American cemeteries represented independence and reinforced African American identity. Folklorist John Michael Vlach emphasizes, "For black Americans the cemetery has long had special significance. Beyond its association with the fear and awe of death, which all humans share, the graveyard was, in the past, one of the few places in America where an overt black identity could be asserted and maintained" (Vlach 1991: 109). The cemetery, Vlach concluded, also enhanced African Americans "sense of ethnicity as well as satisfying their personal need to communicate with their deceased family members" (Vlach 1991: 112). The establishment of a separate cemetery showed African-American autonomy, while also displaying the racist beliefs that often surfaced in the white community that relegated different races to separate facilities. No matter Emancipation, Reconstruction, no matter the Civil Rights Movement of the twentieth century, cemeteries remained segregated because many whites believed that even in death, minorities represented "a dirty, vile, degraded, unredeemable humanity" (Sloane 1991: 187).

Over the past 15 years, the author has investigated scores of historic cemeteries in Tennessee as part of a larger project to document the landscape of rural and small town African American churches in the state. More recently the study has expanded into Alabama's Black Belt as part of a larger effort to landmark significant African American properties associated with the U.S. Public Health Service's Syphilis Study in Macon County, Alabama, from 1930 to 1976. (More on this subject below.) Hundreds of cemeteries share the characteristics noted in the following case studies—located on the margins of the larger dominant culture, but situated deep into the cultural heart of African American communities (Fig. 33.1). Larger ornate gravemarkers typical of the Victorian era are found only in the urban segregated cemeteries; otherwise these cemeteries provide more understated yet compelling testimony of identity, faith, and family.

Fig. 33.1 Indian Hill Missionary Baptist Church Cemetery, Giles County, Tennessee, is located only miles from where the KKK first organized. It is a representative example of how these rural black cemeteries can be located on the margins of the landscape (Photo by Carroll West)

33.2 Golden Hill Cemetery, 1863

The Golden Hill Cemetery on Seven Mile Ferry Road in Clarksville, Montgomery County, Tennessee, began in 1863, when Stephen Cole, a former slave who purchased his freedom in 1859, bought land for a cemetery (Allison and West 2001; West 1998). Both Stephen Cole and his son, Edward, are interred in Golden Hill. Edward Cole was one of several members of the Civil War-era 101st Regiment, United States Colored Troops who is buried at Golden Hill. The fact that Cole started the cemetery soon after Emancipation in the middle of the war also shows the impact of Union occupation on Clarksville's tiny free black population and its much larger population of emancipated and contraband residents. Federal occupation helped to give local African-Americans the freedom to establish their own burial ground.

The occupying Union forces left Clarksville in September 1865, and from that time to 1869, racial conflict resulting in riots and Ku Klux Klan activities occurred frequently. According to historian Howard Winn, "Racial conflicts persisted in the 1870s, and black businesses were especially hurt by the great arson fire of 1878 that destroyed 15 acres of the central business district" (West 1998: 175). During the 1880s, Stephen Cole sold the cemetery to John W. Page. One of the most prominent members of Clarksville's African-American community during the 1880s and 1890s, John W. Page served his community as a grocery store owner, real estate investor, and Sunday school superintendent at St. Peter's African Methodist Episcopal Church. Like his other ventures, Page operated the cemetery as a profit-making business and he incorporated it, selling shares in the cemetery to raise money for its development and naming it Golden Hill. It became the first black-owned corporation in Clarksville (Beech 1964). Although it is impossible to give Page the credit for the circular road system and the planting of ornamental trees in the fashion of the Rural Cemetery Movement, oral tradition is that these improvements to the cemetery took place under his guidance.

The Rural Cemetery Movement began in large antebellum era urban cemeteries. The Hollywood Cemetery (1849) in Richmond, Virginia, is an important southern example. The designs of these new cemeteries of the industrial age created park-like environments with ornamental plantings, cemetery furniture, and circular drives and walkways. This type of cemetery design became more prevalent in smaller southern places after the Civil War when towns, like Huntsville, Alabama, in 1870, passed ordinances creating separate cemeteries for whites and blacks. Many of these new Reconstruction-era cemeteries took on design attributes from the Rural Cemetery Movement (Sloane 1991).

Judging by the dates of gravemarkers, the installation of larger, more ornate monuments on the hilltop in the middle of the cemetery dates to Page's management. Not only was the re-invigorated cemetery a statement of rising African American middle class pride, it was a political statement by Page, who in 1892 was elected as city councilman, the first African American to serve in political office in Clarksville. Page served in the City Council for 7 years, before the elimination of ward voting by whites who wanted to keep African-American power in city government to a minimum.

By the turn of the century, patterns of Jim Crow residential segregation were increasingly imprinted into the Clarksville landscape, leaving African American places like Golden Hill Cemetery separated from the booming town. For example, when city fathers worked with developers to launch a streetcar line, the trolley extended to the city-supported and designated white only Greenwood Cemetery. The trolley line, however, stopped at the white cemetery gates and African Americans would leave the line and walk the remaining 400 m or so to Golden Hill Cemetery on the farthest outskirts of town.

In the early twentieth century, Golden Hill was part of DuBois's African American world behind the veil, as he discussed in *The Souls of Black Folks*. The African American middle class of Clarksville could express their success and taste through the monuments of their leaders, with the designs coming from an African American carver named Hiram Johnson. According to the Montgomery County census, in 1880, Hiram Johnson was a 24-year-old black male who was living in the household of Samuel Hodgson, a white Clarksville merchant. Johnson's occupation is not marked on the census record. However, having lived in the household of Samuel Hodgson, merchant and owner of the Clarksville Marble Works, Johnson likely learned the art of stonemasonry with the Hodgson family. Established in 1852, the Clarksville Marble Works imported marble from Italy and red and gray granite from Scotland, giving Johnson a wide variety of stone types to work with. According to Ursula Beach, a Montgomery County historian, "Monuments and statuary were carved upon order by Samuel Hodgson and his artisans or executed by craftsmen in Italy" (Beech 1964: 275). No further indication of Hiram Johnson's occupation was found. He was not listed among other African-American stonemasons during the late nineteenth and early twentieth centuries. However, linking him to the Hodgson family substantiates oral histories that say he was a master stonemason that worked in both the black and white cemeteries.

The work of Hiram Johnson at Golden Hill Cemetery is physical evidence of class similarities between the white and black middle class of Tennessee towns in the late nineteenth century. Both groups embraced Victorian iconography in their homes, their interior furnishings, and in this case, in the gravestones that marked their places of burials. Similarly embellished Victorian era markers are found at the white Greenwood Cemetery. The difference between the two cemeteries lies in the relatively few large, ornate markers at Golden Hill Cemetery compared to the many Victorian era markers at Greenwood Cemetery. These artifacts of material culture confirm what census data and tax lists from those years point out—there was a black and a white middle class in turn-of-the-century Clarksville, but the white middle class was much larger in numbers and more wealthy in possessions than the African-American middle class.

One of the most elaborate monuments attributed to Johnson is in his family plot, with a full size depiction of his wife, Lena, who died in 1899. Other significant Johnson-carved monuments are the Dabney family monument (Charles Dabney was the first African American constable in Clarksville), which shows a full size winged cherub holding flowers; the Wheeler family monument, which consists of an angel holding a cross chiseled into the stone; and the Buck family monument, which is tall obelisk topped with a classical urn.

33.3 Shiloh Cemetery, c. 1870

Shiloh Missionary Baptist Church Cemetery, located south of the town of Notasulga, Macon County, Alabama, is a wholly different place than the formal Rural Cemetery Movement design of Golden Hill Cemetery. Shiloh cemetery is located immediately west of Alabama Highway 7 approximately a third of a mile south of the Shiloh Missionary Baptist Church and Rosenwald School. Placed on the rise of a small hill, the cemetery has a roughly rectangular shape, measuring approximately 36.5 m in length on its north side, 21.3 m on its west side, 36.5 m on its south side, and 21.3 m on its east side. Although scattered ornamental plantings exist, and a dirt and gravel drive circles the cemetery, there was no intent here to design a park-like place. Like the majority of rural African-American cemeteries, Shiloh has a rough-edge linear appearance, with rows of graves separated by grass walking paths (Fig. 33.2).

Fig. 33.2 Shiloh Missionary Baptist Church Cemetery, Macon County, Alabama (Photo by Carroll West)

The location of the Shiloh's community cemetery on a rise on the historic road halfway between Notasulga and Tuskegee, the seat of Macon County, reflects a Reconstruction-era African American settlement trait of locating rural communities/neighborhoods away from established white communities, but in close enough proximity that residents could walk or travel by horseback to employment in the towns. This section of rolling hills and thin, rocky soil also marked land of marginal agricultural opportunity but land that African American farmers could afford to purchase in the initial decades after emancipation. The cemetery's hillside location and roadway presence also made it a visible landmark within the larger agrarian landscape of the county.

The community insists that burials date to the time when the nearby Shiloh Missionary Baptist Church was established c. 1870; 1881 is the earliest identified marker date. The markers are arranged north-to-south, with headstones facing east. Several large pine trees are centered on the crest of the hill; a few ornamental bush plantings and other smaller pine trees are scattered throughout the cemetery. The estimated number of all burials, determined by a count of headstones and of rectangular-shaped depressed areas, is approximately 300.

The grave markers vary in materials and design. Materials used to make markers include pottery, metal (iron and aluminum), limestone, sandstone, granite, river stone, and marble. The majority of graves have been covered by concrete slabs, installed from c. 1940 to c. 2000, that help to protect individual graves from erosion. Two unmarked graves exhibit traditional African American burial practices in that these graves have been covered by broken pieces of stone, concrete, and pottery.

A metal container in the shape of a coffin covers one unmarked grave. Many family groupings exist, but only that of the Potts family (built c. 1956–1958) have been marked by any sort of barrier, in this case a concrete block wall, faced with bricks on the top of the wall, that also protect these two graves from erosion.

The artistry of various markers ranges from formal to hand engraved. Themes are religious; symbols include the cross, book outlines representing the Bible, and hands clasped in prayer. Hand-blocked or hand carved markers are found throughout the cemetery. Examples include Mrs. Louisa Tolbert (d. 1947) and the unique stoneware jar marker for Willie J. Coming (b. 1880–d. 1881). This jar is signed "made by J. C. Humphries" and contains the hand-carved verse, "A voice so sweet/ now is still/ awaken+[unintelligible]/no one can fill." Almost all markers are rectangular or roughly square in shape, except for the diamond-shaped marble marker for "Jo Ana/ wife of Rev. I. M. Pollard" (d. 1924) (Fig. 33.3).

The cemetery documents the impact of two different avenues of opportunity for African Americans from Reconstruction era of the late nineteenth century through the Jim Crow era of the mid-twentieth century: (1) membership in fraternal organizations and (2) service in the U.S. military. African American fraternal organizations, points out historian Jacqueline Moore, "functioned largely as mutual benefit associations. Members paid dues and in turn were entitled to burial benefits or help for the family in times of illness. These organizations also provided opportunities not found in the larger society for blacks to practice leadership skills, as well as places where they could meet socially in a friendly environment" (Moore 2003: 6). Liberty Chamber 2076 of Notasulga, Alabama, provided standardized markers for

Fig. 33.3 Charlie Pollard was a victim of the U.S. Public Health Service Syphilis Study of the mid-twentieth century and is buried at Shiloh Cemetery (Photo by Carroll West)

Fig. 33.4 Cloverdale Cemetery, Milan, Tennessee. Military veterans are often highlighted in rural African American cemeteries (Photo by Carroll West)

Viola Rowell (d. 1915), Mary C. Pearson (d. 1917), Ellen Conner (d. 1926), and Fannie Gibson (d. 1927), among others. Liberty Chamber also provided a marker for one male, Sam Moss (d. 1918), who was minister at Shiloh Missionary Baptist Church for 15 years. The Zion Traveler Chamber 2813 provided a standardized marker for Rufus Bentley (d. 1917).

During and after the Civil War, the U.S. army provided opportunities, although placed in segregated units, to serve the nation and gain respect among their fellow black citizens for their military service, a trend documented by the World War I through the Vietnam War veterans buried in the cemetery. Military burials include those for World War I veterans Joseph H. Holliday (d. 1957) and Eugene Hart (d. 1952) (Fig. 33.4).

Shiloh Cemetery is significantly associated with the victims of the Macon County Syphilis Study carried out by the U.S. Public Health Service from 1930 to the mid-1970s. Better known as the Tuskegee Syphilis Study, federal officials for four decades deceived African American men into believing that they were treating the disease; in reality the physicians were treating the men as living specimens in a federally funded study that designed to chart the impact that untreated syphilis would have over the long term. "For more than a quarter century now," medical historian Susan M. Reverby concluded, "the images conjured up by the words 'Tuskegee Syphilis Study' or 'bad blood' have haunted our cultural landscape" (Reverby 2000: 1). Identified victims of the syphilis study buried at Shiloh Cemetery include: Charles (Charlie) Pollard, Will Pollard, Frank Cooper, Mont Pollard, and Albert Robinson. Of this group, Charles Pollard had the highest public profile. Pollard was one of the interviewees for an influential *New York Times* expose of the program in July 1972. Pollard later testified to the Senate in 1973 about the abusive program. He also attended the ceremony at the White House on May 16, 1997, when President Bill Clinton, on behalf of the federal government, offered a formal national apology.

33.4 Pierce-Bond Cemetery, 1882

The Pierce-Bond Cemetery, nestled in the Appalachian foothills of northeast Tennessee, is far removed from the Black Belt setting of Shiloh Cemetery. Pierce-Bond Cemetery lies at the junction of two rural roads, Seaver Road and Horse Creek Road, in Sullivan County. Placed on the top of a small hill that overlooks the surrounding countryside, the cemetery has a roughly rectangular shape, measuring approximately 59.4 m in length on its north side, 42.6 m on its west side, 58.5 m on its south side, and 46 m on its east side. Historic tree lines designate the boundaries of all four sides. A historic wire fence (c. 1940), supported by wooden posts spaced at regular intervals, runs along the north, east, and south sides of the cemetery.

The cemetery is the highest point, and the oldest historical resource, associated with the place name Butterfly, a post office that existed there from 1883 to 1905, and the place name Shinbone, which is a name known to current African-American residents of Sullivan County. The cemetery was the burial ground for local African-American residents and members of the Pierce Chapel African Methodist Episcopal Church. The majority of the approximately 150 historic grave markers (out of an estimated 250 burial sites) in the cemetery are clustered along the brow of the hill, bisected by a line of large mature oak trees. A large evergreen bush, c. 1930, is located near the center of this concentration of grave markers. The markers are arranged in north-to-south rows, with headstones facing east. The great majority of extant grave markers date between 1882 (the earliest marker that had a death date listed) and 1950. The cemetery remains in use as the only rural African-American cemetery in Sullivan County (Fig. 33.5).

When the author prepared a National Register of Historic Places nomination for the cemetery in the late 1990s, Theresa Dykes was the oldest surviving member of the congregation. According to Mrs. Dykes, many graves no longer have their markers. Some markers have been lost to vandalism; simple wooden crosses marked other graves and these have disappeared over the decades. According to oral traditions passed on to Mrs. Dykes, the earliest graves had small triangular-shaped rocks as headstones. Approximately 20 of these still exist in the center of the cemetery. The estimated number of all burials, determined by a count of headstones and of rectangular-shaped depressed areas, is approximately 250 burials.

The cemetery contains no elaborate examples of grave markers. The existing markers are small and often unadorned, and made from concrete, limestone, and granite. Among the more architecturally distinctive markers are those of Nancy Russell (died 1882), which contains a carved hand with an extended finger pointing skyward (a traditional grave marker symbol); Frances Brown (died 1900), which includes carved script; and Sarah Amanda Bachman (died 1919), which has a carved

Fig. 33.5 Early twenty-first century memorial stones now mark once unidentified graves at the Pierce-Bond Cemetery in the Appalachian community of Sullivan County, Tennessee (Photo by Carroll West)

rose bush. Extant markers indicate, however, that the cemetery was not a place of indiscriminate burial—most burials are arranged in family groupings.

The exact date of the founding of the cemetery and church is unknown. According to the author's interviews with Theresa Dykes, Jack Pierce, Anna Coley, Virginia Leeper, and Orvel Bond, the cemetery began in the antebellum "slave days" and contains unmarked burials from those years. The small triangular headstones found at the cemetery are similar in size and design to slave graves the author earlier documented in the National Register of Historic Places nomination of the Bailey Graveyard, established circa 1840, in Wilson County, Tennessee.

The cemetery may well have an association with Emancipation and the Civil War due to white and black history of the immediate rural neighborhood. Throughout the Civil War decade, the Bay Mountain area of rural Sullivan County was known as a Unionist stronghold in an otherwise Confederate county. William S. DePew, for example, had established the nearby DePew Chapel Methodist Church before the war; after the fighting began, DePew served in K company of the 8th Calvary of the U.S. Army of Tennessee as a chaplain. After the war (or at least after the 1860 census) two comrades in arms joined DePew in the neighborhood. Benjamin F. Hood had served in K company of the 8th Calvary; he became a minister at the white church. Their neighbor John N. Dolan was a quartermaster sergeant in B company of the 4th Tennessee Calvary of the Union Army of Tennessee. It is a tradition among African-American residents of Kingsport that this group of Union veterans included Jerome Pierce, who had been born a slave, but was the mulatto son of a

Fig. 33.6 The Pierce family marker at the Pierce-Bond Cemetery (Photo by Carroll West)

white farmer named John Pierce. The older members of the congregation date the history of the church and cemetery as beginning with Jerome Pierce; a gravemarker at the cemetery records that Jerome Pierce died in 1942 (no birth date is listed) (Spoden 1976) (Fig. 33.6).

However, no listing for a Jerome Piece exists in the National Park Service's database of individual United States Colored Troops soldier names. According to past community resident Anna R. Coley in an interview with the author, Jerome Pierce joined the Union army near Roanoke, Virginia, when he was about the age of 16. He was assigned work with a cavalry unit, where he worked with horses (both DePew and Hood served in the 8th Calvary). After the war, he returned to Sullivan County to live with his mother. Her oral testimony suggests that Jerome Pierce may be best described as a volunteer, who assisted Union soldiers without officially enlisting.

The presence of white Unionists and a small community of African Americans is telling—here in the hills of Appalachia was a rural safe haven, which nurtured African American culture and identity for over 100 years. Here at one spot they established three key community institutions: a church, a school, and a cemetery, which persisted operating until the early 1960s. "On Memorial Day the families would gather at the cemetery with lunches to clean the graves of their loved ones with buttermilk and baking soda," Anna Coley wrote in a letter to the author. The third Sunday in September was Homecoming Day, continued Coley, when "people came from all over the state for the services and to fellowship with their friends. They would put up tents and serve all kinds of foods. This all took place on the grounds of the church and cemetery." Descendants of the original congregation members also continue to have annual Homecomings, but now the event has shifted from the third Sunday in September to the fourth Sunday in July.

33.5 Alexandria Cemetery, 1869

One of the most starkly segregated cemetery landscapes in the South is in the small country town of Alexandria, DeKalb County, Tennessee. High on a hill overlooking the town are two historic cemeteries, with a hand-laid stone wall between, clearly differing the white world from the black world in this rural community. East Lawn Cemetery is the segregated white cemetery, a beautiful Rural Cemetery Movement property. On the other side of the fence is what is known as the Alexandria Black Cemetery, or sometimes the Seay Chapel Cemetery, named for the nearby Seay Chapel United Methodist Church. Here, according to local stories, is where slave owners buried their dead slaves and where in May 1869, the local government officially designated the former slave graveyard as the community cemetery for African Americans.

The Alexandria Cemetery began with a plot of three acres for a cemetery, church, and school given to a group of African-American trustees and the Methodist church. There is no mention in this deed transaction of an earlier existing cemetery. More likely, the town deeded this largely worthless property of limestone outcroppings and cedar breaks because recently approved state laws commanded counties to create public schools for whites and blacks. (Obviously, by being listed in the deed, the Methodist church was sponsoring both the school and church as a mission). Yet, the transaction is clear. The graveyard was the primary objective; the school and church was to receive the remainder of the plot. Over 130 years later, the cemetery is the only extant active property of the three. The African-American cemetery is not known to have ever had a proper name. It is known to this day as the "cemetery on the hill," according to an interview with Carrie Helen Smith of the black community, and the only place that a black person could be buried in Alexandria between Reconstruction through the era of Jim Crow segregation. This piece of property is the oldest known African-American property in Alexandria.

When the author assessed the property for possible listing in the National Register of Historic Places in 2000, it appeared at first glance to have little to say about the African-American community that developed around these three institutions in the Reconstruction era and the late nineteenth century. There are only 56 extant gravemarkers. Yet the large number of other graves within the cemetery—an estimated 500—testify to the size of the black community in Alexandria in the years after the Civil War and before World War II. Most of the graves originally either had no grave markers, fieldstone markers, or merely wooden ones, which rotted away, especially between 1980 and 2000 when the cemetery was not maintained. The patterns documented in the survey indicate that the Alexandria African-American cemetery was like other black cemeteries documented in small towns across Middle Tennessee. Families tend to be grouped together, often with a rough rectangle of stones marking a family plot. There was only one extant example of a cast-iron fence marking a family plot. Two gravestones were of veterans: one World War I veteran, Hurshel Williams (died 1937) and World War II veteran, Henry Clay Floyd (1914–1948). The range of cemetery artwork, from the clasped hands of Annie Dowell (1909) to the hand pointing skyward of Fannie Dowell (1890), and to the four-leaf clover of Emma Philips (1897), shows how African Americans embraced and adapted Victorian era symbolism to their final place of rest. That most of the burials had either small triangle-shaped fieldstone markers, or now missing wooden markers, is a testament to the general poverty of the black community during the years of the cemetery's greatest use.

The best evidence of the nature of the African American neighborhood that developed around the cemetery comes from an unexpected source, the writings of William E. B. DuBois. During the two summers of 1886 and 1887, while an undergraduate at Fisk University, W.E.B. Du Bois taught at the Wheeler School, a seasonal "colored school" (no longer extant) in neighboring Wilson County, and he attended church at the Methodist Church in Alexandria. He brought to the community his passionate commitment to education for the black race. He returned to Alexandria in 1897 after being the first African-American to receive his Ph.D. from Harvard in order to experience the change in an area that "would remain in his memory bank for a lifetime" (Lewis 1993: 68).

During his sojourn in Alexandria, Du Bois was left with vivid impressions of the importance of religion in rural African-American communities, as well as the power of the local minister, themes he explored in *The Souls of Black Folks*. In this seminal book, DuBois described the Alexandria African American landscape:

Cuddled on the hill to the north was the village of colored folks, who lived in three- or four-room unpainted cottages, some neat and homelike, and some dirty. The dwellings were scattered rather aimlessly, but they centered about the twin temples of the hamlet, the Methodist, and the Hard-Shell Baptist churches. These, in turn, leaned gingerly on a sad-colored schoolhouse. Hither my little world wended its crooked way on Sunday to meet other worlds, and gossip, and wonder, and make the weekly sacrifice with frenzied priest at the alter of the "old-time religion." The soft melody and the mighty cadences of Negro song fluttered and thundered. I have called my tiny community a world, and so its isolation made it; and yet there was among us but a half-awakened common consciousness, sprung from common joy and grief, a burial, birth, or wedding; from a common hardship in poverty, poor land, and low wages; and above all, from the sight of the veil that hung between us and Opportunity (DuBois 2007: 102).

As one approaches the cemetery today, one may experience a landscape different than that recalled by Du Bois in *The Souls of Black Folks* as "a little world. . . cuddled on a hill north of town." The homes that once surrounded the church and school are gone and forgotten as the rugged landscape of limestone outcroppings and scrub trees is increasingly reclaiming the hilltop. It was poor land, a perfect spot in the antebellum era for a town's white cemetery and a perfect spot to place the freedmen's church/school after the Civil War.

The year 1869 is key—a time still when the experiment of Reconstruction held promise—because it was then that the town fathers chose to give three acres of this rocky, worthless hilltop to its freedmen, and left them the challenge of building a community. They met the challenge, and the community thrived until World War II when new opportunities elsewhere began a process of slow community decline. By the 1980s the congregation of Seay Chapel was mostly elderly and poor, and the cemetery began to be reclaimed. By 1998 the congregation was so small that the Methodist Church closed the church. Members today moved to the Dowelltown Methodist Church, also in DeKalb County. Two years later, black and white together in Alexandria began the process of renewing these resources, and remembering this African-American community. The cemetery was cleared; the church was partially restored, cleaned, and opened for a Christmas 2000 celebration.

The Alexandria Cemetery is like the others explored in this chapter and countless additional historic segregated southern landscapes. It is significant because it reflects the history of African Americans as they moved from slavery to freedom, from economic dependence to self-sufficiency, from exclusion by whites to forming their own institutions, as they struggled to overcome white imposed legal and social restrictions, and as they, all the while, continued to build and expand their communities. Segregated cemeteries are more than remnants of the Jim Crow South; they also are sacred community space speaking to identity and culture for African Americans today.

33.6 Discussion

These four cemeteries are representative of hundreds the author has surveyed in Tennessee, Alabama, and Mississippi. Two are located in rural, isolated areas of the Alabama Black Belt or the southern Appalachia Mountains. One is in a small town and the fourth is in what is now a city of over 100,000 but once was an average southern county seat. According to oral testimony, one began in the age of slavery; the other three more clearly began as Emancipation and Reconstruction-era institutions, between 1863 and 1870.

The four properties share many traits. Hundreds of burials lack grave markers of any sort. Most of the missing markers were probably wooden or, later in the mid-twentieth century, small metal markers installed by funeral homes that have been lost as cemeteries are kept clear by large, motorized mowers. Many other graves are marked only by small triangular bits of stone, a type of early grave marker that older African American residents believe to date to the era of slavery. The two town cemeteries have the more elaborate Victorian-styled tombstones; although when compared to any of the segregated white counterparts in the region, the African American cemeteries clearly differ in the number, scale, and degree of ornament found in their larger designed tombstones compared to those found in white cemeteries.

33 Sacred, Separate Places: African American Cemeteries in the Jim Crow South

But the major difference between white and black cemeteries in the rural South is location. Historic white cemeteries lie often within the boundaries of the local historic district; they are prominent, even dominant landmarks within the city, town, or village. No better example exists than in Selma, Alabama. During recent fieldwork within the city for a study of Civil Rights Movement landmarks, the author found abundant literature about the National Register-listed Live Oak Cemetery, which situated between a section of the town's residential historic district and its public parks and country club. City and state officials and local and state historians all knew where that cemetery was located; they did not know the same about the historic Elmwood Cemetery, the once-segregated historic cemetery for African Americans in Selma.

Elmwood was easy enough to locate, however, once you become accustomed to the arrangement of space in southern communities. Elmwood exists on the opposite end of town, on the far eastern boundary, on Race Street. No public park or country club marks its boundaries. Rather a few scattered homes face its entrance; traffic on a modern four-lane bypass highway roars past the backside of the cemetery. Yet exploring the spaces and grave markers of Elmwood Cemetery brings forth the same themes of pride, accomplishment, and African American identity found in countless other black cemeteries. Officialdom might not recognize the location of the cemetery but clearly African American families do; it may be located on the margins of the dominant world but it is central to the African American sense of heritage, identity, and culture. These distinctive, significant properties help us to delineate that "physical color-line" between the black and white worlds of the South noted by W.E.B. DuBois over 100 years ago (Fig. 33.7).

Fig. 33.7 Elmwood Cemetery, Selma, Alabama (Photo by Carroll West)

33.7 Conclusion

Like other culturally-rich resources, every African American cemetery has its own stories and distinctive design to convey, be that embedded in its location, design, or date of establishment. This research into the creation, location, meaning, and preservation of African American cemeteries really has no end, until we reach that time that scholars, preservationists, and officials look at these properties with the same reverent eye they gaze toward the "beautiful" dominant white cemeteries of their communities. As professionals, we, inexplicably to my mind, still do not have rural African American cemeteries "on our radar." We feel empathy about their disheveled condition or the deteriorated condition of the surrounding neighborhood but we do not step inside that world and pull back the curtain on the meaning of the landscape underneath the overgrowth.

Studying and appreciating these unassuming properties, however, can contribute to (1) locating the historic boundaries of post-Emancipation African American communities; (2) identifying the historic institutional center of post-Emancipation African American communities since the cemetery is often adjacent to or shares property with the first schools, lodges, and churches established in the community; and, (3) finding the names and symbols of generations who were marginalized or even wiped from the pages of our shared history during the worst decades of Jim Crow America. Too often I hear of my colleagues remarking that if only they could find "more" African American history, they could better incorporate those narratives into their work. Those histories are there, unrecognized yet still powerful in what a careful study of southern African American cemeteries says about the past, and present.

References

Aiken, C. (1998). *The cotton plantation South since the Civil War*. Baltimore: Johns Hopkins University Press.

Allison, T., & West, C. V. (2001). *Golden Hill Cemetery, Montgomery County, Tennessee, National Register of Historic Places nomination*. Nashville: Tennessee Historical Commission.

Beech, U. (1964). *Along the Warioto: A history of Montgomery County, Tennessee*. Nashville: McQuiddy Press.

Butchart, R. (1980). *Northern schools, southern blacks and reconstruction: Freemen's education, 1862–1875*. Westport: Greenwood Press.

DuBois, W. (2007). *The souls of black folk*. New York: Oxford University Press.

Foner, E. (2002). *Reconstruction: America's unfinished revolution, 1863–1877*. New York: HarperCollins.

Lewis, D. (1993). *W. E. B. Du Bois: Biography of a race, 1868–1919*. New York: Henry Holt and Company.

Litwack, L. (1998). *Trouble in mind: Black southerners in the age of Jim Crow*. New York: Knopf.

Montgomery, W. (1993). *Under their own vine and fig tree: The African-American church in the south, 1865–1900*. Baton Rouge: Louisiana State University Press.

Moore, J. (2003). *Booker T. Washington, W.E.B. DuBois, and the struggle for racial equality*. Wilmington: Scholarly Resources.

Reverby, S. (2000). *Tuskegee truths: Rethinking the Tuskegee syphilis study*. Chapel Hill: University of North Carolina Press.

Sloane, D. (1991). *The last great necessity: Cemeteries in American history*. Baltimore: Johns Hopkins University Press.

Spoden, M. (1976). *Historic sites of Sullivan County*. Kingsport: Kingsport Press.

Vlach, J. M. (1991). *By the work of their hands: Studies in Afro-American folklife*. Ann Arbor: University of Michigan Press.

West, C. V. (1998). *Tennessee encyclopedia of history and culture*. Nashville: Tennessee Historical Society.

West, C. V. (2008). Sacred spaces of faith, community, and resistance: Rural African American churches in Jim Crow Tennessee. In A. Nieves & L. Alexander (Eds.), *"We shall independent be:" African American place making and the struggle to claim space in the United States* (pp. 439–462). Boulder: University Press of Colorado.

Printed by Books on Demand, Germany